COATED GRAINS

Edited by
Tadeusz M. Peryt

With 359 Figures

Springer-Verlag
Berlin Heidelberg New York Tokyo 1983

Dr. TADEUSZ M. PERYT
Instytut Geologiczny
ul. Rakowiecka 4
00-975 Warszawa/Poland

ISBN 3-540-12071-8 Springer-Verlag Berlin Heidelberg New York Tokyo
ISBN 0-387-12071-8 Springer-Verlag New York Heidelberg Berlin Tokyo

Library of Congress Cataloging in Publication Data. Main entry under title: Coated grains. Includes indexes. 1. Rocks, Carbonate. 2. Concretions. I. Peryt, Tadeusz Marek. QE471.15.C3C6. 1983. 552'.5. 82-19449.

Typesetting: K + V Fotosatz, Beerfelden.
Offsetprinting and bookbinding: Konrad Triltsch, Graphischer Betrieb, Würzburg.
2132/3130-543210

Preface

Coated grains have always attracted attention, at first of naturalists, and later of geologists, and the interest in these peculiar bodies was related both to their intriguing form and their significance in facies interpretation and sedimentology and to their relevance to accumulations of hydrocarbons and other mineral deposits. This resulted in numerous publications on this subject, and the intention of this volume is to summarize the present state of knowledge on coated grains. The idea of the book was to unite some general papers with papers reporting case studies of both recent and ancient coated grains. The organization of the book follows this intention.

The papers presented in this volume have been invited by the editor; the theme of the book merits a few words of personal history.

The development of studies of coated grains during the last two decades has not only resulted in a great increase in knowledge of recent and ancient environments of coated grain formation, but also numerous important and controversial questions of classification, environmental significance, mineralogical composition etc. of ancient coated grains have arisen. To answer these questions, in 1978 I started the study of many ancient and recent occurrences of coated grains at the Institut für Geologie, Ruhr-Universität Bochum, following the invitation of Hans Füchtbauer and sponsored by the Alexander von Humboldt-Stiftung. While setting together a part of my results in spring 1980 at the Geobiological Laboratory of Oldenburg University, I came to the conclusion that by reason of the time needed I would better serve the scientific community as editor of papers dealing mainly with case histories with which I am familiar thanks to earlier samples supplied by many colleagues. The idea of the volume as presented here found the support of Dieter Hohm of Springer-Verlag and of most of the friends and colleagues invited to contribute to the volume, who patiently suffered my (sometimes radical) intervention in their papers. Accordingly, my gratitude is extended to all friends and colleagues who collaborated on this book, as well as to those who did not contribute but had earlier furnished me with numerous samples of coated grains for comparative studies; to the Alexander von Humboldt-Stiftung for offering me a research fellowship for the study of coated grains and to the staff of Springer-

Verlag for their kind collaboration. Special thanks are addressed to Hans Füchtbauer, whose invitation to me to Bochum initiated this volume, and to Danusia for her constant understanding throughout.

Warszawa, Spring 1983 TADEUSZ M. PERYT

Contents

I Approaches

I.1 Classification of Coated Grains
T. M. Peryt 3

I.2 Classification of Coated Grains: Discussion
D. K. Richter 7

I.3 Origin of Coated Grains: Trace Element Constraints
U. Brand and J. Veizer (With 1 Figure) 9

I.4 Carbon and Oxygen Isotope Composition of Recent and Ancient Coated Grains
M. Magaritz (With 4 Figures) 27

I.5 Neritic Macroid Genesis, an Ecological Approach
L. Hottinger (With 30 Figures) 38

I.6 Accretionary Lapilli in Volcanic Ash Falls: Physical Factors Governing Their Formation
T. O. Reimer (With 5 Figures) 56

II Ooids

II.1 Calcareous Ooids: A Synopsis
D. K. Richter (With 4 Figures) 71

II.2 Ferriferous Ooids
M. M. Kimberley (With 3 Figures) 100

II.3 Cortical Fabrics in Calcite and Aragonite Ooids
D. A. Medwedeff and B. H. Wilkinson (With 2 Figures) .. 109

II.4 Relict Sand Bodies and Bedforms of the Northern Bahamas: Evidence of Extensive Early Holocene Sand Transport
A. C. Hine (With 13 Figures) 116

II.5 The Joulters Ooid Shoal, Great Bahama Bank
P.M. Harris (With 7 Figures) 132

II.6 Holocene Lacustrine Ooids from Pyramid Lake, Nevada
B.N. Popp and B.H. Wilkinson (With 4 Figures) 142

II.7 Subrecent High-Sr Aragonitic Ooids from Hot Springs Near Tekke Ilica (Turkey)
D.K. Richter and H. Besenecker (With 6 Figures) 154

II.8 Coated Grains Along the Dead Sea Shore
R.A. Garber and G.M. Friedman (With 4 Figures) 163

II.9 Structural and Geochemical Features of Jurassic Oolitic Limestones in the Veneto Region (NE Italy)
L. Scudeler Baccelle (With 4 Figures) 169

II.10 Sedimentary Characteristics of Oolitic Carbonates from the Jialing-Jiang Formation [Lower Triassic (T_1^2J1)], South Sichuan Basin, China
Yun Fu Zeng, Nan Hao Lee, and Yang Zhou Huang (With 8 Figures) 176

II.11 Ooid Zonation as Indication for Environmental Conditions in a Givetian-Frasnian Carbonate Shelf-Slope Transition
T.J.A. Reijers and A.H.M. ten Have (With 10 Figures) . 188

II.12 Ooids and Oolites of the Proterophytic Boomplaas Formation, Transvaal Supergroup, Griqualand West, South Africa
N.J. Beukes (With 7 Figures) 199

III Rhodoids

III.1 Description and Classification of Rhodoliths (Rhodoids, Rhodolites)
D.W.J. Bosence (With 4 Figures) 217

III.2 The Occurrence and Ecology of Recent Rhodoliths – A Review
D.W.J. Bosence (With 5 Figures) 225

III.3 Rhodoids in Temperate Carbonates from the Cenozoic of New Zealand
C.J. Burgess and J.M. Anderson (With 10 Figures) 243

III.4 Early Permian Coated Grains from a Lagoonal Environment, Laborcita Formation, Sacramento Mountains, Southcentral New Mexico, U.S.A.
D.F. Toomey (With 6 Figures) 259

IV Oncoids

IV.1 Oncoids: Comment to Recent Developments
T.M. Peryt ... 273

IV.2 Cyanoliths (Cyanoids): Oncoids Formed by Calcified Cyanophytes
R. Riding (With 5 Figures) ... 276

IV.3 Algal Micro-Reefs – Coated Grains from Freshwater Environments
J. Schneider, H.G. Schröder, and Th. Le Campion-Alsumard (With 7 Figures) ... 284

IV.4 Brackish-Water Oncoids Composed of Blue-Green and Red Algae from a Pleistocene Terrace Near Corinth, Greece
D.K. Richter and R. Sedat (With 4 Figures) ... 299

IV.5 Environmental Significance of Freshwater Oncoids, Eocene Guarga Formation, Southern Pyrenees, Spain
E. Nickel (With 15 Figures) ... 308

IV.6 Depositional Environment, Diagenesis and Reservoir Properties of Oncolitic Packstones, Macaé Formation (Albian-Cenomanian), Campos Basin, Offshore Rio de Janeiro, Brazil
A.V. Carozzi, F.U.H. Falkenhein, and M.R. Franke (With 7 Figures) ... 330

IV.7 Albian Pelagic Phosphate-Rich Macrooncoids from the Tatra Mts (Poland)
K.P. Krajewski (With 10 Figures) ... 344

IV.8 Oncoids and Stromatolites in the Rosso Ammonitico Sequences (Middle-Upper Jurassic) of the Venetian Alps, Italy
F. Massari (With 6 Figures) ... 358

IV.9 Pelagic Oncoids and Ooids in the Middle-Upper Jurassic of Eastern Sardinia
F. Massari and I. Dieni (With 6 Figures) ... 367

IV.10 Depositional Environments of Some Upper Jurassic Oncoids
K. Dahanayake (With 8 Figures) ... 377

IV.11 Carnian Oncolites of the Northern Calcareous Alps and the Drau Range (Austria)
G. Tichy (With 4 Figures) ... 386

IV.12 *Girvanella* Oncoids From Middle to Upper Triassic Allochthonous Boulders of the Dolomite Alps, Northern Italy
K. T. Biddle (With 6 Figures) 390

IV.13 Triassic Oncoids from Central Balkanides (Bulgaria)
G. A. Čatalov (With 7 Figures) 398

IV.14 The Genesis of Algal Nodule Limestones from the Upper Carboniferous (San Emiliano Formation) of N. W. Spain
M. B. J. Bowman (With 6 Figures) 409

IV.15 Morphogenesis of Oncoids in the Lower Carboniferous Llanelly Formation of South Wales
V. P. Wright (With 7 Figures) 424

V Vadoids

V.1 Vadoids
T. M. Peryt (With 7 Figures) 437

V.2 Recent Travertine Pisoliths (Pisoids) from Southeastern Idaho, U.S.A.
H. S. Chafetz and J. C. Meredith (With 4 Figures) 450

V.3 Pisoids in the Caliche Profiles of Tarragona (N. E. Spain)
F. Calvet and R. Julià (With 7 Figures) 456

V.4 Pisoliths (Pisoids) in Quaternary Travertines of Tivoli, Italy
R. L. Folk and H. S. Chafetz (With 17 Figures) 474

V.5 Quaternary Pisoids of Lau, Fiji
C. J. Burgess (With 11 Figures) 488

V.6 Pisoids and Pisolite Facies (Permian), Guadalupe Mountains, New Mexico and West Texas
M. Esteban and L. C. Pray (With 11 Figures) 503

V.7 Vadose Pisolites of the Tongying Formation (Upper Sinian System) in Southwest China
Cao Ruiji and Xue Yaosong (With 7 Figures) 538

V.8 A Proterozoic Calcrete in the Amos Formation, McArthur Group, Northern Territory, Australia
M. D. Muir (With 12 Figures) 548

VI Contrasted Occurrences

VI.1 Coated Grains from the Great Barrier Reef
P.G. Flood (With 4 Figures) 561

VI.2 Coated Grains Facies in the Lower Cretaceous of the Outer Dinarides (Yugoslavia)
J. Tišljar (With 6 Figures) 566

VI.3 Coated Grains in Contrasted Environmental Situations: Norian and Lower Liassic of Northern Apennines
G. Ciarapica and L. Passeri (With 10 Figures) 576

VI.4 Coated Grains from the Zechstein Limestone (Upper Permian) of Western Poland
T.M. Peryt (With 7 Figures) 587

VI.5 Unusual Algal-Crystalline Carbonate Coated Grains from the Capitan Reef (Permian, Guadalupian), New Mexico, USA
S.J. Mazzullo and J.M. Cys (With 4 Figures) 599

VI.6 Contrasted Occurrence of Lower Devonian Coated Grains, Northeastern Armorican Massif, France
J. Poncet (With 7 Figures) 609

VI.7 Accretionary Lapilli and Other Spheroidal Rocks from the Archaean Swaziland Supergroup, Barberton Mountain Land, South Africa
T.O. Reimer (With 10 Figures) 619

Author Index 635

Subject Index 647

Contributors

You will find the addresses at the beginning of the respective contributions

Anderson, J.M. 243
Besenecker, H. 154
Beukes, N.J. 199
Biddle, K.T. 390
Bosence, D.W.J. 217, 225
Bowman, M.B.J. 409
Brand, U. 9
Burgess, C.J. 243, 488
Calvet, F. 456
Cao Ruiji 538
Carozzi, A.V. 330
Čatalov, G.A. 398
Chafetz, H.S. 450, 474
Ciarapica, G. 576
Cys, J.M. 599
Dahanayake, K. 377
Dieni, I. 367
Esteban, M. 503
Falkenhein, F.U.H. 330
Flood, P.G. 561
Folk, R.L. 474
Franke, M.R. 330
Friedman, G.M. 163
Garber, R.A. 163
Harris, P.M. 132
Hine, A.C. 116
Hottinger, L. 38
Julià, R. 456
Kimberley, M.M. 100
Krajewski, K.P. 344
Le Campion-Alsumard, Th. 284
Magaritz, M. 27
Massari, F. 358, 367
Mazzullo, S.J. 599
Medwedeff, D.A. 109
Meredith, J.C. 450
Muir, M.D. 548
Nan Hao Lee 176
Nickel, E. 308
Passeri, L. 576
Peryt, T.M. 3, 273, 437, 587
Poncet, J. 609
Popp, B.N. 142
Pray, L.C. 503
Reijers, T.J.A. 188
Reimer, T.O. 56, 619
Richter, D.K. 7, 71, 154, 299
Riding, R. 276
Schneider, J. 284
Schröder, H.G. 284
Scudeler Baccelle, L. 169
Sedat, R. 299
ten Have, A.H.M. 188
Tichy, G. 386
Tišljar, J. 566
Toomey, D.F. 259
Veizer, J. 9
Wilkinson, B.H. 109, 142
Wright, V.P. 424
Xue Yaosong 538
Yang Zhou Huang 176
Yun Fu Zeng 176

I Approaches

I.1 Classification of Coated Grains

Tadeusz M. Peryt [1]

The term "coated grains" was coined by Wolf (1960) as a substitute for Folk's (1959) "oolites", "to include other concentrically formed materials such as pisolites" (Wolf 1960, p. 1415). Among the different terms proposed for specific kinds of these grains the term "ooid" has unquestionably the longest tradition in geological literature, going back to the 18th century, although the orthography as used here is of later date, having been introduced by Kalkowsky (1908). As ooids, "small egg-like grains, resembling the roe of a fish, each of which has usually a small fragment of sand as a nucleus, around which concentric layers of calcareous matter have accumulated" (Lyell 1855, p. 12) have been named. In 1916 Heim introduced a new term: "oncoid" (in original German orthography: Onkoid; "onchos" in Greek means "nodules"), for denoting a type of coated grain found by him in Mesozoic deposits of the eastern Swiss Alps which differed from an ooid in that the coated grain called oncoid possessed (Heim 1916, p. 566): (1) non-concentric overlapping of separate envelopes, (2) not exclusively rounded form, (3) compact structure, with or without nucleus, and (4) sharp or indistinct boundaries. As noted by Heim (1916), transitions between ooids and oncoids exist, and this is expressed in encrustation of ooids by irregular calcareous envelopes. This has been also noted by other authors (e.g., Fabricius 1967, Radwański 1968, Schuler 1968, cf. Richter, this vol.), but it simply means that the ooids formed nuclei of oncoids, just as the oncoids or their fragments could form nuclei of ooids (e.g., Radwański 1968), and therefore the conclusion that there is no clear limit between ooids and oncoids, as expressed by some workers, is ungrounded.

The term "oncoid" quickly became of common use in continental Europe for denoting all coated grains which were not ooids, as schematically shown in Table 1, and with some rather small modifications, the subdivision of coated grains into two morphological groups, ooids and oncoids, is preferred by numerous workers (e.g., Füchtbauer 1974, Flügel 1978, Purser 1980, Richter, this vol.). When used only descriptionally, this subdivision is very practical; however, unfortunately the oncoids were very early related to algae. It should be stressed that Heim (1916) – although suspecting responsibility of microorganisms for oncoid formation – did not define the type of microorganisms involved, and only after being distinguished by Pia (1927, p. 37) as a group "Oncolithi" within the

1 Instytut Geologiczny, Rakowiecka 4, 00-975 Warszawa/Poland

Coated Grains (ed. by T.M. Peryt)

Table 1. Comparison of classifications of coated grains

<table>
<tr><td colspan="6">Classification:</td></tr>
<tr><td>Continental</td><td colspan="2">Ooids</td><td colspan="3">Oncoids</td></tr>
<tr><td>Proposed</td><td>Ooids</td><td colspan="2">Vadoids</td><td>Oncoids</td><td>Rhodoids</td></tr>
<tr><td>Anglo-Saxon</td><td colspan="2">Ooids</td><td colspan="2">Pisoids</td><td>Rhodoids</td></tr>
</table>

Spongiostromata did the algal origin of oncoids become commonly accepted. Accordingly, a clearly different group of coated grains – rhodoids ["rhodolites" in the original orthography of Bosellini and Ginsburg (1971)] – has been incorporated into oncoids (e.g., Flügel 1978, p. 113; Purser 1980, p. 31). Perhaps even more incorrect was the theory that vadose deposits have been also incorporated into oncoids and were thought to be the algal deposits (see Peryt, this vol.). It should be noted that the term "oncoid" might be regarded as purely descriptional, like the term "stromatolites" as defined by Semikhatov et al. (1979), but evidently the algal origin of oncoids became very deeply rooted in geologists' minds.

In Anglo-Saxon literature, on the other hand, the term "oncoid" was rarely used except in the last two decades. Instead, the term "pisoid" has been widely used for denoting grains similar to the ooids "but less regular in form (commonly crenulated), and generally 2 mm or more in diameter" (Leighton and Pendexter 1962, p. 38) (Table 1). However, as noted by Richter (this vol.) and many earlier workers, there is no reasonable ground to base the classification of coated grains on the magic 2 mm boundary. In Anglo-Saxon literature the oncoids have been variously called: algal nodules, algal balls, algal-encrusted grains (Leighton and Pendexter 1962, p. 38), and Wolf (1965) distinguished a number of varieties within the grains which he called "sedimentary products of calcareous algae". In 1971, Bosellini and Ginsburg distinguished a new group of coated grains – "nodules and detached branch growths with a nodular form composed principally of coralline algae" – which they called "rhodoids" (rhodolites in the original definition, later changed into rhodoliths).

Both in continental and Anglo-Saxon literature the term ooids has been used for particles of clearly different origin (marine, lacustrine, fluvial, cave and artificial). This was strongly criticized by Fabricius (1977), who proposed reserving the term ooids for particles of marine origin; other coated grains with a radial crystalline cortex he placed in "spheroids". However, such subdivision was based on Fabricius' (1977) conviction that marine ooids are only primarily concentric forms (p. 37, 39), but as discussed by Richter (this vol.), radial ooids occur also in marine settings. Besides, as already stressed by Fabricius (1977, p. 34), gradual transitions between marine and lagoonal ooid-bearing environments exist. Nevertheless, it is of great importance that Fabricius (1977) tried to relate the distinguished types of coated grains with specific environments, although he related environments and structures and textures of coated grains too closely. The intention of the classification presented in Table 2 is to relate the environments with specific types of coated grains, and although the structures and tex-

Table 2. Classification of coated grains

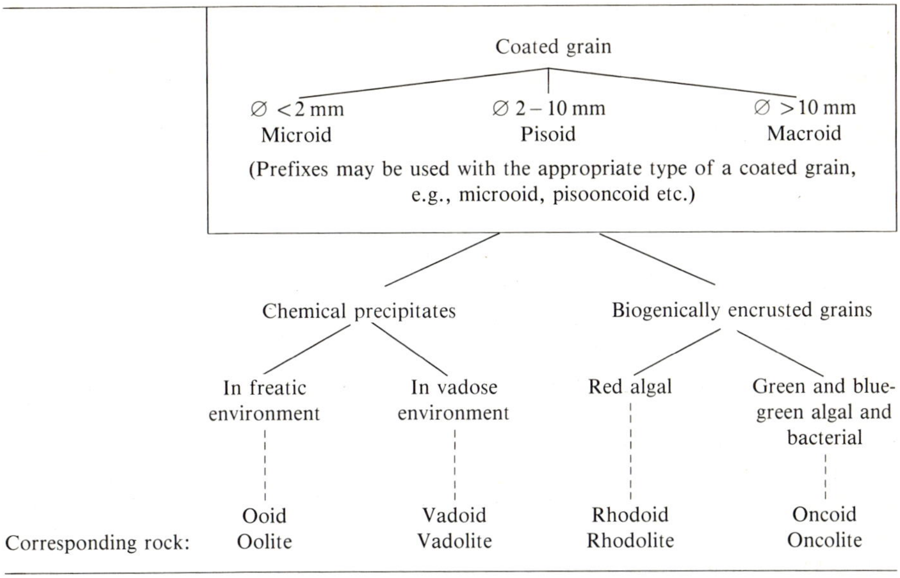

tures of coated grains are considered, they are thought to be nondiagnostic. This in turn is related to the fact that the same structures and textures of coated grains can originate in quite different geological settings.

Considering the origin of the cortex, coated grains may be subdivided into two groups: chemical precipitates and biogenically encrusted grains. Chemical precipitates are represented by ooids (which originate in the phreatic zone) and vadoids (which originate in the vadose zone). Biogenically encrusted grains are formed by algae, cyanobacteria and bacteria, and sporadically by other organisms which very often co-occur with algae and cyanobacteria [e.g., foraminiferal-algal oncoids – see Peryt (1977) and Catalov (this vol.)]. Biogenically encrusted grains are most frequently represented by rhodoids and oncoids. Rhodoids are formed by red algae; the classification of rhodoids has been presented by Bosence (this vol.). Oncoids comprise grains of suspected or inferred (non red-)algal, cyanobacterial and bacterial origin. There are two main groups of oncoids: spongiostromate oncoids and porostromate oncoids (Peryt 1981). When the oncoids are formed by cyanobacteria, they should be termed cyanoids, as suggested by Riding (1982). Oncoids originating as a result of the decay of organic matter are called tuberoids (Fritz 1958) and putroids (Matyja 1978).

When compared to the other classifications (Table 1), the proposed new one (Table 2) differs in narrowing the meaning of ooids. The use of this term is proposed to be restricted to (?bio)chemically precipitated coated grains in phreatic environments. The forms, which are often similar to some ooids but which originated in vadose environments, such as "cave pearls", "diagenetical ooids" and others, are included into vadoids. To vadoids belong also "vadose pisoids", "caliche pisoids", as well as hot spring ooids and pisoids and travertine pisoids (Peryt, this vol.).

The proposed meaning of the terms and the construction of classification stress the environmental approach to the classification. The distinguished types of coated grains are closely related to particular environments and therefore the classification itself may be regarded as genetical, but the particular types of coated grains may be easily distinguished during field work, although certainly explicitness as to the variety of any given type of coated grain and hence detailed environmental conclusions are only possible after laboratory examination of rock samples.

Acknowledgements. This study was supported by Alexander von Humboldt-Stiftung. Discussions with Hans Füchtbauer, Jean-Paul Loreau and Detlev Richter, and many other colleagues, are highly appreciated.

References

Bosellini A, Ginsburg RN (1971) Form and internal structure of recent algal nodules (Rhodolites) from Bermuda. J Geol 79:669 – 682

Bosence DWJ (this vol.) Description and classification of rhodoliths (rhodoids, rhodolites), p. 217 – 224

Catalov G (this vol.) Triassic oncoids from Central Balkanides (Bulgaria), p. 398 – 408

Fabricius F (1967) Die Rät- und Lias-Oolite der nordwestlichen Kalkalpen. Geol Rundschau 56:140 – 170

Fabricius F (1977) Origin of marine ooids and grapestones. Contr Sedimentology 7:1 – 113

Flügel E (1978) Mikrofazielle Untersuchungsmethoden von Kalken. Springer, Berlin Heidelberg New York

Folk RL (1959) Practical petrographic classification of limestones. Amer Assoc Petrol Geol Bull 43:1 – 38

Fritz GK (1958) Schwammstotzen, Tuberolithe und Schuttbrekzien im Weissen Jura der Schwäbischen Alb. Arb geol Inst TH Stuttgart, NF 13:1 – 118

Füchtbauer H (1974) Sediments and sedimentary rocks. Schweizerbart, Stuttgart

Heim A (1916) Monographie der Churfürsten-Mattstock-Gruppe. 3. Lithogenesis. Beitr Geol Karte Schweiz, NF 20:369 – 662

Kalkowsky E (1908) Oolith and Stromatolith im norddeutschen Buntsandstein. Z dt geol Ges 60:68 – 125

Leighton MW, Pendexter C (1962) Carbonate rock types. Amer Assoc Petrol Geol Mem 1:33 – 61

Lyell Ch (1855) A manual of elementary geology. 5th edition, London

Matyja BA (1978) Struktury wczesnodiagenetyczne w wapiennych utworach oksfordu. Prz geol 26:156 – 158

Peryt TM (1977) Environmental significance of foraminiferal-algal oncolites. In: Flügel E (ed) Fossil Algae. Springer, Berlin Heidelberg New York, p. 61 – 65

Peryt TM (1981) Phanerozoic oncoids – an overview. Facies 4:197 – 214

Peryt TM (this vol.) Vadoids, p. 437 – 449

Pia J (1927) Thallophyta. In: Hirmer M (ed) Handbuch der Paläobotanik, 1:31 – 136. Oldenbourg, München Berlin

Purser BH (1980) Sédimentation et diagenèse des carbonates néritiques récents. Editions Technip, Paris

Radwański A (1968) Studium petrograficzne i sedymentologiczne retyku wierchowego Tatr. Studia Geol Polonica 25:1 – 146

Richter DK (this vol.) Calcareous ooids: A synopsis, p. 71 – 99

Riding R (this vol.) Cyanoliths (cyanoids): oncoids formed by calcified cyanophytes, p. 276 – 283

Semikhatov MA, Gebelein CD, Cloud P, Awramik SM, Benmore WC (1979) Stromatolite morphogenesis – progress and problems. Can J Earth Sci 16:992 – 1015

Schuler G (1968) Lithofazielle, sedimentologische und paläogeographische Untersuchungen in der Raibler Schichten zwischen Inn und Salzach (Nördliche Kalkalpen). Erlanger geol Abh 71:1 – 60

Wolf KH (1960) Simplified limestone classification. Amer Assoc Petrol Geol Bull 44:1414 – 1416

Wolf KH (1965) Gradational sedimentary products of calcareous algae. Sedimentology 5:1 – 37

I.2 Classification of Coated Grains: Discussion

Detlev K. Richter [1]

In his paper on *Classification of coated grains*, Peryt (this vol.) subdivides the coated grains in ooids, vadoids, oncoids, and rhodoids. In my opinion, such a classification mixing descriptive (ooids, oncoids) with genetic terms of the same rank (vadoids, rhodoids) is inconvenient. Moreover, Peryt considers the "hot spring ooids" of Richter and Besenecker (this vol.) and some of the ooids discussed by Richter (this vol.) – thermal ooids, cave ooids, aqueduct ooids, and power plant ooids – as vadoids, though this is not justified in my opinion. This was the reason for preparing the following comment.

Ooids (from Greek oon = egg) and oncoids (from Greek onkos = nodule) differ, as is well known, mainly by a regular and irregular layering of the cortex, respectively. Moreover, in ooids no organisms are visible as participants in the construction of the ooidic laminae, though the ooid formation may be controlled by microorganisms. The latter is suggested for the Bahama-type ooids by the distribution and composition of organic matter and the uniform shape and size of nannograins and rods in these ooids (e.g. Richter, this vol.). This makes the classification of all ooids as "chemical precipitates" (Peryt, this vol., Table 2) questionable.

The introduction of the term "vadoid" (from Latin vadosus = shallow) may be meaningful for nodules in pedocretes, since the vadose zone is, compared with the phreatic zone, above the sea-, lake-, or groundwater-level (e.g. Land 1970, Schroeder 1973, Scholle and Kinsman 1974, Richter 1979). Considering the etymology of the term "vadoid", also the shallow-water ooids should be vadoids, which is of course not acceptable. Consequently, Peryt (this vol.) confines "vadoids" to coated grains of the vadose zone, in contrast with phreatic coated grains (Peryt, this vol., Table 2). Cave pearls, hot spring ooids and other coated grains of the phreatic environment (!), however, should then not be termed "vadoids". As descriptive nomenclature, even Sorby (1879) referred to the structural similarity of the hot spring ooids of Karlsbad (CSSR) and the Bahama ooids.

I prefer the descriptive subdivision of coated grains into ooids and oncoids, because it implies less mistakes and is more convenient in the field than the genetic classification. Moreover, both terms do not imply a genetical meaning.

1 Geologisches Institut, Ruhruniversität Bochum, Universitätsstraße 150, D-4630 Bochum-Querenburg, FRG

Coated Grains (ed. by T.M. Peryt)

Though Heim (1916) attributed the formation of oncoids to bacteria and many authors confine the term "oncoid" to coated grains of "(non red-) algal, cyanobacterial and bacterial origin" (Peryt, this vol.), I suggest a more general definition of oncoids, as Flügel (1978) did in his book. This definition would also include irregularly layered coated grains without organic structures such as spongiostromate oncoids sensu Peryt (1981), nodules in the vadose environment (e.g. pedocretes), and diagenetically altered other oncoids (rhodoids, bryoids, see below). Also for ooids, Richter (this vol.) suggested a descriptive subdivision based on structure and mineralogy, because similar ooids occur in rather different environments. Although a descriptive subdivision of the oncoids may be possible, a genetic subdivision is preferred for traditional and practical reasons (cf. Flügel 1978, p. 113). Accordingly, the oncoids may be subdivided as follows: cyanoids (= bacterial, blue-green algal oncoids; Riding, this vol.), chloroids (= green algal oncoids), rhodoids (= red algal oncoids, Peryt, this vol.; "rhodolites" of Bosellini and Ginsburg 1971), bryoids (= bryozoan oncoids, e.g. in Pleistocene rocks of the Channel of Corinthos/Greece), vadoids (Peryt, this vol. = irregularly laminated nodules of the vadose environment, e.g. in pedocretes), and others, though I do not favour these new terms.

Based on the descriptive ooid classification (Richter, this vol.) and the genetic oncoid classification discussed above, regularly layered coated grains in pedocretes would belong to the "ooids", while irregularly layered grains in the same environment would be termed "vadoids". In my opinion the vadoids should be included in the oncoid group, inasmuch as some of the vadoids contain Microcodium, a typical biogenic structure in calcretes (Klappa 1978). Such vadoids are common in Quaternary calcretes at different localities near Corinthos (Greece). These coated grains are in fact "biogenically encrusted", as required by Peryt (this vol., Table 1) for oncoids.

References

Bosellini A, Ginsburg RN (1971) Form and internal structure of recent algal nodules (rhodolites) from Bermuda. J Geol 79:669–682

Flügel E (1978) Mikrofazielle Untersuchungsmethoden von Kalken, 454 p. Springer, Berlin Heidelberg New York

Heim A (1916) Monographie der Churfürsten-Mattstock-Gruppe. 3. Teil: Lithogenesis. Beitr. Geol. Karte Schweiz, N. F. 20, Basel

Klappa CF (1978) Biolithogenesis of Microcodium: elucidation. Sedimentology 25:489–522

Land LS (1970) Phreatic versus vadose meteoric diagenesis of limestones: evidence from a fossil water table. Sedimentology 14:175–185

Peryt T (1981) Phanerozoic oncoids – an overview. Facies 4:197–214

Richter DK (1979) Die Stufen der meteorisch-vadosen Umwandlung von Mg-Calcit in Calcit in rezenten bis pliozänen Biogenen Griechenlands. N Jb Geol Paläont Abh 158:277–333

Scholle PA, Kinsman DJJ (1974) Aragonitic and high-Mg calcite caliche from the Persian Gulf – a modern analog for the Permian of Texas and New Mexico. J Sed Petrol 44:904–916

Schroeder JH (1973) Submarine and vadose cements in Pleistocene Bermuda reef rock. Sediment Geol 10:179–204

Sorby HC (1879) On the structure and origin of limestones. Quart J Geol Soc London XXXV:56–95

I.3 Origin of Coated Grains: Trace Element Constraints

Uwe Brand[1] and Ján Veizer[2]

1 Introduction

Coated grains, which may have skeletal cores, are non-skeletal carbonate and non-carbonate grains formed in marine as well as non-marine environments. These grains are classified according to their structure and environmental parameters (Peryt, this vol.). Peloids are micritic grains without recognizable structures and are represented by individual grains and aggregates. Ooids, oncoids, rhodoids, and vadoids are coated grains with distinct structural features.

In general, these grains consist of calcium carbonate. However, the halotites of the Dead Sea, which consist of sodium chloride (Weiler et al. 1974) are an exception. Non-marine coated grains can consist also of iron oxide (e.g., Pedro et al. 1978). Only coated grains consisting of aragonite, high-Mg calcite, low-Mg calcite and diagenetic low-Mg calcite will be discussed in this paper.

Carbonate-coated grains are composed of silicate and carbonate portions, or are entirely carbonate in composition. Major, minor and trace elements are contributed by both fractions. However, only carbonate-bound elements are diagnostic of original mineralogy, facies, and the pathway and degree of diagenetic alteration. Consequently, only these diagnostic elements will be discussed in the following presentation.

2 Incorporation of Minor and Trace Elements into Carbonate Minerals

Minor and trace elements can be incorporated into carbonate minerals in the following ways (McIntire 1963, Zemann 1969):

1. substitute for Ca^{2+} in the $CaCO_3$ lattice;
2. be incorporated interstitially between lattice planes;
3. occupy lattice positions which are free due to defects in the structure;
4. be adsorbed due to remanent ionic charges.

1 Department of Geological Sciences, Brock University, St. Catharines, Ontario, L2S 3A1, Canada
2 Department of Geology, University of Ottawa, Ottawa, Ontario K1N 6N5, Canada

Coated Grains (ed. by T.M. Peryt)

Rhombohedral calcite owing to its smaller lattice dimensions than orthorhombic aragonite incorporates preferentially cations with ionic radii smaller than that of Ca^{2+} (0.99 Å). Conversely, orthorhombic aragonite shows preference for larger cations (Graf 1960). Thus Fe^{2+}, Mg^{2+}, Mn^{2+} and Co^{2+} (0.66 – 0.80 Å) are more concentrated in calcite while Sr^{2+}, Pb^{2+} and Ba^{2+} (1.12 – 1.34 Å) are more concentrated in the aragonite polymorph of $CaCO_3$ (Zemann 1969).

The actual elemental concentration in $CaCO_3$ is controlled by a partition coefficient between solution and the particular mineralogical polymorph. This partition coefficient, k, is defined as follows (McIntire 1963, Kinsman 1969):

$$\left(\frac{m_T}{m_{Ca}}\right)_S = k\left(\frac{m_T}{m_{Ca}}\right)_L ,$$

where m indicates the molar ratio, T = tracer, S = solid (in this case $CaCO_3$) and L = liquid.

The understanding of elemental partitioning is at present incomplete and subject to considerable controversy (cf. Bathurst 1975, p. 265, Marrow and Mayers 1978, Angus et al. 1979, Land 1980, Lahann and Siebert 1981). This controversy is partly due to inconsistencies and differences in experimental techniques utilised for determination of these partition coefficients (Veizer 1978a). Additional difficulties arise from variations in physical and chemical properties in both experimental and natural environments. For instance, a fast crystal-growth rate results in entrapment of larger quantities of tracers in the $CaCO_3$ phase (Kitano et al. 1971). Since this growth rate is frequently a function of the physical and chemical properties of the environment, it can result in spurious correlations between temperature or salinity and the concentrations of tracers. These difficulties are compounded by the fact that the measured elemental concentrations are not necessarily those of tracers bound strictly in the $CaCO_3$ lattice positions. This is especially true for Na^+, which can also be present as NaCl in fluid inclusions or in entrapped aluminosilicate particles (Fritz and Katz 1972, Land and Hoops 1973, Veizer et al. 1977). Similarly, Fe^{2+} is usually derived in considerable amount from entrapped iron oxide coatings (Brand and Veizer 1980).

The above difficulties notwithstanding, it is useful to define probable theoretical concentrations of tracers for $CaCO_3$ in equilibrium with seawater at near-surface temperatures. Utilizing the partition coefficients of Kinsman (1969) and Kinsman and Holland (1969), calcite in equilibrium with seawater should contain about 1100 ppm Sr^{2+}, and aragonite about 8500 ppm Sr^{2+}. Conversely, using the partition coefficients of Katz et al. (1972) and Ichikuni (1973), the postulated Sr^{2+} content of calcite would be about 1/3 – 2 times that discussed above. However, natural materials, as will be also seen in the subsequent sections, do not support the validity of these k's (Veizer 1978b). The equilibrium concentration for Mg^{2+} should be about 2.5% – 7.5% (Winland 1969, Richter and Füchtbauer 1978), suggesting that high-Mg calcite is the stable mineralogical species in seawater. Predicted equilibrium concentrations in calcites for other trace elements of interest are: Na^+ ~150 – 300 ppm (Möller et al. 1976, White 1978), Fe^{2+} ~1 to $\lesssim$70 ppm (Veizer 1974, Richter and Füchtbauer 1978), Mn^{2+} ~30 ppm (Bodine et al. 1965, Michard 1968). The partition coefficients for aragonite (except for Na^+) are not known. Considering the lattice dimensions of aragonite, tracers such as Mg^{2+}, Fe^{2+} and Mn^{2+} should be present at concentration levels lower than those in calcites. The Na^+ equilibrium value for aragonite is about 2000 ppm (White 1977). Although k's for Zn^{2+}, Cd^{2+}, Cu^{2+}, Co^{2+}, F^{2-}, UO_2^{2+} and Ba^{2+} were also established (see Kitano et al. 1979, Ohde and Kitano 1980, Lorens 1981 for references), they will not be considered because no comparative data are known for coated grains.

Utilizing elemental concentrations in the *average* continental water (Turekian 1972, Table 6-3) and the above partition coefficients, the average compositions for fresh water calcites can also be calculated. Theoretically, freshwater calcites should contain about 260 ppm Sr^{2+}, 2200 – 6500 ppm Mg^{2+}, 3 – 5 ppm Na^+, 1.7% – 3.5% Fe^{2+} and about 3200 ppm Mn^{2+}. The Fe^{2+}, and probably the Mn^{2+}, values may be too high, possibly a consequence of an overestimate of the dissolved Fe^{2+} and Mn^{2+} in natural waters.

The above theoretical estimates provide a comparative basis for the subsequent discussion of natural constituents. The predicted concentrations are for equilibrium conditions that are not distorted by biogenic fractionation effects. If these effects were involved in formation of coated grains, they would add further complications to those already discussed at the beginning of this section.

3 Holocene Coated Grains

Coated grains, as mentioned in the introduction, are classified on the basis of their internal structure, or its absence, into (1) Ooids, (2) Vadoids, (3) Peloids, (4) Oncoids, and (5) Rhodoids. Chemical data are, however, available only for the first four coated grain types.

3.1 Ooids

Structurally, ooids are divided into several types such as (a) radial, (b) micritic, and (c) tangential. Some of them are known from both marine and non-marine environments.

Radial ooids are composed of crystals with their long axes arranged radially relative to the outer surface. Tangential ooids can be either high-Mg calcite (e.g., Baffin Bay, Texas) or aragonite (Umm Said, Quatar) (Table 1). The crystallites in micritic ooids, composed generally of aragonite, have no preferred orientation. In tangential ooids, which are similar in mineralogy to micritic ooids, the long axes of crystals are aligned parallel to the outer surface. However, most ooids reported in the literature include no descriptions of their structures.

3.1.1 Marine Ooids

A survey of the chemistry of marine ooids (Table 1) shows clearly a mineralogical control on the trace element distribution. Aragonitic ooids contain about

Table 1. Chemical composition of Holocene ooids

Type / Mineralogy / *Locality*	Structure	(%) Ca^{2+}	$\bar{x} \pm s$ (ppm) Mg^{2+}	Sr^{2+}	Na^{+}	Fe^{2+}	Mn^{2+}
Marine							
Aragonite							
Bahamas							
Hogsty Reef (7)		39.7	1100	10050 ± 70			
Bimini Lagoon (8)		38.0 ± 0.2	5970 ± 1030	9280 ± 390		10 ± 2	10 ± 4
Joulters, Sand Cays (9)			640 ± 110	9890 ± 480			
Bahama Bank (10)				5510 ± 3140*			
Bahama Bank (15)		39.9	1100			910	
Bahama Bank (22)		38.2	800	9900	2400	300	10
Inagua Bank (22)			3200	9400	4400	350	10
Joulters Cays (1)		38.0	1300	9600	1600		
Bahama Bank (3)		38.5	2050		2000	490	
Bahama Bank (4)			780	10570		60	10
Bahama Bank (6)		39.5 ± 0.3	2320 ± 350	9760 ± 150			
Great Bahama Bank (22)		37.8	500		3400		
Florida							
Florida Reef (15)		39.7				2940	
Texas							
Baffin Bay (2)	Tangential		870	10400			
Baffin Bay (2)	Micritic		870	10400			
Baffin Bay (2)			1230 ± 360	10050 ± 60			

Persian Gulf							
Gulf (9)			1330 ± 620	9590 ± 500			
Abu Dhabi (6)		38.0 ± 1.2	4290 ± 1490	9100 ± 800			
Sabkha Matti (6)		37.2 ± 1.6		9060 ± 800			
Umm Said (6)	Radial	38.1		8820			
India							
Continental Shelf (19)			1520 ± 420				
Average		38.6	1760	9720	2760	720	10
High-Mg Calcite							
Texas							
Baffin Bay (2)	Radial		37660 ± 3320	2020 ± 160			
Australia							
Great Barrier Reef (13, 23)			43250	600 – 750			
Average			40460	2020			
Non-marine							
Great Salt Lake (4)			7240	4230		770	20
Argus Range (5)		33.5 ± 4.6	3080 ± 1520		2460 ± 800	1760 ± 870	
Lake Chad (20)		1.2	7840			1000**	2000
Experimental (14)		36.5 ± 1.4	2420 ± 940	2910 ± 260			

Data from: (1) Halley and Harris 1979, (2) Land et al. 1979, (3) Newell et al. 1960, (4) Zeller and Wray 1956, (5) Hay and Wiggins 1980, (6) Loreau and Purser 1973, (7) Milliman 1967, (8) Till 1970, (9) Kinsman 1969, (10) Kahle 1965, (13) Marshall and Davies 1975, (14) Suess and Fütterer 1972, (15) Illing (1954), (19) Rao 1978, (20) Dupont and Lemoalle 1971, (22) Milliman 1974, (23) Richter 1979. The strontium value marked by asterisk (Kahle 1965) is excluded from the calculated average. The Fe concentration of the Lake Chad ooids (two asterisks) is for the carbonate part only

8000 – 10,000 ppm Sr^{2+}, which is in reasonably good accord with theoretical prediction. The slightly higher observed than predicted concentrations are partly a consequence of variable analytical techniques and partly, perhaps, due to a high rate of crystallization (Kitano et al. 1971). Thus marine ooids are precipitated in approximate inorganic equilibrium with ambient seawater. The only exception are the ooids from the Bahamas reported by Kahle (1965). His measured concentrations of about 5510 ppm Sr^{2+} (Table 1) are at variance with the other observations from the same general area. The cause of this discrepancy is either analytical error or partial recrystallization of the ooids. Similarly, high-Mg calcite ooids contain somewhat higher Sr^{2+} concentrations than predicted. This can be perhaps attributed to causes similar to those postulated for the aragonitic ooids. This discrepancy is slightly more disturbing (almost factor of 2) than in the case of aragonite ooids. On the other hand Sr^{2+} contents of 2000 – 4000 ppm are common for marine biogenic high-Mg calcites (see Dodd 1967, Milliman 1974, p. 144) and this may suggest that the measured values may not be atypical for marine high-Mg calcites. However, this remark should not be construed as a proof of biogenic influence on the formation of high-Mg calcite ooids.

The Mg^{2+} content of high-Mg calcite ooids is within the predicted equilibrium range. This is consistent with the postulated mode of direct inorganic precipitation of these ooids from ambient seawater. The observed Mg^{2+} content of aragonitic ooids is, in agreement with theory, lower than that predicted for calcitic ooids. However, due to our ignorance of k, no further discussion concerning their origin is advisable. Accepting that the aragonitic ooids were precipitated in chemical equilibrium with seawater, that their Mg^{2+} content is about 2500 ppm (range 800 – 6000), and that Mg^{2+} and Ca^{2+} contents of seawater are 1290 and 411 ppm, respectively (Turekian 1972, Table 6-3), then the calculated $k^{Mg}_{aragonite}$ is about 2×10^{-3} (range $0.6 - 4.8 \times 10^{-3}$).

The observed Na^{+} concentrations in the aragonitic ooids are similar to, or slightly higher than, those predicted from partitioning theory. The higher values may be due to the presence of sodium in non-lattice positions, for instance in fluid inclusions.

The Mn^{2+} contents observed in the aragonitic ooids are about 1/2 – 1/3 of those predicted for marine calcite. Since the aragonite structure discriminates against manganese, these concentrations are well within the expected theoretical range. The results suggest, therefore, that $k^{Mn}_{aragonite}$ is about 1/3 to 1/2 of that for calcite. However, considering analytical and other problems, this value must be regarded as tentative only.

The large variation in Fe^{2+} concentrations of aragonitic ooids (10 – 3000 ppm) shows that its distribution is controlled by several factors. The lower values are within the order of magnitude expected for pure $CaCO_3$. Conversely, the higher concentrations reflect probably iron which is present in iron oxide coatings and/or associated with aluminosilicate phases.

3.1.2 Origin of Marine Ooids

Summarizing the above discussion, one can accept that the trace element composition of marine ooids is consistent with their direct inorganic precipitation from ambient seawater. The data do not necessarily exclude all organic roles in their formation, but any such influence should not appreciably disturb the inorganic partitioning equilibria between ambient water and the precipitated $CaCO_3$ phase. At this time we have sufficient data only for Sr^{2+} and Mg^{2+} partitioning by different phyla (see Dodd 1967, for review). These results suggest that only the lower phyla (e.g. algae) can *perhaps* play some role in the formation of marine ooids. However, it must be stressed again that such a role, while broadly admissible, is not required by the chemistry of marine ooids.

In the present state of the art it is not clear what processes control the nucleation and the subsequent growth of ooids and what the factors are controlling their mineralogy. The mechanical "snowball" theory, with accumulation of aragonite and mud-needles around a nucleus (Sorby 1879), was negated by observations in natural environments (Kahle 1974, Sandberg 1975, Land et al. 1979) and by experimental work (e.g., Suess and Fütterer 1972). The alternative, algal, origin also appears to be at variance with the trace element data. Red algae precipitate high-Mg calcite with higher Na^+ concentrations than those observed in the ooids (~5000 vs. ~2000 ppm) (cf. Milliman 1974, Table 15; Table 1, this vol.). The aragonite precipitated by green algae is, in general, depleted by about 10% in Sr^{2+} relative to ooid Sr^{2+} content (cf. Milliman 1974, Table 18; Table 1, this vol.). Petrographic and isotopic observations cast additional doubt on the postulated algal origin of the ooid coatings (Bathurst 1975, Chap. 7; Land et al. 1979).

The theories invoked to explain the mineralogy of ooids also appear unsatisfactory. It has been proposed that (a) changes in Mg/Ca ratio of seawater, either locally or through geologic time (Sandberg 1975, Mackenzie and Pigott 1981), (b) substrate or (c) salinity controlled the mineralogy and texture of ooid coatings. The existence of both high-Mg calcitic and aragonitic ooids in marine environments with similar Mg/Ca ratios and with analogous substrates argues against alternatives (a) and (b) (Davies and Martin 1976, Land et al. 1979). The discovery of ooids with radial texture and with both mineralogies in normal marine as well as hypersaline environments (Davies and Martin 1976, Loreau and Purser 1973, Marshall and Davies 1975, Land et al. 1979) argues against alternative (c) as the major controlling factor.

An indirect biological control, probably by humic and other organic acids, can perhaps account for the mineralogical mode of the ooid coatings (Kahle 1974, Land et al. 1979). Their texture and possibly trace element composition can be, in turn, a function of the nucleation rate; the latter being dependent on the agitation of the ambient water. The formation of radial and micritic ooids tends to be favoured by quiet water conditions. Conversely, tangentially coated ooids tend to form in agitated waters (Loreau and Purser 1973, Kahle 1974, Land et al. 1979). Finally, a fast nucleation rate may explain also the observed 10% enrichment in the Sr^{2+} content of Holocene ooids (Kitano et al. 1971).

3.1.3 Non-Marine Ooids

Only few chemical results are available for non-marine ooids. These ooids consist of either calcium carbonate or goethite (McGannon 1975). Due to the absence of data for ambient waters it is not possible to ascertain whether the aragonitic ooids of the Great Salt Lake (Table 1) are formed by an equilibrium chemical process. Similarly, the chemical compositions of Argus Range and Lake Chad ooids (Table 1) are presented only for the sake of completeness. No interpretation is possible due to the lack of fundamental data.

3.1.4 Experimental Ooids

Experimentally precipitated ooids from supersaturated artificial seawater resemble naturally formed ooids in mineralogy and major element chemistry (Table 1). Their low Sr^{2+} content – by a factor of 3 – can be related to compositional differences between seawater and the artificial seawater of Suess and Fütterer (1972). Although low in Sr^{2+} content, the experimental ooids, according to Suess and Fütterer (1972, Table 1), were precipitated in chemical equilibrium with the utilized artificial solution.

3.2 Vadoids

3.2.1 Caliche Vadoids

Caliche vadoids are non-skeletal coated grains which form in hypersaline beachrock environments. According to Scholle and Kinsman (1974) the vadoids of the Persian Gulf form at a subaerially exposed sediment surface in the vadose and phreatic zones of the sabkhas with pore water close to halite saturation. The vadoids consist of high-Mg calcite and/or aragonite and their mineralogy is controlled by differences in the pore water chemistry.

The chemistry of caliche vadoids (Table 2) reflects the fact that they are a mixture of aragonite and high-Mg calcite. However, the calculated values for the aragonite and high-Mg calcite end-members are chemically identical to those of their ooid counterparts. This suggests that vadoids, similar to ooids, precipitat-

Table 2. Chemical composition of Holocene vadoids

Type Locality	Mineralogy	(%)	$\bar{x} \pm s$ (ppm)				
		Ca^{2+}	Mg^{2+}	Sr^{2+}	Na^{+}	Fe^{2+}	Mn^{2+}
Caliche							
Persian Gulf (11)	HMC, A	35.7 ± 1.1	33130 ± 6100	5000 ± 990	1380 ± 280		
Persian Gulf (11)	A, HMC	34.9 ± 4.7	8860 ± 3890	8520 ± 1010	2810 ± 190		
Persian Gulf (12)	A, HMC		28830	9040			
Cave							
Missouri (4)	A		180	210		20	< 10
Missouri (4)	HMC		60310	90		20	< 10

Data from: (4) Zeller and Wray 1956, (11) Scholle and Kinsman 1974, (12) Purser and Loreau 1973. Mineralogy: A = aragonite, HMC = high-Mg calcite, LMC = low-Mg calcite, dLMC = diagenetic low-Mg calcite, D = dolomite ed in inorganic chemical equilibrium with ambient seawater or with little altered brines of seawater derivation.

3.2.2 Cave Vadoids

Cave vadoids, although precipitating from $CaCO_3$ supersaturated solutions, require somewhat specific physical conditions for their formation. They need detrital material to act as nuclei for calcium carbonate precipitation and require a splash cup which keeps the vadoid from vacating the growth-intensive environment (Donahue 1969). In addition, agitation is required to prevent cave vadoids from becoming stalagmites and at the same time to expose the grain surface intermittently to the growth-promoting solution. Murray (1954) demonstrated that high temperatures, high Mg^{2+} (and possibly Sr^{2+}), and low HCO_3^- contents favour the precipitation of aragonitic cave vadoids.

The measured Mg^{2+}, and to some extent Sr^{2+}, contents of the cave vadoids (Table 2) are direct reflections of their mineralogy. The low Sr^{2+} content of these vadoids shows that they precipitated from solutions with $^{m}Sr/^{m}Ca$ of $2.7-7.3\times10^{-4}$. Such solutions can be formed by groundwater dissolution of limestones with 240 – 640 ppm Sr^{2+}; this range is frequently encountered in old stabilized limestones of low-Mg calcite mineralogy (cf. Veizer 1978b). Such a conclusion for cave vadoids is not surprising; indeed it is almost mandatory.

3.3 Peloids

3.3.1 Marine Peloids

Two competing hypotheses attempt to explain the origin of marine peloids. These are (a) aragonite mud pressed and bonded into a round shape by passage through the gut of mud-feeding organisms (Milliman 1974) and (b) micritization of skeletal debris by endolithic algae (Bathurst 1975). However, the latter alternative is not conducive to formation of well-rounded peloids.

The studied marine peloids, individuals and aggregates, from the Bahamas and the Persian Gulf consist of aragonite. Their chemical composition (Table 3), apart from a slightly higher Mg^{2+} content, is similar to that of the marine aragonite ooids. This suggests a direct precipitation from ambient seawater for the marine peloids; a view supported also by stable isotope data (Lowenstam and Epstein 1957). These results are in favour of the alternative (a) for the origin of at least those peloids which were studied chemically.

3.3.2 Non-Marine Peloids

Nontronite peloids are forming in sediments of Lake Chad (Pedro et al. 1978). Formation of these peloids is a two-stage process. In the first stage, iron precipitates in the interior of the grain and proceeds outward. In the second stage, silica migrates into the peloid and recombines with the iron to form nontronite.

Table 3. Chemical composition of Holocene peloids

Type Locality	Mineralogy	(%)	$\bar{x} \pm s$ (ppm)				
		Ca^{2+}	Mg^{2+}	Sr^{2+}	Na^{+}	Fe^{2+}	Mn^{2+}
Marine							
Bimini Lagoon, Bahamas (8)	A	37.4	4410	8710		30	30
Hog Cay, Bahama Bank (9)	A	38.1	4180	9480 ± 300			
Hogsty Reef, Bahamas (7)	A	38.3 ± 0.1	3900 ± 710	9520 ± 600			
Great Bahama Bank (22)	A	36.7 ± 2.6	3000 ± 1560	9000 ± 1400	2300	780 ± 720	50 ± 50
Bahamas (22)	A	36.8 ± 2.1	5150 ± 1630	9450 ± 350	3550 ± 490	290	50
Persian Gulf (6)	A	34.5 ± 2.4	9050 ± 4740	9750 ± 330			
Persian Gulf (9)	A			9390 ± 500			
	Average	36.9	4950	9330	2930	370	40
Non-Marine							
Argus Range (5)	LMC	35.0 ± 3.8	5750 ± 610		2240 ± 1300		
Lake Chad (16)	Fe oxide	1.9 ± 0.3	6780 ± 2920		1010 ± 290	279890 ± 86500	

Data from: (5) Hay and Wiggins 1980, (6) Loreau and Purser 1973, (8) Till 1970, (9) Kinsman 1969, (7) Milliman 1967, (16) Pedro et al. 1978, (22) Milliman 1974. Mineralogy as in Table 2

Table 4. Chemical composition of Holocene oncoids

Locality	Mineralogy	(%)	$\bar{x} \pm s$ (ppm)	
		Ca^{2+}	Mg^{2+}	Sr^{2+}
Lake Constance (18) (Germany)	LMC	35.9 ± 3.6	2780 ± 320	830 ± 160

Data from: (18) Müller 1968. Mineralogy as in Table 2

The absence of fundamental data on water chemistry, partition coefficients, etc., precludes any comment on the chemistry (Table 3) and origin of fresh water low-Mg calcite and nontronite peloids.

3.4 Oncoids

3.4.1 Marine Oncoids

Marine oncoids are disrupted algal mats shaped by the agitation of water. They have a low ratio of carbonate sediment to organic matter (Kinsman 1969, Scholle and Kinsman 1974), and the trapped sediment is usually similar to the high-Mg calcite or aragonite substrate. Consequently, their chemistry should be similar to that of the inorganically precipitated aragonite (and/or high-mg calcite) lime-muds (Milliman 1974). No data are presently available to verify this assertion.

3.4.2 Non-Marine Oncoids

The non-marine oncoids from Lake Constance (Germany) consist of low-Mg calcite (Table 4). The lake carbonate sand, which is derived from diminuted alga skeletal particles, is also low-Mg calcite. In contrast, pelecypods and gastropods consist of aragonite (Müller 1968). This suggests that the trapped material of Lake Constance oncoids was also derived from the above algal source.

The chemistry of the oncoids (Table 4) is consistent either with direct precipitation or with a biologically mediated precipitation without distortion of inorganic partitioning of tracers from Lake Constance water. The high Sr^{2+} content of the lake oncoids (830 ppm) (Table 4) is due to the high Sr^{2+} content of lake water. The source for this is the unusually high strontium content of the Alpenrhein water (Müller 1968, Table 3). These oncoids cannot be derived from diminuted molluscan debris (gastropods, $\bar{x}$ = 1140 ppm Sr^{2+}; pelecypods, $\bar{x}$ = 1730 ppm Sr^{2+}). Instead, the mineralogy and Sr^{2+} content suggest a freshwater algal source (~920 ppm Sr^{2+}; Müller 1968) for the calcitic Lake Constance oncoids.

4 Diagenetic Repartition of Minor and Trace Elements in Carbonates

Diagenetic alteration of coated grains, and of all other carbonate constituents, proceeds through simultaneous changes in texture and mineralogy as well as in elemental and isotopic chemistry (Brand and Veizer 1980, 1981). These changes are accompanied by diagenetic alteration, compaction and cementation of their enclosing matrix (e.g., Meyers 1980).

A $CaCO_3$-metastable phase (aragonite, high-Mg calcite) precipitated originally in equilibrium with contemporaneous seawater will eventually be exposed to meteoric water in which it will dissolve, equilibrate (partially or fully) and reprecipitate as a stable low-Mg calcite (dLMC) or dolomite. Both, the dissolving phase and the intervening meteoric water contribute ions to the localized microenvironment (cf. Brand and Veizer 1980) in which this mineralogical stabilization occurs. The relative contribution of ions from both sources depends on the water/rock ratio of the diagenetic system; high ratios define open systems while low ratios typify closed systems. However, the same diagenetic system (microenvironment) can be relatively open for some elements, and at the same time relatively closed for others (cf. Pingitore 1982). This depends on the respective proportions of the element derived from the dissolving metastable phase(s) and from the intervening meteoric water. Due to (a) difference in chemical compositions of sea vs. meteoric water, and (b) trace element partitioning coefficients, the above described process will lead to repartitioning of tracers in the precipitated diagenetic low-Mg calcite (dolomite). In general, Mn^{2+} and Fe^{2+} are higher and Sr^{2+}, Na^{+} and Mg^{2+} lower in meteoric than in seawater (cf. Turekian 1972, Table 6-3). In addition, the first group of trace elements has partition coefficients >1 and the second group <1. Because of (a) and (b) post-depositional recrystallization (at whatever stage) will cause an increase for the first group and decrease for the second group of tracers in the newly precipitated phase(s). However, Mg^{2+} shows a diagenetic depletion only if the original mineralogical phase was high-Mg calcite. If, on the other hand, the original mineralogical phase was aragonite diagenetic stabilization may lead to an increase in the Mg^{2+} content of the successor low-Mg calcite.

In general, the bigger the sea vs. meteoric water compositional gradient, the larger the deviation of a partition coefficient from unity, and the higher the water/rock ratio the more pronounced the tracer displacement will be. In a completely open diagenetic system the elemental content should approach chemical composition predicted for a non-marine calcite (cf. Sect. 2). This is rarely the case, since the precursor phase generally exerts some degree of control on the chemistry of the diagenetic microenvironment. In addition, the rate of approach towards meteoric equilibrium appears to be a function of the thermodynamic stability of the original mineralogical phase(s) (Brand and Veizer 1980). Thus, all diagenetically modified carbonate components should display the advocated Mn^{2+} and Fe^{2+} increases and Sr^{2+}, Na^{+} and possibly Mg^{2+} decreases, but the magnitude of this shift will depend on the previously discussed factors and on their original mineralogy (see Pingitore 1978, 1982; Brand and Veizer 1980 for further details). As an example, strontium and manganese are associated mostly with the carbonate lattice and have widely divergent partition coefficients of opposing signs (0.14 vs. ~17). In addition, their concentration ratios to Ca^{2+} in sea and meteoric water are different by orders of magnitude. Consequently, they should serve as particularly suitable tracers for delineating the path and degree of diagenetic stabilization. The composition of the diagenetically stabilized pre-Holocene low-Mg calcitic coated grains must be evaluated, therefore, with the above considerations and constraints in mind.

5 Ancient Coated Grains

Geochemical data for ancient oolites are available from formations covering most of the Phanerozoic. However, these are mostly Sr^{2+} concentrations, with Mn^{2+}, Mg^{2+} and Ca^{2+} values reported only sporadically (Table 5). This table and the accompanying Fig. 1 show that oolites depict diagenetic chemical signatures similar to those of other carbonate constituents. For instance, there is a sharp drop in the strontium content between Holocene ooids and Pleistocene and older oolites (Fig. 1). Apart from a few exceptions, the older oolites contain, on average, about 300 ppm Sr^{2+} (Table 5), with no overall Sr^{2+} decrease with increasing age. This shows that a carbonate component, once stabilized, is not usually involved in perpetual dissolutions and recrystallizations (cf. Bathurst 1975, p. 425; Veizer 1977); a statement contrary to the generally held consensus by the geologic fraternity.

Of special interest are Sr^{2+} highs in the Pleistocene, Eocene and the Pennsylvanian. The higher Sr^{2+} content of the Pleistocene oolites is related to a partial preservation of the original metastable aragonite. However, this explanation is not applicable to their Eocene and Pennsylvanian counterparts. Similar Sr^{2+} enrichments were also noted by Kahle (1965) for diagenetically altered and by

Table 5. Chemical composition of ancient oolites

Age	Mineralogy	(%)	$\bar{x} \pm s$ (ppm)		
		Ca^{2+}	Mg^{2+}	Sr^{2+}	Mn^{2+}
Pleistocene (17)	A, dLMC		2000	4300	10
Pleistocene (17)	dLMC		550 ± 300	1100 ± 310	10 ± 2
Pleistocene (10)	dLMC			280	
Eocene (10)	dLMC, D			1340 ± 1000	
Cretaceous (10)	dLMC			310 ± 50	
Jurassic (10)	dLMC			310 ± 30	
Jurassic (21)	dLMC	26.79	49450	260	
Pennsylvanian (10)	dLMC			950 ± 360	
Mississippian (10)	dLMC			330 ± 60	
Silurian (10)	dLMC			270 ± 30	
Cambrian (10)	dLMC, D			250 ± 20	

Data from: (10) Kahle 1965, (17) Stehli and Hower 1961, (21) Bordine 1965. Mineralogy as in Table 2

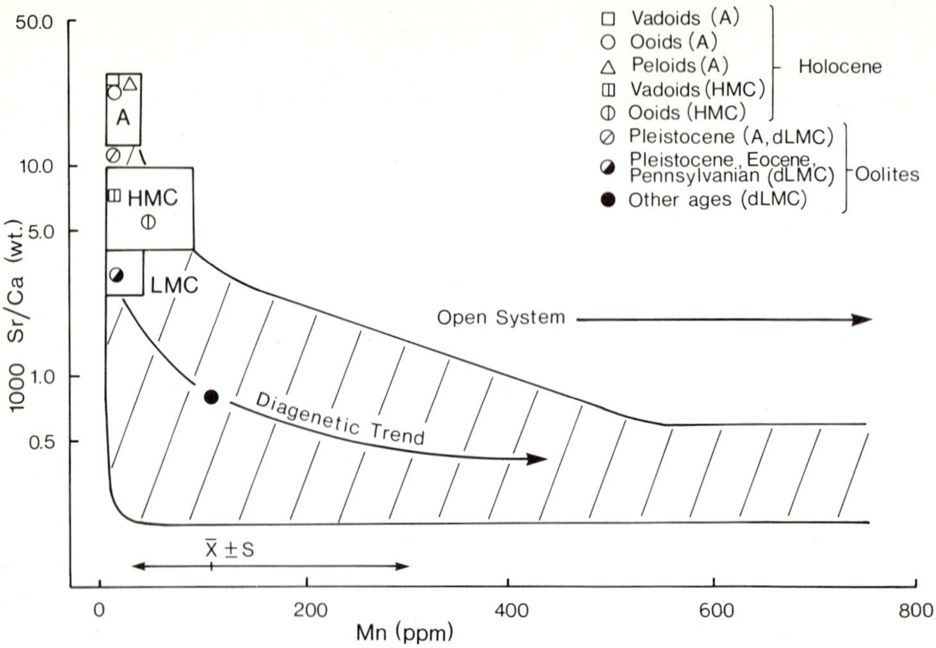

Fig. 1. Summary diagram of the diagenetic chemical trend for coated grains. (Modified from Brand and Veizer 1980). Manganese values for pre-Pleistocene oolites are assumed to be comparable to those of ancient carbonate components ($\bar{x} \pm s$ in Brand and Veizer 1980). Mineralogy as in Table 2

Brand (1982) for unaltered Pennsylvanian fossil components. In the present state of the art it is difficult to decide whether this Sr^{2+} enrichment is a consequence of better post-depositional preservation of their chemistry, related to some original feature (e.g., mineralogy, water chemistry), or a combination of both. The discussion of Holocene examples demonstrated that ooids are either aragonite or high-Mg calcite; both eminently unstable phases. In view of $k^{Sr}_{calcite} \sim 0.14$, it is unlikely that originally high-Mg calcite ooids could retain – after conversion into diagenetic low-Mg calcite in a meteoric or mixed water environment – $\gtrsim 50\%$ of their original Sr content. In fact, the observed concentrations in these oolites (950–1340 ppm Sr^{2+}) are still high enough to be comparable to the equilibrium value for calcite precipitated from seawater. For these reasons we believe that the high-Sr oolites had to start with Sr^{2+} content much higher than that of the high-Mg calcite (1200–2000 ppm) and were, therefore, of original aragonitic mineralogy ($\sim$9000 ppm Sr^{2+}).

The low Sr^{2+} oolite group ($\sim$300 ppm) can be interpreted either as (a) originally aragonitic ooids, which underwent extensive post-depositional alteration, or (b) originally high-Mg calcite ooids with post-depositional alteration similar to that of the high-Sr^{2+} group. The advocated original calcitic mineralogy for most of the Paleozoic oolites (ooids), regardless whether caused by secular changes in Mg/Ca ratio of seawater (Sandberg 1975, Mackenzie and Pigott 1980) or by other factors, would be consistent with the alternative (b). However, in the absence of distinctive data, we prefer to defer our judgement on the above options.

6 Conclusions

1. Holocene marine ooids are composed of aragonite or high-Mg calcite. Their trace element composition is consistent with direct inorganic precipation from ambient seawater. The slight enrichment (~10%) in strontium may be a consequence of kinetic effects; possibly a high rate of nucleation. An indirect biologic influence (e.g., the role of humic acids) on the formation of ooids can neither be confirmed not discounted by the presently available data.
2. Caliche vadoids and peloids of marine origin have trace element composition similar to that of the marine ooids. This suggests a comparable chemical mode of formation.
3. Cave vadoids contain strontium concentrations typical of calcium carbonate precipitated from meteoric waters in equilibrium with mineralogically stabilized ancient limestones.
4. Freshwater oncoids have trace element composition consistent with inorganic (or organically mediated, but without distortion of trace element partitioning) equilibrium with the ambient lake water. Their chemistry and mineralogy suggest an algal source for the entrapped sediments of oncoids.
5. The diagenetic mineralogical transformation of ooids into low-Mg calcite oolites results in a progressive decrease in their strontium content. This decrease follows the previously described (Brand and Veizer 1980) diagenetic trends for allochems and matrix in carbonate sediments.
6. Exceptionally high strontium concentrations (950–1340 ppm) in some ancient oolites indicate that they were originally aragonitic. However, most oolites have strontium concentrations of about 300 ppm. The data are insufficient to decide whether this low Sr content is a consequence of a more prevasive post-depositional alteration or of their original high-Mg calcitic mineralogy.
7. Many of the questions discussed in the previous contribution have been recently summarized in the compendium of Loreau (1982). This contribution was not available to us at the time of writing and the reader is advised to consult his findings.

Acknowledgements. We thank R. C. G. Bathurst for the review of the manuscript, F. Gombos for drafting the figure, and H. J. De Gouffe for typing the manuscript. Financial support was provided by the President's Grant of Brock University (to U. Brand), and by the National Sciences and Engineering Research Council of Canada (to J. Veizer).

References

Angus JG, Raynor JB, Robson M (1979) Reliability of experimental partition coefficients in carbonate systems: evidence for inhomogeneous distribution of impurity cations. Chem Geol 27:181–205

Bathurst RCG (1975) Carbonate sediments and their diagenesis. Elsevier, Amsterdam, 658 p

Bodine MW, Holland HD, Borcsik, M (1965) Coprecipitation of manganese and strontium with calcite. In: Štemprock M, Rieder M (eds) Symposium on problems of postmagmatic ore deposition, vol II. ČSAV, Prague, pp 401–406

Bordine BW (1965) Paleoecologic implications of strontium, calcium, and magnesium in Jurassic rocks near Thistle, Utah. Brigham Young Univ. Geol Stud 12:91 – 120

Brand U (1981) Mineralogy and chemistry of the lower Pennsylvanian Kendrick fauna, eastern Kentucky, 1: trace elements. Chem Geol 32:1 – 16

Brand U, Veizer J (1980) Chemical diagenesis of a multicomponent carbonate system – 1: trace elements. J Sediment Petrol 50:1219 – 1236

Brand U, Veizer J (1981) Chemical diagenesis of a multicomponent carbonate system – 2: stable isotopes. J Sediment Petrol 51:987 – 997

Davies PJ, Martin K (1976) Radial aragonite ooids, Lizard Island, Great Barrier Reef, Queensland, Australia. Geology 4:120 – 122

Dodd JR (1967) Magnesium and strontium in calcareous skeletons – a review. J Paleontol 41:1313 – 1329

Donahue J (1969) Genesis of oolite and pisolite grains: an energy index. J Sediment Petrol 39:1399 – 1411

Dupont B, Lemoalle J (1971) Les oolithies et pseudo-oolithes ferrugineuses du Lac Tchad. Soc Geol Fr C R 6:309 – 310

Fritz P, Katz A (1972) The sodium distribution of dolomite crystals. Chem Geol 10:237 – 244

Graf DL (1960) Geochemistry of carbonate sediments and sedimentary carbonate rocks – 1. Ill State Geol Surv Circ 297:1 – 39

Halley RB, Harris PM (1979) Fresh-water cementation of a 1,000-year-old oolite. J Sediment Petrol 49:969 – 988

Hay RL, Wiggins B (1980) Pellets, ooids, sepiolite and silica in three calcretes of the southwestern United States. Sedimentology 27:559 – 576

Ichikuni M (1973) Partition of strontium between calcite and solution: effect of substitution by manganese. Chem Geol 11:315 – 319

Illing LV (1954) Bahaman calcareous sands. Am Assoc Petrol Geol Bull 38:1 – 95

Kahle CF (1965) Strontium in oolitic limestones. J Sediment Petrol 35:846 – 856

Kahle CF (1974) Ooids from Great Salt Lake, Utah, as an analogue for the genesis and diagenesis of ooids in marine limestones. J Sediment Petrol 44:30 – 39

Katz A, Sass E, Starinsky A, Holland HD (1972) Strontium behaviour in the aragonite-calcite transformation: an experimental study at 40 – 98 °C. Geochim Cosmochim Acta 36:481 – 496

Kinsman DJJ (1969) Interpretation of Sr^{2+} concentrations in carbonate minerals and rocks. J Sediment Petrol 39:486 – 508

Kinsman DJJ, Holland HD (1969) The co-precipitation of cations with $CaCO_3$ – IV. The co-precipitation of Sr^{2+} with aragonite between 16 and 96 °C. Geochim Cosmochim Acta 33:1 – 18

Kitano Y, Kanamori N, Oomori T (1971) Measurements of distribution coefficients of strontium and barium between carbonate precipitate and solution – Abnormally high Sr values of distribution coefficients measured at early stages of carbonate formation. Geochem J 4:183 – 206

Kitano Y, Tokuyama A, Arakaki T (1979) Magnesian calcite synthesis from calcium bicarbonate solution containing magnesium and barium ions. Geochem J 13:181 – 185

Lahann RW, Siebert RM (1981) A kinetic model for distribution coefficients and application to Mg-calcite. Abstr Geol Soc Can Annu Meet, Calgary, p A-33

Land LS (1980) The isotopic and trace element geochemistry of dolomite: the state of the art. SEPM Spec Publ 28:87 – 110

Land LS, Hoops GK (1973) Sodium in carbonate sediments and rocks: a possible index to the salinity of diagenetic solutions. J Sediment Petrol 43:614 – 617

Land LS, Behrens EW, Frishman SA (1979) The ooids of Baffin Bay, Texas. J Sediment Petrol 49:1269 – 1278

Loreau J-P (1982) Sediments aragonitiques et leur genèse. Mémoires du Muséum National d'Histoire Naturelle, Série C, 47, 312 pp

Loreau J-P, Purser BH (1973) Distribution and ultrastructure of Holocene ooids in the Persian Gulf. In: Purser BH (ed) The Persian Gulf. Springer, Berlin Heidelberg New York, pp 279 – 328

Lorens RB (1981) Sr, Cd, Mn and Co distribution coefficients in calcite as a function of calcite precipitation rate. Geochim Cosmochim Acta 45:553 – 561

Lowenstam H, Epstein S (1957) On the origin of sedimentary aragonite needles of the Great Bahama Bank. J Geol 65:364 – 375

Mackenzie FT, Pigott JD (1981) Tectonic controls of Phanerozoic sedimentary rock cycling. J Geol Soc London 138:183–196
Marshall JF, Davies PJ (1975) High-magnesium calcite ooids from the Great Barrier Reef. J Sediment Petrol 45:285–291
McGannon Jr DE (1975) Primary fluvial oolites. J Sediment Petrol 45:719–727
McIntire WL (1963) Trace element partition coefficients – a review of theory and applications to geology. Geochim Cosmochim Acta 27:1209–1264
Meyers WJ (1980) Compaction in Mississippian skeletal limestones, southwestern New Mexico. J Sediment Petrol 50:457–474
Michard G (1968) Co-precipitation de l'ion manganeaux avec le carbonat de calcium. C R Ser D 267:1685–1688
Milliman JD (1967) Carbonate sedimentation on Hogsty Reef, a Bahamian Atoll. J Sediment Petrol 37:658–676
Milliman JD (1974) Marine carbonates. Springer, Berlin Heidelberg New York, 375 p
Möller P, Rajagopolan G, Germann K (1976) A geochemical model for dolomitization based on material balance; part II. Geol Jahrb B-20:57–76
Morrow DW, Mayers IR (1978) Simulation of limestone diagenesis – a model based on strontium depletion. Can J Earth Sci 15:376–396
Müller G (1968) Exceptionally high strontium concentrations in fresh-water oncolites and mollusc shells of Lake Constance. In: Müller G, Friedman GM (eds) Recent developments in carbonate sedimentology in central Europe. Springer, Berlin Heidelberg New York, pp 116–135
Murray JW (1954) The deposition of calcite and aragonite in caves. J Geol 62:481–492
Newell ND, Purdy EG, Imbrie J (1960) Bahamian oolitic sands. J Geol 68:481–497
Ohde S, Kitano Y (1980) Incorporation of fluoride into Ca-Mg carbonate. Geochem J 14:321
Pedro G, Carmouse JP, Velde B (1978) Peloidal nontronite formation in Recent sediments of Lake Chad. Chem Geol 23:139–149
Pigott J, Mackenzie FT (1980) Global tectonic control of Phanerozoic long-term cycling of carbon. Abstr 26th Int Geol Congr, Paris, 506 p
Pingitore NE Jr (1978) The behaviour of Zn^{2+} and Mn^{2+} during carbonate diagenesis: theory and applications. J Sediment Petrol 48:799–814
Pingitore NE (1982) The role of diffusion during carbonate diagenesis. J Sediment Petrol 52:27–39
Purser BH, Loreau J-P (1973) Aragonitic, supratidal encrustations on the Trucial Coast, Persian Gulf. In: Purser BH (ed) The Persian Gulf. Springer, Berlin Heidelberg New York, pp 343–376
Rao CM (1978) Distribution of $CaCO_3$, Ca^{2+} and Mg^{2+} in sediments of the northern half of western continental shelf of India. Indian J Mar Sci 7:151–154
Richter DK (1979) Die Stufen der meteorisch-vadosen Umwandlung von Mg-Calcit in Calcit in rezenten bis pliozänen Biogenen Griechenlands. Neues Jahrb Geol Palaeontol Abh 158:277–333
Richter DK, Füchtbauer H (1978) Ferroan calcite replacement indicates former magnesium calcite skeletons. Sedimentology 25:843–860
Sandberg PA (1975) New interpretations of Great Salt Lake ooids and of ancient non-skeletal carbonate mineralogy. Sedimentology 22:497–537
Scholle PA, Kinsman DJJ (1974) Aragonitic and high-Mg calcite caliche from the Persian Gulf – A modern analog for the Permian of Texas and New Mexico. J Sediment Petrol 44:904–916
Sorby HC (1879) The structure and origin of limestones. Proc Geol Soc London 35:56–95
Stehli FG, Hower J (1961) Mineralogy and early diagenesis of carbonate sediments. J Sediment Petrol 31:358–371
Suess E, Fütterer D (1972) Aragonitic ooids: experimental precipitation from seawater in the presence of humic acid. Sedimentology 19:129–139
Till R (1970) The relationship between environment and sediment composition (geochemistry and petrology) in the Bimini Lagoon, Bahamas. J Sediment Petrol 40:367–385
Turekian KK (1972) Chemistry of the Earth. Rinehart and Winston, New York, 131 p
Veizer J (1974) Chemical diagenesis of belemnite shells and possible consequences for paleotemperature determinations. Neues Jahrb Geol Palaeontol Abh 147:91–111
Veizer J (1977) Diagenesis of pre-Quaternary carbonates as indicated by tracer studies. J Sediment Petrol 47:565–581
Veizer J (1978a) Simulation of limestone diagenesis – a model based on strontium depletion: Discussion. Can J Earth Sci 15:1683–1685

Veizer J (1978b) Strontium: abundance in common sediments and sedimentary rock types. In: Wedepohl KH (ed) Handbook of geochemistry, vol II – 5. Springer, Berlin Heidelberg New York, K1 – K13

Veizer J, Lemieux J, Jones B, Gibling MR, Savelle J (1977) Sodium: paleosalinity indicator in ancient carbonate rocks. Geology 5:177 – 179

Weiler Y, Sass E, Zak I (1974) Halite oolites and ripples in the Dead Sea, Israel. Sedimentology 21:623 – 632

White AF (1977) Sodium and potassium co-precipitation in aragonite. Geochim Cosmochim Acta 41:613 – 625

White AF (1978) Sodium co-precipitation in calcite and dolomite. Chem Geol 23:65 – 72

Winland HD (1969) Stability of calcium carbonate polymorphs in warm, shallow sea water. J Sediment Petrol 39:1579 – 1587

Zeller EJ, Wray JL (1956) Factors influencing precipitation of calcium carbonate. Am Assoc Petrol Geol Bull 40:140 – 152

Zemann J (1969) Crystal chemistry. In: Wedepohl KH (ed) Handbook of geochemistry, vol I. Springer, Berlin Heidelberg New York, pp 12 – 36

I.4 Carbon and Oxygen Isotope Composition of Recent and Ancient Coated Grains

Mordeckai Magaritz[1]

1 Introduction

Coated grains form in various environments, some of which have been characterized by their carbon and oxygen isotopic composition. The origin and history of ancient carbonate rocks may be inferred by studying these compositions. The carbon and oxygen isotopes in carbonate minerals have both been used extensively in the studies of environment of deposition, diagenesis and burial history (Gross 1964, Land 1970, Tan and Hudson 1974, Weber et al. 1976, Hudson 1977, Campos and Hallam 1979).

The various environments in which it has been suggested that coated grains form have their characteristic isotopic record. The question of chemical precipitation vs. biological precipitation of carbonate minerals may be elucidated using the stable isotope record of the carbonate. The diagenetic environment may have different carbon isotope composition from the environment of deposition in certain cases. But the main advantage in the stable isotope usage lies in the ability to distinguish between the marine and continental environment of formation of the carbonate phases (Keith and Weber 1964).

It is the purpose of this review to indicate the isotopic variation of carbon and oxygen in natural environments; and to present case studies where isotopic measurements have been used to study the various types of coated grains.

2 Methods and Isotopic Fractionation

The natural variation of carbon and oxygen isotopes is measured on CO_2 gas generated or equilibrated with the sample using a double collector mass spectrometer. The isotopic composition is conventionally given as the parts per thousands (per mil) deviation from a given in standard. The standards are SMOW (Standard Mean Ocean Water) for water (Craig 1961a) and PDB (Peede formation Belemnite, Urey et al. 1951) for carbon and oxygen in carbonate phases. The carbonate rocks are reacted with phosphoric acid (1.80 specific gravity, McCrea 1950)

1 Incumbent of Graham Rhana Beck Development Chair, Isotope Department, The Weizmann Institute of Science, 76100 Rehovot, Israel

Coated Grains (ed. by T. M. Peryt)

The commonly used notation is the δ-notation, where

$$\delta^{18}O = \left(\frac{(^{18}O/^{16}O)_{sample} - (^{18}O/^{16}O)_{stand.}}{(^{18}O/^{16}O)_{stand.}} \right) \times 1000 \,.$$

Typical analytical errors range between 0.2 to 0.05 for both carbon and oxygen.

The isotopic partition between various phases in the $CO_2 - H_2O - CaCO_3$ system had been studied both theoretically and experimentally. The isotopic fractionation factors between the various phases depend on temperature only. The oxygen and carbon isotopic fractionation was recently summarized by Friedman and O'Neil (1977).

The main reservoirs of water in the world differ from carbonate minerals by about 30‰ in $\delta^{18}O$ values. For the carbon isotope two large reservoirs exist, the organic matter is depleted by 20‰ – 25‰ from the carbonate system.

Within each large reservoir isotopic differences exist due to the various modes of formation (Fig. 1). The significant difference between continental and marine carbonate is that the meteoric water from which the former arises is depleted in $\delta^{18}O$ compared to seawater (Craig 1961b). The dissolved carbonate in meteoric water is also depleted in ^{13}C due to the huge influence of biologically derived CO_2 (Deines et al. 1974), while in the ocean the dissolved carbonate is only slightly influenced by organic derived carbon. On the other hand, in many supertidal environments, both the water and the carbonate precipitate are enriched in $\delta^{18}O$ values due to evaporation (Lloyd 1966).

When $CaCO_3$ precipitation is biologically influenced , one can observe deviations from chemical and isotopic equilibrium condition depending on the biological system: such is the case for example, of $CaCO_3$ precipitated by corals (Weber

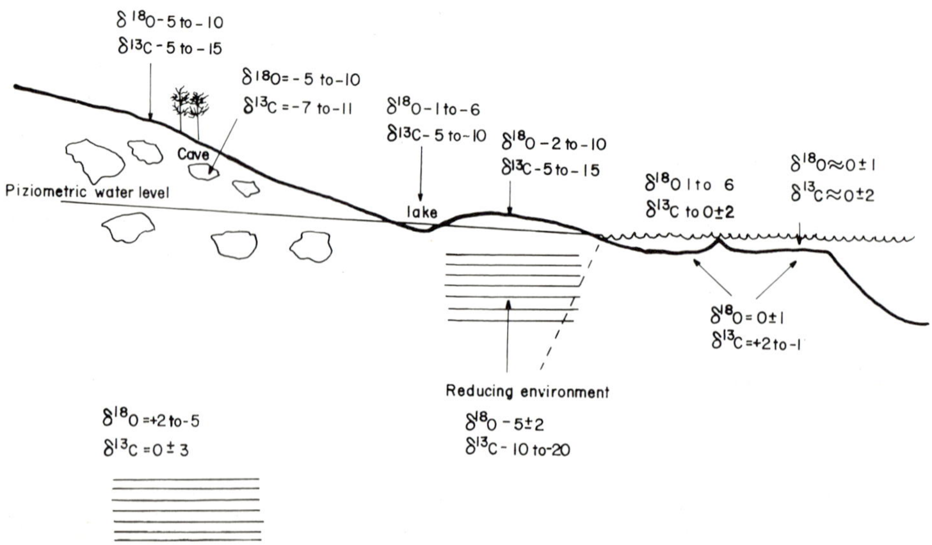

Fig. 1. Range of carbon and oxygen isotopic composition of carbonate forming in various modern environments reported in literature

et al. 1976). Here, the differences in the isotopic composition between biologically and chemically precipitated carbonates may be detected (Weber 1967).

3 Discussion

The following discussion will be separated into three groups of samples; the isotopic record of recent coated grains from the marine environment, that from the continental environment and the isotopic record of ancient coated grains.

3.1 Recent Coated Grain

3.1.1 Marine Environment

The first study using stable isotopes to study the origin of aragonitic ooids was done on samples from the Great Bahama Bank (Lowenstam and Epstein 1957). The ooids studied had concentric layers in the outer zone and a cryptocrystalline core. Both $\delta^{13}C$ and $\delta^{18}O$ values were found to be enriched in the heavy isotope in the ooids compared to both the biogenic (algal) aragonite and to the aragonitic needle mud. The values also show a narrower range than do the other data of algal aragonite. Lowenstam and Epstein (1957) concluded that "it would be difficult to attribute the carbonate of the ooids to a direct algal origin". The accretion of needles from the mud also was ruled out based on the dissimilarity in the isotopic record. The temperature calculated for the formation of the ooids base on the measured $\delta^{18}O$ of the water was found to be 24° – 25.7 °C, which is similar to the mean water temperature (Lowenstam and Epstein 1957). Cloud (1962) argues that the relation between salinity and ^{18}O used by the above should be corrected and the real temperature range is 22.8° – 23.7 °C. This temperature range indicates that the ooids formed during winter months only.

The second environment from which ooids were studied is Baffin Bay, Texas (Land et al. 1979). In Baffin Bay, some laminae of the ooids are built of aragonite, others of high-Mg calcite. The mineralogy and morphology of these ooids are similar to those reported from other aragonitic environments. In this case the ooids show no large isotopic differences which may be due to mineralogy (Land et al. 1979). The isotopic record (shown in Fig. 2) had been interpreted by Land et al. (1979) as equilibrium precipitation of the $CaCO_3$ with CO_2 derived almost totally from dissolved carbonate. The $\delta^{18}O$ values indicate equilibrium growth during the summer months. The organic mats derived from blue-green algae have no influence on the ooids isotopic composition, because their $\delta^{13}C$ values are highly depleted in ^{13}C (– 13‰ to – 17‰, Behrens and Frishman 1971).

3.1.2 Continental Environment

Coated grains are found in the continental environment mainly in semiarid soils and in swamp sediments. Most of these do not commonly have concentric or radial texture. A central grain or fragment is commonly observed with a micritic rim surrounding it.

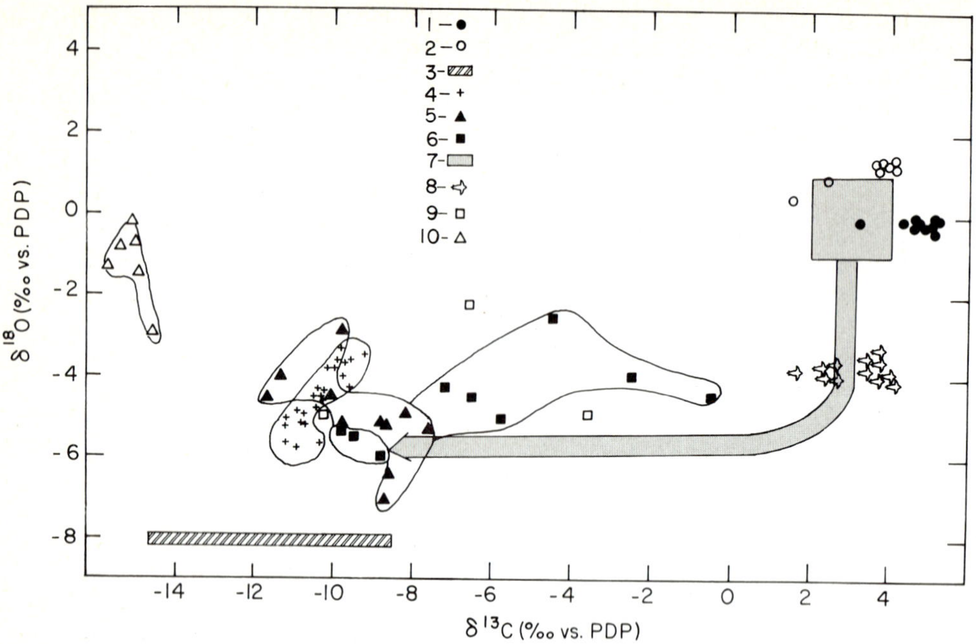

Fig. 2. The carbon and oxygen isotopic composition of various coated grains appearing in the literature, listed in the order of the text discussion. *1* Recent oolites from the Great Bahama Bank (Lowenstam and Epstein 1957); *2* Recent ooids from Baffin Bay, Texas (Land et al. 1979); *3* Swamp nodules from Louisiana (Whelam and Roberts 1973); *4* Vadoids from caliche layers, Israel (Magaritz 1974); *5* Average of Recent and Pleistocene soil nodules from the Coastal Plain of Israel (Magaritz et al. 1981); *6* Average of calcrete carbonates from various countries (Salomons et al. 1978); *7* The behaviour of carbon and oxygen isotopes in ancient pelleted limestone from Israel (Magaritz 1975); *8* Ooids from the Smackover Formation (Jurassic) (Moore and Druckman 1981); *9* The nodules from Jurassic rocks in Scotland (Tan and Hudson 1974); *10* The nodules from Jurassic rocks of the North Sea (Irwin et al. 1977)

Macroids of this type were reported from swamp environments by Whelam and Roberts (1973). In two cores from a clay rich freshwater swamp near Port Basse, Louisiana, $CaCO_3$ macroids were observed. The $CaCO_3$ occur in a spectrum of nodular shapes and sizes. Some of the macroids show recrystallization features and some are found to be surrounded rootlets. Unfortunately, only carbon isotope data were reported (Fig. 2). The range in $\delta^{13}C$ values, $-9.5‰$ to $-19‰$ is large. A distinct difference in $\delta^{13}C$ values is observed in the two cores; the one from the well-drained environment shows only a minor fluctuation with depth. On the other hand the macroids from the poorly drained reducing environment show up to 10‰ depletion in ^{13}C at the core bottom. These data are within the range of pedogenically influenced CO_2, where part of the carbon in the well drained environment may be derived from the dissolution of older carbonate, while in the poorly drained core the CO_2 is derived from the decay of organic material in the reducing environment.

The second type of continental macroids are those found in soil profiles. The carbonate in soil can have either of two origins: detrital or pedogenic. In many cases the detrital grain becomes the coated core of the vadoids.

Only in specific cases does the vadoid represent pedogenic carbonate phases only. Such a case was studied in soil-coated grains from the Coastal Plain of Israel (Magaritz et al. 1981). Low Mg-calcite vadoids of various sizes are found in buried soils profiles and show no concentric texture. They were formed in the burial position in a decalcified soil horizon by carbonate elluviation from the overlying horizon. In such a case Magaritz et al. (1981) suggested that all the carbonate in the nodules is pedogenic and that the isotopic record and ^{14}C concentration represent the conditions of formation. The vadoids studied could be divided into two groups: soft vadoids of Holocene age which are found around rootlets and on ped surfaces, and hard vadoids of Upper Pleistocene age. The mean $\delta^{18}O$ and $\delta^{13}C$ values of the vadoids population shown in Fig. 2 indicate deposition from unevaporated local rainwater and dissolved carbonate characteristic of this region today. The $\delta^{13}C$ values of soil CO_2 in the region today range from $-16‰$ to $-21‰$.

Another type of pedogenic nodule found in the semi-arid zone is hard crust, the calcrete or caliche, as it is called in the literature. Stable isotopic studies by Salomons et al. (1978) of various calcrete deposites from Europe, Africa and India did not include only the coated grain type and did not eliminate the detrital carbonate component from the samples analyzed. In general the $\delta^{18}O$ values of these carbonates are in most cases more enriched in ^{18}O relative to the calculated value of carbonate in isotopic equilibrium with the oxygen isotope composition of local rain and the average temperature (Fig. 5 in Salomons et al. 1978). This indicates that the evaporation of soil water causes the oversaturation of $CaCO_3$ and its precipitation. In other localities the $\delta^{18}O$ values of the soil carbonate are in isotopic equilibrium with the $\delta^{18}O$ value of local rain, suggesting that the cause of $CaCO_3$ formation is loss of CO_2 from the soil solution.

In a study of vadoids from caliche layers from Israel (Magaritz 1974) separation of the detrital core from the pedogenic phase of the vadoids was possible because of the mineralogical difference. (The core in these cases is dolomitic and the pedogenic phase in the rim is 5 mol% Mg-calcite). Only part of the CO_2 generated from the reaction of the carbonate powder and phosphoric acid after 1 h was collected for analysis. This portion contains mainly the calcitic part of the mixture. The isotopic record (Fig. 2) indicates that the carbonate precipitated from evaporated meteoric water, and that the carbon was derived from a mixture of organic derived carbon and rock carbonate. The evaporated origin of the calcite is suggested from the fact that the calcite which was dated as Holocene is enriched by 2‰ in ^{18}O relative to calcite that should precipitate in isotopic equilibrium from the present day meteoric water at present day temperature.

3.2 Ancient Coated Grains

3.2.1 Isotope Exchange Problem

The differences in isotopic composition between the various coated grains found in the recent environment could not apply directly without coation in the study of ancient samples. In processes such transformations from unstable minerals like aragonite and Mg-calcite (which characterize coated grains from the marine environment) to more stable low Mg-calcite isotopic exchange with the diagenetic solution may take place. Other late processes like recrystallization, crystal growth and late cementation also may cause isotopic changes in the rock. These changes were observed even within the Pleistocene sequence (Gross 1964, Gavish and Friedman 1969, Magaritz et al. 1979). But continuous isotopic studies of ancient carbonate sequences have helped to clarify the relationship between the stable isotope record and the various stages in the history of the rock (Choquette 1968, Tan and Hudson 1974, Magaritz 1975, Hudson 1977, Walls et al. 1979).

A study of a pelleted limestone unit of Turonian age from Northern Israel (Magaritz 1975) may be used as an example of post-depositional isotopic change. Samples from a layer rich in pelleted sediments, very similar to that reported by Land and Goreau (1970) were analyzed for their $\delta^{18}O$ and $\delta^{13}C$ values. A drastic difference in $\delta^{18}O$ values were observed between the Turonian rocks and their modern analog. The average $\delta^{18}O$ value of $-5.5‰$ in these marine rocks (as in the case in Fig. 3) is very similar to the values of cave and soil carbonate in that region (Magaritz 1975). These values indicate a post-depositional exchange of oxygen isotopes with meteoric water. This exchange may have occurred during lithification or later. The fact that the $\delta^{13}C$ values in the petrographically non-recrystallized samples from this unit have the same range (0‰ to 2‰) as the modern pellet mud, suggests that the carbon isotope record in carbonate rocks behaves differently from the oxygen isotope record. The negative $\delta^{13}C$ values occur only in samples showing petrographical evidence for recrystallization. The sample with the most depleted $\delta^{13}C$ value is a totally recrystallized sample. This type of study indicates that two types of isotopic exchange occur in carbonate rocks. In the first of these no petrographical evidence is seen under light microscope, but the mineralogical change from unstable phases to the stable phase may be reflected. Under this condition the $^{18}O/^{16}O$ ratio changes dramatically from its original value. The second is a recrystallization event which can be seen petrographically and correlates with a carbon isotope change. These changes (Fig. 2) can in some cases cause coated grains of marine origin to have an isotopic composition very similar to a continental one.

The distinct difference between carbon and oxygen isotope behaviour in ancient carbonate rocks may be understood if one uses the material balance approach for the system fluid + solid as it is done for hydrothermal systems (Taylor 1978).

A material balance calculation can be formulated

$$W\delta^{i}_{Solution} + R\delta^{i}_{Rock} = W\delta^{f}_{Solution} + R\delta^{f}_{Rock},$$

where i = initial value, f = final value observed, W = atom per cent of oxygen or carbon in the solution and R = atom percent in the rock. If the final isotopic composition in the fluid phase is determined by the equilibrium value (Δ) at a specified temperature then

$$W/R = \frac{\delta^{f}_{R} - \delta^{i}_{R}}{\delta^{i}_{F} - \delta^{f}_{F} + \Delta}.$$

In the case of carbonate rocks since large quantities of fluid usually go through as in an aquifer, one uses the open system solution (Taylor 1978). In such a system, each increment of fluid makes a single pass through the system, then

$$W/R = \log_e \left[\frac{\delta_F^i + \Delta - \delta_R^i}{\delta_F^i - \delta_R^f + \Delta} \right].$$

Using this formulation one gets the atom wt%. Converting these values to weight units show the difference between carbon and oxygen. While a typical carbonate rock contains 48 wt% oxygen and 12 wt% carbon, any fluid contains about 89 wt% oxygen but commonly less than 0.0019% carbon (equivalent to 1000 ppm HCO_3). The results of using the above convertion factors between these two units (atom % to weight) are 0.5 for oxygen and 6300 for carbon.

In Fig. 3 a calculation based on this model is presented. The initial conditions are a carbonate rock with $\delta^{18}O = +31‰$ vs. SMOW ($= +0.5‰$ vs. PDB) and $\delta^{13}C = +2‰$ interacting at 50 °C with fluid having $\delta^{18}O = -5‰$ SMOW and $\delta^{13}C_{CO_2} = -20‰$. Note the difference in the amount of water needed to change the oxygen in the rock from that of the carbon. Since these values are *minimal* values (assuming exchange occurs with every atom in the fluid) one should realize that the very large number for carbon indicates that such exchange cannot occur

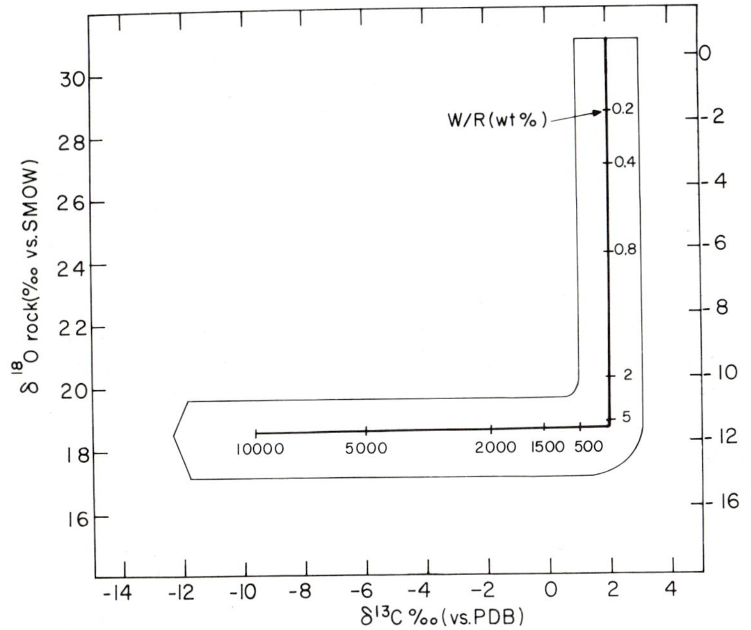

Fig. 3. Plot of calculated values of $\delta^{18}O$ and $\delta^{13}C$ that would be obtained during a hypothetical meteoric water alteration assuming the initial condition in the text. Note that $\delta^{18}O$ values are in the SMOW scale and not in PDB as is the rest of the figures. A conversion formula is $\delta^{18}O_{SMOW} = 1.0305\ \delta^{18}O_{PDB} + 30.5‰$

in normal conditions. Whereas the interaction between the water and the oxygen in the rocks are easily achieved as was shown for the case of pelleted limestone mentioned above. A similar calculation can be performed for different initial conditions and temperatures. It is independent of the type of reaction that cause the isotopic exchange.

3.2.2 Case Studies

Ooids from the upper Jurassic Smackover formation in the Gulf of Mexico were suggested by Moore and Druckman (1981), to have been originally composed of aragonite. These show depletion in ^{18}O (Fig. 2) compared to recent marine carbonate. The values indicate post formational exchange of oxygen isotope, probably in the transformation stage from aragonite to calcite. Moore and Druckman (1981) suggest that the isotope exchange process took place with pore fluid during burial at temperature lower than 90 °C. The preserved carbon isotope data clearly indicate that the oxygen isotope exchange (mineralogical transformation) took place in a dominant rock system. In such a system no carbon changes occur.

Coated grains from the so-called "algal" beds in ancient rocks create problems when one tries to use the stable isotope method to detect their origin. This may be because their equivalent has not been found and studied isotopically in the recent environment. These are coated grains, which were suggested in many studies to form in the interstitial diagenetic environment. Such is the case in beds containing coated grain from Great Estuarine series of Jurassic age in Scotland (Tan and Hudson 1974). These coated grains appear in beds which contain evidence of evaporitic condition (pseudomorphs after gypsum). The $\delta^{13}C$ values are highly variable but usually more negative than −1.1‰ (up to −14‰). Their average oxygen isotopic values, shown in Fig. 2, are significantly different from their modern analog. Tan and Hudson (1974) suggest that the depletion in ^{13}C in the carbonate is the result of incorporation of metabolic CO_2. But, as was indicated in the recent model from Baffin Bay, no evidence for such incorporation was observed. One can only suggest that the formation of the coated grains occurred either in the interstitial water where algal mats decay or in a swamp environment.

Such an interstitial model was suggested for the origin of macroids in clay rich units of Kimmeridgian age from the North Sea (Irwin et al. 1977). These macroids have normal marine $\delta^{18}O$ values (0‰ to −3‰), but they are highly depleted in ^{13}C (up to −15‰). Irwin et al. (1977) suggested that the carbonate in the solution from which the macroids grow derived mainly from sulphate-reducing bacteria source (Raiswell 1976).

In a case where two generations of coated grains are reported, the use of stable isotope may be of great help in studying the history of the sequence. In a sequence of marine rocks of Turonian age from northern Israel where marine ooids are the dominant phase ($\delta^{13}C$ values about 0‰, Fig. 4), two layers of carbonate rich in iron staining were found. Petrographically those rocks contain vadoids which in the centers contain rock fragments in comparison to biogenic fragment of the marine ooids. Based on the drastic difference in $\delta^{13}C$ values

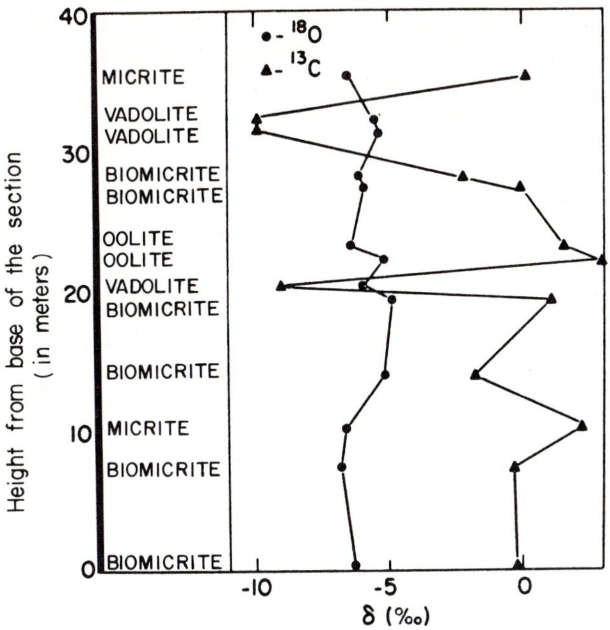

Fig. 4. $\delta^{18}O$ and $\delta^{13}C$ isotopic variation in the Turonian sequence from northern Israel

(−8 to −11‰) (Fig. 4). Magaritz (1973) indicated that these units represent fossil caliche layers within marine ooid-bearing sequence.

4 Conclusions

Stable isotope of carbon and oxygen can be of great help in identifying the environment of deposition and diagenesis of coated grains. In cases of recent coated grains the isotopic study may indicate the season, the exact environment and the biogenic influence on the formation of these carbonates. On the other hand in ancient cases one can use the isotopic record, more to differentiate between various types of carbonate forms in the rocks than to deduce the exact origin of the coated grain. Hopefully more isotopic studies of ancient coated grain may help clarify the various environments where those carbonate forms had been formed.

Acknowledgements. I thank Dr. Kaufman for his critical suggestions. This report was written during a visit of the author in California Institute of Technology. Supported by a grant from the Minna-James-Heinemann-Stiftung.

References

Behrens EW, Frishman SA (1971) Stable carbon isotope in blue green algal mats. J Geol 79:94 – 100

Campos HS, Hallam A (1979) Diagenesis of English Lower Jurassic limestones as inferred from oxygen and carbon isotope analysis. Earth Planet Sci Lett 45:23 – 31

Choquette PW (1968) Marine diagenesis of shallow marine lime-mud sediments: insights from $\delta^{18}O$ and $\delta^{13}C$ data. Science 161:1130 – 1132

Cloud Jr PE (1962) Environment of calcium carbonate deposition west of Andros Island, Bahams. US Geol Surv Prof Pap 350:1 – 138

Craig H (1961a) Standard for reporting concentrations of deuterium and oxygen-18 in natural water. Science 133:1833 – 1834

Craig H (1961b) Isotopic variation in meteoric waters. Science 133:1702 – 1703

Deines P, Langmuir D, Harmon RS (1974) Stable carbon isotope ratios and the existance of a gas phase in the evolution of carbonate ground waters. Geochim Cosmochim Acta 38:1147 – 1164

Friedman I, O'Neil JR (1977) Compilation of stable isotope fractionation factors of geochemical interest. US Geol Surv Prof Pap 440 KK

Gavish A, Friedman GM (1969) Progressive diagenesis in Quaternary to Late Tertiary carbonate sediments: sequence and time scale. J Sediment Petrol 39:980 – 1006

Gross MG (1964) Variation in the $^{18}O/^{16}O$ and $^{13}C/^{12}C$ ratios of diagenetically altered limestone in Bermuda Island. J Geol 72:170 – 194

Hudson JD (1977) Stable isotopes and limestone lithification. J Geol Soc London 133:637 – 660

Irwin H, Cartis C, Coleman M (1977) Isotopic evidence for source of diagenetic carbonate formed during burial of organic rich sediments. Nature (London) 269:209 – 213

Keith ML, Weber JN (1964) Carbon and oxygen isotopic composition of selected limestones and fossils. Geochim Cosmochim Acta 28:1787 – 1816

Land LS (1970) Phreatic versus vadose meteoric diagenesis of limestones: evidence from a fossil water-table. Sedimentology 14:175 – 185

Land LS, Goreau TF (1970) Submarine cementation of Jamaican reefs. J Sediment Petrol 40:457 – 462

Land LS, Behrens EW, Frishman SA (1979) The ooids of Baffin Bay Texas. J Sediment Petrol 49:1269 – 1278

Lloyd RM (1966) Oxygen isotope enrichment of sea water by evaporation. Geochim Cosmochim Acta 30:801 – 814

Lowenstam HA, Epstein S (1957) On the origin of sedimentary aragonite needles of the Great Bahama Bank. J Geol 65:364 – 375

Magaritz M (1973) Carbon and oxygen isotopic composition of some carbonate rocks from Israel (in Hebrew). Ph D thesis, Weizmann Inst Sci, Rehovot, Israel

Magaritz M (1974) ^{14}C ages, $^{18}O/^{16}O$ and $^{13}C/^{13}C$ ratios in calcareous crusts (caliche) from Israel. Geol Soc Am Abstr Prog 12:1046 – 1047

Magaritz M (1975) Sparitization of a pelleted limestone: a case study of carbon and oxygen isotope composition. J Sediment Petrol 45:599 – 603

Magaritz M, Gavish E, Bakler N, Kafri U (1979) Carbon and oxygen isotope composition – indicators of cementation environment in Recent, Holocene and Pleistocene sediments along the coast of Israel. J Sediment Petrol 49:401 – 412

Magaritz M, Kaufman A, Yaalon DH (1981) Calcium carbonate nodules in soils: $^{18}O/^{16}O$ and $^{13}C/^{12}C$ ratios and ^{14}C content. Geoderma 25:157 – 172

McCrea JM (1950) On the isotopic chemistry of carbonates and a paleotemperature scale. J Chem Phys 18:849 – 857

Moore CH, Druckman Y (1981) Burial diagenesis and porosity evolution, upper Jurassic Smackover, Arkansas and Louisiana. Am Assoc Petrol Geol 65:597 – 628

Raiswell R (1976) The microbiological formation of carbonate concretions in the Upper Lias of NE England. Chem Geol 18:227 – 244

Salomons W, Goudie A, Mook WG (1978) Isotopic composition of calcrete deposits from Europe, Africa and India. Earth Surf Proc 3:43 – 57

Tan FC, Hudson JD (1974) Isotopic studies on the paleoecology and diagenesis of the Great Estuarine Series (Jurassic) of Scotland. Scotl J Geol 10:91 – 128

Taylor HP Jr (1978) Water/rock interactions and the origin of H_2O in granitic batholiths. J Geol Soc London 133:509–558

Urey HC, Lowenstam HA, Epstein S, McKinney CR (1951) Measurements of paleotemperatures and temperatures of the Upper Cretaceous of England, Denmark and the southeastern United States. Geol Soc Am Bull 62:399–416

Walls RA, Mountjoy EW, Fritz P (1979) Isotopic composition and diagenetic history of carbonate cements in Devonian Golden Spike Reef, Alberta, Canada. Geol Soc Am Bull 90:963–982

Weber JN (1967) Factor affecting the carbon and oxygen isotopic composition of marine carbonate sediments. I. Bermuda. Am J Sci 265:586–608

Weber JN, Deines P, Weber PH, Baker PA (1976) Depth related changes in the $^{13}C/^{12}C$ ratio of skeletal carbonate deposited by the Caribean reef-frame building coral *Montastraea annulasis:* further implications of a model for stable isotope fractionation by scleractinian corals. Geochim Cosmochim Acta 40:31–40

Whelam TH, Roberts HH (1973) Carbon isotope composition of diagenetic carbonate nodules from freshwater swamp sediments. J Sediment Petrol 43:54–58

I.5 Neritic Macroid Genesis, an Ecological Approach

Lukas Hottinger [1]

1 Introduction

Neritic coated grains of macroid size grow below the tidal zone over restricted surfaces of the sea floor reflecting thus the narrowness of the equilibria conditioning their genesis. As the growth of the grains is a time dependent process, the abiotic and biotic factors involved in the genesis of coated grains must be time-dependent also. The equilibria between the factors are therefore dynamic.

As the surfaces where neritic coated macroids grow are restricted, "blind", punctual measurements in space and time from ships navigating on the surface of the sea are insufficient to study the sensitive equilibria responsible for the formation of coated macroids. As the equilibria are time-dependent, understanding of the equilibria and of the factors involved must be based mainly on circumstantial evidence: Repetitive events and/or slow processes are recorded in the sea by sedimentation and/or by the succession of benthic communities. Insight into such processes and repetitive events is based best on direct observation by scuba diving. Repetitive dives to the same spot over many years furnished the necessary understanding of the indirect evidence recording events or processes over long periods of time. Methods and results of such observations are presented here in form of three case histories, two from the Mediterranean, one from the Red Sea. They represent exclusively subtidal, neritic, comparatively simple environments where the atmosphere does not take part in the system as an environmental factor. Moreover, the case histories are restricted to the euphotic zone where light is available for algal growth.

The present study is part of research project 301976 on functional morphology of sediment producing organisms supported by the Swiss National Science Foundation.

1 Geologisch-Paläontologisches Institut der Universität Basel, Bernoullistraße 32, CH-4056 Basel, Switzerland

Coated Grains (ed. by T.M. Peryt)

2 Rhodoids from Punta Galera, Elba Island, Italy

2.1 Methods

On the South coast of Elba, at Porto Azzurro, a Swiss diving center directed by R. Galster and W. Guggenbühl operates from Barbarossa Bay. With the gratefully acknowledged help of this center, over 8 consecutive years, observations by changing groups of about 20 biology and geology students were carried out at Punta Galera (Fig. 1) under the direction of D. Senn and the author, during 10 or 14 days in early summer or autumn.

A topographical map of the underwater area at Punta Galera (Fig. 1) was established be measuring distances and directions under water from key points marked by a buoy to the next one with ropes of defined length, compass and the divers depth gauge. The key point buoys were located by measuring their distances over the sea surface by triangulation. On this topographic map printed on water-resistant Synthosihl paper all distributional observations where marked in situ by the diver.

In order to estimate life span and growth rate of gorgonians, several dozens of colonies were marked by fastening to their foot a strip of white plastic tablecloth numbered with black, water-resistant ink marker. Maximum length and number of branches were recorded in situ. Detailed results of this study will be published elsewhere, but the general results will be used in this paper to estimate the turnover rate of rhodoids on which gorgonians grow.

Rhodoids where fixed in situ by pouring colored epoxy-resin onto a macroid covered surface delimited by a frame. After polymerisation of the resin (for 24 h), the macroids fixed in their original position can be collected by the diver to be sectioned in the laboratory as if they were sedimentary rocks.

2.2 General Setting

The map of Punta Galera (Fig. 1) shows parallel relief elements running North-South as a major topographic feature. This is due to faults sectioning the rock bottom into long slabs of rock resisting differently to erosion. The prevailing currents enter the system from the North-East (Hottinger 1981). During stormy weather with rain brought in from the South, the current system is reversed for a comparatively short time. Depending on weather conditions, current velocities reach over 1 m s^{-1} at the surface in both directions. Wave action is felt down to 20 m depth during summer by the diver and produces ripple marks in the coarse sand accumulated at the foot of the boulder-covered areas. In the stormy winter season, wave action reaches probably greater depths (as was observed in other parts of the Mediterranean) but at Punta Galera wave ripples have not been found below 20 m depth so far.

The thermocline is located during the summer months between 20 and 35 m depth, depending on the stability of weather conditions.

2.3 Sedimentation Process

Based on in situ observation of the nature of sediment grains, their distribution in different seasons and by the effects of single high-energy events, the following mechanisms of sedimentation condition the deposits off Punta-Galera: The input of terrestrial sediment into the system is very small and gets insignificant below 25 m depth where the production of skeletons, their coating by incrusting organisms or their breakdown by boring organisms determines the size of the coarser fraction.

The psammitic and pelitic fractions fall out of the suspension in the water column and accumulate during the calm periods of the summer season, submerging slowly the coarser biogenous grains. This material is obviously resuspended and transported towards the open sea during winter (Fig. 3). Uniform muds sampled 1 km off the coast in 100 m depth yield, as a kind of tracer, light fragments of bryozoan skeletons derived from hard bottom environments near the coast.

Exceptional high energy events observed once in 8 consecutive years and corresponding to a 30-year-storm event on land, interrupt the normal sedimentation process (Fig. 2) covering the normal sediment with comparatively well sorted storm deposits (Fig. 1).

Observing in early summer the results of each winter period under water, one must conclude to the overriding importance of the yearly cycles in the system considered. The discontinuous energy regime is best described by the periodic sinking and rising of the erosion base level for different grain size fractions during the seasonal cycle (Fig. 2). Water energy in this system is not measurable directly. Linear and turbulent currents together with wave action immediately above the bottom produce the sedimentary effects. No measurement gear is available at a reasonable price to explore and record turbulence over the necessary time in such an exposed part of the sea.

2.4 Rhodoid Distribution

At Punta Galera, rhodoids (Figs. 4–7) occur frequently on mobile, soft substrate in a belt along the South coast. Its upper limit corresponds approximately to the 30 m isobath (Fig. 1). To the South-West of the mapped area where the outermost rocky relief is known as "mine reef", rhodoids were found down to 60 m. Rhodoid depth distribution corresponds exactly to the one of typical "coralligène" vegetation cover (Pérès and Picard 1964, p. 79) on permanent hard substrate exposed to light. Depth distribution of rhodoids does not correspond to

→

Fig. 1. Pta. Galera, Elba Island, Italy. Distribution of rhodoids and selected substrates

Fig. 2. Depth of base level for coarse sand, fine sand and mud off Pta. Galera during the seasonal cycle. Schematic. *WB* Wave base as recorded by wave ripples in coarse sand. *Arrow* Rare high energy event as recorded by storm deposits

Fig. 3. Process of sedimentation off Pta. Galera between 20 and 50 m depth. Schematic. *A* "Coralligène" vegetation cover on permanent hard substrate as main producer of sediment. *B* Zone of accumulation of biogenous sediment at the foot of hard substrate. *C* Current-exposed zone of low sedimentation rates where rhodoids grow. *D* Zone of rhodoid accumulation. *Arrows* Transport direction of the fine-grained fractions of the sediment

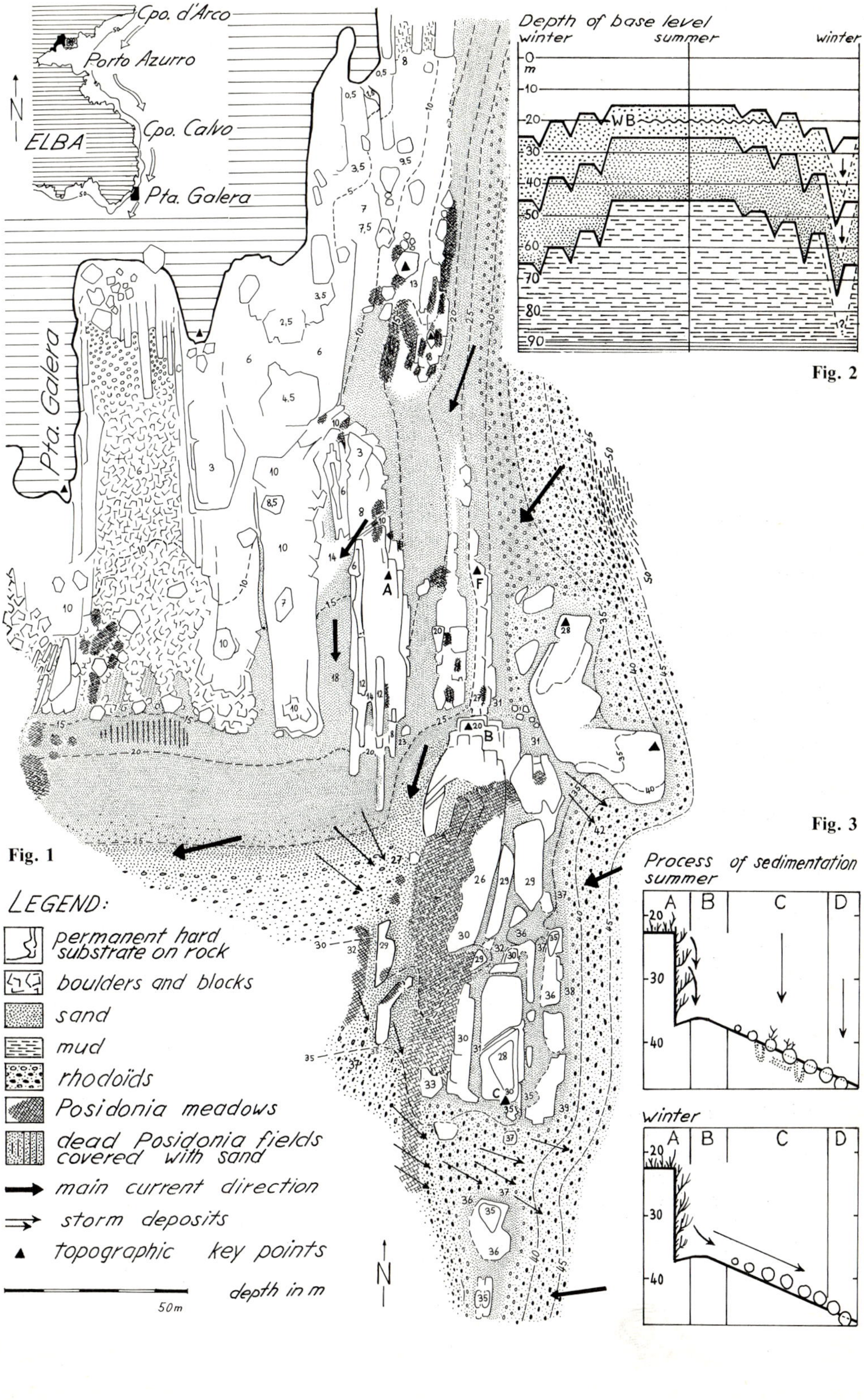

Fig. 2

Fig. 1

Fig. 3

the distribution of "coralligenous" incrustations on *Posidonia* "roots" or in caves which may occur in much shallower but shaded environments, well above the thermocline. Rhodoid formation is therefore dependent on light quantities and qualities rather than on temperature.

In the depth zone between 30 and 60 m, rhodoids are not found on hard substrates nor where sediments are deposited locally in a significant way. In particular, they are not found at the base of current-exposed walls where biogenous debris expelled from the "coralligène" vegetation cover by as yet unidentified endobionts accumulate. Rhodoids are thus restricted to current-swept softbottom surfaces where the sedimentation rate is extremely low or nil.

3 Estimations of Time Conditioning the Process of Rhodoid Formation

Gogònian growth rates are an important tool to estimate the time periods conditioning the sedimentary regime and rhodoid formation (cf. Riedl 1966, p. 56, 281). Most white gorgonians (*Eunicella stricta*) grow and disappear after one year whereas yellow gorgonians (*Eunicella cavolinii*), producing about 6 – 7 new branches per year (resulting in 5 cm additional length, average values for 20 measurements) reach at least 20 years of age as estimated from maximum sizes.

Red gorgonians (*Paramuricea chamaeleon*) produce about 20 new branches (10 cm additional length) and may reach therefore an age of 30 years as a conservative estimate. They grow largest at the base of current exposed walls of the underwater relief. The position of the top of the sediment accumulating below the walls does not change significantly during a full life span of red gorgonians (at least 30 years).

At some distance from the wall, the sedimentation rate is even lower. The slow linear accumulation of the sediment below the wall being distributed over large surfaces, the net accumulation rate integrated over the yearly cycle must be insignificant, varying around zero according to the yearly weather variations.

Figs. 4 – 10. Rhodoids

Fig. 4. Exceptionally large rhodoid from 35 m off Pta. Galera, Elba, with 2 – 3-year old *Eunicella cavolinii*, formed by an advanced coating community of red algae, serpulids and bryozoans

Fig. 5. Artificial hard substrate planted at 35 m near key point C off Pta. Galera, showing the same advanced pioneer community after 3 years of exposure

Fig. 6. Coarse grain of transparent crystalline rock deposited by exceptional storm event in 35 m depth SW of key point B, coated during 1 year and fixed in situ. Thin section showing rhodoid coat on its light-exposed upper side, overgrowing a small bryozoan colony. Lower side with thin bryozoan coat only. Frame corresponding to Fig. 7

Fig. 7. Detail of Fig. 6 as indicated by frame. *B* Bryozoan colony. *G* Grain. *M* Artificial matrix fixing the grains in their original position (*arrow*)

Fig. 8. Rhodoids from Ras Muhammed abraded reef platform growing in "suspension" by wave action. Gulf of Elat, Red Sea

Fig. 9. Thin section of rhodoid as presented in Fig. 8 showing comparatively regular concentric structure

Fig. 10. Similar red algae growing on permanently fixed substrate in a sheltered hole of the reef platform. Maledivas

4

1cm

5

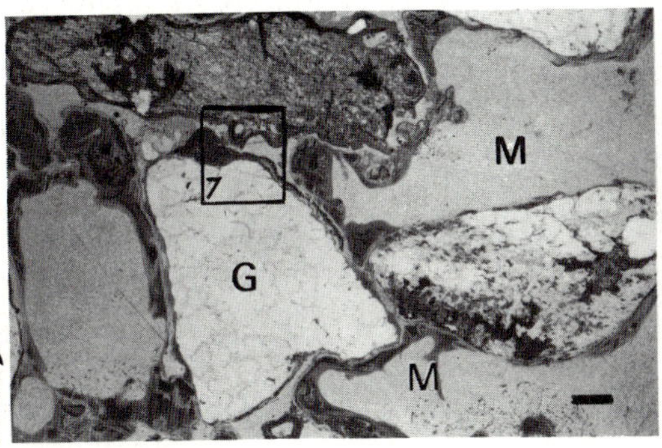

6

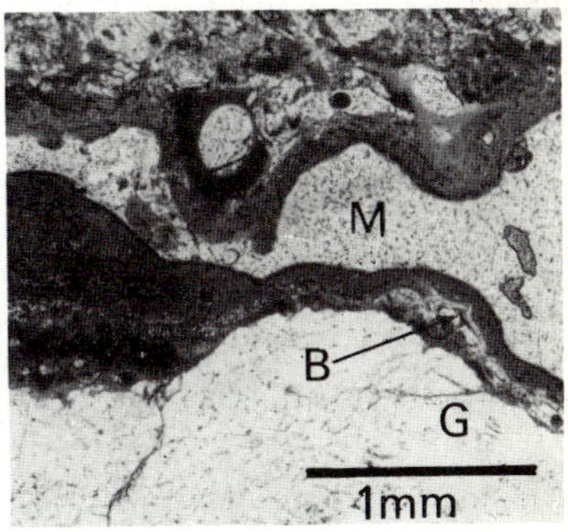

7

8

9

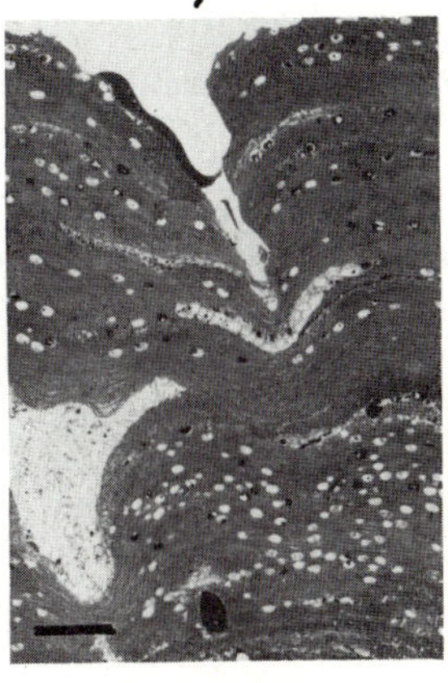

10

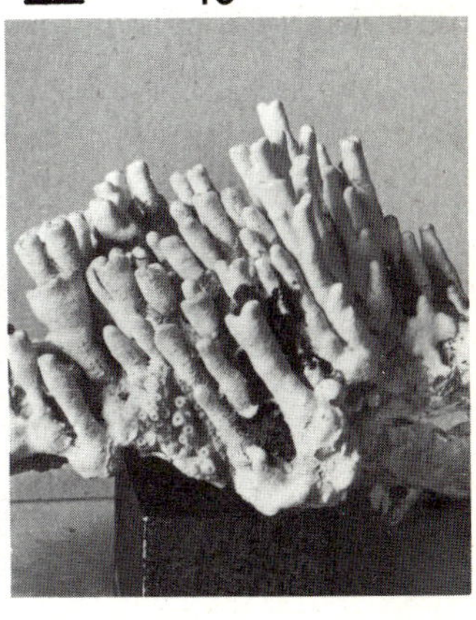

This almost perfect equilibrium is kept up to the moment when exceptional weather events in decennial cycles disrupt the equilibrium for a short time and stop the process of rhodoid growth.

Red and yellow gorgonians are rarely growing on rhodoids although small colonies have been observed to occur on large macroids in depths over 35 m (Fig. 4). These rare occurrences prove that macroids of large size may not be turned over in every winter season. On many other macroids between 30 and 35 m white gorgonians are growing frequently, in particular on macroids of the soft substrate slope fully exposed to the prevailing currents North of key point F in Fig. 1. This proves that the macroids are not turned over during the growth period of white gorgonians in the summer season. Therefore, the living algal incrustation found on the exposed side of a rhodoid in situ represents in most cases one year's one-sided accretion of less than 1 mm thickness.

This is in reasonable accordance with rhodoid growth rates from similar depth on tropical fore reefs off Bermuda (Focke and Gebelein 1978).

3.1 Conclusions on Case History 1

At Punta Galera, rhodoid formation is restricted to open surfaces of soft bottom where accumulation of sediment is so slow that each larger grain may act as temporary hard substrate and may get coated by the pioneer settlers of the "coralligène" environment. The latter is conditioned by light. Zones of rhodoid formation will therefore extend parallel to depth lines. Moreover, the water energy must rise periodically to a level high enough to clean the sea bottom from mud falling out of the suspension but low enough that the grains in process of coating are turned over without being transported out of the environment of their coating organisms. Light penetration and exposure of the investigated area are such that the necessary level of water energy is available at depths where light conditions favour the growth of coralline algae. The periodicity of the seasonal cycle conditioning the yearly rise and fall of the water energy is such that the coralline algae have time to cover substantial parts of the surface offered by the coarser grains.

4 Foraminiferal Macroids from Dahab (Gulf of Elat, Red Sea)

The second case history is presented here to illustrate biological processes involving macroid formation and in particular the pioneer character of the community of organisms coating the macroids.

4.1 Methods

The map of Dahab bay (Fig. 11) was established by controlling the detailed admiralty map by series of scuba dives with the divers depth gauge during two

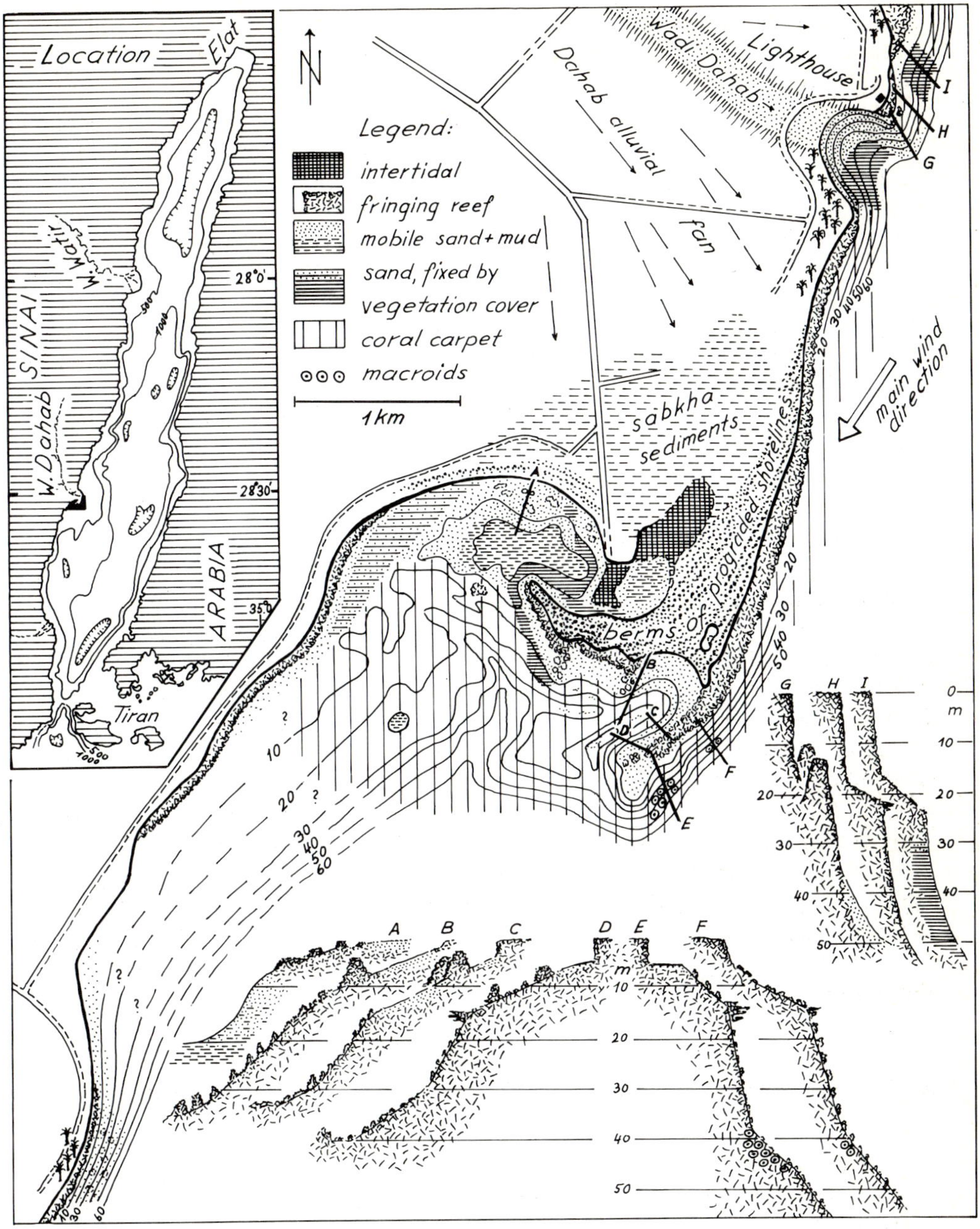

Fig. 11. Sketch map of Dahab bay and schematic cross-sections of fringing reefs in the Gulf of Elat, Red Sea. Location of *Acervulina* macroids

seasons of work in 1971 and 1973 (Hottinger 1977, Fig. 18, p. 71). Long-term observations by repetitive dives to the same spots are not available in this region. In other sections of the fringing reefs of the Gulf of Elat, particularly in front of the biological laboratory of Elat (Hottinger 1977, Fig. 16, Sect. 2) large nodules of the same type occur on the current-exposed terrace between 40 and 50 m depth (Fig. 17). The general environmental conditions in the Gulf of Elat are characterized by an exceptional transparency of the water column, an insignificant thermocline and low nutrient levels (Klinker et al. 1976, 1978; Reiss 1977).

4.2 Occurrence and Composition of the Macroids

Macroids occur on current-exposed parts of the foot along the fringing reefs in discontinuous belts between 40 and 60 m depth. Their density varies from several to hundreds of macroids m^{-2}. The macroids are generated mainly by the incrusting foraminifer *Acervulina inhaerens* Schulze 1845 (Hottinger and Levinson 1978) in competition with coralline algae and other organisms of minor importance (bryozoans, incrusting smaller foraminifera, *Sorites* sp. etc.) coating mainly coral debris (Figs. 12 – 17).

The distribution of these foraminiferal macroids corresponds to the general depth zone where the reef is cemented by the same unicellular organism incrusting larger coral fragments incorporating them to the reef body (Hottinger and Levinson 1978). *Acervulina inhaerens* is thus in this depth zone the most successful settler on new hard substrate as offered in the lower parts of the reef covered by carpets of living corals. The latter succeed the pioneers by settling on the cemented coral fragments.

The mechanism and the frequence of the periodical turnover of coral debris and macroids in these environments is not known. However, temporary very strong longshore currents have been observed in front of the biological laboratory of Elat. They must be responsible for winnowing out the pelitic and finer sandy fractions of the sediment which are absent even at the foot of the coral growth at 60 – 70 m depth. I suppose that such currents are responsible also for turning over the macroids. A yearly cycle of significant changes in water energy would fit the general observations about the life-span of large sized foraminifera measured to be one year in *Amphisorus* and *Sorites* for instance (Zohary et al. 1980).

On reef platforms of the same area, under similar conditions of salinity and temperature, other types of macroids are generated exclusively by coralline algae

Figs. 12 – 17. Macroids formed by the foraminifer *Acervulina inhaerens* from the Gulf of Elat, Red Sea

Fig. 12. Particularly spherical macroids from Dahab

Fig. 13. Section of macroid showing composite nucleus and concentric coats built up by *Acervulina* shells. Dahab

Fig. 14. Detail of sectioned coat showing four successive *Acervulina* shells

Figs. 15, 16. Sectioned spherical *Acervulina* macroids from Dahab

Fig. 17. Section of large ovoid *Acervulina* macroid from Elat, off the Biological laboratory, 40 m

Figs. 18, 19. *Balanus* macroid, viewed from opposite sides. Ile Maire, Marseille, 40 m

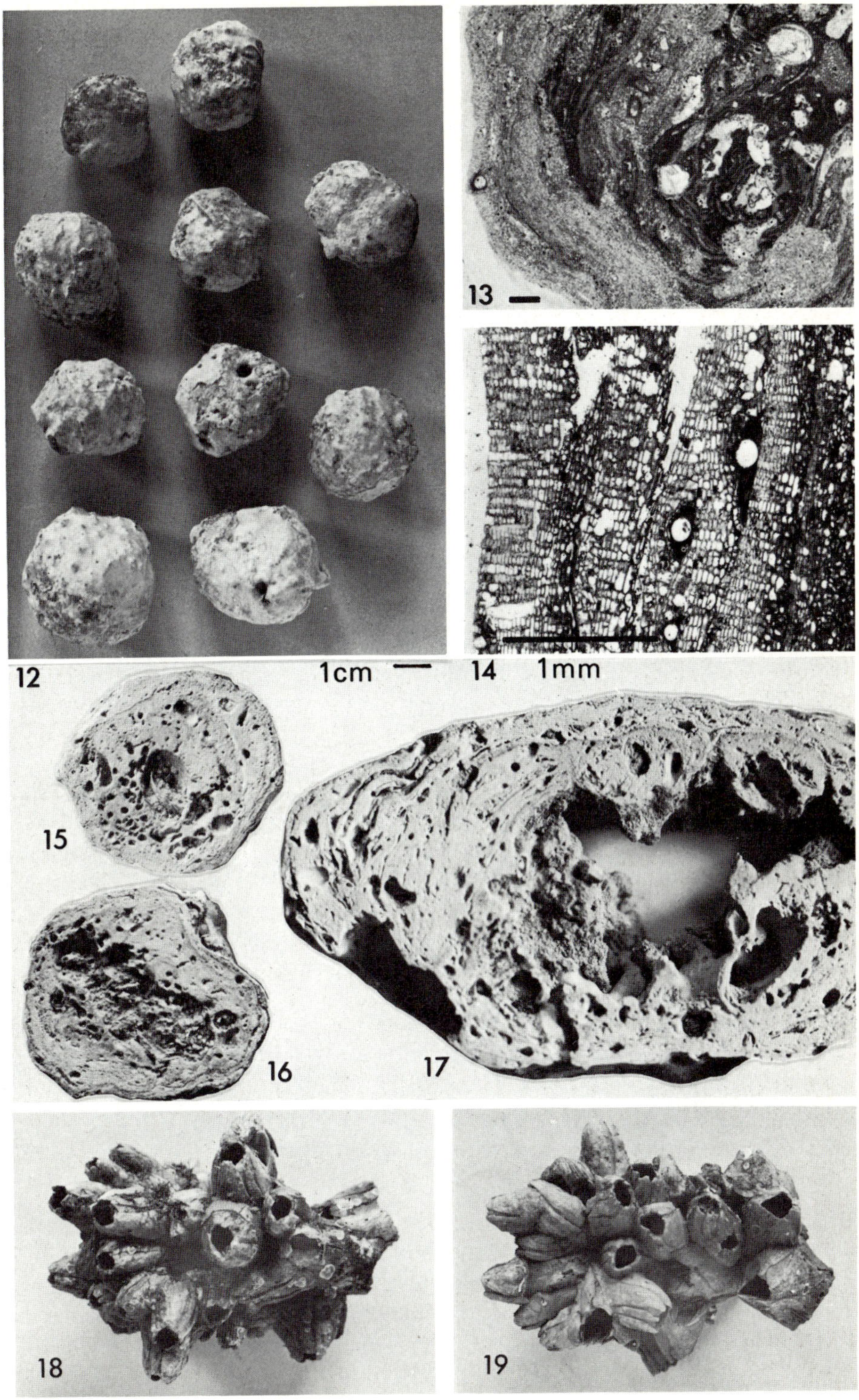
13
12
1cm
14
1mm
15
16
17
18
19

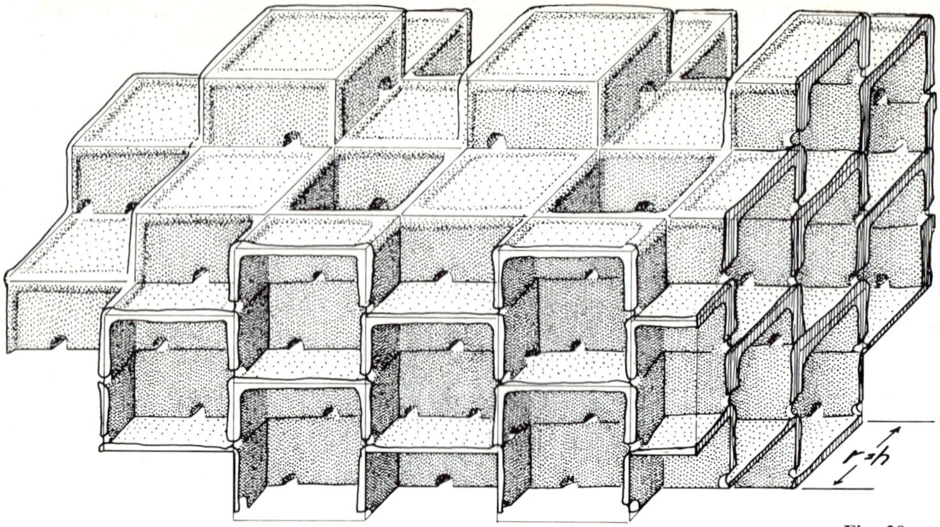

Fig. 20

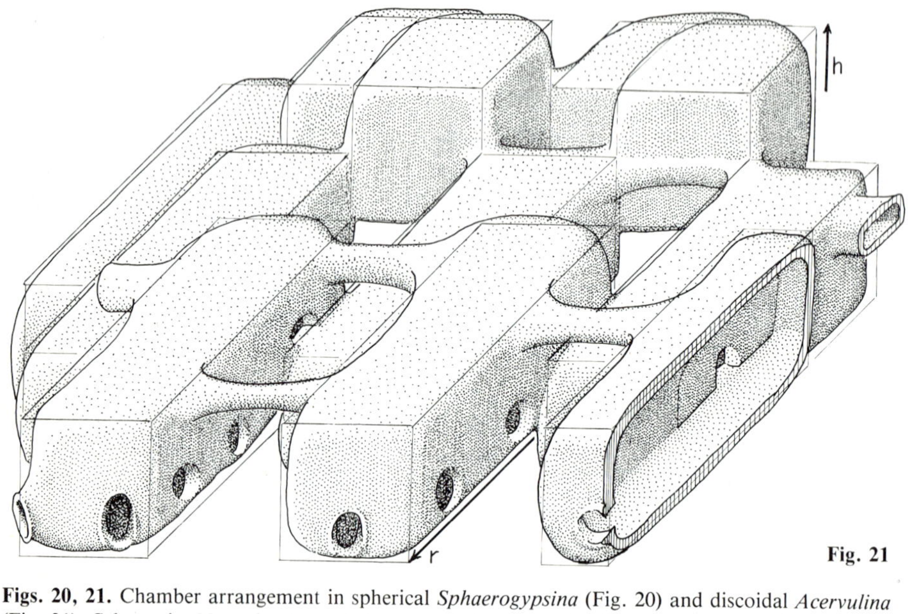

Fig. 21

Figs. 20, 21. Chamber arrangement in spherical *Sphaerogypsina* (Fig. 20) and discoidal *Acervulina* (Fig. 21). Schematic. Note the 4:1 relation of radial (*r*) to axial (*h*) growth rate in *Acervulina*, a device to accelerate radial growth

without *Acervulina* (Figs. 8 and 9). These rhodoids are generated on the Ras Muhammed reef platform (Gvirtzmann and Buchbinder 1978, p. 176 in between stop 1 and 2) under the following particular ecologic conditions: This reef platform is slowly uplifted and therefore continually abraded by wave action. The mechanism of erosion of such platforms can be studied best at Ras Um Sid (Gvirtzmann and Buchbinder 1978, p. 184, stop 12) where left-overs from old reefs are still standing on the new abrasion surface. Under these erosional circumstances, water energy is extremely high in permanence and suppresses the usual succession of communities on hard substrate. No coral can grow in such environments, only thin layers of coralline algae are coating the exposed surface. In erosional lows and hollows, rhodoids are growing. They are kept in more or less continual suspension by wave action and grow concentrically. Their build is exceptionally massive and their specific weight higher than in any other type of macroid observed so far (Figs. 8 and 9). This type of macroid is mentioned here to point out once more the ecological mechanisms selecting the pioneering organisms settling on hard substrate. In particular environments, the selected organisms on permanently fixed hardground (Fig. 10) and on macroids are the same, when the succession of the community is suppressed not only on the moving grain but also on the permanent hard substrate.

4.3 A Growth Strategy to Compete for Space (Figs. 20 – 28)

Free hard substrate is scarce. The pioneer settlers on hard substrate are particularly adapted to occupy as much surface as possible by rapid growth (Figs. 22 – 24). In the particular environment discussed in this section, the unicellular *Acervulina inhaerens* is the most successful organism (Fig. 25). It has developed a particular strategy for rapid growth to compete with coralline algae: As a member of the foraminiferal family *Acervulinidae, A. inhaerens* is basically programmed for growth by multiple chamberlet formation in each growth step. The free-living relatives *Gypsina, Planogypsina* and *Sphaerogypsina* generate their more or less globular shells by proliferation of the chamberlets in all directions. The attached *A. inhaerens* generates successive layers of chamberlets with each growth step. In order to minimize the volume of new protoplasm necessary to fill new chamberlet layers, each growth step generates not a continuous layer of chamberlets but a more or less regular network. Each new chamberlet layer fills up the gaps between the previous chamberlets (Figs. 20, 21, 25). Moreover, the front of the chamberlet network can be detached from the substrate and raise freely into the medium when the periphery of the large shell enters in contact with a thallus front of coralline algae (Figs. 27 and 28). This is an example of how the generation of macroids follows the processes of pioneering communities on hard substrate (cf. Pérès 1961, p. 215) competing for space. The succession of communities on the surface of a macroid is of course restricted in time by the periodicity of the turnover mechanism so that the coating organisms on the macroids surface represent a kind of perpetuated pioneer community on hard substrate.

5 Balanid Macroids from Marseille (Mediterranean)

The third case history is reported here in order to demonstrate the relation between periodicity of water energy changes and growth rate of coating organisms and to indicate the importance of residence time of macroids in their original environment. It is founded on observations during a few reconnaissance skin dives in May 1980 and 1981 on Maire Island, on both sides of Les Farillons (Fig. 29). The Southern tip of Ile Maire is characterized by vertical rock walls falling to depths between 30 and 40 m. These walls, large boulders accumulating at their foot and several ship wrecks creating artificial hard substrates in some distance from the walls are covered with a community dominated by *Balanus* down to at least 45 m depth. The horizontal orientation of large parts of the gorgonian community growing on vertical substrates and ripple marks on patches of coarse sand off the wall down to at least 40 m depth indicate an exceptionally deep wave base (cf. Riedl 1966, p. 362). Vertical movements in the water column are thus more important than lateral currents. This particular regime of the water energy seems to be responsible for the presence of dense balanid colonies in this particular environment down to unusual depths.

On mobile substrate, off the walls, macroids generated exclusively by *Balanus* were collected in depths between 35 and 45 m. The macroids show rough coats by balanid colonies. Individuals growing closely together in the same direction have usually a similar size and can therefore be distinguished as a generation from larger or smaller sized groups of individuals growing in opposite directions. On all macroids collected, the balanid shells were dead, the opercula missing.

In order to be sure that the collected macroids were generated on soft substrate and would not represent dead colonies fallen from the vertical walls near by, the nearest balanid colonies growing on permanently fixed substrate were collected also and exposed to 3 months of natural weathering in the garden in order to remove the dense organic overgrowth. On all balanid colonies from permanent hard substrate, a part of the organic overgrowth resists weathering, in particular the calcareous serpulid tubes, whereas the macroids collected on mobile substrate are absolutely clean (Figs. 18 and 19). In this particular environment, *Balanus* is obviously the most important first settler on hard substrate. Whereas on permanent substrate the community is diversified by subsequent successor organisms (serpulids first, bryozoans and coralline algae next, gorgonians probably last), the balanid macroids must be turned over frequently enough so that no successor organism nor any significant amount of boring organisms can alter the original macroid.

Figs. 22–28. Competition on space between coating organisms. Scanning electron micrographs
Fig. 22. Red algal thallus overgrowing living colony of spiny bryozoans. Pta. Galera, Elba
Fig. 23. Detail of Fig. 22 showing thallus margin growing around the spines
Fig. 24. Red algal thallus versus bryozoan colony, both coating a macroid. To the left of the point of equilibrium (*arrow*), the algal thallus grows over the bryozoan, to the right, the relationship is inverse. Pta. Galera
Fig. 25. Shell margin of *Acervulina inhaerens* coating a macroid. Ultimate and penultimate chamber. Compare Fig. 20. *Arrows* Position of marginal apertures. Dahab, Gulf of Elat. 40 m
Fig. 26. Red algal thallus overgrowing dead bryozoan colony and a dead arborescent foraminifer (*Miniacina*). Dahab
Fig. 27. The same algal thallus as in Fig. 26 overgrown on its opposite side by *Acervulina* with its ultimate and penultimate chambers forming bridge over the thallus margin
Fig. 28. Detail of ultimate chamber of *Acervulina inhaerens* bridging the gaps in a dead bryozoan colony. Dahab

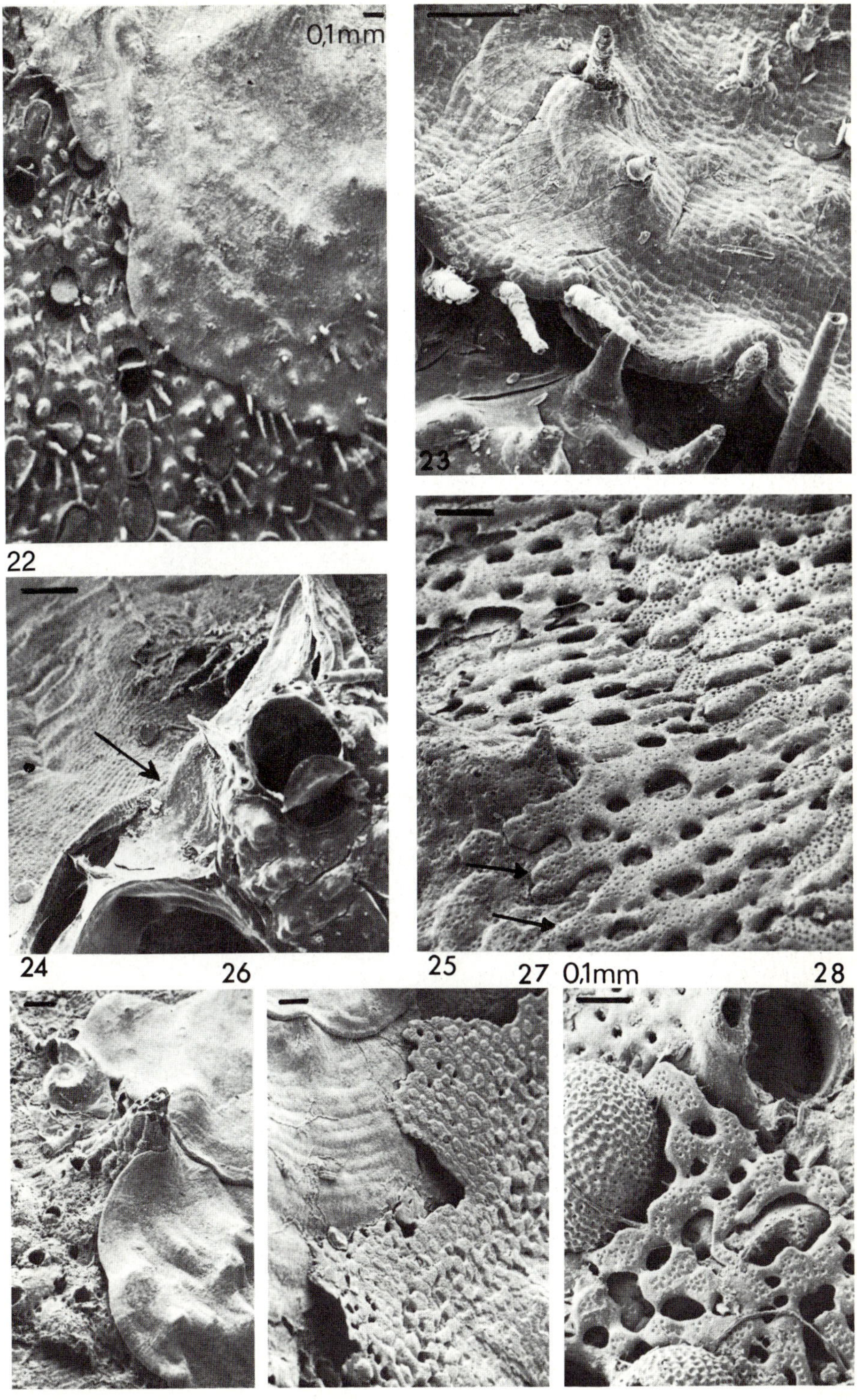
0,1mm
23
22
24
26
25
27
0,1mm
28

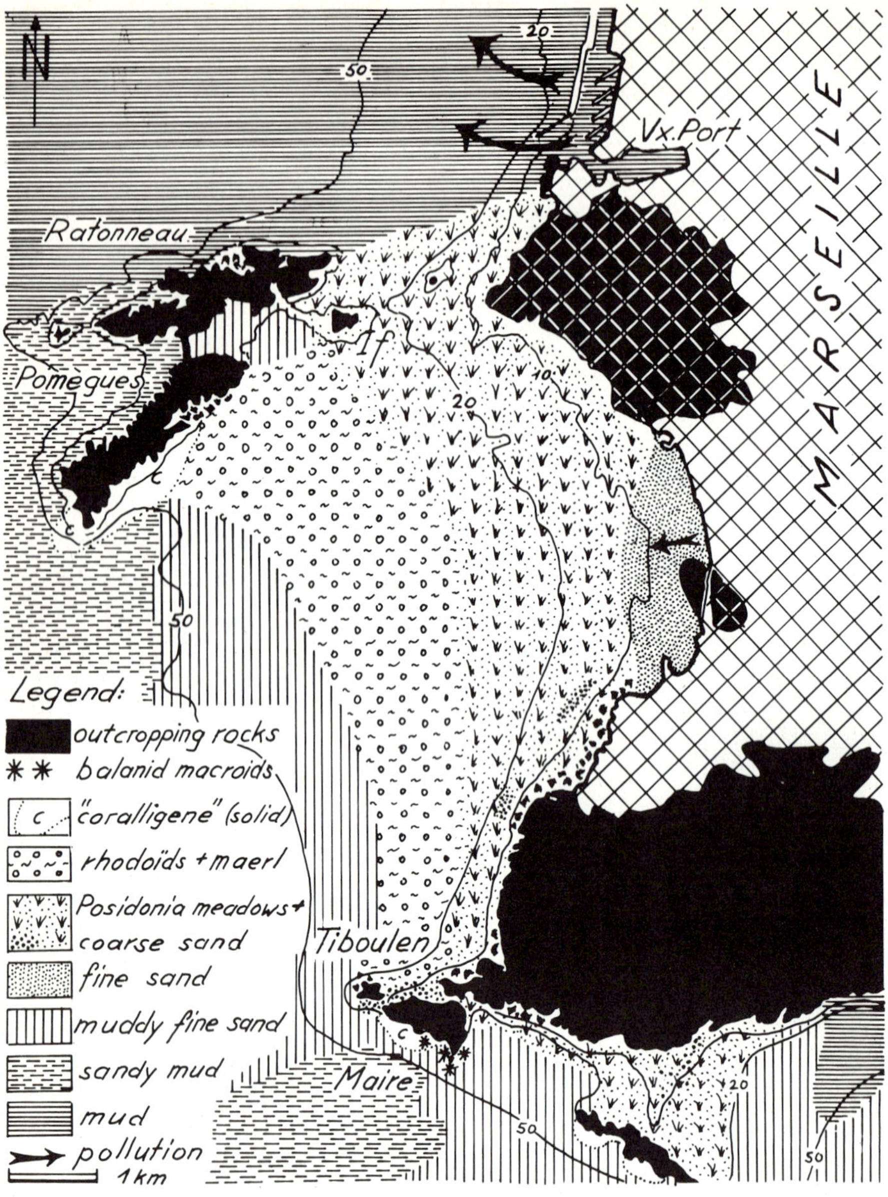

Fig. 29. Sketch map of Ile Maire and its surroundings near Marseille, France, based on the Geological Map of France, 1:50,000, sheet Aubagne – Marseille, 2d. ed. 1969, and corrected around Ile Maire as far as personal diving experience permits. Location of *Balanus* macroids (*asteriscs*)

Some species of *Balanus* are known to exhibit growth rates exceeding 1 cm per year. This fits the idea that in our case history one generation of 2 to 5 mm long balanid shells represents only some months of time in between successive turnover processes. This again is in accordance with the particular water energy regime in this environment where with each heavy storm large energies are turning up the water column down to considerable depths.

Comparing the Marseille macroids with the ones generated in similar depths and under similar, north-Mediterranean conditions of water temperature, light penetration and nutrient supply in Elba, we must conclude that the turnover-periodicity due to water energy changes is in some equilibrium with the growth rates of the coating organisms. The shorter the turnover periods, the more strictly the coating organisms are selected by their growth rate.

6 Conclusions

Considering the three case histories presented here we may formulate the following generalized hypothesis as to the genesis of neritic macroids below the tidal zone including all kinds of organisms involved in their coating (Fig. 30).

Macroids are generated where water energy changes periodically, leaving the grain in place on its mobile substrate during a certain time and turning it over periodically. During the immobile period, the sedimentation rate must be low enough not to submerge the grains. During the mobile period, the water energy

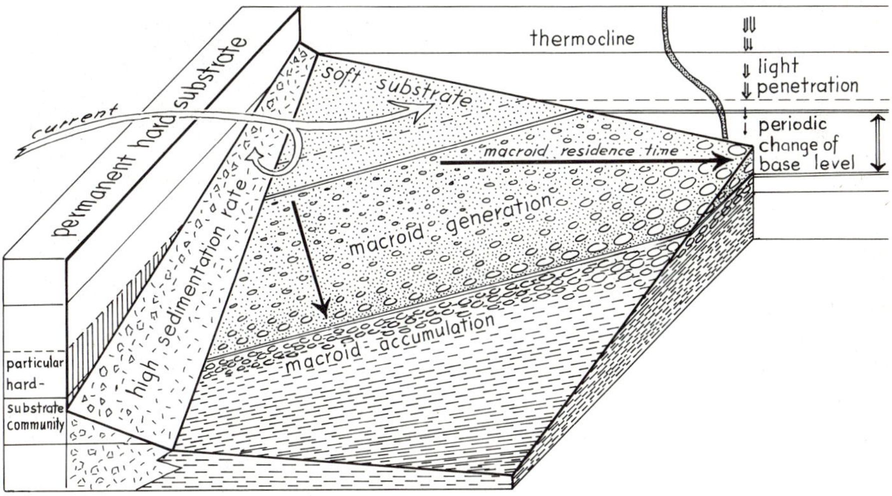

Fig. 30. Mechanisms of macroid generation in the photic and below the tidal zone. Schematic. Note the delimitation of macroid growth parallel to depth lines conditioned by light penetration and periodic change of base level. Note also the relation between the direction of slow transport of the macroids across the zone of their generation and the residence time conditioning macroid size in the moment of final deposition in the accumulation zone. The latter is likeliest to be definitively buried and fossilized

turning over the coated macroids must be high enough to winnow out the finer sediment fractions but low enough to guarantee a residence time of the coated grain in its original environment of sufficient length to permit reasonable growth of the grain coat.

These conditions form a complex equilibrium between biotic and abiotic factors: The periodicity of the changes in water energy must be equilibrated with the sedimentation rate and the growth rate of the coating organism.

The residence time of the grain in its original environment determines the size of the macroid and the selection of coating organisms. The residence time of the coated grain depends on the direction and velocity of movement in respect to the limits of its original environment or to the limits of equilibria conditions as outlined above. These again depend on the topographical configuration, particularly the inclination of the substrate.

The composition of the coating community corresponds to the first or the first few pioneer organisms on hard substrate as available in a particular environment. They are selected mainly according to their growth rates in relation to the turnover time and the residence time.

This hypothesis involving multiple-sided equilibria between abiotic and biotic factors is restricted to macroids generated by coating organisms needing hard substrate as a support and growing as patches in radial direction on a larger surface. It does not take into account the complications arising when different coating organisms compete not only for space but depend in the same time on the upper- or underside of the grain (light dependence may be one of the reasons). It does neither take into account biological mechanisms of grain turnover (by infaunal burrowing activity, for instance).

This hypothesis does not apply to grains growing concentrically in a kind of suspension, as seems to be the case of the deep-water, bacteria-induced nodules or the rhodoids suspended by continuous wave action as mentioned above.

The complex equilibria conditioning the genesis of neritic macroids discussed here indicates the difficulties to expect when fossil occurrences of such macroids have to be interpreted as paleoenvironmental indicators. The sedimentological characteristics of the matrix deposited in between the fossilized coated grains and of the overlaying strata must provide us first with an answer to the question of if we are in the presence of a fossilized zone of more or less continuous accumulation of dead macroids immediately or far below their zone of generation, or if the surface of macroid generation itself was buried by a single event of storm deposits. The community of coating and of all other organisms fossilized in the particular occurrence must provide us next with insight into the general climatic conditions in which the fossil macroids were formed. In the framework of these interpretations only, an analysis of the coating organisms, the grain size distribution and the sphericity of the grains may give us further indications on the paleoenvironment, their periodicity of water energy and the bathymetry of the deposit. Although the interpretation of fossil neritic macroids may be complex, the presence of such macroids indicates always periodic changes in water energy and very low sedimentation rates on soft substrates in the zones where the macroids were generated.

References

Focke JW, Gebelein CD (1978) Marine lithification of reef rock and rhodolites at a fore-reef slope locality (−50) off Bermuda. Geol Mijnboukwkd 57:163−171

Gvirtzmann G, Buchbinder B (1978) Recent and pleistocene coral reefs and coastal sediments of the Gulf of Elat. In: 10th Int Congr Sedimentol Guideb, vol II. Int Assoc Sedimentol, Jerusalem, pp 161−191

Hottinger L (1977) Distribution of larger peneroplidae, Borelis and nummulitidae in the Gulf of Elat, Red Sea. Utrecht Micropaleontol B15:35−109

Hottinger L (1981) L'énergie de l'eau, son action sur les substrats et la répartition d'organismes benthiques choisis à Punta Galera, Ile d'Elbe, Italie. XXVIIe Congrès-Assemblée plénière CIESM. Rapp PR Reun Cons Int Explor Mer 27(5):181−182

Hottinger L, Levinson G (1978) Cementation of reefs in the Gulf of Elat (Red Sea) by Acervulinid Foraminifera. In: Friedman GM (ed) Abstr Lect 10th Int Congr Sedimentol, vol I, Jerusalem, p 315

Klinker J, Reiss Z, Kropach C, Levanon I, Harpaz H, Halicz E (1976) Observations on the circulation pattern in the Gulf of Elat, Red Sea. Isr J Earth-Sci 25:85−103

Klinker J, Reiss Z, Kropach C, Levanon I, Harpaz H, Shapiro Y (1978) Nutrients and biomass distribution in the Gulf of Aqaba (Elat), Red Sea. Mar Biol 45:53−64

Pérès JM (1961) Océanographie biologique et biologie marine. T1. La vie benthique. Presses Univ, Paris, pp 1−541

Pérès JM, Picard J (1964) Nouveau manuel de bionomie benthique de la Mer Méditerranée. Recl Trav Stn Mar Endoume-Marseille Fasc Hors Ser Suppl 31/47:5−138

Reiss Z (1977) Foraminiferal research in the Gulf of Elat-Aqaba − A review. Utrecht Micropaleontol B15:7−34

Riedl R (1966) Biologie der Meereshöhlen. Parey, Hamburg Berlin, pp 1−636

Zohary T, Reiss Z, Hottinger L (1980) Population dynamics of Amphisorus hemprichii (Foraminifera) in the Gulf of Elat (Aqaba), Red Sea. Eclogae Geol Helv 73/3:1071−1094

I.6 Accretionary Lapilli in Volcanic Ash Falls: Physical Factors Governing Their Formation

Thomas O. Reimer[1]

1 Introduction

"It cannot give a better description of its (the cloud's) shape than by comparing it with that of a pine tree; it rushed to a great height, straight and smooth as the trunk of a tree which divides into branches at the top." This is the description of the initial ash cloud of the eruption of Mount Vesuvius on August 24, 79 A. D. as given by C. Plinius (the younger) in a letter to the historian P. C. Tacitus. Had he shown the same scientific interest as his uncle C. Plinius (the elder) who died during the eruption, probably from respirational problems, he might have discovered in the ash layers peculiar pellets frequently exhibiting concentric textures. Such bodies are now generally referred to as accretionary lapilli, in preference over terms as volcanic hailstones, mud balls, etc.

Their possible paleo-environmental significance lies in the suggestions by Lowe and Knauth (1978, 1979) that their occurrence in Archaean rocks could be used as indicators for the composition of the contemporaneous atmosphere, a view that was questioned by Self and Sparks (1979). The present paper attempts a classification of the various types of accretionary lapilli described in the literature. It furthermore discusses the physical and mechanical factors governing the formation of these peculiar bodies in order to assess their presumed indicative value.

2 Classification

While accretionary lapilli have so far been rarely described from Precambrian rocks, they are a well-documented feature in the Phanerozoic and their formation has been observed during a number of historic eruptions in various parts of the world (Table 1). Little work has been done on the internal texture of such recent specimens, but from the information available three types can be recognized.

The first group (type A) are spheres or balls of mud or irregular-shaped aggregates (Fig. 1) which can attain sizes of up to 15 cm. They form either through agglomeration of moist ash in an eruptive cloud or around rain drops falling

1 c/o Dyckerhoff Engineering GmbH, P.O. Box 2247, D-6200 Wiesbaden-Biebrich, FRG

Coated Grains (ed. by T.M. Peryt)

Table 1. Observations of accretionary lapilli falls during historic eruptions

Volcano	Date	Types			Petrology of ash	
		A	B	C	basalt.	andes/rhyol.
Vesuvius	79	——	?	——	×	
Kilauea (Hawaii)	1790	×			×	
Vesuvius	1822			×	×	
Krakatao	1883	×				×
Sta. Maria (Guatemala)	1902	——	?	——		×
Mont Pelé	1902	×		×		×
Vesuvius	1906	×			×	
Taal (Philippines)	1911	×			×	
Kilauea	1924	×	×		×	
Paricutin	1943	——	?	——	×	
Capelinhos (Azores)	1957/58	×			×	
Taal	1965	——	?	——	×	
St. Helens	1980	×				×

Sources: References in Moore and Peck (1962) and Self and Sparks (1979) St. Helens after M. Walsh (pers. comm. 1982).
Type A: Mostly round bodies of mud, partly plastic and flattened on impact, usually several millimeters to centimeters in diameter, maximum 15 cm.
Type B: "Brown, compact, dry pisolites", recent samples only 1 mm in diameter, fossil ones up to 2 cm.
Type C: Formed around rain drops and moist mineral or lithic fragments falling onto and rolling on fresh ash, usually several millimeters in diameters, also roller-shaped.
(Note: all three types can be expected to be hollow and moisture-filled upon formation)

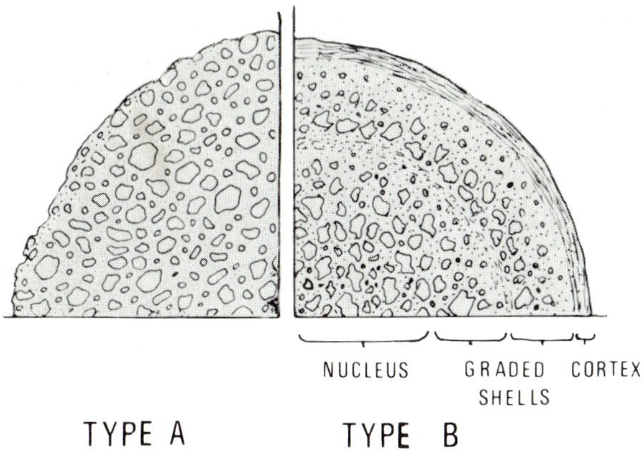

Fig. 1. Schematic drawing of types of accretionary lapilli

through ash clouds. If the moisture increases beyond a certain value mud rains will develop as observed in Jakarta during the 1883 eruption of the Krakatao or during the 1980 eruption of Mount St. Helens in the western USA. On impact on the ground this type of accretionary lapilli is frequently deformed and flattened into mud cakes. Their formation can carry on for up to 2 days as observed during the 1906 eruption of Vesuvius (Jaggar 1921, in Moore and Peck 1962), their occurrence usually being limited to the vicinity of the volcano.

The formation of the second type of accretionary lapilli (type B) has so far been observed only once, during the 1924 eruption of the Kilauea in Hawaii when a "shower of dry, brown, compact pisolites about 1 mm in diameter" fell about 2 h after an eruption (Stearns 1925, in Moore and Peck 1962). Even this short reference suggests that this type is notably different from the type A spheroids, the reference to their dry nature being especially important. Although no data are available on their internal texture it is tentatively suggested here that this type actually respresents the accretionary lapilli most frequently observed in ancient rocks. Due to their usually concentrically banded outer zone (Fig. 1) the term "coated accretionary lapilli" is proposed for them. Their brown surface could be due to an outer enrichment of ferrous minerals which has also been observed in fossil examples (Moore and Peck 1962). Individual fossil beds are known to extend over distances of up to 140 km in New Zealand (Self and Sparks 1979). Trendall (1965) found accretionary lapilli within the early Proterozoic Fortescue Group of the Pilbara Block of Western Australia over a lateral extent of up to 500 km.

A third type of accretionary lapilli (type C) forms on the ground when rain drops, larger moist lithic or mineral fragments as well as other accretionary lapilli fall onto beds of freshly deposited fine-grained ash. When these objects roll on the fresh ash fine particles stick to the moist surface thereby increasing the size of these spheroids. Some of them are roller-shaped, consisting of a continuous rolled-up layer. Their frequency of formation appears to be the same as that of the type-A ones, also usually being restricted to the vicinity of the volcano.

Depending on the meteorological conditions apparently even accretionary lapilli consisting of tephra and ice can form. Thorarinsson (1969) described hail showers from the 1963 eruption of Surtsey, a grain of tephra forming the core of each hailstone.

3 Physical Forces of Granulation

Accretionary lapilli are a well-documented phenomenon accompanying ash eruptions, but there is a notable lack of information on the processes which lead to their formation in the air. A relatively high moisture content of the parent ash cloud is usually thought to be a prerequisite (Moore and Peck 1962). In contrast to this, there is a vast amount of data in the technical literature on comparable processes which are collectively known as "agglomeration" or "granulation".

In the technology of fine-grained materials granulation entails the formation of larger, usually spherical bodies through the combination or recombination of finer-grained materials. It can be carried out at different temperatures in layers of materials, in liquids or in air or other gases. A comprehensive survey of the physical factors governing these processes has recently been given by Schubert (1979). For the present investigation five processes are of main interest, the respective forces exerted by them are shown in Fig. 2.

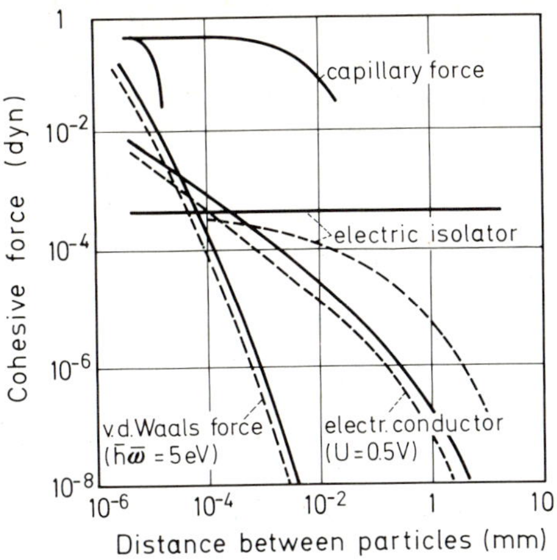

Fig. 2. Cohesive forces between airborne particles. (*Solid line:* 10 μm sphere against flat plate. *Broken line:* between 2 spheres of 10 μm ∅. (Modified after Schubert 1979)

3.1 Electrostatic Attraction

This force is especially effective between non-conductive materials (isolators) such as the majority of the ash-forming minerals and glasses. Technically it is extensively used in the removal of dust form waste gases as, for example, in the cement industry and in power plants. The omnipresence of electrostatic potentials during volcanic eruptions is witnessed by the frequent occurrence of flashes of lightning within ash clouds. Anderson et al. (1965) calculated concentrations of 10^5 to 10^6 elementary charges per cm^3 in the cloud originating from the Surtsey eruptions 1964/1965.

It has been found that electrostatic attraction is most effective for bringing small charged particles into contact with each other over distances which are large in comparison with the particle diameter.

3.2 Capillary Pressure of Pore Fluids

In partly moisture-filled pores of fine-grained granulated materials a strong cohesive force is exerted by the surface tension of the respective liquids. Figure 2 presents a calculated example for the forces acting between a plate and a sphere. The lower curve is for a small moisture content, the upper one for a higher content. For a model of two spheres the calculation is complicated by the size of the liquid bridges and the surface tension of the liquids used.

Moisture is a consistent component of ash clouds which represent highly turbulent aerosols. During their ascent they pull in large amounts of atmospheric air from the sides, heating it at the same time. The resulting decrease in density contributes to the rise of the thermal plume. The moisture contained in the air together with the one inherited from the initial ash cloud eventually condensate, thereby supplying abundant water droplets. In the presence of non-hygroscopic nuclei as represented by the ash particles condensation only occurs above 100% relative humidity while below this value the particles are already covered by a thin film of moisture (Meldau 1964). It is thus not surprising that freshly fallen ashes are frequently moist even if not accompanied by rain falls (E. Dimroth 1980, pers. comm.).

While electrostatic attraction is the main force for bringing the particles into contact, it is actually the capillary pressure which keeps the resulting granules together. However, it will only remain active as long as not all of the pores are filled with liquid. Limestone dust granules investigated by Schubert (1979) began to disintegrate once the moisture content of the pore volume grew to above 90%. Figure 2 shows that the capillary cohesion is effective only over short distances and collapses beyond about 2.5 μm.

3.3 van-der-Waals Forces

These forces are caused by the mutual interaction of the dipole moments of atoms and molecules and are of great importance for the processes discussed here. Figure 2 shows that they can exert cohesive forces close to those due to capillary pressure of pore fluids. They are especially effective in the granulation in dry aerosols.

3.4 Crystallisation of Dissolved Matter

While the forces described so far are to some extent reversible, non-reversible contacts can also be established. In industry a large variety of granules are produced with the specific addition or in the presence of moisture. This moisture either dissolves some of the material to be granulated or contains already certain dissolved substances on introduction into the process. When the resulting granules are dried, the dissolved matter will become concentrated and eventually precipitated at the site of the original liquid bridges, thereby establishing solid crystal bridges which greatly increase the stability of the granules. If the precipitated material is again water-soluble the granules will disintegrate when immersed in water, a process that is applied in a large number of "instant"-soluble products of everyday life. In experiments it has been found that such solid bridges can increase the stability of technically formed granules by up to two orders of magnitude (Knacke and Pohl 1959).

In ash clouds this process is also possible. The condensated warm moisture within accretionary lapilli will contain some of the original volcanic gases in a dissolved form and thus be chemically rather aggressive. The ash particles are to a large extent warm glass shards and thus highly susceptible to dissolution. If these accretionary lapilli are losing their moisture, e.g., when falling through warm dry air zones, the dissolved matter will start to crystallize and form material bridges which will have a large stabilizing effect on them.

3.5 Sintering at High Temperatures

In industry this process is rarely used although the formation of clinker nodules from raw meal in a rotary cement kiln could be compared to this process. In nature a comparable process would be welding in ignimbrites which can take place at temperatures as low as 500° – 600 °C (Füchtbauer and Müller 1970). In industry it has been noted that for sintering to occur readily the temperature has to reach at least 60% of the melting temperature of the materials concerned. At lower temperatures the reaction would be too slow to be of importance. This process might make some contribution to the formation of the cores of accretionary lapilli. However, considering the great rapidity with which water condensates in ash clouds, thereby indicating temperatures below 100 °C, it is unlikely that it is of great importance for the formation of accretionary lapilli.

3.6 Reactions in Fluidized Beds

Applying the above processes on an industrial scale, it is possible to produce granules similar to accretionary lapilli by the "fluidized bed" principle in its various modifications. This process basically entails the introduction of fine powders and finely sprayed fluids or finely sprayed fluids containing the material to be granulated into a reaction vessel at a narrowly controlled rate. A fluidized bed of particles is established by blowing air, which in some instances is heated, into the vessel from below. The granulation of the particles takes place initially through electrostatic attraction and the cohesion of the resulting granules is stabilized by the capillary pressure. The moisture is eventually usually evaporated and permanent material bridges are established at or near the original grain contacts. Alternatively van-der-Waals forces keep the dry granules together.

The size of the resulting granules is governed mainly by the gas velocity and limited to a maximum of about 5 mm due to technical constraints. A narrow grain size spectrum is usually obtained. The granules shown in Fig. 3 with a median diameter of 0.8 mm were produced at a gas velocity of about 3.8 m/s (Grunert et al. 1980). Depending on the materials used spherical granules can be obtained while concentric textures are the result of variations in the grain size and/or type of the materials fed into the process. Under certain conditions it is even possible to produce granules around a nucleus of liquids which will become hollow on drying.

In nature the fluidized-bed principle would be represented by the ascent of the hot turbulent ash cloud. However, are the particle concentrations and the distances between the particles in the two environments such that the comparison is really valid? No information on these factors in ash clouds could be found, and a technical example is used as a starting point for theoretical considerations.

The thick gray to white smoke belching from the stack of an unfiltered cement kiln contains a particle concentration of about 50 g/m^3. Assuming that this smoke consists of spheres of uniform size of 7 μm the average distance between them will be about 0.25 mm (Fig. 4). Considering the volumes deposited from volcanic ash clouds this concentration appears to be rather low. Tazieff

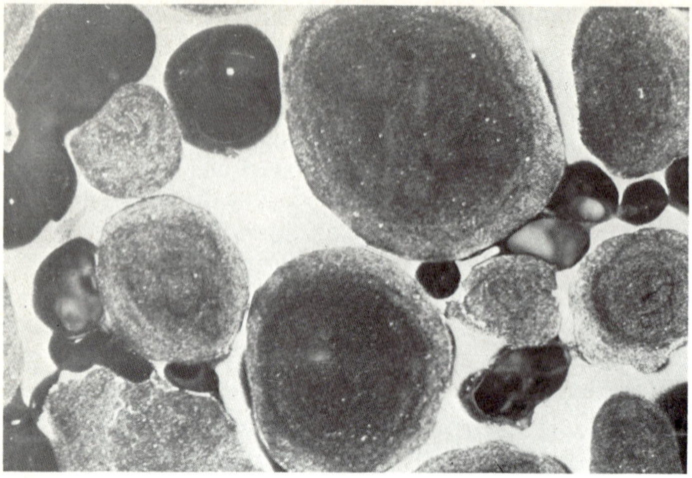

Fig. 3. Granules of bariumferrite produced in fluidized-bed reactor (Grunert et al. 1980, diameter of largest granule 0.8 mm)

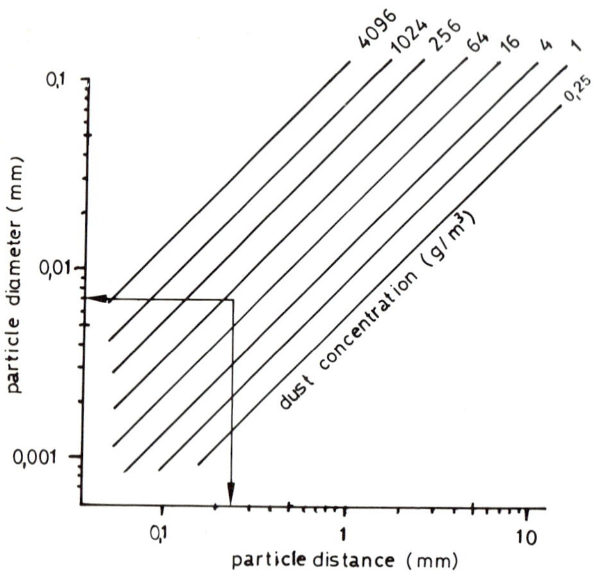

Fig. 4. Relation between particle concentration and distance in aerosols

(1968) presented a model calculation in which the vaporization of 900 kg of water, equivalent to about 1800 m^3 of vapour, is able to pulverize 1 m^3 of basaltic lava. This would correspond to a dust load of about 1.5 kg/m^3, a rather high value which most probably is only realized in the lower parts of the ash cloud shortly after leaving the vent. A more realistic example would be a cloud 500 m thick and depositing 5 cm of ash, without considering the porosity of the material as deposited. The initial dust load would then have been 250 g/m^3 and the median distance for spheres of 10 μm diameter would have been about 0.2 mm. Thus within volcanic ash clouds the average particle intervals would be around 0.1 – 0.5 mm and consequently within the range of the forces described above. To go back to a technical example – in the investigations of Grunert et al. (1980) dust loads were 14 – 70 g/m^3. This results in particles distances of 0.3 – 0.5 mm for 10-μm spheres. It appears that for the formation of accretionary lapilli medium to low dust loads are more suitable. In denser clouds the mutual attrition of the ash particles could probably counteract the small forces of granulation.

4 Formation of Accretionary Lapilli

While accretionary lapilli, both fossil and recent, are frequently described in literature, one important feature so far apparently has escaped notice. The fossil examples are in their majority different from the ones which have been observed during recent ash falls, the fossil ones usually being coated (type B) while the recent ones are mostly of the uncoated type A (Fig. 1). Any theory of the origin of these spheroids will therefore have to explain two modes of formation and at the same time explain why the ones most frequently forming are the ones least likely to be preserved.

Using the above-described technical factors, a conceptual model was developed for the explanation of the processes governing the formation of accretionary lapilli. It might still leave a number of features unexplained, but should at least serve as an impetus for further work. Experiments on the formation of accretionary lapilli in a fluidized-bed reactor are presently under way.

The formation of the two types mentioned is shown schematically in Fig. 5. It is difficult to ascertain when the cores of accretionary lapilli are forming, but considering the weak forces in question it is unlikely that they originate in the lower to middle parts of the ascending stem of an ash cloud where velocities well in excess of 100 km/h are reached (Thorarinsson 1968) and where particle concentrations and thus mutual attrition are high. Once velocities have decreased and moisture starts to condensate, in the lower part of the laterally spreading cloud, ash particles and droplets of moisture will be drawn together by electrostatic attraction. Mineral and lithic fragments as well as larger drops of moisture can act as nuclei.

The continued cohesion of the resulting speroids is effected mainly by the capillary pressure developing from the enclosed droplets of moisture. At the same time the warm water starts to dissolve the surfaces of the ash particles due to its content of acids inherited from the gases of the original ash cloud. Initially there should be little size grading of the particles in the resulting spheroids as they and the enclosing ash are moving in the same direction being carried by the gas stream.

The forces of attraction here are not counteracted by large differences in the direction of movement. Once they have grown to a size, the free fall velocity of which is larger than the upward velocity of the ash cloud, the spheroids will start to descend. For a sphere of 2 mm diameter this would represent a velocity of 10 m/s. Thus in a highly turbulent ash cloud larger nuclei can form than in a cloud with "quieter" flow conditions.

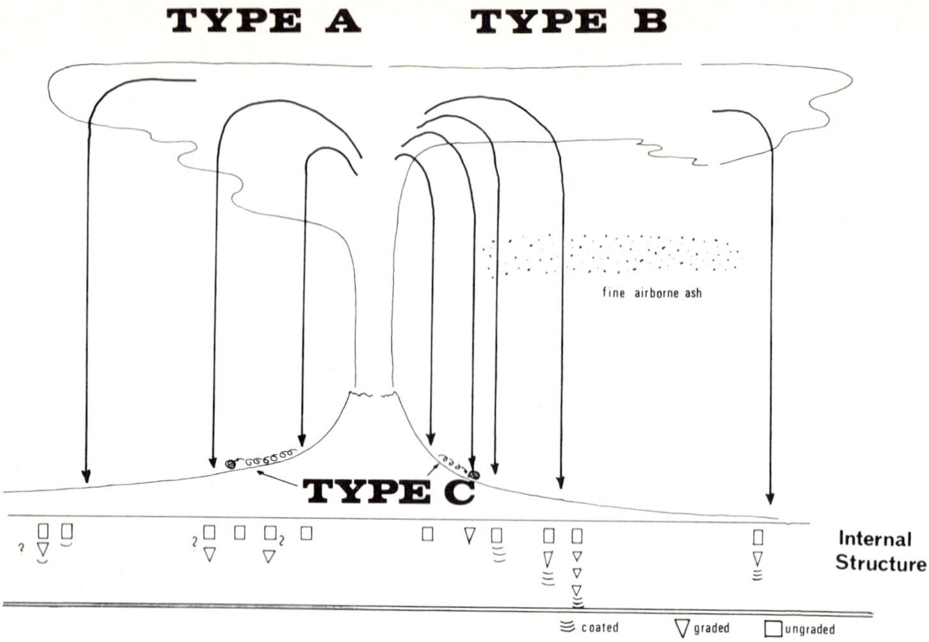

Fig. 5. Conceptional model for formation of accretionary lapilli

From this stage onwards it becomes necessary to assume two different paths of developments for the two types of accretionary lapilli forming in the air (Fig. 4). In the highly turbulent ash cloud a rather high content of moisture and ash particles is required. Once the spheroids start to descend through the cloud the size of the particles added to them becomes successively smaller due to the difference in the direction of movement of spheroids and ash particles. Size grading can develop. It is a well-established fact that the median grain size of particles in accretionary lapilli is usually up to 3 phi-units smaller than that of the embedding ash bed from which they most probably formed while in the air (Walker 1971). However, no data exist on the internal texture of recent accretionary lapilli.

The spheroids will now have grown to a size of several millimeters and contain a fair amount of moisture. As they are falling with increasing speed they collide with further ash particles and drops of moisture and thereby grow to considerable sizes by plain mechanical gathering, even forming irregular aggregates up to 15 cm across. Some coating with very fine-grained ash could develop if they fell through drier zones of fine-grained ash. The essential point is, however, that they retain most of their moisture until impact on the ground, their falls at times even being accompanied by rains. Their formation would take place over a time interval of probably less than 5 min and they usually fall within the vicinity of the volcanic vent. These type A accretionary lapilli are the ones observed falling most frequently.

For a typical specimen of the coated accretionary lapilli (type B) the following sequence of events is visualized: Around a descending nucleus size grading starts to develop in the ash cloud. Should it continue to fall outside ash clouds no grading will develop, while multiple grading would be the result of fall in a turbulent environment with repeated upward and downward movement. Moore and Peck (1962) observed that the grading in the vast majority of samples is unidirectional from the inside outwards. Reverse grading is rare and double nuclei can develop. Eventually they fall through rather dry zones containing abundant fine ash, probably the result of previous eruptions. Their pore moisture will now gradually start to evaporate. Minerals crystallizing from the solution will establish stable material bridges at or near the original grain contacts.

With the continued fall it will become increasingly difficult to attach larger particles to the spheroid as due to its small diameter it does not have the same gathering effect as those of the type A. In contrast to this the attraction of very fine ash particles is effected by electrostatic forces, aided by the potential of the cloud traversed and by the friction between the airborne ash and the falling spheroids. Some fracturing of spheroids can occur. A sand-blasting effect of the ash particles against the failling spheriod (Moore and Peck 1962) should be rather small due to the very fine size of the ash, the forces of attraction and cohesion exerted by the electrostatic and van-der-Waals forces being larger than the erosive effect of the free ash particles. Platy particles might become attached preferentially due to the electrostatic principles discussed above. The various laminae of the coating could represent different compositions, grain sizes, and concentrations of ash within the different zones of the ash cloud traversed. Irregular growth rings can also develop.

The result of the fall through a dry cloud of fine ash will be a fine-grained dense coating of great cohesion and probably low permeability which should give the resulting accretionary lapilli a pearl-like appearance most likely as the "dry compact pisolites" of Stearns (1925). Some of the coatings appear to have been still unstable and flexible as indicated by broken and deformed portions. It is possible that certain specimens of this type containing amygdal-like cores formed originally around droplets of moisture and were filled shortly after deposition.

A notable feature of this type of accretionary lapilli is their wide distribution, identifiable individual beds having been traced for up to 140 km in New Zealand (Self and Sparks 1979). For a bed identified as accretionary lapilli by Lowe and Knauth (1978) in the Archaean Fig Tree Group of South Africa the author previously had reconstructed an original extent of at least 40 × 100 km (Reimer 1975). Trendall (1965) described their occurrence over a lateral extent of about 500 km.

This wide extent of coated accretionary lapilli poses the question as to the reasons for the distribution of nuclei over such large areas. It is not yet clear whether the falls occur simultaneously over the whole area or whether a front of nucleation moves laterally through the cloud or whether the nuclei are transported by strong winds away from the vent and start to grow and fall only farther away. Their formation itself will only take a few minutes and even the 140 km mentioned above can be traversed by the ash cloud within about 2 h using the travel velocities of volcanic ash clouds reconstructed by Thorarinsson (in Walker 1971). As the coated accretionary lapilli (type B) apparently only rarely form, the respective ideal conditions of grain size, moisture, turbulence, and electric charge in and around ash clouds are not met easily. If, however, they do occur, they can spread over large areas, as shown by the fossil examples.

5 Preservation of Accretionary Lapilli

The question remains why the type least frequently observed in recent falls is the one predominantly preserved in the fossil record and vice versa. The model proposed above offers an explanation for this apparent contradiction. It has been pointed out that for the accretionary lapilli of type A the continued presence of capillary pressure from pore fluids is the essential stabilizing factor. This, however, will collapse if these bodies are deposited in water and they will consequently disintegrate (Moore and Peck 1962). Only those deposited subaerially will have a chance of survival, although also a large number of them will be destroyed by rains accompanying or following ash falls. As on the other hand subaerial volcanic deposits have only a low potential of preservation, this also applies to the accretionary lapilli contained in them. It appears furthermore possible that even when preserved in tuffs of similar mineralogical composition their outlines can be easily obliterated by diagenesis. Walker (1971) suggested that air bubbles preserved in certain tuffs might represent the nuclei of originally water-filled accretionary lapilli that formed around water droplets rolling on freshly deposited ash. The same considerations apply to the preservation of the type C accretionary lapilli.

In contrast to these two types, the coated accretionary lapilli form less frequently but due to their dense fine-grained cortex and the stabilizing material bridges at the grain contacts they will not disintegrate as easily when deposited in water. Their spherical outline will furthermore aid their preservation even during subsequent sedimentary transport. An additional factor aiding their preservation is their frequently wide primary dispersal and deposition also in non-volcanic environments such as in deeper water geosynclinal sediments (Bateson 1965). Their predominance in the fossil record thus is mainly the result of their potential for preservation.

6 Conclusions

The "apparent abundance" of accretionary lapilli in Precambrian rocks of South Africa and Australia led Lowe and Knauth (1978) to suggest the existence of "an abnormally wet early atmosphere" at that time. However, before one can talk about abundances or relative abundances of accretionary lapilli at all it will be necessary to carry out a systematic survey of their fossil occurrences, describing the various types present, their sizes, chemical compositions, enclosing sedimentary environments, lateral extent, etc. in detail.

The small number of observations in Precambrian rocks could be accidental, occurrences in other rocks of this age having been overlooked due to their usually small size or even been misinterpreted as oolites. A brief survey of the literature on the Archaean of Zimbabwe furnished already three questionable reports of "oolites" from non-carbonate environments (Reimer 1983) Several bands of accretionary lapilli have recently been observed by the author in upper Archaean (pre-2.7 b.y.) rocks of the Lopian Suite of the Baltic Shield northwest of Petrozavodsk (Karelia/USSR). Others might have been mistaken for ferruginous granules in banded iron formations.

The indicative value of accretionary lapilli for atmospheric conditions and especially for atmospheric moisture content is rather limited as they form during processes which themselves introduce large amounts of moisture into the atmosphere. Furthermore, the evaluation of the physical processes governing their formation shows that if higher moisture contents had prevailed during the Precambrian, type-A accretionary lapilli would have formed preferentially which, however, have only a low potential for preservation. The lack of this type in the fossil record is thus not in itself an indicator for dry atmospheric conditions.

On the other hand, coated accretionary lapilli only form if dry air zones are present around the parent ash cloud. Their widespread formation over large areas indicates that such conditions were not unusual in the Precambrian and could not have been just "dry holes" in an otherwise continuous cloud cover.

It appears therefore reasonable to assume that the distribution of moisture in the atmosphere since the early Archaean has not deviated much from present conditions and that at least by about 3.5 b.y. ago the interaction between atmosphere and hydrosphere had stabilized at or near modern levels. This does not exclude the possibilities of higher ambient temperatures in the Archaean.

Further factors militating against an abnormally wet Precambrian atmosphere are Archaean subaerial evaporites from 3.55 b.y. old sediments of the Pilbara Block of Western Australia (Dunlop 1978) and the occurrence of mudcracked shales in tidal and fluviatile sediments of the 3.3 b.y. old Moodies Group of South Africa (Visser 1956) and numerous other Precambrian formations.

Consequently, if one wants to use accretionary lapilli as paleoenvironmental indicators they would rather point to the existence of dry atmospheric zones around volcanic ash clouds. Their importance lies more in the fact that they can represent excellent time markers in fossil-free sedimentary and volcanic formations. It is hoped that this paper will stimulate further investigations of fossil and recent examples of these peculiar structures.

Acknowledgements. This paper presents a contribution to IGCP-Project 160. Its presentation at the meeting of IGCP-projects 158 and 160 at Mexico City in January 1982 was made possible through a travel grant of the Deutsche Forschungsgemeinschaft. The permission of VEB Verlag Technik (Berlin, DDR) to publish Fig. 3 is gratefully acknowledged.

References

Anderson R, Björnsson S, Blanchard DC, Gathman S, Hughes J, Jonasson S, Moore CB, Survilas HJ, Vonnegut B (1965) Electricity in volcanic clouds. Science 148:1179–1189

Bateson JH (1965) Accretionary lapilli in a geosynclinal environment. Geol Mag 102:1–8

Dunlop JSR (1978) Shallow water sedimentation at North Pole Pilbara Block, Western Australia. In: Glover JE, Groves DI (editors) Archaean cherty metasediments: their sedimentology, micropaleontology, biogeochemistry, and significance to mineralization, vol II. Publ Geol Dep Extensions, Univ. West Aust, pp 30–38

Füchtbauer H, Müller G (1970) Sedimente und Sedimentgesteine, Schweitzerbart, Stuttgart
Grunert L, Schöps W, Lübke M (1980) Granulierung von Materialien der technischen Keramik. Silikattechnik 31:235 – 237
Knacke O, Pohl H (1959) Experimenteller Beitrag zu den Grundlagen des Pelletisierens. Chem-Ing Tech 31:50 – 54
Lowe DR, Knauth PR (1978) The oldest marine carbonate ooids reinterpreted as accretionary lapilli, Onverwacht Group, South Africa. J Sediment Petrol 48:709 – 722
Lowe DR, Knauth PR (1979) The oldest marine carbonate ooids reinterpreted as accretionary lapilli, Onverwacht Group, South Africa: Reply. J Sediment Petrol 49:664 – 666
Meldau R (1964) Handbuch der Staubtechnik, vol II. VDI, Düsseldorf
Moore JG, Peck DL (1962) Accretionary lapilli in volcanic rocks of the western continental United States. J Geol 70:182 – 193
Reimer TO (1975) Paleogeographic significance of the oldest known oolite pebbles in the Archaean Swaziland Supergroup (South Africa). Sediment Geol 14:123 – 133
Reimer TO (this vol.) Accretionary lapilli and other spheroidal rocks in the Swaziland Supergroup of the Barberton Mountain Land (South Africa), pp 619 – 634
Schubert H (1979) Grundlagen des Agglomerierens. Chem-Ing Tech 51(4):266 – 277
Self S, Sparks RSJ (1979) The oldest marine carbonate ooids reinterpreted as accretionary lapilli: Discussion. J Sediment Petrol 49:661 – 663
Tazieff H (1968) Sur le mécanisme des éruptions basaltiques sous-marines à faibles profondeurs et la génese d'hyaloclastites associés. Geol Rundsch 57:955 – 966
Thorarinsson S (1968) On the rate of lava and tephra production and the upward migration of magma in four Icelandic eruptions. Geol Rundsch 57:705 – 718
Thorarinsson S (1969) Surtsey – The new island in the North Atlantic. Cassell, London
Trendall AF (1965) Pisolitic tuffs in Western Australia. Geol Surv West Aust Annu Rep 1964:105 – 109
Visser DJL (1956) The geology of the Barberton area. S Afr Geol Surv Spec Publ 15
Walker GPL (1971) Grain-size characteristics of pyroclastic deposits. J Geol 79:697 – 714

II Ooids

II.1 Calcareous Ooids: A Synopsis

Detlev K. Richter [1]

1 Introduction

In 1855, Lyell gave an appropriate definition of "oolitic grains" (p. 12): "The variety of limestone called 'oolite' is composed of numerous small egg-like grains, resembling the roe of a fish, each of which has usually a small fragment of sand as a nucleus, around which concentric layers of calcareous matter have accumulated." Also Sorby's detailed description of calcareous "oolitic grains" from 1879 is still modern. He distinguished between the following three types: (1) "Concentric structure" characterizes the aragonitic oolitic grains of the hot springs at Karlsbad (CSSR) and of the modern sediments in the Bahamas. "Concentric" means, according to Sorby, layers of tangentially oriented aragonite crystals. (2) "Radiate structure" refers to the radial-fibrous structure of many calcitic fossil oolitic grains. Sorby recognized this structure as primary. This was neglected, at least for marine ooids, from Cayeux (1935) to Shearman et al. (1970) and has been reestablished by Simone (1974), Sandberg (1975) and Wilkinson and Landing (1978) only in the last ten years. (3) "Recrystallized structure" is developed in calcitized former aragonitic oolitic grains, in which "the larger irregular crystals formed in them have either no special orientation or are arranged with the principal axis converging towards the centre".

Kalkowsky (1908), in his publication on the oolites of the Lower Buntsandstein of Middle Germany, distinguished between the rock ("oolite") and the particles ("ooids"). Since then, "ooid" has replaced "oolitic grains" [Oolithkörner, e.g. Frantzen (1887) in German publications]. In the English language, the distinction of "oolite" and "ooid" was not established until Teichert's discussion in 1970; in earlier publications, "oolith" or "oolite" was used also for the particles.

In the sixties, electron microscopy provided a closer view of single ooids. Thus, the aragonite crystals of the Bahama-type ooids turned out to be micron-sized needles (Shoji and Folk 1964), rods or batons (Fabricius and Klingele 1970, Loreau 1970). The latter shape, which is otherwise only known from biocrystallites, was used by Fabricius (1977) to suggest an organic participation in the formation of Bahama-type ooids.

---

1 Geologisches Institut, Ruhruniversität Bochum, Universitätsstraße 150, D-4630 Bochum-Querenburg, FRG

Coated Grains (ed. by T.M. Peryt)

In 1975, marine ooids of radial-fibrous high magnesian calcite were found by Marshall and Davies in the Great Barrier Reef and by Milliman and Barretto on the Amazon shelf. These observations are most important, since already Sorby (1879) considered the radial-fibrous texture of fossil calcitic ooids primary, which, however, was substantiated in modern environments only in spelaeothems and mines (Hess 1929, Hahne et al. 1968). A primary radial-calcitic composition of Paleozoic and Mesozoic ooids, in contrast to the prevailing tangential-aragonitic ooids in the modern sea, has been attributed to increasing Mg/Ca ratio in the seawater since the Paleozoic, by Sandberg (1975) and Wilkinson (1979). However, well-preserved ancient radial-fibrous ooids (the author uses the continental classification of coated grains and consequently the paper discusses also (chapter 3) vadoids) with ferroan calcite composition indicate an original magnesian calcite which was replaced in situ by ferroan calcite in a reducing environment rich in ferrous iron (Richter and Füchtbauer 1978 and, based on a larger number of ooids, Richter 1980). The chances are that all marine radial-calcitic ooids in the sedimentary record consisted primarily of high magnesian calcite, like the two modern occurrences mentioned above. Therefore the ooids cannot be used as an indication of a changing Mg/Ca ratio in seawater since Paleozoic times.

The intention of this paper is to provide geologists with all the information on ooids necessary for thin section analyses. The paper is based on literature as well as on the author's observations. Important facts on ooids such as sorting, hydraulic energy, special shape, and experimental formation are not or only briefly discussed, because excellent reviews have been published recently by Simone (1981) and Flügel (1978, p. 114 – 140) which include these subjects.

2 Definition and General Properties of Ooids

In all definitions of ooids from Lyell (1855) to Flügel (1978), a concentric arrangement of carbonate layers around a core is emphasized. This applies to ooids with spherical, but not to those with angular, elongate and skeletal cores, in which during ooidic encrustation a more and more spherical shape is attained (Fig. 2A). This *increasing sphericity with growth,* as described for marine ooids by Carozzi (1960, p. 240) and Newell et al. (1960, p. 490) and for hot spring ooids by Richter and Besenecker (this vol.) is probably the best criterion of ooids.

Though the *thickness of the laminae* in the cortex is important with respect to the size variance in the different occurrences, it cannot be used for the definition of ooids, because in many oolitic rocks the real thickness of laminae cannot be recognized. Whereas the micron-sized laminae of the tangential aragonite ooids of the Bahama-type (Newell et al. 1960, plate 1 and 2) and of hot spring occurrences (Richter and Besenecker, this vol.) are visible under the microscope, the lamination of radial-calcitic and "superficial ooids" (Carozzi 1960, p. 238) can often be recognized only be cathodoluminescence (Fig. 2F), provided that the laminae differ with respect to Mn and/or Fe. If such differences in composition of radial-aragonitic, radial-calcitic, and micritic ooids are lacking, cathodoluminescence will not show any lamination in the cortex. In such a case one has to depend on inclusion rims which, however, do not necessarily correspond to the thickness of the original laminae.

Though most ooids are smaller than 2 mm (Flügel 1978, Fig. 17), it is not meaningful to postulate a *grain size* boundary between "ooids" and "pisoids" at a diameter of 2 mm as suggested by Choquette (1978) and Donahue (1978), based on older usage, because the ooids of several occurrences exceed this limit without structural and mineralogical changes. Selected occurrences with ooid populations covering a range from < 2 mm to > 2 mm are listed below.

1. Radial-fibrous dolomite ooids up to 10 mm diameter from the Precambrian of Spitsbergen (Radwanski and Birkenmajer 1977).
2. Radial-fibrous and micritic calcite ooids up to 5 mm from the Precambrian Boomplaas Formation of Griqualand West, South Africa (Beukes, this vol.).
3. Radial-fibrous ferroan calcite ooids up to 5 mm from the Dinantian near Velbert, Nordrhein-Westfalen (Richter 1980).
4. Radial-fibrous calcite ooids up to 7 mm of the Lower Triassic "Rogenstein" near Aderstedt/Bernburg and Eggerode/Thale, Harz, DDR (Kalkowsky 1908).
5. Microcrystalline calcite ooids up to 8 mm with indications of in-situ transformation (aragonite → calcite) from the Lower Anisian (Triassic) near Episkopi, island of Hydra, Greece (see 16 in Table 2).
6. Tangential aragonite ooids up to 3,5 mm of a Quaternary marine terrace north of Kira Wrissi, Isthmus of Corinthos, Greece (thin section 4.408, Geolog. Inst. Bochum; other thin sections of calcitized rocks of the same locality contain ooids up to 5 mm).
7. Radial calcitic ooids up to more than 10 mm from the mine at Guggenbach, Steiermark, Austria (Hahne et al. 1968).
8. Tangential aragonite ooids of the "Sprudelstein"-deposits in the hot spring of Karlsbad, CSSR (Sorby 1879), – up to more than 5 mm in diameter.
9. Tangential aragonite ooids up to 50 mm of the hot springs near Tekke Ilica, Turkey (Richter and Besenecker, this vol.).

Because in many oolitic rocks the *primary composition* of the ooids cannot be deduced from their present internal structure, it is impossible to include the mineralogy in a general definition of ooids. Calcareous ooids were primarily composed of aragonite, calcite or magnesian calcite. Even the dolomite ooids in the Precambrian of Spitsbergen, though they were considered primary dolomite by Radwanski and Birkenmajer (1977), were most probably (?Mg-) calcitic or aragonite and became dolomitized during early diagenesis (well preserved concentric structure), because dolomite precipitation at the surface is unlikely; an original aragonite mineralogy is excluded by their well-preserved structure. In diagenetically mature oolitic limestones, the cortices of the ooids consist either of calcite or of ferroan calcite. In the case of calcite, preservation of the original structure indicates a previous calcitic or Mg-calcitic composition, while previous aragonite is indicated by sparry calcite (see below). In micritic ooids even this distinction may be impossible. Fe-calcitic ooids cannot be primary, because they require an anoxic environment; in cases of well-preserved structure, Mg-calcitic precursors have to be assumed.

The *orientation of crystallites* in the ooid cortices and their possible relations to the depositional environment can also not be used as general criteria for calcareous ooids. Though in modern marine environments tangential-aragonitic ooids predominate, radial-aragonitic ooids occur as well (Davies and Martin 1976, and others). On the other hand, Sorby (1879) described tangential-aragonitic ooids from hot springs, and Kahle (1974) documented radial-aragonitic ooids from the hypersaline Great Salt Lake. These primary structures are destroyed during diagenesis, because aragonite is dissolved to form oomoldic porosity which may be filled with cement, or the aragonite is calcitized "in situ", as shown by relics of the original laminae. Anyway, the same diagenetic fabrics develop from tangential- and radial-aragonitic ooids.

Organisms contributing to the lamination should not be visible in ooids; otherwise such coated grains should be named "oncoids" (but see reservation f on p. 75). The endolithic microorganisms described by Newell et al. (1960) and many other authors must be considered destructive for the ooid lamination, though a constructive participation of microorganisms is also suggested by the rod-shaped micron-sized aragonite crystals in the Bahama ooids (see Chap. 3.1; Fabricius 1977). Mitterer (1968, 1972) is even inclined to argue for an organic induction of carbonate precipitation in ooids. He investigated the organic matrix of modern ooids from the Bahamas, the Persian Gulf, the Red Sea, and the Great Salt Lake, as well as Pleistocene ooids from the Atlantic shelf off North Carolina and found great similarities in the composition of the organic matrix of ooids and skeletal carbonates. Finally, Davies et al. (1978) and Ferguson et al. (1978) were able to synthesize different ooid types at different turbulence by adding organic compounds. The significance of the latter, however, is considered quite differently in these publications: (a) Davies et al. (1978, p. 703): "Ooids exhibiting a tangential orientation of battenlike crystals have been synthesized under conditions of agitation, supersaturation and without the intervention of organic processes during the precipitation", (b) Ferguson et al. (1978, p. 295): "The morphology, mineralogy and rate of formation of the carbonate precipitates were all strongly dependent on the type of organic matter added to the solution." Although an organic participation may be proven for some of the ooid types, this cannot be part of a general ooid definition, all the more because this cannot be identified in fossil ooids.

Thus, the *definition of ooids* must be completely descriptive. Due to the variety of ooid types, it should be restricted to the following general properties:

1. Calcareous ooids are formed by a cortex and a nucleus variable in composition and size.
2. The cortex is smoothly laminated.
3. The laminae are either concentric or they are thinner on points of stronger curvature of the nucleus, and vice-versa, thus increasing the sphericity of the ooid during its growth.
4. Constructive biogenic structures are lacking.

But even these few statements require a number of reservations:

a) In oomoldic rocks, cortex and lamination are of course lacking.

STRUCTURAL TYPE	PRIMARY			SECONDARY
	ARAGONITE	Mg-CALCITE	CALCITE	CALCITE
1  TANGENTIAL	marine lacustrine/hypersaline thermal	?	caliche	?
2 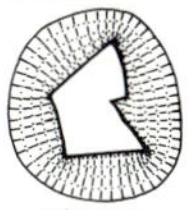 RADIAL	marine lacustrine/hypersaline	marine	cave pearls lacustrine fluvial	Mg-calcite → calcite
3  RANDOM	marine	?	lacustrine caliche	Mg-calcite → calcite or aragonite → calcite
4  OOMOLDIC				aragonite ↘ dissolution ↘ calcite cement
5 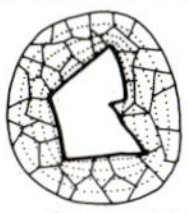 "IN SITU" CALCITIZED				aragonite → calcite

Fig. 1. Classification of the main types of calcareous ooids

b) This general definition does not allow for environmental implications. They require a specification of ooid types given below.

c) Diagenesis may have obliterated constructive biogenic structures, so that they cannot be excluded in some ooid types. It is not a great mistake, however, to regard such particles as ooids, because, in Bahama-type ooids, the carbonate is presumably in part precipitated by blue-green algae, thus making the difference between such ooids and oncoids rather small.

d) In micritic or microsparitic coated grains only the regularity of lamination can be used for the distinction of ooids and spongiostromate oncoids, though there is certainly no clear limit between these grain types (see c). Transitions and mixtures have been mentioned by Fabricius (1967, p. 150) and Schuler (1968, p. 34), respectively. They are frequent and possibly important in Upper Permian dolomites of northwestern Germany and differ only in shape, not in size (Füchtbauer 1964, p. 491; Sannemann et al. 1978, p. 43).

e) Special ooid types, e.g., cerebroid ooids and asymmetrical ooids, are discussed in Sect. 5.

f) Encrusters are possible within and on ooidic cortices. The case that foraminifera encrust radial ooids is quite common, e.g., in the Zechstein Limestone of the basinal facies in N and W Poland or in the Raibler Beds, and in some

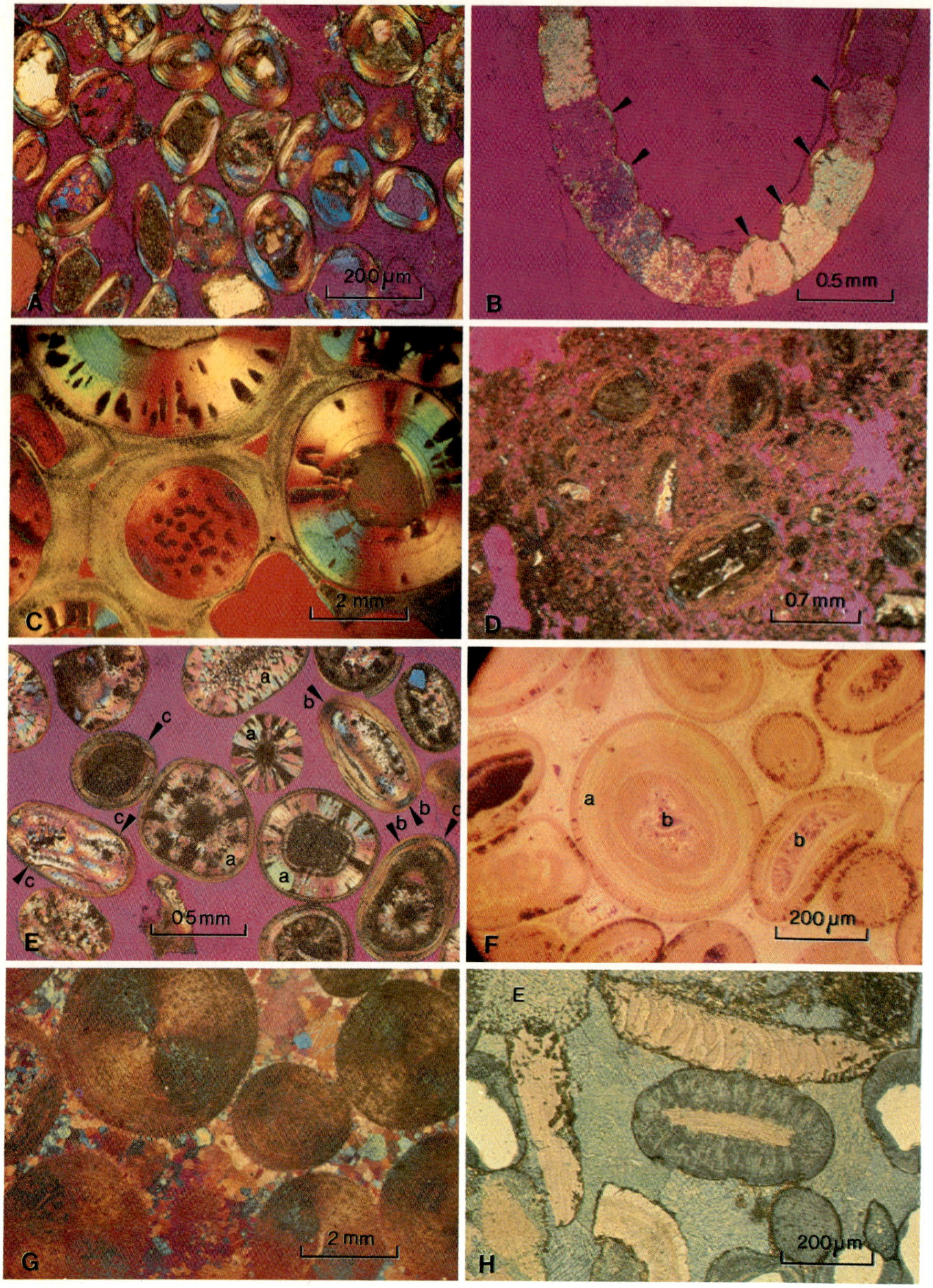

Fig. 2A – H. Different types of aragonitic and calcitic ooids; **A – E** and **G** with crossed nicols and gypsum plate (red 1st order; direction of vibration = NE – SW in figures); thickness of the thin sections: **A – F** and **H** 20 – 30 µm, G = <5 µm.

A Tangential-aragonitic ooids (anomalous birefringence) of the Bahama-type; note the increasing sphericity of many ooids with increasing thickness of the cortex; Holocene oolite of Tolon near Nauplion (Peloponnesus/Greece – see Richter 1976).

cases foraminifera form envelopes between successive ooid envelopes (T. Peryt, pers. communication).

We are dealing with a variety of mineralogical and structural *types of ooids*. Only the latter were used by Kahle (1974, Fig. 1) in his scheme. In the following classification, both criteria as well as the diagenetic modifications were considered (Fig. 1).

3 Primary Ooid Types

3.1 Tangential-Aragonitic Ooids

3.1.1 Marine Occurrences

The famous Bahama-type ooids consist of *aragonite rods* without crystallographic terminations. They are oriented tangentially, i.e., parallel to the ooid lamination (Fig. 2A). According to Fabricius (1977, p. 19), the average length of these rods is 1 micron, their maximum length 3 micron, and their diameter ranges between 0.1 and 0.3 micron. Granules of 0.1 – 0.3 micron in diameter can also occur (Fabricius 1977), though rods are prevailing, according to the literature and to observations by the author.

The *spongy structure* of these ooids is reflected by a microporosity of about 20% as measured by Oti and Müller (1979). This porosity figure may still be too low, because the method included only pores above 75 Å. A striking characteristic of Bahama-type aragonite ooids is their low apparent birefringence, i.e., only 0.03 – 0.035 instead of 0.155 (Illing 1954, Loreau and Purser 1973). Only due to such special conditions the orientation of the crystallites can de determined in normal thin sections (20 – 30 micron in thickness) by means of a gypsum slide (1st order). According to these authors, the low apparent birefringence is caused

◄

B Tangential-aragonitic crusts (anomalous birefringence) in the interior of a sea urchin (test of the species Echinocyamus pusillus); Holocene oolite near Neapolis (South-Peloponnesus/Greece – see Richter 1976).
C Tangential-aragonitic ooids (anomalous birefringence) cemented by radial fibrous crystals of aragonite (nearly normal birefringence); sample from the hot springs of Karlsbad (CSSR).
D Tangential-calcitic ooids (anomalous birefringence) from a caliche crust at the top of a marine terrace of Tyrrhenian age (northeast of Cape Hiräon/Peninsula of Perachora near Corinth/Greece).
E Aragonitic ooids of the Great Salt Lake (Utah): (*a*) radial with normal birefringence, (*b*) radial with anomalous birefringence, (*c*) tangential with anomalous birefringence.
F Cathodoluminescence photograph of Upper Jurassic radial-calcitic ooids (Holzen near the Ith mountains/Northern Germany; thin section No. 4.42, Bochum); multilayered cortices (*a*) become visible due to different ratios of Mn/Fe in the calcitic layers (primarily varying Mg-content?); molds of formerly aragonitic particles were cemented by calcite with luminescence zonation (*b*).
G Radial-calcitic ooids with normal birefringence (thickness of thin section: < 5 μm); "Kohlenkalk" of Dinantian age (Velbert/Nordrhein-Westfalen; No. 1 in Table 1).
H Radial-fibrous Fe-calcitic ooids (blue-violet); fragments of an echinoderm (E, blue-violet) and of primarily calcitic molluscs (red); stained thin section; Middle Jurassic oolite of the well Anzing 3 (E of Munich; No. 4.403 in Table 1)

by the microporosity mentioned above and by the different orientation of the crystallites within and relative to the laminae ("statistically tangential") as demonstrated by calculations (Loreau in Loreau and Purser 1973).

The origin of the *tangential orientation* of the crystallites is not fully understood. Shearman et al. (1970) suggested that the close alternation of laminae of organic mucilage and aragonite needles is instrumental in the tangential orientation and that the agitation of the ooids in a high-energy environment has an additional influence. The latter mechanism, however, is difficult to understand for ooids with angular nuclei where the tangential orientation follows the reentrants in the nuclei. Moreover, aragonitic crusts forming at the sediment surface with a tangential crystal orientation have been found in the Trucial Coast area by Evamy (1973) and Purser and Loreau (1973). Richter (1976) found such crusts in the interior of Echinocyamus pusillus, a sea urchin of less than 1 cm diameter, in a Holocene oolitic sediment on the southern coast of Laconia (S. Peloponnesus, Greece). These crusts consist of tangentially oriented aragonite crystals and oocur preferentially on projections (Fig. 2B; better access of light?) or they drape the sea urchin's interior at constant thickness. Random orientation occurs only in the intraskeletal pores (e.g., ambulacral pores). It is suggested that the tiny organic envelopes of the Bahama-type ooids allow only a statistically tangential growth of the rods (Fabricius 1977) and that the same applies to the interior of echinoids.

The *strontium* content of Bahama-type ooids of 8,300 – 10,600 ppm (Kinsman 1969, Loreau and Purser 1973, Milliman 1974, Brand and Veizer 1982) compares well with the high Sr figures of marine aragonite cement (see compilation by Richter 1979, Table 9), thus indicating their dependency on seawater composition.

A *recent formation* of ooids is not yet proven. However, it is likely, because well-preserved mucilage is still present in the cortex of "modern" ooids, as shown by SEM photographs (Fabricius 1977, Plate 3 B). C^{14} determinations (Newell et al. 1960) yielded an age of 225 $\pm$ 100 years for the outer layers and 2350 $\pm$ 100 years for the inner layers of the ooids from Cat Cay (Great Bahama Bank). A very slow formation of ooids has thus been established.

The *occurrence* of modern Bahama-type ooids is restricted to warm areas of low latitude, e.g., Great Bahama Bank (Sorby 1879, and others), Gulf of Suez, Persian Gulf (Loreau and Purser 1973, and others), Yucatán shelf, Mexico (Logan et al. 1969), and Shark Bay, western Australia (Davies 1970). In general they are considered as sediments of very shallow marine environments. Comprehensive modern investigations of the structure of these ooids have been conducted by Loreau and Purser (1973) and Fabricius (1977).

3.1.2 Lacustrine-Hypersaline Occurrences

In the prevailing radial-aragonitic ooids of the hypersaline Great Salt Lake, tangential-aragonitic laminae occur. According to Halley (1977), the crystals are either rod-shaped or show crystallographic terminations. A few ooids of this occurrence are composed of alternating layers of tangentially and randomly oriented crystals (author's observation). They cannot be distinguished – either by structure or by composition – from many marine ooids, since Sorby (1879) described such ooids from the Bahamas.

The tangential-aragonitic laminae of the Great Salt Lake ooids are characterized by a low birefringence (Fig. 2E) as shown also by the ooids from the Persian Gulf (Loreau and Purser 1973) and from the hot springs of Tekke Ilica, Turkey (Richter and Besenecker, this vol.). Since SEM photographs of the tangential-aragonitic layers indicate a loose packing (Halley 1977, Fig. 2a), a high microporosity and an only statistically tangential orientation is responsible for this phenomenon (see last paragraph).

3.1.3 Hot Springs

The tangential orientation of the aragonite crystals in the ooids of the hot springs at Karlsbad (CSSR) has already been observed by Sorby (1879), who began his ooid chapter with the description of this occurrence. Similar ooids with micron-sized aragonite crystals and rods, occur near the hot springs of Tekke-Ilica, Turkey (Richter and Besenecker, this vol.). Again, both occurrences show reduced birefringence (Fig. 2C). Lacroix (1898, cited in Illing 1954) suggested a new mineral "klypeite" with a birefringence of 0.02 only, but SEM photographs of the Turkish occurrence demonstrated the high microporosity and the type of orientation discussed above.

Because only two occurrences of tangential-aragonitic ooids are known from hot springs, a comparison with the numerous equivalent marine Holocene occurrence may be inadequate. Three differences, however, are remarkable: (1) The ooids from hot springs can become very big (> 5 mm in Karlsbad; up to 50 mm in Tekke Ilica, see Fig. 4 in Richter and Besenecker, this vol.). (2) The Sr content may differ considerably from marine occurrences (8300 – 10,000 ppm), since 25,000 ppm were registered in Tekke Ilica ooids (Besenecker and Richter, this vol.) and 2500 – 3500 in Karlsbad. (3) The Tekke Ilica ooids show frequently erosional "unconformities" of the lamination, unlike Bahama-type ooids (see Fig. 3 in Richter and Besenecker, this vol.).

3.2 Tangential-Calcitic Ooids

Coated grains are not rare in calcretes. Some of them meet the requirements of ooids, i.e., their sphericity increases with growing cortex (Fig. 8c in Hay and Wiggins 1980; see also James 1972, Siesser 1973, Read 1974). A tangential orientation has been described by James (1972) for calcite crystals 9 – 30 microns long and 1 – 2 microns in diameter, whose c-axis coincides with the crystal elongation. Ooids with tangential low magnesium calcite crystals are frequent in calcretes from marine terraces in the Quaternary of the Isthmus of Corinth and on the Perachora Peninsula (north of Corinth, Greece; Fig. 2D). Generally they are associated with tangential-calcitic crusts. This association has been also described from Barbados by James (1972).

Two properties of the calcrete ooids seem to be most significant, because they occur also in the tangential-aragonitic Bahama-type ooids: (1) The tangential orientation of the crystallites in the cortex of calcrete ooids and crusts proves an independence from water agitation. (2) The calcitic layers in the cortex of the calcrete ooids and in the crusts are characterized by an apparently low birefringence indicating high microporosity and only statistically tangential orientation, though SEM photographs are not yet available. These observations suggest that an organic matrix may be instrumental not only in the formation of Bahama-type ooids but also in the formation of calcrete ooids.

Ooids with tangential crystals of high magnesian calcite have not been described, though they are conceivable in calcretes of special environments, e.g., in calcareous serpentinite conglomerates.

3.3 Radial-Aragonitic Ooids

3.3.1 Marine Occurrences

Radial-aragonitic layers have been described frequently in marine ooids (Logan et al. 1969, p. 58; Davies 1970, p. 134; Fabricius and Klingele 1970, Fig. 7; Fabricius 1977, Plate 1/Fig. F), but completely radial-aragonitic ooids in marine environments are mentioned only from the Persian Gulf (Loreau and Purser 1973), from the Gulf of Aqaba (Friedman et al. 1973 – sea marginal hypersaline pool) and from the Great Barrier Reef (Davies and Martin 1976). Based on field observations, the radial texture of marine aragonite ooids is generally explained by low water agitation during their formation. This has been confirmed by Davies et al. (1978) in seawater experiments. They were able to "synthesize" radial ooids under quiet-water conditions and tangential ooids in agitated water.

The density in the radial-aragonitic layers in marine ooids has not yet been investigated systematically. In thin sections of many Quaternary oolites of the Isthmus of Corinthos (Greece), I found in radial – as well as in tangential – aragonitic layers the same abnormally low birefringence as in the Bahama-type ooids. This may be caused by a high microporosity and by the only statistically radial texture of the aragonite crystals. Such a spongy packing of statistically radial aragonite crystals is described for the Persian Gulf ooids by Loreau and Purser (1973) and is also shown on the SEM photographs of Fabricius (1977, Plate 1, Fig. F). Comparable observations are missing for the radial-aragonitic ooids of the Gulf of Aqaba (Friedman et al. 1973) and the Great Barrier Reef (Davies and Martin 1976).

Layered radial-aragonitic ooids have been synthesized by Suess and Fütterer (1972), Ferguson et al. (1978) and Davies et al. (1978) in seawater experiments only by adding organic compounds, which indicates their importance in the ooid formation.

3.3.2 Lacustrine-Hypersaline Occurrences

The hypersaline Great Salt Lake (Utah) is certainly the classical occurrence of radial-aragonitic ooids. They were first mentioned by Eardley (1938, p. 1384): "Since the new aragonite-calcite test of Leitmeier and Feigl indicates that some of the rays may be aragonite, the secondary development in radial structure may be due in some oolites to recrystallization without inversion". According to Fabricius and Klingele (1970), Kahle (1974), Sandberg (1975) and Halley (1977), the radial orientation of the aragonite crystals is primary. It is by far the dominant structure in the Great Salt Lake ooids; individual layers with randomly or tangentially oriented nannograins are relatively rare (Halley 1977). Three types of "radial" aragonite layers can be distinguished:

1. Layers with radial aragonite rays 100 – 200 micron in size; their layering is due to inclusions. Groups of radial rays are embedded in a matrix of clay and aragonitic micrite forming radial structures, in thin sections (Fig. 2E) (see Sandberg 1975, Fig. 3; Halley 1977, Fig. 4).
2. Layers with interfering spherulitic fans of aragonite crystals (Sandberg 1975, Fig. 5).
3. Layers of radially oriented needle- or rod-shaped nannograins; these layers are very thin (microns; Halley 1977, Fig. 2c).

Type (1) predominates in the Great Salt Lake ooids. In thin sections the aragonite shows normal birefringence. The low microporosity of the Great Salt Lake ooids (1.4% – 6.9%; Oti and Müller 1979) is in accord with that birefringence. Type (3) is relatively rare and is characterized by an abnormally low birefringence (Fig. 2E). A high microporosity and a statistically radial orientation are therefore suggested. The latter is indicated in Fig. 2c of Halley (1977).

Popp and Wilkinson (1982) described radial-aragonitic layers in ooids composed of aragonite and low magnesian calcite, from the brackish-lacustrine Pyramid Lake (Nevada). Some radial-aragonitic layers occur in the tangential-aragonitic hot spring ooids of Karlsbad (CSSR) and of Tekke Ilica (Turkey).

3.4 Radial (Mg) Calcitic Ooids

3.4.1 Marine Occurrences

The ooids of the marine-hypersaline Baffin Bay (Texas) contain tangential-aragonitic, micritic, and radial layers (Rusnak 1960). The latter are composed of magnesian calcite (Frishman and Behrens 1969) with 12.4 – 14.4 mol percent $MgCO_3$ (Land et al. 1979), while, according to the latter authors, tangentially and randomly oriented laminae consist of aragonite. The anhedral magnesian calcite crystals are ranging between 1 and 1.5 microns in diameter at a length of tens of microns (Land et al. 1979, p. 1272); information on their apparent birefringence is missing.

Ooids completely composed of radial magnesian calcite are reported from two occurrences only:

a) Late Pleistocene ooids from the Brazil-Guayana shelf north of the Amazon mouth consist of magnesian calcite with 11.7 (Moguedet 1973) or 11 – 13 mol percent $MgCO_3$ (Milliman and Barretto 1975). The latter authors distinguished, in non-micritized ooids, between those with concentric laminations separated by organic inclusions and those with diagenetic alterations resulting in merged layers of radial magnesian calcite.

b) Holocene ooids from the Great Barrier Reef (Australia) consist or radial magnesian calcite with 15 (Marshall and Davies 1975) or 15 – 17 mol percent $MgCO_3$ and 600 – 750 ppm Sr (Richter 1979, Table 9). In thin sections (Fig. 4C) I observed normal birefringence, indicating tightly packed and nearly parallel oriented crystals.

3.4.2 Non-Marine Occurrences

Primary ooids consisting of radial calcite poor in, or free of, magnesium indicate a non-marine environment. Mg-calcitic radial ooids are conceivable only in saline lakes and in brackish environments, though such ooids have not yet been described. The cave pearls from dripstone caves (Hess 1929, and others) and from mines (Hahne et al. 1968, including many references) are the classical radial-calcitic ooids. Artificial radial-calcitic ooids are also formed in the water system of power plants (Knatz 1966) and in water treatment plants (Popp and Wilkinson 1982, Fig. 5B). Richter (1980) mentioned such ooids from the calcareous sinter in a Roman aquaeduct near Kreuzweingarten SW of Cologne (Fig. 4D). Late Pleistocene radial-calcitic ooids in the fluvial environment have been described by McGannon (1975). Isolated radial-calcitic layers occur in the aragonitic-calcitic ooids of the brackish-lacustrine Pyramid Lake, Nevada (Popp and Wilkinson 1982, Fig. 3D).

The calcite fibres of the ooids can constitute the whole ooid or single layers only. A fan structure of these fibres is shown by Fig. 15 of Hahne et al. (1968). Sphericity increases with growth of the cave pearls (Mackin and Coombs 1945), Hahne et al. 1968) and of fluvial ooids (McGannon 1975, Fig. 8A). Early diagenetic alterations of radial-calcitic ooids such as secondary development of larger radial fibres have been mentioned especially by Hahne et al. (1968, p. 19).

The birefringence of these ooids has not been described. According to my inspections of cave pearls and the artificial ooids mentioned above (Fig. 4D), the calcitic cortices are massive.

3.5 Micritic to Microsparitic Ooids with Randomly Oriented Crystals

3.5.1 Marine Occurrences

The Bahama-type ooids frequently contain layers of cryptocrystalline randomly oriented crystals (Sorby 1879, Loreau and Purser 1973). In most cases, however, it is not clear whether the structure of these micritic layers is primary or secondary. Newell et al. (1960) described endolithic microorganisms from such layers. According to Fabricius (1977, p. 56 – 58), micritic ooids with "random orientation and large differences in grain size" of the crystals are formed by "micritization" with biogenic influences. Micritization is also suggested by Marshall and Davies (1975) for the radial magnesian calcitic micritic ooids from Australia.

Marine micritic ooids with primarily randomly oriented crystals are conceivable though they have not yet been described. Systematic investigations of the composition and origin of such marine micritic ooids are missing. Only for aragonitic and magnesian calcitic ooids of the Baffin Bay Land et al. (1979) concluded that an aragonitic composition of the micritic layers with random orientation was most likely.

3.5.2 Non-Marine Occurrences

Layers with randomly oriented aragonitic crystals occur in the mainly radial-aragonitic ooids of the Great Salt Lake (Sandberg 1975, Halley 1977). The crystal size varies between 0.1 × 1.0 and 1.0 × 5.0 microns (Halley 1977, p. 1103).

Completely micritic to microsparitic ooids with random crystal orientation have been found only with calcitic composition in the nonmarine realm: (a) calcrete ooids from western Australia (Read 1974, p. 256), (b) lacustrine ooids in marls of a lake in northern central Michigan (Wilkinson et al. 1980).

The water turbulence in dripstone caves seem to influence the structure of the ooids (Donahue 1969); cave pearls of the agitated environment (splash cup) are radial-calcitic, whereas those of the non-agitated environment (rimstone dam, pool) are composed of random equant calcite crystals. However, splash cup experiments have only yielded calcitic ooids with randomly oriented crystals (Donahue 1965).

4 Secondary Ooid Types

4.1 Radial-Calcitic Ooids

Radial-calcitic ooids are one of the most important secondary ooid types, although this has been only recently recognized. For this reason, they will be treated in some detail.

The radial-fibrous structure of many calcitic ooids in the geological record has already been recognized as primary, i.e., as the original structure, by Sorby (1879). This was not rediscovered before Simone (1974) and Sandberg (1975). According to Wilkinson and Landing (1978), ooids with aragonitic nuclei collapsed during burial and compaction, whereas those with calcitic nuclei remained unaffected. In these Jurassic ooids, textural evidence suggested that both the fabric and mineralogy of the radial-calcitic cortex was primary. Ooids with calcite "fans" (Simone 1981, Fig. 16) and fibrous calcite "cones" (Kalkowsky 1908, Usdowski 1962) are to be included in the radial-calcitic ooids.

Although these ooids, due to the described properties, could be primary as well, they were most probably composed of magnesian calcite and then altered to calcite. Otherwise, the seawater must have been poor in magnesium in the past. This, however, has been suggested by Sandberg (1975) and Wilkinson (1979), also based on the observation that radial-calcitic ooids dominate in Paleozoic rocks while originally aragonitic ooids predominate in Cenozoic rocks. They concluded an increase of the Mg/Ca ratio in the seawater from about 2:1 to 5:1 since Paleozoic time.

Investigations of Quaternary coastal terraces with different stages of diagenesis showed that certain calcitic skeletons with excellently preserved structure (Echinoidea, Corallinacea et al.) consisted originally of magnesian calcite, because both calcite and magnesian calcite still occur in the same skeleton (Richter 1974, 1979, with many references). In most of the fossil – especially pre-Tertiary – occurrences a Mg-calcitic composition of the original skeleton or cement can only be identified by special methods presuming distinct diagenetic processes:

a) If quartz is precipitated in primary Mg-calcitic skeletons before these are completely turned into stable calcite, the calcite inclusions in the quartz crystals contain more Mg than the surrounding unsilicified skeleton (Richter 1972).
b) In reducing, iron-bearing environments, Mg-calcitic skeletons are turned into ferroan calcite with preservation of the skeletal structures, whereas calcitic skeletons are not replaced (Richter and Füchtbauer 1978).
c) A variation in the Mn/Fe ratio of the interstitial fluids during the magnesian calcite-calcite transformation produces blotchy cathodoluminescence which can be observed especially in the large crystals of echinoderms (Richter and Zinkernagel 1981).

(b) is the most convenient method if combined with staining of the thin section, by which ferroan calcite is covered by a violet or blue film, depending on the iron content (Evamy and Shearman 1962, Dickson 1966). Richter and Füchtbauer (1978) mentioned radial Fe-calcitic ooids from the Jurassic of well Anzing 3 near Munich (Fig. 2H) and concluded that they consisted originally of magnesian calcite. Using the same method, Richter (1980) was able to prove originally Mg-calcitic ooids in five other occurrences of Carboniferous through Jurassic age. Based on these observations he argued against Sandberg's (1975) and Wilkinson's (1979) interpretation of a changing oceanic Mg/Ca ratio.

In Table 1, 11 marine occurrences and three brackish and restricted occurrences of radial ferroan calcite ooids are listed. The stained thin sections reveal the following facts:

1. The magnesian calcite-ferroan calcite transformation occurs early in the ooid diagenesis, because the Fe-content of the ooids corresponds mostly with that of an early cement generation (especially nos. 2, 5 – 9, 11, 12, in Table 1).

2. As in Quaternary coastal terraces (Gavish and Friedman 1969, Richter 1979), the aragonite particles are dissolved frequently after the transformation of magnesian calcite to calcite or ferroan calcite. This is shown by the observation that the moulds of the dissolved aragonitic particles are frequently filled with a late cement generation (especially nos. 2, 5 – 9, 11, 12, 14, in Table 1). Sample no. 13 is particularly impressive, containing radial Fe-calcitic ooids enclosing gastropods which are preserved as aragonite.

3. The radial-calcitic ooids of marine rocks consisted originally of high-magnesian calcite, as shown by their relatively high iron content in many rocks. In order to understand this, we will discuss the iron stain of different parts of echinoderm skeletons, in thin section 4.404 (no. 7 in Table 1) and compare it with the anal-

Table 1. Fourteen oolites of Lower Carboniferous to Quaternary age with ooids consisting of layers of "radial" Fe-calcite (samples 1–3 and 5–12 are from marine, 13–14 from brackish, and 4 from restricted brackish environments). Interparticle cement generations are indicated ("calcite" = <0.3 wt.% FeO: red; FeI calcite = 0.3–1.5 wt.% FeO: red violet; FeII calcite = 1.5–2,5 wt.% FeO: blue violet; FeIII calcite = >2.5 wt.% FeO: blue, according to Lindholm and Finkelman 1972; this relationship between staining and FeO-content was confirmed by microprobe or chemical analyses in 1, 3, 4, and 13). The composition of the ooids is connected with the corresponding cement generation by vertical arrows. Horizontal lines indicate the composition of the cement replacing former aragonitic skeletons; if there is more than one cement generation present, this is shown by horizontal arrows. Both the lines and the arrows are connected with the interparticle cements. Table 1 of Richter (1980), completed by additional samples

(1) DINANTIAN / Tournaisian ("Kohlenkalk" - Velbert / Nordrhein-Westfalen./ Germany) No. 4.375

↓ Fe II calcite → Fe I calcite → calcite → Fe II calcite

(2) MISSISSIPPIAN ("Newman Formation" - Pike Co. / Kentucky / USA) No. 4.95

↓ Fe I calcite → calcite

(3) MISSISSIPPIAN ("Blue River Group" - Indiana 64 / Harrison Co. / Indiana / USA) No. 4.97

calcite → Fe I calcite → calcite

(4) LOWER TRIASSIC ("Rogenstein" - Vienenburger Sattel / SE HANNOVER / Germany) No. 4.8

calcite → Fe I calcite → calcite

(5) MIDDLE TRIASSIC ("Trochitenkalk" - Quarry NW Haueda / 25 km NW Kassel / Germany) No. 4.401

↓ Fe I calcite

(6) MIDDLE JURASSIC / Bajocian-Bathon. (Well Waibling 1 / SE Regensburg / Germany) No. 4.402

↓ Fe II calcite → Fe III calcite → Fe II calcite

(7) MIDDLE JURASSIC / Bajocian-Bathon. (Well Anzing 3 / E München / Germany) No. 4.403

↓ (FeII) Fe III calcite

(8) UPPER JURASSIC / Oxfordian (Osmington Mills/ SW England) No. 4.390

↓ Fe II calcite → Fe I calcite

(9) UPPER JURASSIC / Kimmeridgian (6 km N Ricla / Province Saragossa / Spain) No. 4.404

↓ Fe II calcite → Fe III calcite

(10) UPPER JURASSIC / Kimmeridgian (Quarry Holzmühle / Osterwald / S Hannover / Germany) No. 4.13

calcite → Fe I calcite → calcite

(11) UPPER JURASSIC / Portlandian ("Eimbeckhäuser Plattenkalk" - Holzen / Ith / S Hannover / Ger.) No. 4.40

↓ Fe I calcite → Fe II calcite

(12) UPPER JURASSIC / Purbeckian ("Serpulit" - Deister / SW Hannover / Germany) No. 4.63

↓ Fe I calcite → Fe II calcite

(13) UPPER PLIOCENE (Top of the Corinthian Marl - Channel of Corinthos / Greece) No. 4.405

↓ Fe I calcite (skeletal aragonite preserved)

(14) QUATERNARY (1 cycle before Paleotyrrhenian - Channel of Corinthos / Greece) No. 4.406

↓ Fe I calcite → calcite

ogues in modern sediments. While in the latter, the spines mostly contain less Mg than the other echinoderm parts (Chave 1954), the spines are generally Fe-free in the above-mentioned Jurassic thin section, except for the intraparticle cement, while the other echinoderm parts are Fe-calcitic. The reason is that the echinoid spines are more stable diagenetically because of their low Mg content compared with the other parts. They resist therefore transformation into low-Mg calcite and thus have no chance to take up ferroan iron during diagenesis. Alternatively, their transformation may have occurred later, in an Fe-poor pore solution. Coming now back to the ooids, their iron stain corresponds to those parts of the echinoderms originally consisting of high-Mg calcite. It is concluded that their chemical instability, i.e., the Mg content before diagenesis, was comparable with those echinoderm parts.

4. In the radial-calcitic ooids of several limestones, a primary layering is indicated by alternating strong and weak iron staining. It is caused either by primary differences in the Mg content or by intraparticle cementation replacing pigment layers during later diagenesis, as shown by the following examples.

a) In the "Rogenstein" oolites of the lowermost Triassic north of the Harz Mts. (no. 4 in Table 1, Fig. 4B) the most unstable Mg-rich ooid layers were the first to transform into calcite, when the pore solutions were still low in iron. The earliest cements are correspondingly low in iron. The less Mg-bearing layers transformed later, when the Fe-content in the solutions was increased and thus show a stronger iron stain than do later cements (Richter 1980, Richter and Zinkernagel 1981).

b) A similar sequence of diagenesis has been found in radial-calcitic ooids of Bajocian/Bathonian age from the well Anzing 3 east of Munich. They show thin calcitic layers with strong orange cathodoluminescence (because of a low Fe/Mn ratio) embedded in the Fe-calcitic ooids (Richter and Zinkernagel 1981). Since the calcite layers are only a few microns thick, it cannot be excluded that they are thin cement rims, replacing pigment layers in an interstitial water free in iron. Moreover, this sample (no. 17 in Richter and Füchtbauer 1978) contains also very different ooids which consist of low-iron calcite; they are possibly reworked and were therefore not included in Table 1.

5. Fossil radial ooids of originally Mg-calcitic composition occur in marine as well as in brackish or restricted environments. Marine Mg-calcitic ooids have occurred at least since Cambrian time. The first occurrence is still earlier, if one considers the radial dolomite ooids of the Precambrian of Spitsbergen (Radwanski and Birkenmajer 1977) as Mg-calcitic ooids which were dolomitized with preservation of the structures.

4.2 Micritic to Microsparitic Ooids

Micritic layers of randomly oriented aragonite crystals in tangential-aragonitic ooids coincide, according to Newell et al. (1960) with endolithic microorganisms. This structure is therefore considered early-diagenetic. The micritic layers are "at least in part demonstrably a product of recrystallization, but it may in part result from interstitial precipitation" (Newell et al. 1960, p. 481). Dravis (1979, p. 198) mentioned a "complete micritization of ooids" from the Eleuthera Bank (Bahamas) and thus introduced a diagenetic formation of peloids. This secondary aragonitic ooid type was not included in Fig. 3, because it is geologically insignificant. Ooids with "micritized" (Mg) calcitic rims can be observed in the secondary types of "oomoldic" and "in-situ calcitized" ooids.

Fossil micritic to microsparitic ooids of the marine environment can develop from micritic to microsparitic Mg-calcitic ooids or from aragonitic ooids. Because micritic to microsparitic Mg-calcitic ooids with random orientation of the crystals have not been found in modern sediments, I prefer the aragonitic interpretation. Fabricius and Klingele (1970, p. 603) described a calcitization of separate layers in the aragonitic ooids of strongly cemented Pleistocene oolites in the Isthmus of Corinth (Greece). They explain this by replacement or internal precipitation of calcite cement layers. I found this particularly well developed in the Pleistocene oolite south of Kira Wrissi (Isthmus of Corinth), where all transitions occur from tangential-aragonitic ooids to micritic/microsparitic, partially calcitized (Fig. 3E) and completely calcitized ooids (with preserved layering, Fig. 3F). The same transformation type has been described by Richter and Besenecker (1982) from the subrecent tangential-aragonitic hot spring ooids of Tekke Ilica (Turkey). A micritic to microsparitic instead of a sparitic calcitization of tangential-aragonitic ooids is possibly due to a high microporosity, in which the calcite precipitation is initiated. This is particularly well conceivable in the meteoric-vadose domain of semiarid regions, in which the highly porous ooids are exposed to Mg-poor capillary water for long periods of time.

The significance of the microporosity of tangential-aragonitic ooids during diagenesis is also reflected by the reaction of different aragonitic particles in meteoric-vadose water. While in the above-mentioned Pleistocene oolite south of Kira Wrissi the ooids are dissolved or more or less transformed into micritic/microsparitic calcite, the gastropods are preserved as aragonite with exceptional dissolution or calcitic replacement (nos. 4.407 and 4.408).

Micritic/microsparitic calcite ooids are common in the geological record; part of them may be calcitized aragonite ooids. In the Zechstein formation of northern Germany, in the Lower Triassic of Hydra (Greece) and in the Lower Muschelkalk of the Germanic Triassic, such micritic/microsparitic ooids are associated with "oomoldic" and "in-situ calcitized" ooids, which both are diagenetically transformed from aragonite (see below).

4.3 Oomolds

Dissolved aragonitic ooids are common in subaerially exposed Quaternary coastal terraces. Friedman (1964) introduced the term "oomoldic porosity"; in his Fig. 53 he presented a Paleozoic example with calcitic cementation. After dissolution of the ooid cortices the non-aragonitic nuclei were excentrically displaced (Fig. 3A, B), normally toward the bottom of the mould as shown by Knewtson and Hubert (1969) also for Carboniferous ooids. Mazzullo (1977) described the same texture from Upper Cambrian ooids and explains those geopetal ooids by recrystallization in a vadose environment.

Oomolds are known from rocks at least as old as Cambrian; they can be traced through the geological periods (Table 2) and testify for originally aragonitic ooids. In general it is not clear whether the aragonite crystals were arranged tangentially, radially, or random. The moulds are normally filled with blocky calcite cement or with coarse-fibrous calcite cement growing preferentially from the periphery towards the nucleus (Sorby 1879).

4.4 "In-Situ" Calcitized Ooids

Aragonitic skeletons are calcitized either by biogenic micritization, or by cementation after dissolution of the aragonite, or by a micro- to macrocrystalline "in-situ"-calcitization. In the latter case there was only a solution film between the aragonite and the replacing calcite. This thin-film alteration allowed a preservation of certain properties of the original wall structure, for instance of pigment layers which form inclusion-rich zones in the calcitized skeletons. In-situ calcitization is proven if the inclusions penetrate individual calcite crystals. The orientation of the aragonite crystals, however, is normally not preserved. This has been well known since Sorby (1879, Fig. 3) and is documented for instance by Bathurst (1971, p. 488) and Füchtbauer (1974, Fig. 5-68). Of course, transitions are common between micritic/microsparitic ooids and "in-situ" calcitized ooids (e.g., Richter and Besenecker 1982, Fig. 3 B).

This ooid type as well as the oomoldic type occur at least since Paleozoic time (Table 2, Fig. 3 C, D). They have been found in the Upper Permian Zechstein fm. of northern Germany and in the Lower Triassic of Hydra (Greece). "In-situ" calcitization in an iron-bearing reducing environment is indicated by Fe-calcitic ooids in the Silurian "Burksvik Oolite" of Gotland (Sweden) and in the Triassic oolites from Episkopi (Hydra; no. 16 – 4.397 in Table 2).

Fig. 3 A – H. Ooids affected by diagenesis.
A Oomolds in part with excentrically displaced nuclei (Pleistocene oolite/Corinth/Greece).
B Oomolds cemented by calcite (*arrows* excentrically displaced nuclei); oolite of Lower Anisian age (Episkopi/Island of Hydra/Greece; No. 16a in Table 2).
C "In-situ" calcitized ooids of Pleistocene age (terrace near Corinth/Greece).
D_1 "In situ" calcitized ooids of Upper Permian age (Niedermarsberg/Rheinisches Schiefergebirge; No. 14 in Table 2); left ooid is part of a complex ooid.
D_2 "In situ" calcitized ooids of Lower Anisian age (Episkopi/Island of Hydra/Greece; No. 16b in Table 2).
E Tangential-aragonitic ooid partially affected by micritic to microsparitic calcitization (Pleistocene oolite south of Kira Wrissi/Isthmus of Corinth/Greece; No. 4.408, Bochum).
F Ooid (formerly aragonite) affected by micritic to microsparitic calcitization and by dissolution; note the gravitational calcite cement (*arrow*); same thin section as **E**.
G Half moon ooids of Lower Anisian age (Episkopi/Island of Hydra/Greece; No. 4.396a, Bochum).
H Distorted ooids of Middle Triassic age ("Unterer Muschelkalk" near Heidelberg; like sample No. 17 in Table 2, but another part of the rock)

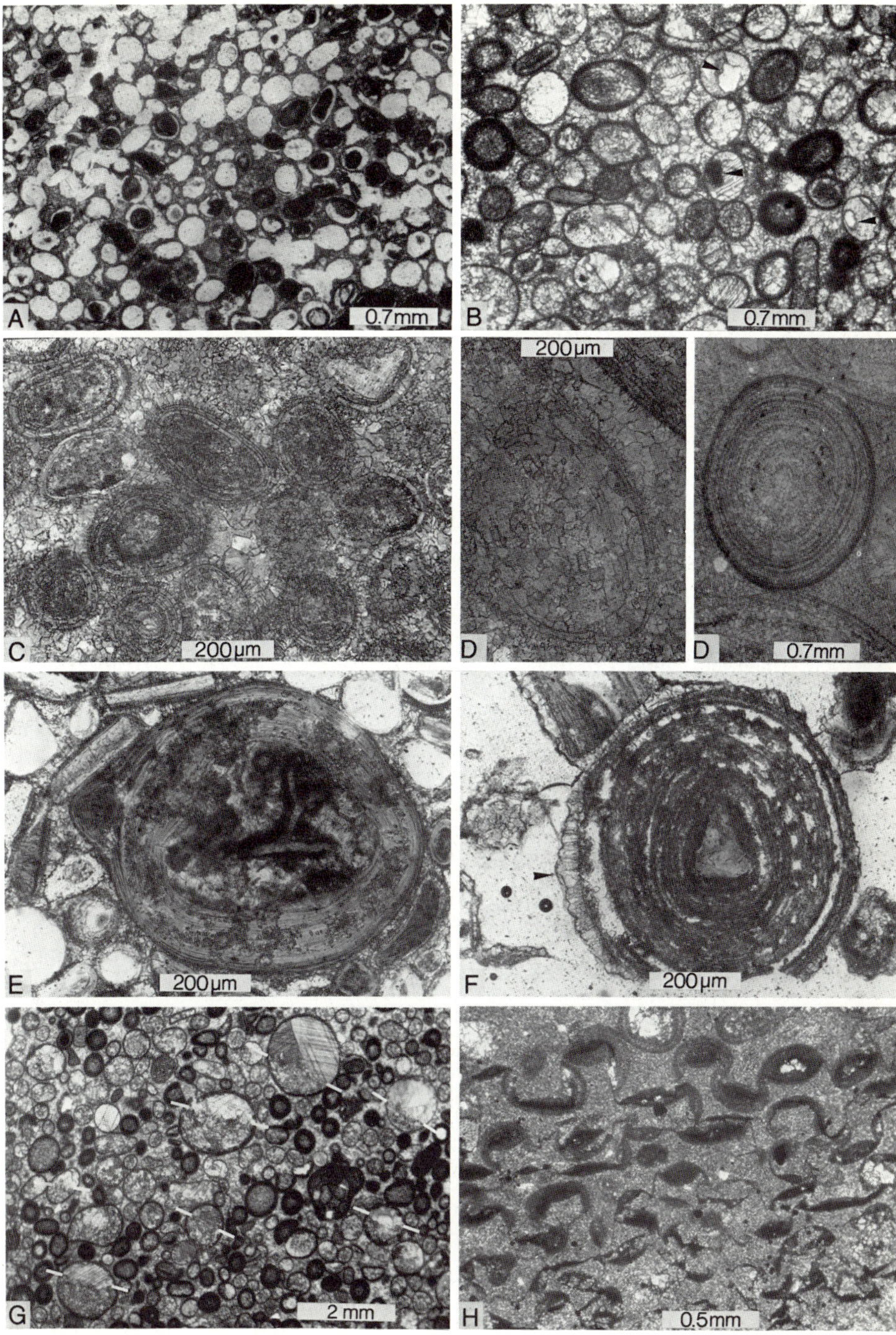
A
0.7mm
B
0.7mm
C
200µm
200µm
D
D
0.7mm
E
200µm
F
200µm
G
2 mm
H
0.5mm

Table 2. Former aragonitic ooids of Cambrian to Upper Jurassic age. The examples from the literature are in part reinterpreted based on the cited figures. The numbers 4.392 – 4.400 indicate thin sections of the Geological Institute of Bochum

Nr.	Age	Provenance	Type of Ooids		Reference
			"oomoldic"	"in situ-calcitized"	
1	Cambrian	Pennsylvania/U.S.A.	+ (Fig. 123)		Bathurst (1971)
2	Cambrian	Tyrone/Pennsylvania/U.S.A.	+ (Fig. 4-47)		Pettijohn (1975)
3	Upper Cambrian	"Gailor Formation"/ Mohawk Valley/ New York/U.S.A.	+ (Fig. 1)		Mazzullo (1977)
4	Cambro-Ordovician	"Durness Limestone"/ NW Scotland	+ (Fig. 2C)	+ (Fig. 2B)	Swett (1965)
5	Lower Ordovician	"Shakopee Formation"/ Southeastern Minnesota/ U.S.A.	+ (Fig. 4)		Davis (1966)
6	Ordovician	"Nittany Dolomite"/ Centre County/ Pennsylvania/U.S.A.	+ (Fig. 6)		Folk and Pittman (1971)
7	Ordovician	"Shakopee Dolomite"/ Wyalusing/Grant County/ Wisconsin/U.S.A.	+ (Plate III, Fig. 2)		Carozzi and Textoris (1965)
8	Silurian (Ludlow)	"Burksvik Oolite"/ Gotland/Sweden		+	4.392
9	Carboniferous (Mississippian)	"Ste. Genevieve Limestone"/Missouri/U.S.A.	+ (Fig. 14)		Knewtson and Hubert (1969)
10	Carboniferous (Pennsylvanian)	"Bethany Falls Limestone"/Missouri/U.S.A.	+ (Fig. 6)		Payton (1966)
11	Carboniferous (Pennsylvanian)	"Winchell Limestone"/ Palo Pinto/Texas/U.S.A.	+ (Fig. 53)		Friedman (1964)
12	Upper Permian	"Z 2-Carbonate"/ Denmark	+ (Fig. 1B)		Clark (1980)
13	Upper Permian	"Z 2-Limestone"/ Niedermarsberg/Rhein. Schiefergebirge/F.R.G.	+		4.394
14	Upper Permian	"Z 3-Limestone"/ Niedermarsberg/Rhein. Schiefergebirge/F.R.G.		+	4.395
15	Lower Triassic	"Jialingitang Formation"/ T_1^2J1/South Sichnan Basin/China	+ (Plate I, Fig. C/F)		Zeng Yun-Fu et al. (1982)
16	Triassic (Anisian)	Episkopi/Hydra/Greece	+		4.396
				+	4.397
17	Triassic (Anisian)	"Unterer Muschelkalk"/ Quarry Nußloch/ S Heidelberg/F.R.G.	+		4.398
18	Triassic (Anisian)	"Unterer Muschelkalk"/ Elm (E Braunschweig)/ F.R.G.	+		4.399
19	Rhaeto-Liassic	Siria/Prov. Soria/Spain	+		4.400
20	Rhaeto-Liassic	"Calcari grigi"/S. Massenza and Loppio/Trento/ Italy	+ (Plate II, Fig. 4 – 8)		Castellarin and Sartoni (1973)
21	Jurassic	Arkansas/U.S.A.	+ (Fig. 124)		Bathurst (1971)
22	Jurassic	"Kelloways Rock"/ Scarborough/England	+		Sorby (1879)
23	Upper Jurassic	"Smackover Formation"/ NE Texas/U.S.A.	+		Dickinson (1969)

5 Special Ooid Types

5.1 Broken and Regenerated Ooids

Broken ooids which are healed by new ooid laminae are frequent in fossil radial-calcitic oolites (Kalkowsky 1908, Table V, Fig. 2; Carozzi 1961 a). However, they occur also in aragonitic oolites as far as they are not yet transformed to calcite. Examples are the radial-aragonitic ooids of the Great Salt Lake (Carozzi 1961 a, Fig. 8) and the tangential-aragonitic hot spring ooids of Tekke Ilica, Turkey (Richter and Besenecker 1982, Fig. 3 A). This type has been termed "hiatus Ooids" by Berg (1944).

5.2 Complex Ooids

These include multiple ooids as described by Kalkowsky (1908) from the "Rogenstein" (roestone) in the Bunter Sandstone (Lower Triassic) of northern Germany, by Carozzi (1964) from Triassic lake sediments in Virginia (U.S.A.), and by Radwanski and Birkenmajer (1977, Fig. 6) from Precambrian dolomites of Spitsbergen. While the latter are radial-dolomitic, the others are radial-calcitic. Complex ooids, however, can also be tangential-aragonitic (e.g., the Pleistocene oolite south of Kira Wrissi, Isthmus of Corinth; thin section 4.407 and 4.408, Bochum; Fig. 4 A), and even calcitized aragonitic (e.g., in the Lower Triassic of Episkopi (Hydra; Table 2, no. 16, 4.397).

5.3 Asymmetrical Ooids

In cases of very little or no agitation during ooid growth, the latter is preferably directed upward. Such asymmetrical ooids containing a primarily eccentric nucleus are formed only in areas of low turbulence, in the Laguna Madre (Texas), according to Freeman (1962). They should not be confused with diagenetic types such as half moon ooids and oomolds with eccentric nuclei. After diagenesis, asymmetrical ooids can correctly be identified only in primarily (Mg-) calcitic oolites.

5.4 Half Moon Ooids

In this ooid type part of the ooid layers was dissolved, while the others, as well as the nucleus, were plastically deformed and settled, thus forming a geopetal structure with a lower dense micritic sediment and an upper coarse sparry mosaic (Carozzi 1963). Half moon ooids were first described from the Upper Cambrian Allentown Dolomite (Northampton County, Pennsylvania) by Wherry (1916), who interpreted them as originally aragonitic ooids with carbonaceous particles of probable organic origin which, after the dissolution of the aragonite, settled

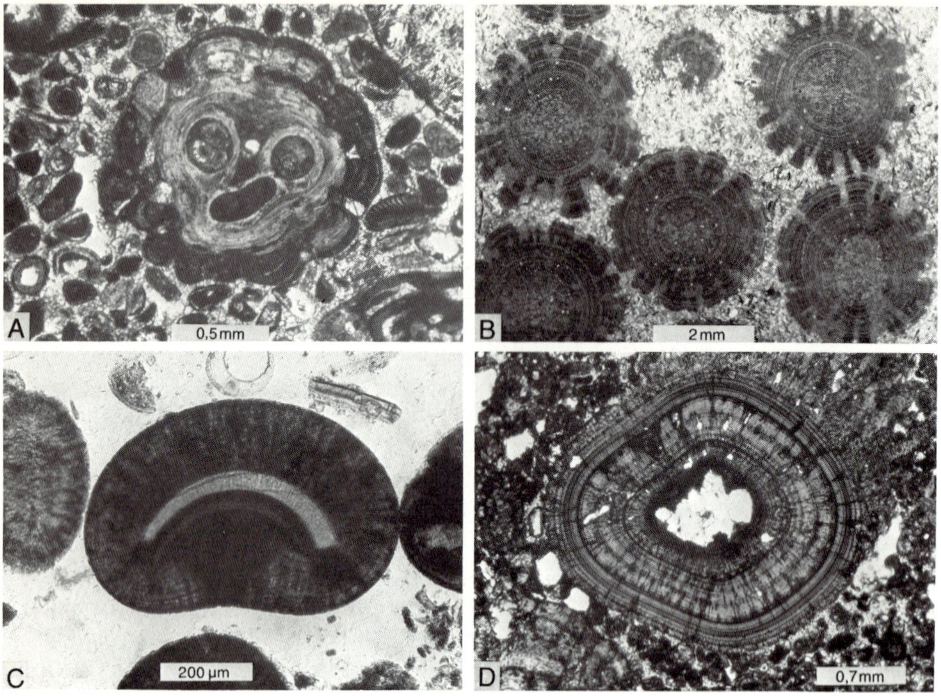

Fig. 4. A Complex ooid with cerebroid structure at the surface; preferentially tangential-aragonitic, radial-aragonitic layers occur only in the domes (orientation visible by using a gypsum plate); Pleistocene oolite south of Kira Wrissi (Isthmus of Corinth/Greece; No. 4.409, Bochum). **B** Radial-calcitic ooids with laminated cones ("Kegelstruktur"); "Rogenstein" of Lower Triassic age from the northern borderland of the Harz (No. 4 in Table 1). **C** Radial-fibrous ooids composed of Mg_{15-17}-calcite Great Barrier Reef; sample of J. F. Marshall/Canberra). **D** Radial-fibrous ooid composed of Mg_0-calcite (Roman aquaeduct near Kreuzweingarten/SW of Cologne)

with the nucleus at the bottom of the oomolds. Carozzi (1963) investigated an adjacent occurrence of the same formation. He observed "the settling and deformation of the core and of sets of concentric rings toward the bottom of the outer shell of the oolites" and concluded (p. 633): "These diagenetically modified bodies are considered to have been originally normal oolites consisting of a calcite core surrounded by alternating layers of calcite and anhydrite (or gypsum). Immediately after deposition and before completion of compaction, the sulfate layers were dissolved generating the internal deformation and settling of the insoluble parts." The same explanation was adopted by Radwanski and Birkenmajer (1977, p. 11 – 13) for dolomitic half moon ooids in the Precambrian of Spitsbergen. Choquette (1955) mentioned aragonite and calcite as dissolved minerals in the "bipartite oolites" (= half moon ooids) of the silicified Upper Cambrian rocks near State College (Pennsylvania). Dissolution mechanisms are also stressed by Conley (1977, p. 558) for half moon ooids in the Pennsylvanian Plattsburg limestone (Wilson County, Kansas). The genetically close relationship between half moon ooids and oomolds is shown by transitional stages (Castellarin and Sartoni 1973, Plate II, Fig. 4 – 8). The half moon ooids and the coexistent oomolds in the Lower Triassic of Episkopi (Hydra, Greece; thin section 4.396a; Fig. 3G) are presumably due to dissolution of aragonitic laminae.

5.5 Distorted Ooids

They are, according to Conley (1977, p. 554) "characterized by structures such as notched and stretched grains, grains connected by narrow apophyses, and series of grains linked in long zigzag chains". Explanations given to such structures include fracturing of "soft ooliths" in a turbulent environment ("reciprocal impacts") during deposition (Carozzi 1961b, Sarkar 1973), sediment sliding (Bachmann 1973, Fig. 61), compaction before cementation (Kettenbrink and Manger 1971), and compaction after a first interparticle cementation and particle dissolution of the ooids (Conley 1977). The latter interpretation is presumably in most cases applicable, though attention has to be given to the question whether nucleus, cortex, or both are dissolved. If only the nucleus is dissolved, e.g., in radial-calcitic ooids with aragonitic nuclei, distortion is possible even without interparticle cementation. Beautiful zigzag chains are illustrated by Cayeux (1935, Plate XV, Fig. 55), Kettenbrink and Manger (1971, Fig. 3F), Conley (1977, Fig. 5), and already by Frantzen (1887, Table III, Fig. 2/3) from the Lower Muschelkalk (Triassic) (see also Fig. 3H).

The term "distorted ooids" includes also the pitted ooids and cracked ooids which are formed by pressure solution and mechanical distortion, respectively (Radwanski and Birkenmajer 1977). Pressure solution between calcitic and Fe-calcitic ooids led to indentations in the latter because of their higher solubility. This has been observed in radial-fibrous ooids of Bajocian/Bathonian age from the well Anzing 3 east of Munich (No. 17 in Richter and Füchtbauer 1978). Crushed or cracked radial-calcitic ooids are frequently formed by compaction after dissolution of aragonitic nuclei ("eggshell diagenesis", Wilkinson and Landing 1978). Intensive stress results in elongations parallel to the schistosity planes (Cloos 1947, Nissen 1964).

5.6 Cerebroid Ooids

They were defined by Carozzi (1962, p. 241 f.) based on ooids of the Great Salt Lake (Utah): "Mottled and speckled spherical body resulting from numerous light-colored and well-crystallized bulging areas separated by darker and concave zones with crypto-crystalline structure. In section such an oolite displays a festooned or cerebroid aspect representing the completed stage. These bodies have been described by Eardley (1938, pp. 1380–1384, Plate 8, 10a, 12b) under the name of 'mottled oolites'". According to Carozzi (1962), the cerebroid structure was formed in the radial-fibrous and micritic Ste. Genevieve ooids (Mississippian of Illinois) by "recrystallization" and in the Great Salt Lake ooids during the aragonite-calcite transformation. Kahle (1974), however, was able to prove an aragonitic composition also for cerebroid ooids of the Great Salt Lake and ascribed the formation of micrite in the concave parts of the ooids to dissolution and precipitation by bacteria. In many cerebroid ooids, traces of lamination are visible also in the micritic areas; these laminae are concave, suggesting an inhibition of ooid growth, possibly by microorganisms. Green and blue-green algae have been found in pits of the Bahama ooids by Newell et al. (1960, p. 490). According to Bachmann (1973, Fig. 34), micritic borings in radial-calcitic ooids of the Upper Muschelkalk (SW-Germany) project into the nucleus. In the present author's opinion, the micritic areas with random orientation in the tangential-aragonitic thermal ooids from Karlsbad (Fig. 2C) can also be explained by local growth inhibition due to microorganisms.

Two phenotypes can be distinguished in cerebroid ooids. The concave areas can be circular (Kahle 1974, Fig. 4B) or reticular (Kahle 1974, Fig. 3E, and "Rogenstein" ooids in the Lower Buntsandstein of northern Germany; Kalkowsky 1908, Usdowski 1962).

Cerebroid ooids occur in nonmarine (Great Salt Lake) and marine environments (Ste. Genevieve formation, Mississippian of Illinois, Carozzi 1962). This author mentioned cerebroid structures from radial-fibrous ooids only. However, they occur also in tangential-aragonitic marine ooids (Fig. 4A). Radial-aragonitic layers have only been observed in the domal areas of diagenetically fresh ooids with a strongly developed "oncoid-like" cerebroid structure (Fig. 4A) from the Pleistocene oolite south of Kira Wrissi (Isthmus of Corinth).

A primary cerebroid-type morphology can also be observed in artificial ooids (Oti and Müller 1979, Fig. 7). It may have been caused by a fan-like growth of the calcite fibres.

6 Discussion and Conclusions

1. A large variety of ooid types, including tangential-aragonitic, radial-aragonitic, radial-Mg-calcitic and random-aragonitic ooids occur in recent and subrecent marine sediments. During diagenesis they are transformed and modified to form additional ooid types. The same variety of types occurs in non-marine oolites. For these reasons, a broad definition of the ooids as a subgroup of the coated grains is recommended, which includes specific types such as cave pearls, caliche ooids, and centimeter-sized thermal ooids.

2. Based on the structure of the cortex, five ooid types can be distinguished, (a) tangential, (b) radial, (c) random, (d) oomoldic, (e) in-situ calcitized. A combination of structure and mineralogy (aragonite, magnesian calcite, calcite) and a consideration of the depositional environment results in the classification shown in Fig. 1. The environment is critical, for instance in the distinction of primary radial-calcitic cave pearls and secondary radial-calcitic marine ooids.

3. It is impossible to differentiate a marine from a non-marine depositional environment on the basis of ooid structure and composition. This is in agreement with the conclusions of Land et al. (1979, p. 1275), using the diagenetically unaltered ooids from Great Salt Lake, Red Sea, Persian Gulf, Baffin Bay, Brazilian shelf, and Lizard Island, which are considered also in the present paper: "Of the

seven examples of Holocene radially coated ooids, four are associated with hypersaline conditions (Great Salt Lake, Red Sea, Persian Gulf, and Baffin Bay), whereas three are not (Brazilian shelf and two examples from the Australian Barrier Reef). Four examples are aragonite (Great Salt Lake, Red Sea, Persian Gulf, and Lizard Island) whereas three are Mg-calcite (Baffin Bay, Brazil, and Australian Barrier Reef shelves). The correlation between salinity, mineralogy and texture is imperfect and we therefore reject salinity as a clear-cut controlling variable on oolite texture or mineralogy."

4. Especially modern marine ooids are rich in organic material (e.g. Mitterer 1968, 1972). According to seawater experiments of Davies et al. (1978) and Ferguson et al. (1978), coated grains of different structure and composition are formed if different organic substances are added. This suggests a participation of organic substances in the formation of carbonate ooids. Up to now, however, it is not clear how, for instance, the aragonite rods in the Bahama-type ooids are formed. If microorganisms participate in the ooid formation, no genetic difference exists between ooids and onkoids.

5. An abnormally low birefringence which is conspicuous in thin sections of diagenetically unaltered ooid occurrences listed below can be explained by an only statistically predominant orientation of the carbonate crystals and a high microporosity. The phenomenon can be observed in tangential-aragonitic ooids of marine, thermal, and lacustrine-hypersaline origin, in radial, aragonitic ooids of marine, lacustrine-hypersaline and thermal origin, and in tangential-calcitic caliche ooids.

6. During diagenesis, aragonitic ooids are (a) dissolved or replaced by calcite either (b) in-situ or (c) by micrite to microsparite (cement in part). oomolds (a) and in-situ calcitized ooids (b) are known from rocks as old as Paleozoic (Table 1). Micritic/microsparitic ooids (c) are well developed in Pleistocene oolites on the Isthmus of Corinth (Greece): All transitions between unaltered aragonitic ooids and completely micritic/microsparitic ooids occur. The crystal size in the latter is smaller than in calcitized aragonitic skeletons (e.g., gastropods), presumably because of the finely dispersed micropores in the original ooids. At least part of the numerous occurrences of Paleozoic and Mesozoic micritic/microsparitic ooids were presumably originally aragonitic.

7. Marine radial-calcitic ooids consisted originally of magnesian calcite, as demonstrated by the occurrence of radial Fe-calcitic ooids even in Paleozoic rocks (Table 2). Mg-calcitic particles are transformed into ferroan calcite with preservation of the original structure (skeletal or ooid), in reducing, iron-bearing pore solutions (Richter and Füchtbauer 1978). Special observations discussed earlier suggest an original high-magnesian calcite composition of fossil radial-calcitic ooids in marine limestones.

8. Observations indicating a more or less stable Mg/Ca-ratio in the seawater since Paleozoic time include (a) the dolomite/calcite ratio in Mississippian cements of New Mexico points to a closed-system diagenesis of a $Mg_{0.13}Ca_{0.87}CO_3$ cement (Meyers and Lohmann 1978, p 481) which, according to experiments by

Füchtbauer and Hardie (1980, Fig. 2), is compatible with modern seawater. (b) Originally aragonitic and Mg-calcitic ooids were formed in marine environment at least since Paleozoic times (Tables 1, 2), as they are still formed. It remains, however, an interesting problem, why the distribution of Mg-calcitic and aragonitic ooids is so different from place to place and from period to period. In the Germanic Triassic for example, aragonitic ooids were formed predominantly in the Lower Muschelkalk, Mg-calcitic ooids in the Upper Muschelkalk (mo_1).

Acknowledgements. For discussions and assistance with the English version I am indebted to H. Füchtbauer (Bochum) and Ph. Sandberg (Urbana/Illinois).

References

Bachmann GH (1973) Die karbonatischen Bestandteile des Oberen Muschelkalkes (Mittlere Trias) in Südwest-Deutschland und ihre Diagenese. Arb Inst Geol Palaeontol, Univ Stuttgart, NF 68:1 – 99

Bathurst RGC (1971) Carbonate sediments and their diagenesis. Dev Sedimentol 12:620

Berg G (1944) Vergleichende Petrographie oolithischer Eisenerze. Arch Lagerstättenforsch (Berlin) 76:1 – 128

Beukes NJ (this vol.) Ooids and oolites of the proterophytic Boomplaas Formation, Transvaal Supergroup, Griqualand West, South Africa, pp 199 – 214

Brand U, Veizer J (this vol.) Origin of coated grains: trace element constraints, pp 9 – 26

Carozzi AV (1960) Microscopic sedimentary petrography. John Wiley, New York, 485 p

Carozzi AV (1961a) Oolithes remaniées, brisées et régénérées dans le Mississippien des chaînes frontales, Alberta Central, Canada. Arch Sci (Genève) 14 Fasc. 2:281 – 296

Carozzi AV (1961b) Distorted oolites and pseudoolites. J Sediment Petrol 31:262 – 274

Carozzi AV (1962) Cerebroid oolites. Trans Illinois State Acad Sci 55:238 – 249

Carozzi AV (1963) Half-moon oolites. J Sediment Petrol 33:633 – 645

Carozzi AV (1964) Complex ooids from Triassic lake deposit, Virginia. Am J Sci 262:231 – 241

Carozzi AV, Textoris DA (1967) Paleozoic carbonate microfacies of the Eastern Stable Interior (U.S.A.). Int Sediment Petrogr Ser XI:41

Castellarin A, Sartoni R (1973) Desication shrinkage and leaching vugs in the Calcari grigi Infraliassic tidal flats (S. Massenza and Loppio, Trento, Italy). Eclogae Geol Helvet 66:339 – 343

Cayeux L (1935) Les roches sédimentaires de France. Roches carbonatées. Masson, Paris, 447 p

Cayeux L (1970) Carbonate rocks. Hafner, Darien (Conn), 506 p (Transl)

Chave KE (1954) Aspects of the biogeochemistry of magnesium. – 1 Calcareous marine organisms. J Geol 62:266 – 283

Choquette PW (1955) A petrographic study of the "State College" siliceous oolite. J Geol 63:337 – 347

Choquette PW (1978) Oolite. In: Fairbridge Rh W, Bourgeois J (ed) The encyclopedia of sedimentology. Stroudsburg, Pennsylvania, pp 510 – 515

Clark DN (1980) The diagenesis of Zechstein carbonate sediments. Contr Sedimentol 9:167 – 203

Cloos E (1947) Oolite deformation in the South Mountain Fold, Maryland. Bull Geol Soc Am 58:843 – 918

Conley CD (1977) Origin of distorted ooliths and pisoliths. J Sediment Petrol 47:554 – 564

Davies GR (1970) Carbonate bank sedimentation, eastern Shark Bay, western Australia. Am Assoc. Petrol Geol Mem 13:85 – 168

Davies PJ, Martin K (1978) Radial aragonite ooids, Lizar Island, Great Barrier Reef, Queensland, Australia. Geology 4:120 – 122

Davies PJ, Bubela B, Ferguson J (1978) The formation of ooids. Sedimentology 25:703 – 730

Davis RA (1966) Quiet water oolites from the Ordovician of Minnesota. J Sediment Petrol 36:813 – 818

Dickinson KA (1969) Upper Jurassic carbonate rocks in northeastern Texas and adjoining parts of Arkansas and Louisiana. Trans. Gulf-Cst. Assoc Geol Soc 19:175 – 187

Dickinson JAD (1966) Carbonate identification and genesis as revealed by staining. J Sediment Petrol 36:491 – 505

Donahue J (1965) Laboratory growth of pisolite grains. J Sediment Petrol. 35:251 – 256

Donahue J (1969) Genesis of oolite and pisolite grains: an energy index. J Sediment Petrol 39:1399 – 1411

Donahue J (1978) Pisolite. In: Fairbridge Rh W, Bourgeois J (eds) The encyclopedia of sedimentology. Stroudsburg, Pennsylvania, pp 582 – 583

Dravis J (1979) Rapid and widespread generation of Recent oolitic hardgrounds on a high energy Bahamian platform, Eleuthera Bank, Bahamas. J Sediment Petrol 49:195 – 208

Eardley AJ (1938) Sediments of Great Salt Lake, Utah. Am Assoc Petrol Geol Bull 22:1359 – 1387

Evamy BD (1973) The precipitation of aragonite and its alteration to calcite on the Trucial Coast of the Persian Gulf. In: Purser BH (ed) The Persian Gulf. Springer, Berlin Heidelberg New York, pp 329 – 341

Evamy BD, Shearman DJ (1962) The application of chemical staining techniques to the study of diagenesis in limestones. Proc Geol Soc London 1599:102

Fabricius FH (1967) Die Rät- und Lias-Oolithe der nordwestlichen Kalkalpen. Geol Rundsch 56:140 – 170

Fabricius FH (1977) Origin of marine ooids and grapestones. Contrib Sedimentol 7:1 – 113

Fabricius FH, Klingele H (1970) Ultrastrukturen von Ooiden und Ooilithen: zur Genese und Diagenese quartärer Flachwasserkarbonate des Mittelmeeres. Verh Geol Bundesanst Jg 1970:594 – 617

Ferguson J, Bubela B, Davies PJ (1978) Synthesis and possible mechanism of formation of radial carbonat ooids. Chem Geo 22:285 – 308

Flügel E (1978) Mikrofazielle Untersuchungsmethoden von Kalken. Springer, Berlin Heidelberg New York, 454 p

Folk RL, Pittman JS (1971) Length-slow chalcedony: A new testament for vanished evaporites. J Sediment Petrol 41:1045 – 1058

Frantzen W (1887) Untersuchungen über die Gliederung des unteren Muschelkalkes in einem Theile von Thüringen und Hessen und über die Natur der Oolithkörner in diesen Gebirgsschichten. Jahrb K Preuss Geol Landesanst Berlin Jg 1887:1 – 93

Freeman T (1962) Quiet water oolites from Laguna Madre, Texas. J Sediment Petrol 32:475 – 483

Friedman GM (1964) Early diagenesis and lithification in carbonate sediments. J Sediment Petrol 34:777 – 813

Friedman GM, Amiel AJ, Braun M, Miller DS (1973) Generation of carbonate particles and laminates in algal mats – example from sea-marginal hypersaline pool, Gulf of Aqaba, Red Sea Am Assoc Petrol Geol Bull 57:541 – 557

Frishman SA, Behrens EW (1969) Geochemistry of oolites, Baffin Bay, Texas. Geol Soc Am Abstr 7:71

Füchtbauer H (1964) Fazies, Porosität und Gasinhalt der Karbonatgesteine des norddeutschen Zechsteins. Z Dtsch Geol Ges. 114:484 – 531

Füchtbauer H (1974) Sedimentary petrology, part II. Sediments and sedimentary rocks 1. Schweizerbart, Stuttgart, 464 p

Füchtbauer H, Hardie LA (1980) Comparison of experimental and natural magnesian calcites. Int Assoc Sedimentol 1st Eur Meet Abstr, pp 167 – 169

Gavish E, Friedman GM (1969) Progressive diagenesis in Quaternary to Late Tertiary carbonate sediments: sequence and time scale. J Sediment Petrol 39:980 – 1006

Hahne C, Kirchmayer M, Ottemann J (1968) "Höhlenperlen" (Cave Pearls), besonders aus Bergwerken des Ruhrgebietes. Modellfälle zum Studium diagenetischer Vorgänge an Einzelooiden. Neues Jahrb Geol Palaeontol Abh 130:1 – 46

Halley RB (1977) Ooid fabric and fracture in the Great Salt Lake and the geologic record. J Sediment Petrol 47:1099 – 1120

Hay RL, Wiggins B (1980) Pellets, ooids, sepiolite and silica in three calcretes of the southwestern United States. Sedimentology 27:559 – 576

Hess FL (1929) Oolites or cave pearls in the Carlsbad caverns. Proc US Natl Mus 76:1 – 5

Illing LV (1954) Bahaman calcareous sands. Bull Am Assoc Petrol Geol 38:1 – 95

James NP (1972) Holocene and Pleistocene calcareous crust (caliche) profiles: criteria for subaerial exposure. J Sediment Petrol 42:817 – 836

Kahle CF (1974) Ooids from Great Salt Lake, Utah, as an analogue for the genesis and diagenesis of ooids in marine limestones. J Sediment Petrol 44:30–39
Kalkowsky E (1908) Oolith and Stromatolith im norddeutschen Buntsandstein. Z Dtsch Geol Ges 60:68–125
Kettenbrinck EC, Manger WL (1971) A deformed marine pisolite from the Plattsburg Limestone (Upper Pennsylvanian) of southeastern Kansas. J Sediment Petrol 41:435–443
Kinsman DJJ (1969) Interpretation of Sr^{2+}-concentrations in carbonate minerals and rocks. J Sediment Petrol 39:486–508
Knatz H (1966) Zur Bildung "Künstlicher Ooide" in Kraftwerken. Leitz Mitt Wiss Techn 3:176–178
Knewtson SL, Hubert JF (1969) Dispersal patterns and diagenesis of oolitic calcarenites in the Ste. Genevieve Limestone (Mississipian), Missouri. J Sediment Petrol 39:954–968
Land LS, Behrens EW, Frishman SA (1979) The ooids of Baffin Bay, Texas. J Sediment Petrol 49:1269–1278
Lindholm RC, Finkelman RB (1972) Calcite staining: semiquantitative determination of ferrous iron. J Sediment Petrol 42:239–242
Logan BW, Harding JL, Ahr WM, Williams JD, Snead RG (1969) Carbonate Sediments and Reefs, Yucatán Shelf, Mexico. Am Assoc Petrol Geol Mem 11:1–198
Loreau JP (1970) Ultrastructure de la phase carbonatée des oolithes marines actuelles. CR Acad Sci 271:816–819
Loreau JP, Purser BH (1973) Distribution and Ultrstructure of Holocene Ooids in the Persian Gulf. In: Purser, BH (ed) The Persian Gulf. Springer, Berlin Heidelberg New York, pp 279–328
Lyell CH (1855) A manual of elementary geology, 5th edn, London, 655 p
Mackin JH, Coombs HA (1945) An occurrence of "Cave Pearls" in a mine in Idaho. J Geol 53:58–65
Marshall JF, Davies PJ (1975) High-magnesium calcite ooids from the Great Barrier Reef. J Sediment Petrol 45:285–291
Mazzullo SJ (1977) Shrunken (geopetal) ooids: evidence of origin unrelated to carbonate-evaporite diagenesis. J Sediment Petrol 47:392–397
McGannon DE (1975) Primary fluvial oolites. J Sediment Petrol 45:719–727
Meyers WJ, Lohmann KC (1978) Microdolomite-rich syntaxial cements: proposed meteoric-marine mixing zone phreatic cements from Mississipian limestones, New Mexico. J Sediment Petrol 48:475–488
Milliman JD (1974) Marine carbonates. Recent sedimentary carbonates, part I. Springer, Berlin Heidelberg New York, p 375
Milliman JD, Barretto HT (1975) Relict magnesian calcite oolite and subsidence of the Amazon shelf. Sedimentology 22:137–145
Mitterer RM (1968) Amino acid composition of organic matrix in calcareous oolites. Science 162:1498–1499
Mitterer RM (1972) Biogeochemistry of aragonite mud and oolites. Geochim Cosmochim Acta 36:1407–1412
Moguedet G (1973) Contribution à l'étude des sédiments superficiels du plateau continental de la Guyane Française. Unpubl Ph D thesis Univ Nantes, p 143
Newell ND, Purdy EG, Imbrie J (1960) Bahamian oolitic sand. J Geol 68:481–497
Nissen HU (1964) Calcite fabric analysis of deformed oolites from the South Mountain Fold, Maryland. Am J Sci 262:892–903
Oti M, Müller G (1979) Recent ooids from different environments. Erdöl Kohle Erdgas Petrochem 32:107–115
Payton CE (1966) Petrology of the carbonate members of the Swope and Dennis Formations (Pennsylvanian), Missouri and Iowa. J Sediment Petrol 36:576–601
Pettijohn FJ (1975) Sedimentary rocks, 3rd edn. Harper & Row, New York, 628 p
Popp BN, Wilkinson BH (1982) Holocene lacustrine ooids from Pyramid Lake, Nevada. (this vol)
Purser BH, Loreau JP (1973) Aragonitic, Supratidal Encrustations on the Trucial Coast, Persian Gulf. In: Purser BH (ed) The Persian Gulf. Springer, Berlin Heidelberg New York, pp 343–376
Radwanski A, Birkenmayjer K (1977) Oolitic/pisolitic dolostones from the Late Precambrian of south Spitsbergen: their sedimentary environment and diagenesis. Acta Geol Polon 27:1–39
Read JF (1974) Calcrete Deposits and Quaternary Sediments, Edel Province, Shark Bay, Western Australia. Am Assoc Petrol Geol Mem 22:250–282

Richter DK (1972) Authigenic quartz preserving skeletal material. Sedimentology 19:211 – 218
Richter DK (1974) Zur subaerischen Diagenese von Echinidenskeletten und das relative Alter pleistozäner Karbonatterrassen bei Korinth (Griechenland). Neues Jahrb Geol Palaeontol Abh. 146:51 – 77
Richter DK (1976) Gravitativer Meniskuszement in einem holozänen Oolith bei Neapolis (Süd-Peloponnes, Griechenland). Neues Jahrb Geol Palaeontol Abh. 151:192 – 223
Richter DK (1979) Die Stufen der meteorisch-vadosen Umwandlung von Mg-Calcit in Calcit in rezenten bis pliozänen Biogenen Griechenlands. Neues Jahrb Geol Palaeontol Abh 158:277 – 333
Richter DK (1980) Former magnesian-calcite ooids in Paleozoic to Recent sediments. Int Assoc Sedimentol 1st Eur Meet Abstr, pp 215 – 218
Richter DK, Besenecker H (this vol.) Subrecent thermal ooids with tangentially oriented high-Sr aragonite (Tekke Ilica/Turkey), pp 154 – 162
Richter DK, Füchtbauer H (1978) Ferroan calcite replacement indicates former magnesian calcite skeletons. Sedimentology 25:843 – 860
Richter DK, Zinkernagel, U (1981) Zur Anwendung der Kathodolumineszenz in der Karbonatpetrographie. Geol Rundsch 70:1276 – 1302
Rusnak GA (1960) Some observations of recent oolites. J Sediment Petrol 30:471 – 480
Sandberg PA (1975) New interpretations of Great Salt Lake ooids and ancient nonskeletal carbonate mineralogy. Sedimentology 22:497 – 537
Sannemann D, Zimdars J, Plein F (1978) Der basale Zechstein (A 2 – T 1) zwischen Weser und Ems. Z Dtsch Geol Ges 129:33 – 69
Sarkar B (1973) Deformed oolites in the Pre-Cambrian Bhander limestones, India. J Sediment Petrol 43:636 – 643
Schuler G (1968) Lithofazielle, sedimentologische und paläogeographische Untersuchungen in den Raibler Schichten zwischen Inn und Salzach (Nördliche Kalkalpen). Erlanger Geol Abh 71:1 – 60
Shearman DJ, Twyman J, Zand Karimin M (1970) The genesis and diagenesis of oolites. Proc Geol Assoc London 81:561 – 575
Shoji R, Folk RL (1964) Surface morphology of some limestone types as revealed by electron microscope. J Sediment Petrol 34:144 – 155
Siesser WG (1973) Diagenetically formed ooids and intraclasts in South African calcretes. Sedimentology 20:539 – 551
Simone L (1974) Genesi e significato ambientale degli ooidi a struttura fibroso-raggiata di alcuni depositi mesozocici dell'area Appennino-dinarica e delle Bahamas merdionali. Boll Soc Geol Ital 93:513 – 545
Simone L (1981) Ooids: a review. Earth Sci Rev 16:319 – 355
Sorby HC (1879) On the structure and origin of limestones. Q J Geol Soc London XXXV:56 – 95
Suess E, Fütterer D (1972) Aragonitic ooids: experimental precipitation from seawater in the presence of humic acids. Sedimentology 19:129 – 139
Swett K (1965) Dolomitization, silification and calcitization patterns in Cambro-Ordovizian oolites from northwest Scotland. J. Sediment Petrol 35:928 – 938
Teichert C (1970) Oolite, oolith, ooid: discussion. Am Assoc Petrol Geol Bull 54:1748 – 1749
Usdowski HE (1962) Die Entstehung der kalkoolithchen Fazies des norddeutschen Unteren Buntsandsteins. Beitr Mineral Petrogr 8:141 – 179
Walther J (1888) Die Korallenriffe der Sinaihalbinsel. Abh Math Phys Cl Kgl Sächs Ges Wissensch 24:437 – 505
Wherry ET (1916) A peculiar oolite from Bethlehem, Pennsylvania. Proc US Natl Mus 49:153 – 156
Wilkinson BH (1979) Biomineralization, Paleooceanography, and the evolution of calcareous marine organisms. Geology 7:524 – 527
Wilkinson BH, Landing E (1978) "Eggshell diagenesis" and primary radial fabric in calcite ooids. J. Sediment Petrol 48:1129 – 1138
Wilkinson BH, Popp BN, Owen RM (1980) Nearshore ooid formation in a modern temperate region marl lake. J. Geol 88:697 – 104
Zeng Yun-Fu, Lee Nan Hao, Huang Yang-Zohn (this vol.) Sedimentary characteristics of oolitic carbonates of Jialingjtang Formation of the Lower Triassic (T_1^2J1) in South Sichuan Basin, China, pp 176 – 187

II.2 Ferriferous Ooids

Michael M. Kimberley [1]

1 Mineralogy and Primary Morphology

Ferriferous ooids are a common feature in several banded cherty iron formations as well as in chert-poor iron formations (Kimberley 1978). Ooids in both cherty and chert-poor iron formations are similar morphologically to Recent calcareous ooids (Markun and Randazzo 1980, Rohrlich 1974). Ferriferous oolitic layers are always concentric and do not display the pervasive radial structure that cuts across concentric layers in many ancient calcitic ooids. Ancient calcitic ooids with just a small proportion of iron minerals also display well-preserved concentric layering (Figs. 1 A, 2 C).

Ferriferous ooids have formed on shallow continental platforms, on shallow volcanic platforms, and within inland seas (Kimberley 1978). On platforms, ooids have formed preferentially along high-energy margins (Goodwin 1956, Dimroth 1976, Button 1976). In pre-metazoan iron formations, this environment also produced abundant algal stromatolite (Fig. 1 B).

Primary morphology generally is preserved better in ferriferous than in calcitic ooids of comparable age. The best-preserved Archean calcitic ooids (Fig. 1 A) probably owe their preservation to a small proportion of iron silicates. Ferriferous ooids commonly display more than one set of oolitic layers due to asymmetric secondary growth (Fig. 1 C) or to secondary growth around a broken ooid (Figs. 1 E, 2 A). Asymmetric growth causes secondary oolitic layers to abut against primary layers (Fig. 1 C; Nahon et al. 1980, Fig. 8). In calcareous ooids, this feature is characteristic of ooid growth in quiet water (Freeman 1962). In all iron-formation ooids, new layering has formed only on surfaces of ooids, as on modern calcareous ooids. By contrast, new concentric layering commonly forms within and cross-cutting old layering in the weathering environment, e.g., in formation of bauxite spherules. Moreover, ferriferous ooids appear to have been sorted in a water body, unlike spherules developed in soil. There is no known morphological evidence for a weathering origin of ferriferous ooids in any iron formation (cf. Nahon et al. 1980, Siehl and Thein 1978).

The number of discernable layers in ferriferous ooids varies from dozens (Fig. 1 D), each a few microns thick as in Recent calcareous ooids (Bathurst 1975)

1 Department of Marine, Earth, and Atmospheric Sciences, North Carolina State University, Raleigh, NC 27650, USA

Coated Grains (ed. by T. M. Peryt)

to just a few layers in diagenetically altered ooids (Fig. 3 F). The thickness of the entire oolitic sheath may vary from the bulk of the ooids (Fig. 1 D) to a thin coating on fossil fragments (Schoen 1962, p. 37). Preferred orientation of ferrous silicate crystals within ooids is discernable by extinction of crystals parallel to directions of light polarization (Fig. 1 D).

Broken ooids with secondary oolitic growth constitute a larger proportion of ironstone ooids than of most limestone ooids (Fig. 1 E). Halley (1977) has shown that ooid breakage occurs most readily in an evaporitic environment. Independent evidence of an evaporitic environment is found in length-slow chalcedony and gypsum casts accompanying broken ooids (Dimroth and Chauvel 1973, Barghoorn et al. 1977, Markun and Randazzo 1980).

All minerals that occur in ironstone are found within ooids. Preservation of oolitic layering is best achieved by chamosite and greenalite, both ferrous silicates, and by fine-grained goethite, hematite, and magnetite. Thin layers of oxides usually are less continuous than thin layers of silicates (Fig. 2 A, C versus Fig. 3 A, C). Siderite invariably preserves oolitic texture less well than do chamosite or greenalite coexisting with siderite (Figs. 1 E, 2 A).

Ooids completely replaced by siderite are virtually the only monomineralic ferriferous ooids. The mineralogical heterogeneity of ooids is due either to growth in a variable chemical environment or to the influence of enclosed organic matter on diagenetic mineral growth. Gruner (1922) argued that interlamination of ferrous and ferric minerals precludes a replacement origin. However, calcareous ooids contain carbonaceous layers that could affect the oxidation state of iron during ferruginization (Shearman et al. 1970).

Algal boring is obvious in a few ferriferous ooids where tubes about 10 microns in diameter spiral inward (Hayes 1915). However, most algal and fungal tubes are much smaller and intertwine to produce structureless micritic rims on allochems (Bathurst 1975). These rims cannot be identified unequivocably in most ferriferous ooids because of a mineralogical difference between the rim and the rest of the ooid. For example, the goethitic rim on the ooid in Fig. 2 C probably has been bored, given that the rim cuts across oolitic layering.

2 Origin of Ferriferous Ooids

Chamosite and greenalite typically form the thinnest and most continuous oolitic layers, when present (Fig. 3 A, B). There is no petrographic evidence for an origin of these ferrous silicates by diagenetic alteration of observed oxides (cf. Bubenicek 1971). Ferrous silicates occur in ooids fragmented by waves (Fig. 1 E) and in ooids concentrated by metazoans within burrows (Fig. 1 F). However, ferrous silicates are unstable in the presence of molecular oxygen (e.g., Eugster and Chou 1973). Kimberley (1979) therefore modeled an origin of ferriferous ooids by early diagenetic ferruginization of calcareous ooids.

Gygi (1981) rejects the carbonate-replacement model for inland-sea iron formations in Europe because chamositic ooids could grow under intermittent cover by marine mud. Goethitic layers would aggregate onto the ooids during periods

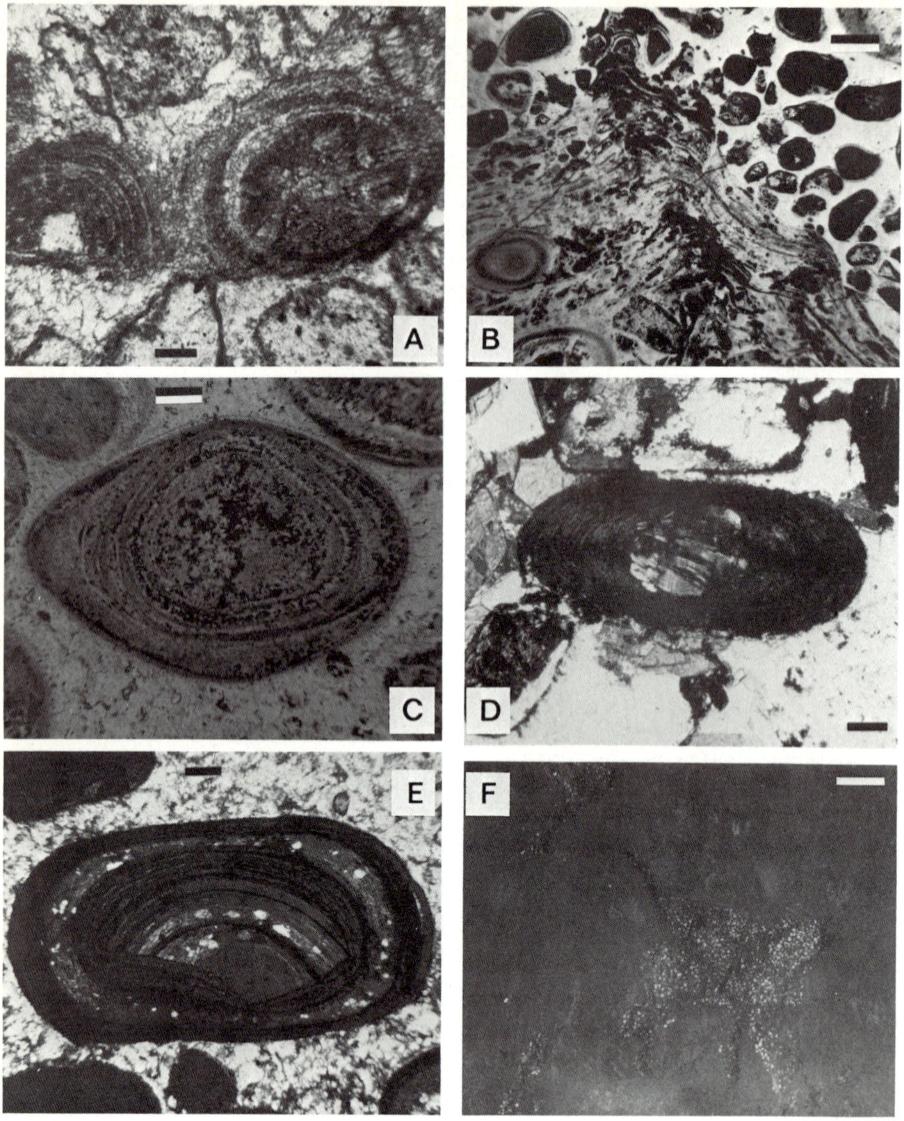

Fig. 1A–F. Primary structures in ferriferous ooids.
A An ooid and ooid fragment largely composed of calcite (*light gray*) with minor iron silicate (*black*) defining oolitic structure. Surrounding grains are mostly calcite. Archean Outerring iron formation, Back River area of Canada. Plane polarization. Scale bar is 0.2 mm.
B Algal stromatolite with ooids both within stromatolitic layers and surrounding the algal structure. *White areas* are chert. Variable proportions of greenalite appear as *gray* to *black*. Greenalite is concentrated in the ooids with respect to the matrix. Middle Precambrian Gunflint iron formation, Ontario, Canada. Plane polarization. Scale bar is 0.5 mm.
C Asymmetrical ooid, largely composed of chert (*light gray*) with minor greenalite (*medium gray*) in oolitic layers. Hematite (*black*) occurs preferentially within outermost and innermost oolitic layers and in ooid core. Middle Precambrian Gunflint iron formation, Ontario, Canada. Plane polarization. Scale bar is 0.1 mm.

of exposure. Petrographic evidence for a primary versus replacement origin is commonly ambiguous and paleoecological evidence like that of Gygi (1981) is helpful.

Several petrographic characteristics of ferriferous ooids have been cited in discussions of ironstone genesis (e.g., Kimberley 1979). Ferriferous ooids are commonly equal in size or larger than intermixed quartz grains (Fig. 3E). This may indicate a smaller settling velocity than that of quartz. A small settling velocity could be due to less dense primary calcareous mineralogy or simply to a high water content in primary ferriferous ooids. Diagenetic alteration of ooids is obvious where ooid cracks have controlled mineralization (Fig. 2A) but this does not prove that the ferruginization itself was diagenetic.

The occurrence of partially ferruginized ooids in limestone and dolostone of the Jurassic Green River Formation is probably due to diagenetic introduction of iron. As first noted by Bradley (1929, p. 221), hematitic ooids accompany dolomitic ooids and iron is concentrated in ooids relative to the matrix (Fig. 2B). This concentration of iron within ooids and the occurrence of ooid fragments as cores of ooids (Fig. 2B) are both characteristic features of ironstone (Fig. 1B, E). Compaction of several dolomitic ooids in Fig. 2B probably is due to preferential dissolution of calcareous ooid cores (Wilkinson and Landing 1978).

The Jurassic Marlstone Rock Bed and Korallenoolith are extensive examples of a lithology intermediate between oolitic limestone and oolitic ironstone. Figure 2C illustrates a partially ferriferous ooid from the Marlstone Rock Bed. Bottke et al. (1969, p. 385, Fig. 1) illustrate a similar ooid in the Korallenoolith, surrounded by more ferriferous ooids. If ferriferous ooids are primary precipitates, the environment of ironstone sedimentation must be potentially compatible with that of calcareous sedimentation. Fossil shell fragments in the Marlstone Rock Bed locally are replaced by siderite, goethite, or apatite (Edmonds et al. 1965). At least some diagenetic mineralization is therefore apparent. A limestone bed about a metre above the Marlstone Rock Bed contains typical calcitic ooids with both concentric and cross-cutting radial structure (Edmonds et al. 1965, p. 78). The lack of pervasive radial structure in partially ferriferous ooids (Fig. 2C) may indicate that radial structure preferentially developed in nonferriferous ooids diagenetically. Cross-cutting radial structure could develop before compactive deformation (cf. Wilkinson and Landing 1978).

Diagenetic ferruginization of calcareous ooids can occur on at least a small scale, given a Quaternary example on Andros Island, Bahamas (Fig. 2D). Oolitic layering has been accentuated during this ferruginization but subsequent expo-

◄

D Chamositic ooid with subordinate goethite and hematite. The horizontal black band through the middle of the ooid is an extinction cross that reveals alignment of chamosite crystallites. Ooid is surrounded by siderite. Eocene Paz de Rio iron formation in Colombia. Crossed polarization. Scale bar is 0.1 mm.

E Ooid fragment in core of ooid. Variable shades of *gray* are due to variable proportions of hematite and goethite. *White areas* within and surrounding ooid are siderite. Chamosite (*gray*) occurs with siderite in matrix. Eocene Paz de Rio iron formation in Colombia. Crossed polarization. Combined transmitted and reflected light. Scale bar is 0.05 mm.

F Burrows in sandstone immediately underlying the Paz de Rio iron formation. Burrows are filled with chamositic ooids (*white*). The top of the core is to the *left*. Eocene Paz de Rio iron formation in Colombia. Scale bar is 1 cm

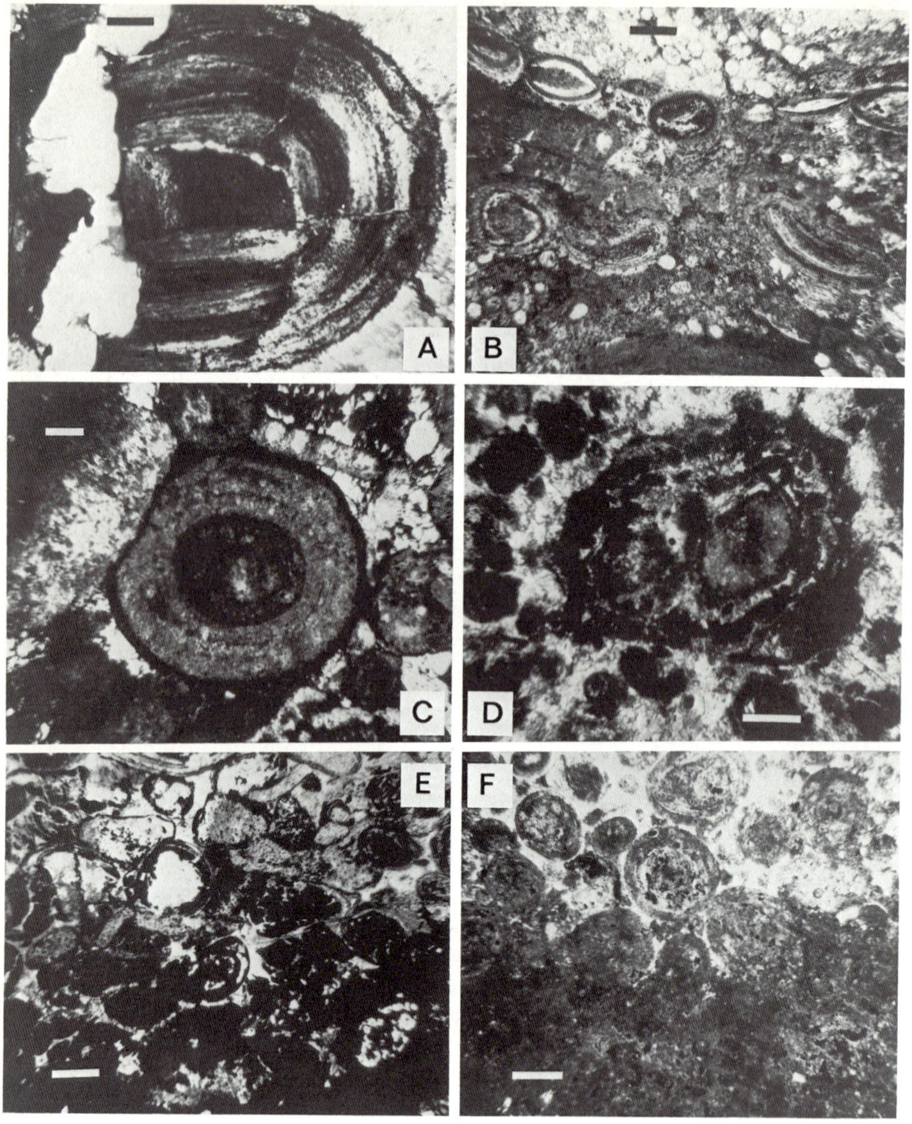

Fig. 2A – F. Features of genetic significance in ooids.

A Cracked ooid traversed by quartz veins (*white*) and surrounded by cherty cement. Color variation in the ooid is controlled by chamosite-rich (*light*) versus hematite-goethite-rich (*dark*) areas. Oxidation proceeded inward from cracks. Eocene Paz de Rio iron formation in Colombia. Plane polarization. Scale bar is 0.1 mm.

B Oolitic and pelletal dolostone. The dark ooid in the *upper central* area contains hematite (*black*). The ooid to its *left* displays solution collapse. An ooid to the *lower right* contains an ooid fragment as its core. Lamination in the rock is algal; the hand sample is stromatolitic. Eocene Green River Formation in Wyoming. Plane polarized light. Scale bar is 0.3 mm.

C Ooid composed of calcite (*gray*) with subordinate geothite (*black*). Surrounding grains are calcite. Minor amounts of goethite within the ooid enhance oolitic layering but the goethitic rim cuts across outer layers in one corner. Jurassic Marlstone Rock Bed, England. Plane polarized light. Scale bar is 0.1 mm.

sure has caused oxidation and mottling. Banding also has developed diagenetically in Quaternary ferriferous oolite (Fig. 2E) and some banding in Precambrian oolitic ironstone appears to be a diagentic modification of massively oolitic sediment (Fig. 2F).

3 Deformation of Ferriferous Ooids

Predominantly chamositic ooids generally are more deformed by compaction than other types of ooids (Fig. 3A). Chamositic ooids surrounded by quartz grains may be compacted until the ratio of length/width exceeds 10. Oxide-rich and sideritic ooids that have experienced the same conditions of burial display little or no deformation (Figs. 1E, 2A). Dissolution of ooid cores also has caused collapse during compaction (Fig. 3B). The hooked form of Figure 3B commonly is attached to an adjacent hooked ooid (Chanda et al. 1977, Fig. 5). A continuous row of attached, S-shaped ooids probably represents a plane of solution collapse along which there has been small lateral movement (Conley 1977).

Diagenetic loss of oolitic texture is difficult to prove except in cases like Fig. 3C where remnants of the sheath are scattered around the ooid core. Remnant oolitic segments also occur without a preserved core as in Fig. 3D. Diagenetic growth of siderite may force ooid layers outward (Fig. 3D) or inward to form bizarre ooid outlines (Fig. 3E). Iron oxides are also affected by siderite growth within ooids and form irregular rings (Fig. 3F).

4 Conclusion

Primary morphology of ooids is independent of deposit age. Primary features of ferriferous ooids more closely resemble those of calcareous ooid than those of concentric spherules that form during weathering in soil. Correspondence of compactive features to mineralogy indicates that iron minerals were present before compaction. Most genetic models for ferriferous ooids invoke direct precipitation of dissolved iron or agglomeration of particulate iron minerals at the sediment-water interface.

D Quaternary calcareous-ferriferous ooid with two ooids forming its core. Calcite and intergrown amorphous silica are white. Silica constitutes about 15% of the rock and iron about 2%. Authigenic iron minerals (*black*) accentuate oolitic layering. Quaternary oolite near Red Bay, Andros Island, Bahamas. Crossed polarization. Scale bar is 0.1 mm.

E Middle Precambrian ferriferous oolite. Diagenetic lamination (*banding*) has partially destroyed oolitic texture. Upper band is siderite (*light gray*) and chert (*white*). Lower band is partially oxidized greenalite (*black*) and siderite. Gunflint iron formation, Ontario, Canada. Plane polarization. Scale bar is 0.3 mm.

F Quaternary calcareous oolite exposed on Andros Island, Bahamas. Diagenetic lamination (*banding*) has partially destroyed oolitic texture. Diagenetic lamination is parallel to stratification. Indistinguishable fine-grained silica permeates this limestone; average SiO_2 is 15%. Scale bar is 0.3 mm

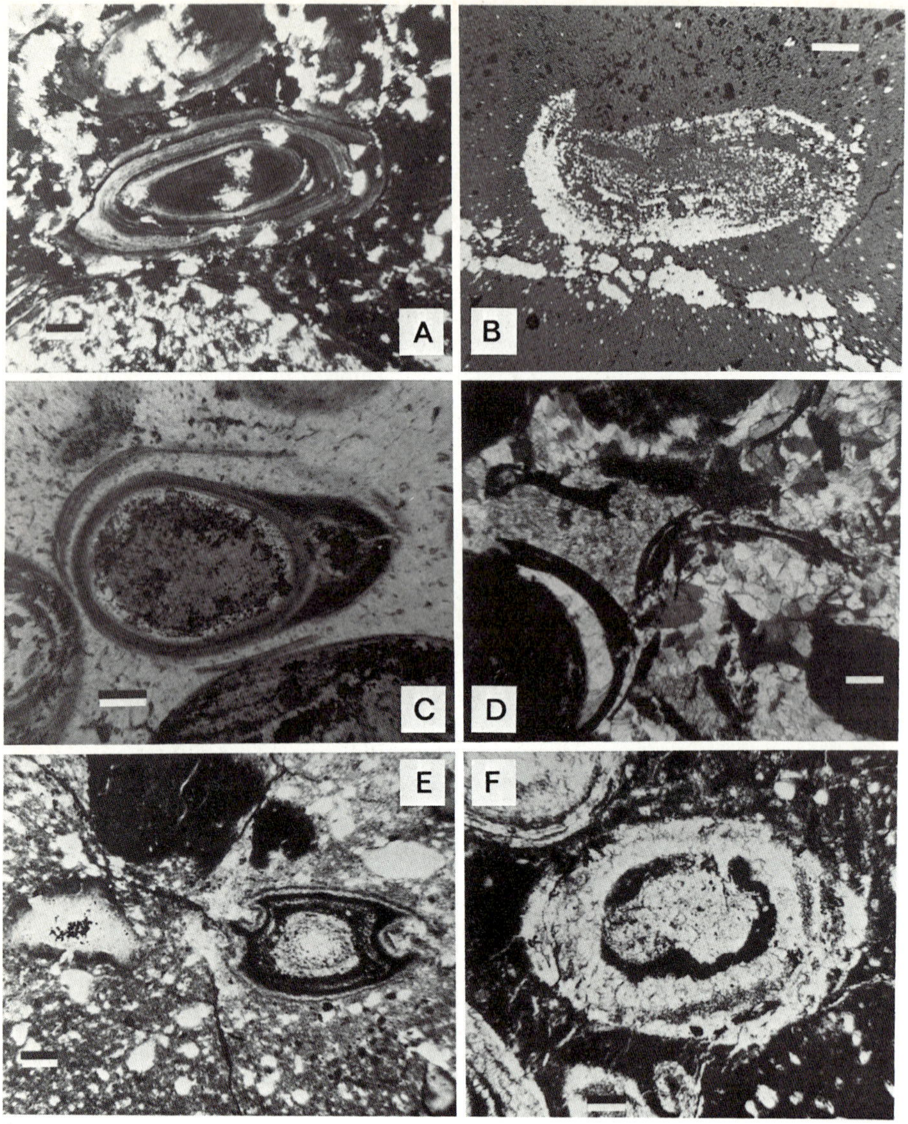

Fig. 3A – F. Secondary structures in ferriferous ooids.

A Chamositic ooids with subordinate layers of dark goethite and patches of siderite. Matrix is chamosite with patches of siderite. Two ooids have been compacted against each other and display a common grain boundary for about 0.2 mm in the *upper right corner.* Eocene Paz de Rio iron formation in Colombia. Plane polarization. Scale bar is 0.1 mm.

B Deformed ooid composed of pyrite (*white*) and chert (*gray*) surrounded by kaolinite (*gray*) and pyrite (*white*). Holes in the polished thin section are *black.* Eocene Paz de Rio iron formation in Colombia. Reflected light. Scale bar is 0.05 mm.

C Ooid largely composed of chert (*light to medium gray*) with subordinate greenalite (*dark gray*) in oolitic layers and hematite (*black*) in the ooid core. Oolitic layers are partially replaced by chert. Chert surrounds the central ooid and adjacent ooids. Middle Precambrian Gunflint iron formation, Ontario, Canada. Plane polarization. Scale bar is 0.1 mm.

References

Barghoorn ES, Knoll AH, Dembicki AH (1977) Variation in stable carbon isotopes in organic matter from the Gunflint iron formation. Geochim Cosmochim Acta 41:425 – 430

Bathurst RGC (1975) Carbonate sediments and their diagenesis, 2nd edn. Springer, Berlin Heidelberg New York

Bottke H, Dengler H, Finkenwirth A, et al. (1969) Sammelwerk Deutsche Eisenerzlagerstätten II. Eisenerze im Deckgebirge (Post varistikum) 1. Die marin-sedimentären Eisenerze des Jura in Nordwestdeutschland. Beih Geol Jahrb Heft 79, Hannover

Bradley WH (1929) Algal reefs and oolites of the Green River Formation. US Geol Surv Profess Pap 154:203 – 223

Bubenicek L (1971) Géologie du gisement de fer de Lorraine. Bull Centre Rech Pau SNPA 5:223 – 320

Button A (1976) Transvaal and Hamersley Basins – Review of basin development and mineral deposits. Miner Sci Eng 8:262 – 293

Chanda SK, Bhattacharyya A, Sarkar S (1977) Deformation of ooids by compaction in the Precambrian Bhander Limestone, India: implications for lithification. Geol Soc Am Bull 88:1577 – 1585

Conley CD (1977) Origin of distorted ooliths and pisoliths. J Sediment Petrol 47:554 – 564

Dimroth E (1976) Aspects of the sedimentary petrology of cherty iron-formation. In: Wolf KH (ed) Handbook of strata-bound and stratiform ore deposits 7:203 – 254

Dimroth E, Chauvel J-J (1973) Petrography of the Sokoman iron formation in part of the central Labrador trough, Quebec, Canada. Geol Soc Am Bull 84:111 – 134

Edmonds EA, Poole EG, Wilson V (1965) Geology of the country around Banbury and Edge Hill. G B Geol Surv Mem, p 137

Eugster HD, Chou IM (1973) The depositional environments of Precambrian banded iron-formation. Econ Geol 68:1144 – 1168

Freeman T (1962) Quiet water oölites from Laguna Madre, Texas. J Sediment Petrol 32:475 – 483

Goodwin AM (1956) Facies relations in the Gunflint iron formation. Econ Geol 51:565 – 595

Gruner JW (1922) The origin of sedimentary iron formations. Econ Geol 17:407 – 460

Gygi RA (1981) Oolitic iron formations: marine or not marine? Eclogae Geol Helv 74:233 – 254

Halley RB (1977) Ooid fabric and fracture in the Great Salt Lake and geologic record. J. Sediment Petrol 47:1099 – 1120

Hallimond AF (1925) Iron ores: bedded ores of England and Wales, petrography and chemistry. G B Geol Sur, Spec Miner Resour GB 29:129

Hayes AO (1915) Wabana iron ore of Newfoundland. Can Geol Surv Mem 78:163

Kalliokoski J (1966) Diagenetic pyritization in three sedimentary rocks. Econ Geol 61:872 – 885

Kimberley MM (1978) Paleoenvironmental classification of iron formations. Econ Geol 78:215 – 229

Kimberley MM (1979) Origin of oolitic iron formations. J Sediment Petrol 49:111 – 131

Markun CD, Randazzo AF (1980) Sedimentary structures in the Gunflint iron formation, Schreiber Beach, Ontario. Precambrian Res 12:287 – 310

Nahon D, Carozzi AV, Parron C (1980) Lateritic weathering as a mechanism for the generation of ferruginous ooids. J Sediment Petrol 50:1287 – 1298

◄———

D Two stages of siderite growth, an early fine-grained and late coarse-grained stage. Hematitic ooids have been partially replaced and outer oolitic layers have expanded due to siderite growth. Eocene Paz de Rio iron formation in Colombia. Crossed polarization. Scale bar is 0.1 mm.
E Highly deformed ooid with rim of siderite (*gray*), chamositic layers (*gray*), and goethitic layers (*black*). Adjacent pellets are composed of chamosite with a variable degree of oxidation to goethite, either concentrated in the pellet rim (*left* of ooid) or as partial oxidation throughout pellet (*above* ooid). Ooids and pellets are surrounded by kaolinite (*medium gray*) and silt- to sand-sized quartz (*white*). Eocene Paz de Rio iron formation in Colombia. Plane polarization. Scale bar is 0.1 mm.
F Siderite-replaced ooid with two remnant layers rich in hematite. The inner layer is most hematitic. Clear siderite also has precipitated around one corner of the central ooid and around all other allochems. The matrix is finely intergrown chamosite, siderite, and chert. Eocene Paz de Rio iron formation in Colombia. Plane polarization. Scale bar is 0.1 mm

Rohrlich V (1974) Microstructure and microchemistry of iron ooliths. Miner Depos 9:133 – 142

Schoen R (1962) Petrology of iron-bearing rocks of the Clinton Group in New York State. D thesis, Harvard Univ, 151 p

Sherman DJ, Twyman J, Karimi MZ (1970) The genesis and diagenesis of oolites. Proc Geol Assoc (London) 81:561 – 575

Siehl A, Thein J (1978) Geochemische Trends in der Minette (Jura, Luxemburg/Lothringen). Geol Rundsch 67:1052 – 1077

Wilkinson BH, Landing E (1978) "Eggshell diagenesis" and primary radial fabric in calcite ooids. J Sediment Petrol 48:1129 – 1138

II.3 Cortical Fabrics in Calcite and Aragonite Ooids

Donald A. Medwedeff [1,2] and Bruce H. Wilkinson [1]

1 Introduction

Calcareous ooids from modern and ancient marine and non-marine settings are highly variable owing to differences in initial cortical composition and fabric. Compositionally, modern ooids consist of low-magnesian calcite (0% – 5% $MgCO_3$), high-magnesian calcite (10% – 18% $MgCO_3$) and aragonite with more than one phase commonly occurring within an individual cortex (e.g, Land et al. 1979). The primary mineralogy of ancient (now low-magnesian calcite) cortices was also variable (e.g., Kettenbrink and Manger 1971), although the initial magnesium content of ancient calcite ooids is undocumented and the relative abundance of aragonite versus calcite phases has not been constant throughout Phanerozoic time (e.g., MacKenzie and Pigott 1981).

Cortical fabric in modern coatings is defined primarily through differences in the size, shape, and orientation of constituent crystals. In addition, the frequency of changes with respect to one or more of these features defines the density of cortical banding within an individual cortex. Modern ooid fabrics may be classified as radial, tangential or random, with random fabrics arising either when elongate crystals lack a preferred orientation, or when cortical laminae are composed of equant crystals. As with composition, more than one fabric may occur within an individual cortex (e.g., Land et al. 1979). Fabrics of ancient ooids may be even more variable owing to the development of neomorphic fabrics during the calcitization of aragonite grains (e.g., Kettenbrink and Manger 1971, Sandberg 1975).

Despite a voluminous literature with respect to oolitic carbonates, little is known concerning the relationships between cortical composition and cortical fabric. To this end, we have examined aragonite and calcite oolitic sands in order to document the relationships between composition and fabric in these coated grains.

1 Department of Geological Sciences, The University of Michigan, Ann Arbor, MI 48109, USA
2 Present address: Department of Geological Sciences, Queens University, Kingston, K7L-3N6, Canada

Coated Grains (ed. by T.M. Peryt)

2 Calcite Ooids

The Twin Creek formation in northwestern Wyoming contains numerous thin oolite beds which record normal marine sedimentation on a shallow open Jurassic shelf (Implay 1952, Wright 1971). These units are typically well-sorted medium sand with rare intraclasts, pellets, and uncoated crinoid debris. Ooid nuclei consist of foraminifera, crinoidal and molluscan debris, ostarcode valves, and quartz sand. Composite ooids, consisting of aggregates of two to five grains, are rare.

Early diagenetic dissolution of aragonitic molluscan nuclei within some of these units resulted in the development of hollow elliptical cortical shells which were subsequently crushed during burial and compaction. Ooids with calcitic nuclei, on the other hand, were unaffected during this early dissolution and burial event. These relationships demonstrate that the mineralogy of these calcite cortices is a primary feature (Wilkinson and Landing 1978). Because non-spelian calcite cortices are exceedingly rare in modern settings, the Twin Creek grains afford an excellent opportunity to document fabric variations in calcite ooids.

The gross fabric of Twin Creek ooids is typically radial, with calcite crystals oriented normal to grain surfaces (Fig. 1). Within this framework, cortices exhibit two general types of fabric which are directly related to ooid size; radial cortices and banded radial cortices (Fig. 1). Unbanded coatings characterize smaller grains and thc inner (older) coatings on larger grains. Radial crystals are 30 to 50 μm in length and up to 7 μm in width, and commonly increase in width from the nucleus, ending in blunt rounded terminations. These radial crystals rarely exhibit banding parallel to ooid exteriors (Fig. 2A).

Banded radial fabric occurs in the younger (outer) portions of larger ooids (Fig. 2C) with a gradational boundary between radial and banded radial fabrics occurring at an average diameter of 0.12 mm. Banded radial cortices consist of radially oriented fine laths of calcite, 2 to 3 μm in width and 5 to 15 μm in length. Owing to significantly larger width/length ratios for these crystals (compared to

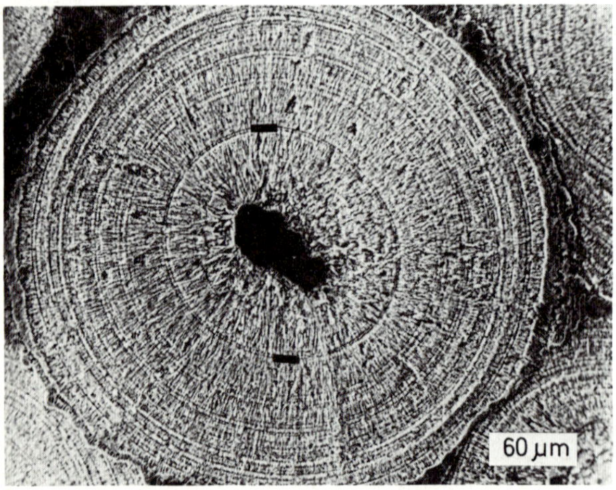

Fig. 1. SEM photomicrograph of a typical ooid from the Twin Creek Formation. Note the quartz nucleus, the radial arrangement of cortical crystals, the coarse calcite cement between grains, and the pronounced increase in banding density toward the ooid exterior. The *two dark dashes* mark the concentric division between radial (interior) and banded radial (exterior) cortical fabrics

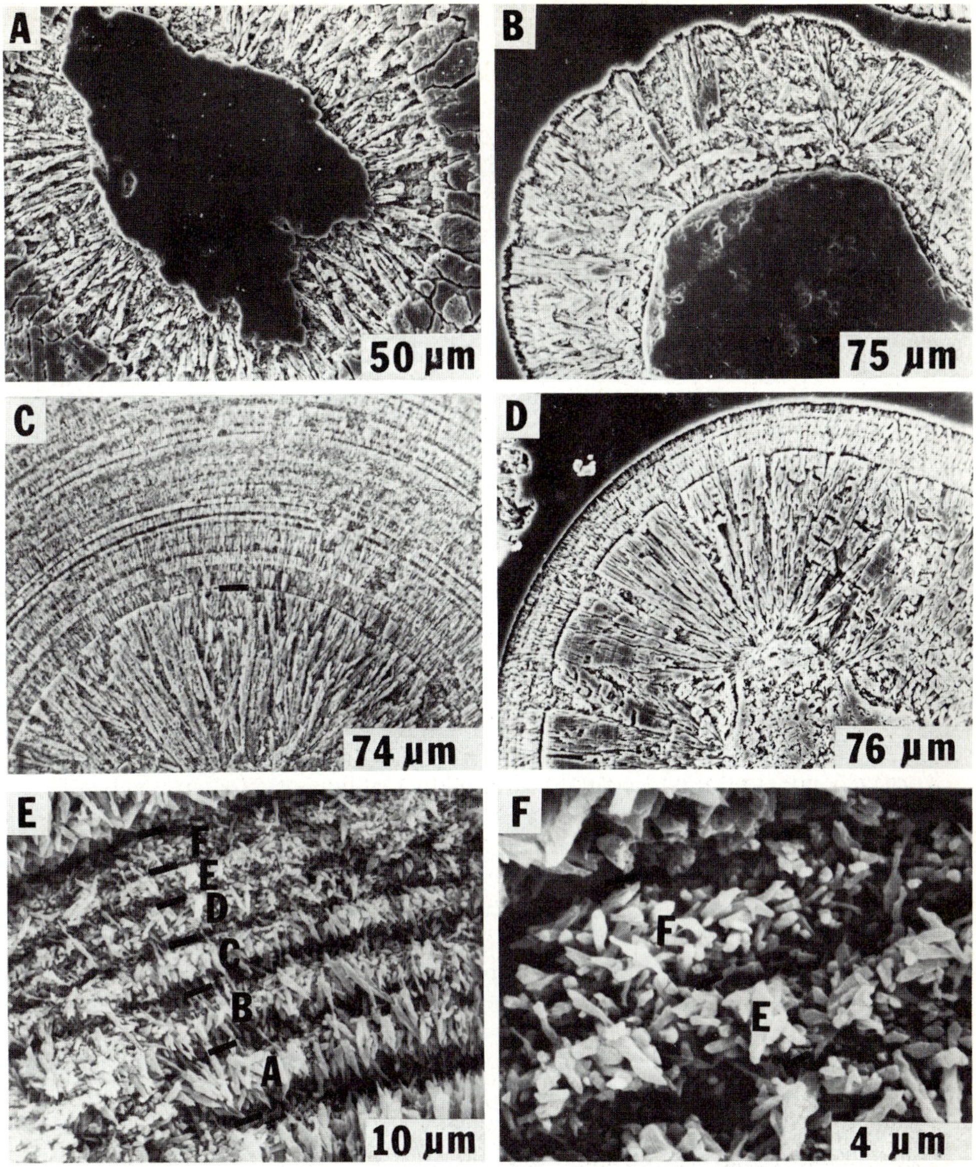

Fig. 2A – F. SEM photomicrographs of Twin Creek and Great Salt Lake ooids. All samples were sectioned, polished, and etched for 30 s in 50% glacial acetic acid: **A** Small calcite ooid (0.19 mm dia.) from the Twin Creek Formation with a cortex exhibiting radial fabric. Note that large crystals become wider away from the quartz grain which serves as the ooid nucleus. **B** Small aragonite ooid (0.28 mm dia.) from the Great Salt Lake with a cortex exhibiting radial fabric. Note the large radial crystals which become wider away from substrates, and the similarity between this grain and the calcite ooid in **A**. **C** Large calcite ooid (0.43 mm dia.) from the Twin Creek Formation with a radial inner cortex (0.23 mm dia.) and a banded radial outer cortex. The *bar* marks the boundary between these two fabric types. **D** Large aragonite ooid (0.37 mm dia.) from the Great Salt Lake with a radial inner cortex (0.32 mm dia.) and a banded radial outer cortex. **E** Densely laminated cortical aragonite in the banded radial portion of a large ooid from the Great Salt Lake. The six complete laminae within this field of view (*A* to *F*) are 6.4, 5.2, 4.4, 3.6, 2.8, and 2.4 µm in thicknesses. Crystals in **A**, **B**, and **C** are radial, crystals in **D** and **E** are random, and crystals in **F** are tangential. **F** Higher magnification of laminae *E* and *F* of **E**. Note the lack of radial orientations within these thin laminae those in cortices with radial fabrics), crystals sides are nearly parallel; crystals do not become wider toward ooid exteriors. The most striking aspect of this type of fabric is the presence of concentric bands which truncate or cross-cut cortical crystals. These discontinuities occur at 5 to 15 μm intervals; the distance between bands decreases with increasing distance from ooid interiors. Where banding is most dense, with spacings of 4 to 8 μm, cortical crystals are not distinctly radial. Rather, individual thin laminae frequently consist of 3 to 6 μm diameter equant calcite nannograins which exhibit a random fabric.

3 Aragonite Ooids

Radial aragonite ooids occur in a variety of modern marine and hypersaline settings. Because ooids from the Great Salt Lake, Utah, have been studied in considerable detail, samples collected from the south shore of this lake were examined in order to document textural variation in aragonite corties. Nuclei of ooids commonly consist of ovoid fecal pellets and, rarely, of quartz and igneous rock fragments. Mean ooid size ranges from 0.1 to 1.0 mm among different localities. Grains examined during this study have an average diameter of 0.4 mm and are very well sorted.

The fabric of Great Salt Lake cortices, described in detail by Halley (1977), Kahlc (1974), and Sandberg (1975), consist of tangentially, randomly, and radially oriented elongate laths. Radial crystals are by far the most common type and also the largest, ranging up to 150 μm in length.

As with Twin Creek ooids, two general types of fabrics characterize Great Salt Lake ooids. Smaller grains have cortices exhibiting a radial fabric which is similar to that of small ooids from the Twin Creek limestones (Fig. 2B). Radial crystals are more variable in size, however, and may deviate by as much as 30° from a truly radial orientation. Like Twin Creek grains, smaller Great Salt Lake ooids are less spherical, more closely reflecting the shapes of their respective nuclei (Fig. 2B). Larger Great Salt Lake ooids commonly consist of older (inner) cortices exhibiting a radial fabric, surrounded by younger laminae with significantly denser banding (Fig. 2D). Densely banded cortices occur on grains larger than about 0.23 mm in diameter, and are characterized by radial, random, and tangential orientations of constituent cortical crystals. In addition, there is a strong correlation between laminae thicknesses (banding density) and crystal orientation. Laminae thicker than 4 μm typically exhibit radial fabrics, laminae between 4 and 2.5 μm most commonly exhibit crystals with random orientations, and crystals within laminae thinner than 2.5 μm are oriented tangential to ooid exteriors. Also, there is a general trend in Great Salt Lake cortices for banding to be more dense toward the exteriors of larger grains. As a result, the predominance of radial fabrics decreases from grain interiors.

4 Discussion

The general significance of cortical fabrics in oolitic carbonates has been discussed at length elsewhere (see Halley 1977, Simone 1981, for excellent reviews), and will not be belabored here. Rather, we comment on several aspects of cortical fabric which may be drawn from examination of Twin Creek and Great Salt Lake ooids.

1. Twin Creek cortices were precipitated as calcite of an unknown composition, although their fabric bears a striking resemblence to Holocene radial high-magnesian calcite ooids from Baffin Bay, Texas. The fabric of Twin Creek ooids is also nearly identical to that of calcite ooids from the upper Cambrian Warrior Formation of Pennsylvania (Heller et al. 1980), and may be typical of calcite ooids which predominated as coated grains throughout much of the Phanerozoic (Sandberg 1975). Within both the Warrior and Twin Creek Formations, cortical fabric of smaller ooids is typically radial whereas cortical exteriors of larger ooids are progressively more densely laminated. In banded radial cortices from both formations there is a progression in crystal shape from large radial laths to stubby radial laths to equant nannograins with increasing banding density. It is clear that there is a direct relationship between ooid size, banding density, component crystal shape, and component crystal orientation in calcite ooids. As first suggested by Heller et al. (1980), it is likely that as ooids grew larger through cortex accretion, they came into contact with other grains more frequently and with greater force. In other words, the relative role of agitation and abrasion during cortex accretion is directly related to ooid size, and has significantly modified cortical fabric during the progressive growth of individual grains.

Great Salt Lake ooids exhibit nearly identical relationships with respect to ooid size, banding density, and orientation of cortical crystals. In these grains there is a common progression in cortical crystal size and orientation from large radial crystals to small radial crystals to small elongate randomly oriented crystals to small elongate tangentially oriented crystals with increased banding density. As such, these ooids record a relationship between grains size, banding density, component crystal shape, and component crystal orientation in aragonite ooids. Like calcite ooids, aragonite grains reflect the significance of agitation during cortex accretion with respect to cortical fabric.

2. Smaller calcite and aragonite ooids exhibit similar textural relationships suggesting that primary cortical mineralogy is of limited significance in predicating cortical fabric when ooid mass is small. Both exhibit unbanded radial fabrics which are similar in many respects. Under conditions of high abrasion, on the other hand, cortical mineralogy may play a significant role in predicating cortical fabric. Increased mass and the development of thin laminae as densely banded cortices in calcite ooids is accompanied by a progressive change in crystal shape from elongate laths to equant nannograins. Analogous transitions in aragonite ooids, on the other hand, are accompanied by a progressive change in crystal orientation from radial to random to tangential. This difference is significant, particularly in the context of cortical fabrics in ooids from other settings. Tangential fabric in calcite ooids is unreported from any modern or ancient set-

tings, marine or otherwise. It appears that, for some as yet undocumented reason, calcite crystals are incapable of assuming or growing in tangential orientations on ooids in natural systems. Similarly, the fabric of aragonite ooids from modern marine settings is predominantly tangential. Radial fabrics such as those in ooids from the Great Salt Lake are rare. It seems that aragonite crystals are readily rotated to and/or grow in tangential orientations on grain exteriors in natural systems.

3. The significance of cortical fabric in oolitic carbonates is predicated by both primary mineralogy and the effectiveness of physical energy. While it has frequently been suggested that radial fabrics record accretion in low energy settings, this relationship may only be valid for aragonite grains. Because aragonite cortical fabric is almost completely obliterated during diagenetic transformations of aragonite to calcite, and because radial fabrics predominate in all calcite cortices, radial fabrics in ancient ooids probably have little to do with energy levels in ancient carbonate environments.

4. The processes by which ooids originate are complexly interrelated. While we have pointed out some features that can be readily observed in many cortical coatings, it should be emphasized that the nature of the processes which give rise to such features are very poorly understood. Water chemistry, for example, predicates cortex composition, which in turn is related to crystal shape, size, and orientation. It is also true that environmental water chemistry largely determines if, and at what rate, cortex accretion takes place. The role of physical agitation and abrasion, which also relates to ooid size, banding density, and cortical fabric, is probably not an independent variable, and to some degree must relate to the rate at which component crystals grow. Heller et al. (1980), for example, observed that radial to banded radial fabric transitions in Warrior Formation ooids occur at about 0.6 mm diameters. Similar transitions occur in Twin Creek and Great Salt Lake ooids at 0.12 and 0.23 mm diameters respectively. These differences in size, all of which record the onset of significant abrasion on ooid exteriors, may reflect different energy levels in depositional environments. It is equaly likely, however, that these differences record variable ooid growth rates from locality to locality. Slower growth rates in concert with low abrasion could well give rise to fabrics similar to those in grains which grew rapidly in high energy settings.

Acknowledgements. We thank the staff at The University of Michigan Scanning Electron Microscope Laboratory for assistance during this study. Research on recent lacustrine carbonates at The University of Michigan is supported by the National Science Foundation, NSF grant EAR 78-93634.

References

Halley HB (1977) Ooid fabric and fracture in the Great Salt Lake and the geologic record. J Sediment Petrol 47:1099–1120

Heller PL, Komar PD, Pevear DR (1980) Transport processes in ooid genesis. J Sediment Petrol 50:943–952

Imlay RW (1952) Correlation of the Jurassic formations of North America exclusive of Canada. Geol Soc Am Bull 63:953–992

Kahle CF (1974) Oooids from the Great Salt Lake, Utah, as an analogue for the genesis and diagenesis of ooids in marine limestones. J Sediment Petrol 44:30 – 39

Kettenbrink EC, Manger WL (1971) A deformed marine pisolite from the Plattsburg Limestone (Upper Pennsylvanian) of southeastern Kansas. J Sediment Petrol 39:486 – 508

Land LS, Behrens SW, Frishman SA (1979) The ooids of Baffin Bay, Texas. J Sediment Petrol 49:1269 – 1278

MacKenzie FT, Pigott JD (1981) Tectonic controls of Phanerozoic sedimentary rock cycling. J Geol Soc 138:183 – 196

Sandberg PA (1975) New interpretations of Great Salt Lake ooids and of ancient nonskeletal carbonate mineralogy. Sedimentology 25:673 – 702

Simone L (1981) Ooids: A review. Earth Sci Rev 16:319 – 355

Wilkinson BH, Landing (ed) (1978) "Eggshell Diagenesis" and primary radial fabric in calcite ooids. J Sediment Petrol 48:129 – 138

Wright RP (1971) The Marine Jurassic of Wyoming and South Dakota: its paleoenvironments and paleobiogeography. Unpubl PhD diss, Univ Mich 167 p

II.4 Relict Sand Bodies and Bedforms of the Northern Bahamas: Evidence of Extensive Early Holocene Sand Transport

A.C. Hine[1]

1 Introduction

Recent studies along the shallow margins of the northern Bahama Banks have strongly indicated that sand transport has been and continues to be one of the dominant processes determining the character of these platform edges and distal lagoons (Hubbard et al. 1976, Hine 1977, Hine and Neumann 1977, Harris 1979, Palmer 1979, Hine et al. 1981a, b, Wilber 1981). The recent acquisition and analysis of high quality space and aerial imagery (see acknowledgements) show extensive fields of relict, large-scale bedforms. These large bedforms (maximum size: spacing – 1 km, length – 10 km, height – 5 m) are generally covered and stabilized with sparse to dense benthic flora and are thus termed "relict" – that is, the original sediments and depositional topography are no longer in equilibrium with the modern environment. This suggests that sand movement was more vigorous and pervasive earlier in the Holocene on the bank tops and margins than at present.

This paper intends to demonstrate the extent and variability of these relict and semi-relict features and to relate them to the late Holocene sea-level flooding history of the banks and ultimately to oolite-grainstone cycles seen in ancient settings.

2 Previous Work

Gebelein (1974), by analyzing surficial sediment data from the Bahama banks (Illing 1954, Purdy 1963, Ball 1967, Winland and Matthews 1974), determined that widespread blankets of oolite (now mostly micritized) and grapestone extended across much of the bank tops (Enos 1974, his Fig. 3) (Fig. 1). These sediments formed in response to "cross-bank currents" which were more vigorous earlier in the Holocene (Gebelein 1974, p. 56). Although Gebelein does not mention it, large bedforms, indicative of significant sand transport, lie within the ooid and grapestone facies.

1 Department of Marine Science, University of South Florida, 140 Seventh Avenue South, St. Petersburg, FL 33701, USA

Coated Grains (ed. by T.M. Peryt)

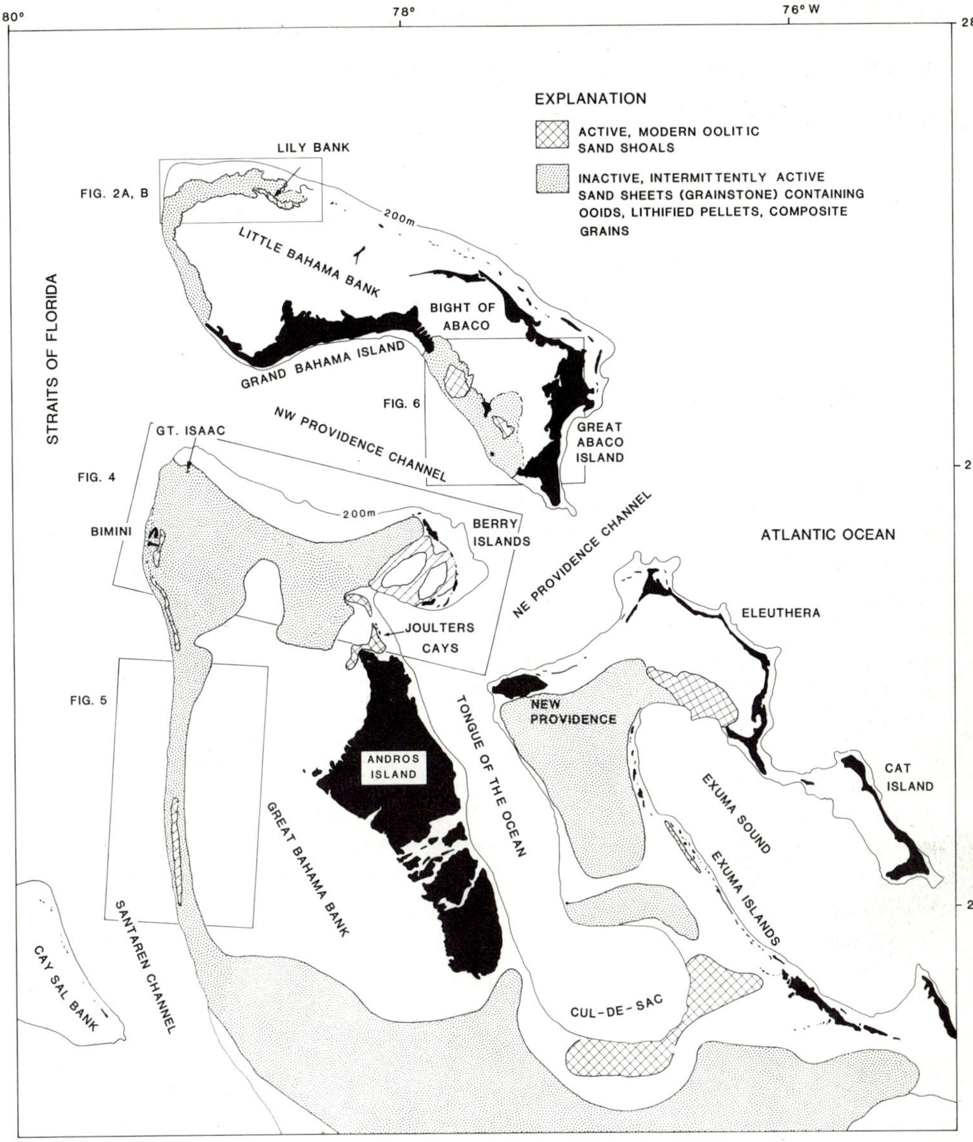

Fig. 1. Location map of northern Bahamas. Distribution of modern, active oolite shoals and relict sand bodies sometimes dominated by large bedforms are shown. Relict sand bodies consist of micritized ooid, peloid, and composite or grapestone grains. North and east facing margins are windward. West facing margins are leeward. Map composed from data of Purdy (1963), Enos (1974), Gebelein (1974), Palmer (1979), Harris (1979), Hine et al. (1981a)

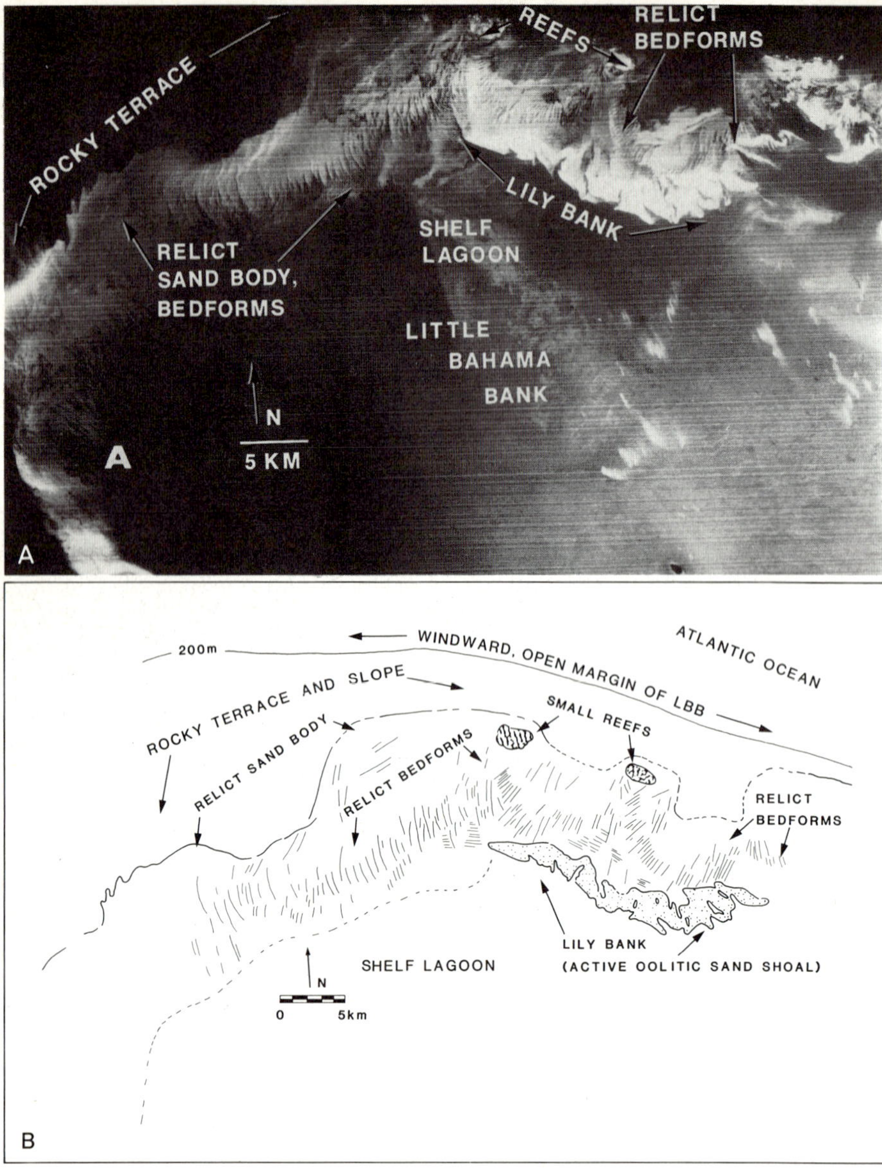

Fig. 2. A Detail of Landsat image of north-northwestern Little Bahama Bank. This is a windward, open margin. **B** Map of Landsat detail shown in **A**. Note the two orientations of relict bedforms. Also note small size of Lily Bank – the present-day active oolite shoal compared to the size of the relict sand body

Fig. 3. a Detail of Landsat image of northeastern Great Bahama Bank. Note the different orientations of the relict bedforms. **b** Detail of Landsat image of northwestern Great Bahama Bank. Note the large field of sand waves indicating bedload transport in the offbank (west) direction. The western edge is a leeward, open margin

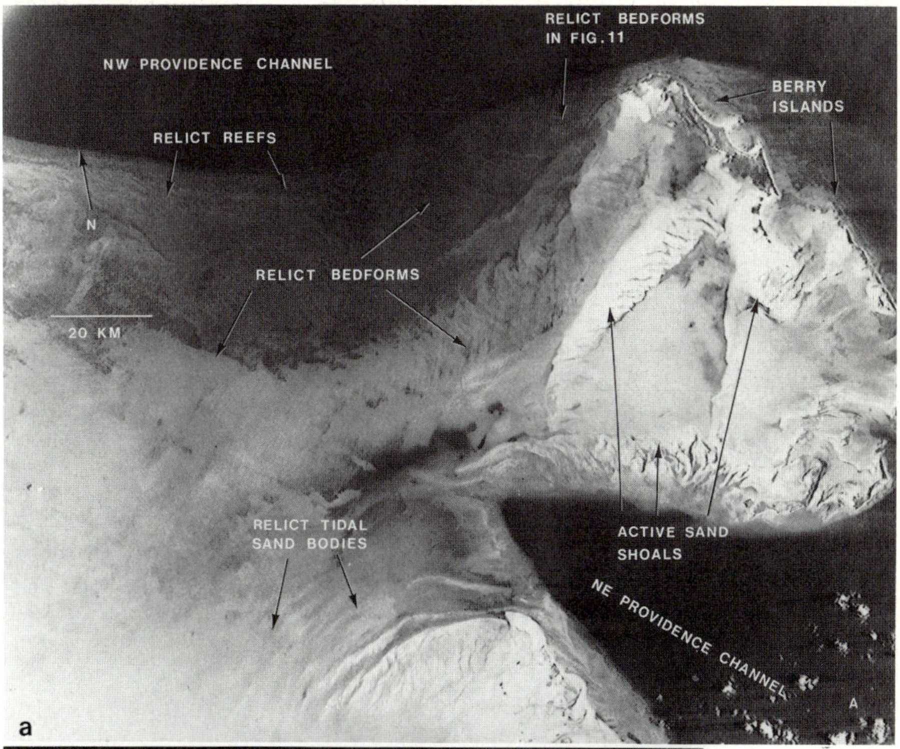
RELICT BEDFORMS
IN FIG. 11
NW PROVIDENCE CHANNEL
BERRY
ISLANDS
RELICT REEFS
N
RELICT BEDFORMS
20 KM
RELICT TIDAL
SAND BODIES
ACTIVE SAND
SHOALS
NE PROVIDENCE CHANNEL
A
a

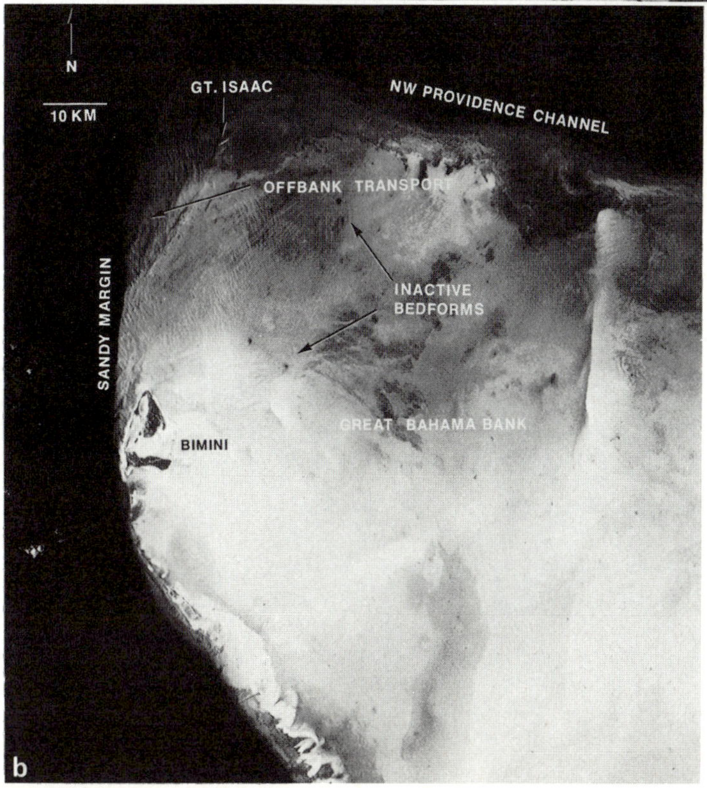
N
GT. ISAAC
NW PROVIDENCE CHANNEL
10 KM
OFFBANK TRANSPORT
SANDY MARGIN
INACTIVE
BEDFORMS
GREAT BAHAMA BANK
BIMINI
b

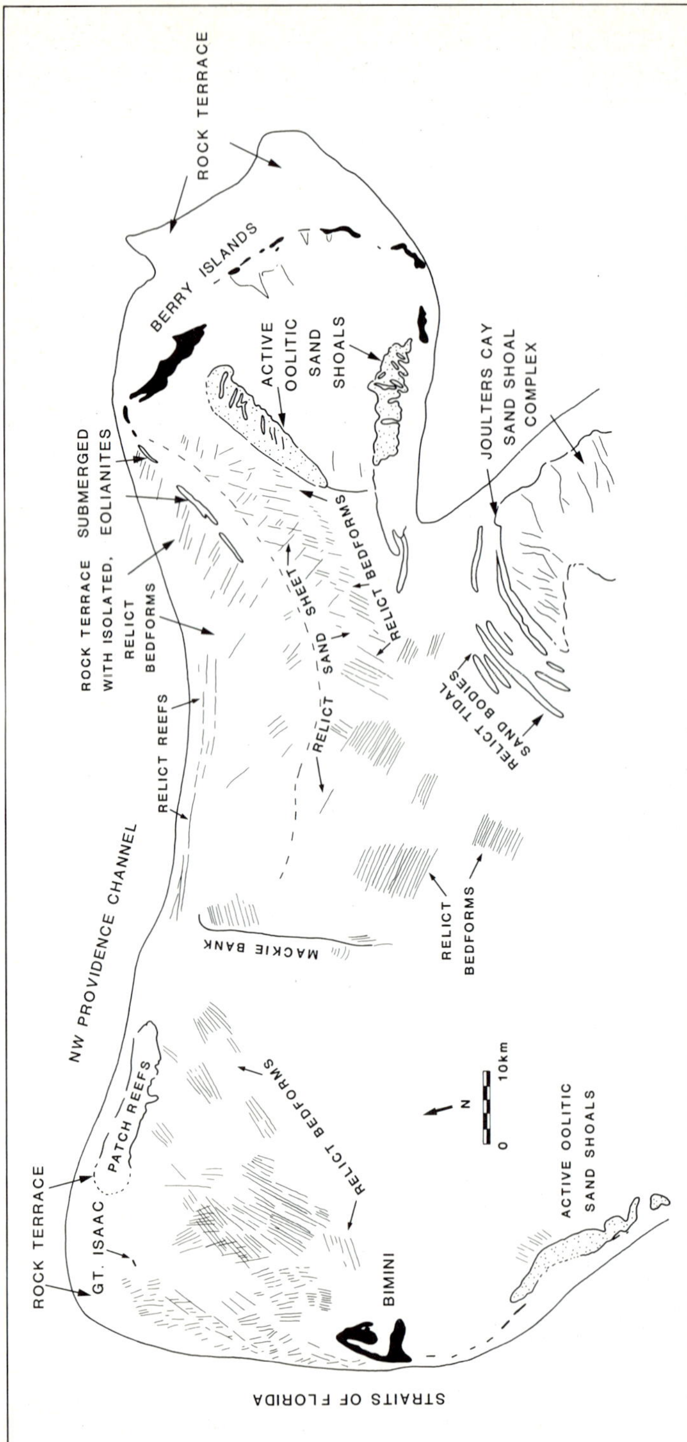

Fig. 4. Map of relict bedforms shown in Fig. 3A, B. Note inactive tidal sand bodies bankward of Joulters Cay sand shoal complex. Also note the few, narrow zones of modern, active oolitic-sand shoals

For reasons not well understood, these blankets of higher energy sands were unable to keep pace with the continuing Holocene sea-level rise. Stabilization processes such as algal-mat formation, incipient cementation, and sea-grass colonization began to retard sand transport (Taft et al. 1968, Neumann et al. 1970, Scoffin 1970, Winland and Matthews 1974, Dravis 1979). Ultimately, shelf-lagoonal sedimentary conditions prevailed.

It is only the visibility of large grass and algal-covered bedforms and lobate sand bodies that indicate the presence of the formerly active shoals in the absence of extensive coring data. In some areas, this benthic cover is thin and sparse which indicates that intermittent sand transport still occurs (Hine et al. 1981b). However, other sand bodies are covered with such a thick, dense benthic growth that only the large, dim outline of the huge, rhythmic bedforms belies the earlier sedimentary activity. Since these organisms are extremely difficult to erode and grow back slowly (Scoffin 1970, Patriquin 1975), it is doubtful that even major hurricanes have a significant effect on these vegetated zones (Perkins and Enos 1968; Hine 1977, pp. 1573–1575).

Ground studies to be used as detailed verification of the overhead imagery are meager. However, in a few areas, there are coring, drilling, and seismic data to

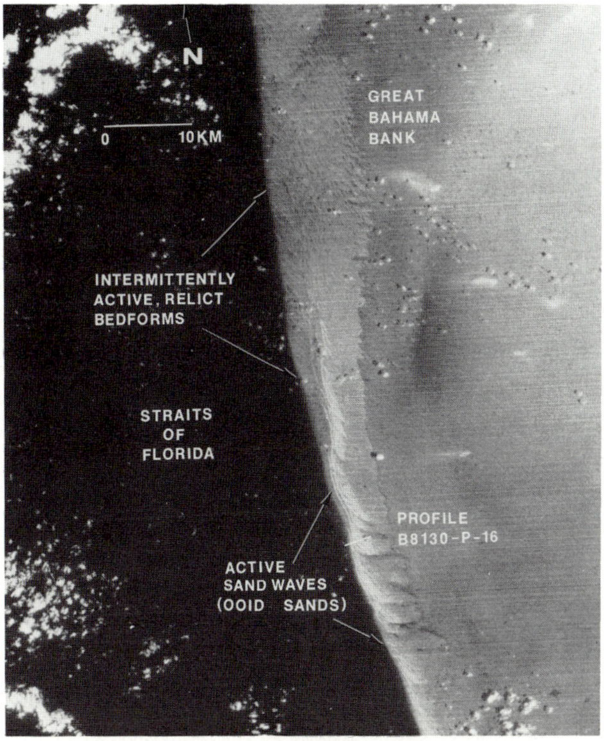

Fig. 5. Detail of Landsat image of western Great Bahama Bank – a leeward, open margin dominated by active and intermittently active sand shoals. Note extensive, linear sand belt dominating bank margin. Note also the location of seismic line B8130-P-16 seen in Fig. 12

corroborate conclusions drawn from the image analysis (Newell and Rigby 1957, Purdy 1963, Buchanan 1970, Boardman 1976, Hine 1977, Hine and Neumann 1977, Dravis 1979, Harris 1979, Palmer 1979, Locker 1980, Hine et al. 1981a, b, Wilber 1981).

3 Bank Margin Relict Sand Bodies and Bedforms

3.1 General Setting of Sand Bodies

In the northern Bahamas (Little Bahama Bank – LBB, and Great Bahama Bank – GBB) bank margins have been classified on the basis of physical energy flux and antecedent topography (Ginsburg and Shinn 1964, Hine and Neumann 1977, Hine et al. 1981a). The largest and most extensive zones of relict sand bodies (belts and sheets) occur along the windward-open (Figs. 2A, B, 3A, B, 4) and leeward-open margins (Figs. 3B, 4, 5). Both margins presently feature an unrestricted exchange of water flow on and off the banks with net sand transport occurring in a bankward direction along the windward margins (east-north facing) and a net sand transport occurring off the bank along the leeward margins (west facing).

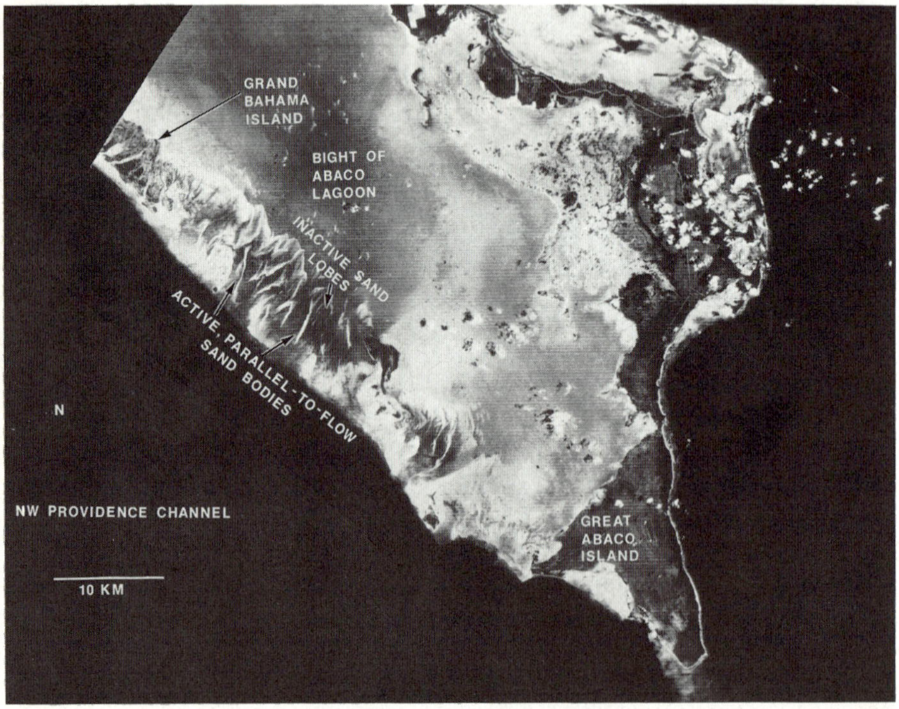

Fig. 6. Detail of Landsat image of open bank margin between Great Bahama Bank and Great Abaco Island. Note large sand lobes penetrating the Bight of Abaco. These are mostly relict. Linear sand bodies just seaward and normal to the bank edge are active zones of sand transport. This open margin was probably once a zone of more dominant tidal currents

Other areas where extensive sand shoals occur are located on the tide-dominated margins such as the cul-de-sac of the Tongue of the Ocean and Exuma Sound (Fig. 1) (Newell and Rigby 1957, Off 1963, Ball 1967, Dravis 1979, Halley et al., in press). These sand shoals are still very active. However, there are two tide-dominated zones that are apparently less active now than earlier: (1) the south edge of LBB between Grand Bahama Island and Abaco Island (Figs. 1, 6) and (2) portions of the Joulters Cay sand shoal complex (Figs. 1, 3A).

The bedforms are almost always confined to these bank margin sand belts/sheets which range in width from 10–30 km and in thickness from 1–25 m (Figs. 2A, B; 3A, B). A deeper, seaward, rocky terrace occasionally fronts the sand belt. Behind (bankward) the sand bodies, an open, shelf-lagoon environment exists. Few bedforms, either active or relict are found in this platform-interior environment.

Within each sand body lies a much smaller active belt dominated by bedforms responding to the daily tides and wave-generated currents. Along northern LBB, this active area is Lily Bank (Fig. 2A, B). Along GBB, this active zone lies behind the northern Berry Islands (Figs. 3A, 4). Along southern LBB, between Grand Bahama Island and Great Abaco Island a zone of inactive sand lobes lies bankward of a zone of active sand transport (Fig. 6). Coated grain formation is most likely still occurring within all these zones (Purdy 1963, Hine 1977, Hine et al. 1981a).

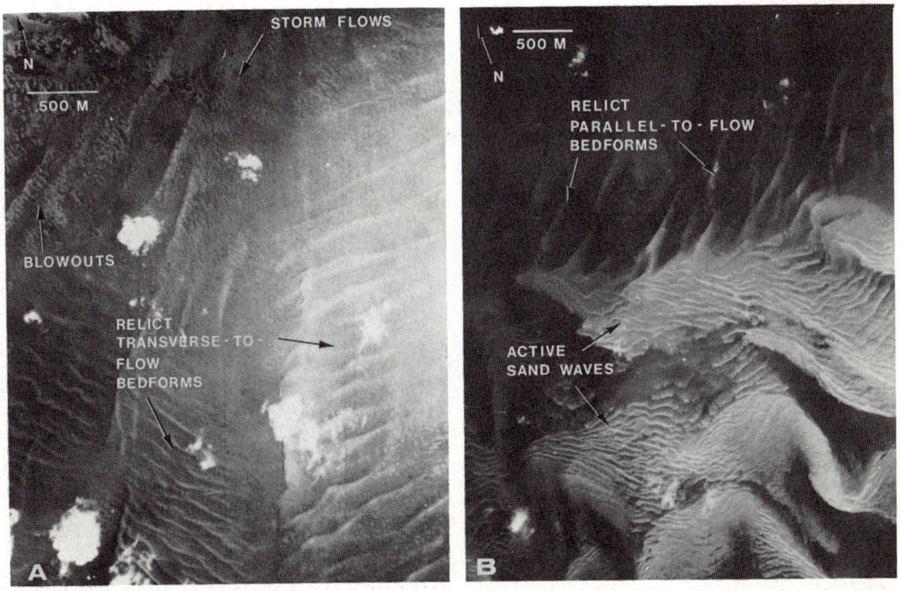

Fig. 7. A Aerial photograph of relict bedform field just to north (seaward) of Lily Bank, Little Bahama Bank. Note two bedform orientations: those parallel-to-flow (strom flows) and those transverse-to-flow. Extensive field of blowouts (erosional cuts in sea grass cover) is also present. **B** Aerial photograph of an active portion of the modern Lily Bank oolite shoal. Photograph shows the large, relict, parallel-to-flow features as well as the much smaller (in spacing) active transverse-to-flow (sand waves). These latter bedforms respond to the daily tidal currents and are not stabilized. DMA photo

3.2 Bedform Character

Two basic types of relict bedforms are apparent: (1) those that were formed transverse-to-flow and (2) those that formed more parallel-to-flow. These two types of relict bedforms are well illustrated in the Lily Bank LBB area (Fig. 7A, B). Here, it was speculated that higher velocity, bankward-directed flows generated the parallel-to-flow features nearer to the bank edge. As the water velocities decreased crossing the bank top due to frictional effects, the transverse-to-flow forms were generated, thus creating the distinct segregation of the two types (Hine 1977).

Fig. 8. Aerial photograph of intersecting set of two bedform types on northwestern Great Bahama Bank. DMA photo

Fig. 9. Aerial photograph of very even, parallel, symmetrical bedforms on Great Bahama Bank northeast of Bimini. Note "tuning fork" junctures. DMA photo

This relationship can be seen in a number of other examples as well (Fig. 8). In the northwestern corner of GBB, where the open, windward margin grades into an open, leeward margin, a huge field of each bedform type is shown with a transition zone where both bedforms occur simultaneously (Figs. 3B, 4). This suggests that the flows necessary to generate these features came from the northeast and travelled to the southwest. Some of these parallel-to-flow bedforms are very symmetrical (Fig. 9, as seen through a stereoscope). Others appear to be distinctly asymmetric (Fig. 10), have one side predominately stabilized by the benthic flora, and have gently curving axial trends. The orientation of the "blowout" structures (Patriquin 1975, Wanless 1981) indicate that the present dominant flows are oblique to the bedform axis suggesting that these features may have responded to changing flow conditions and direction through time (Fig. 10).

Bedform fields changing through time can generate complex, overlapping patterns. This is well illustrated along the relict sand body just to the west of the Berry Islands on northern GBB (Figs. 3A, 4). Here, two intersecting sets of large-scale bedforms can be seen. The earlier orientation appears to be the one where the bedform crests run NW-SE. These can be seen existing alone out on the deeper, rock terrace fronting the sand body. In places, these bedforms represent the only sediment resting on the rock surface (Fig. 11). Further in, and higher up on the sand body itself, this trend continues, but seems to be overprinted by a stronger NE-SW lineation. Beyond this zone where both orientations co-exist,

Fig. 10. Aerial photograph of large, parallel-to-flow bedforms on Little Bahama Bank west of Lily Bank. These features are asymmetric in profile, have benthic vegetation growing on the more gently dipping slope, and have curving axial trends. A blowout field indicates that more recent flows are oblique to the general axial trend of these bedforms indicating changing flow conditions

the NE-SW trend dominates and the NW-SE trend disappears. Finally, well in on the bank top, a new, more E-W trend appears. These bedforms are more closely spaced and appear to be transverse-to-flow type sand waves. The other two probably are the parallel-to-flow linear sand ridges.

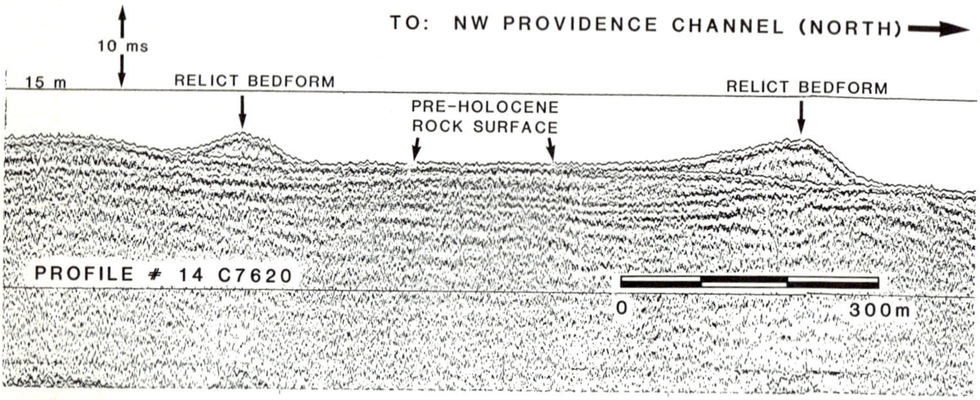

Fig. 11. Section of a high resolution seismic reflection profile (#14-C7620) along northern Great Bahama Bank. Isolated relict bedforms consisting of coarse, skeletal sands, lie directly on a pre-Holocene rock surface. Location of seismic section is shown in Fig. 3A

3.3 Other Evidence of Early Holocene Sand Transport

3.3.1 Leeward-Open Margins

Seismic data, drilling, coring, and aerial photos clearly show that Holocene sands, up to 25 m thick, lie on top of earlier Holocene fringing reefs (Fig. 12). This indicates, without doubt, that significant sand transport has occurred in the past and probably continues today during major storms (Hine et al. 1981b).

The presence of relatively modified, unvegetated, fresh appearing sand waves along portions of these leeward margins indicates that these features are still activated by passing storms. However, deeper (15 – 20 m) more heavily vegetated bedforms off Great Isaac Island (NW GBB) (Figs. 3B, 13) and the impressive volume and thickness of this leeward sand belt suggest that sand transport earlier in the Holocene along these leeward margins was vigorous.

Palmer (1979) suggests that fine peloid and ooid-rich sands accumulated on the bank edge early on during the Holocene sea level rise. With continual rise, ooid formation and accumulation shifted to a more bankward position, thus shielding the outer edge from normal current motions. As a result, the outer, bank-margin slope became covered with pellet-rich silty sands (Palmer 1979, pp. 131 – 132). This also suggests that much of the sand (mixed peloid, pellet packstone to grainstone and ooid grainstone; Palmer, his Fig. 37) along the outer portions of these leeward margins was deposited soon after flooding of the bank tops, thus indicating vigorous, early sediment transport.

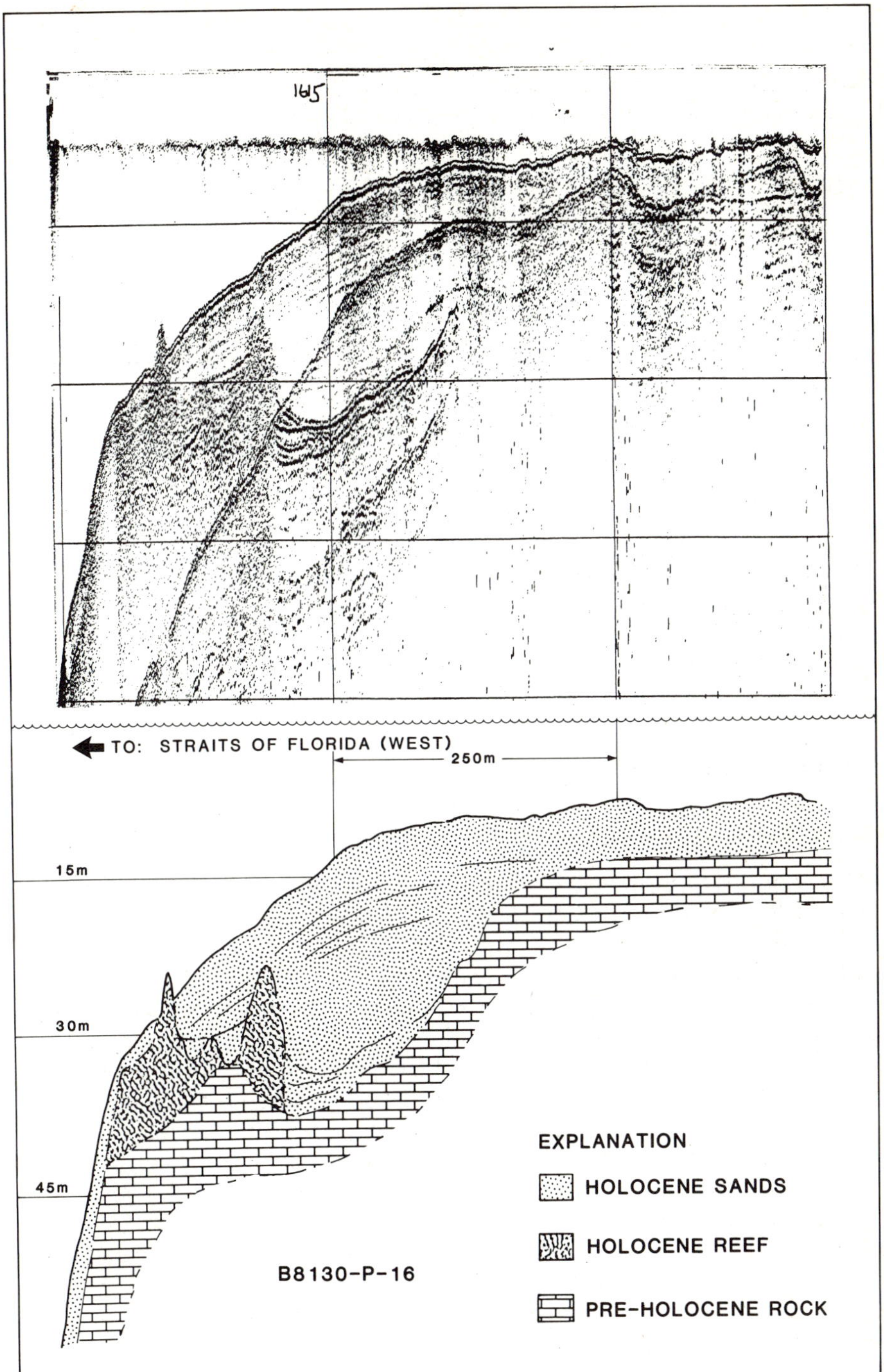

Fig. 12. Detail of high resolution seismic trace profile (B8130-P-16) across leeward, open margin seen in Fig. 5. Thick Holocene sands now bury earlier Holocene fringing reefs. The presence of such thick sands indicates vigorous sand transport from the distal lagoon to bank edge

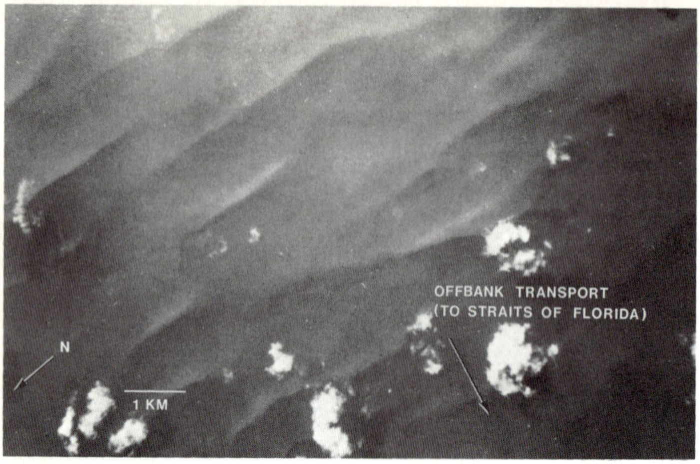

Fig. 13. Aerial photograph of deep (>15 m), mostly vegetated, relict sand waves west of Great Isaac Island on NW Great Bahama Bank. Bedforms are asymmetric and indicate sand transport in offbank (W) direction. DMA photo

3.3.2 Inactive Tide-Dominated Zones

There are two zones which appear to have been tide-dominated margins that have since become less active. They are the Joulter's Cay area (Harris 1979) on GBB and the open bank margin between the east end of Grand Bahama Island and the southern tip of Great Abaco Island (Hine et al. 1981a). Harris, in a very extensive study, has shown that an active broad, mixed ooid and ooid sand sheet probably dominated by tidal bars and channels extended up to 35 km in from the bank edge soon after the bank top in this area had been flooded (Fig. 3A). Ultimately, water motions were restricted due to the size and shallowness of the sand body. Also, islands formed on the bank margin further cutting off tidal currents. However, at one time, a huge, extensive zone was dominated by active sand movement and mixed, fine peloid and ooid formation.

The area between Grand Bahama Island and Great Abaco Island is covered with large, inactive bank-oriented sand lobes/sheets and active linear, parallel-to-flow sand bodies (Fig. 6). For the most part, the large sand lobes are presently devoid of large bedforms – at least none are visible from the air and a seismic/PDR line nearby indicated no rhythmic topographic features. The sand lobes appear to be mostly stabilized by algal mats and sparse benthic vegetation. Obviously, there must have been more vigorous current activity in the past to have formed these large, lagoon-penetrating sandy features.

4 Discussion

The broad sheets of grapestone and micritized oolitic sediments, the large, relict fields of bedforms, thick, marginal sand bodies, and stabilized sand lobes all indicate that there was a period of elevated, more intense sand transport on the Bahama Banks earlier in the Holocene than what exists in the present.

It is difficult to establish the timing of this high energy, high sediment transport interval. However there are a few studies which suggest that open marine conditions and early sand shoal formation began within the 3000 – 4700 yr B.P. time interval. Boardman (1976) and Locker (1980) both studying lagoons on LBB indicate that open, marine conditions appeared approximately at 3640 and 4700 yr B.P. respectively. Hine (1977) suggests that the now-relict bedforms surrounding Lily Bank oolite shoal were active around 4000 yr B.P. Harris (1979) indicates that widespread sand shoal formation in the Joulter's Cay area began between 3000 – 4000 yr B.P.

During this time (3000 – 4700 yr B.P.) the bank tops were covered with a minimum of 2 m (4700 yr B.P.) and a maximum of 6 m (3000 yr B.P.) of water. [These values were obtained using Neumann's sea level curve (1971), assuming an average depth to limestone of 8 m below present-day sea level, and a sediment accumulation rate of 0.5 m/1000 yr.]

When water depths were quite shallow (<1 m), wave and tidal activity were minimal. Certainly some winnowing and sediment movement did occur. However, sedimentation consisted mostly of pellet and skeletal rich muds. By the time 2 – 5 m of water covered the banks, a more vigorous active exchange between the bank tops and adjacent deep water had been established. Both tidal currents and storm-generated flows could more easily traverse the margin and penetrate into the bank. It was during this time that the major fields of bedforms and sheets of oolitic and grapestone sand formed. It was also during this time that the rate of sea-level rise slowed, thus prolonging the period of relatively high bottom turbulence.

As the seas rose the last 2 – 4 m to their present level, the added depth caused a reduction in the intensity and duration of sediment motion on the bottom. As the water deepened, surface currents generated by wind stress extended to the sea floor less frequently. Also, bottom oscillatory currents from waves generated on the banks became less vigorous. This energy reduction was gradual as it slowly allowed benthic organisms to begin to colonize and stabilize the sea floor. Consequently, many of the active sand transport zones narrowed to the relatively limited belts seen today.

5 Geological Significance

The relative sudden appearance, development, and waning of this widespread oolitic, grapestone sand sheet mantled by bedforms indicates that water depth, and therefore sea level, plays a critical role in the formation of this type of lithosome. As sea level sufficiently covers a wide, broad, flat surface, a transgressive sheet of diagnostic sediments forms over an extensive area (Gebelein 1974, p. 56). This could be a bank top such as the Bahamas or an epicontinental sea. If sea level rose rapidly and flooded the shallow basin quickly only a thin deposit would form. If sea level rose very slowly allowing sand accumulation to keep pace, then a much thicker unit would develop.

Additionally, if sea level fluctuated as little as several meters, the changes in energy levels at the sea floor could be large enough to form widespread and different sedimentary facies. This is a possible mechanism to explain the apparent high frequency cyclicity seen in the ancient oolite-grainstone units (Wilson 1975, pp. 283 – 297).

Along leeward-open margins, the amount of sand transported off to the deep also is a function of the level and rate of flooding. As developed by Hine et al. (1981b), the supply of shallow-water peri-platform sands to the deep flanks surrounding the banks, particularly along leeward, open margins may occur during brief, intense intervals when the banks are newly flooded or when they are nearly exposed during the ensuing sea level fall. In between, when water levels are high, offbank sand transport is retarded. During these periods of high stands, stabilized, shelf lagoon conditions exist all the way to the bank edge. Obviously, during periods of low stands, the banks are exposed and few shallow water sediments are formed.

6 Conclusions

1. Large, broad sheets and lobes of oolitic (may be micritic) and grapestone sands as well as large, relict bedforms indicate that a period of elevated sand transport occurred earlier in the Holocene and affected extensive zones of the tops of the shallow platforms.
2. Since that high energy interval (approximately 3000 – 4000 yr B.P.) the zones of active, constant sand movement and ooid formation have narrowed to distinct tidal-current dominated zones.
3. This suggests that broad, sheets of sands of this type form only during periods when there is an appropriate balance between the rate of sea level rise/fall, the regional level of physical energy, and the ability of the sand shoal to produce more sediments. If flooding is rapid, broad, thin oolitic sheets are formed. When flooding is slow or stable, wide, thick sand bodies form. When sea level slowly outpaces sand production, as in the present case on the northern Bahama Banks, a wide, thin sheet is initially formed followed by narrow zones of continuing sand movement and production.
4. The export of peri-platform sands to the deep flanks responds similarly to this balance between benthic turbulence, water depth, and the rate of sea level rise.

Acknowledgements. Funds to conduct much of this work were obtained from a President's Research Council Grant (University of South Florida). This paper also drew heavily upon other published research which in turn was supported by the National Science Foundation – grants OCE 76-04330 and OCE 77-24639 awarded to A. C. Neumann and A. C. Hine. Mr. James Hammack of the Defense Mapping Agency provided numerous, excellent space and aerial imagery that contributed significantly to this manuscript. The EROS Data Center provided the Landsat imagery. Dr. H. T. Mullins provided numerous helpful comments in his review.

References

Ball MM (1967) Carbonate sand bodies of Florida and the Bahamas. J Sediment Petrol 37:556–591

Boardman MR (1976) Lime mud deposition in a tropical island lagoon, Bight of Abaco, Bahamas. Unpubl thesis, Univ N Ca, Chapel Hill, NC

Buchanan H (1970) Environmental stratigraphy of Holocene carbonate sediments near Frazers Hog Cay, British West Indies. Unpubl thesis, Col Univ, 229 p

Dravis J (1979) Rapid and widespread generation of recent oolite hardgrounds on a high energy Bahamian platform, Eleuthera Bank, Bahamas. J Sediment Petrol 49:195–208

Enos P (1974) Surface facies map of the Florida-Bahamas Plateau. Geol Soc Am map Suppl statem

Gebelein CD (1974) Modern Bahamian platform environments, field trip guide. Bermuda Biol Stn, Bermuda

Ginsburg RN, Shinn EA (1964) Distribution of the reef-building community in Florida and the Bahamas. Am Assoc Petrol Geol Bull 48:527

Halley RB, Harris PM, Hine AC (in press) Bank margin sand bodies. Am Assoc Petrol Geol Mem

Harris PM (1979) Facies anatomy and diagenesis of a Bahamian ooid shoal. Sedimenta vii: Comp Sediment Lab, Univ Miami, Miami Fla

Hine AC (1977) Lily Bank, Bahamas: history of an active oolite sand shoal. J Sediment Petrol 47:1554–1581

Hine AC, Neumann AC (1977) Shallow carbonate bank margin growth and structure, Little Bahama Bank, Bahamas. Am Assoc Petrol Geol Bull 61:376–406

Hine AC, Wilber RJ, Neumann AC (1981a) Carbonate sand bodies along contrasting shallow bank margins facing open seaways in northern Bahamas. Am Assoc Petrol Geol Bull 65:261–290

Hine AC, Wilber RJ, Bane JM, Neumann AC, Lorenson KR (1981b) Offbank transport of carbonate sands along open, leeward margins, northern Bahamas. Mar Geol 42:327–348

Hubbard DK, Ward LG, FitzGerald DM, Hine AC (1976) Bank margin morphology and sedimentation, Lucaya, Grand Bahama Island. Univ SC Dep Geol Tech Rep 7-CRD:36

Illing LV (1954) Bahamian calcareous sands. Am Assoc Petrol Geol Bull 38:1–95

Locker S (1980) Holocene depositional history of Abaco Sound, Bahamas. Unpubl thesis, Univ North Ca, Chapel Hill NC

Neumann AC (1971) Quaternary sea level data from Bermuda. Quaternaria XIV Roma, pp 41–43

Neumann AC, Gebelein CD, Scoffin TP (1970) The composition, structure, and erodability of subtidal mats, Abaco, Bahamas. J Sediment Petrol 40:274–297

Newell ND, Rigby JK (1957) Geological studies on Great Bahama Bank. In: Le Blanc RJ, Breeding JG (eds) Regional aspects of carbonate deposition. Soc Econ Paleontol Miner Spec Publ 5, Tulsa OK, pp 15–72

Off T (1963) Rhythmic linear sand bodies caused by tidal currents. Am Assoc Petrol Geol Bull 47:324–341

Palmer MS (1979) Holocene facies geometry of the leeward bank margin, Tongue of the Ocean, Bahamas. Unpubl thesis, Univ Miami Fla 199 p

Patriquin DG (1975) "Migration" of blowouts in seagrass beds at Barbados Capriacou, West Indies, and its ecological and geological implications. Aquat Bot 1:163–189

Perkins RD, Enos P (1968) Hurricane Betsy in the Florida-Bahama area: geologic effects and comparison with Hurricane Donna. J Geol 76:710–717

Purdy EG (1963) Recent calcium carbonate facies of the Great Bahama Bank, 2. sedimentary facies. J Geol 71:472–497

Scoffin TP (1970) The trapping and binding of subtidal carbonate sediments by marine vegetation in Bimini Lagoon, Bahamas. J Sediment Petrol 40:249–273

Taft WH, Arrington F, Haimovitz A, MacDonald C, Woolheater C (1968) Lithification of modern carbonate sediments at Yellow Bank, Bahamas. Bull Mar Sci Gulf Carib 18:762–828

Wanless HR (1981) Fining-upwards sedimentary sequences generated in seagrass beds. J Sediment Petrol 51:445–454

Wilber RJ (1981) Late Quaternary history of a leeward carbonate bank margin, Little Bahama Bank. Unpubl thesis, Univ North Ca, Chapel Hill NC

Wilson JR (1975) Carbonate facies in geologic history. Springer, Berlin Heidelberg New York

Winland HD, Matthews RK (1974) Origin and significance of grapestone, Bahama Islands. J Sediment Petrol 44:921–927

II.5 The Joulters Ooid Shoal, Great Bahama Bank

Paul M. Harris [1]

1 Introduction

Although Recent (Holocene) carbonate sands have been deposited during an extremely brief epoch of geologic time, their subsurface facies relationships may be complex. The development of ooid shoals in the Bahamas has been examined by coring and seismic profiling by Ball (1967), Buchanan (1970), Hine (1977), Harris (1979), Palmer (1979) and Hine et al. (1981). The studies document facies, sedimentary structures, vertical sequences and geometry of bank-margin ooid accumulations. It is these sedimentary features that are critical in our interpretation of ancient oolite deposits seen in outcrop or borehole cores.

Fig. 1. The Joulters ooid shoal is an accumulation of Holocene sands that is located on Great Bahama Bank immediately north of Pleistocene Andros Island

1 Gulf Oil Exploration and Production Company, P.O. Box 36506, Houston, TX 77236, USA

Coated Grains (ed. by T.M. Peryt)

The Joulters ooid shoal on Great Bahama Bank (Fig. 1), described in detail by Harris (1979), is important because it displays a variety of subenvironments in which ooid sands can accumulate, some of which are quite different from the environment in which the ooids are generated. Facies relationships illustrate one possible evolutionary path of the bar and channel pattern commonly seen elsewhere: the channels have been filled in to form a vast sand flat of ooid sands that have been mixed with other sediments by burrowing.

2 Depositional Environments

The Joulters ooid shoal today is not a series of bars or shoals built up to sea level with deeper channels between them, but is instead a 400 km^2 sand flat fringed on the ocean-facing borders by mobile sands (Fig. 2). The sand flat, nearly intertidal and stabilized by sea grasses and algae, is penetrated partially by numerous tidal channels. The narrow rim of mobile ooid sands is intertidal, 1 to 2 km wide, 25 km long, and trends northwest parallel to the shelf break. The name of the shoal is derived from the Joulters Cays, islands up to 5 m above sea level, that formed along the rim of mobile sands.

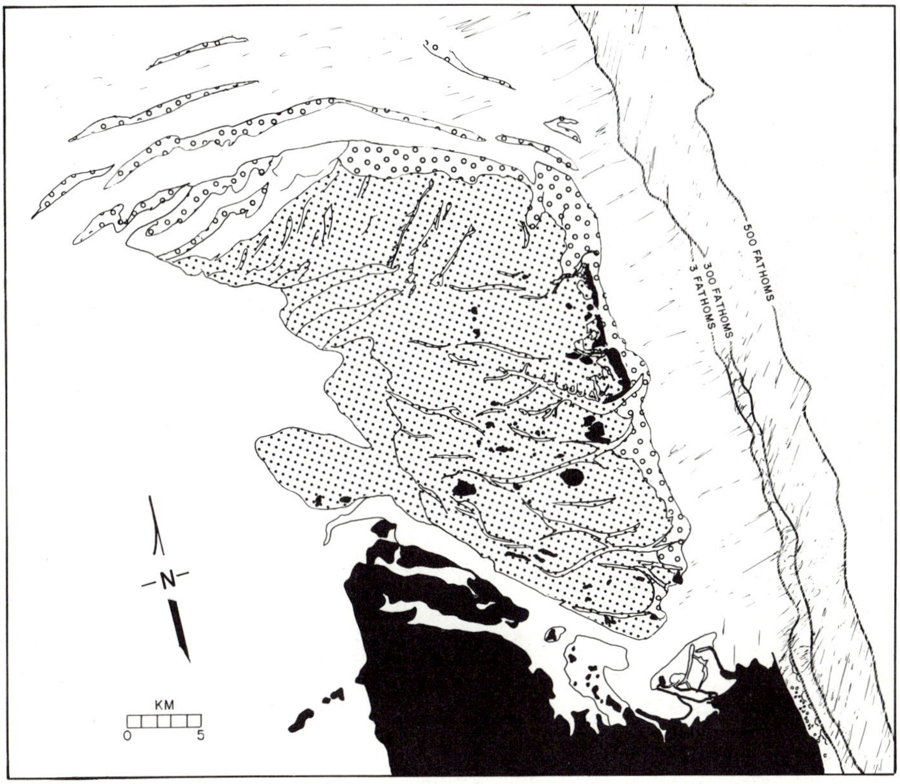

Fig. 2. Mobile sands of the Joulters ooid shoal (*open circle pattern*) rim a sand flat (*dot pattern*) that is penetrated by tidal channels

The shallow bank surrounding the shoal is variable. A 4- to 6-km-wide shelf lies east and in abrupt contact with the shoal. Skeletal sands on the shelf are commonly less than 0.5 m thick. An open shelf to the north of the shoal comprises a tidal bar belt (Ball 1967) with curving bars, 0.5 km wide and 1 to 3 m thick, that trend east-west for 12 km and then northeast-southwest for 15 km. The shoal changes gradually to the west into a mounded, grass-covered bottom of peloidal sands and muds.

3 Ooids

Ooids from the Joulters shoal average 200 to 500 μm in diameter. The nuclei, generally 100 to 300 μm in diameter, are commonly peloids. Ooid size reflects agitation (Illing 1954): small nuclei that are easily put into motion quickly develop thick coatings, while large nuclei have only a superficial coating. Typically, the coatings on fresh, shiny ooids from agitated bottoms are alternating micron-thick layers of oriented and less-oriented fabric. Within an oriented layer, the individual acicular aragonite crystals, 1 to 4 μm long and a few tenths of a micron wide, lie with their C-axis nearly tangential to the plane of the layer. In a less-oriented layer, similar aragonite crystals are more randomly positioned.

Ooids from stabilized bottoms characteristically do not retain their microstructure. During diagenetic alteration, such as micritization, which is common on stabilized bottoms, the concentric coatings may be disrupted and the preferred tangential orientation of aragonite crystals may be destroyed. Such ooids are a major unidentified contributor to the grains of aragonite matrix of Illing (1954) and the cryptocrystalline grains of Purdy (1963) and Buchanan (1970). Winland (1971) outlined criteria for recognizing altered ooids in thin section. The coatings on altered ooids may contain remnants of original concentric structure in the form of thin voids or streaks of wide crystal spacing. In completely altered ooids, the coatings are more coarsely crystalline than is the nucleus. The coatings appear transparent and grayish in plane light in contrast with the nucleus that appears opaque and brownish; in cross-polarized light the altered coatings are much brighter than is the nucleus.

4 Subsurface Facies Relations

Although coring of the Joulters ooid shoal has revealed six facies, the actual thickness of the shoal relative to the surrounding sea floor is primarily the result of contributions by three facies in differing amounts throughout the area. Ooid grainstone (Fig. 3) is present as a narrow belt along the ocean-facing borders of the shoal, where ooid accumulation coincides with formation. The most widespread facies, ooid packstone (Fig. 4) which is exposed on the sand flat and fine-peloid packstone (Fig. 5) which is exposed in the bank interior west of the shoal, are the result of mixing of ooids with other grain types. The significant contribu-

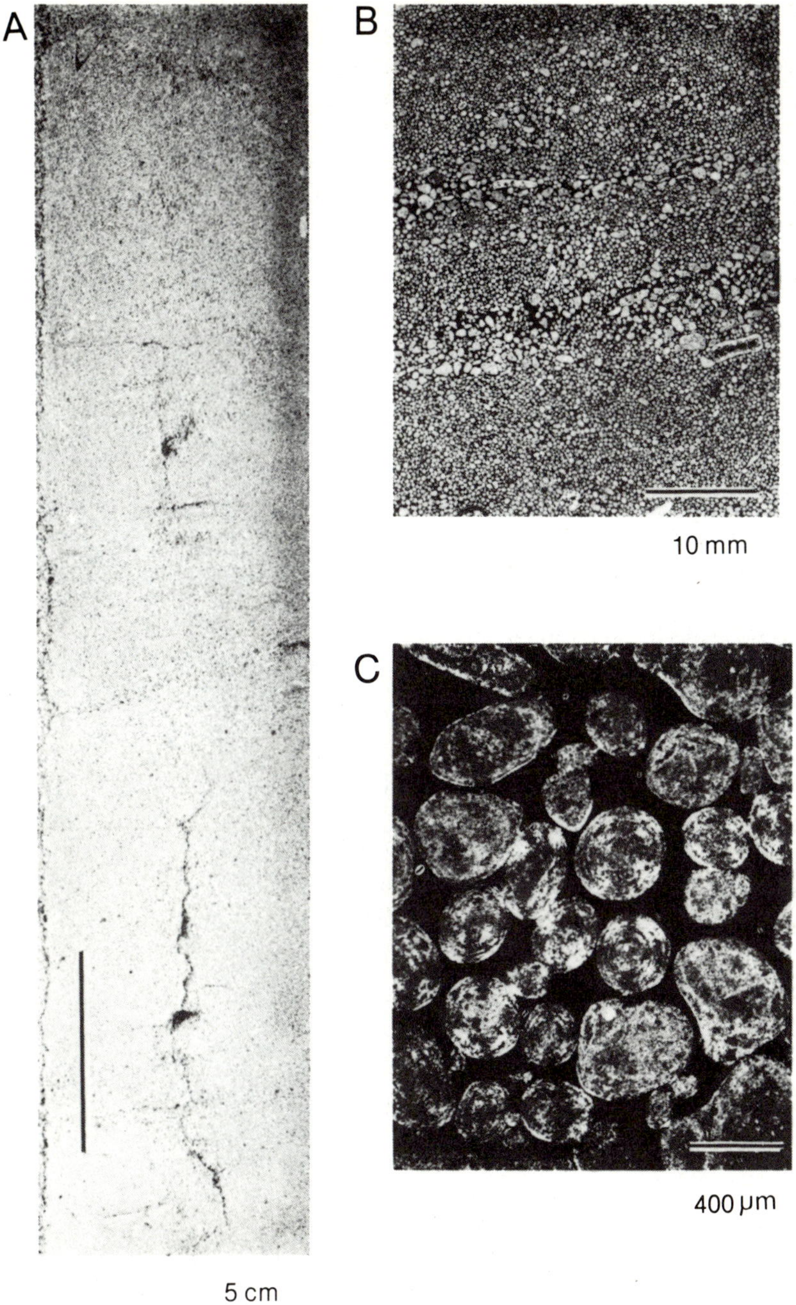

Figs. 3 – 5. A is a slab of plastic-impregnated core; **B** is a negative print of a thin section from the core slab; **C** is a photomicrograph in cross-polarized light of a portion of that thin section

Fig. 3 A – C. Ooid grainstone: **A** Inclined laminae of ooids (*top*) and laminae to thin beds of coarse grains (middle and right of scale). **B** Well to very well sorted, fine to medium ooids are laminated with fine-peloids (*top middle*) and with coarser aggregates and skeletal grains (*middle*). **C** Fresh ooids, peloids and coarser aggregates

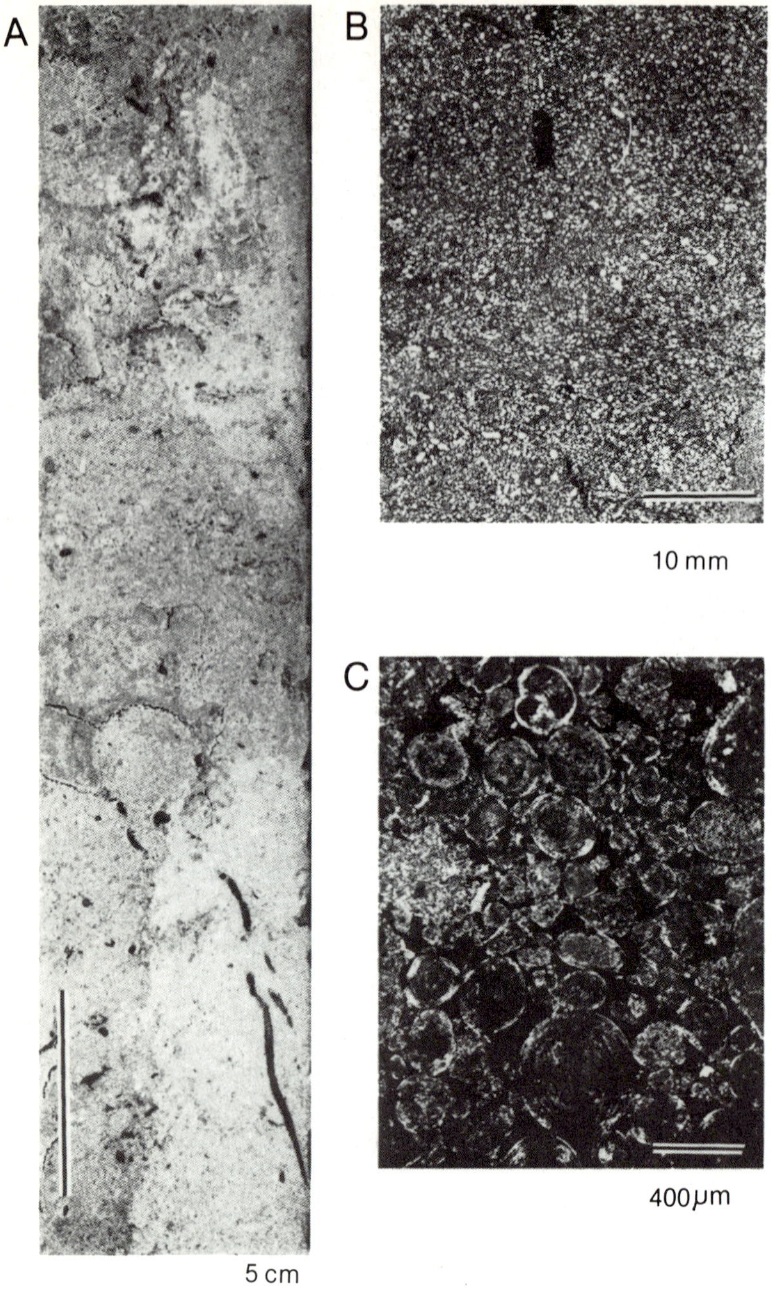

Fig. 4A – C. Ooid packstone: **A** Roots (*top* and *bottom*), bioturbated sand (*middle*), distinct large burrows (*top* and *middle*). **B** Very poorly to moderately sorted, bioturbated fine to coarse sand and mud with irregular, discontinuous laminae (*middle* to *bottom*) and roots (*top center*). **C** Altered ooids and fine-peloids with fewer pellets, coarser aggregates, foraminifers and mud

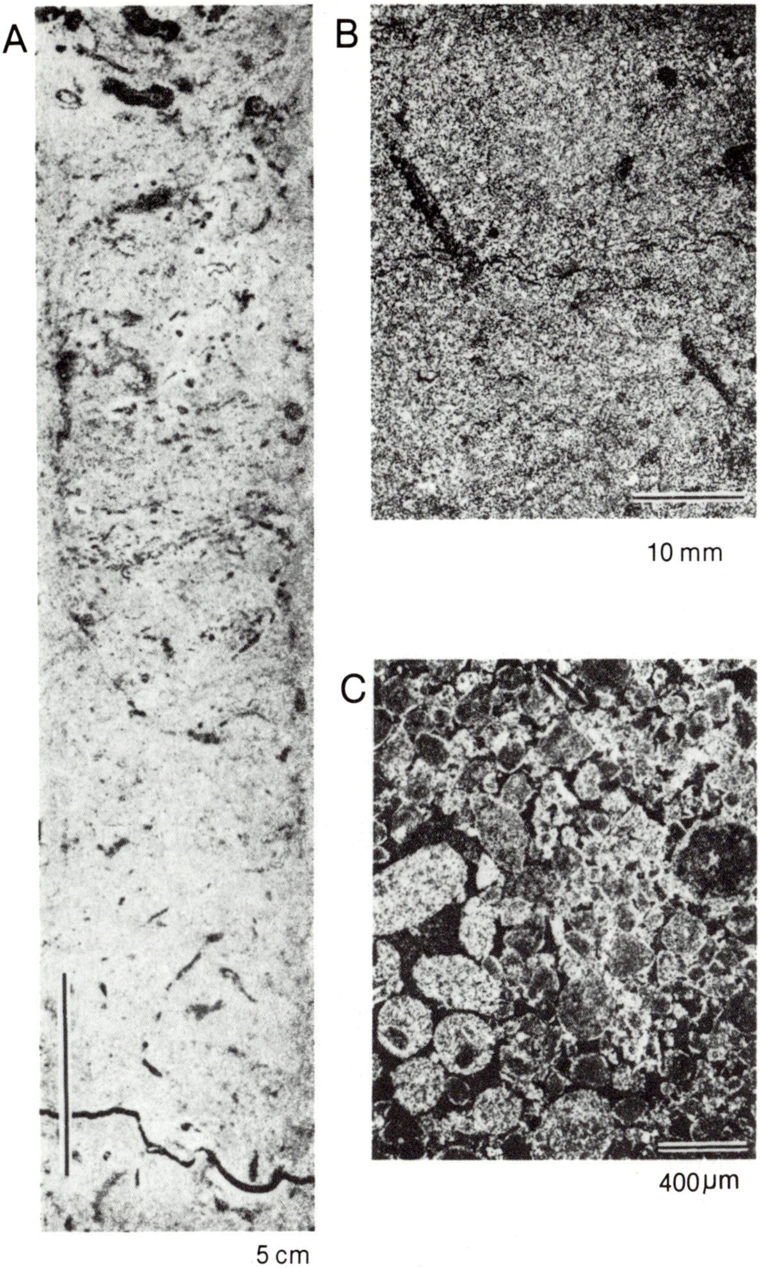

Fig. 5A – C. Fine-peloid packstone: **A** Discontinuous laminae of coarser grains, bioturbated sand, distinct unlined and mud-lined burrows with ooid sand infills (*top* and *right* of scale). **B** Moderately to very poorly sorted, very fine to coarse sand, with unlined burrows (*middle* to *bottom*) and roots (*top left* and *lower right*). **C** Fine-peloids and pellets with altered ooids, skeletal grains and mud. Fine-peloids are very fine sand and silt-size, subangular to subrounded micritic grains that are not recognizable broken fragments of larger grains. Garrett (1971) suggests that they are polygenetic: either broken lithified pellets, most likely from worms; altered skeletal or non-skeletal grain fragments; or accretions of mud-size aragonite tions of facies other than ooid grainstone indicate that the shoal is compound and not the product of a single depositional environment or process.

The basic facies suite is a fringe of ooid grainstone bordering a shoal composed of opposing wedges of ooid packstone and fine-peloid packstone (Fig. 6). Ooid grainstone up to 4.5 m thick directly overlies limestone toward the seaward limit of the facies, but it overlies in part and interfingers with packstones at its bankward margin. Ooid packstone up to 3 m thick forms an upper bankward-thinning wedge within the shoal. The facies is thickest near the seaward edge of the sand flat. It thins bankward as the underlying facies thickens and also thins to the south over a shallowing limestone floor. Ooid packstone is also present as tidal bars to the north of the shoal. Muddier fine-peloid packstone up to 5.5 m thick forms the underlying seaward-thinning wedge within the shoal. The facies is thickest in the western part of the sand flat. It thins seaward as ooid packstone thickens and thins to the south as the limestone floor shallows. Fine-peloid packstone thins to the north and west where it forms an inter-platform sheet. The vertical sequence of the facies in a core shows distinct trends from bottom to top: increasing grain size, better sorting, more ooids, abundant stratification, and a more grain-supported fabric.

5 Growth of the Shoal

The growth through time of the Joulters ooid shoal shows that the resulting facies are not exclusively a product of today's depositional environments, but that they have resulted from changing depositional patterns in response to rising sea level. Another reason that patterns of deposition changed is that there was topographic relief on the underlying floor of Pleistocene limestones; the shoal overlies and wraps around a high projecting from Andros Island.

Growth of the shoal occurred in three stages: bank flooding, shoal formation and shoal development (Fig. 7). During bank flooding when rising sea level was still about 5 m below present level, about half of the platform underlying today's shoal was submerged. Muddy sands of fine-peloids and pellets accumulated in protected lows around the high on the limestone floor. At −3 m sea level, the shoal began to form. As the muddy sands extended seaward over the now-submerged high, ooid production began where bottom agitation was most pronounced. During shoal development, which occurred in the last 3000 years, as determined from radiocarbon ages of core samples, production and dispersal of ooid sands established the present size and physiography of the shoal and changed the general nature of bank-margin sediments from muddy peloidal sands to ooid sands.

Ooid sands formed in a belt of bars and channels along the seaward edge of the limestone high on the seafloor and as bars curving spit-like around its northern edge. As the belt of active bars increased in width, ooid sands were transported farther bankward. During the last 1500 years, the exchange of water between the seaward and bankward sides of the shoal was increasingly restricted by widespread sediment build-up nearly to sea level, restriction of tidal channel flow,

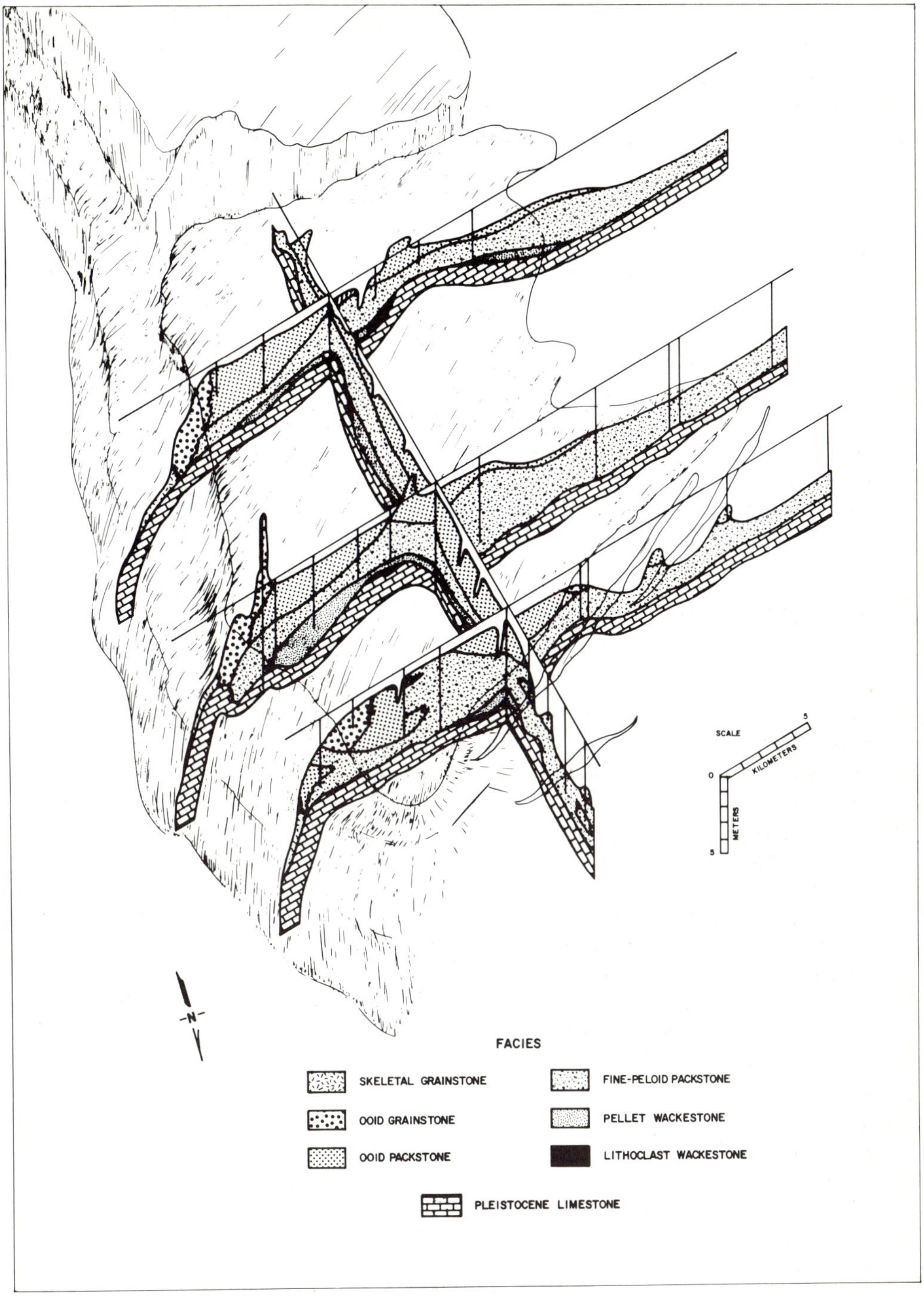

Fig. 6. Fence diagram of the Joulters ooid shoal based on 60 cores and hundreds of measurements of sediment thickness. Three facies – ooid grainstone, ooid packstone and fine-peloid packstone – are responsible for the build-up of the shoal

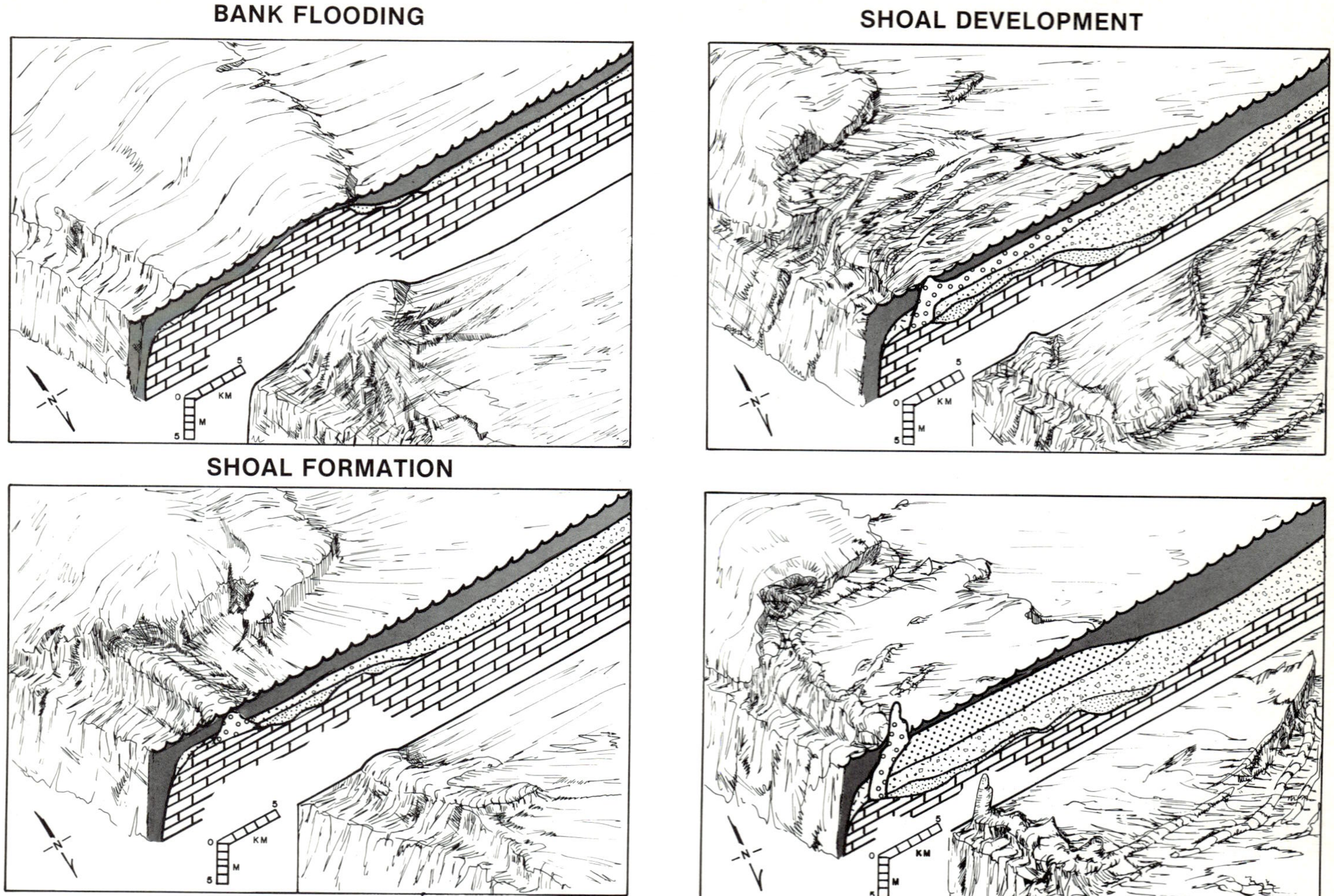

Fig. 7. Block diagrams showing bank flooding, shoal formation and stages in the growth of the Joulters ooid shoal. Lithological symbols are the same as those

and island formation along the shoal's ocean-facing margin. What began as a series of bars and channels became an intertidal sand flat, and ooid formation was increasingly restricted to a narrow rim of mobile sands at its seaward margin. Sand flat sediments are a burrowed mixture of ooid and peloid sands. They overlie and grade bankward into muddier sands of the bank interior. The tidal channels that partially penetrate the present-day sand flat are probably remnants of the original channels of the active sand belt. The channels still function as pathways for the bankward transport of ooid sands, but deposition is now localized as spill-overs and lobe-shaped fans at their ends.

References

Ball MM (1967) Carbonate sand bodies of Florida and the Bahamas. J Sediment Petrol 37:556–591

Buchanan H (1970) Environmental stratigraphy of Holocene carbonate sediments near Frazers Hog Cay, British West Indies. Unpubl PhD thesis, Col Univ, NY

Garrett P (1971) The sedimentary record of life on a modern tropical carbonate tidal flat, Andros Island, Bahamas. Unpubl PhD thesis, Johns Hopkins Univ, Baltimore, Maryland

Harris PM (1979) Facies anatomy and diagenesis of a Bahamian ooid shoal, Sedimenta VII. Univ Miami, Fla

Hine AC (1977) Lily Bank, Bahamas: History of an active oolite sand shoal. J Sediment Petrol 47:1554–1581

Hine AC, Wilber RJ, Neumann AC (1981) Carbonate sand bodies along contrasting shallow bank margins facing open seaways in Northern Bahamas. Am Assoc Petrol Geol Bull 65:261–290

Illing LV (1954) Bahamian calcareous sands. Am Assoc Petrol Geol Bull 38:1–95

Palmer MS (1979) Holocene facies geometry of the leeward bank margin, Tongue of the Ocean, Bahamas. Unpubl MS thesis, Univ Miami, Fla

Purdy EG (1963) Recent calcium carbonate facies of the Great Bahama Bank – Petrography and reaction groups. J Geol 71:334–355

Winland HD (1971) Diagenesis of carbonate grains in marine and meteoric waters. Unpubl PhD thesis, Brown Univ, Providence, Rhode Island

II.6 Holocene Lacustrine Ooids from Pyramid Lake, Nevada

Brian N. Popp[1] and Bruce H. Wilkinson[2]

1 Introduction

Since the pioneering work of Sorby (1879), considerable attention has been directed toward the examination of compositions and fabrics exhibited by modern and ancient ooids. Although volumetrically important in modern settings and in the rock record, marine ooids show little variation from locality to locality, and therefore represent a small sample of compositions and fabrics possessed by coated grains. Although volumetrically insignificant, non-marine ooids from both high- and low-salinity systems are important in understanding the physical and chemical processes which give rise to variations in coated grains.

Holocene ooids occur along the northwestern shore of Pyramid Lake, a brackish ensialic lake in northwestern Nevada. Although ooids from several other modern lacustrine systems have been described in some detail (e.g., Great Salt Lake, Utah; Kahle 1974, Sandberg 1975, Halley 1977; Lake Malinou, U.S.S.R.; Vital 1948, and Higgins Lake, Michigan; Wilkinson et al. 1980), none exhibit the spectrum of fabrics and compositions present in Pyramid Lake cortices. Because of the relative rarity of lacustrine ooids in modern settings, and because Pyramid Lake ooids are composed of both aragonite and low-magnesian calcite, examination of these grains was undertaken. This paper describes Pyramid Lake ooid compositions, fabrics, and environmental setting, and relates these features to the spectrum of compositions and fabrics reported for modern and ancient ooids as a whole.

1.1 General Setting

Pyramid Lake lies within a north-trending block-faulted and tilted Tertiary basin in the Basin and Range province of western North America. The lake is located in the Carson Desert of western Nevada (Fig. 1), and is the largest and deepest remnant of Pleistocene Lake Lahontan (Russell 1885). In response to climatic change, lake levels have fluctuated greatly over the past 20,000 years, ranging from a high at 1330 m to an undetermined low below the modern elevation of

1 Department of Geology, University of Illinois, Urbana, IL 61801, USA
2 Department of Geological Sciences, The University of Michigan, Ann Arbor, MI 48109, USA

Coated Grains (ed. by T. M. Peryt)

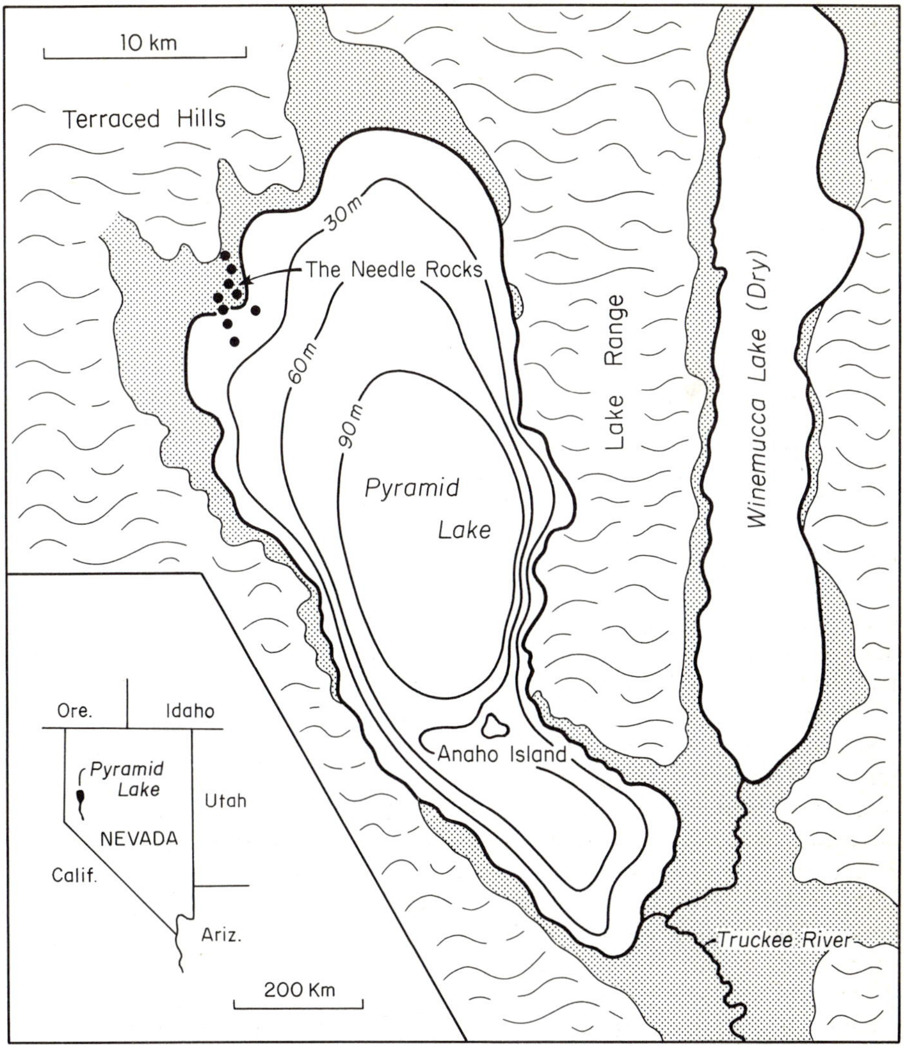

Fig. 1. Generalized map of Pyramid Lake, Nevada, showing bathymetry, the distribution of basins and ranges, and the aerial extent of the Lahontan Lake bottom (*stippled*). Note the location of The Needle Rocks in the northwestern part of the basin

the lake at 1156 m (Broecker and Kaufman 1965). However, from about 3000 years ago until the late 19th century, lake level rose to approximately 1179 m, and then began dropping rapidly, primarily in response to agricultural practices (Benson 1978).

At present, Pyramid Lake is approximately 50 km in length, 20 km in width, and attains a maximum depth of about 100 meters. The basin has no surface outlet and is fed primarily by the Truckee River which enters the lake from the south (Fig. 1). Mean annual rainfall into the lake basin is 10 to 20 cm, with an evaporation/precipitation ratio of 12. The lake is oxygenated to the bottom, and experiences an incomplete stratification cycle with one long overturn annually. The pH of surface water is 9.2, decreasing to 8.9 in the deep basin center. At present, waters are slightly brackish with total undissolved solids at 4.7‰ (Born 1972).

Extensive algal carbonates occur throughout the lake basin as thick massive blankets on now-exposed bedrock surfaces and as spectacular tufa pinnacles which reach heights of 60 m. These towers are most striking along the northwest lake margin where a roughly linear assemblage called "The Needle Rocks" have formed over springs which emanate from a north-trending fracture zone (Fig. 1).

Oolitic sands, first reported by Jones (1925), occur along high abandoned beaches in the area of the Needles and are actively forming along the modern shoreline in the same area. The age of the oldest (highest) ooids are imprecisely known, but Broecker and Kaufman (1965) radiocarbon dated a single oolite sample collected at 1177 m (about 20 m above present lake level) at 1000 ± 200 years. Almost certainly, ooid deposits in the Needles area are younger than 3000 years in age.

As suggested by Jones (1925), ooid formation in the Needles area appears to be directly related to the mixing of calcium-rich spring waters and bicarbonate-rich lake water (Table 1) in nearshore wave-agitated environments. During June of 1978, for example, water from a small spring, flowing approximately 50 l/min, was entering the lake near the Needles. Immediately surrounding the point of entry, a whitish semicircular turbid cloud of carbonate extended lakeward some 50 m in all directions. Because actively forming ooids were only found along the shoreline in this immediate area, and because the tufa pinnacles (The Needles) owe their very existence to spring emanations along this portion of the lake basin, we believe that mixing of the two waters has been responsible for ooid formation in this restricted area over the past several thousand years.

Table 1. Water chemistry data in ppm except as noted

	Surface water		Spring water
	(1)	(2)	(3)
T	–	17°	48° – 63°
pH	–	8.9	–
Ca^{2+}	18	7.2	272
Mg^{2+}	80	110	5
Na^{2+}	1173	1610	1312
K^+	77	99	349
SO_4^-	180	280	255
Cl^-	1429	1920	2693
SiO_2	41	1.3	147
CO_2	500	1166	78
Total solids	3.5‰	4.7‰	5.2‰
[a]Mg^{2+}/Ca^{2+}	4.5	15	0.08

Data from (1) Russell 1885, (2) Whitehead and Feth 1961, (3) Jones 1925.
[a] Mg^{2+}/Ca^{2+} ratios determined during this study: surface water = 21, spring water = 0.0089

2 Pyramid Lake Ooids

Unconsolidated oolitic sands were collected along the modern beach, as were weakly-cemented and unconsolidated samples from higher shorelines. Both populations are well sorted, but grains from the modern beach are considerably smaller, averaging 1 mm in diameter, whereas those from higher shorelines average 5 mm in diameter due to significantly thicker cortices. In general, all ooids are spherical to ovate in form (Fig. 2A). Ooid nuclei are variable, consisting of shell debris, tufa fragments, and terrigenous quartz or volcaniclastic rock fragments. Commonly, compound ooids and cortical debris also act as nuclei, demonstrating some degree of reworking during ooid growth. Reworking and grain abrasion has also been seen in sections of larger ooids as angular unconformities among growth laminae.

2.1 Ooid Mineralogy and Composition

The mineralogy and composition of cortical layers on the larger grains was determined both by X-ray diffraction and by staining. The staining method, described by Schneidermann and Sandberg (1971), consists of treating sectioned and polished ooid cortices with Feigl's solution, and then examining the treated surface by SEM. Aragonite, owing to its higher solubility, reacts rapidly with the solution, causing a granular precipitate to form on aragonite surfaces. This technique allows for the discrimination between calcite and aragonite on a fine scale, even where individual laminae were less than 10 μm thick.

In addition to being stained, cortical layers were physically separated from nuclei for X-ray diffraction analyses. Powdered bulk cortical samples were run, and their peak heights were evaluated relative to those from a standard curve for the X-ray diffractometer used during this study. Pyramid Lake cortices range from 34% to 42% aragonite and 66% to 58% calcite, values consistent with visual estimates made on stained ooids. The magnesium concentration in calcite phases was determined by measuring the relative shift in the principal calcite peak (3.035 Å) against the principal peak of an internal standard, quartz (3.343 Å). The relative shift corresponds to a magnesium content of 3 to 4% $MgCO_3$ in the calcite (Goldsmith et al. 1961).

2.2 Ooid Fabric

Not only do cortical layers of larger Pyramid Lake ooids have differing mineralogy, individual laminae also exhibit characteristic fabrics which reflect habits of the mineral phases present. Aragonite always occurs as short, radial fibrous laths to needles, approximately 5 to 10 μm in length and 1 to 2 μm in width. Calcite crystals exhibit a spectrum of habits ranging from isopachous laminae to equant, hexagonally-zoned crystals to large radial laths and blades (Fig. 2B–F). Typically, radial calcite laths are 10 to 20 μm long and 4 to 8 μm wide. Hexagonal

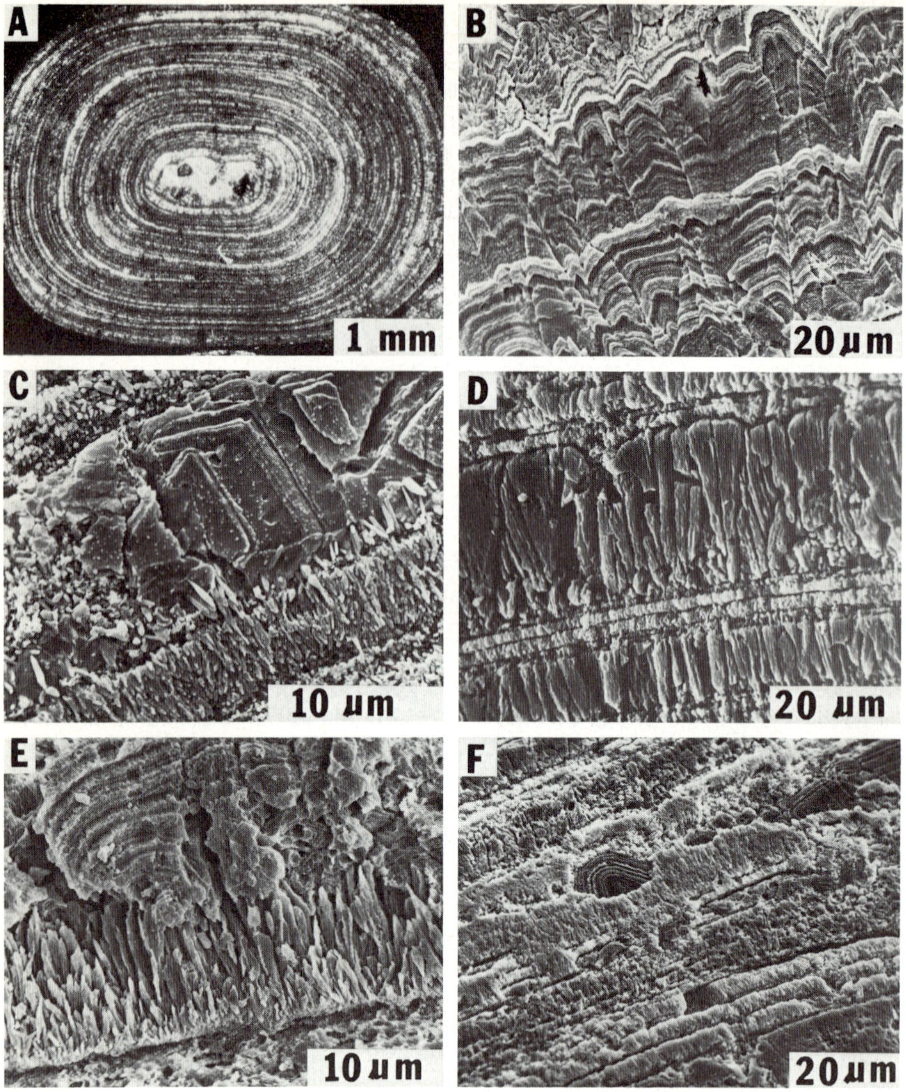

Fig. 2 A–F. Pyramid Lake ooids. All figures are SEM photomicrographs of sectioned ooids which were polished and etched with formic acid, except **A** which is a thin-section photomicrograph, crossed polars. **A** Thickly coated ovoid ooid exhibiting a faint pseudouniaxial cross. The nucleus of this grain is a fragment of calcareous tufa. The alternation in laminae color (light gray vs. dark gray) in part reflects the sizes of cortical crystals. The calcitic lamina (coarsely crystalline) appear as lighter layers. **B** Isopachous layer of coalesced calcite crystals. Note the fine zoning and subhedral crystal terminations. A younger layer of bladed calcite crystals is just visible in the *upper left* of the photograph. **C** Hexagonally zoned equant calcite crystals directly overlying an older lamina of acicular aragonite needles. Note that in several instances calcite surrounds and incorporates the more distal portions of some aragonite crystals. **D** Lamina of elongate calcite laths. This thick layer directly over lies two thin (*lighter*) laminae of aragonite, which in turn overlie a moderately thick lamina composed of calcite. **E** Acicular needles of aragonite overgrown by an isopachous lamina of zoned calcite similar to that in Fig. 3B. **F** Typical portion of a Pyramid Lake ooid cortex. Laminae of acicular aragonite (*light gray*) alternate with laminae of equant and bladed calcite (*dark gray*). Note the zoned equant calcite crystal in the center of the photo

zoned calcite crystals, which have diameters frequently greater than 30 μm, are the largest cortical crystals. Each discrete lamina is typically uniform in thickness and mineralogy, and almost always surrounds the entire circumference of a grain. Although cortical laminae of low-magnesian calcite and aragonite alternate, there is no regular or systematic relationship between lamellar thickness and either mineralogy or position within the ooid.

Thick cortices composed of lamellae of alternating mineralogies are only found in older ooids collected from high abandoned beaches. The modern ooids have thin coatings, usually less than 20 μm, which exhibit 3 to 5 laminae, all of radial aragonite needles. The lack of bimineralic compositions and the thinness of cortices on these grains probably reflects a lack of fluctuating water chemistries during the very short time period in which the modern beach has been an area of carbonate precipitation.

2.3 Interpretation

The restriction of Holocene ooid occurrence to the Needles area, and the formation of modern ooids only in areas of spring discharge, demonstrate an intimate association between hot spring activity and ooid formation within this system. However, owing to insufficient chemical data on the spring waters, we were unable to determine whether or not the mixing model proposed by Jones (1925) is the only process responsible for ooid formation. Conversely, the presence of turbid carbonate-rich water observed in the mixing zone, strongly suggests that this mechanism results in local carbonate supersaturation.

Analyses of hot spring waters and lake water indicate strikingly different chemistry, expecially with respect to magnesium and calcium concentrations (Table 1). In general, Mg/Ca ratios (by weight) of lake water average 13.7, whereas spring waters have ratios of about 0.01. Many authors (e.g., Müller et al. 1972, Folk 1974) have related Mg/Ca ratios to the carbonate phases precipitated from various solutions. Qualitatively, when Mg/Ca ratios are greater than about 2, aragonite precipitates; when ratios are less, calcite forms. Within the context of bimineralic Pyramid Lake ooids, then, it is probable that relative volumetric dominance of one of the two waters in the nearshore mixing zone at any given time has determined the phase precipitated. Calcitic laminae, which make up the greater volume of cortical layers, probably formed in the mixing zone when Mg/Ca ratios were low (spring waters were dominant), and aragonite was precipitated when Mg/Ca ratios were high (when lake water predominated).

The general fabric of individual monomineralic laminae in Pyramid Lake ooids is directly related to characteristic habits exhibited by calcite and aragonite crystals, a relationship also seen in cements and in ooids from other settings. Although the precise processes by which environmental variables relate to crystal habit are poorly understood (cf. Folk 1974 and Lahann 1978), it generally has been concluded by many authors that low-magnesian calcites are typically equant to slightly elongate, that magnesian calcites are characteristically elongate, and that aragonite is typically acicular to fibrous. The habit of Pyramid Lake cortical crystals is consistent with these generalizations.

3 Discussion

Holocene ooids from Pyramid Lake are particularly significant in that they represent one of very few occurrences of low-magnesian calcite in cortices of modern non-spelean ooids. More importantly, however, the coexistence of aragonite and calcite from this, admittedly unusual, geological setting serves to illuminate those processes which give rise to variations in the composition and fabric of coated grains in general.

3.1 Ooid Mineralogy and Composition

Modern ooids exhibit three primary compositions and three primary fabrics, although some ooids such as those from Pyramid Lake and Baffin Bay, Texas (Land et al. 1979), may possess more than one composition and/or fabric as alternating cortical laminae. Modern ooids are composed of calcite, magnesium calcite, or aragonite. Virtually all modern normal and hypersaline marine ooids are composed of aragonite (e.g., Illing 1954, Newell et al. 1960, Purdy 1963, Friedman et al. 1973, Loreau and Purser 1973), and aragonite has also been reported from numerous hypersaline lacustrine settings (e.g., Kahle 1974, Sandberg 1975, Halley 1977). Some typical aragonite ooids are shown in Fig. 3.

To our knowledge, aragonite in Pyramid Lake laminae represents the first reported occurrence of this phase in ooids from a system with salinities less than that of sea water (35‰). It is significant to note, however, that all aragonite laminae occur in waters with elevated Mg/Ca ratio, suggesting that concentrations of these cations, rather than concentrations of total dissolved solids, are primarily responsible for aragonite precipitation during ooid formation.

Calcite ooids are much less abundant both volumetrically and aerially in modern environments. Magnesian calcite ooids are known from the Amazon shelf (Milliman and Barretto 1975), the Great Barrier Reef (Marshall and Davies 1975), and Baffin Bay on the south Texas coast (Freeman 1962, Rusnak 1960, Land et al. 1979). Samples from the Amazon shelf (85 to 205 m depth) and those from Australia (100 to 120 m depth) are relict, however, and may have formed in environments similar to Baffin Bay, but during the Holocene transgression. Low-magnesian calcite ooids are largely restricted to spelean settings (e.g., Hess 1929, Davidson and McKinstry 1931, Baker and Frostick 1947), and have only recently been reported from a low-salinity lake (Wilkinson et al. 1980). Figure 4 shows several modern and ancient calcite cortices.

The coexistence of low-magnesian calcite and aragonite in Pyramid Lake ooids is, as far as we know, a unique occurrence, and reflects the widely fluctuating setting in which the ooids originate. Although atypical of coated grains in general, Pyramid Lake ooids demonstrate the intimate relationship which exists between the chemistry of the waters in which ooids form and the mineralogy and composition of constituent cortical crystals.

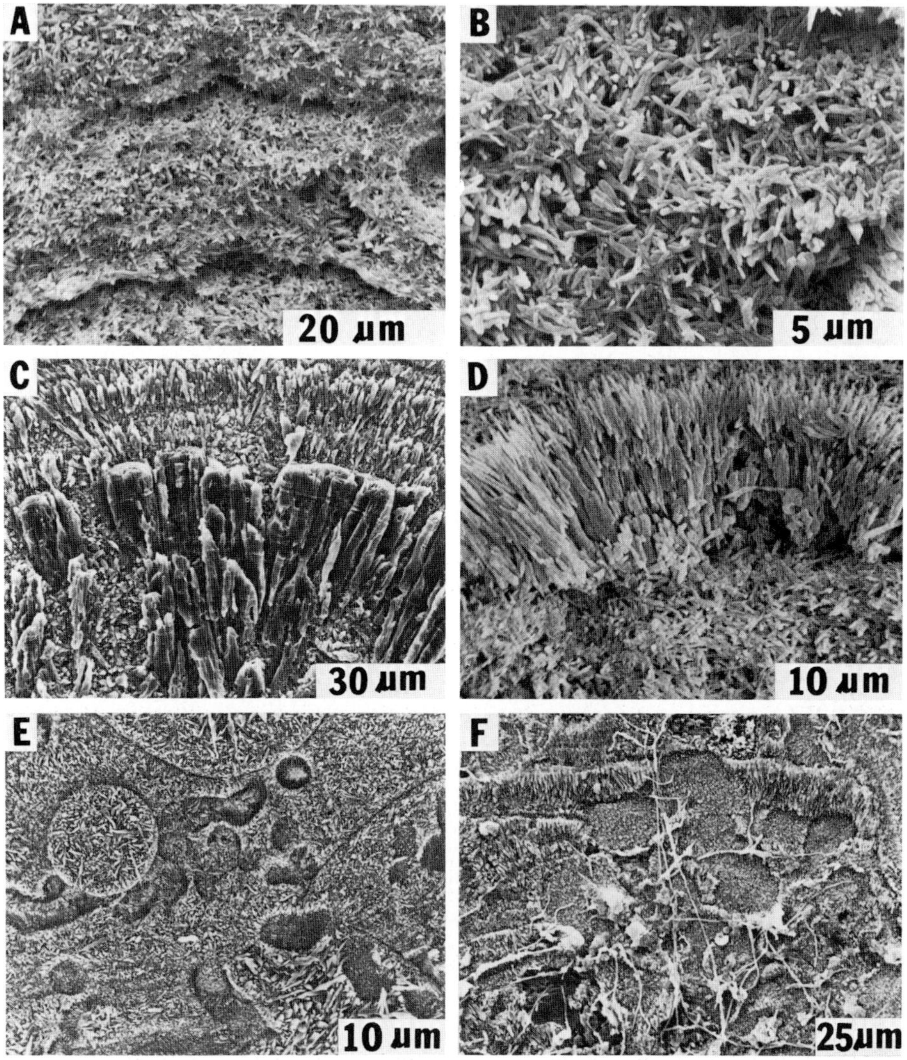

Fig. 3A – F. Fabrics in modern aragonite ooids. All are SEM photomicrographs. **A** Statistically tangential orientation of aragonite in a normal marine Bahamian ooid from Cat Cay. Note two discontinuities between three successive laminae. This fabric is typical of modern marine ooids. **B** Detail of tangential aragonite in **A**. Note that although highly variable, most laths are oriented roughly normal to the photograph of a surface which passes through the ooid center. **C** Radial aragonite laths and needles typical of ooids from the hypersaline Great Salt Lake, Utah. Note growth bands in the large laths in the lower portion of the photo. **D** Radial aragonite lens from an unaltered portion of the Pleistocene Miami oolite from Florida. This fabric, while rare in normal marine ooids, occurs in marine cortices from the Persian Gulf, Australia, Florida, and the Bahamas, and probably records ooid accretion during periods of little agitation by waves. Fabrics similar to this are characteristic of aragonite laminae in Pyramid Lake cortices. **E** Random orientation of aragonite needles characterizes this Bahamian ooid from Cat Cay, which has been heavily bored by microendolithic organisms. Secondary infilling of these borings results in a typical "micritized" cortex of unoriented crystals. This ooid probably once had a primary fabric similar to that in **A** and **B**. **F** Partial micritization of a high-magnesian calcite ooid from Baffin Bay, Texas. Note the remnants of radial high-magnesian calcite laminae, the filled borings of randomly oriented aragonite, and the (now-desiccated) filaments which lie across the etched surface

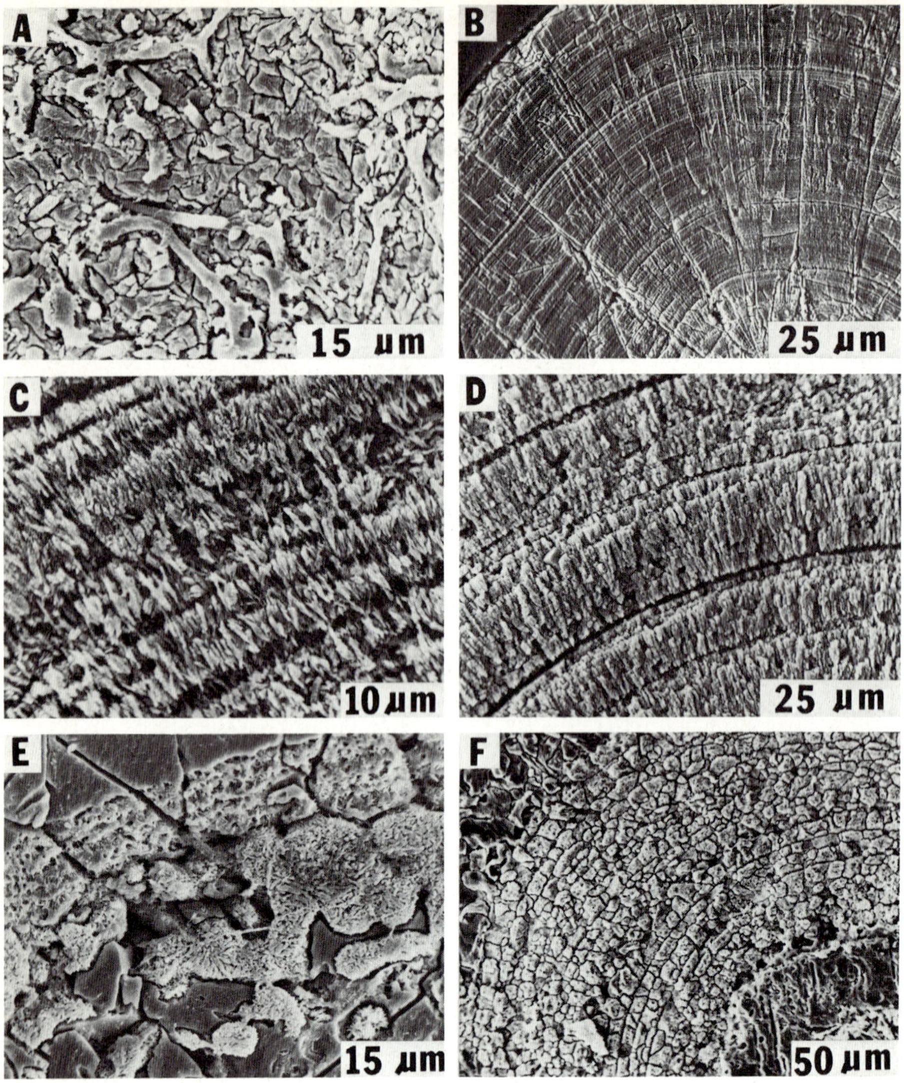

Fig. 4A–F. Fabrics in modern and ancient calcite ooids. All are SEM photomicrographs. **A** Cortex of equant anhedral crystals of low-magnesian calcite in an ooid from Higgins Lake, a low-salinity temperate-region marl lake in north-central Michigan. The cyanophytes within this cortex are passive endoliths which cross-cut primary equant calcite crystals. **B** Radial fabric exhibited by ooids from a water treatment plant in Key Largo, Florida. These artificial ooids, spelean grains, and some laminae in the Pyramid Lake ooids are all composed of radial low-magnesian calcite. **C** Radial needles of magnesian calcite in an ooid from Baffin Bay, Texas. **D** Radial laths of low-magnesian calcite in an ooid from the normal marine Jurassic Twin Creek formation of northwestern Wyoming. Such a fabric is typical of many, if not most, Paleozoic and Mesozoic marine ooids. Note the similarity in fabric to **C**. Although now low-magnesian calcite, these Jurassic crystals may once have had a composition similar to those in Baffin Bay ooids. **E** Partial neomorphism of aragonite cortical carbonate in the Pleistocene Miami Oolite from Florida. Note: (1) partially neomorphosed ooid laminae exhibiting included aragonite laths along the ooid exterior to the *upper left*; (2) the crystallographic continuity between the spar cement and the inverted exterior laminae; (3) the darker gray inclusion-free calcite which has replaced parts of the ooid; (4) unreplaced microendolith borings filled with randomly oriented aragonite. **F** Equant blocky low-magnesian calcite, typical of neomorphic marine aragonite. This ooid is from the Miocene Mishan Formation of southeastern Iran. Note equant calcite spar cement to the *upper left*, relict laminations within the cortex, and the complete obliteration of primary fabric which was originally similar to that in Fig. 3A, B

3.2 Ooid Fabric

The orientation of constituent cortical crystals with respect to grain exteriors may be radial, tangential, or random in coated grains (Figs. 3 and 4). These fabric variations are chiefly a function of primary cortex mineralogy, and to a lesser extent to the degree of agitation experienced by individual grains during cortex accretion (Rusnak 1960, Halley 1977).

Radial fabric predominates in magnesian calcite ooids (Marshall and Davies 1975, Milliman and Barretto 1975, Land et al. 1979), in some hypersaline lacustrine aragonite ooids (e.g., Kahle 1974, Sandberg 1975, Halley 1977), in most spelean ooids (e.g., Davidson and McKinstry 1931), and rarely in marine aragonite ooids (e.g., Loreau and Purser 1973, Davies and Martin 1976). Tangential aragonite is largely restricted to high-energy normal marine ooids (see Bathurst 1976), although these form the vast majority of all modern oolitic sands.

Ooids which lack a preferred orientation of constituent crystals arise through primary fabric obliteration by microendolithic organisms (micritization), through cortex accretion of equant micritic nannograins, or where development of preferred orientation is precluded because cortical layers are composed of large crystals of equant low-magnesian calcite. Lack of preferred orientation is present in Pyramid Lake ooids where large zoned low-magnesian calcite crystals form discrete layers (Fig. 2B, C) or single crystals within a layer (Fig. 2F). These crystals, resembling equant meteoric calcite cements, probably formed when Mg/Ca ratios were low. A similar lack of preferred orientation has been observed in other material to result from neomorphic replacement of aragonite. Although the crystals of such neomorphic mosaics are generally irregular in shape and variable in size (Sandberg 1975), there are exceptions in which crystals appear equant (Fig. 4F).

Fabrics in Pyramid Lake ooids bear a striking similarity to those seen in monomineralic ooids from other settings. Radial aragonite is nearly identical to that which occurs rarely in normal marine ooids and commonly in hypersaline lacustrine ooids, whereas equant low-magnesian calcite crystals are analogous to those in cortices from other low-salinity lacustrine systems and from some calcitized aragonite precursor ooids (Fig. 4F). The presence of radial low-magnesian calcite, although uncommon from other modern natural non-spelean systems, is similar to that reported from caves, and demonstrates that such fabrics may be expected in other natural non-marine low-salinity settings.

4 Summary

The fabrics and compositions of Pyramid Lake ooids fit well into the spectrum of known primary and secondary cortical fabrics and compositions exhibited by modern and ancient ooid cortices. Calcite cortical laminae exhibit equant to bladed habits, analogous to radial fabrics in spelean calcite ooids or to random fabrics in other primary equant calcite ooids and some calcitized aragonite grains from Cenozoic marine settings. Similarly, radial fibrous aragonite from Pyramid

Lake is analogous to radial aragonite in ooids from a few normal marine settings and from hypersaline lakes and lagoons.

Observation of Pyramid Lake coated grains serves to strengthen generalizations concerning the relationships between environmental variables and the mineralogy, composition, and habit of cortical crystals. Land et al. (1979) have pointed out that most ooids precipitate in near inorganic equilibrium with environmental waters, and should be viewed as subaqueous cements, centrifugally deposited on nuclei. The major features of Pyramid Lake coated grains strongly support such a point of view.

Acknowledgements. We would like to thank Gerald R. Smith and Krystyna Swirdczuk who assisted during field work. Many individuals at The University of Michigan have contributed significantly to ooid research over the past several years through examination of samples from many areas. We acknowledge the considerable contribution of Bernard Coakley, Donald Medwedeff, and Mary Robert in particular. Philip A. Sandberg (University of Illinois), Robert L. Folk and Lynton S. Land (University of Texas at Austin) read the original manuscript and offered helpful suggestions. Drafting was done by Derwin Bell, Department of Geological Sciences, The University of Michigan. We would like to thank those individuals at both the Center for Electron Microscopy, University of Illinois, and The University of Michigan scanning Electron Microscope Laboratory for assistance during this study. Research on recent lacustrine carbonates at The University of Michigan is supported by the National Science Foundation, NSF grant EAR 78-93634.

References

Baker G, Frostick AC (1947) Pisoliths, ooliths and calcareous growths in limestone caves at Port Campbell, Victoria, Australia. J Sediment Petrol 21:85 – 104

Bathurst RGC (1976) Carbonate sediments and their diagenesis. Elsevier, Amsterdam

Benson LV (1978) Fluctuation in the level of Pluvial Lake Lahontan during the last 40,000 years. Q Res 9:300 – 318

Born SM (1972) Late Quaternary history, deltaic sedimentation and Mudlump Formation at Pyramid Lake, Nevada. Nev Univ Center Water Resour Res, pp 1 – 97

Broecker WS, Kaufman A (1965) Radiocarbon chronology of Lake Lahontan and Lake Bonneville II, Great Basin. Bull Geol Soc Am 76:537 – 566

Davidson SC, McKinstry HE (1931) "Cave Pearls", oolites and isolated inclusions in veins. Econ Geol Monogr 26:289 – 294

Davies PJ, Martin K (1976) Radial aragonite ooids, Lizard Island, Great Barrier Reef. Geology 4:120 – 122

Folk RL (1974) The natural history of crystalline calcium carbonate: effect of magnesium content and salinity. J Sediment Petrol 44:40 – 53

Freeman T (1962) Quiet water oolites from Laguna Madre, Texas. J Sediment Petrol 32:475 – 483

Friedman GM, Amiel AJ, Braun M, Miller DS (1973) Generation of carbonate particles and laminates in algal mats. Examples from a sea marginal hypersaline pool, Gulf of Aqaba, Red Sea. Am Assoc Petrol Geol Bull 57:551 – 557

Goldsmith JR, Graf DL, Heard HC (1961) Lattice constants of the calcium-magnesium carbonates. Am Miner 46:453 – 457

Halley HB (1977) Ooid fabric and fracture in the Great Salt Lake and the geologic record. J Sediment Petrol 47:1099 – 1120

Hess FL (1929) Oolites or cave pearls in the Carlsbad Caverns. US Natl Mus Proc 75:1 – 5

Illing CV (1954) Bahamian calcareous sands. Am Assoc Petrol Geol Bull 8:1 – 95

Jones JC (1925) The geological history of Lake Lahontan. Carnegie Inst Wash Publ 325:3 – 50

Kahle CF (1974) Ooids from Great Salt Lake, Utah, as an analogue for the genesis and diagenesis of ooids in marine limestones. J Sediment Petrol 44:30 – 39

Lahann RW (1978) A chemical model for calcite crystal growth and morphology control. J Sediment Petrol 48:337–344
Land LS, Behrens EW, Frishman SA (1979) The ooids of Baffin Bay, Texas. J Sediment Petrol 49:1269–1278
Loreau J, Purser BH (1973) Distribution and ultrastructure of Holocene ooids in the Persian Gulf. In: Purser BH (ed) The Persian Gulf. Springer, Berlin Heidelberg New York, pp 279–328
Marshall JF, Davies PJ (1975) High-magnesium calcite ooids from the Great Barrier Reef. J Sediment Petrol 45:285–291
Milliman JD, Barretto HT (1975) Relict magnesian calcite oolite and subsidence of the Amazon Shelf. Sedimentology 22:137–145
Müller G, Iron G, Forstner U (1972) Formation and diagenesis of inorganic Ca-Mg carbonates in the lacustrine environment. Naturwissenschaften 59:158–164
Newell ND, Purdy EG, Imbrie J (1960) Bahamian oolitic sand. J Geol 68:481–497
Purdy EG (1963) Recent calcium carbonate facies of the Great Bahama Bank. J Geol 71:472–497
Rusnak GA (1960) Some observations of recent oolites. J Sediment Petrol 30:471–480
Russell IC (1885) Geologic History of Lake Lahontan. US Geol Surv Mon 11:287 p
Sandberg PA (1975) New interpretations of Great Salt Lake ooids and nonskeletal carbonate mineralogy. Sedimentology 22:497–537
Schneidermann N, Sandberg PA (1971) Calcite-aragonite differentiation by selective staining and scanning electron microscopy. Gulf Coast Assoc Geol Soc Trans 21:349–352
Sorby HC (1879) On the structure and origin of limestones. The anniversary address of the president. Q J Geol Soc London Proc 35:56–95
Vital DA (1948) Modern lime-magnesium concretions and oolites in the lakes of the Kulindin Steppe, Altai Region U.S.S.R. Moscow Soc Naturalists 23:83–100
Whitehead HC, Feth JH (1961) Recent chemical analyses of waters from several closed-basin lakes and their tributaries in the western United States. Geol Soc Am Bull 72:1421–1426
Wilkinson BH, Popp BN, Owen RM (1980) Nearshore oolite formation in a modern temperate-region marl lake. J Geol 88:697–704

II.7 Subrecent High-Sr Aragonitic Ooids from Hot Springs Near Tekke Ilica (Turkey)

Detlev K. Richter [1] and Horst Besenecker [2]

1 Introduction

Aragonitic coated grains <1 mm in diameter with a layered cortex formed by crystals of tangential orientation are characteristic of ooids from the Bahamas (Illing 1954, Newell et al. 1960). This structure is normal for recent ooids from marine environments (for exceptions see Land et al. 1979). The µm-sized aragonite crystals are rod- or baton-shaped (Fabricius and Klingele 1970, Loreau 1970).

Nonmarine ooids showing tangential orientation of aragonite crystals in the layered cortex have only been described by Sorby (1879) from the "Sprudelstein" of Karlsbad (CSSR), which is of thermal origin. In addition, the author mentioned a similar structure for many layers of the Bahama ooids.

In this paper we present petrographic details about preserved as well as calcitized thermal ooids from Turkey, which may help to understand the genesis and diagenesis of ooids. The field work was carried out in spring 1973. We are indebted to Dr. J. Becker-Platen (Hannover) for the locality. Furthermore, we are grateful to Dr. G. Bijvank (Bochum) for providing the SEM close-ups and to Dr. Mook (Groningen) for isotopic investigations.

2 Locality

In the Menderes Valley, about 20 km west of Denizli, close to the gravel road south of the Menderes River at Tekke Ilica hot springs occur at the boundary between calcareous nonmarine sediments of Neogene age and the valley fill of the Menderes River (Fig. 1). Closely east of the ruins of the bathing-establishment (Tekke Hamam) the hot water (60° – 80 °C) pours out of fissures in the bedrock into several pools being a couple of meters in diameter (Fig. 2A). Most of the pools are coated by calcareous crusts reaching up to 20 cm of thickness. Ooids

1 Geologisches Institut, Ruhruniversität Bochum, Universitätsstraße 150, D-4630 Bochum-Querenburg, FRG

2 Bundesanstalt für Geowissenschaften und Rohstoffe, Alfred-Bentz-Haus, D-3000 Hannover 51, FRG

Coated Grains (ed. by T. M. Peryt)

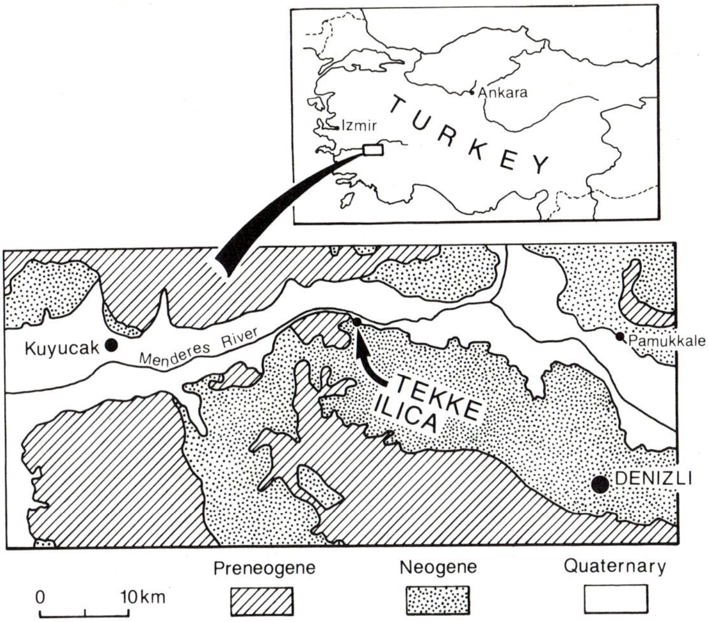

Fig. 1. Location of the thermal occurrence of Tekke Ilica

occur in the crusts and at the bottom of the pools. The occurrence of both calcareous crusts and ooids up to several meters above the water surface of the recent pools may be an indicator for a subrecent age of the carbonate precipitation. We took samples from outside the recent pools.

The hot springs are accompanied by fumarole activity (smell of H_2S). Exhalation minerals like sulphur, anhydrite and (or) gypsum are common in fractures of the Neogene limestones which are exposed in a little quarry south of the gravel road across the pools.

3 Size and Distribution of the Ooids

The grain-size distribution of the ooids is varying between <1 mm and 5 cm with a maximum between 0.5 cm and 3 cm (Fig. 2B – E). As ooids of different size always occur in separate layers, their sorting in a single layer is excellent (Fig. 3). Grading appears to be normal and inverse, possibly due to an decreasing or increasing activity of the springs. Intercalations of calcareous crusts between the oolitic layers show the same mineralogical and chemical composition as the cortices of the ooids. Calcareous crusts composed of bundles of aragonite needles (Fig. 4G) prevail in the recent surface outside the oolitic deposits. These crusts, which may contain ooids, show upwarpings and are locally shaped like cascades. Probably the crusts are grown on topographic higher positions than the ooids. Water turbulence must have been low in these regions which are now affected by erosion.

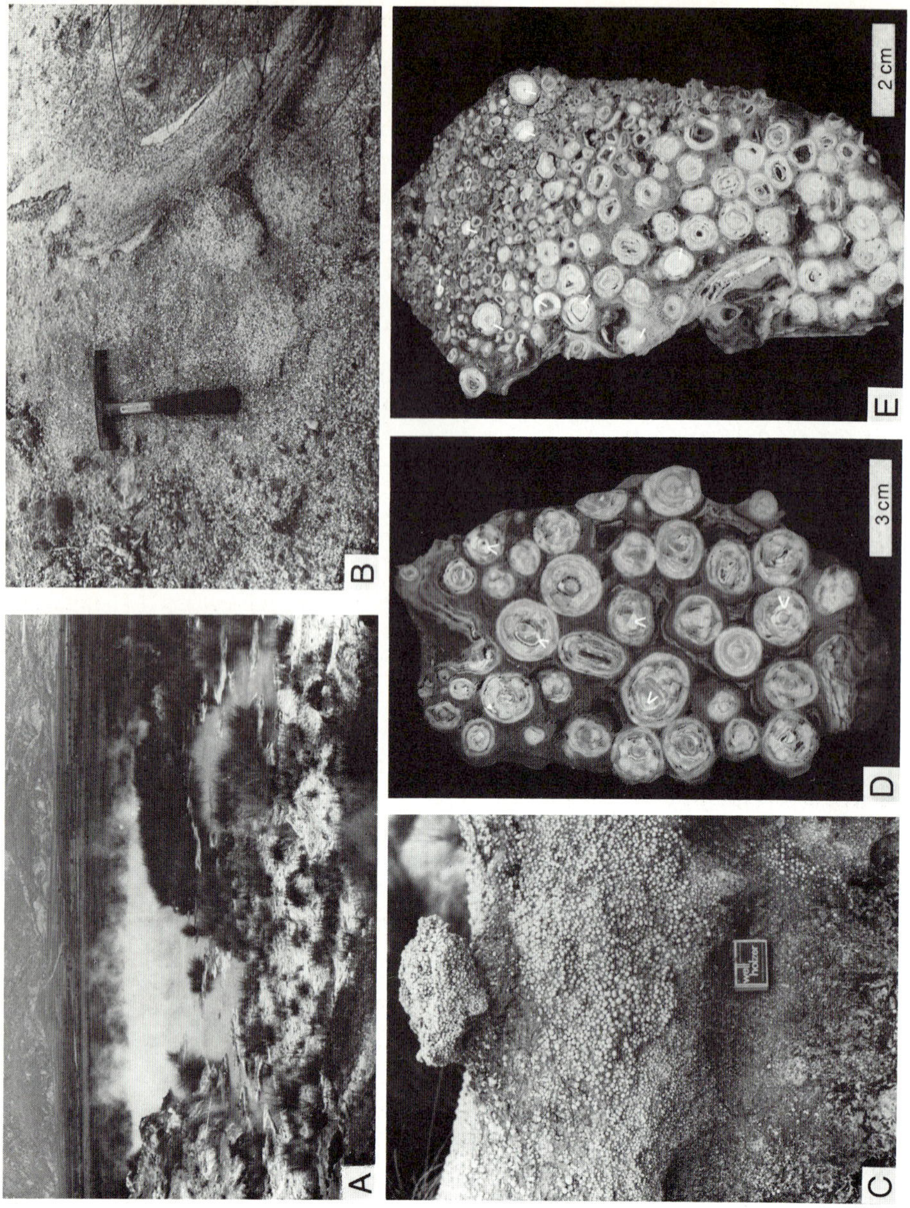

Fig. 2. A View from the subrecent calcareous sinter crusts over a part of the recent thermal pools to the plain of the Menderes River in northern direction. **B** Outcrop of partly laminated oolite layers at the rim of a subrecent pool. **C** Section through a subrecent pool showing ooid-containing calcareous sinter crusts at the base, fine-grained oolite in the central part, and coarse-grained oolite at the top. **D** Very coarse-grained oolite cemented by calcite with a sinter crust at the top; the *black areas* in the ooids are pores caused by dissolution, the white parts (*arrows*) are relics of aragonite in the calcitized cortices. **E** Normally graded oolite cemented by calcite showing mainly calcitized ooids; white parts (*arrows*) relics of aragonite. **D** and **E** are polished sections

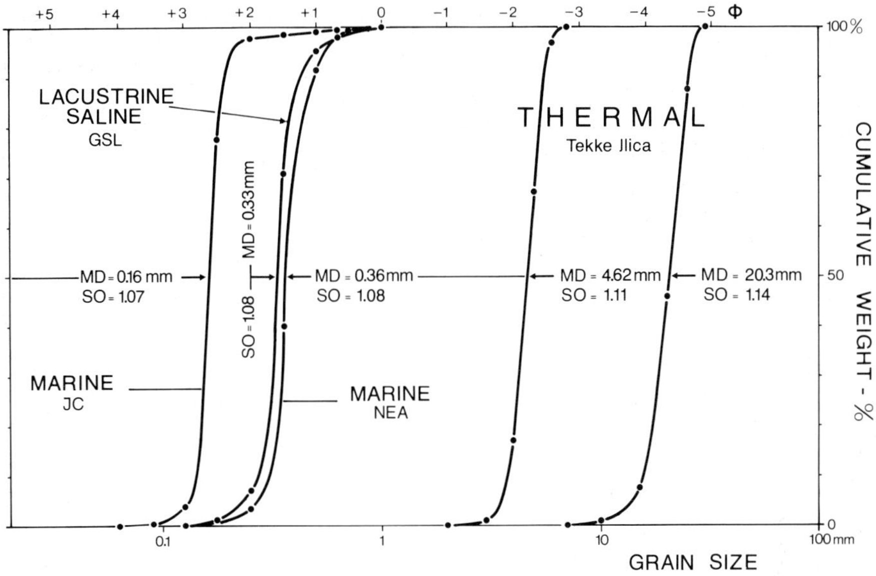

Fig. 3. Grain-size distribution of ooids of two layers from Tekke Ilica compared with Holocene marine occurrences (*JC* Joulters Cay, Oti and Müller 1979; *NEA* Neapolis/South Peloponnesus/Greece, Richter 1976) and with a sample of the hypersaline Great Salt Lake/U.S.A. (*GSL*, Oti and Müller 1979). *MD* median, *SO* sorting (after Trask)

4 Internal Structure

The ooids show a layered structure of aragonitic laminae being only a few μm thick. A proof for a tangential orientation of the aragonite crystals is obtained by observing thin-sections in transmitted light: using crossed nicols and a gypsum plate with red 1st order, the pseudo-uniaxial image appears to be positive. Radial-aragonitic layers occur only seldom (Fig. 4E). The aragonite crystals are 0.5 to 2 μm long and 0.2 to 0.5 μm in diameter which is the same as in recent marine ooids of aragonitic composition. Crystal faces are commonly visible at high magnification using a SEM (Fig. 4E). Thus, aragonitic cortices with imperfect crystals like rods or batons without crystal faces seem to be typical only for marine ooids (see Fabricius and Klingele 1970, Loreau 1970). Both thermal and marine ooids show the same loose pattern of aragonite crystals as shown by SEM pictures (Fig. 4E), which may indicate a high porosity. A microporosity of about 20% was obtained for ooids from the Bahamas by Oti and Müller (1979). High porosity and an only statistically tangential orientation of the crystallites in marine ooids (Loreau and Purser 1973) are shown even in normal (i.e., 20 μm in thickness) thin sections of the present ooids by a low apparent birefringence (2nd order blue parallel to the vibration of a gypsum plate with red of 1st order). Ooids with radial orientation of fan-like fibres of aragonite (major part of Great Salt Lake ooids) or calcite (e.g., "Rogenstein" of the Lower Triassic from the northern borderland of the Harz) show the same apparent birefringence in thin

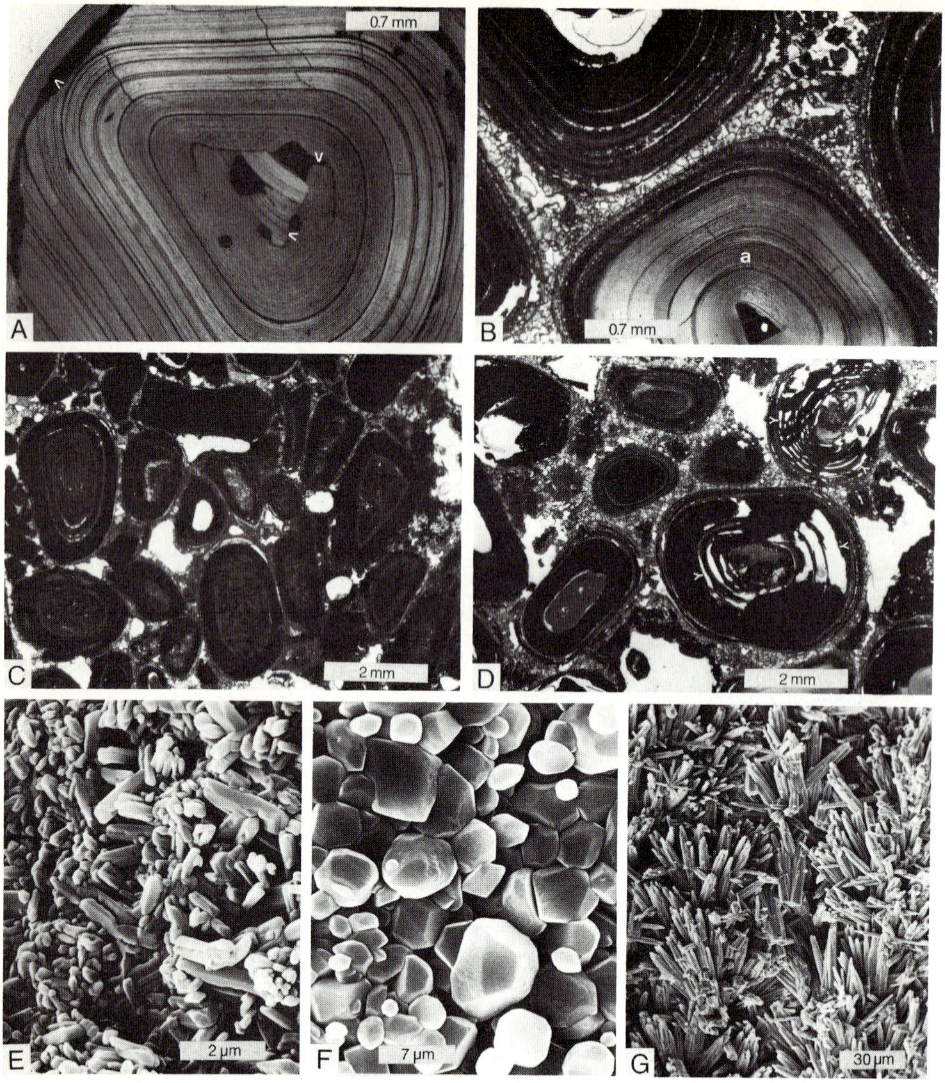

Fig. 4. A Aragonitic ooid showing three erosive phases (*arrows*) during growth. **B** Extensive preservation of layered structure despite of micritic to microsparitic calcitization of the ooids; a preservation of aragonite in the cortex of the lower ooid; *arrow*: pigmented line in the calcite crystal explains a calcitization of aragonite in situ. **C** Completely calcitized aragonitic ooids showing layered structure. **D** Micritic and microsparitic calcitized aragonitic ooids partly showing pores caused by dissolution (*arrows*). **E** Aragonite crystals in the cortex of an ooid not affected by diagenesis; SEM photography; laminae in vertical direction; note the two radial-aragonitic laminae. **F** Calcite crystals in the cortex of completely calcitized ooid; SEM photography. **G** Bundles of aragonite crystals in a diagenetically unaltered calcareous crust of a subrecent pool; SEM photography

sections below 5 μm thick – of course a negative pseudo-uniaxial figure (Richter, this vol.).

Ooids with spheric cores show concentric laminae while those with angular or elongate cores (e.g., broken older ooids) are either coated concentrically or become more and more spherical with increasing thickness of the cortex. The latter was mentioned also by Newell et al. (1960, p. 490) from the Bahamas; according to Carozzi (1960, p. 240) this is a typical feature of ooids. The layered structure of the cortices of thermal ooids from Tekke Ilica often shows disruptions caused by erosional processes. A new coating after an erosional phase always tries to smoothen the surface (Fig. 4A).

Alteration of primary structures of ooids by calcitization or dissolution is very common, resulting in a more or less porous micritic to microsparitic fabric which is either confined to individual layers or penetrates the aragonitic laminae or alters the ooids completely (Fig. 4B). The calcitized parts are composed of subhedral crystals (Fig. 4F) with the primary layering still visible (Fig. 4C, D). The preservation of the lamination may be due to an irregular distribution of the original microporosity forming distinct pathways for the calcitizing solutions. The boundaries of calcite spots crossing the laminae often coincide with fissures in the cortex. In such ooids the replacement results in an irregular net of thin calcitized zones (Fig. 4D).

5 Mineralogical and Chemical Composition

Ooids from Tekke Ilica not affected by diagenesis are made up by aragonite like marine ooids from the Bahamas. Both, however, show deviating $d_{(111)}$-values as determined by XRD with an internal quartz standard when compared to the 3.396 Å major reflexion of stoichiometric aragonite (Swanson and Fuyat 1953, cited in Swanson et al. 1954). They indicate an expansion of the crystal lattice which is correlated with the concentration of Sr as determined by spectral photometry and XRF (Fig. 5). The correlation between $d_{(111)}$-value and weight-% seems to be a straight line connecting aragonite (3.396 Å) and strontianite (3.535 Å).

The concentration of Sr in aragonitic ooids and crusts of thermal origin from Tekke Ilica is in the range between 1.8 and 2.5 wt.% and thus two to three times higher than for the Bahama ooids. According to Kinsman (1969), the distribution coefficient for Sr is $k_{Sr\ (70\,°C)}^{aragonite} = 0.92$. The solution equilibrium with the thermal ooids containing 20,000 ppm Sr (molar Sr/Ca rate = 0.024) would therefore have a molar Sr/Ca rate of 0.026.

The concentration of Mg increases during calcitization, while Sr decreases (Fig. 6). The dots with 1 wt.% Sr (molar Sr/Ca ratio: 0.012) correspond to nearly pure calcite, without aragonite. According to Kinsman's distribution coefficient ($k_{Sr\ (70\,°C)}^{calcite} = 0.1$), this calcite requires a solution with a molar Sr/Ca ratio of 0.12, which clearly exceeds the Sr/Ca in the solution responsible for the aragonite precipitation discussed above (0.026). The high content of Sr in the calcitic samples is probably due to traces of Sr-bearing minerals like strontianite or celestite. A tiny peak (<5 mm) at the position of the major reflexion of strontianite in the XRD diagrams is not a clear proof for the occurrence of strontianite.

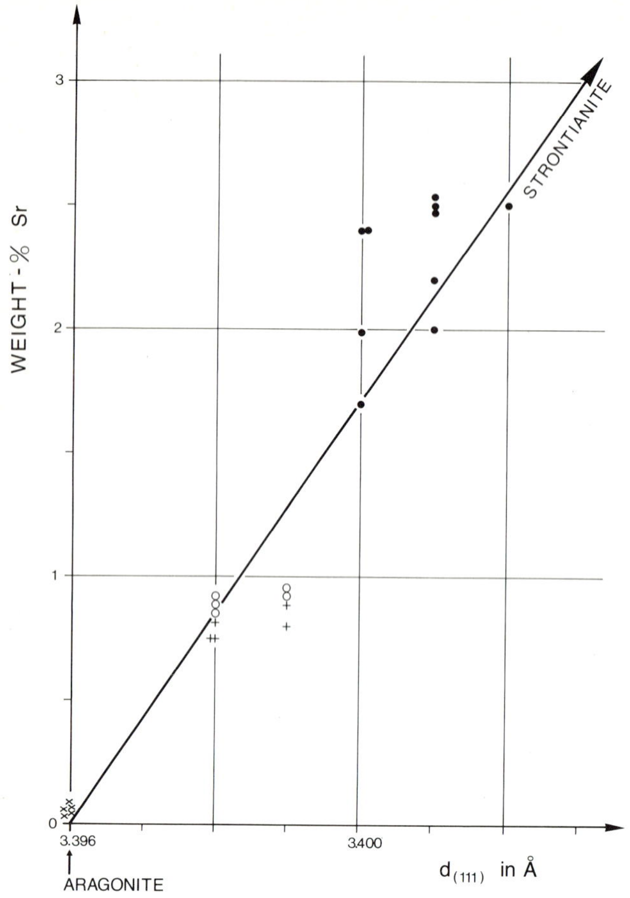

Fig. 5. Wt.% Sr and $d_{(111)}$-values of different aragonites. ● aragonitic ooids and crusts from the thermal occurrence of Tekke Ilica; ○ holocene ooids from Tolon near Nauplion (Peloponnesus/Greece, see Richter 1976); + recent corals from the Red Sea near Port Sudan; × aragonite crystals from a weathered zone of the ophiolite complex east of Loutraki (north of Corinth/Greece)

The lack of gravitational fabrics (e.g., cement) in the partly calcitized areas and the traces of primary pigment laminations in microcrystalline ooids suggest that calcitization has taken place in the phreatic zone (Fig. 3B). Measurements of C and O isotopes (PDB) illustrate the similarity in composition of the solutions in which aragonite and calcite were precipitated; the values are characteristic of thermal meteoric water:

$$\text{aragonite: } \delta C^{13} = +2.49, \quad \delta O^{18} = -19.17$$
$$\text{calcite: } \delta C^{13} = +2.29, \quad \delta O^{18} = -15.70.$$

A reason for the higher O^{18} value of calcite may be the lower temperature of the solution in which calcitization took place compared with the formation of aragonitic ooids.

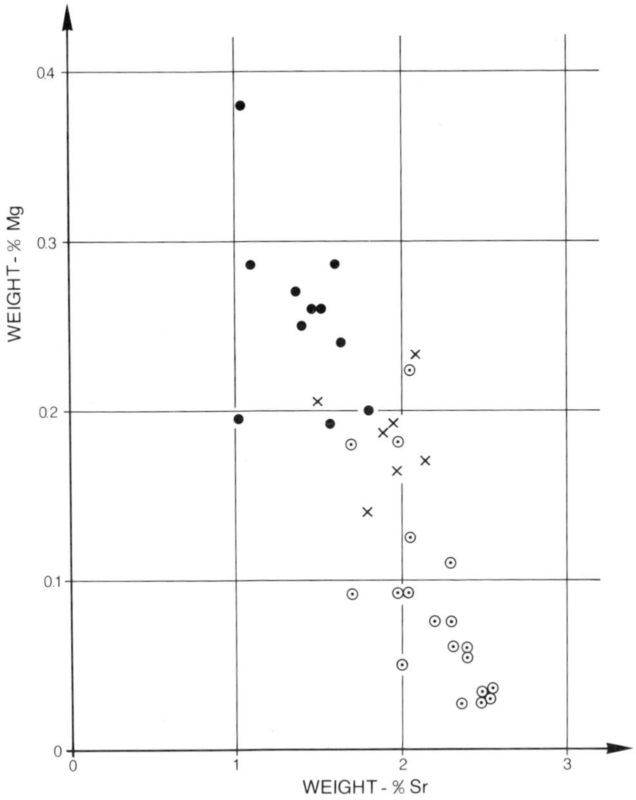

Fig. 6. Wt.% Sr and wt.% Mg of more or less calcitized ooids from Tekke Ilica: ⊙ = >90% aragonite; × = 10% – 90% aragonite; ● < 10% aragonite; the remainder is always calcite

6 Conclusions

1. There is no phenomenological difference between the subrecent aragonitic thermal ooids from Tekke Ilica and marine Bahamian ooids except for their larger size and the absence of rods in the thermal ooids. The ooids of both occurrences consist of micron-size laminae of aragonite crystals showing a statistically tangential (rarely radial) orientation and become more and more spherical with increasing size.
2. According to recent compilations by Choquette (1978) and Donahue (1978), the grain-size distribution (<1 mm to 50 mm in diameter) would include ooids (<2 mm in diameter) as well as pisoliths (or pisoids; >2 mm in diameter). Such a distinction, however, does not make sense, because there is no difference in structure between small and large ooids from Tekke Ilica and because Sorby (1879) has already described similar large ooids from Karlsbad (CSSR) as "oolitic grains".
3. The high content in Sr ranging between 1.8 and 2.5 wt.% is fixed to aragonite as identified by XRD investigations (d_{111} > 3.396 Å). If Sr in aragonite exceeds 1 wt.% (= marine aragonite), we suggest the term "strontian aragonite".

4. The aragonitic thermal ooids have been altered diagenetically by dissolution or (and) micritic to microsparitic calcitization in the warm phreatic zone. Lamination may be preserved when calcitized. These observations are important for the interpretation of fossil ooids, because many occurrences of ooids from the past, e.g., in the Lower Muschelkalk (Triassic) in Central Germany (Frantzen 1887), show a micritic to microsparitic lamination and thus presumably are calcitized aragonitic ooids.

References

Carozzi AV (1960) Microscopic sedimentary petrography. John Wiley, New York, 485 p

Choquette PHW (1978) Oolite. In: Fairbridge RHW, Bourgeois J (eds) The encyclopedia of sedimentology. Stroudsburg, Pennsylvania, pp 510 – 515

Donahue J (1978) Pisolite. In: Fairbridge RHW, Bourgeois J (eds) The encyclopedia of sedimentology. Stroudsburg, Pennsylvania, pp 582 – 583

Fabricius F, Klingele H (1970) Ultrastrukturen von Ooiden und Oolithen: zur Genese und Diagenese quartärer Flachwasserkarbonate des Mittelmeeres. Verh Geol Bundesanst Jahrg 1970:594 – 617

Frantzen W (1887) Untersuchungen über die Gliederung des unteren Muschelkalkes in einem Theile von Thüringen und Hessen und über die Natur der Oolithkörner in diesen Gebirgsschichten. Jahrb K Preuss geol Landesanst Berlin Jahrg 1887:1 – 93

Illing LV (1954) Bahaman calcareous sands. Bull Am Assoc Petrol Geol 38:1 – 95

Kinsman DJJ (1969) Interpretation of Sr^{2+} concentrations in carbonate minerals and rocks. J Sediment Petrol 39:486 – 508

Land LS, Behrens EW, Frishman SA (1979) The ooids of Baffin Bay, Texas. J Sediment Petrol 49:1269 – 1278

Loreau JP (1970) Ultrastructure de la phase carbonatée des oolithes marines actuelles. C R Acad Sci 271:816 – 819

Loreau JP, Purser BH (1973) Distribution and Ultrastructure of Holocene Ooids in the Persian Gulf. In: Purser BH (ed) The Persian Gulf. Springer, Berlin Heidelberg New York, pp 279 – 328

Newell ND, Purdy EG, Imbrie J (1960) Bahamian oolitic sand. J Geol 68:481 – 497

Oti M, Müller G (1979) Recent ooids from different environments. Erdöl Kohle Erdgas Petrochem 32:107 – 115

Richter DK (1976) Gravitativer Meniskuszement in einem holozänen Oolith bei Neapolis (Süd-Peloponnes, Griechenland). Neues Jahrb Geol Palaeont Abh 151:192 – 223

Richter DK (this vol.) Calcareous ooids: a synopsis, pp 71 – 99

Sorby HC (1879) On the structure and origin of limestones. Q J Geol Soc London XXXV:56 – 95

Swanson HE, Fuyat RK, Ugrinic GM (1954) Standard X-ray diffraction powder patterns. Data for 34 inorganic substances. National bureau of standards circular 539, vol III. Washington DC, pp 1 – 73

II.8 Coated Grains Along the Dead Sea Shore

R. A. Garber and G. M. Friedman[1]

1 Introduction

Coated grains form in a number of environments. In the modern marine environment, the cortex of coated grains contains individual aragonite crystals with their long axes oriented tangential to the nucleus. In the cortex of some coated grains, the long axes or aragonite (and calcite) crystals are normal to the rims forming ooids with a radial fabric. Modern coated grains having this fabric occur in hypersaline waters such as the Great Salt Lake of Utah; in Baffin Bay, off Laguna Madre, Texas; in lakes in the Kulundin Steppe; or in thick accumulations of algal mats, as found in sea-marginal pools of the modern Red Sea, and along the west coast of Australia (Vital 1948, Friedman et al. 1973, Kahle 1974, Sandberg 1975, Davies and Martin 1976, Halley 1977, Friedman and Sanders 1978, p. 52–53).

Coated grains of a non-radial fabric form in the Dead Sea, unlike many other occurrences from hypersaline environments. This report describes this occurrence and compares these coated grains with others from hypersaline environments and coated grains formed under normal marine conditions.

2 Setting

The Dead Sea is a hypersaline lake situated in the lowermost depression of the Jordan-Arava Rift Valley (Fig. 1). The salinity of the water at the present day is approximately 340 g/l. A comprehensive description of water chemistry is given by Neev and Emery (1967), Nissenbaum (1977) and Beyth (1980).

The coated grains were found as a distinct internal layer within a finely varved aragonitic crust that covers pebbles and boulders along the Dead Sea margin (Fig. 2). The aragonitic crust has an irregular surface in which sand-size particles are collecting today. The aragonitic pavement may also contain sand to pebble-size terrigenous particles in addition to the coated grains.

The wave energy is quite low during the summer months. During the winter, however, there may be several violent storms at which time the waves may reach

1 Department of Geology, Rensselaer Polytechnic Institute, Troy, NY 12181, USA

Coated Grains (ed. by T. M. Peryt)

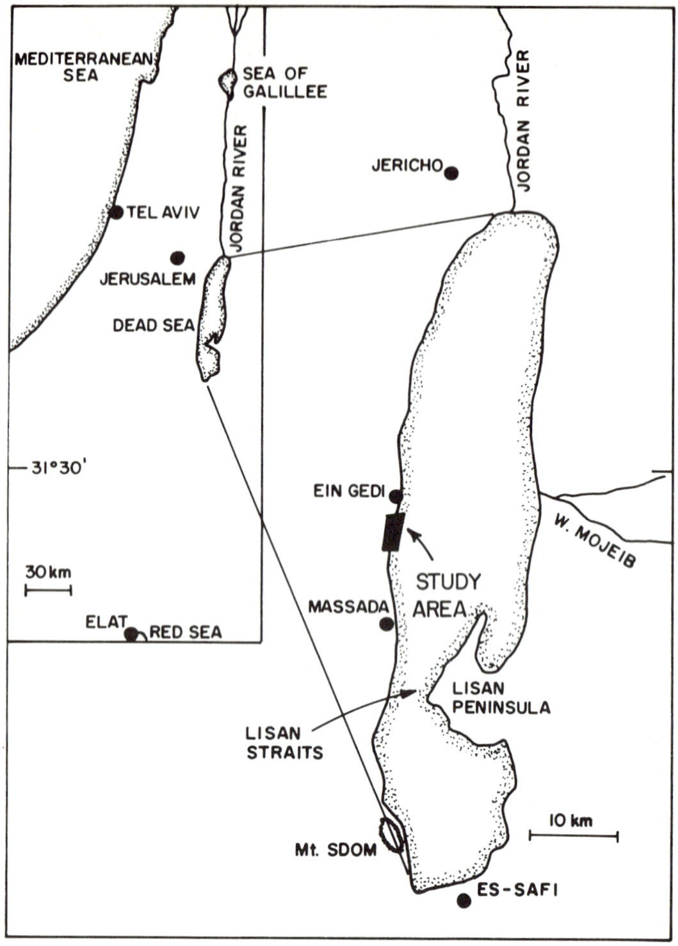

Fig. 1. Index map of the Dead Sea showing the location of coated grain occurrences

up to 2 m in height. Sand trapped in pockets in the crust are rippled in addition to relic ripple marks preserved in the crusts, which attest to high-energy conditions.

3 Description

The coated grains possess nuclei of recycled particles of dolostone, limestone, chert, quartz and feldspar, derived from exposed bedrock. These particles, which are irregular in shape, are surrounded by up to 20 concentric layers of aragonite. The coated grains range in size from fractions of a mm up to 0.5 mm in diameter. They are fairly well sorted and show no signs of breakage.

With the aid of a petrographic microscope, it is possible to discern both light and dark laminae concentric to the nucleus (Fig. 3). The total thickness of the

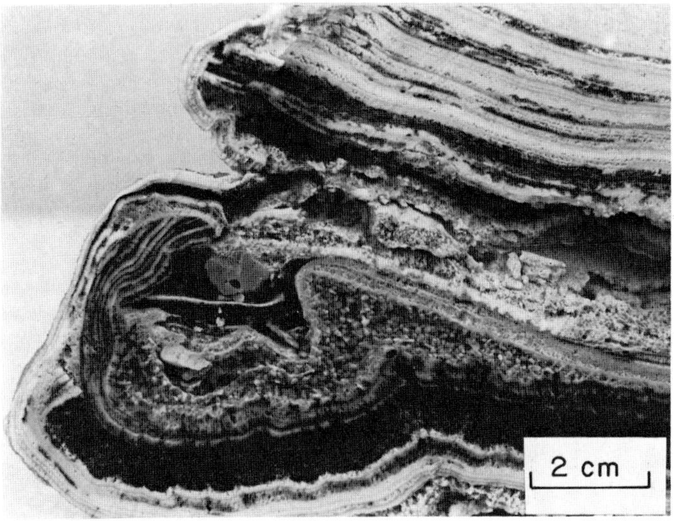

Fig. 2. Cross-section through aragonitic pavement. Central sandy area contains coated grains

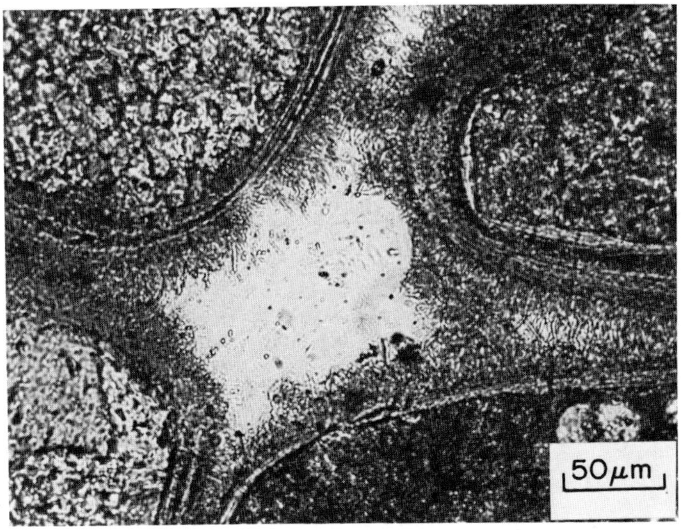

Fig. 3. Photomicrograph of a thin section showing several coated grains. Note the variety of nuclei present. Fibrous aragonite partially occludes pore space. Plane polarized light

cortex is as much as 50 μm. The individual laminae contain equant crystals of aragonite (0.5 μm in diameter) arranged in thin (4 μm) layers as revealed in SEM (Fig. 4). They may be separated by ultra-thin layers (1 μm) of randomly oriented needle-like crystals of aragonite (average needle length = 2 μ).

The coated grains are lightly cemented together by a fibrous aragonite cement in which the individual needle-like crystals are oriented perpendicular to the surface thus the pore space between the coated grains is partially occluded.

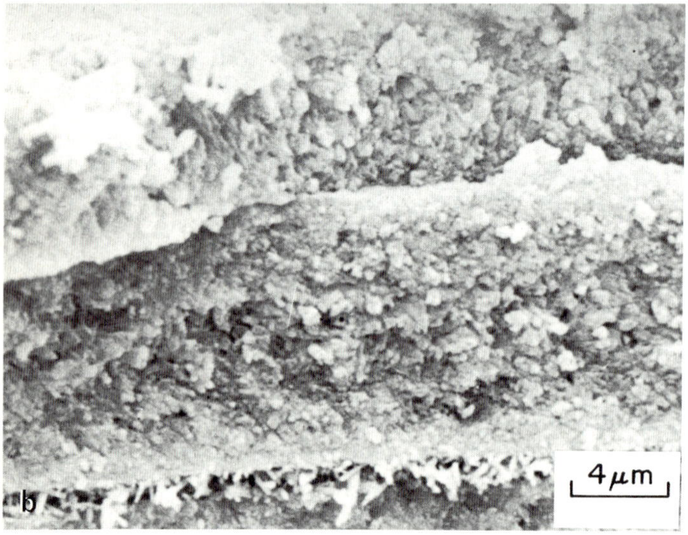

Fig. 4. a SEM photomicrograph of a cross-section through a coated grain. Note the smooth surface of the nucleus. Large, randomly oriented crystals of aragonite are exposed on the surface of the grain. Thin laminae of the cortex are exposed in this section. **b** SEM photomicrograph of a cross-section through a coated grain (close-up). The thicker layers contain micritic aragonite crystals, whereas fine layers contain randomly oriented aragonite needles

4 Discussion

The coated grains from the Dead Sea possess a non-radial texture consisting of equant crystals of aragonite surrounding a detrital nucleus. Similarly, coated grains from the normal marine environment possess a cortex of rods (tangentially oriented) and granules arranged concentrically around the nucleus (Fabricius 1977). The coated grains from the Dead Sea differ from their marine counterpart in two respects: (1) they do not contain rod-shaped aragonite crystals except for minor layers of unoriented needles found between the concentric layers of equant crystals, (2) there is no evidence of borings by algae which is common to coated grains in the marine environment (Fabricius 1977). These two points may help to differentiate between coated grains from the marine and hypersaline environments in the rock record, assuming diagenesis has not obscured the original texture.

Coated grains from the hypersaline environment up until now were thought to consist of primary radial crystals (Friedman et al. 1973, Kahle 1974, Fabricius 1977, Halley 1977). Coated grains of freshwater origin also possess a crystalline structure but their crystals tend to be more equant than fibrous (Loreau and Purser 1973). Several investigators have concluded from the few examples of coated grains from hypersaline settings that they are always of radial crystalline texture (above references). Land et al. (1979) list occurrences of present-day radial aragonite coated grains from non-hypersaline environments as well as radial Mg-calcite coated grains from hypersaline environments. They conclude that the correlation among salinity, mineralogy and texture is imperfect and thus salinity is not a controlling factor for texture in coated grains. This study further demonstrates that coated grains from a hypersaline environment do not invariably possess a radial texture.

Acknowledgements. This research was supported by a grant from the United States-Israel Binational Science Foundation (BSF), Jerusalem, Israel. The writers express their thanks to Dr. Joel Gat for help in carrying out the field work.

References

Beyth M (1980) Recent evolution and present stage of Dead Sea brines. In: Nissenbaum A (ed) hypersaline brines and evaporitic environments. Elsevier North Holland Biomedical Press, Amsterdam New York, pp 155 – 166

Davies PJ, Martin K (1976) Radial aragonite ooids, Lizard Island, Great Barrier Reef, Queensland, Australia. Geology 4:120 – 122

Eardley AJ (1938) Sediments from the Great Salt Lake, Utah. Am Assoc Petrol Geol 22:1305 – 1411

Fabricius FH (1977) Origin of marine ooids and grapestones. Contrib Sedimentol 7:1 – 113

Friedman GM, Sanders JE (1978) Principles of Sedimentology. John Wiley, New York

Friedman GM, Amiel AJ, Braun M, Miller DS (1973) Generation of carbonate particles and laminites in algal mats-example from sea-marginal hypersaline pool, Gulf of Aqaba, Red Sea. Am Assoc Petrol Geol 57:541 – 557

Halley RB (1977) Ooid fabric and fracture in the Great Salt Lake and the geologic record. J Sediment Petrol 47:1099 – 1120

Kahle CF (1974) Ooids from Great Salt Lake, Utah, as an analogue for the genesis and diagenesis of ooids in marine limestones. J Sediment Petrol 44:30–39

Land LS, Behrens EW, Frishman SA (1979) The ooids of Baffin Bay, Texas. J Sediment Petrol 49:1269–1278

Loreau JP, Purser BH (1973) Distribution and ultrastructure of Holocene ooids in the Persian Gulf. In: Purser BH (ed). The Persian Gulf. Springer, Berlin Heidelberg New York, pp 279–328

Neev D, Emery KO (1967) The Dead Sea, depositional processes and environments of evaporites. Geol Surv Isr Bull 41:1–147

Nissenbaum A (1977) Minor and trace elements in Dead Sea water. Chem Geol 19:99–111

Sandberg PA (1975) New interpretations of Great Salt Lake ooids and of ancient non-skeletal carbonate mineralogy. Sedimentology 22:497–537

Vital DA (1948) Modern lime-magnesian concretions and oolites in lakes of the Kulundin Steppe, Altai region, USSR. Moscow Soc Nat 23:83–100

II.9 Structural and Geochemical Features of Jurassic Oolitic Limestones in the Veneto Region (NE Italy)

L. Scudeler Baccelle[1]

1 Introduction

The Liassic carbonate platform in the Veneto is rich enough in oolitic rock-types to justify the fact that one of its components, an Upper Lias carbonate formation, is called the S. Vigilio Oolite.

The sedimentological features and macroscopic aspects of the oolitic beds (i.e., textural and structural features) are generally very similar both in the S. Vigilio Oolite and in the underlying Calcari Grigi Formation (grey limestones of Lower and Middle Liassic age). However, in detail the ooids show features which suggest different genetic mechanisms and/or micro-environments.

Representative samples from both formations have been studied. They were collected from one of the better-exposed rock sequences, cropping out for a total of about 900 m in the Rio Bissole Valley (province of Verona) (Fig. 1). This area corresponds to the western margin of the Liassic carbonate platform of the Veneto (Trento Plateau according to Winterer and Bosellini 1980).

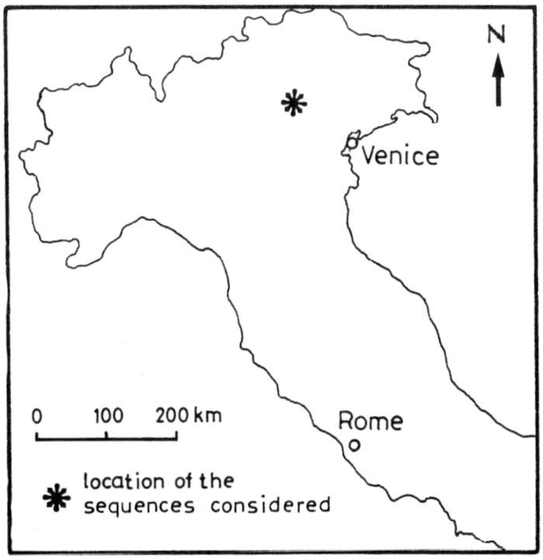

Fig. 1. Sampling station of oolitic Liassic sequences in the Veneto (NE Italy)

1 Istituto di Mineralogia e Petrologia, Università di Padova, Corso Garibaldi, 37, 35100 Padova, Italy

Coated Grains (ed. by T.M. Peryt)

Isolated ooids and total rocks were analyzed using different methods. The main aim was the study of morphological features and mineral and chemical composition (Mg, Sr, and amino acid contents), both in whole rocks and in single grains. The interpretation of results was based on a comparison with data collected from Bahamian ooids (from Joulters Key and other localities) and from some Jurassic oolitic rocks from the Tethys basin, and with data from the literature.

2 Petrographic Features

2.1 Structural Features

The ooids are sometimes irregular in shape, being polygonal, elliptical, lobate, or with disrupted outer layers. Only regular forms are considered here, as the others resulted from deformation due to physical or biological events (impact, compaction, cementation, fractures due to synaeresis, perforations or micritization by vegetal and/or animal organisms).

The structures which may be recognized are shown in Fig. 2.

The nannograins are needle-like, bladed, equant, rhombohedral, anhedral or pseudoeuhedral. The needle- or blade-shaped nannograins, often showing competitive growth, occur in structures a′ and a″, while micritic ooids (a‴, b′ and b″), laminar or not, seem to be composed of equant, anhedral or pseudoeuhedral grains of differing size.

As regards the cement, it may be observed that the micritic grains (b′ and b″) are always encrusted with a more or less dense border of fibrous-radial calcite. The other laminar-structure grains (a′, a″ and a‴) sometimes lack a border and, when it is present, it may be either in fibrous-radial or rhombohedral calcite.

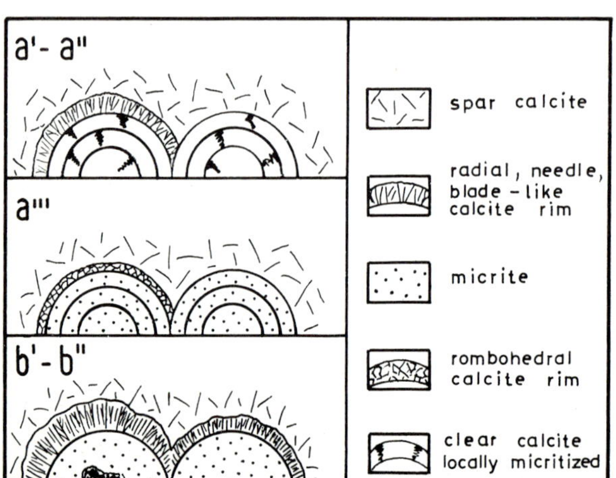

Fig. 2. Grains and cements structures. Laminar: *a′* with rare, clear and porous laminae; *a″* with dense, clear, locally micritized laminae; *a‴* with dense, micritic laminae, alternately more or less dense. Non-laminar: *b′* micritic with *Nubecularia* in the nucleus; *b″* micritic. Radial rim: *a′*, *b′* and *b″*. Rhombohedral rim: *a‴*

2.2 Mineral and Chemical Composition

In their outermost layer, the Liassic grains consist of a low-Mg calcite (mean $MgCO_3$ content: 2.6 mol%, corresponding to 2.1 wt%; N = 4; s = 0.89), while the corresponding rocks have the following average chemical contents, in wt.%:

CaO: 55.10 (N = 5; s = 0.2145), corresponding to $CaCO_3$ 98.392;
MgO: 0.43 (N = 5; s = 0.0693), corresponding to $MgCO_3$ 0.908.

Therefore, the $MgCO_3$ contents, which are in any case very low in the rocks (0.908 wt.%), are preferentially concentrated in the grains (2.1%) rather than in the cement. This location is also confirmed by microprobe analyses, which generally indicate low, scattered MgO content in the grains with local significant increases in the organogenic inclusions (Scudeler Baccelle 1980).

Among the minor elements, Sr was also analysed by X-ray fluorescence and atomic absorption spectrometry. The modern Bahamian aragonitic ooids turn out to have 9274 ± 30 ppm of Sr, while the Liassic oolitic limestones show significantly lower Sr contents (160 ppm; N = 5; s = 18.4), with homogeneous distribution between grains and cement.

The Joulters Key ooids show Sr contents which are consistent with the high values reported in the literature for modern ooids from other Bahamian localities and from the Persian Gulf (9500 ± 500 ppm).

On the other hand, as sketched in Fig. 3, the low Sr content found in the analyzed Liassic samples of Veneto (1) fall into the range of Sr values reported in the literature concerning other Jurassic oolitic limestones of Tuscany (2), Lazio (3), Sicily (4), Sardinia (5) and Western Carpathians (6), (Manetti and Turi 1973, Fornaseri and Grandi 1963, 1968, Veizer and Demovič 1973).

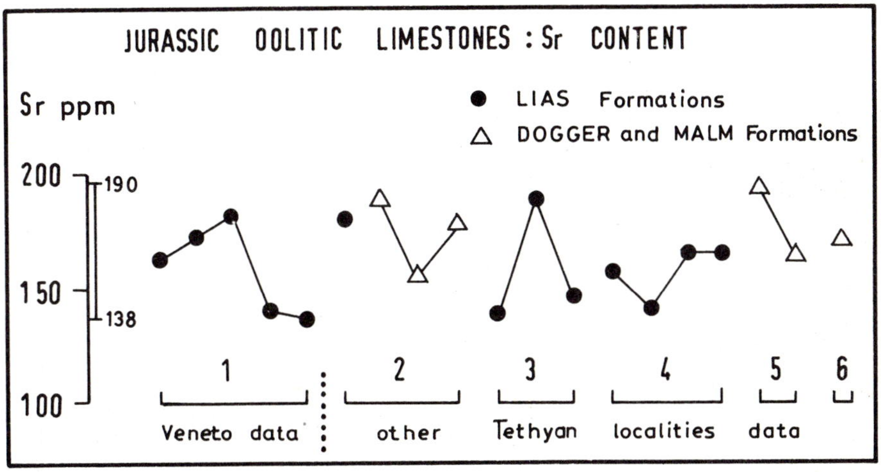

Fig. 3. Comparison of Sr content of Jurassic oolitic limestones from Veneto (*1*), Tuscany (*2*), Lazio (*3*), Sicily (*4*), Sardinia (*5*) and Western Carpathians (*6*)

3 Organic Matter

The organic mucilage which occurs in small quantities in the Liassic rocks and in larger quantities in the Joulters Key ooids was isolated by means of HCl 1 N leaching and analysed using an amino acid autoanalyser. It turned out to be completely composed of basic, neutral and acid amino acids.

The total amino acid content averages 0.1470 μM per gram of ooids (N = 4; s = 0.087) in the Liassic samples. In contrast, the Bahamian ooids of Joulters Key contain 5.3502 μM amino acids per gram of ooids. The absolute values in the old ooids are thus definitely lower than those in the modern ones; the modern ooids analysed show total amino acid contents (mainly aspartic acid) which are only slightly higher than the values reported by Mitterer (1968, 1969, 1971, 1972) for ooids from the Great Salt Lake, Atlantic Shelf, Persian Gulf, Red Sea and localities in the Bahamas (4.50 μM per gram of ooids; N = 7; s = 0.885). Mean values from my analyses and the maximum and minimum values from the literature are shown in Table 1.

With regard to the quantitative ratios among the various amino acids, it is clear that the sum of the neutral and basic amino acids (78.63%) clearly prevails over the aspartic and glutammic acids (21.36%).

Table 1. % amino acid composition of the ooids

	Jurassic ooids from Veneto			Modern ooids from Joulters Key	Other modern ooids taken from Mitterer	
		N	s		max.	min.
Aspartic acid	8.7050	4	0.7739	37.21148	30.2	16.8
Threonine	4.9026	4	0.4657	2.74007	7.0	4.2
Serine	11.6033	4	0.7117	5.48387	6.3	4.6
Glutamic acid	12.9012	4	0.3352	12.16957	16.8	8.6
Proline	3.5963	4	0.8371	2.49896	4.9	2.8
Glycine	19.5045	4	0.8553	12.32657	15.5	12.3
Alanine	8.2883	4	0.4458	6.70999	9	11.5
Cystine ($\frac{1}{2}$)		4		2.15878	1.6	0.2
Valine	3.8579	4	0.4813	4.71008	7.9	4.6
Methionine	0.7879	4	0.4310	0.41923	6	0
Isoleucine	3.5532	4	0.4305	2.81483	5.1	2.8
Leucine	6.7370	4	0.7763	3.72139	6.6	4.5
Tyrosine	2.7225	4	0.1285	0.31027	1.3	0.3
Phenylalanine	2.7359	4	0.0953	0.93828	5.5	2
Ornithine		4		0.79996	1.7	0.1
Lysine	7.5124	4	1.4282	3.60358	3.6	1.4
Histidine		4		0.61493	0.8	0.2
Arginine	3.7109	3	1.9220	0.76819	3.7	0.7
Total	101.1189					
Total aspartic and glut. acid	21.6062 i.e., 21.36%					
Total remaining amino acids	79.5127 i.e., 78.63%					

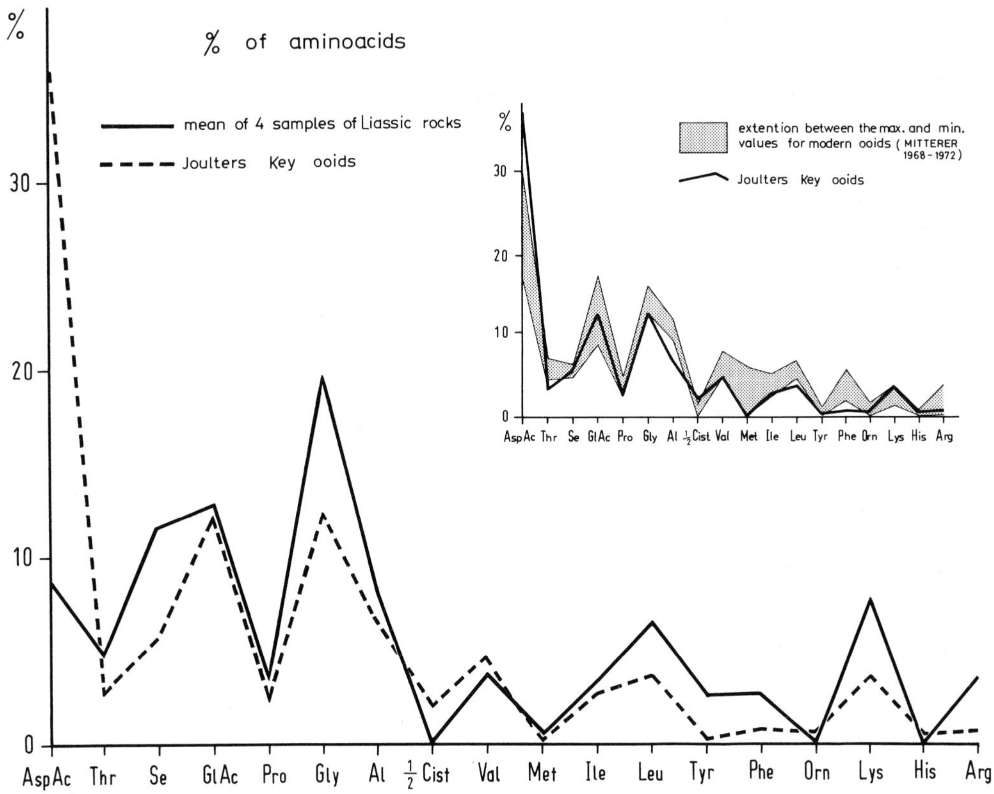

Fig. 4. Amino acid ratios: *lower sketch* compares the modern and fossil data, *upper sketch* compares literature and Joulters Key analyzed ooids data

It may also be observed that the ratios among the various amino acids (excluding proline and serine) may be compared with values found in the membranes of the modern blue-green alga *Anacystis nidulans* (Trichet 1968). The diagrams in Fig. 4 compare my analytical data and the values found in the literature.

It is clear that the nature of the organic fraction of the oolitic deposits does not depend on the age or geographic distribution of the deposits themselves; it is instead linked to the presence of blue-green algae and the products of their decomposition. Confirmation of this seems to be given also by the finding of recent ooids (Abu Dhabi-Persian Gulf, Schooner Cays-Bahamas) of light hydrocarbons, considered to be syngenetic products of the transformation of filamentous algae, bacteria and fungi (Ferguson and Ibe 1981).

4 Discussion and Conclusions

The morphological and mineralogical data suggest that:

1. According to recent studies (Folk 1974, Assereto and Folk 1976), due to the variety of needle, bladed and rhombohedral nannograins, the Jurassic ooids may be considered of original calcitic and aragonitic composition, similar to the Pleistocene ooids of the Bahamas (Bourrouilh 1976). These carbonate mineral phases were altered into low-Mg calcite during diagenesis.
2. The films of organic matter, which are still preserved in them rhythmically alternating with the carbonate layers, show that each grain grew episodically.
3. The organic matter in these ooids consists largely of basic and neutral amino acids (78.63%) with subordinate aspartic and glutammic acids (21.36%).

Considering the above points, the following conclusions can be drawn:

a) The organic matter may have played a double role, even within the ambit of the history of each single grain, so that phases of aggregation or capture of pre-existing particles (e.g., Mg-rich clastic grains or microskeletal fragments) may have been alternated with phases of carbonate precipitation. Processes of this type have recently been reproduced by Suess and Fütterer 1972, Ferguson et al. 1978)

b) Two possible carbonate mineral phase – aragonite and calcite – are not completely unrelated to the type and quantity of available organic matter (Jackson and Bischoff 1971). The latter may also control the change in the Ca^{2+}/CO_3^{2-} ratio, which is in turn responsible (Lippmann 1973) for the radial or tangential ultrastructure of the inorganic $CaCO_3$. In particular, diagenesis enhances the role of the organic matter; it is well known that the basic and neutral amino acids assist the alteration of aragonite into calcite, while the aspartic and glutammic acids inhibit it (Jackson and Bischoff 1971). The neutral and basic amino acids which still occur in ooids studied, as the main components of their organic fraction, doubtless allowed the alteration of aragonite into calcite even during the early diagenetic stage preceding cementation. In this way circulating fluids, being able to remove the large quantities of Sr leached from the aragonite, greatly depleted the originally high Sr content of the rocks (9,000/10,000 ppm). Judging from the uniformity of the data, this mechanism seems to be common not only in the Liassic oolitic limestones from Veneto, but also in the other oolitic formations of the same and younger age (Dogger-Malm) from the Apennines, Sicily, Sardinia and the Western Carpathians. Therefore, a similar diagenetic and epigenetic evolution may be assumed for all these oolitic formations.

c) Lastly the early diagenetic depletion of Sr from the oolitic sediments also may explain the Sr contents, which are commonly lower in oolitic limestones (Kahle 1965, Scudeler Baccelle 1980) than in other limestones having different textural features (presence of a matrix, variety of grains and differences in the matrix-grains relationship).

Acknowledgements. I would like to express my sincere thanks to Tadeusz Peryt and Chriss Burgess for their help and the critical reading of may article

References

Assereto R, Folk RL (1976) Brick-like texure and radial rays in triassic pisolites of Lombardy, Italy: a clue to distinguish ancient aragonitic pisolites. Sediment Geol 16:205–222

Bourrouilh F (1976) Les particules du Plèistocène bahamien et de sa croute: sédimentologie et diagenèse. Sediment Geol 17/34:247–284

Ferguson J, Ibe AC (1981) Origin of light hydrocarbons in carbonate oolites. J Petrol Geol 4/1:103–107

Ferguson J, Bubela B, Davies PJ (1978) Synthesis and possible mechanism of formation of radial carbonate ooids. Chem Geol 22:285–308

Folk RL (1974) The natural history of crystalline calcium carbonate: effect of magnesium content and salinity. J Sediment Petrol 44/2:40–53

Fornaseri M, Grandi L (1963) Contenuto in Stronzio di serie calcaree italiane. G Geol Bologna Ser II 31:171–198

Fornaseri M, Grandi L (1968) Nuovi dati sul contenuto in Stronzio di serie calcaree italiane. Period Miner XVII:733–776

Jackson TA, Bischoff JL (1971) The influence of amino acids on the kinetics of the recrystallization of aragonite to calcite. J Geol 79:493–497

Kahle CF (1965) Strontium in oolitic limestones. J Sediment Petrol 35/4:846–856

Lippmann F (1973) Sedimentary carbonate minerals. Springer, Berlin Heidelberg New York, p 228

Manetti P, Turi A (1973) Ricerche chimiche sulle rocce mesozoiche nella Val di Lima (Lucca). Period Miner XIII 1/2:295–325

Mitterer RM (1968) Amino acid composition of organic matrix in calcareous oolites. Science 162:1498–1499

Mitterer RM (1969) The origin of calcareous oolites. Geol Soc Am Progr S E Sect 4:54–55

Mitterer RM (1971) Influence of natural organic matter on $CaCO_3$ precipitation. In: Bricker P (ed) Carbonate Cements, The Johns Hopkins University Studies in Geology N° 19. Johns Hopkins Press, Baltimora London, p 376

Mitterer RM (1972) Biogeochemistry of aragonite mud and oolites. Geochim Cosmochim Acta 36:1407–1422

Scudeler Baccelle L (1980) Mineralogia e Geochimica degli ovoidi carbonatici della Piattaforma Liassica Veneta. Miner Petrol Acta 24:17–34

Süess E, Fütterer D (1972) Aragonitic ooids: experimental precipitation from seawater in the presence of humic acid. Sedimentology 19:129–139

Trichet J (1968) Étude de la composition de la fraction organique des oolites. Comparaison avec celle des membranes des bactéries et des Cyanophycées. C R Acad Sci 267:1492–1494

Veizer J, Demovič R (1973) Environmental and climatic controlled fractionation of elements in the mesozoic carbonate sequences of the Western Carpathians. J Sediment Petrol 43/1:258–271

Winterer EL, Bosellini A (1980) Subsidence and Sedimentation on Jurassic Passive Continental Margin, Southern Alps, Italy. Am Assoc Petrol Geol Bull 65/3:394–421

II.10 Sedimentary Characteristics of Oolitic Carbonates from the Jialing-Jiang Formation [Lower Triassic (T_1^2J1)], South Sichuan Basin, China

Yun Fu Zeng[1], Nan Hao Lee[2], and Yang Zhou Huang[2]

1 Introduction

Sichuan Basin was an epicontinental sea in the late (Aurenig) of the Early Triassic. The sea was bounded to the north and west by land masses, and to the south by a submarine high, while to the east lay the open sea (Fig. 1). These paleogeographic features controlled sedimentation in the basin.

The area of study, located at the southern extension of the central submarine uplift zone, stretches over the western and eastern depressions, and records a long history of tidal sedimentation. It is therefore a region favourable for the formation and accumulation of oil and gas, since successive transgressions and regressions resulted in polycyclic depositional facies in the strata of the Jialing-Jiang Formation of the Lower Triassic (Fig. 2).

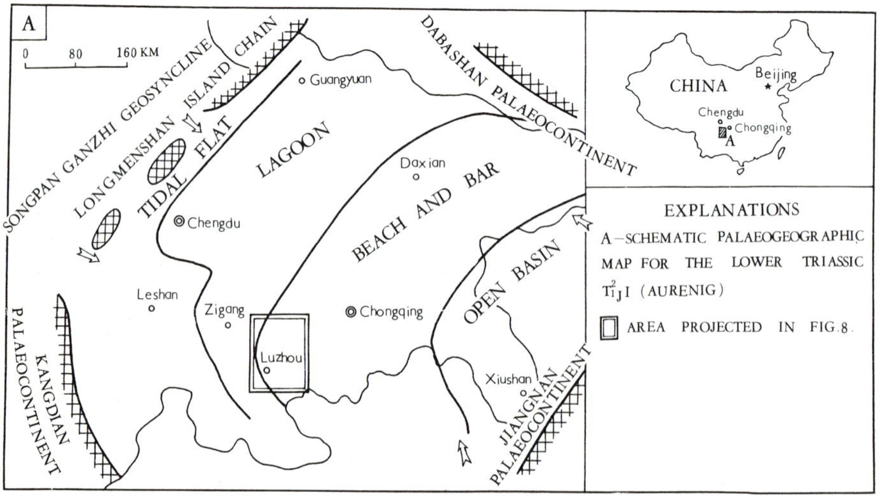

Fig. 1. A paleogeographic map of the area studied in this paper

1 Geological College of Chengdu, Department of Geology, Sichuan, China
2 Geological College of Chengdu, Department of Petroleum Geology, Sichuan, China

Coated Grains (ed. by T.M. Peryt)

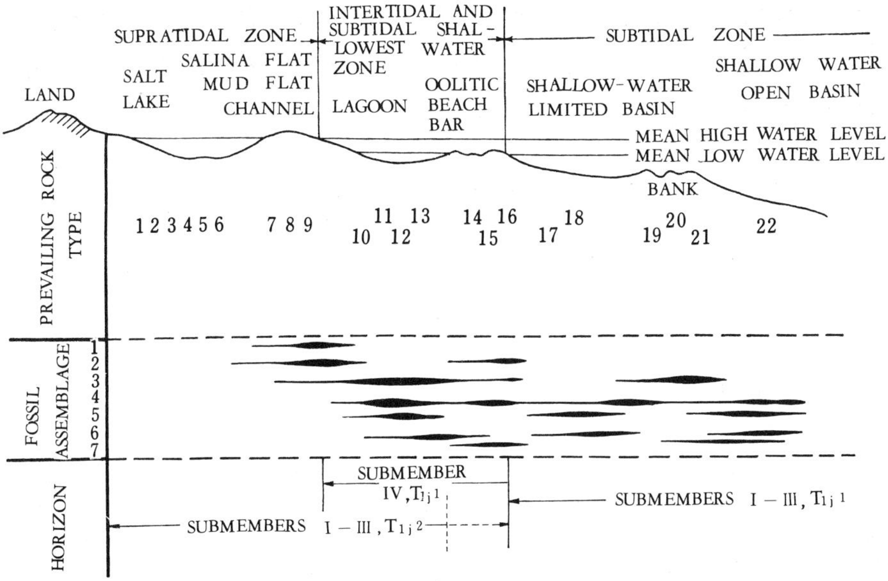

Fig. 2. Depositional pattern of members I and II, Jialing-Jiang Formation, in the Lower Triassic of South Sichuan Basin.
Prevailing rock types: *1* Mudstone, *2* anhydritic dolomite (mudstone), *3* dolomite, *4* anhydrite, *5* rock salt, anhydritic dolomite, *6* dolomite, *7* anhydritic clayey dolomite, *8* hydromica mudstone, *9* slightly pyritized fine-grained quartz-sandstone, *10* slightly clayey dolomite, *11* slightly clayey dolomitic limestone, *12* laminated micritic limestone, *13* pelletal micritic limestone and dolomitic limestone, *14* sparry oolitic limestone (dolomite), leached oolitic limestone, *15* oolitic dolomite, micritic (sparry) bioclastic limestone, *16* green algal micritic (dolomitic) limestone, *17* bioclastic intraclastic limestone, *18* slightly clayey micritic limestone, *19* slightly bioclastic intraclastic micritic limestone, *20* sparry oolitic limestone, *21* sparry (micritic) intraclastic limestone, *22* micritic limestone.
Fossil assemblages: *1* Stromatolites, *2* phylloid algae, *3* ostracods, *4* foraminifers, *5* gastropods, *6* pelecypods, *7* crinoids

The strata consist chiefly of a carbonate-evaporite cycle (Fig. 3) reflecting the gradual change from a shallow water open basin at the start of the cycle to a supratidal saline lake at the end (see Fig. 2). This gradual regressive sequence constitutes a complete suite of source-reservoir and capping horizons for oil and gas.

Oolitic carbonates are chiefly developed in the intertidal and shallow subtidal facies, and now comprise leached oolitic limestone and dolomite with high porosity and permeability. These rock types are chiefly situated in the strata of submember IV (T_1^2J1) in the area of ancient uplift. They are divided, from top to bottom, into intervals C, E and G (interpreted as beach facies) with intercalated strata D and F (lagoon facies) (Fig. 4). Lithologically, the strata D and F consist mainly of slightly clayey micritic limestone and dolomite, intercalated with layers and bands of skeletal limestone. The strata C, E and G are composed mainly of oolitic limestone and dolomite (Fig. 4).

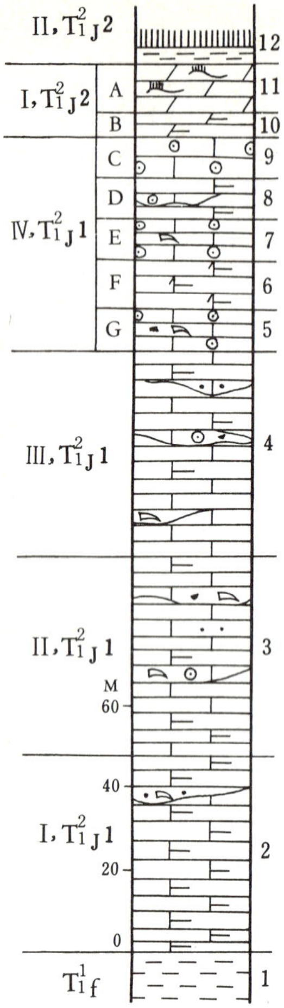

Fig. 3. Schematic columnar section for the Lower Triassic T_1^2J1 (Aurenig). *1* Violet-red mudstone (Indian), *2* slightly clayey limestone, occasionally intercalated with thin bedded grainstone about 42 – 65 m, *3* slightly clayey limestone, intercalated with thin bedded grainstone about 45 – 72 m, *4* micritic limestone, intercalated with thin bedded or lenticular intraclast-oolitic limestone about 40 – 58 m, *5* oolitic limestone and micritic limestone about 5 – 10 m, *6* slightly clayey-dolomitic limestone about 7 – 12 m, *7* oolitic and micritic limestone about 4 – 10 m, *8* slightly clayey limestone or dolomite, intercalated with skeletal-oolitic limestone about 3 – 9 m, *9* oolitic limestone about 2 – 9 m, *10* slightly clayey micritic dolomite about 1 – 5 m, *11* algal-detrital dolomite about 4 – 13 m, *12* mudstone, anhydritic rock and dolomite

2 Lithology

2.1 Oolitic Limestone

Ooids make up more than 25% of the rock. They are mostly superficial ooids, with subsidiary leached ooids, normal ooids and distorted ooids. Compound ooids occur only occasionally. The micritic outer envelopes or cortices of all the ooids possess concentrically lamellar structure and occasionally radial structure. The size of ooids ranges from 0.18 to 0.38 mm, but is generally about 0.22 – 0.28 mm. The largest superficial and normal ooids are mostly found in stratum C.

2.1.1 Normal Ooids. The nuclei are in most cases, composed of a pellet or intraclast of microcrystalline or micritic calcite and in a few cases comprise shelly debris or foraminifers. The number of concentric envelopes in the cortex is generally small, being only three fo four. Ooids with more than ten envelopes are found only in stratum C of well 12 (Fig. 5A). The thickness of the cortex makes up more than half of an ooid raidus.

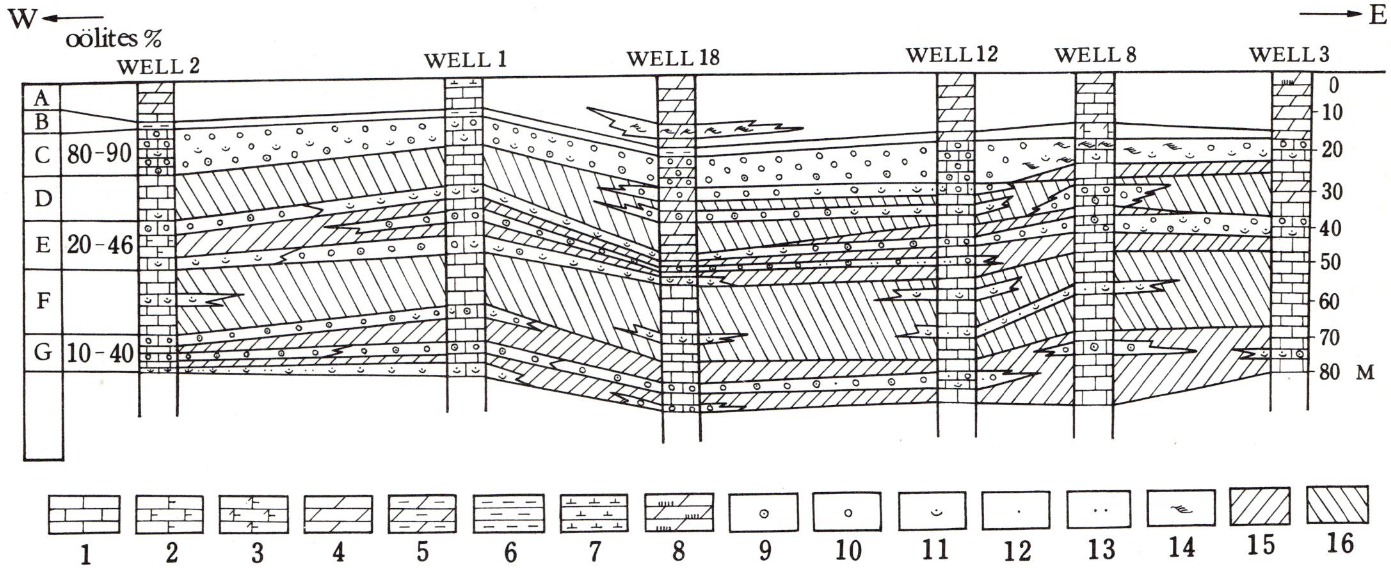

Fig. 4. Cross section through the submembers IV in T_1^2J1. *1* Limestone, *2* slightly clayey limestone, *3* slightly clayey dolomitic limestone, *4* dolomite, *5* clayey dolomite, *6* mudstone, *7* calcitic mudstone, *8* anhydritic dolomite, *9* ooids, *10* leached ooids, *11* shell debris, *12* sand-sized detritus, *13* pellet, *14* algal debris, *15* micritic limestones and dolomites (beach), *16* micritic clayey limestones (lagoon)

2.1.2 Superficial Ooids. The nuclei are mainly composed of pellet or sand-sized detritus of micritic calcite, and sometimes contain a fragment of foraminifer or shell. They were generally coated with only one envelope of micritic calcite, which is 0.01 – 0.02 mm in thickness (Fig. 5 D), but sometimes 3 or 4 envelopes are found, with total thickness of 0.038 – 0.08 mm (2/3 – 1/10 of the ooid radius). Superficial ooids which contain several envelopes of concentric laminae in the cortex occur mostly in strata E and C.

2.1.3 Distorted Ooids. These were generally deformed at an early diagenetic stage (Fig. 5 B). The ooids were deformed by compaction, being somewhat cracked first, then rent completely from each other, and finally drawn into a tadpole shape. The sparry calcite of the first generation cement around the ooid surface was broken, but that is not so with the second generation, which was precipitated directly on the fractured surface (Fig. 5 B). So it is possible that the distortion of the ooids was subsequent to the deposition of tiny crystals of first generation cement during earlier diagenesis and prior to the deposition of larger crystals of second generation cement at a later diagenetic stage.

2.1.4 Leached Ooids. Leached ooids have secondary intragranular pores whose edges are irregular. If the ooids are totally dissolved, only the outermost envelopes remain, forming oomoldic pore spaces (Fig. 5 C, F). When the nucleus is made up of a foraminifer, the pore space is offset from center, with the nucleus itself left alone (Fig. 5 F). If the pore space is completely filled by calcite, the ooids are termed filled leached-ooids, or filled ooids. In some cases, filled ooids cannot easily be distinguished from recrystallized ooids.

2.1.5 Peloids. Peloids occur in strata G, E and C in different quantities, have no internal structure and are composed of micritic or microcrystalline calcite (Fig. 5 D). They may be considered as well-rounded intraclasts or pellets (less than 0.1 mm in diameter).

2.1.6 Associated Grains. The ooids are associated mainly with bioclasts, mostly of sand-and-gravel-sized fragments of pelecypods and gastropods (generally less than 10%), foraminifers, and ostracods, as well as with quantities of green algal and echinodermal debris. The intraclast content varies between 10% and 25%.

Fig. 5 A – H. Oolitic limestone. **A** Normal ooid with 7 – 8 envelopes of cortex. The dissolved intra-ooid pore spaces are filled entirely by calcite with irregular edges. Stratum C, well 12-1424. **B** Distorted ooids. Stratum C, well 1-1423. **C** Sparry leached oolitic limestone, whose dissolved intra-ooid pores were subsequently enlarged and are connected to each other. Statum E, well 13-75. **D** Peloids and superficial oolitic limestone. Stratum E well 14-125. **E** Sparry oolitic dolomite whose intergranular pore spaces are filled with freshwater dolomite and dark bitumen. Stratum C, well 3-1379. **F** Sparry leached oolitic limestone whose intra-ooid pore space shows the remains of a foraminifera nucleus. Stratum E, well 3-1379. **G** Silliceous oolitic dolomite, with intergranular pore spaces filled with granular authigenic quartz. The ooid surface shows pressure-solution pits due to compaction (the *upper right corner*). Stratum C, well 13-63. **H** Intercrystalline pores are well-developed in the ooids (*lower left* and *right corners*) and cements. Intergranular dissolved pores can also be seen (*upper left part*). Stratum D

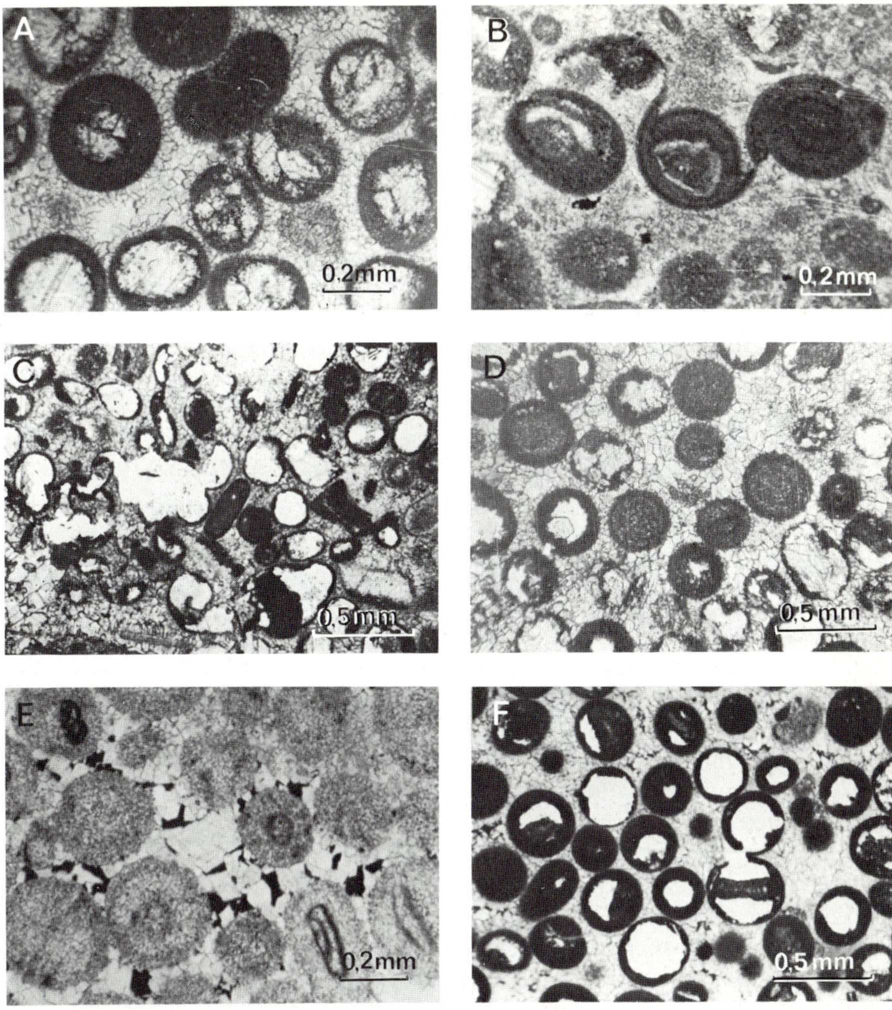
A
0.2mm
B
0,2mm
C
0,5mm
D
0,5mm
E
0,2mm
F
0,5mm
G
0,2mm
H
20μ

2.2 Oolitic Dolomite

In same cases ooids have been completely replaced by microcrystalline dolomite. Normal ooids, superficial ooids and very common peloids were recognized. The nucleii of normal and superficial ooids comprise pellets or a sand-sized intraclast. Multiple envelopes may be seen within the cortex.

The cements are of various types. Around the edges of the ooids, in the intergranular pore space, very fine crystalline rhombohedra are found replacing the dolomite of the first generation. The cements in some pore spaces are single coarsely crystalline calcite (with crystal stocks) or finely crystalline rhombohedral sparry dolomite (Fig. 5E). Sometimes each pore space is partially or wholly filled with anhydrite, celestite or occasionally with secondary quartz (Fig. 5G).

The number of ooids (including peloids of supposed initial ooid origin) is generally high, ranging from 70% to 90%. The compaction process is obvious, for the ooids have been pressed flat or even show pressure-solution pits (Fig. 5G). In the oolitic dolomite the intergranular pore space is generally well developed, with only a few vugs in the ooids which are relatively small and sometimes filled with secondary quartz.

3 Diagenesis

The oolitic carbonates in this district formed in intertidal and shallow subtidal environments. Because of hydraulic controls, the inner parts of the beach were separated into microenvironments consisting of an oolitic bar, shoal and ridge. During the formation of the oolitic beach, the shoal, bar and ridge were exposed at several metres above sea level due to uplift or eustatic sea level changes. This allowed the percolation of meteoric water, with resulting dissolution of aragonitic ooids and cementation by two generations of calcites in the active freshwater phreatic zone (Longman 1980). The intra-oolitic leached pore spaces were filled partially with freshwater euhedral calcite and quartz. In the mixed saline-freshwater zone, the undissolved ooids were preserved and the inter-granular pore spaces cemented with sparry calcite. When the oolitic beach was buried, the subsurface freshwater level began to rise. The leached ooids in the freshwater zone were preserved, but those in the upper part of the mixed zones were further cemented with calcite. Finally, only the ooids in the upper part of the section remain leached. In the middle part, they occur in the form of "filled ooids" of single or multiple crystalline calcite, while in the lower part they remained undissolved. When the beach was buried below sea level, it was overlain by low energy conditions which formed another bed of lime mud which preserved the sequence described in the above section (Fig. 6).

Oolitic dolomite occurs in the ridge of beach C, which is limited on either side by the higher and larger bar and shoal of NNE strike. When the bar and shoal were exposed above seal level, the ridge was still submerged below the water level of the tidal channels, where connection with the open sea was poor, water circulation restricted and the evaporating volume rather high under conditions of shallow water and arid climate. The salinity and Mg/Ca ratio increased gradually

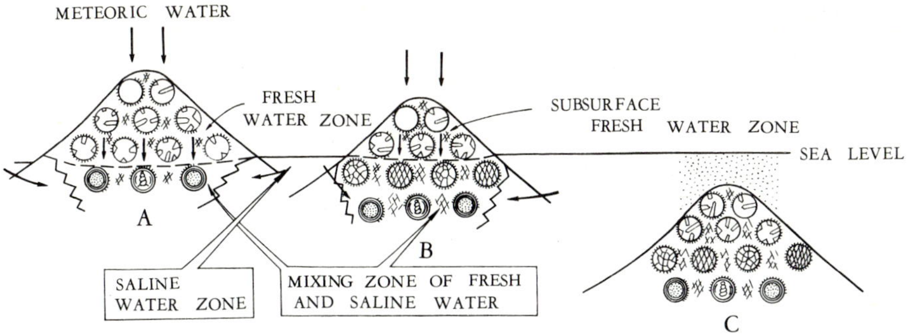

Fig. 6 A – C. Schematic section of the diagenetic evolution of the oolitic bar. **A** The oolitic bar is uplifted and exposed above sea level, resulting in leaching and dissolution of aragonitic ooids, by meteoric water. In the freshwater zone: intra-oolitic water creates pores which are filled partially by euhedral calcite and quartz after cementation in intergranular pores by first and second generation calcite in the active freshwater phreatic zone. In the mixing zone: undissolved ooids are preserved. The intergranular pores are cemented with sparry calcite. **B** The oolitic beach is buried. In the freshwater zone: leached ooids are preserved. In the mixing zone: in the upper part, the leached ooids, filled by euhedral quartz and calcite, are completely cemented with sparry calcite. In the lower part, undissolved ooids are also preserved. **C** The oolitic beach is buried below the sea level, and the beach is covered with lime mud. In the buried section below the lime mud the following types of ooids occur: (a) In the upper part: leached oolitic limestone. (b) In the middle part: filled single- or multiple-crystalline oolitic limestone. (c) In the lower part: undissolved oolitic limestone

so that the initial aragonitic ooids in the ridge as well as the first generation high magnesian calcite cement were subjected at an early diagenetic stage to thorough dolomitization forming dolomitic ooids and very fine crystalline dolomite replacement cement. At the same time, the surface of the ridge was covered with dolomitic lime mud. Later, when the bar and shoal on either side of the ridge were uplifted again and exposed above sea level, dissolution took place. The lime mud deposits on the ridge surface acted then as a barrier, preventing meteoric water from directly percolating downward. In the northern extension of the ridge, which was nearer to the bar and shoal, the intergranular pore spaces, influenced by the lateral percolation of subsurface freshwater from both sides, where cemented by freshwater calcite and dolomite at a later diagenetic stage and filled locally with secondary quartz (Fig. 7B) (Folk 1974). In the southern end of the ridge which was farther away from the bar and shoal, saline water percolated into the intergranular pore spaces which were then cemented successively with precipitated fine to coarse crystalline dolomite, anhydrite and celestite (Fig. 7C). These minerals, however, were unable to totally fill the intergranular pore spaces because of the limited supply. This has resulted in good remnant porosity and permeability in the rock.

4 Depositional Environment

The oolitic carbonates were originally oolitic limestones, and the varieties have resulted from diagenetic transformation. They were chiefly formed in intertidal

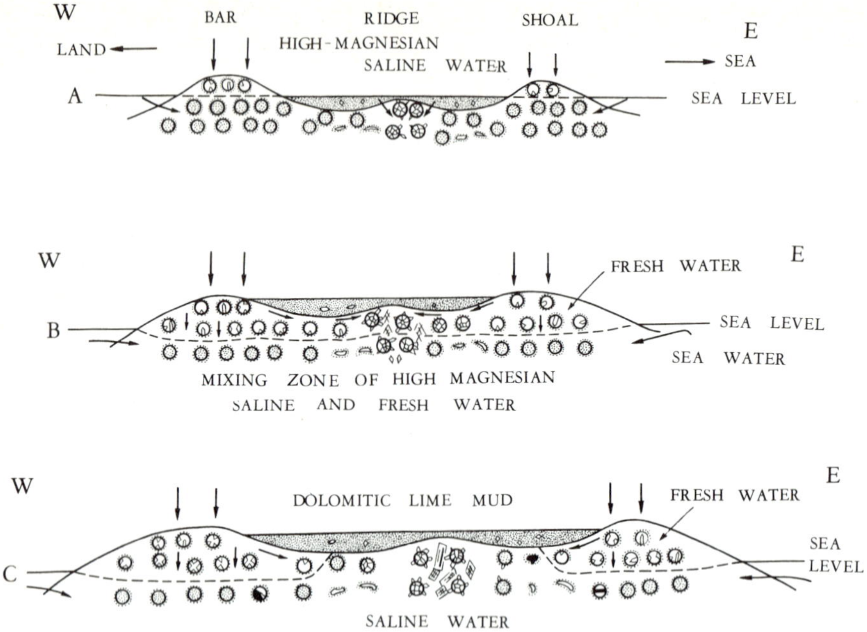

Fig. 7 A – C. Schematic section of the diagenetic evolution of oolitic dolomite. **A** The ridge was submerged and aragonitic ooids as well as first generation high magnesian calcite cements were dolomitized. **B** With the lowering of sea level the ridge was covered by lime mud and its northern end was influenced by the lateral penetration of subsurface freshwater from both the bar and shoal. This water provided a source of calcite and dolomite cements. In the high magnesian saline and freshwater mixed zone dolomite cements were formed. **C** With the continual lowering of sea level, the ridge became covered with lime mud and its southern end was removed from the influence of freshwater. The saline water in pore spaces was supersaturated, yielding precipitates of dolomite, anhydrite and celestite as cementing materials

and shallow subtidal environments, in a beach complex separated from the coast by a lagoon on the western side and adjacent to a shallow water basin on the eastern side.

The oolitic carbonate features are described below:

1. The ooids occur in globular or ellipsoidal shape.
2. The nulei are composed of pellets, rounded sand-sized detritus and foraminifers and occasionally shelly debris.
3. Most of the ooids are superficial ooids, the thickness of whose cortex is 1/10 – 2/5 of the radius. These ooids make up 30% – 49% of the rock. Only in beaches E and C do normal ooids which form 5% – 26% of the rock, occur.
4. Where the beach facies is thickest, 60% – 75% of the content is allochems, and of this ooids make up 60% – 100% of the content in the bar, 70% – 90% in the ridge, 61% – 84% in the shoal, 60% – 78% in beach G and 86 – 91% in overlying facies.

5. The average diameter of ooids is 0.22 – 0.33 mm, in beach C the maximum being 0.38 mm and in beach E 0.21 – 0.28 mm. The minimum diameter in beach G is 0.20 – 0.24 mm.
6. The standard deviation (B_1) of the oolites is 0.32 – 0.53, and the grains are well-sorted.
7. 10% – 25% of the rock content is sand-gravel sized intraclasts which are mostly angular to subangular in shape and are liable to plastic deformation. The exception is beach C where the intraclasts are subrounded to rounded.
8. The bioclasts are mainly of pelecypods and gastropods, forming several to 20% of the content, with echinodermal debris forming 1% to 2%.
9. Most of the oolitic limestones are cemented with sparry calcite which makes up 25% – 40% of the rock.
10. Leached oolitic limestone is developed in the shoal and bar as well as in beach G and oolitic dolomite prevails in the submarine oolitic ridge.

These features have led us to believe that beach G (Fig. 8) in this district originated in the shallowest subtidal zone. In the light of the studies of recent oolitic depositional environments (Bathurst 1971), it is believed that the depth of water was about 2 – 5 m. In the shallow water zone, evaporation was great due to the arid climate, and the water was often supersaturated with $CaCO_3$ and subjected to the influence of high and low tides and shoreward wind and waves. Because the energy of the high tidal race is greater than that of the low tidal race, the pellets, foraminifers, etc. on the beach were moved to and fro intermittently, resulting in the formation of a thin cortex around the grains. During the formation of beaches E and C (Fig. 8), the oolitic beach was moved into the intertidal zone because of a gradual regional regression, while the beach margin and the major tidal channel which was approximately of EW strike still lay in the subtidal shallowest water zone. Consequently, the microrelief of the oolitic beach was dissected. In the vicinity of the major tidal channel, the energy was so great that ooids were relatively concentrated, resulting in the formation of oolitic bars and shoals, while in the vicinity of the minor channel oolitic ridges and small shoals occur. The oolitic content of the bars, shoals and ridges is as high as 60% – 90% of the grains. The energy in the channel interior and the beach margin was comparatively small, so that mixed sediments of ooids, shelly and skeletal debris and sand-gravel sized intraclasts occur. Since the tidal race was subjected to the influence of submarine friction when passing through the shallow water basin in the east, the total energy of the tidal water reaching the beach was not great enough to transport the grains very far within the beach drainage system. Therefore the grains are mostly coated with thin cortex. However, during the formation of beach C when the tidal water advanced and regressed into the beach time and again because of the frequent occurrences of slight regional elevation and subsidence, the energy was relatively high and yielded some normal ooids with a thicker cortex, with sand-gravel sized debris with a greater degree of roundness. The bars, shoals and beach G were exposed above sea level during this time and subjected to erosion by meteoric water, forming several layers of leached oolites, while the ridge in beach C was protected on either side by a bar and shoal so that

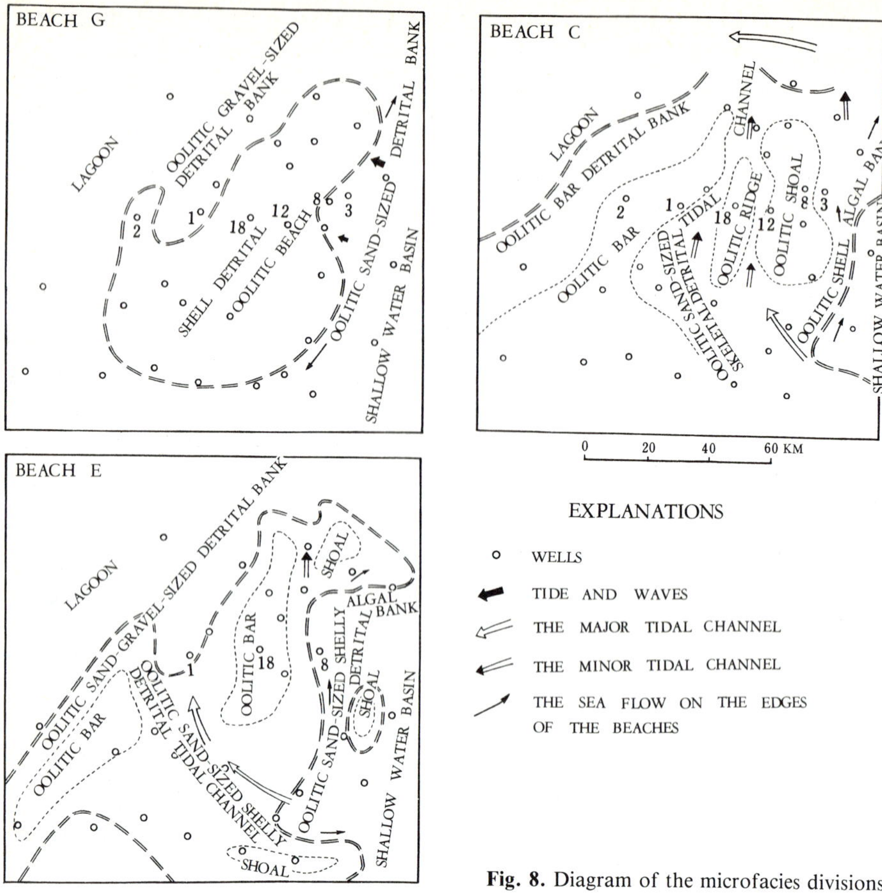

Fig. 8. Diagram of the microfacies divisions

high-magnesian saline water from the lower channels percolated down into the ridge with resulting dolomitization of ooids.

The oolitic limestones of this district are very similar in character to the sediments of mixed ooid-pellet-bioclastic grains with an oolitic content of 10% – 90%, which occur at depths of 2 – 5 m within the axial channel and oolitic beaches on the outer edges of the oolitic tidal delta along the Trucial Coast of the southern Persian Gulf (Loreau and Purser 1973).

5 Summary

1. The Lower Triassie oolitic carbonates of South Sichuan Basin occur mainly in the middle and upper parts of a carbonate-evaporite series of regressive nature.
2. The oolitic carbonates which include mainly oolitc limestone and a small quantity of oolitic dolomite were formed in intertidal and shallow subtidal en-

vironments located between a lagoon and a shallow water basin, controlled by the tectonic uplift.

3. During a regression, oolitic beaches were formed and exposed above sea level from time to time. Shoals and bars situated in the upper part of the section were frequently subjected to leaching by meteoric water ragonitic ooids were dissolved preferentially, resulting in extensive leached oolitic limestone with intragranular pore space.

The ridge in the central part of the beach were flanked and limited by a bar and shoals during exposure and were submerged in the saline water of channels, resulting in dolomitization of the ooids and the formation of oolitic dolomite with intergranular pore space. The leached oolitic limestone and dolomite are the main reservoir rocks in the formations of submember IV, T_1^2J1 in this district (see Cheng Tsung-lu et al. 1977), and they are controlled by the distribution of bars, shoals and ridges. This indicates that oolitic beaches confined to uplifted districts in subsiding basins of an epicontinental sea are promising zones for the discovery of high-yield oil and gas reserves.

Acknowledgements. The authors wish to acknowledge the helpful suggestions of Chriss Burgess and Tad Peryt who read the manuscript, and to express thanks to the Sichuan Administrative Bureau of Petroleum for the help in the research work.

References

Bathurst RGC (1971) Carbonate sediments and their diagenesis, Dev Sedimentology 12

Cheng Tsung-lu et al. (1977) Sedimentary characteristics and reservior behaviour of oolitic carbonates of submember IV, member I, formation II of the lower Triassic (T_1J1) in the ancient uplift district of Nanhua Basin, Peking, China

Folk RL (1974) The nature history of crystalline calcium carbonates: effect of magnesium content and salinity. J Sediment Petrol 44/1:40 – 50

Longman NW (1980) Carbonate diagenetic textures from near surface diagenetic environment. AAPG Bull 64:461 – 487

Loreau JP, Purser BH (1973) Distribution and ultrastructure of holocene ooids in the Persian Gulf. In: Purser BH (ed), The Persian Gulf. Springer, Berlin Heidelberg New York, pp 279 – 328

II.11 Ooid Zonation as Indication for Environmental Conditions in a Givetian-Frasnian Carbonate Shelf-Slope Transition

T. J. A. Reijers [1] and A. H. M. ten Have [2]

1 Introduction

The Givetian-Frasnian Portilla Limestone Formation is well exposed in the Cantabrian Mountains in N.W. Spain (Fig. 1). It is sandwiched between the siliciclastic underlying Huergas and overlying Nocedo Formations. The member subdivision is area-dependable (Fig. 2). Six facies have been recognised (Reijers 1972); all reflect deposition in a shallow marine environment on an extensive carbonate platform and its slope. One facies is siliciclastic. It predominates in one member (C in the central and eastern and B in the western part of the study area; Fig. 2) and straddles the Givetian-Frasnian boundary. This unit reflects a change in the sedimentation. It is the result of a Givetian-Frasnian pulse of reactivation of the hinterland supply area. Neither this member nor this facies will further be considered in the present paper.

Members B and D are mainly composed of organic framework carbonates (Fig. 3a). Detailed studies of selected sections and cross sections (Reijers 1980) allowed recognition of sedimentary mechanisms in these units. These reflect selective sea-level or sea-bottom changes. The sedimentation patterns of facies in these members can be described in space and time with the aid of a sedimentation model (Reijers 1972) and with the recognised sedimentary mechanisms (Reijers 1980).

Fundamental fault lines delineate the carbonate platform. They control the areal distribution of facies-groups. The León-line, for example, separates a northerly area of non-deposition (the Asturo-Leonese geanticlinal land area) from the southerly carbonate shelf.

In the same area the Sabero-Gordón fault-zone acts as a hinge-line south of which rapid subsidence took place causing the growth of juxtaposed reefoid carbonate bodies (Reijers 1974).

These suggestions are supported by observations and interpretations made in the organic framework carbonates in members B and D (see also Reijers 1972, 1974, 1980). However, it is still problematic to understand what environmental conditions triggered carbonate sedimentation. Neither is it self-evident that such environmental settings as those prevailing in members B and D were present right from the beginning of Portilla carbonate production.

1 Shell Internationale Petroleum Maatschappij, The Hague, The Netherlands
2 Koninklijke/Shell Exploratie en Produktie Laboratorium, Rijswijk, The Netherlands

Coated Grains (ed. by T. M. Peryt)

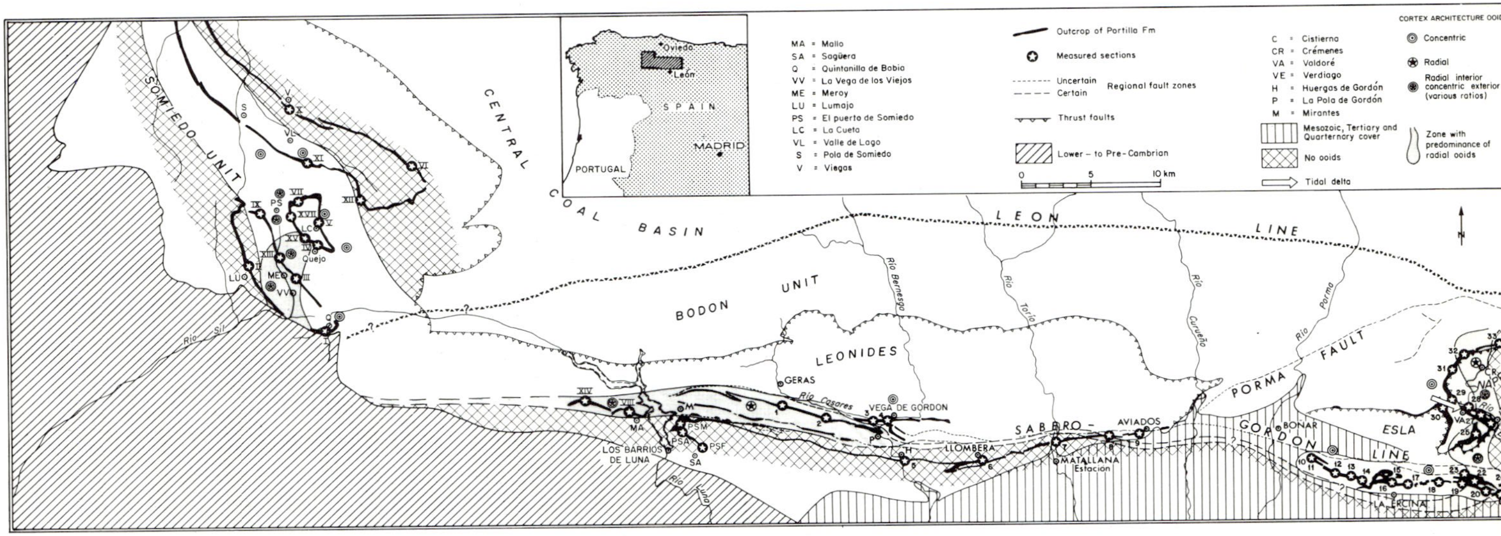

Fig. 1. Outcrop and index map of area studied, showing zonation of ooids in carbonate-clastic lower beds of Portilla Limestone Formation

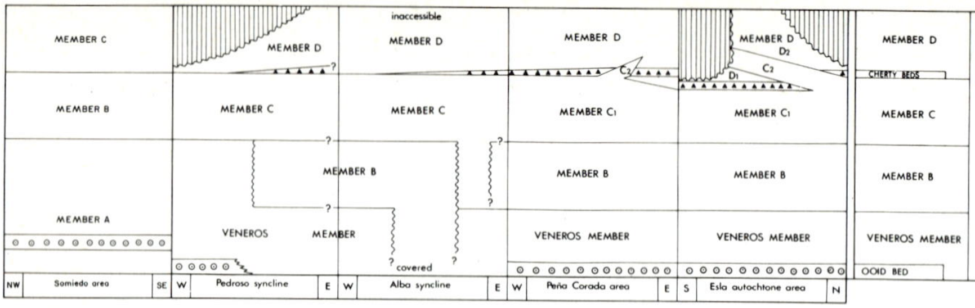

Fig. 2. Lithostatigraphic correlation of recognised units in study area

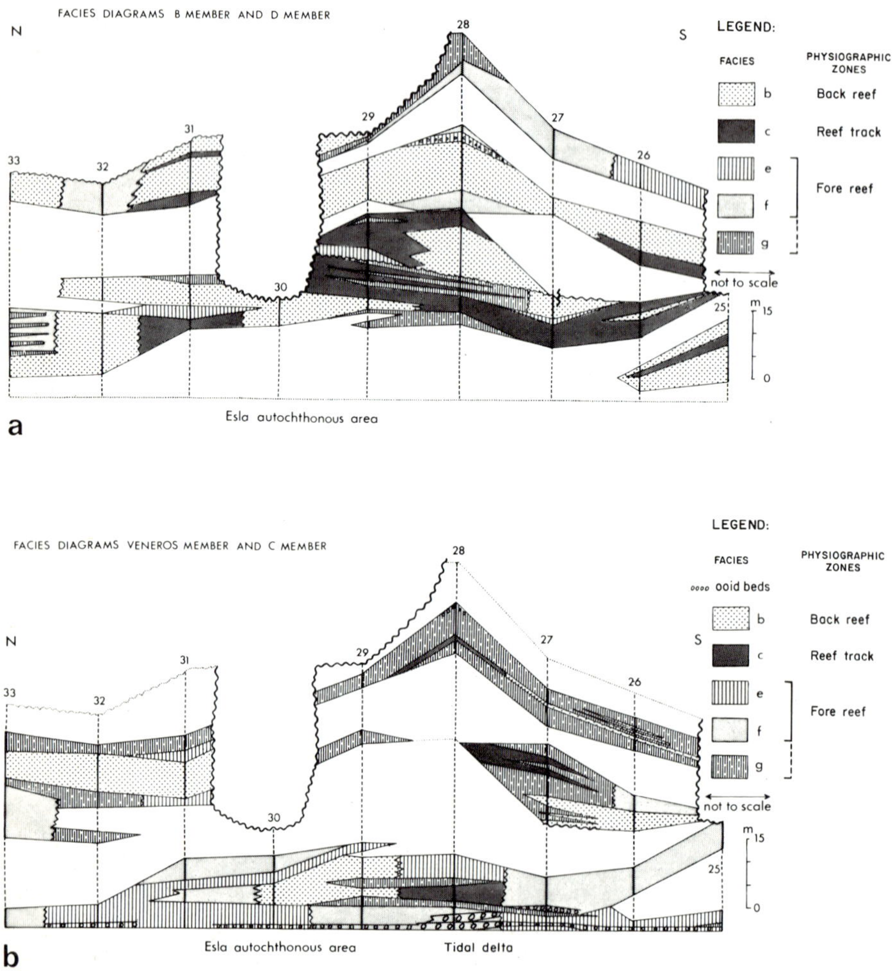

Fig. 3. a Lateral correlation of facies recognised in Members B and D in the Esla area. **b** Lateral correlation of facies recognised in the Veneros Member and Member C in the Esla area

The solution of these uncertainties is hidden within the ostensibly monotonous carbonate-clastic lowermost unit of the Portilla Formation (i.e., the Veneros member or the lower half of member A; Fig. 2). In several places the first beds of this unit are oolitic. In the Somiedo area, however (Figs. 1 and 2), cross-bedded bioclastic grainstones precede oolites. These basal beds herald the change in sedimentary regime and to a lesser degree in depositional environment. They will be studied in the present paper with the object of deriving an answer to the question: what was the environmental setting at the transition Huergas-Portilla sedimentation and what environmental condition prevailed?

2 The Carbonate-Clastic Member

The lowermost unit of the Portilla Formation (the Veneros Member and its equivalent lower half of member A in the Somiedo area) is dominated by encrinal bryozoan grainstones (facies e; Reijers 1972) and by marly encrinal brachiopodal packstones (facies f) as can be seen in Fig. 3b. Both facies are carbonate-clastic. In addition oolites occur in this unit.

Facies e reflects an open shallow marine depositional environment in which crinoid-bryozoan meadows occurred with prolific organic growth. Desintegration products of these organisms were commonly swept together in cross-bedded encrinite grainstone bars. In particular in the Somiedo area these are a conspicuous element in the blanket deposits. In several places this deposit is further characterised by the presence of channels and lateral pinch-outs. The crinoid ossicles are commonly of various size. The size classes occur together, suggesting transport as bed load for the coarser particles, and in suspension for the finer particles in areas where currents were strong enough to entrain such particles.

Facies f reflects an open marine depositional environment that is slightly deeper than the one in which facies e was deposited. The average hydraulic energy here was lower. As a consequence the depositional texture of the carbonate is muddy but it remains overall grain supporting. Fan-shaped bryozoans act as sediment binders. The faunal habitat in places reveals some depth restriction. Siliciclastic material occurs both admixed and interbedded in this facies. It is the cause of the wavy, irregular bedding aspect and the slight nodularity of the facies and it gives the carbonate commonly a marly aspect.

Ooids are concentrated in oolite beds. Near Valdore the oolite reaches a thickness of 2.8 m (Fig. 3b) and near Quintanilla de Babia 2.5 m. The oolite occurs exclusively in facies e and f but does not show any preference for either facies. Locally it has a bimodal size distribution.

Oolites are cross-bedded and slightly wavy and well developed bedding planes and channels are commonly present (Figs. 4 and 5). The intergranular space in oolite beds is filled with a drusy and blocky cement generation (Fig. 6). Colloform, clotty haematite is frequently concentrated between the various enveloping layers making up the cortex of the ooid. It then produces "dark haloes" within the ooids, contrasting with the "light haloes" immediately surrounding the ooids as a result of the presence of drusy, blade-shaped crystals. The nuclei of ooids

Fig. 4. Section near Valdoré. Note intraformational unconformity (*arrows*) below which, in the lower unit of the Portille Formation, a thick, cross-bedded oolite is present. Channels and cross-bedding structures are common

Fig. 5. Detail of Fig. 4. Edge of a channel in which minor slumping features have been seen

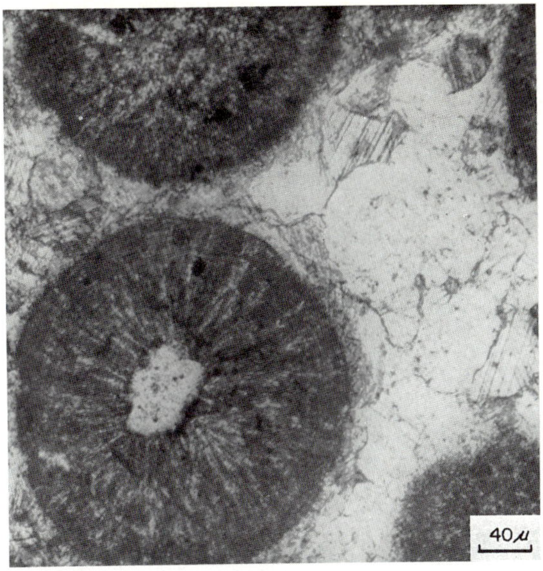

Fig. 6. Ooids with radical cortex architecture. Note the faint onset of concentric crystal orientation at the exterior. The ooid is surrounded by irregular drusy cement. The remainder of the intergranular space is taken by blocky cement. Plane light

Fig. 7. Ooids with concentric cortex architecture. Note the thin inner part of the cortex in which radial crystal orientation prevails. Note the irregular outline of the ooid, giving it a pisoid-like appearance, and the dark layers due to haematite clots. Plane light

consist either of skeletal fragments or of quartz grains. The cortex organisation of the ooids is commonly normal or superficial (with only one cortex layer); occasionally no distinct cortex is present. When normal, several cortex layers are present in the ooids. Ooids have radial (Fig. 6), concentric (Fig. 7) or composite cortexes (Fig. 8). Composite cortexes are scarce in the eastern and central area but common in the west. Where present the radial cortex layers occur at the inside and the concentric cortex layer at the outside (Fig. 8). Ooids with these internal cortex fabrics occur in distinct zones throughout the study area (Fig. 1).

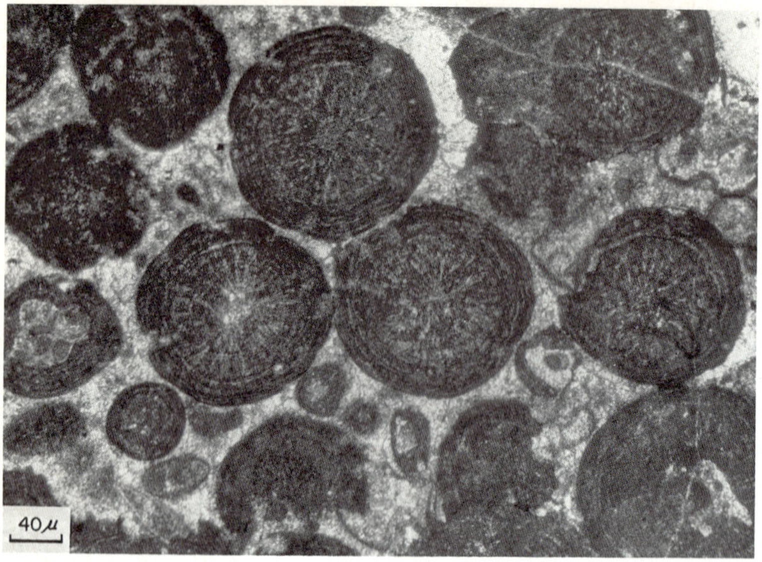

Fig. 8. Ooids with mixed radial/concentric cortex architecture. Note that in all cases the radial coating occupies the inner part, whereas the concentric coating forms the outer part of the cortex. Plane light

3 Environmental Interpretation of Ooids

In an agitated shallow marine warm environment with water saturated with respect to $CaCO_3$, slow precipitation around nuclei results in the formation of cortexes with a concentric internal structure. This has been observed in several recent environments and in laboratory test as have recently been summarised by Heller et al. (1980). Similar environmental conclusions have been derived from such cortex fabrics in ancient coated grains (see Heller et al. 1980).

In slightly deeper settings further away from the coast and down the slope of the carbonate platform, the sea may be weakly agitated but could still be saturated with respect to $CaCO_3$. Here, slow precipitation around nuclei results in radial fabrics in ooid cortexes. Various authors (Rusnak 1960, Loreau and Purser 1973, Heller et al. 1980) have observed similar ooids in comparable environments and settings both in the recent and in the past. Davies et al. (1978) reproduced such ooid-fabrics in the laboratory.

Another ooid-type has a mixed cortex. The interior layers have a radial crystal growth fabric, the exterior ones a concentric one. Heller et al. (1980) interpret such composite ooids in terms of variable hydrodynamic behavior. Radial crystal growth reflects a period of weak agitation in which the comparatively small ooid resides predominantly in suspension and radial crystal growth occurred. Transport occurs rarely. With increasing agitation (perhaps as a result of a slight regressive tendency) the grain could roll about and concentric layers with tangentially oriented crystals are added. Thus the grain grows and this larger grain can participate in sediment transport as bed load. The result is a mixed end-product with bimodal grains having radial or composite cortex fabrics reflecting a shallow marine agitated depositional environment. This is commonly present in the Portilla oolite. The pure radial ooid in contrast reflects a low-energy slightly deeper depositional environment, is less common and is restricted to a distinct [illegible] in the study area (Fig. 1).

In the Portilla Formation all three ooid-types occur in the oolite at or near its base. This base is progressively younger away from the landmass formed by the Asturo-Leonese geanticline and the Porma high. (For paleontological evidence see Reijers 1972.)

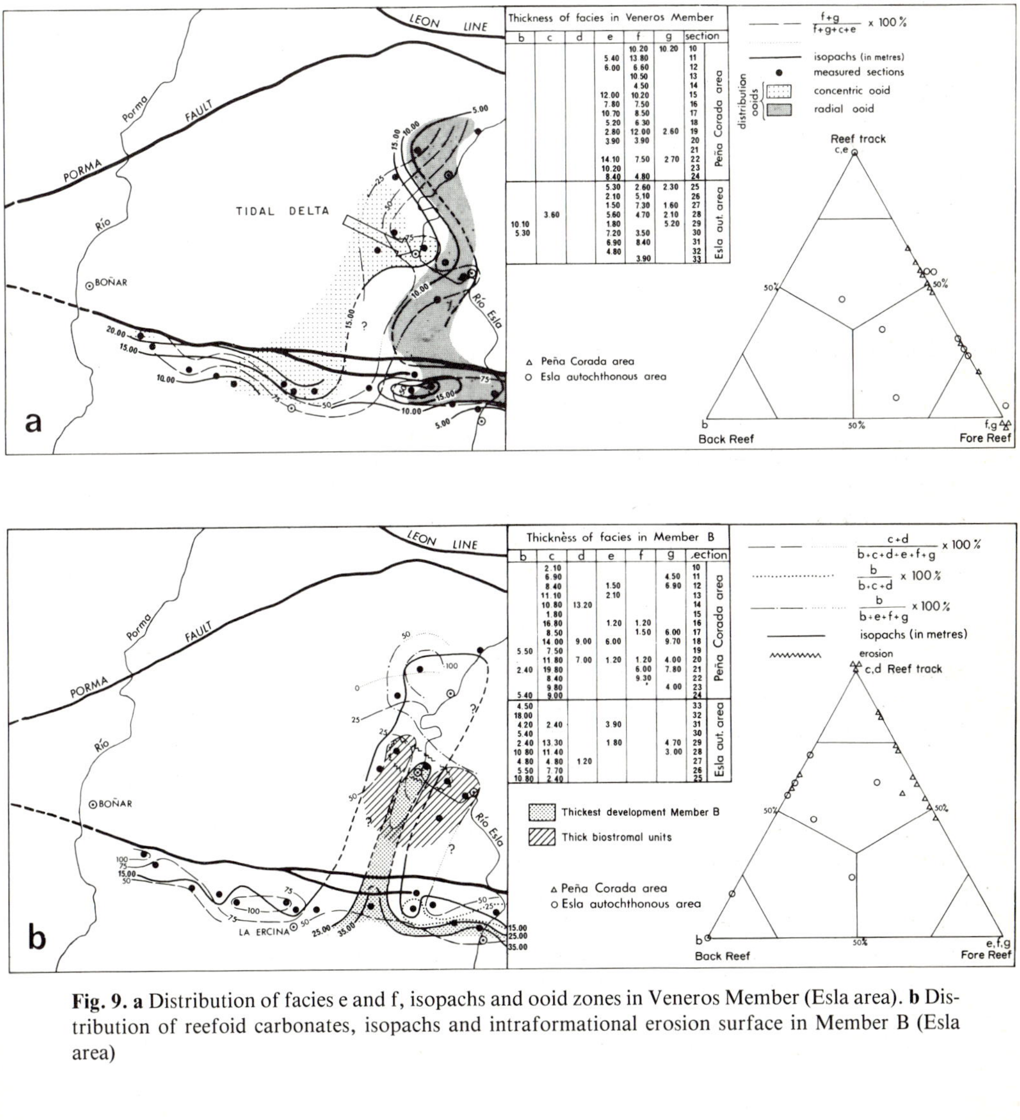

Fig. 9. a Distribution of facies e and f, isopachs and ooid zones in Veneros Member (Esla area). **b** Distribution of reefoid carbonates, isopachs and intraformational erosion surface in Member B (Esla area)

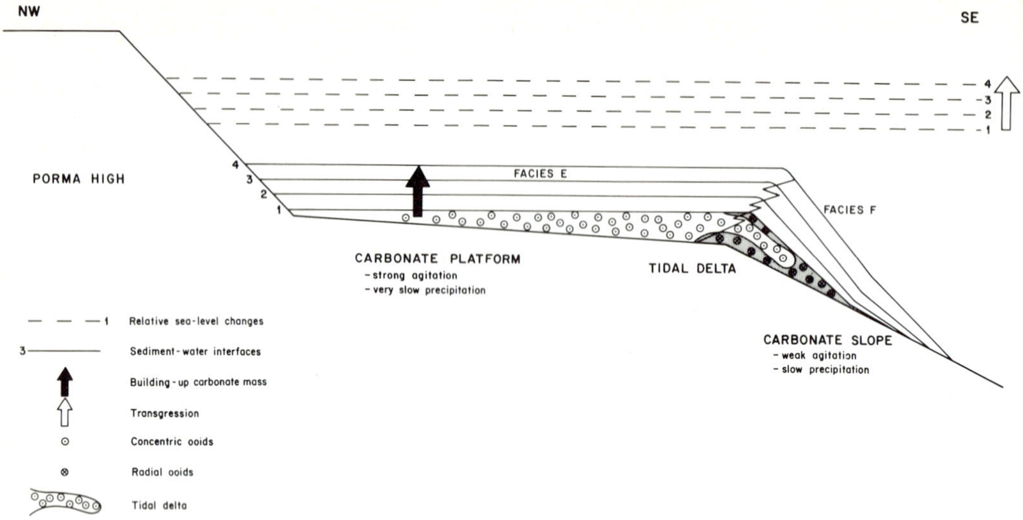

Fig. 10. Hypothetical cross-section through carbonate platform on which lowermost beds of Veneros Member east of Porma high are deposited. Note tidal delta in Valdore

A number of additional observations in the carbonate clastic unit of the Portilla Formation help to further reconstruct the environmental conditions prevailing during the deposition of the initial beds.

Moderate to high hydraulic energy deposits (facies e) laterally give way to moderate to low hydraulic energy deposits of facies f. In the same direction isopach values decrease (example Esla area Fig. 9a). Thus the thickest part of this unit is closest to land and consists of encrinal bryozoan grainstones reflecting a vast shallow platform with prolific crinoid-bryozoan growth. The thinnest part of this unit is farther away from land and reflects a platform slope in which siliciclastics are admixed and brachiopods are more common. In the former part of the Veneros Member the ooids are concentric or composite, in the latter part radial. Thus the initial beds of the carbonate clastic unit of the Portilla Formation reflect already environmental differences which persist throughout this formation, as will be discussed below.

In the smooth isopach contours (Fig. 9a) of the Veneros member in the Easla area an irregularity coincides with an unusual thick oolite interval near Valdore. The ooids here have concentric cortexes in a location where radial ooids would be expected. Channels and cross beddings (Figs. 4 and 5) are clearly recognisable and indicate high hydraulic energy conditions. It is speculated that this "bulge" in contours indicates the site of a tidal delta of concentric ooids sands spilling over the edge of carbonate platform towards the east (Fig. 10) where they interfere with the belt in which radial ooid occur. Similar tidal delta ooid accumulations are known to exist in the present-day Persian Gulf and have extensively been described by Purser and Evans (1973). The tidal delta disturbs locally the smooth ooid-belt arrangement that appear to describe contour-wise the slope of the seabottom going down towards the southeast.

After having established environmental differences that are reflected in the basal beds of the Portilla Formation, it is here attempted to correlate these environmental differences with those reflected in reefoid members. This correlation links the environmental response of fundamentally different carbonate deposits, carbonate-clastic ones and organic-framework ones, and hints at events that influence the environment locally but concistantly.

4 The Reefoid Members

Members B and D are predominantly composed of frame-building organisms. The (im)balance between the growth potential of organisms and the relative sea-level fluctuations trigger sedimentation mechanisms in these units. The results of such (im)balances are carbonate masses that are building-in (grading towards the coastline), building-out and building-up (Reijers 1980). Building-in and building-out sedimentary mechanisms are the principle ones that govern facies distribution. The resulting facies mosaics readily demonstrate the local and temporal presence of reefoid bodies laterally grading into facies of the back-reef or of the fore-reef belts.

At first glance the distribution of facies over the shallow carbonate platforms appears random. However, lumping the facies into groups that reflect the main physiographic zones results in three end-members for a 100% facies triangle. The back-reef zone is situated north of the reef track; the fore-reef to the south.

In the Esla area, e.g., reef-track sediments are present from Valdoré to Cistierna and from there they follow, with some interruptions, the Sabero-Gordón fault-zone (Fig. 9b). From Valdoré towards the south the reef-track is biostromal (Figs. 3 and 9b) and composed of sediment-trapping branching tabulate corals. These groups combined characterise the stage of colonisation of an organic carbonate body in the model of Alberstadt et al. (1974). Along the Sabero-Gordón fault-zone building-up bioherms predominate and the colonisation stage mentioned above is frequently followed by a stage of domination, as is expressed by the presence of additional great amounts of massive, frame-building tabulate and rugose corals and stromatoporoids.

5 Organic Framework Units

The organic framework carbonates present in member B show a number of features that support the environmental conclusion reached in the carbonate-clastic lowermost unit of the Portilla. The main conclusion is that the seabottom slopes down towards the east and the south, and that variations in environmental conditions as a result of this are reflected in facies groups occurring in specific physiographic zones.

As frame-building carbonates tend to accentuate and exaggerate pre-existing reliefs the thickest development of member B is expected to coincide with the break in the slope (Fig. 9b). It is also expected that here biostromal (facies c) carbonates reach their maximum thickness. The frame-builders occasionally reached the sealevel and were locally and temporarily exposed. Mild epeirogenetic bottom movements might have triggered this. Local intraformational unconformities are a result (Fig. 4). The belt of frame-building carbonates in Member B strikes approximately parallel to the belts of ooids in the Veneros Member. Thus they too demonstrate the presence of a slightly sloping seabottom. The upslope predominance of lagoonal deposits and the downslope occurrence of forereef facies further support this conclusion.

6 Conclusions

In a shallow marine setting in which carbonate-clastic deposits are followed by frame-building units a number of environmental trends aid in establishing a sloping seabottom The lithofacies reflect shelf and slope settings. The ooids have developed cortex fabrics that appear to be a remarkably sensitive hydraulic energy gauge. Consequently the distribution of ooids in belts follows narrowly the palaeocontours of the slope of the depositional basin. The reasonableness of this conclusion has been checked with the facies distribution of the overlying organic framework unit in which the facies distribution reflects a similar sloping seabottom. The facies sensitivity of ooid cortexes might prove to be more common than in this case alone.

Thus, a criterium has been established to recognise shelf-slope transitions.

Acknowledgement. The authors are indebted to Shell for permission to publish this paper.

References

Alberstadt LP, Walker KR, Zurawski RP (1974) Patch reef in the Carters Limestone (Middle Ordovician) in Tennesse and vertical zonation in Ordovician reefs. Geol Soc Am Bull 85:1171 – 1182

Davies PJ, Bubela B, Ferguson J (1978) Transport processes in ooid genesis. J Sediment Petrol 50 3:943 – 952

Heller PL, Komar PD, Pevear DR (1980) Transport processes in ooid genesis. J Sediment Petrol 50 3:943 – 952

Loreau JP, Purser BH (1973) Distribution and ultrastructure of holocene ooids in the Persian Gulf. In: Purser BH (ed) The Persian Gulf-holocene carbonate sedimentation and diagenesis in a shallow epicontinental sea. Springer, Berlin Heidelberg New York, pp 279 – 328

Purser BH, Evans G (1973) Regional sedimentation along the Trucial coast, SE. Persian Gulf. In: Purser BH (ed) The Persian Gulf-holocene carbonate sedimentation and diagenesis in a shallow epicontinental sea. Springer, Berlin Heidelberg New York, pp 211 – 231

Reijers TJA (1972) Facies and diagenesis of the Devonian Portilla Limestone Formation between the river Esla and the Embalse de la Luna, Cantabrian Mountains, Spain. Leidse Geol Meded 42 2:163 – 249

Reijers TJA (1974) Hinge movements influencing deposition during the Upper Devonian in the Esla area of the Cantabrian Mountains, Spain. Geol Mijnbouw 53 1:13 – 21

Reijers TJA (1980) Sedimentary mechanisms in Spanish Devonian Carbonates. Geol Mijnbouw 59 1:87 – 96

Rusnak GA (1960) Some observations of recent oolites. J Sediment Petrol 30:471 – 480

II.12 Ooids and Oolites of the Proterophytic Boomplaas Formation, Transvaal Supergroup, Griqualand West, South Africa

N. J. Beukes [1]

1 Introduction

The Boomplaas Formation represents a newly recognized carbonate platform deposit (Beukes 1978) below the Campbellrand carbonate platform sequence in the lower part of the Transvaal Supergroup on the Kaapvaal craton in Griqualand West, South Africa (Fig. 1A). Deposition of the Transvaal Supergroup commenced more than 2300 m.y. ago (Button 1976) making the Boomplaas Formation one of the oldest known carbonate platform deposits in the world.

Oolite rock units composed of concentrically laminated ooids make up an important part of the Boomplaas Formation, the stratigraphy of which is described by Visser and Grobler (1972) and Beukes (1979). This paper supplies information regarding the character of the ooids and oolite rock units, their distribution, environments of formation, deposition and diagenesis.

2 Stratigraphic Setting

The Boomplaas Formation is uncorformably underlain by argillite and arenite of the Vryburg Formation and overlain by shale of the Lokammona Formation. It reaches a maximum known thickness of 100 m along the Griquatown fault zone southwards of which it interfingers with argillite of the Vryburg Formation in the Niekerkshoop basin (Fig. 1B, C). The formation thins northwards onto the interior of the Kaapvaal craton and eventually pinches out by interfingering with arenites of the Vryburg Formation on the Ganyesa dome (Fig. 1C). The Ganyesa dome acted as an internal source of siliciclastic material during deposition of the Boomplaas Formation. Similarly an external siliciclastic source, which led to the formation of quartz arenite in the southern facies of the Vryburg Formation, was present towards the south of the Doringberg fault zone along the extreme southwestern margin of the Kaapvaal craton (Fig. 1B). All these features suggest that the Boomplaas carbonate platform deposits prograded from north-east, on the Kaapvaal craton, towards the south-west into the Niekerkshoop basin.

1 Rand Afrikaans University, Department of Geology, P.O. Box 524, Johannesburg 2000, South Africa

Coated Grains (ed. by T.M. Peryt)

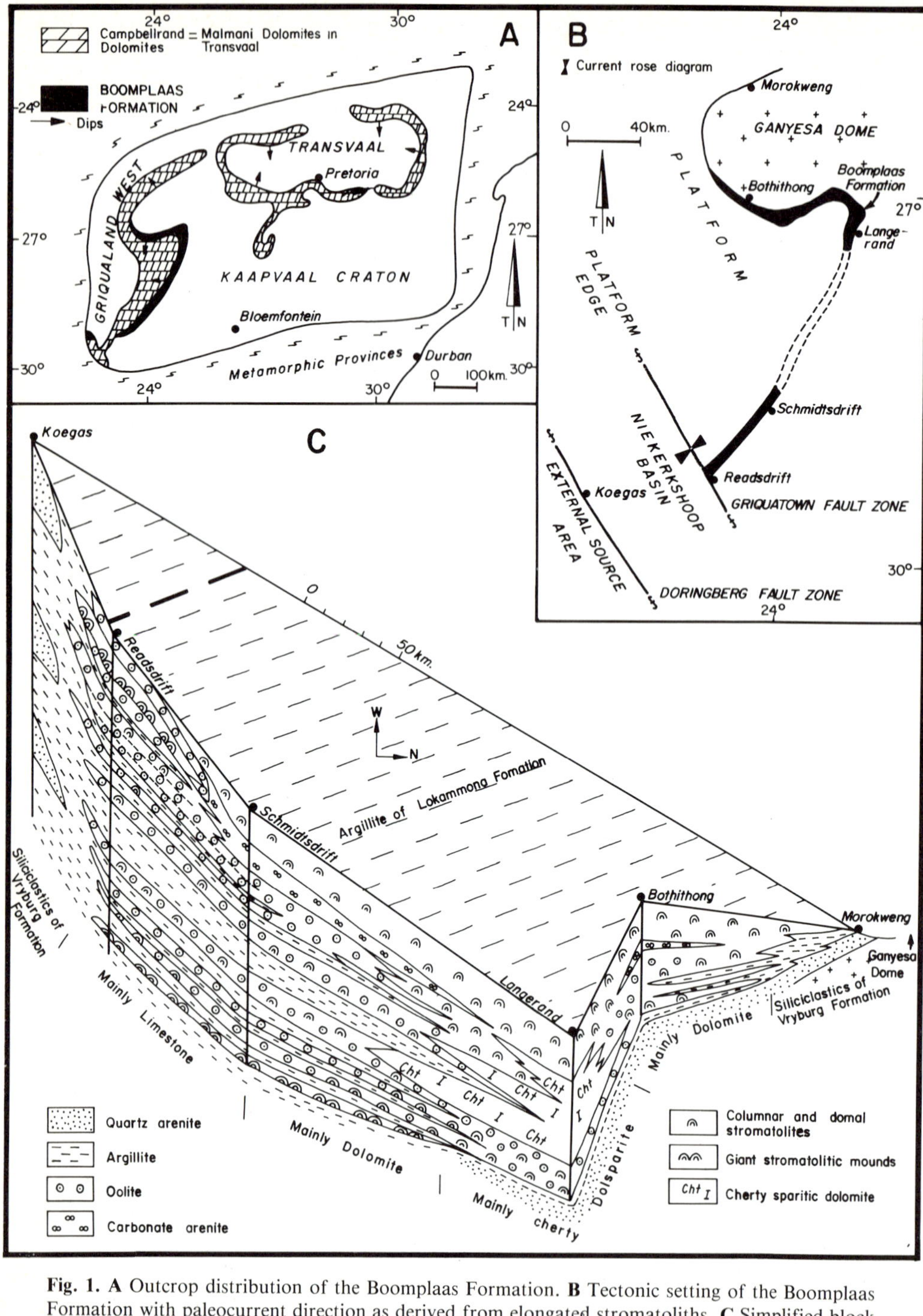

Fig. 1. A Outcrop distribution of the Boomplaas Formation. **B** Tectonic setting of the Boomplaas Formation with paleocurrent direction as derived from elongated stromatoliths. **C** Simplified block diagram of the Boomplaas Formation indicating interfingering with siliciclastics of the Vryburg Formation

Oolite rock units are best developed in the south-west along the Griquatown fault zone near Readsdrift. In this area the oolites are on average between 1 m and 3 m thick with one unit reaching a thickness of 6 m. Northwards, onto the platform interior, the oolite units thin to a thickness of less than 1 m on average. In this north-eastern part of the outcrop area supermature quartz grains, derived from the Ganyesa dome, are present in the oolite units (Beukes 1977).

Along the platform edge, in the south-west, the oolite units are interbedded with cryptalgal laminated giant stromatoliths (mounds), cross-laminated carbonate arenites, edgewise carbonate conglomerates and argillite. On the platform interior most of these units pinch out and the oolites are associated with columnar and domal stromatoliths and cherty sparitic dolomite, the latter especially well developed at Langerand (Fig. 1C). The carbonate rocks consists of limestone with minor dolomite in the south-west and only of dolomite in the north-east (Fig. 1C).

The oolite units of the Boomplaas Formation are well sorted and represent oosparites (Fig. 2A) containing hydrodynamic structures such as cross-lamination and horizontal lamination. This is an indication that the ooids were deposited as free-rolling particles in a high-energy environment capable of sorting them. The environment of formation of such ooids may differ from their environment of deposition (Loreau and Purser 1973). The physical character of individual ooids are to some extent diagnostic of their environment of formation, whereas, the character of oolite rock units and their relationship to other lithofacies types would be more diagnostic of their environment of deposition. These aspects are discussed separately in this paper.

3 Physical Character of the Ooids and Their Possible Environments of Formation

The ooids of the Boomplaas Formation range in size from microoids (<2 mm in diameter) to pisooids of up to 5 mm in diameter. They display differences in physical character, some of which are the result of secondary recrystallization, silicification, dolomitization and compaction. However, in many ooids an idea of their original structure and texture can still be obtained and these are to be concentrated on in this section. Such ooids are spherical to nearly spherical and irrespective of size consist of a nucleus surrounded by a concentrically laminated cortex (Fig. 2A).

The nuclei invariably consist of monocrystalline or polycrystalline carbonate grains suggesting that the ooids formed in an environment remote from siliciclastic influx. Even where quartz grains are present in the oolite, as at Bothithong, the ooids have nuclei of carbonate (Fig. 2B), indicating that they became mixed with the quartz grains only after formation (Beukes 1977). Loreau and Purser (1973) found that those ooids in the Persian Gulf forming in close proximity to terrigenous influx have nuclei of siliciclastic grains whereas those forming in areas remote from such influx have autochthonous carbonate nuclei.

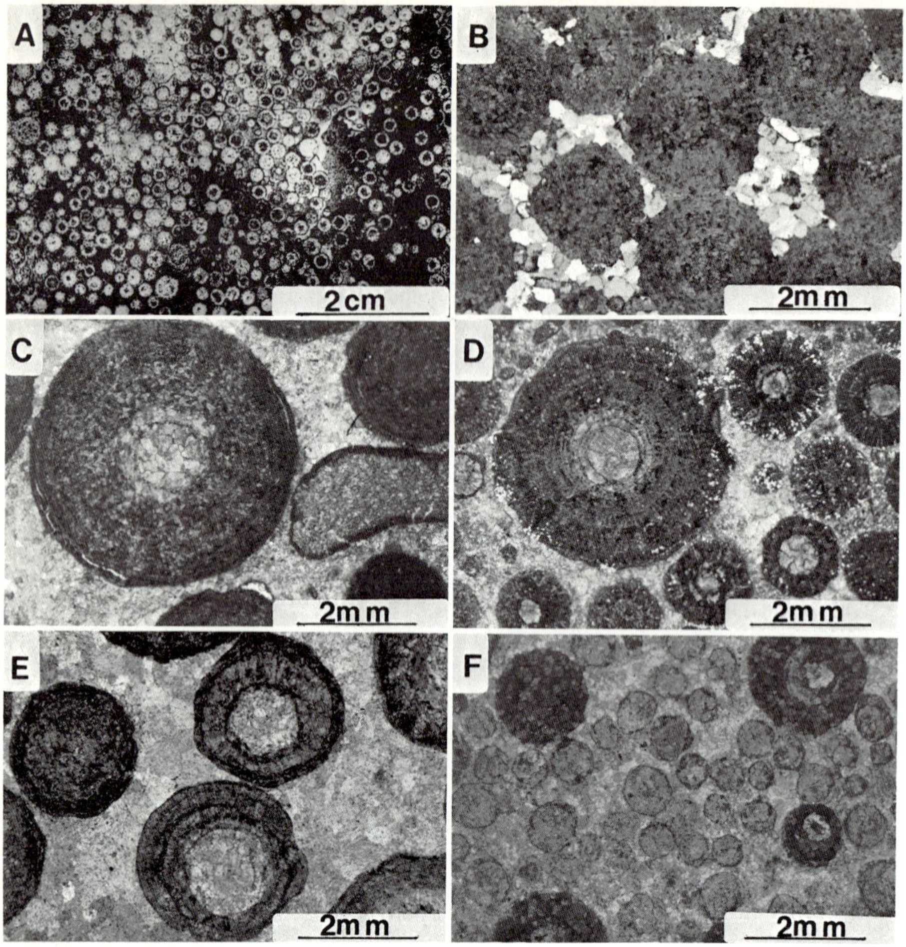

Fig. 2. A Print of a stained acetate peel of a partly dolomitized oosparite. *Black* is calcite and *white* is dolomite. Note that some ooids are completely dolomitized whilst in others the nucleus consists of calcite and the cortex of dolomite. Sample stained with Alizarin Red-S and potassium ferricyanide after the method of Davies and Roger (1968). **B** Quartz-free dolomitic ooids in a matrix of quartz arenite. Note that primary structure of ooids has been obliterated almost completely by dolomitization. **C** Pisooid with a concentric laminated cortex surrounding a polycrystalline carbonate nucleus. Note coated intraclast on *right*. **D** Radial calcite crystals replacing the concentric laminated cortexes of ooids. Borders of ooids and the radial calcite crystals are in turn replaced by white quartz. **E** Radial-concentric laminated ooids. Note asymmetry. **F** Infant limestone ooids mixed with larger dolomitized more mature ooids

Two types of "primary" cortexes are recognizable. The first type is semi-opaque and cryptocrystalline (micritic) and consists of regular, smooth, very thin (<0.01 mm) concentric laminae (Fig. 2C). Such cortexes are comparable to those of present day ooids which form in high-energy agitated environments such as tidal deltas and tidal channels of the Persian Gulf (Loreau and Purser 1973), oolite shoals on the Bahama carbonate bank (Illing 1954, Newell et al. 1960,

Shoji and Folk 1964, Fabricius and Klingele 1970), and the coastlines of the Gulf of Suez (Fabricius and Klingele 1970) and Lagune Madre (Rusnak 1960). In all these Recent examples, aragonite needles tend to be tangentially orientated to the surfaces of the ooids and most probably it is this texture which is still preserved in some of the calcitic and/or dolomitic ooids of the Boomplaas Formation. In some concentric-textured ooid cortexes the laminae are transversed by radial-textured calcite or dolomite crystals (Fig. 2D) developed during recrystallization. Such transverse radial structures are common amongst partly recrystallized calcite-bearing aragonite ooids of Salt Lake (Eardley 1938) and amongst recrystallized ancient ooids (Greensmith 1978).

The second type of cortexes consist of relatively thick (0.05 – 0.4 mm) sparitic radial-textured laminae alternating with very thin cryptocrystalline (micritic) semi-opaque micro-laminae (Fig. 2E). It is possible that the radial laminae formed in more protected environments such as for example below wave base in deeper water; lagoons or tidal flats. Concentric laminae only formed when the ooid was moved to a high energy environment. An analogous process of formation is suggested by Loreau and Purser (1973) for radial-textured Recent ooids of the Persian Gulf. Some radial-concentric Boomplaas ooids are asymmetrical with the nucleus off-centre and the cortex thicker on one side (Fig. 2E). Similar asymmetrical ooids are described by Freeman (1962) from Lagune Madre. The asymmetry may have developed through preferential upward growth whilst the ooid was at rest in a protected environment.

Infant ooids with only one coat surrounding well-rounded carbonate sand particles (Fig. 2F) are also present in the Boomplaas Formation. Such infant ooids may be mixed with more mature ooids in bimodal oolite units (Fig. 2F) or may form the base of upward-coarsening microoolite (Fig. 3A) to pisoolite rock units. The sizes of nuclei of infant ooids corresponds to that of nuclei of the mature ooids suggesting a link in origin. The size of nuclei varies between samples but maximum diameters observed in infant ooids are in the order of 0.5 – 0.75 mm. Smaller nuclei are sometimes not coated suggesting some optimum minimum size necessary for ooid formation. This is however difficult to prove because in thin section the observed nucleus to cortex ratios and nucleus diameters depend upon the position of the section in the ooid.

Giant Boomplaas ooids with diameters between 2 and 5 mm have nuclei of up to 2 mm in diameter (Fig. 3B). These nuclei are coarser than the average maximum grain size of about 1 mm which recent high energy ooids reach (Bathurst 1975). According to Bathurst (1975) this maximum size limit is probably controlled by equilibrium between inorganic precipitation of $CaCO_3$ on, and abrasion of the larger ooids. However, the Boomplaas giant ooids are concentrically textured (Fig. 3B), suggesting formation in a high-energy environment which could mean that their formation was controlled by some other process – most probably. The semi-opaque, micritic character of the cortexes of the giant ooids is probably the result of organic inclusions. It is well known that many Recent ooids are covered by micro-organisms, but whether these have a direct, indirect or no influence on ooid growth is still debated (Bathurst 1975). Organic coatings like blue-green algal mats could, however, have protected the Boomplaas ooids against abrasion, in this way allowing growth to giant sizes. At the same time the algae could

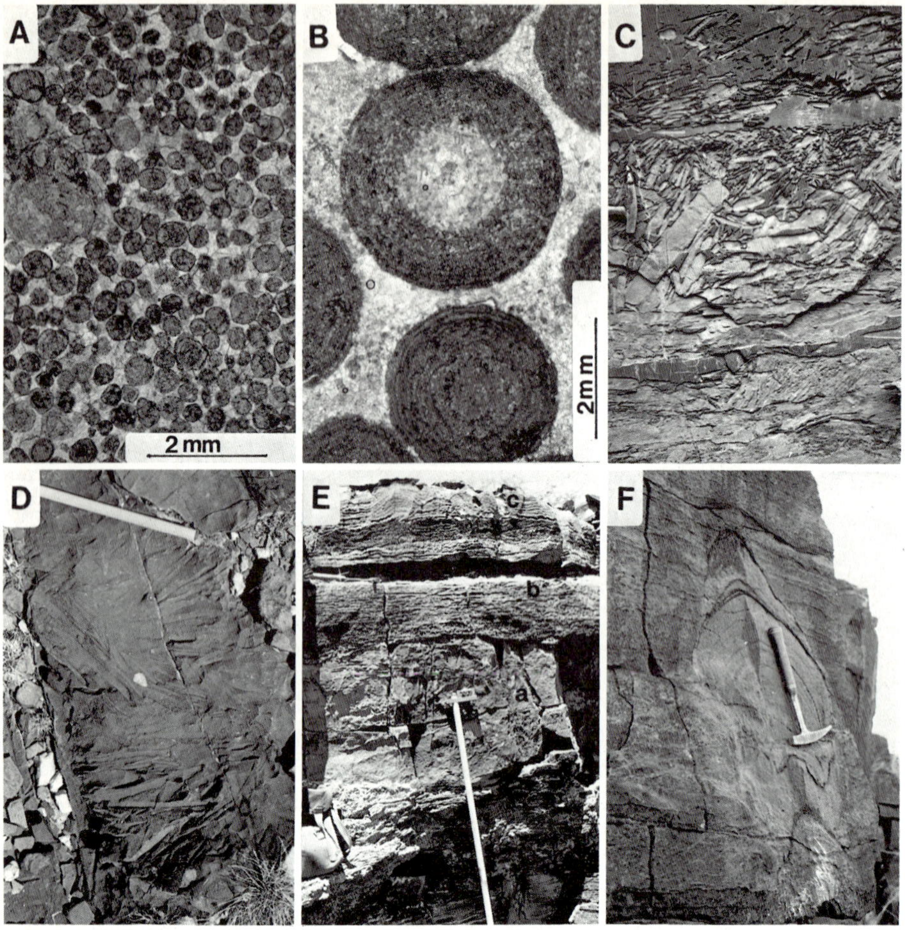

Fig. 3. **A** Microolite. Note that many microoids have only one coat and the concretionary coated grain (vadoid) on *left*. **B** Pisooid with a large nucleus. Note somewhat wrinkled laminae in cortex of lower dark ooid. **C** Edgewise conglomerate in carbonaceous mudstone. Note imbricated structure. **D** Edgewise conglomerate in a channel between stromatolitic columns. Matrix of conglomerate consists of oolite. **E** Upward coarsening microolite (*a*) → pisoolite (*b*) cycle. Coarse grained pisooids on top is overlain by LLH-C stromatoliths (*c*). **F** Large single stromatolitic dome in oolite

have contributed to the precipitation of carbonate so that it would not be unrealistic to think of the Boomplaas giant ooids as highly spherical, regular coated oncoids. Sections cut through the cortexes of giant ooids display a somewhat wrinkled lamination (Fig. 3B) which could possibly be related to growth of an algal mat.

4 Environment of Deposition

The ooids and their associated oolite rock units form part of three asymmetrical lithofacies cycles types. They are: (1) stromatolite → oolite cycles at the top of the Boomplaas Formation at Readsdrift and near the top of the Vryburg Formation between Readsdrift and Schmidtsdrift, (2) argillite → oolite cycles constituting most of the Boomplaas Formation at Readsdrift and Schmidtsdrift and (3) oolite → stromatolite cycles building the sequence at Bothithong. The arrows point up in the sequence.

At Langerand in the extreme northeast most of the Boomplaas Formation consists of cherty sparitic dolomite (Fig. 1C) containing lenticular oolite units.

In the three lithofacies cycle types and the cherty dolomite it would appear as if the oolites were deposited in shallow water, high energy environments – an interpretation supported by the well-sorted oosparitic composition of the units. Further proof for this interpretation is as follows.

In the stromatolite → oolite cycles, oolite fills steep-sided channels between giant cryptalgal laminated mounds, it covers some of the mounds and is also present in channels between overlying elongated cryptalgal laminated columns and domes (Fig. 4A). The giant stromatolitic mounds have widths of between 1 and 3 m and lengths up to 8 m and are elongated perpendicular to the Griquatown fault zone, i.e., the edge of the Boomplaas carbonate platform (Fig. 1B). The channels between them and also the overlying elongated columns have similar orientation and the mounds are thus interpreted as subtidal stromatolitic reefs elongated parallel to tidal currents along the edge of the carbonate platform. Ooids derived from backreef oolite shoals were washed down into channels between the mounds. The stromatolitic mounds, which are similar in appearance to those described by Eriksson and Truswell (1973) from the overlying Campbellrand – Malmani carbonate sequence, built out onto offshore siliciclastic mudflats (Fig. 4A). Carbonate fragments washed down from the reefs and from the oolite shoals led to the development of interbedded shale and carbonate arenite at the toe of the slope (Fig. 4A).

The argillite → oolite cycles (Fig. 4B) differ from the stromatolite → oolite cycles (Fig. 4A) in that no stromatolitic reefs are developed in between the argillite and oolite. Instead the oolite grades progressively down through flaser-bedded oolite; wavy bedded oolite/carbonate arenite-argillite; lenticular bedded oolite/carbonate arenite-argillite, into argillite. These cycles were probably formed by "open" oolite shoals prograding onto subtidal mudflat areas (Fig. 3B), i.e., conditions similar to some areas of oolite outbuilding as described by Loreau and Purser (1973) from the Persian Gulf.

Edgewise conglomerate in mudstone (Fig. 3C) interpreted as storm wave deposits (Beukes 1978) are developed below the oolite units in the argillite → oolite cycles (Fig. 4B). Some of the clasts were washed up onto the oolite shoals as is indicated by their presence in oolite units and in channels between elongated columns (Fig. 3D). The edgewise fragments originated in a subtidal environment (Beukes 1978) and they are not coated by cryptalgal laminations as those of supratidal flats (Davies 1970) typically are.

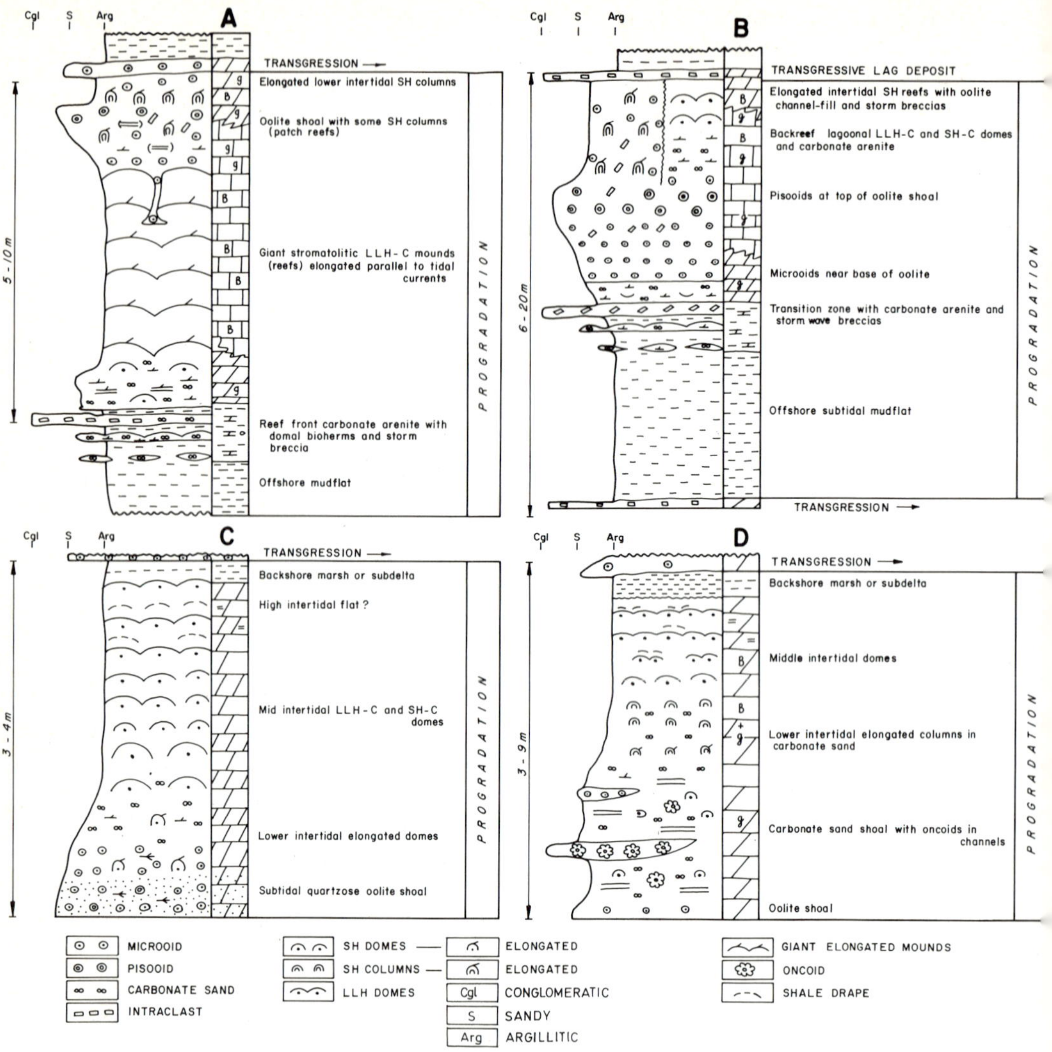

Fig. 4A–D. Ideal lithofacies cycles of the Boomplaas Formation and their interpretation. **A** Stromatolite → oolite cycle from Readsdrift. **B** Argillite → oolite cycle from Readsdrift. **C** Oolite → stromatolite cycle from Bothithong. **D** Oolite → stromatolite cycle from Bothithong with oncolite and carbonate arenite

The oolite units in the argillite → oolite cycles (Fig. 3E) display reversed graded bedding from infant microoids at the base up to mature pisooids on top. Microoids may in turn grade down into well sorted, mature carbonate arenite. These features could be related to the origin of ooids and the progradation of oolite shoals. Bathurst (1975) is of the opinion that ooids of the Browns Cay oolite shoals of the Bahamas start to form in deeper water (a few metres) and are then washed up to the crest of the shoal where the largest ones will remain as a lag deposit, smaller ones being washed back to deeper water. Newell et al. (1960) also found that the largest and most mature ooids (those with the greatest number of concentric laminae) are concentrated immediately below the intertidal zone on the oolite shoals of the Great Bahama Bank. Progradation of these shoals would lead to reversed graded bedding similar to that observed in the Boomplaas Formation.

Cryptalgal laminated domes and columns overlying and interbedded with oolite of the Boomplaas Formation (Fig. 3F) are always directly associated with the coarsest ooids irrespective of the cycle type. This may be an indication that the coarsest parts of the Boomplaas oolite shoals eventually tended to become stabilized with the resulting growth of stromatolite "reefs" and "patch reefs" (discontinuous stromatoliths) on them (Fig. 5A).

Elongated columnar stromatoliths separated by ooid-filled channels (Fig. 5B) are similar to the elongated stromatolitic columns described by Logan (1961) and Logan et al. (1964) from high energy, lower intertidal areas at Shark Bay. That the columns of the argillite → oolite cycles near Readsdrift and Schmidtsdrift were associated with intertidal mudflats connected to a land area, are, however, doubtful. Some of the oolite units are overlain by symmetrical domes of about 1 m in diameter and with a microstructure of fingerlike columns similar to that of the subtidal giant stromatolitic reefs of the stromatolite → oolite cycles. The inference thus is that these domes also formed in a subtidal environment but because of their symmetry it is concluded to have been a low-energy regime. These domes may therefore have grown in a subtidal platform lagoon situated landwards of oolite shoals (Fig. 4B). The elongated lower intertidal stromatolitic reefs and patch reefs mentioned above may have thus grown along the edge of a carbonate platform (Fig. 4B). Such stromatolitic reefs are described by Hagan and Logan (1975) from the Hutchinson Embayment of Hamelin Pool, Shark Bay. Ripple cross-laminated carbonate arenite overlying some of the oolite units at Readsdrift and Schmidtsdrift (Fig. 1C) may represent backshoal and backreef carbonate sand deposits. Such sand deposits are a common feature of present-day areas of carbonate sedimentation (Bathurst 1975).

The platform edge setting of oolite and columnar stromatoliths in the argillite → oolite cycles differs from that encountered in the oolite→ stromatolite cycles described by Beukes (1977) from the Bothithong area. In this area oolite units are overlain by elongated columns and domes separated by oolite-filled channels which grade upwards into linked columns in turn followed by domes and horizontal cryptalgal lamination (Fig. 4C). The cycles are capped by argillite and some of the domes and columns are draped by argillite indicating proximity to a siliciclastic source area which may have been the Ganyesa dome. It is also in this area that supermature quartz grains are present in the oolite units. Beukes (1977) thus interpreted these cycles as being representative of progradational oolite shoals which were connected landwards to intertidal stromatolitic carbonate mudflats in turn flanked by terrestrial siliciclastic muds of backshore marshes and/or subdeltas (Fig. 4C).

The upper part of the Boomplaas Formation at Bothithong is characterized by oolite → stromatolite cycles in which horizontally stream-laminated dolarenite overlies the oolite (Fig. 4D). This bedform is indicative of strong cur-

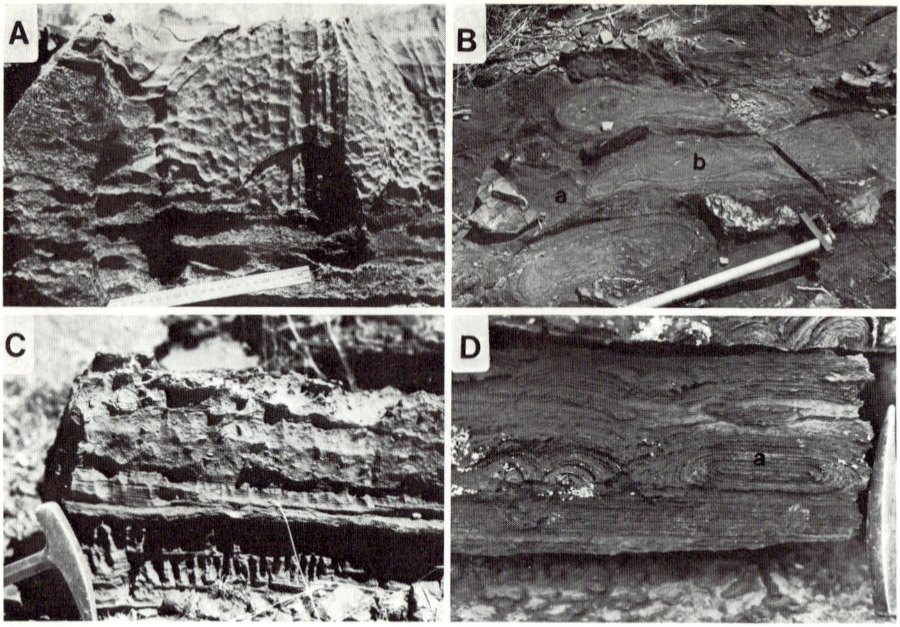

Fig. 5. A Cryptalgal laminated limestone dome overlying coarse grained partly silicified pisoolite. **B** Plan view of oolite (*a*) filling channels between elongated columns (*b*). **C** Oncolite overlying horizontally laminated dolarenite. **D** Algal biscuit (*a*) overlying horizontally laminated dolarenite. Note irregular replacement of LLH-C domes by chert (resistant layers with positive relief)

rent flow, most probably in relatively shallow water, at the transition between lower and upper flow regimes (Simons et al. 1965).

Small single domal stromatoliths of less than 10 cm in diameter are present in the dolarenite. However, their sandy substrata was apparently unstable so that some of them were overturned by the current to form oncoids (Fig. 5C) which became concentrated in channel-like bodies. A similar process is known to be responsible for the formation of modern oncoids on Joulters Cay in the Bahamas (Gebelein 1976).

The lowermost oolite unit of the Boomplaas Formation at Bothithong displays well defined herringbone cross-lamination which indicate tidal current activity (Klein 1970). However, the oolite units in the south near Schmidtsdrift and Readsdrift are poorly bedded displaying low angle cross-lamination which could have been produced by partly washed-out ripples under fairly high energy conditions.

Ooids associated with the cherty dolomite facies association may be partly replaced by chert and occur in channel-like lenticular units. The stratigraphy of this facies association is complex, but cycles are present which consist of lenticular oolite and dolarenite units containing oncoids and algal biscuits (Fig. 5D) overlain by small LLH-C and LLH-S domes → horizontal cryptalgal lamination → encapsulated dolomite breccias. The small domes and horizontal cryptalgal laminations are similar to those described by Logan (1961) from protected intertidal mudflats of Shark Bay and the encapsulated breccias resemble supratidal breccias described by Davies (1970) from Shark Bay. The ooids, oncoids, algal biscuits and dolarenites are thus thought to represent shallow water, tidal channel deposits on intertidal mudflats which were in turn flanked by supratidal carbonate mudflats. A similar intertidal-supratidal facies association is described by Eriksson (1977) from the Malmani carbonate sequence of the Transvaal Supergroup.

5 Regional Depositional Model

Most of the oolite units and their associated progradational sedimentary cycles (Fig. 4) are correlatable over long distances of up to 340 km along strike (Beukes 1978, 1979). The even thickness and sheetlike form of the oolites could probably be best explained by the progradation of oolite shoals similar to those described by Newell and Rigby (1957) from the Great Bahama Bank. Progradation of a complex system of barrier islands and oolitic tidal deltas like that of the Trucial Coast in the Persian Gulf (Loreau and Purser 1973) would probably have led to more variation in thickness and to less continuity along strike.

Subtle facies changes take place within individual lithofacies cycles from the carbonate platform edge in the south onto the Kaapvaal craton in the north. For example, argillite → oolite cycles of the Readsdrift area change northwards to oolite → stromatolite cycles at Bothithong and may in turn be represented by cherty dolsparite at Langerand (Beukes 1978, 1979). An evolution thus took place within a cycle as it prograded southwards towards the edge of the carbonate platform. The infant stage of any one progradational cycle was one of oolite shoals connected to intertidal mud flats (Fig. 6A). The mature stage was represented by subtidal stromatolitic reefs; oolite shoals with platform edge stromatolitic reefs and patch reefs; platform lagoonal subtidal stromatoliths, and platform interior, intertidal mudflats connected to a low lying siliciclastic source area (Fig. 6B). Carbonate sands washed down from the oolite shoals and stromatolitic reefs were deposited in an offshore mudflat environment and also backshore of oolite shoals in the platform lagoon (Fig. 6B).

In the Readsdrift area near the edge of the carbonate platform, the thicknesses of oolite units from their base to the base of overlying lower intertidal stromatolitic columns vary between 1 and 6 m. These thicknesses correspond to the depth of water into which the oolite shoals prograded (Beukes 1977). In the platform interior near Bothithong the water was shallower because the maximum thickness of subtidal oolite shoal units here are only in the order of 1 m. These values are in good agreement with the maximum water depth of 5 m which Loreau and Purser (1973) state for oolite formation in the Persian Gulf. This could in turn mean that the Boomplaas ooids were deposited in close proximity to their environment of formation. Ooids distributed out of their environment of formation could be dispersed to almost any water depth by currents, or onto land by eolian action (Loreau and Purser 1973).

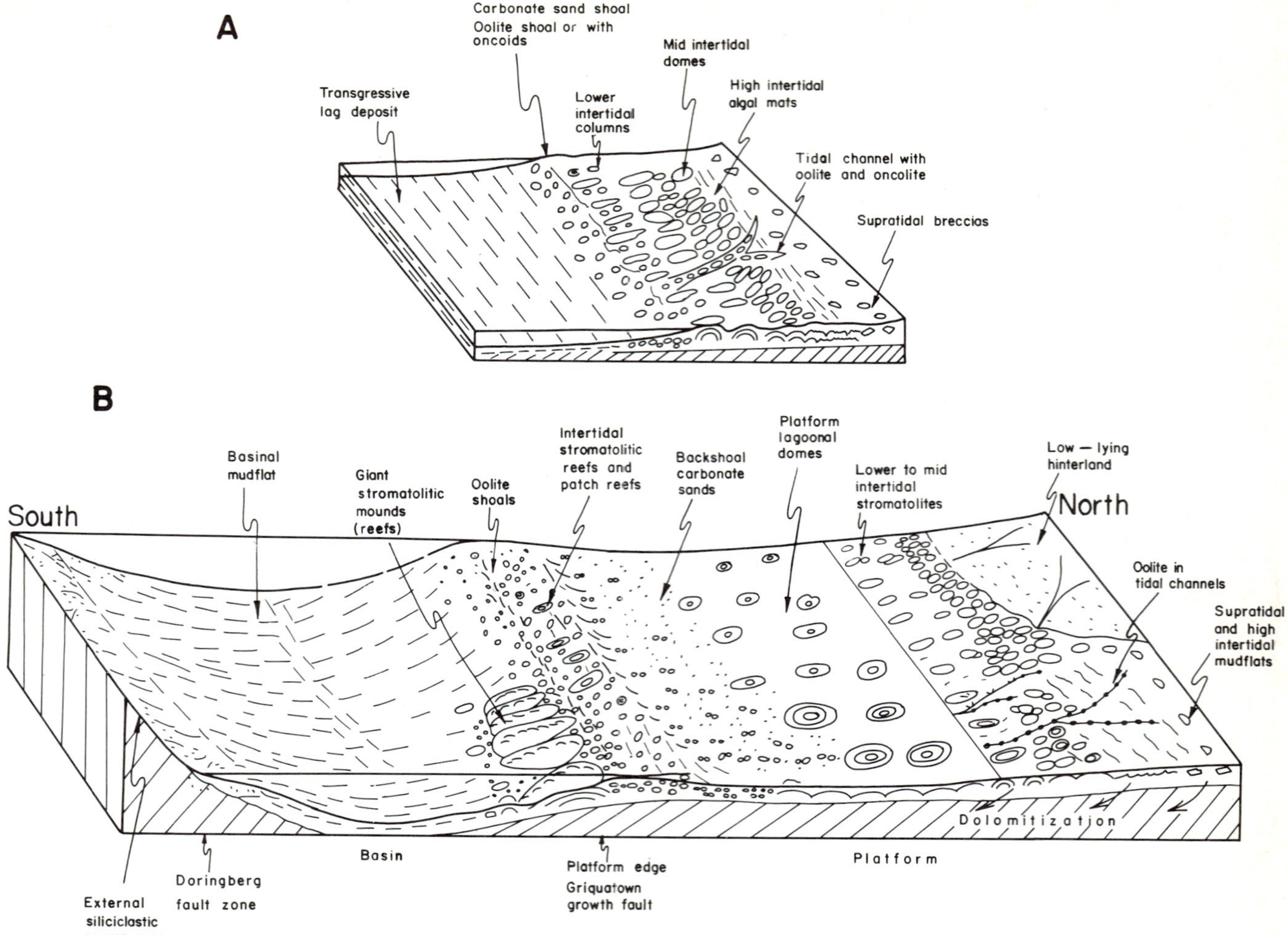

Fig. 6 A, B. Evolution of a progradational carbonate sedimentary increment from an infant stage (**A**) to a mature carbonate platform stage (**B**)

The absence of gravity flow deposits in front of the Boomplaas oolite shoals and stromatolitic reefs is indicative of shallow foreslope dips (Beukes 1978). Total depth of water immediately in front of the platform at Readsdrift calculated from the thickness of subtidal units (Beukes 1977) ranged from 7 to 20 m. Next to the Doringberg fault zone upward coarsening, basinal argillite → shallow marine quartz arenite cycles are developed with a maximum known thickness of 45 m. This figure probably represents the minimum known depth of the basin into which the Boomplaas carbonate platform prograded (Beukes 1978). These water depths are not much unlike that of the Persian Gulf given by Purser and Seibold (1973).

6 Diagenesis

Dolomitization is the most conspicuous diagenetic process that affected the ooids, oolites and other carbonate lithofacies of the Boomplaas Formation. The dolomites are iron-bearing, weathering to a red or reddish brown colour which stands in strong contrast to the dark blue-grey weathered surfaces of the limestones.

On a regional scale the seepage reflux model of dolomitization described by Deffeyes et al. (1965) could explain the preferential dolomitization of the platform interior, intertidal and supratidal deposits of the Boomplaas Formation relative to the platform edge deposits which are only partly dolomitized (Fig. 6B). In the latter area limestone oolites and stromatolites have only been dolomitized where they are in contact with pyritic carbonaceous argillite, suggesting that the dolomitizing fluids were derived locally during compaction from the pore water of the argillites. The FeO content of the dolomites is as high as 10% by weight next to argillite units.

The process of dolomitization is clearly illustrated in the oolite units. Some distance ahead of the dolomitization front only a few ooids consist of dolomite. The number of dolomitized ooids increases gradually as the dolomitization front is approached and eventually all ooids and their interstitial limestone cement material are dolomitized.

On a microscopic scale it is seen that the microcrystalline cortexes of ooids are first dolomitized (Fig. 7A). Dolomitization then starts to affect the sparitic limestone cores of ooids (Fig. 7A) and the surrounding sparite cement. Eventually dolomitization is complete, the ooids recrystallized and their original texture almost totally destroyed (Fig. 2B). Late stage neomorphic calcite aggradation also took place because in some limestone samples, calcite crystals cut across the boundaries between ooids and cement material.

Other diagenetic processes which affected the Boomplaas ooids are silicification, chertification and compaction. Silicification took place on as small scale in platform edge oolite units near Readsdrift. The rims of ooids and radial calcite crystals in them are preferentially silicified (Fig. 2D). The silica may have been derived from devitrified volcanic glass shards present in some of the oolite units (Beukes 1978). Chertification and neomorphic aggradation took place in the

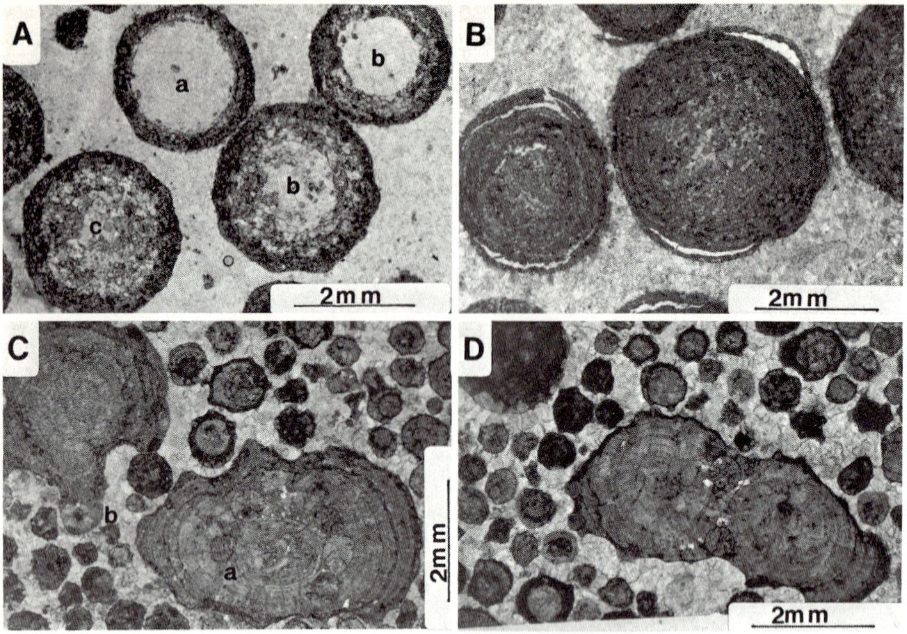

Fig. 7. **A** Preferential dolomitization of ooids. Only cortex of ooid (*a*) is dolomitized; the nuclei of ooids (*b*) are partly dolomitized and ooid (*c*) is totally dolomitized. Calcite (*light grey*) has been dissolved in a weak solution of HCl. **B** Partly detached laminae of pisooids. Detachment probably the result of compaction. **C** Vadoid with microoids incorporated in its cortex (*a*). Note how microoid starts to be incorporated along the boundary of the upper left hand vadoid (*b*). **D** An amalgamated vadoid with microoids incorporated along the boundary between the two original vadoids

intertidal and supratidal cherty dolomite and oolite of the Langerand area. Truswell and Eriksson (1975) ascribed this type of recrystallization and chertification to periodic subaerial exposure and the influence of acid meteoric waters on intertidal and supratidal mudflats.

In some samples close-packed ooids display vague hexagonal or pentagonal outlines due to impregnation by surrounding ooids. In some ooids, laminae became detached from the cortex before cementation took place (Fig. 7B). This feature is ascribed to early compaction in Upper Mississippian ooids of Montana (Bathurst 1975).

In one oolite unit of the Boomplaas Formation some giant irregular-outlined coated grains of 3 – 5 mm in diameter are present. The grains may have nuclei consisting of one or more microoids whilst in others microoids are incorporated in the cortexes (Fig. 7C). The irregular outline of the grains are the result of the incorporation by in situ growth of surrounding ooids (Fig. 7C). Outward in situ growth also resulted in the amalgamation of grains (Fig. 7D). These features indicate that the grains are of a post-depositional concretionary origin and that they may represent vadoids similar to those described by Dunham (1969) from the Capitan reef complex in Texas. The implication is that some of the oolite shoals of the Boomplaas Formation may have been exposed above the water line for rather long periods shortly after deposition.

7 Summary

Concentrically laminated ooids of the Boomplaas Formation formed in high energy shoal environments whereas radial-concentrically textured ones may have originated in more protected platform lagoonal environments. The ooids were deposited in oolite shoals along the edge of the Boomplaas carbonate platform; in tidal channels between subtidal and intertidal stromatolitic reefs; and in tidal channels on intertidal mudflats. In the latter the ooids are associated with oncoids and algal biscuits. The ooids formed remote from clastic influx in water depths varying between 1 and 6 m.

Dolomitization of the ooids took place after deposition with their microcrystalline limestone cortexes first affected and at a later stage their sparitic nuclei and cement material. Secondary silicification, chertification and compaction also took place. Some oolite units became exposed shortly after deposition, as is indicated by the presence of vadoids.

Acknowledgements. This paper represents part of a PhD study which was undertaken at the Rand Afrikaans University, Johannesburg with research grants from the University, the C.S.I.R., ISCOR, Cape Asbestos, Asbestos Investments and The Jim and Gladys Taylor Educational Trust.

References

Bathurst RGC (1975) Carbonate sediments and their diagenesis, 2nd edn. Elsevier, Amsterdam

Beukes NJ (1977) Transition from siliciclastic to carbonate sedimentation near the base of the Transvaal Supergroup, Northern Cape Province, South Africa. Sediment Geol 18:201 – 221

Beukes NJ (1978) Die karbonaatgesteentes en ysterformasies van die Ghaap-Groep van die Transvaal-Supergroep in Noord-Kaapland. Unpubl PhD thesis, Rand Afrikaans Univ, Johannesburg, 580 p

Beukes NJ (1979) Litostratigrafiese onderverdeling van die Schmidsdrif-Subgroep van die Ghaap-Groep, Noord-Kaapland. Trans Geol Soc S Afr 82:313 – 327

Button A (1976) Transvaal and Hamersley basins – review of basin development and mineral deposits. Miner Sci Eng 8:262 – 293

Davies GR (1970) Algal-laminated sediments, Gladstone Embayment, Shark Bay, Western Australia. Am Assoc Petrol Geol Mem 13:169 – 205

Davies PJ, Roger T (1968) Stained dry cellulose peels of ancient and recent impregnated carbonate sediments. J Sediment Petrol 38:234 – 237

Deffeyes KS, Lucia FJ, Weyl PK (1965) Dolomitization of Recent and Plio-Pleistocene sediments by marine evaporite waters on Bonaire, Netherlands Antilles. SEPM Spec Publ 13:71 – 87

Dunham RJ (1969) Vadose pisolite in the Capitan Reef (Permian), New Mexico and Texas. SEPM Spec Publ 14:182 – 191

Eardley AJ (1938) Sediments of Great Salt Lake, Utah. Bull Am Assoc Petrol Geol 22:1305 – 1411

Eriksson KA (1977) Tidal flat and subtidal sedimentation in the 2250 my Malmani Dolomite, Transvaal, South Africa. Sediment Geol 18:223 – 244

Eriksson KA, Truswell JF (1973) High inheritance elongate stromatolitic mounds from the Transvaal dolomite. Palaeontol Afr 15:23 – 28

Fabricius F, Klingele H (1970) Ultrastrukturen von Ooiden und Oolithen: Zur Genese und Diagenese quartärer Flachwasserkarbonate des Mittelmeeres. Verh Geol B A 4:594 – 617

Freeman T (1962) Quiet water oolites from Laguna Madre, Texas. J Sediment Petrol 32:475 – 483

Gebelein CD (1976) Open marine subtidal and intertidal stromatolites (Florida, the Bahamas and Bermuda). In: Walter MR (ed) Stromatolites. Elsevier, Amsterdam, pp 381 – 388

Greensmith JT (1978) Petrology of the sedimentary rocks, 6th edn. Allen & Unwin, London

Hagan GM, Logan BW (1975) Prograding tidal-flat sequences, Hutchinson Embayment, Shark Bay, Western Australia. In: Ginsburg RN (ed) Tidal deposits. Springer, Berlin Heidelberg New York, pp 215 – 222

Illing LV (1954) Bahaman calcareous sands. Bull Am Assoc Petrol Geol 38:1 – 95

Klein G de V (1970) Tidal origin of a Precambrian quartzite (Dalradian) of Islay, Scotland. J Sediment Petrol 40:973 – 985

Logan BW (1961) Cryptozoan and associated stromatolites from the Recent, Shark Bay, Western Australia. J Geol 69:517 – 533

Logan BW, Rezak R, Ginsburg RN (1964) Classification and environmental significance of algal stromatolites. J Geol 72:68 – 83

Loreau JP, Purser BH (1973) Distribution and ultra-structure of Holocene ooids in the Persian Gulf. In: Purser JP (ed) The Persian Gulf. Springer, Berlin Heidelberg New York, pp 279 – 328

Newell ND, Rigby JK (1957) Geological studies on the Great Bahama Bank. SEPM Spec Publ 5:15 – 72

Newell ND, Purdy EG, Imbrie J (1960) Bahamian oolitic sand. J Geol 68:481 – 497

Purser BH, Seibold E (1973) The principal environmental factors influencing Holocene sedimentation and diagenesis in the Persian Gulf. In: Purser BH (ed) The Persian Gulf. Springer, Berlin Heidelberg New York, pp 1 – 9

Rusnak GA (1960) Some observations of Recent oolites. J Sediment Petrol 30:471 – 480

Shoji R, Folk RL (1964) Surphace morphology of some limestone types as revealed by electron microscope. J Sediment Petrol 34:144 – 145

Simons DB, Richardson EV, Mordin CF (1965) Sedimentary structures generated by flow in alluvial channels. SEPM Spec Publ 12:34 – 52

Truswell JF, Eriksson KA (1975) A palaeoenvironmental interpretation of the early Proterozoic Malmani dolomite from Zwartkops, South Africa. Precambrian Res 2:277 – 303

Visser JNJ, Grobler NJ (1972) The transititon beds at the base of the Dolomite Series in the Northern Cape Province. Trans Geol Soc S Afr 75:265 – 274

III Rhodoids

III.1 Description and Classification of Rhodoliths (Rhodoids, Rhodolites)

Daniel W. J. Bosence[1]

1 Introduction

Nodules composed of coralline algae have been described since the latter part of the eighteenth centuary (Pallas 1766, Ellis and Solander 1786). The first extensive descriptive work on free-living corallines was undertaken by Foslie (1894), Lemoine (1910) and later by Cabioch (1966, 1969). Bosellini and Ginsburg (1970) were the first to recognise the palaeoecological significance of rhodoliths and they suggested a descriptive terminology. They proposed the term rhodolite for "... nodules and detached branch growths with a nodular form composed principally of coralline algae ...". Unfortunately the term rhodolite has priority for a variety of garnet (Binda 1973) and therefore "rhodolith" has now become the accepted term (Toomey 1975). More recently "rhodoid" has been suggested (Peryt, this vol.) but this is not preferred by the author because (a) an acceptable term is already in use and (b) its etymology is unsuitable (Rhodolith-rhodophycean or red stone, rhodoid-rhodophycean or red-like).

Despite the clear difference between rhodoliths and oncoliths (blue green algal nodules, see Peryt, this vol.) some authors have confused these terms (McMaster and Conovar 1966, Blanc 1968).

In this paper I propose a taxonomic and morphological classification of rhodoliths together with a suggested scheme for description based on the work of Bosellini and Ginsburg (1970), Bosence (1976), and Bosence and Pedley (1982). This scheme is summarized in Table 1 and from this it can be seen that

Table 1. Descriptive terms for rhodoliths (if brevity is required the abbreviations shown in brackets are recommended)

Monospecific (Mono)	Spheroidal (S)	Laminar (L): concentric (con) or boxwork (box)
Multispecific (Multi)	Ellipsoidal (E)	Branching (B): classes I, II, III, IV
	Discoidal (D)	Columnar (C)

e.g. Multispecific, discoidal, laminar concentric (or Multi. D. L. con) rhodolith
Monospecific, spheroidal, branching III (or Mono. S. B. III) rhodolith

1 Department of Geology, University of London Goldsmiths' College, Rachel McMillan Building, Creek Road, London SE8 3BU, Great Britain

Coated Grains (ed. by T.M. Peryt)

rhodoliths may be described in terms of their morphology (shape, size and structure) and their taxonomic composition (monospecific or multispecific).

2 Classification of Rhodoliths

Early workers recognised the great variety of growth forms present in rhodoliths (Foslie 1894, Lemoine 1910). The classification of these varieties was accomplished by adding form names after the specific name (e.g., *Lithothamnium calcareum* f. *crassa*). There are two drawbacks to this approach if it is extended to cover all rhodoliths. Firstly, many species of unattached corallines (monospecific rhodoliths) show the same sequences of morphological variations (Cabioch 1966). These variations can be quantified as shown below thus resulting in a simpler classification. Secondly, many rhodoliths contain more than one species of coralline (multispecific rhodoliths) and the taxonomic composition of the rhodolith may change with time. In these cases it is again simpler to separate the taxonomy of the corallines from the description of the rhodolith.

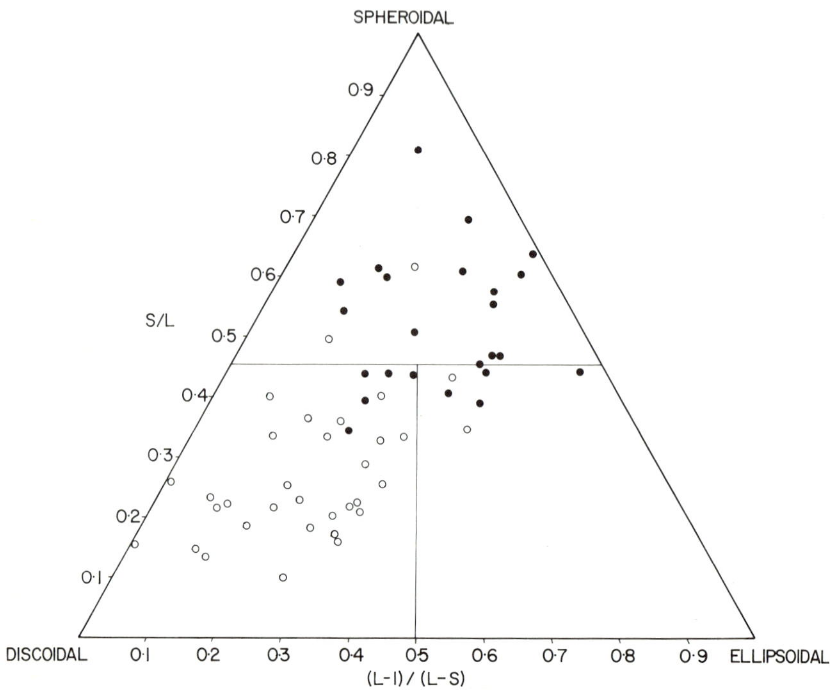

Fig. 1. Shape classification of rhodoliths using Sneed and Folk's (1958) pebble shape diagram. ● Mainly spheroidal rhodoliths from high energy Rhodolith Pavement facies. ○ Mainly discoidal rhodoliths, from low energy Crustose Algal Marl facies, derived from planar crusts (see Fig. 5). Locality 14, Coralline Algal Biostrome, Miocene, Malta. (After Bosence and Pedley 1982)

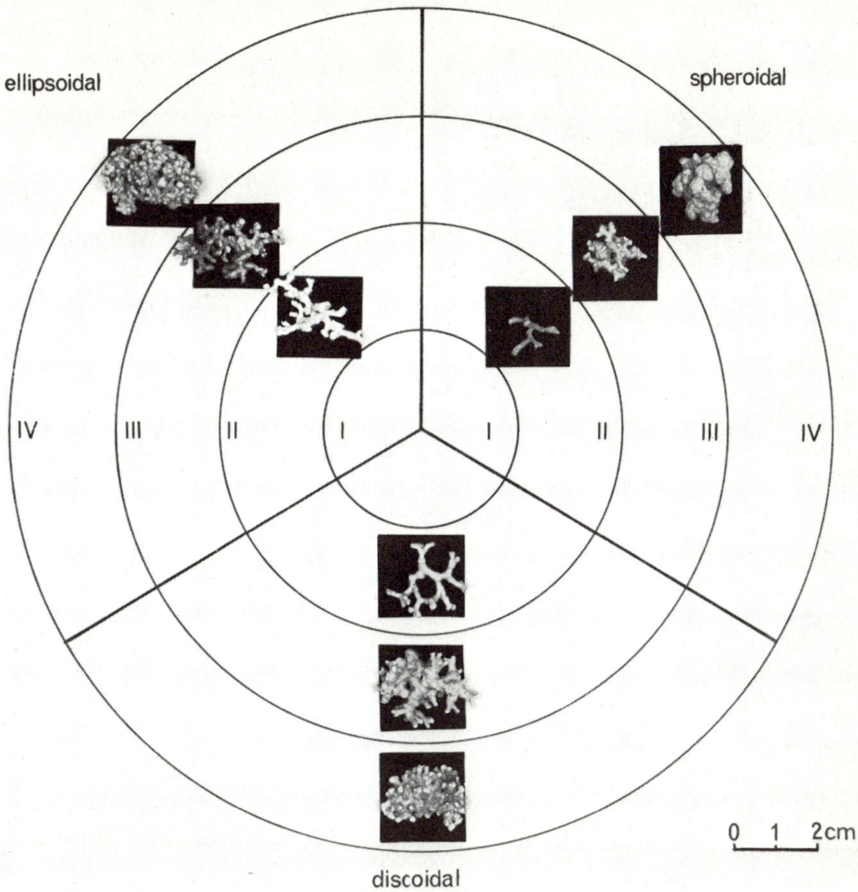

Fig. 2. Morphological classification of monospecific rhodoliths of *Lithothamnium corallioides* illustrating similar ranges of branching density with ellipsoidal, spheroidal and discoidal shaped thalli. (After Bosence 1976)

2.1 Shape

Rhodoliths grow into a number of shapes which can be described both quantitatively and qualitatively. Bosellini and Ginsburg (1970) assigned their Bermudan rhodoliths to four shape classes: spheroidal, elliptical, ameboidal and flat. The sphericity was quantified by measuring the Long, Intermediate and Short axes and applying Sneed and Folk's (1958) maximum projection sphericity formula ($\sqrt[3]{S^2/LI}$). Bosence (1976) and Bosence and Pedley (1982) have further quantified rhodolith shape by measuring the Long, Intermediate and Short axes of the enveloping surface of a rhodolith. This provides data for plotting shapes on Sneed and Folk's (1958) pebble shape diagram (Fig. 1) which defines spheroidal, discoidal and ellipsoidal morphologies. This has been used to classify both recent (Fig. 2) and ancient rhodoliths. In both cases it was shown that rhodolith shapes are characteristic of different sedimentary environments.

2.2 Size

Size may be measured in a number of ways which should be applicable to both recent and fossil material and be independent of the internal structure of the rhodolith. For this reason weight and volume (by displacement) are not recommended. A size measurement based on the Long, Intermediate and Short axes (as outlined above) is convenient. Bosellini and Ginsburg (1970) use the arithmetic mean diameter $(L+I+S/3)$ and Bosence (1976) has used the volume of an ellipsoid $(\sqrt{LIS/4\pi})$.

2.3 Structure

The structure of many rhodoliths is complex and if the details are recorded then they may reflect the changing conditions during the history of a rhodolith. Thin sections and polished surfaces (preferably two at right angles through the long and short axes and the intermediate and short axes) through rhodoliths are required for this information.

Three basic coralline morphologies have been found to be present in rhodoliths: laminar, branching and columnar growths. These may give rise to rhodoliths constructed of one morphology or to rhodoliths constructed of two or all three morphologies.

2.3.1 Laminar Rhodoliths

Simple crusts of corallines are common constructors of rhodoliths. The crusts consist of a basal hypothallus (if present in a taxa) and an upper perithallus (Fig. 3A). Typically these ocur as closely stacked concentric crusts representing successive colonisations and overgrowths of corallines on the rhodolith (Fig. 3B). The thickness of the crust is usually constant for individual species of coralline. In corallines with a more foliaceous growth form (some species of *Lithophyllum* and *Mesophyllum*) crusts may bifurcate and grow over earlier crusts with a leafy habit (Fig. 3D). In extreme cases a non-concentric or boxwork of crusts may result in a very delicate rhodolith. This type of rhodolith constructed by the squamariacean *Peysonnelia* has been described from the Recent of the Mediterranean (Huvé 1954) and leafy crustose rhodoliths forming boxworks constructed by *Mesophyllum* have been described from the Miocene of Malta (Bosence and Pedley 1982). Laminar rhodoliths commonly originate from crusts overgrowing a previous substrate such as a fragment of preexisting rhodolith,

Fig. 3. A Crust of *Lithophyllum albanense* (Lemoine) illustrating lower plumose hypothallus and upper grid-like perithallus. Coralline Algal Biostrome, Miocene, Malta. **B** Spheroidal, laminar concentric rhodolith from Pliocene, Zukawa Formation, Ebizaka, Japan × 1. (Note also serpulids and bryozoans). **C** Discoidal, columnar rhodolith. Conceptacles pick out arched layers of perithallial tissue in columns. Coralline Algal Biostrome, Miocene, Malta. × 1. **D** Discoidal, laminar boxwork rhodolith. Coralline Algal Biostrome, Miocene malta. × 1. **E** Vertical section through branch of *Lithothamnium corallioides* illustrating central medulla with growth layers and outer cortex with smaller cells. Mannin Bay, Ireland. × 64

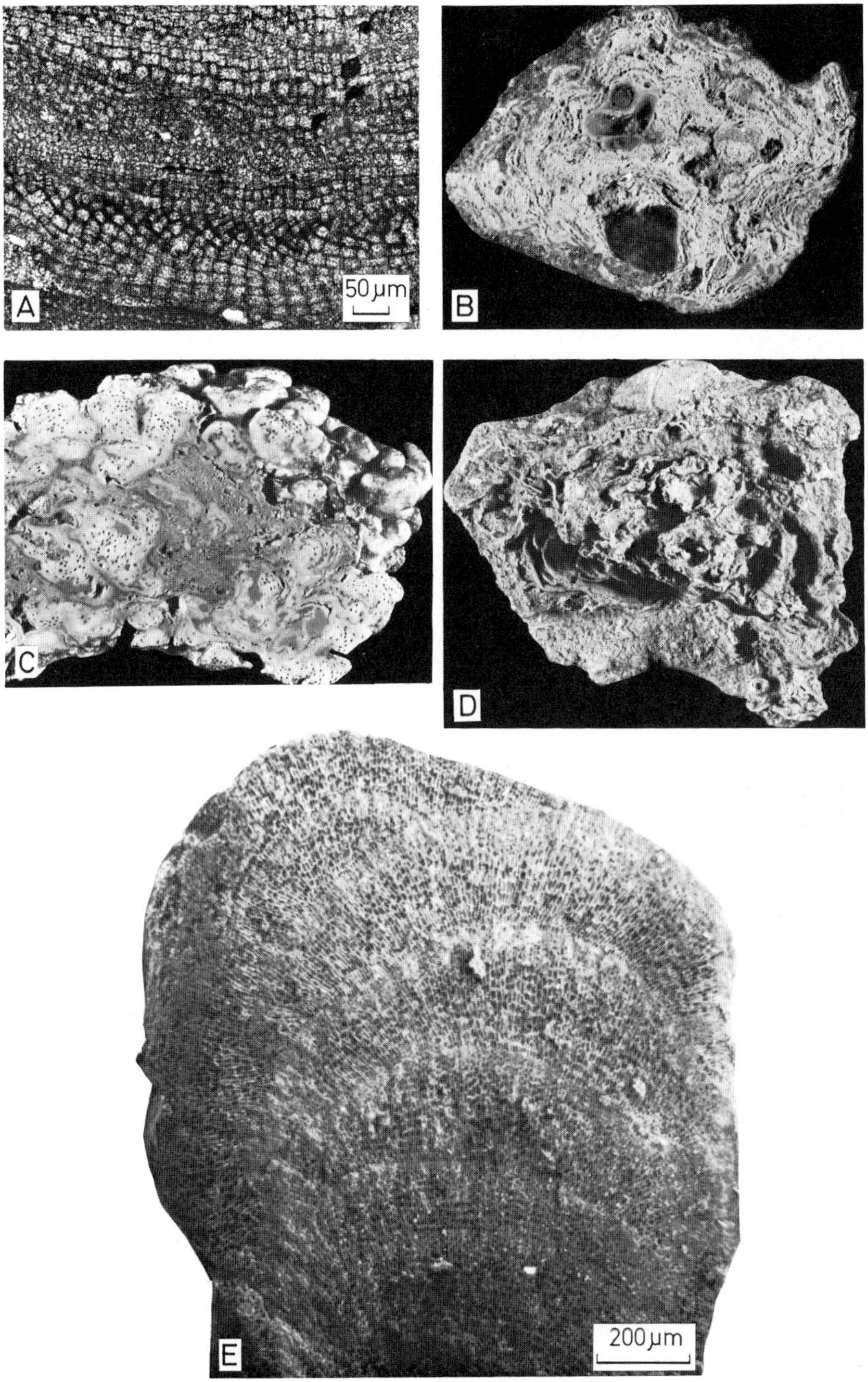
50 µm
A
B
C
D
200 µm
E

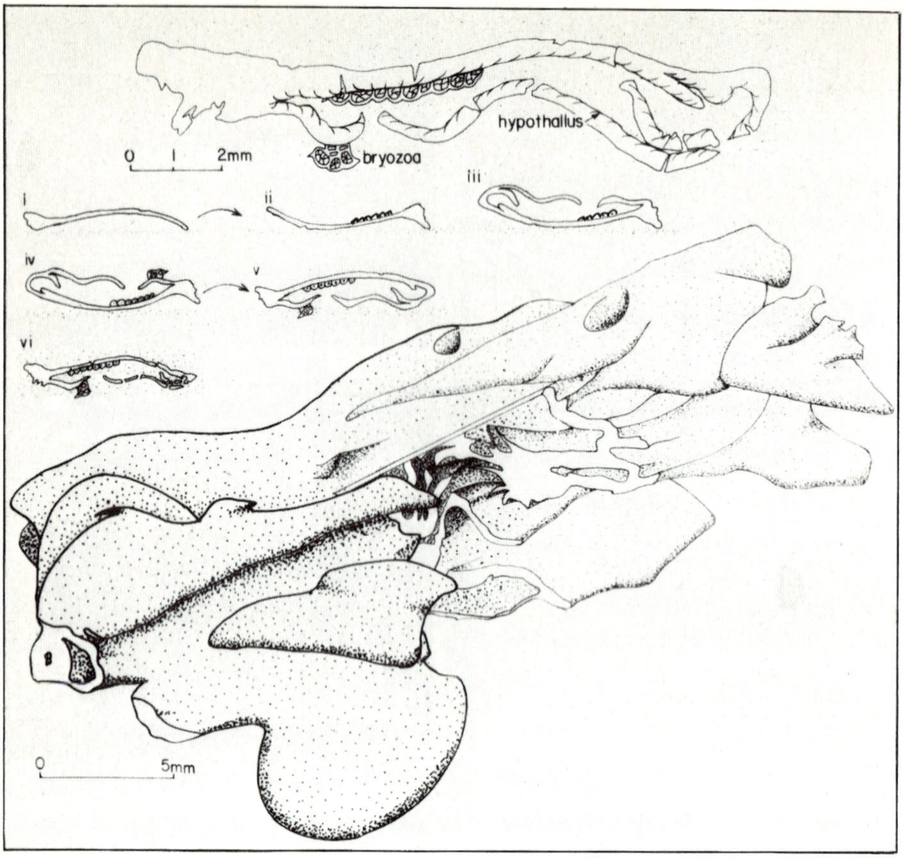

Fig. 4. Monospecific, laminar boxwork rhodoliths constructed by *Mesophyllum commune*. Upper simple discoidal rhodolith shows sequence of turning, overgrowth and encrusting recorded in rhodolith. Coralline Algal Biostrome, Malta. (After Bosence and Pedley 1982)

intraclast, terrigenous pebble, shell etc. However, occasionally they originate from a single isolated crust which is turned by currents or organisms (Fig. 4).

2.3.2 Branching Rhodoliths

Branches arise from crusts through continued growth of perithallial filaments (Fig. 3E). In the branches of many corallines a central medulla (or secondary hypothallus of Cabioch 1972) of filaments with longer cells may be present. These filaments bifurcate and grow outwards to a position normal to the side of the branches. These areas often have smaller cells and are referred to as the cortex. The width of the branch is normally constant for any species. Branching may be either apical and dichotomous, resulting in divergence of the axial filaments of a branch, or branches may arise from rejuvenation of lateral (cortical) filaments to form new medullary and cortical tissue.

The main variation found in branching rhodoliths is the density of branching (Cabioch 1966, Bosence 1976). Figure 2 illustrates varying branch density for monospecific rhodoliths of *Lithothamnium corallioides*. Similar ranges are found in *Phymatolithon calcareum*. A semi-quantitative division of branching density has been used to describe these rhodoliths which has also been extended to fossil material (Bosence and Pedley 1982):

I. Single branch
II. Few branches
III. Frequent branches
IV. Dense intergrown branches

Density IV is also found to be characterised by flattening of branch apices and lateral branching, possibly resulting from apical abrasion, whereas groups II and II normally show dichotomous branching.

2.3.3 Columnar Rhodoliths

The term "columnar" has been used to describe short stocky projections arising from crusts (1 – 2 cm long and 3 – 10 mm wide (Bosellini and Ginsburg 1970) and is probably synonymous with "mammilated protruberences and short stubby branches" (Johnson 1962) and "mammelonnes" Lemoine (1939). The term is used here and in Bosence and Pedley (1982) to describe short stocky protruberences arising from crusts which are distinct from branches in that they are almost as wide as they are high (Fig. 3C). Some corallines (e.g., *Mesophyllum commune*) show both branches and columns on the same thallus with quite different size ranges thus indicating the distinctness of these two structures. The internal structure of columns may be either arched, layered growths of perithallial tissue or medullary and cortical perithallial tissue of successive dome-shaped crusts each with a hypothallus and perithallus.

As far as I am aware no one has commented on why columnar forms develop as opposed to branches in the same coralline species. However, some taxa may only grow branches and others only columns. Bosellini and Ginsburg (1970) observed that columnar growths occurred only in rhodoliths stabilised in sand. Bosence and Pedley (1982), however, note that columnar growths are modified in the same way as branches by abrasion and transport.

3 Discussion

The descriptive terminology and classification of rhodoliths proposed in this paper (Table 1) relies on a number of criteria to distinguish the various rhodolith morphologies. However, it must be realised that individual rhodoliths may have a long history (100's to 10,000's yrs) and that they frequently show changes in their structure and taxonomic composition through time. Common structural changes are from branching and columnar growths to laminar growths and back to branches or columns. This probably is a response to increased and then decreased hydraulic energy (see Chap. 3.2, this vol.). Compositional changes include ecological succession of corallines from the centre to the outside of rhodoliths (see Chap. 3.2, this vol.). Similarly a branching monospecific branching rhodolith may change to a laminar monospecific rhodolith and thence to a laminar multispecific rhodolith as documented by Bosellini and Ginsburg (1970).

For these reasons it is not considered appropriate to erect new names for the various rhodolith types but to use the adjectives listed in Table 1 and defined in the text. This I have found to be convenient and workable for both recent and fossil rhodoliths.

References

Binda PL (1973) Form and internal structure of recent algal nodules (Rhodolites) from Bermuda: a discussion. J Geol 81:238

Blanc JJ (1968) Sedimentary geology of the Mediterranean Sea. Oceanogr Mar Biol Annu Rev 6:377 – 454

Bosellini A, Ginsburg RN (1970) Form and internal structure of recent algal nodules (Rhodolites) from Bermuda. J Geol 79:669 – 682

Bosence DWJ (1976) Ecological studies on two carbonate sediment producing coralline algae from western Ireland. Palaeontology 19:365 – 395

Bosence DWJ, Pedley HM (1982) Sedimentology and palaeoecology of a Miocene coralline algal bioherm from the Maltese Islands. Palaeoclimatol Palaeoecol Palaeogeogr 38:9 – 43

Cabioch J (1966) Contribution à l'étude morphologique anatomique et systématique des deux Melobesiées: *Lithothamnium calcareum* (Pallas) Areschoug et *Lithothamnium corallioides* Crouan. Bot Mar 9:33 – 35

Cabioch J (1969) Les fonds de maerl de la baie de Morlaix et leur peuplement végétal. Cah Biol Mar 10:139 – 161

Cabioch J (1972) Étude sur les Corallinacees. II La morphogenese consequences systématique et phylogenétique. Cah Biol Mar 13:137 – 288

Ellis and Solander (1786) The natural history of many curious and uncommon zoophytes. London, 208 pp

Foslie MH (1894) The Norwegian forms of Lithothamnion. Dan K Norske Vidensk Selske Skr 2:1 – 203

Huvé H (1954) Contribution à l'étude des fonds a *Peysonnelia polymorpha* Schmidtz de la région Marseille. Rec Trav Stu Mar Endoume 7:119 – 143

Johnson JH (1962) The algal genus *Lithothamnium* and its fossil representatives. Colo Sch Mines 57:111

Lemoine P (1910) Répartition et mode de vie du maerl (*Lithothamnium calcareum*) aux environs de Concarneau (Finistère). Ann Inst Oceanogr Monaco 1:1 – 28

Lemoine P (1939) Les algues calcares fossiles de L'Algerie. Matar Carte Geol Alger 1:1 – 128

McMaster RW, Conovar JT (1966) Recent algal stromatolites from the Canary Islands. J Geol 74:647 – 652

Pallas (1766) Elenchus zoophytorum. Imk, The Hague, 451 pp

Peryt T (1982) Classification of coated grains. This volume

Sneed ED, Folk RL (1958) Pebbles in the lower Colorado river Texas; A study in particle morphogenesis. J Geol 66:114 – 150

Toomey DF (1975) Rhodoliths from the Upper Paleozoic of Kansas and the Recent – a comparison. Neues Jahrb Geol Palaeontol Monatsh 4:242 – 255

III.2 The Occurrence and Ecology of Recent Rhodoliths – A Review

Daniel W. J. Bosence[1]

1 Introduction

1.1 Definitions and General Occurrence

Rhodoliths (= rhodolites, coralline algal nodules) are nodules and unattached branched growths with a nodular form composed principally of coralline algae.

Rhodoliths can be produced whenever crustose or branching corallines either settle on non-cohesive particulate substrates or are detached from existing hard substrates. Coralline growth as concentric laminae or radial branches or columns with occasional turning results in the characteristic radial or concentric structures (Fig. 3). This results in rhodoliths forming in a wide range of non-muddy marine environments, normally with depths less than 100 m. Since the Mesozoic the coralline algae have diversified to inhabit a wide range of environments from arctic to tropical.

This chapter attempts to summarise descriptions of rhodoliths from various environments (Fig. 1) and to describe various ecological aspects of rhodoliths.

1.2 Methods of Study

Traditionally rhodoliths have been obtained by dredging. This method is particularly efficient for gathering deep water rhodoliths, but it has the disadvantage of collecting material from a number of, unknown, subenvironments and little ecological information is obtained. More recently shallow water rhodoliths have been extensively studied by diving (e.g. Adey and McKibbin 1970, Bosellini and Ginsburg 1971, Bosence 1976a). It is these in situ studies based on detailed observations and experiments which have led to a greater understanding of the processes involved in the formation of rhodoliths.

1 Department of Geology, University of London Goldsmiths' College, Rachel McMillan Building, Creek Road, London SE8 3BU, Great Britain

Coated Grains (ed. by T.M. Peryt)

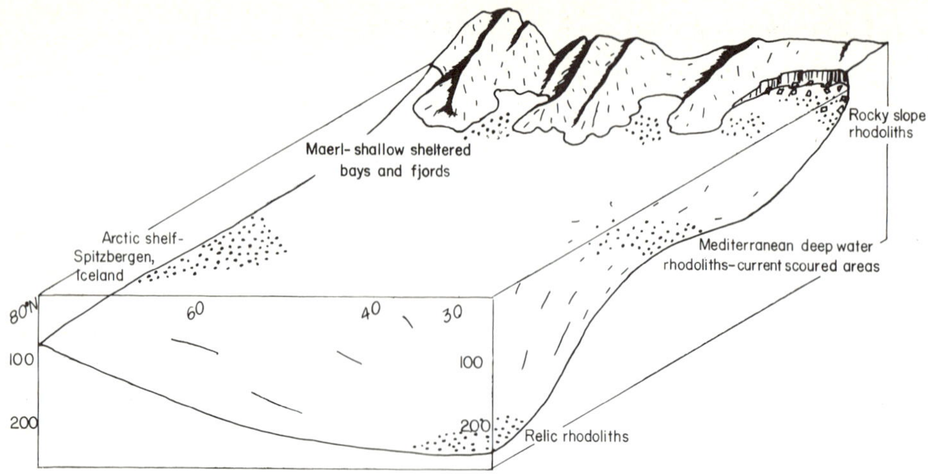

Temperate-Arctic and deep water rhodolith environments

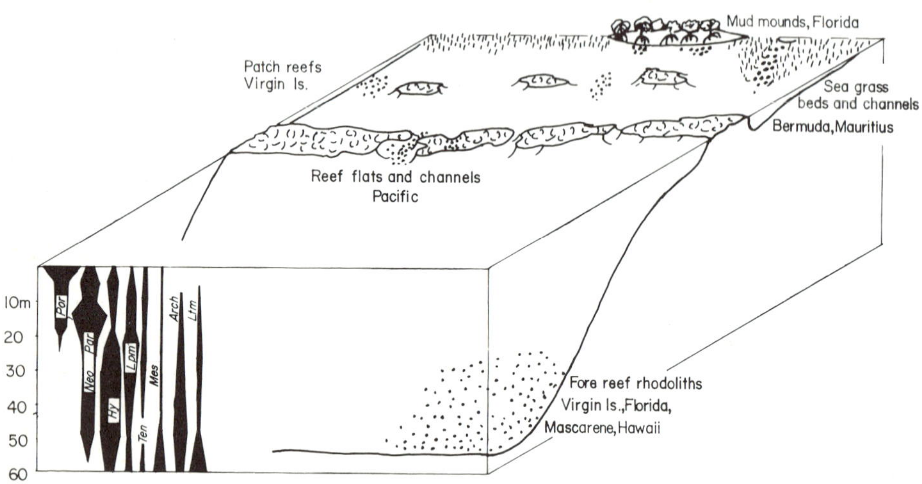

Tropical rhodolith environments

Fig. 1. Environments of rhodolith occurrence. Drawings not to scale. For sources see text. Coralline depth ranges from Adey and Boykins (in press). *Por* Porolithon, *Neo* Neogoniolithon, *Par* Paragoniolithon, *Hy* Hydrolithon, *Lpm* Lithophyllum, *Mes* Mesophyllum, *Arch* Archaeolithothamnium

2 Environments of Rhodolith Occurrence

2.1 Temperate and Polar Rhodoliths

2.1.1 Maerl

Maerl (= Maërl, Lemoine 1910) is a Breton term for the unattached coralline algae and algal gravels found off the northwest coast of France. This useful term has since been extended to cover unattached coralline algae (rhodoliths) and their debris extending from the Mediterranean (Jacquotte 1962) to Arctic waters (Adey and Adey 1973). The species composition of the maerl changes with latitude. Arctic waters are dominated by *Lithothamnium tophiforme* and boreal waters by *L. glaciale* (Adey and Adey 1973). Rhodoliths of *L. glaciale* can be very abundant, covering the sea bed up to 4–5 sq. miles with rhodolith sizes up to 15–20 cm (Kjellman 1883). Similarly, Foslie (1894) describes extensive deposits off the Norwegian coast forming banks up to 3 km long with rhodoliths "... up to nearly 2 feet in diameter ...". Off the western coasts of the British Isles and France the maerl is composed mainly of *Phymatolithon (Lithothamnium) calcareum* and *Lithothamnium corallioides* (Adey and Adey 1973, Bosence 1976a, 1980). Live maerl is found from the lowest intertidal zone down to 27 m, but is most abundant between depths of 1 and 10 m. The environmental conditions required for luxuriant growth are shallow clear waters which are sheltered from excessive wave and tidal currents. Light has been shown to be the major factor for the coralline distribution of these species (Adey and McKibbin 1970, Bosence 1976a) within their probably temperature-controlled geographic range (Adey and Adey 1973). Bosence (1976, 1980) has shown previously that the maerl is composed of a number of sedimentary facies, each with its own distinctive rhodolith morphology and associated invertebrate fauna (Bosence 1979). Both *P. calcareum* and *L. corallioides* can form in situ build-ups 10–20 cm above the surrounding sea floor of branching and intergrown rhodoliths. In more exposed areas algal debris gravels are found with more densely branched rhodoliths. Carbonate production studies on the maerl show that production is equivalent to tropical green algal carbonate production (Bosence 1976b, 1980).

Similar facies occur in south west Britain (Bosence 1980), Brittany (Lemoine 1910, Cabioch 1969) and north west Scotland (Farrow et al. 1979).

The species *P. calcareum* and *L. corallioides* also extend south to the Mediterranean where they form extensive areas of maerl down to depths of 60 m (Jacquotte 1962, Pérès 1967, Caulet 1972). They are joined by an additional maerl species *Lithothamnium valens.*

Further south on the north-west African shelf Summerhayes et al. (1976) record a maerl facies.

2.1.2 Rocky Slope Rhodoliths

Hard substrate slopes throughout the world are encrusted to a greater or lesser extent by corallines. If these crusts break off and continue to grow as they move downslope they may form rhodoliths. Rhodoliths from these environments have not been studied in detail but have been noted by Adey (1966) for crusts of *Lithothamnium* up to 10 cm thick on rocky coasts of the north eastern United States. In semi-exposed areas the crusts are broken off and growth continues on the crusts to form a rhodolith bottom adjacent to the slope. Similarly, Laborel (1961) reports pebbles coated with *Lithophyllum expansum* and *Neogoniolithon mamillosum* accreting downslope off the Ile de Lemnos in the Mediterranean.

2.2 Deep Water and Relic Rhodoliths

Rhodoliths from water depths of 50 to 250 m have been described by many authors including: Thompson (1877) from the Challenger Bank south-east of Bermuda, Parker and Curray (1956) from the north-west shelf of the Gulf of Mexico, McMaster and Conover (1966) and Martin (1971) from the eastern Atlantic, Siesser (1972) from the South African shelf, Logan et al. (1969) from the Yucatan shelf, MacIntyre and Milliman (1970) from the shelf off the south east of North America, Adey and MacIntyre (1973) from the Hawaiin Chain, Toomey (1975) from the West Flower Garden bank of the Gulf of Mexico and Blanc (1968) from the Mediterranean. These rhodoliths vary in size from 5 – 30 cm and are constructed of concentric layers of crustose corallines together with foraminifers, bryozoa and serpulid worms. They are often bored by molluscs, worms, sponges, algae and fungi.

Probably because these rhodoliths are beyond the safe limit for diving no detailed studies on these deeper water rhodoliths has been undertaken. The "coralligène de Plateau" of Blanc (1968) contains abundant rhodoliths ("pralines, oncoliths, pisoliths or algal balls"). These are being constructed by *Neogoniolithon* and occur on current swept submarine plateaus at depths of 80 or 90 m. McMaster and Conover (1966) find that the greatest depths for algal growth on rhodoliths from the Canary islands is between 60 and 70 m. At these depths coralline growth rates would be expected to be slow (Adey and McIntyre 1973) and McMaster and Conover (1966) report an age of 1500 ± 100 for the core of an algal ball. Bosellini and Ginsburg (1970) suggest that the turning required for the concentric growth at these depths is best explained by currents scouring the underlying sand. However, the possibility of turning by benthic organisms cannot be discounted.

Many of the deeper water rhodoliths are dead and radiocarbon dating indicates ages up to 20,000 years. They are, therefore, interpreted as relic grains →

Fig. 2. Locations of rhodoliths referred to in text.
Maerl rhodoliths: *1* Foslie 1894; *2* Farrow et al 1979; *3* Bosence 1976a; *4* Lemoine 1910; *5* Adey and McKibbin 1970; *6* Adey 1966; *7* Jacquotte 1962; *8* Summerhayes et al. 1976. Rocky slope rhodoliths: *1* Adey 1966; *2* Laborel 1961. Deep water live rhodoliths: *1* McMaster and Conovar 1966; *2* Blanc 1968. Relic rhodoliths: *1* Thompson 1877; *2* Parker and Curray 1956; *3* John and Lieberman 1979; *4* Logan et al. 1969; *5* MacIntyre and Milliman 1970; *6* Siesser 1972. Fore reef rhodoliths. *1* Adey and Boykins in press; *2* this chapter; *3* Montaggioni 1979; *4* this chapter. Reef flat and Channel rhodoliths: *1* Weber von Bosse and Foslie 1904; *2* Adey and Boykins in press; *3* Johnson 1961; *4* Konishi 1975; *5* Montaggioni 1979; *6* Flood, this vol. Back reef rhodoliths: *1* Bosellini and Ginsburg 1971; *2* this chapter; *3* Adey 1966; *4* Montaggioni 1979; *5* Logan et al. 1969; *6* Zanevald 1958

■ Maerl rhodoliths
□ Rocky slope rhodoliths
△ Deep water live rhodoliths
▲ Relic rhodoliths
● Fore reef rhodoliths
○ Reef flat & channel rhodoliths
X Back reef rhodoliths

which formed at times of lower sea level stands (Logan et al. 1969, MacIntyre and Milliman 1970, Martin 1971, Siesser 1972, Toomey 1975).

Therefore, it appears that rhodoliths may grow, and be turned to produce their characteristic concentric structure, down to depths of about 90 m. These rhodoliths have a slow growth rate and may be 1000's of yrs old (Growth Rates). The deeper water rhodoliths at 100 – 250 m have core ages up to 20,000 yrs, and these are probably relic. Bioerosion and mineralisation may be used to recognise relic rhodoliths in ancient sequences. Bioerosion in living rhodoliths may be recognised by growth interactions between borer and host (Fig. 3A). The relic algal grains in the Mediterranean are reported to be impregnated with iron and magnesium hydroxides, giving them a red or black colour (Leclaire 1972). This mineralisation is used as evidence for interpreting the rhodoliths as relic. Mineralised dead thalli of *P. calcareum* have also been described from shallow waters in the Firth of Clyde (Deegan et al. 1973).

2.3 Tropical Rhodoliths

2.3.1 Fore Reef

Montaggioni (1979) describes abundant (50 – 200 per m^2) rhodoliths from depths of 25 – 60 m on the fore reef slopes of the Mascarene Is. Diameters vary from 27 – 140 mm and shapes are mainly "almondal". The rhodoliths are constructed mainly by *Lithophyllum incrustans* and *Lithoporella melobesioides* with minor amounts of *Porolithon, Lithothamnium, Dermatolithon, Hydrolithon* and encrusting foraminifers. The coralline thalli are mainly globular and laminar in construction.

Adey and Boykins (in press) mention abundant rhodoliths of *Archaeolithothamnium episoredian, Lithothamnium pulchrum* and *Mesophyllum* in depths of 60 – 90 m off the Hawaiin reefs.

Recently Shinn (1980, pers. comm.) has reported large ares of rhodoliths in 35 – 55 m depths off the front of the Florida reefs. These submersible observations have since been confirmed by dredging by myself and Shinn in 1980. The areas of rhodoliths cover a wide band (1 km × at least 50 km) and have been dredged from similar sites 300 km to the north (Dr. C. Hoskins, pers. comm.). The rhodoliths have a wide variety of size, internal morphologies and structures (Figs. 3B – F) and are to be described in detail in the future. Similar occurrences are known from the Virgin Is in the Caribbean (Shinn 1980, pers. comm.).

Fig. 3. A Growth reaction by *Phymatolithon calcareum* to boring by *Polydora*. Mannin Bay, Ireland. × 100. **B** Branching III rhodolith. 40 m fore reef slope off French Reef, Florida. × 1. **C** Branching IV rhodolith, 40 m, fore reef slope off French Reef, Florida. × 1. **D** Branching IV rhodolith, 40 m, fore reef slope, off French Reef, Florida. × 1. **E** Inner core of rhodolith micritised and replaced by gypsum (now calcite) with outer fresh unaltered coralline. Coralline Algal Biostrome, Miocene Malta. **F** Ellipsoidal laminar rhodolith, 40 m, fore reef slope, off French Reef, Florida. × 1. **G** Discoidal laminar rhodolith, 40 m, fore reef slope, off French Reef Florida. × 1. **H** Apical expansion and lateral branch indicating increased turning and apical abrasion in *Lithophyllum* rhodolith. Coralline Algal Biostrome, Miocene, Malta. × 15

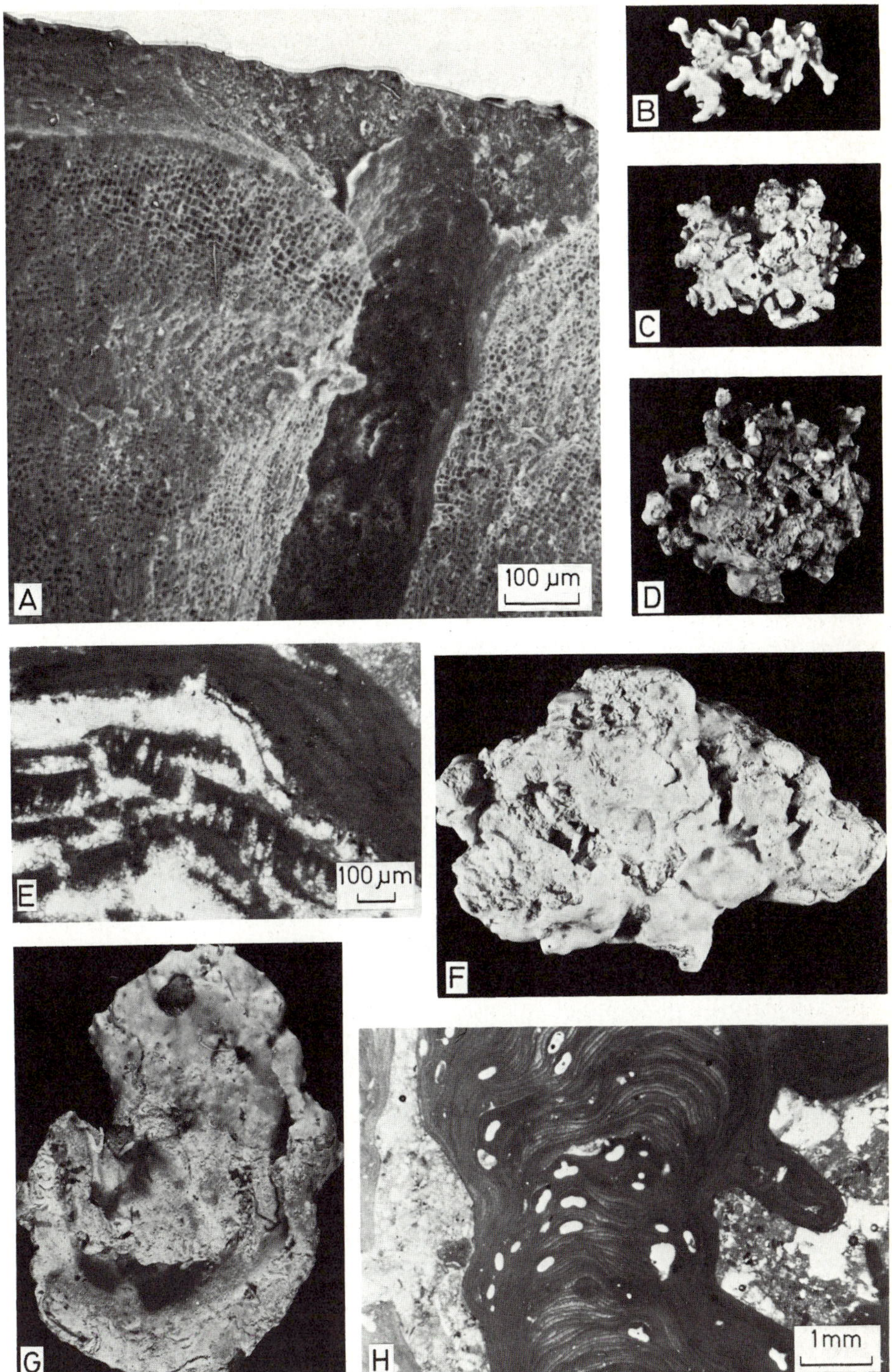
A
100 μm
B
C
D
E
100 μm
F
G
H
1mm

Littler (1973) mentions rounded aggregates of *Hydrolithon breviclavium* which roll or slide down fore reef slopes. They presumably originate from the leafy crusts of *H. breviclavium* found higher up on the fore reef slope.

2.3.2 Reef Flat and Channel Rhodoliths

The intertidal reef flat at Haingsisi (off S. W. Timor) is built up "for as far as the eye can see" of rhodoliths of *Mesophyllum (Lithothamnium) erubescens* (Weber van Bosse and Foslie 1904). The rhodoliths vary in size from 1 – 9 cm in diameter and occur with occasional *Lithothamnium australe, Lithophyllum okamurai* and *Hydrolithon (Goniolithon) reinboldii.*

Pollock (1928), describing the fringing reef flat of Oahu, Hawaiin Islands states that all pools and channels on the flat contain nodular forms of branching corallines. These may also be seen in the raised fossil reefs.

Johnson (1961) also reports reef flat rhodoliths of *Lithophyllum molluccense, L. kotschyanum* and *Neogoniolithon frutescens* from the Saipan Mariana Is.

Rhodoliths from tidal channels between fringing reefs of the Ryukyu Islands have been described by Konishi (1975). The rhodoliths are found covering up to 40% of very coarse sand bottoms. They consist of branching forms of *Lithophyllum* and *Mesophyllum* with a mean size of 45 mm. Most of the rhodoliths are embedded in the sand and are only growing on their upper surfaces. Growth of *Acropora, Pocillopora* and *Favia* on the upper surfaces of the rhodoliths give further evidence of stabilisation.

Reef flats and discharge channels in reefs from the Mascarene Is. are commonly floored with rhodoliths (Montaggioni 1979). The rhodoliths are smaller in Reunion and Mauriti Is. (av. 40 mm) than on Rodriguez (av. 90 mm) and occur as lamellar-columnar, globular or lamellar growth forms. The rhodoliths are constructed from a mixture of corallines (*Hydrolithon, Lithothamnium, Lithoporella* and *Porolithon*) together with encrusting foraminiferans, molluscs, serpulids, bryozoans and sponges. In Mauritius *Lithophyllum* cf. *molluccense* and *Lithothamnium* sp. form monospecific branching rhodoliths. Flood (this volume) reports rhodoliths from reef flats of the Great Barrier Reef, Australia.

2.3.3 Back Reef Rhodoliths

Rhodoliths have been reported from a number of backreef environments as summarised below.

In a now classic study on rhodoliths, Bosellini and Ginsburg (1971) described the growth and occurrence of rhodoliths from sea grass beds and shallow channels in Whalebone Bay, Bermuda.

Branched unattached forms of *Lithothamnium occidentale* grow in the relatively sheltered *Thallasia* beds and may be transported out of these beds into the adjacent sand floored channels. Here they are turned by currents and develope a laminar coating.

On Tavernier Bank (Turmel and Swanson 1976) I have seen *Neogoniolithon (Goniolithon) strictum* rhodoliths in the exposed northern part of the bank. *N. strictum* normally grows as long branches arising from crusts on hard substrates, but these thalli may become detached and continue to grow as rhodoliths as they are transported shorewards.

Adey (1976) reports rhodoliths in sands and gravels between algal and coral patch reefs in Boiler Bay, St. Croix.

In the Indian Ocean rhodoliths are known from *Halodule* and *Halophila* grass beds in muddy bays around Mauritius (Montaggioni 1979). They occur in concentrations of up to 50 m^2 with a mean size of 50 mm and are slightly branched or bumpy on their external surface.

Kornicker and Boyd (1962) and Logan et al (1969) describe a luxuriant growth of unattached corallines from the lagoon of the Alacran reef complex on the Campeche Bank. Similarly a shallow water *"Lithothamnium"* bank is described by Zanevald (1958) formed of *Mesophyllum (Lithothamnium) erubescens* off Bonaire. Adey and MacIntyre (1973) suggest that this may be *Neogoniolithon strictum.*

3 Ecological Aspects of Rhodoliths

3.1 Ecology of Rhodolith Forming Corallines

Field and experimental studies by (Adey 1966, 1970; Adey and McKibbin 1970; Adey and Adey 1973) have shown that the distribution of coralline algae is mainly controlled by temperature and light in marine waters. Most species show a combined effect but others are controlled to a greater extent by light (Fig. 4A) and others by temperature (Fig. 4B).

The biogeographic zones of coralline algae in the northern Atlantic are considered to be determined by water temperature (Adey and Adey 1973).

Adey (1970) has published distribution maps for present-day coralline genera and the table below summarises this work (the taxa are listed in order of percentage abundance within the floras for each province)

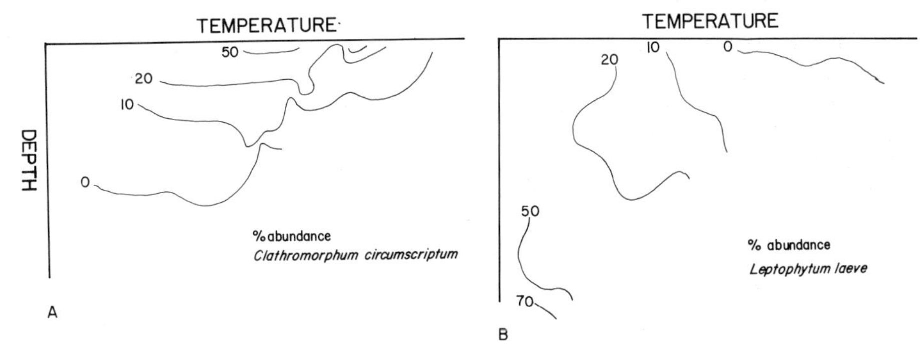

Fig. 4A, B. Examples of **A** dominant light control on distribution of *Clathromorphum circumscriptum* and **B** dominant temperature control on distribution of *Leptophytum laeve.* (After Adey 1966)

	North
Arctic	*Lithothamnium, Clathromorphum, Leptophytum.*
Temperate	*Lithothamnium, Lithophyllum, Leptophytum, Clathromorphum, Phymatolithon Tenarea.*
Tropical	*Porolithon, Neogoniolithon, Tenarea, Hydrolithon, Lithophyllum, Lithoporella, Mesophyllum, Archaeolithothamnium, Lithothamnium, Phymatolithon, Leptophytum, Pseudolithophyllum.*
Temperate	*Mesophyllum, Lithothamnium, Pseudolithophyllum, Clathromorphum.*
Antarctic	*Mesophyllum, Leptophytum, Pseudolithophyllum.*
	South

This table also shows the increased diversity of tropical floras (where genera occur in depth zones, see below) and the differences between northern and southern floras.

Data on the depth distribution of coralline genera is only recently being compiled (Adey and Boykins, in press). Pacific and Caribbean tropical reef coralline genera can be grouped into a number of depth zones as shown in Fig. 1.

The deeper water genera can also occur in cryptic environments in shallow waters.

On a smaller scale rhodoliths dominated by one species may be restricted to similar water temperatures and show equatorial submergence. *Phymatolithon calcareum* commonly forms rhodoliths in depths of 1 to 10 m off the west coast of Ireland, 14 to 25 m in Brittany (Lemoine 1910) and down to 60 m in the Mediterranean (Jacquotte 1962). Similarly in turbid waters corallines may be present in shallower waters than in clearer water areas. Adey and MacIntyre (1973) illustrate mearl rhodoliths at 60 m in clear coastal (Newfoundland, N.W. Norway) waters and at 10 m in the more turbid (Gulf of Maine, British) waters.

Within the above ecological restraints of light and temperature rhodolith occurrence will be controlled by substrate and hydraulic energy. To develope a rhodolith bed suitably sized grains for spore settlement must be present. However, once rhodoliths are established, future recruitment is probably by breakage and overgrowth of rhodoliths. The core of rhodoliths in the maerl beds is frequently found to be an old abraded grain of coralline algae (personal observation). An exposure to moderate hydraulic energy is required to overturn crusts and branched growths of newly settled corallines which results in the concentric (crustose) and radial (branching) growths characteristics of rhodoliths. In one unusual case it has been suggested that the force for turning coralline crusts to produce rhodoliths has been biogenic rather than hydraulic (Bosence and Pedley 1982). If the hydraulic energy is too high, then rhodoliths are broken down into coralline gravels composed of branches and crusts. This erosion can be recorded within the rhodoliths which are smaller, densely branched or crustose with erosion surfaces and fractures (Bosence and Pedley 1982).

The wide range of environments in which the above conditions prevail today is documented in the first section of this paper and includes many intermediate energy shelf environments from the tropics to subpolar regions.

3.2 Size and Shape Variations

The greatest amount of work on rhodolith morphology has been undertaken on the maerl species of the northeast Atlantic. Lemoine (1910) plotted the distribution of growth forms of *Phymatolithon (Lithothamnium) calcareum* form *squarulosa* (= ellipsoidal III) and *crassa* (= spheroidal IV) for an area of southern Brittany. Form *squarulosa* is found throughout the Isles of Brehat between depths of 4 – 18 m and in the outer regions of the Baie de Concarneau at depths of 10 – 15 m. This form is often found on muddy substrates where there is little current activity. She concludes that this branching form is adapted to life in muddy environment. Form *crassa* is found in slightly shallower waters (3 – 15 m) to the north of the Isles de Brehat and south of the islands in the mouth of the Baie de Concarneau. She quotes from Foslie (1894) that the ball-like forms grow under the influence of strong currents.

A systematic description of the growth forms of the maerl species is given by Cabioch (1966) and in 1969 she published a detailed study of the maerl beds of the Baie de Morlaix, showing that the growth forms described were characteristic of different substrates. A coarse shell sand contains all forms of *Lithothamnium corallioides* var. *corallioides* and with *P. calcareum* forms *calcarea* (= ellipsoidal III), *squarulosa* (= spheroidal III) and *compressa* (= discoidal III) are present. In contrast, a fine sand and mud substrate supports *L. corallioides* var. *minima*.

Similar variations in growth form are found in Mannin Bay, Ireland where their morphogenesis and environmental setting have been studied (Bosence 1976a). Both the maerl species show similar variations of growth forms. The commonest shapes are ellipsoidal and spheroidal and these two classes show no obvious pattern of distribution. Discoidal forms are concentrated on the sand-substrates in the south in contrast to the coarse algal gravels of the rest of the bay. The branching density classes illustrate a close correlation with exposure to wave-induced bottom currents as shown by the presence of ripples (72% of stations with more than 5% branching density IV are rippled). Experiments with models and specimens of rhodoliths in wave tanks show that ellipsoidal forms are more easily transported than spheroidal forms and that discoidal forms are the most stable. Open branched forms are found to be more stable than densely branched forms of the same shapes and sizes. In Mannin Bay the more open branched forms are found in the sheltered parts of the bay (cf. Lemoine 1910). This apparent anomaly is explained as follows. When a branching thallus is rolled the apices are abraded and the meristem is damaged. Further growth occurs within the spaces between the preexisting branches as lateral branches. The proportions of dichotomous (apical) to intercalary (lateral) branching in rhodoliths shows that the densely branched forms are typified by lateral branching and the open by dichotomous branching. Therefore growth in a high energy area leads to a densely branched rhodolith which is more easily transported but, because of its dense branching, is a strong thallus.

The discoidal forms growing on the sand substrates are interpreted as a result of growth on an even substrate which does not permit downward growth. Frequent turning will only permit sideways growth, which results in the discoidal growth form. The spheroidal and ellipsoidal forms grow on coarse open algal gravels and these shapes do not appear to be related to any regional parameters but probably relate to their position with respect to the substrate, its nearest neighbours and breakage.

As far as I am aware the only other detailed observations on rhodolith morphology and environment are those from Whalebone Bay, Bermuda by Bosellini and Ginsburg (1971). Here unattached *Lithothamnium occidentale*, which normally grows in sea grass beds, is occasionally washed out into sandy channels. Here, rolling causes branch tips to flatten and join laterally with adjacent branches. Subsequent growth is in the form of laminar crusts. Large rhodoliths which get stabilised in the channels produce columnar growths on their upper surfaces.

In summary, it has now been shown from two environments that densely branched and laminar rhodoliths are the result of frequent turning and high energy conditions. The apical abrasion encurred during turning appears to lead to intercalary branching or lateral filament growth below the apex (Fig. 3H). This results in branches joining laterally. Subsequent growth in this higher energy environment may be recorded by concentric laminar growth. Columnar morphologies are as yet poorly understood. Bosellini and Ginsburg (1971) suggest they form from stabilised rhodoliths but Bosence and Pedley (1982) find columnar growths on rhodoliths that have been turned and the columns may show the same lateral growth in response to apical abrasion.

3.3 Ecological Succession Within Rhodoliths

Several workers have shown that (a) small pebbles or rhodoliths may have a different encrusting flora than larger pebbles or rhodoliths and (b) that the inner parts of the large rhodoliths have different taxa to those on the outside.

Adey (1970) has shown that different pebble weights have different floras and Adey and MacIntyre (1973) point out that the succession of species in rhodoliths may be the result of change in size rather than a change in environment (Fig. 5).

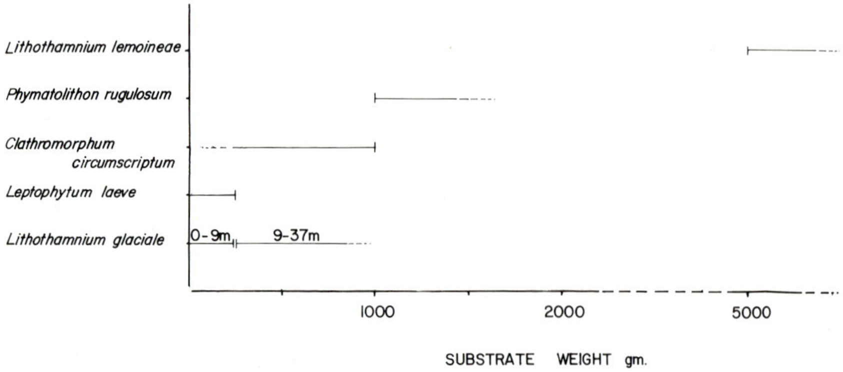

Fig. 5. Relationship between pebble weight and encrusting corallines. (Data from Adey 1966, 1970; Steneck 1978)

Steneck (1978) found similar results in the Gulf of Maine and suggests how corallines may have different strategies for dealing with overturning. High Arctic corallines (*Lithothamnium laeve, Clathromorphum circumscriptum*) may be adapted for periods of long darkness by storing carbohydrates to last through winter (Adey 1973). The same adaptation would favour crusts on overturned rhodoliths.

Similarly coralline crusts with high marginal growth rates are more suitably adapted to substrates which are more frequently turned. No experiments have been attempted to test whether spores preferentially settle on particular sizes or whether they settle ubiquitously and the less tolerant species are removed or die on frequently turned substrates. The latter is suggested in a detailed study by John and Lieberman (1979) on epiphytes on rhodoliths from the coast of Africa. Here seasonal storms remove the more fragile members of the epifloral community which then reestablish themselves during the non-stormy season.

Succession within rhodoliths has not been studied in detail. Logan et al. (1969) and Toomey (1975) describe rhodoliths from the Gulf of Mexico with inner layers of *Lithoporella* and *Lithophyllum* and outer layers of *Lithothamnium* and the foraminifera *Gypsinia*. They suggest that the inner assemblage is more representative of shallower water conditions than the present day outer assemblage and that the change in encrusters may reflect the Holocene sea level rise. An alternative explanation may be that the corallines prefer different substrate sizes and that they accumulated under similar conditions.

Bosence and Pedley (1982) indicate ecological succession within Miocene rhodoliths with the inner cores constructed by *Mesophyllum* and *Lithoporella* and the outer parts by *Lithophyllum, Lithothamnium* and *Mesophyllum* (Table 1).

Further work is required on recent rhodoliths to establish if this succession can be of use in palaeoecology. An important question for palaeontologists is whether this succession represents changing environments or the change in rhodolith size.

3.4 Growth Rates of Rhodoliths

The most detailed work on rhodolith growth rates has been undertaken on the maerl species of the North East Atlantic. Adey and McKibbin (1970) measured rates by measuring, marking, releasing, recapturing and measuring rhodoliths at monthly intervals. Although variable the results show a seasonal growth pattern with maximum growth, characterised by small cells, in the summer and no measurable growth in the winter which is characterised by large cells. This seasonal growth is thought to give rise to seasonal growth rings which were used by Bosence (1980) to establish carbonate production rates from rhodoliths of the maerl species (up to 422 g $CaCO_3$ m^{-2} yr^{-1}). Summed yearly growth rates of branches of 0.105 mm for *Lithothamnium corallioides* and 0.486 mm for *Phymatolithon calcareum* were recorded by Adey and McKibbin (1970).

Konishi (1975), following work by Chave and Wheeler (1965) and Moberley (1968), used magnesium concentrations and growth line methods to locate seasonal growth layers in *Lithophyllum* rhodoliths. Rates of 0.40 to 0.44 mm

Table 1. Frequency of observed transitions in growth form and taxa from centre to exterior of multispecific rhodoliths. Data from Bosence and Pedley (1982)

Rhodolith pavement facies (25 rhodoliths with 63 outward transitions)

Laminar (box)	—13%→	Laminar (con)	—28%→	Branching (IV)
Mesophyllum commune		*Mesophyllum commune*	↘ 33%	
Lithophyllum albanense		*Lithoporella melobesoides*		
Lithoporella melobesoides		*Lithophyllum albanense*		Columnar
		Archaeolithothamnium affine		*Mesophyllum commune*
				Lithophyllum albanense
				Lithophyllum species "a"

Crustose Pavement Facies (14 rhodoliths with 36 outward transitions)

Laminar (box)	—8%→	Laminar (con)	—30%→	Branching IV	→	Laminar (con)
Mesophyllum commune		*Mesophyllum commune*	↘	*Mesophyllum commune*		*M. commune*
Lithoporella melobesoides		*Lithoporella melobesoides*				
		Lithophyllum species "b"		Columnar	→	Laminar (con)
		Lithothamnium praefructiculosum		*Mesophyllum commune*		*M. commune*
				Lithophyllum albanense		*L.* species "b"
				Lithophyllum species "b"		
				Lithothamnium magnum		
				Lithothamnium praefructiculosum		

yr^{-1} were obtained. The use of magnesium concentrations must be used with care as they have been shown to be related to both internal as well as external changes in the corallines (Moberley 1968).

Because rhodoliths are unattached, they are suitable for isolating in aquaria for calcification rate experiments using radioisotopes (Bosence 1976b, Boehm et al. 1978). Calcium^{45} uptake by the corallines can be measured on the sea bed by placing rhodoliths in sealed aquaria for short periods. Boehm et al. (1978) have produced rates of 14 μg $CaCO_3$ hr^{-1} for *Phymatolithon calcareum* in the western Baltic (equivalent to apical branch growth rates of 0.5 – 2.7 mm yr^{-1}).

Tropical rhodoliths have not received so much attention but branch growth rates have been recorded for the Caribbean (non-rhodolith) corallines *Neogoniolithon westindianum* as 22 mm yr^{-1} (Adey and Vassar 1974) and for *Lithophyllum congestum* as up to 80 mm yr^{-1} (Steneck and Adey 1976). These results are up to one order of magnitude higher than rates for temperate corallines.

The rates quoted above can be used for monospecific branching rhodoliths but for multispecific and laminar rhodoliths growth rates are more difficult to measure and interpret. Instantaneous rates will be for concentric marginal crust growth and there may be long periods of time between successive crusts (i.e., "growth discontinuities" of Bosence and Pedley 1982). In additon rhodoliths may be buried and exhumed and recolonised. Bosence and Pedley (1982) describe a Miocene rhodolith in which the core has been bored, micritised and replaced by gypsum, presumably in a littoral or supralittoral zone, prior to colonisation by well preserved corallines (Fig. 3E).

Adey and MacIntyre (1973) consider that with continuous growth deep water rhodoliths up to 30 cm in size could have ages of up to 800 yrs because of slow growth rates. With intermittant growth, which is essential for concentric laminar forms, ages must be much higher. McMaster and Conover (1966) record radiocarbon ages of 400 ± 40 and 1500 ± 100 yrs for crusts and cores of living rhodoliths from 63 – 66 m off the Canary Islands.

These figures seem quite reasonable as it may be calculated that a rhodolith with a diameter of 20 cm with concentric laminar growth (one crust per mm of radius with two growing margins) may be up to 75,000 yrs old with Arctic or Boreal growth rates (0.2 mm/month: Adey and Vassar 1974) or up to 600 yrs old with tropical growth rates (2 mm/month: Adey and Vassar 1974). These calculations are based on continuous growth of crusts and could be much higher with burial and death.

3.5 Associated Fauna and Flora

Rhodolith bottoms present a diversity of niches for benthic plants and animals. Many workers have commented on the rich encrusting fauna and flora of rhodoliths. Detailed descriptions of the benthic faunas associated with the temperate maerls of the north east Atlantic include Cabioch (1968), Keegan (1974) and Bosence (1979). Keegan (1974) and Bosence (1979) recognises distinct faunal communities in the various facies of the maerl. The main ecological groups of organisms inhabiting the maerl substrates are as follows: A vagile epifauna made up principally of crustaceans and gastropods living within the algal branches; sessile epifauna of byssate and cemented bivalves, serpulids, spirorbids, bryozoa, and foraminifera; An infauna of boring bivalves and polychaetes within the algal thalli; and a burrowing infauna within the sediment deposited around the rhodoliths of bivalves, polychaetes and crustacea.

4 Conclusions and Implications for Geological Studies

The wide range of environments where rhodoliths may occur is emphasised in the first part of this chapter. With further investigation of shelf seas this will undoubtedly be enlarged. In the past rhodoliths have been taken to indicate shallow warm waters (see discussion in Adey and MacIntyre 1973) but this is quite clearly incorrect.

The majority of recent coralline genera can be recognised in fossil material and as their present day depth and geographic ranges are documented these genera can be used for palaeoenvironmental interpretation. Of particular use in palaeoecology is the slow rate of evolution of the corallines and their restricted depth ranges within the photic zone (Adey and Boykins, in press).

The external morphology of rhodoliths has been shown to be a useful indicator of hydraulic energy in both recent and fossil examples. The internal sequence of morphologies gives a record of changing hydraulic conditions. Further research on ecological succession within rhodoliths may determine whether this is due to changing external or internal factors.

The realisation that living rhodoliths may be hundreds to tens of thousands of years old means that a relatively long term record of sea floor conditions is preserved compared to the relatively much shorter life span of individual organisms.

In conclusion, it is only in the last ten years that the full palaeoecological significance of rhodoliths has become apparent. They have been shown to preserve information on depth, geography, temperature and prevailing hydraulic energy.

References

Adey WH (1966) Distribution of saxicolous crustose corallines in the northwestern North Atlantic. J Phycol 2:49–54

Adey WH (1970) Some relationships between crustose coralline algae and their substrates. Sci Isl 2:21–25

Adey WH (1973) Temperature control of reproduction and productivity in a subarctic crustose coralline. Phycologia 12:111–118

Adey WH (1976) The algal ridges and coral reefs of St. Croix and their Holocene development. Atoll Res Bull 187:67

Adey WH, Adey PJ (1973) Studies on the biosystematics and ecology of the epilithic crustose corallines of the British Isles. Br Phycol J 8:343–407

Adey WH, Boykins W (in press) The crustose coralline algae of the Hawaiin Archipelago. Smith Contrib Sci

Adey WH, MacIntyre IG (1973) Crustose coralline algae; a re-evaluation in the geological sciences. Bull Geol Soc Am 84:883–903

Adey WH, McKibbin DL (1970) Studies on the maerl species *Phymatolithon calcareum* Pallas (nov. comb.) and *Lithothamnium corallioides* Crouan in the Ria de Vigo. Bot Mar 13:100–106

Adey WH, Vassar JM (1974) Colonisation, succession and growth rates of some tropical crustose coralline algae (Rhodophyta, cryptonemiales). Phycologia 14:55–69

Blanc JJ (1968) The sediments of the Mediterranean Sea. Oceanogr Mar Biol Annu Rev 6:373–454

Boehm L, Schramm W, Rabsch U (1978) Ecological and physiological aspects of some coralline algae from the western Baltic. calcium uptake and skeletal formation in *Phymatolithon calcareum.* Kieler Meeresforsch 4:282–288

Bosellini A, Ginsburg RN (1971) Form and internal structure of recent algal nodules (rhodolites), Bermuda. J Geol 79:669–682

Bosence DWJ (1976a) Ecological studies on two unattached coralline algae from western Ireland. Palaeontology 19:365–395

Bosence DWJ (1976b) Ecological and sedimentological studies on some carbonate sediment producing organisms, Co Galway, Ireland. Unpubl thesis, Univ Reading

Bosence DWJ (1979) Live and dead faunas from coralline algal gravels Co. Galway, Ireland. Palaeontology 22:440–456

Bosence DWJ (1980) Sedimentary facies, production rates and facies models for recent coralline algal gravels, Co. Galway, Ireland. Geol J 15:91–111

Bosence DWJ, Pedley HM (1982) Sedimentology and palaeoecology of a Miocene coralline algal biostome from the Maltese Islands. Palaeoclimatol Palaeoecol Palaeogeogr 38:9–43

Cabioch J (1966) Contribution à l'étude morphologique anatomique et systématique des deux Mélobesiées: *Lithothamnium calcareum* (Pallas) Areschoug et *Lithothamnium corallioides* Crouan. Bot Mar 9:33–55

Cabioch J (1969) Les fonds de maerl de la Baie de Morlaix et leur peuplement végétal. Cah Biol Mar 10:139–161

Cabioch L (1968) Contributions à la connaissance des peuplements benthiques de la Manche occidentale. Cah Biol Mar 9:493–711

Caulet JP (1972) Recent biogenic calcareous sedimentation on the Algerian shelf. In: Stanley DJ (ed) The Mediterranean Sea. Dowden Hutchinson and Ross, pp 261–278

Chave KE, Wheeler BD (1965) Mineralogic changes during the growth of the red alga *Clathromorphum compactum*. Science 147:621

Deegan CE, Kirby R, Rae I, Floyd R (1973) The superficial deposits of the Firth Clyde and its sea lochs. Inst Geol Sci Rep 73/9:42

Farrow G, Scoffin T, Brown B, Cucci M (1979) An underwater television survey of facies variation on the Scottish shelf between Colonsay, Islay and Jura. Scott J Geol 15:13–29

Foslie M (1894) The Norwegian forms of Lithothamnion. D Kgl Norske Vidensk Selske Skr 2:203

Jacquotte P (1962) Etude des fonds de maerl de la Méditerranée. Recl Trav Stn Mar Endoume 26:141–235

John DM, Lieberman DL (1979) Ecology of subtidal algae on seasonally devastated cobble substrates off Ghana. Ecology 60:1151–1161

Johnson JH (1961) Limestone building algae and algal limestones. Colo Sch Mines, 267 pp

Keegan B (1974) The marine fauna of maerl substrates on the west coast of Ireland. Cah Biol Mar 15:513–530

Kjellman FR (1883) The algae of the Arctic sea. K Sven Vetenskaps Akad Handl 20:350

Konishi K (1975) Tidal channel rhodolites in a fringing reef complex of Ishigakijima, southwest Ryukyus. Abs Int Symp Fossil Algae, Erlangen

Kornicker LS, Boyd DW (1962) Shallow water geology and environments of Alacran reef complex, Campeche Bank, Mexico. Am Assoc Petrol Geol Bull 46:640–673

Laborel J (1961) La concrétionnement algal 'conralligène' et son importance géomorphologique en Mediterranée. Recl Trav Stn Mar Endoume Bull 37:37–60

Leclaire L (1972) Aspects on the late Quaternary sedimentation on the Algerian Precontinent and in the adjacent Algiers-Balearic Basin. In: Stanley DJ (ed) The Mediterranean Sea. Dowdon Hutchinson and Ross, 765 pp

Lemoine P (1910) Repartition et mode de vie du maerl (*Lithothamnium calcareum*) aux environs de Concarneau (Finistère). Ann Inst Oceanogr Monaco 1:1–28

Littler M (1973) The distribution, abundance and communities of deepwater Hawaiian crustose corallines (Rhodophyta, Cryptonemiales). Pac Sci 27:281–289

Logan BW, Harding JL, Ahr M, Williams JP, Sleep RG (1969) Late Quaternary carbonate sediments of the Yucatan shelf, Mexico. In: Logan B et al. (eds) Carbonate sediments and reefs, Yucatan shelf Mexico. Am Assoc Petrol Geol Mem 11:5–128

MacIntyre IG, Milliman JD (1970) Physiographic features of the outer shelf and upper Atlantic continental margin, Southeastern United States. Geol Soc Am Bull 81:2577–2598

Martin L (1971) The continental margin from Cape Palmas to Lagos: bottom sediments and submarine morphology. In: Delany FM (ed) The geology of the East Atlantic Continental margin. 4. ICSU/IGS 70/16:79–96

McMaster RW, Conover JP (1966) Recent algal stromatolites from the Canary Islands. J Geol 74:647–652

Moberley R (1968) Microbrobe study of diagenesis in calcareus algae. Sedimentology 14:113–123

Montaggioni LF (1979) Environmental significance of rhodolites from the Mascarene reef province western Indian Ocean. Bull Cent Rech Explor Prod Elf Aquitaine 3:713–723

Parker RH, Curray JR (1956) Fauna and bathymetry of banks on the continental shelf northwest Gulf of Mexico. Am Assoc Petrol Geol Bull 40:2428–2439

Pérès J (1967) The Mediterranean Benthos. Oceanogr Mar Biol Annu Rev 5:449–533

Pollock J (1928) Fringing and fossil coral reefs of Oahu. Bernice P Bishop Mus Bull 55:56

Siesser WG (1972) Relict algal nodules (rhodolites) from the South African continental shelf. J Geol 80:611–616

Steneck R (1978) Factors influencing the distribution of crustose coralline algae in the Damariscotta River estuary, Maine. Unpubl thesis, Univ Maine

Steneck R, Adey WH (1976) The role of environment in control of morphology in *Lithophyllum congestum* a coralline algal ridge builder. Bot Mar 19:197–215

Summerhayes CP, Milliman JD, Briggs SR, Bell AG, Hogan C (1976) The shelf of Morocco and Spanish Sahara. Sedimentation off an arid coast. J Geol 84:277–300

Thompson CW (1877) The voyage of the 'Challenger'. The Atlantic. A preliminary account of the general results of the exploratory voyage of HMS Challenger during 1873 and early part of the year 1876. HMSO, London 423 pp

Toomey D (1975) Rhodoliths from the Upper Palaeozoic of Kansas and the Recent – A comparison. Neues Jahrb Geol Palaeontol Monatsh 4:242–255

Turmel R, Swanson RW (1976) The development of Rodriguez Bank: a Holocene mudbank in the Florida reef tract. J Sediment Petrol 46:497–518

Weber von Bosse A, Foslie M (1904) The coralline algae of the Siboga expedition. Leiden Bull Rep 61:110

Zanevald JS (1958) A *Lithothammnion* bank at Bonaire (Netherlands W. Indies) Blumen Suppl 4:206–219

III.3 Rhodoids in Temperate Carbonates from the Cenozoic of New Zealand

C. J. Burgess[1] and J. M. Anderson[2]

1 Introduction

Some genera of the calcareous red algae (*Phylum rhodphyta*) have an encrusting habit which may produce laminated nodules termed rhodolites by Bosellini and Ginsburg (1971) and which here are termed rhodoids. Rhodoids have been recognized from many modern and ancient environments and increasing use is being made of their paleoecology for paleo-environmental reconstruction (Bosellini and Ginsburg 1971, Adey and MacIntyre 1972, Milliman 1974, 1977, Bosence 1976, Buchbinder 1977, Studencki 1979). The importance of rhodolites in temperate carbonates is being increasingly recognized and this paper describes a number of occurrences from New Zealand Cenozoic sediments deposited under warm temperate conditions at about 45 °S latitude (Kennett et al. 1975, Molnar et al. 1975, Nelson 1978a).

2 General Geological Setting

New Zealand sits astride the active tectonic boundary of the Pacific and Indian plates, a longstanding situation which is widely reflected in its sedimentary record. Periods of relative quiescence have been interspersed with periods of more intense tectonic activity. Carbonate sediments are widely distributed throughout the Cenozoic sequences but are most abundant in the Oligocene, when the influence of Antarctic climatic and circulation changes were becoming more pronounced and surface temperatures were significantly below 20 °C and sedimentation rates became very low (Nelson 1978a). They vary in character from platform edge and peri-platform foraminiferal ooze to bryozoan mollusc and barnacle grainstones and packstones of shelf origin, often displaying massive tabular cross-beds and related traction current features. A general account of these Cenozoic limestones can be found in Nelson (1978a). Algal limestones composed largely of rhodoids form significant but limited accumulations within Cenozoic sequences (Fig. 1).

1 Robertson Research International Limited, Llandudno, North Wales, LL30 1SA, U.K.
2 Shell International Limited, The Hague, Netherlands

Coated Grains (ed. by T. M. Peryt)

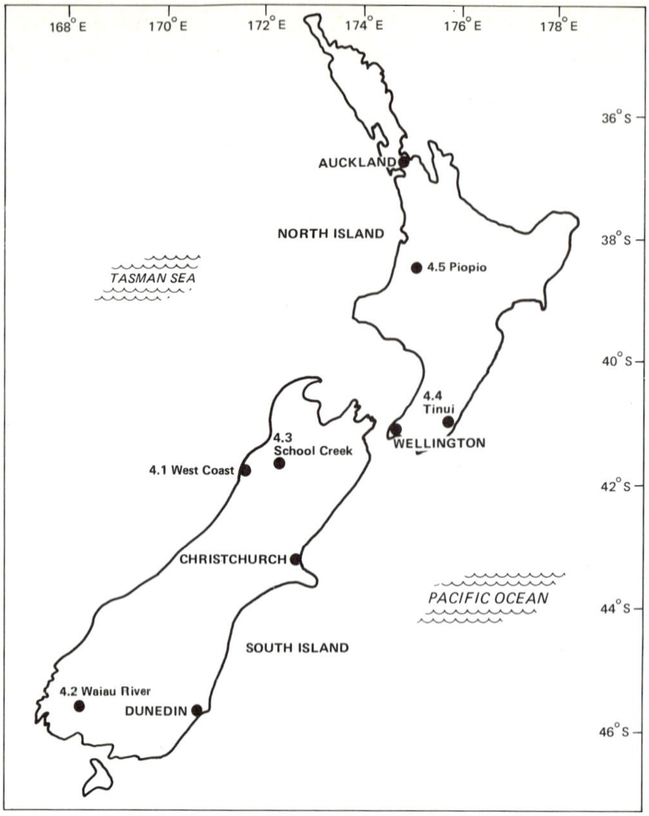

Fig. 1. Distribution of rhodoid localities described in the text

3 Rhodoids

3.1 Generic Composition

The most important genera in New Zealand algal limestones are *Lithothamnium, Lithophyllum* and *Mesophyllum.* Subsidiary contributing genera are *Melobesia* and *Lithoporella.* Milliman (1974) reported *Lithothamnium* in all climates while Johnson (1961) suggested that the diversity of this genus increases in higher latitudes relative to the number of other genera. Adey and MacIntyre (1973) considered *Lithothamnium* primarily a cool water form that lives in conditions of strong to very weak light. This genus is by far the most abundant contributor to rhodoid construction in New Zealand sediments.

Lithophyllum is a dominantly tropical genus but it also occurs in temperate waters (Milliman 1974). Adey and MacIntyre (1973) found that the genus is most prolific in warmer water under conditions of strong to very strong light (intertidal) at shallow depths or where the water is very clear. Although it may live at depths of 100 m in the tropics, it is extremely unlikely to have formed nodules at such a depth in temperate waters.

The genus *Mesophyllum* is a common encrusting form in New Zealand algal sediments. Adey and MacIntyre (1973) reported that this genus is primarily Antarctic in distribution, but Johnson (1957, 1964) recorded the genus from tropical sequences. It appears to fill a similar ecologic niche to *Lithothamnium*.

The genera *Melobesia* and *Lithoporella* are only minor contributors to algal sediments and little information is available on their ecology. From the present New Zealand coastal environment, which is probably similar in climate, wave energy and in some areas morphology to Tertiary marine conditions, Chapman (1974) described over ten species of *Lithothamnium*, two species of *Lithophyllum* and several species of *Melobesia* although not encrusting forms. Rhodoids occur commonly on rocky shores to the south of Wellington although when washed up onto beaches they are abraded and chipped suggesting that formation occurs outside the area of greatest energy. Wave energy on this coast can be great, with the exceptional wave of 8 m height and wave energy of 15 $J\ m^{-2} \times 10^{-3}$. Such waves would turn coated nodules at a depth of 50 m (E. Mathews 1981, pers. comm.).

Rhodoids are not always composed of one genus and a complex generic make-up, with intergrowing encrusting bryozoa is common (see Sect. 3.2).

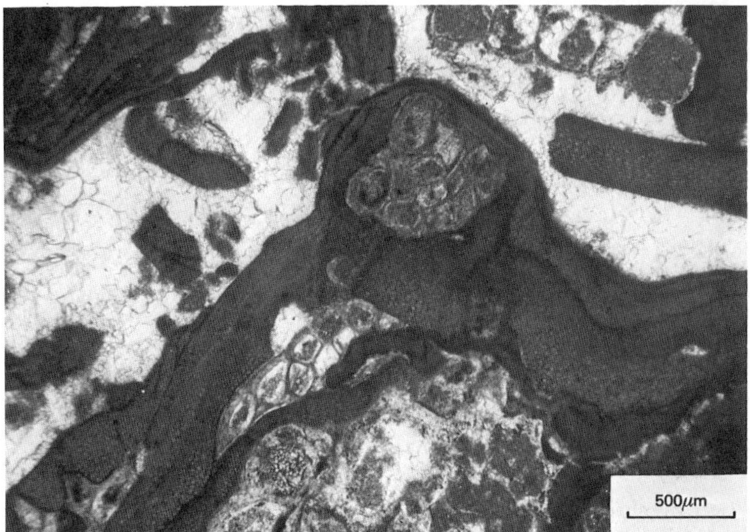

Fig. 2. Encrusting bryozoan in *lithothamnium* layers on the margins of a rhodoid from the West Coast. Detrital algal fragments are also shown

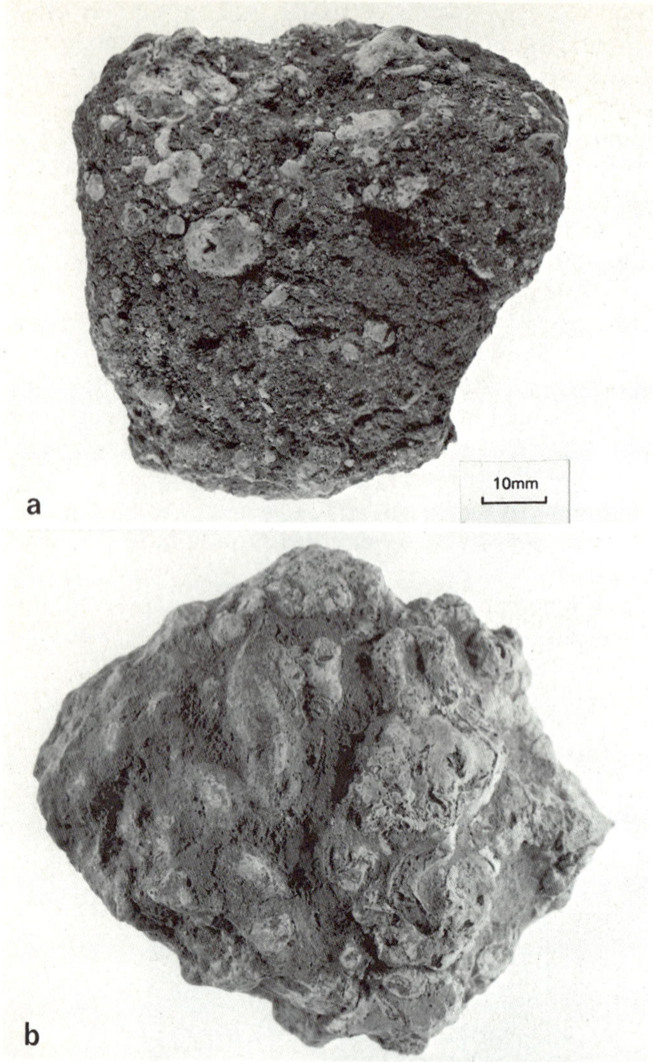

Fig. 3a–d. Examples of rhodoids from differing localities in New Zealand: **a, b** Glauconite-rich rhodoids from Fiordland. **c** Slightly abraded *lithothamnium* nodule from the West Coast, South Island. **d** Section through a *lithothamnium* dominated rhodoid from Tinui, North Island, showing an allochem nucleus and enclosed skeletal muddy sediment

3.2 Associated Allochems and Sediment

A limited number of allochems occur in association with Tertiary rhodoids. The most common of these are benthic foraminifera and Bryozoa, although planktic foraminifera, molluscs, echinoderms, brachiopods and serpulid worm tubes are subsidiary constituents. The benthic foraminifera are dominated by *Amphis-*

c

d

Fig. 3c, d

tegina sp. with rare *Cibicides* sp., Miliolids and Textularia. Non-encrusting, branching coralline algal debris is also a common constituent of associated sediment.

Bryozoans vary in growth form considerably and the variable growth form can be related to different environmental conditions (Stach 1936, Lagaaij and Gautier 1965, Schopf 1969, Wass et al. 1970, Pedley 1976). Identification of growth form is more difficult in a two-dimensional thin section, but a number of forms including membraniporiform and celleporiform were identified. The membraniporiform growth forms are often incorporated into the rhodoid structure (Fig. 2), growing on a dead surface of the algal nodule before being overwhelmed by a further algal layer. Other growth forms are incorporated in a similar way to terrigenous and allochemical fragments, that is by lodgement in a hollow with encasement by algal growth. Allochems are usually fragmented and less than 2 mm in diameter.

3.3 Morphology

The nodules are all macroids commonly varying in size from 20 – 60 mm (Fig. 3). One exceptional rhodoid containing a granite boulder nucleus 150 mm in length surrounded by algal laminae 100 – 120 mm in thickness was recorded from a quarry at School Creek, South Island (Fig. 4). The nodules tend to be abraded and rounded to sub-rounded and occasionally discoidal.

Internal development is very variable. Often an allochem fragment or terrigenous clast acts as a nucleus and the algal laminae grow irregularly coating the nucleus with layers of uneven thickness (Figs. 3 and 5) and incorporating small allochem fragments, terrigenous grains, foraminifera and glauconite. The internal structure reflects the degree of current activity, availability of light and the nature of the substrate as well as species type. Spheroidal form and laminar structure indicate frequent movement and high current activity. If branching forms are subjected to constant movement then the tips become damaged, expand and may connect to form laminar coatings. Flattened form and columnar structure indicate immobility of the rhodoid (Lees et al. 1969).

Fig. 4. Large granite boulder 150 mm in length surrounded by quartz-rich algal layers forming a huge rhodoid. Unfortunately, the true spherical nature of the "nodule" is not preserved. School Creek, South Island. Hammer = 180 mm in length

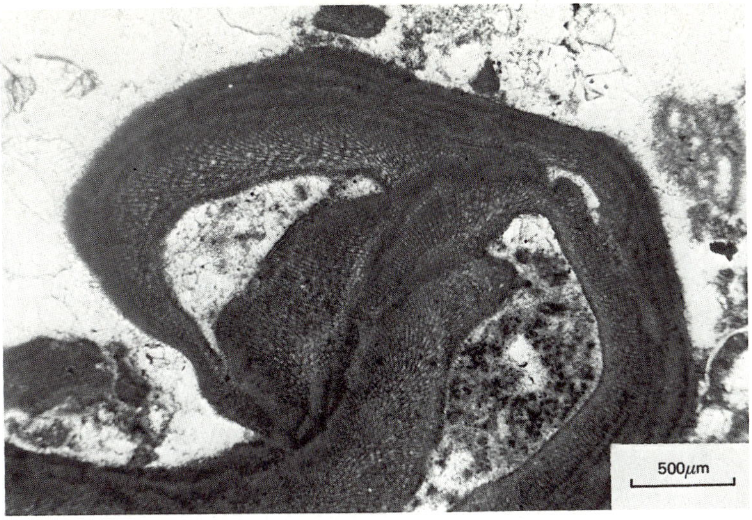

Fig. 5. Grummulose pelletal sediment incorporated into a columnar growth structure in an unabraded rhodoid margin, West Coast, South Island

4 Distribution of Rhodoid Facies

Deposits containing rhodoids are widespread in New Zealand. A brief account of five localities (Fig. 1) is given and factors common to all facies are emphasised.

4.1 South Island, West Coast

Rhodoids up to 150 mm in size occur in a deposit over 10 m in thickness and laterally continuous for well over 100 m. The deposit partially rests on basement and upward and laterally grades into detrital algal lithofacies interbedded with sandstones, mudstones and glauconitic terrigenous-rich limestones. Sedimentary features are obscured by compaction and diagenetic effects but the deposit appears generally unstratified except for some rudimentary localised layering 100 – 200 mm thick (Fig. 6). Sand and mud content are very low, but micrite content (possibly diagenetic in origin) is moderately high. The sediments are poorly sorted probably reflecting trapping of finer grained sediments between the rhodoids. Spheroidal, ellipsoidal and flattened shapes were all seen (Fig. 3c) and internal structure is both laminar and columnar. Associated allochems include benthic foraminifera *Amphistegina,* various bryozoa, brachipoda, echinoidea and non-encrusting red algae.

Interpretation. This deposit probably formed as a bank in fairly shallow water (< 50 m) under moderate to strong hydrodynamic conditions. Sedimentation rate was slow. The variety of shapes and internal structure of the rhodoids suggests

Fig. 6. Rudimentary stratification emphasised by compaction in a rhodolite from the West Coast, South Island. Lens cap is 50 mm in diameter

that they were periodically turned but that some were stable for considerable periods of time. The number of unquantifiable controls on growth rate are such that a frequency of overturning cannot be determined. However, studies of recent rhodoids from tropical seas suggest periods of several weeks or months without turning would be required for some layers (Bosellini and Ginsburg 1971, Adey and MacIntyre 1973). This suggests that some of the rhodoids were formed in environments where the critical velocity for overturning was only reached during periods of storm activity.

This West Coast deposit is one facies of a thick sequence of bryozoan grainstones and packstones which formed in shallow shelf seas (Anderson, in prep.). Exposure and tectonic alteration is such that proximity to an ancient shoreline cannot be determined, although conformably underlying coal measures suggest shallow water conditions.

4.2 Waiau River, Fiordland

Oligocene sedimentation adjacent to the Fiordland Block in the southwest of the South Island was controlled by faulting related to the oblique continental transform, the Alpine Fault (Norris et al. 1978). Deep subsiding basins accumulated over 2000 m of mass-emplaced breccias and muddy flysch adjacent to the rising Fiordland Block, but on the block itself the sediments are arkosic sands, calcareous arkoses and shallow marine sandy limestones. At one locality the regional early Cenozoic coal measures are preserved and separated by a condensed sequence from the overlying Miocene mudstone. This sequence consists of phosphatised-glauconitised fossils and rhodoids (Norris et al. 1978, Carter and

Norris 1978). Rhodoids are spheroidal to biscuit shaped, frequently bored with infillings of glauconite and terrigenous grains (Fig. 3a, b). The thickness of the unit is often only 1 m.

Interpretation. The rhodoids formed on an upraised fault block, probably in fairly shallow water receiving little sediment influx. Accumulation rate was extremely slow suggesting that some material was removed from the site of formation, or that conditions for growth were not good.

4.3 School Creek, Buller, South Island

A rhodolite, lensoid in shape, with maximum thickness of about 30 m, overlies an eroded granitic block. Granite boulders and sand occur within the lower few meters and glauconite is concentrated in a thin layer within the limestone. Vertically and laterally the limestone passes into coarse sandstone, siltstone and mudstone.

Rhodoids within the limestone are rounded to subrounded and large amounts of granite sand occur within the laminae, trapped within growth layers. The average size of nodules is about 30 mm, with rare rhodoids reaching quite exceptional proportions (Fig. 4).

Interpretation. The granite block probably formed a submarine high, swept by strong currents and receiving little or no terrigenous sediment input. Rhodoids formed in the shallow waters overlying the block and were periodically swept off the side to form allochthonous layers in the deeper flank beds.

4.4 Tinui, North Island

An algal limestone containing numerous rounded to elongate, abraded algal rhodoids up to 12 mm across, with layers of severely recrystallized gastropods, occurs interbedded with thick sandstones and minor siltstones (Johnston 1980). Rounded grains of glauconite and small subangular quartz particles are randomly scattered within the matrix and incorporated within the rhodoids. The maximum thickness of the limestone unit is about 150 m. Rhodoids also occur within the associated sandstone but are scattered and probably did not form there. Individual rhodoids (Fig. 3d) tend to be abraded, rounded to well-rounded with irregular internal growth layers with abundant interstitial silt.

Interpretation. The rhodoids were formed in shallow water as algal banks in association with molluscs and benthic foraminifera. Strong currents moved allochem fragments and rhodoids from the site of formation and mixed them with shallow shelf sands. The shallow marine conditions are considered to have been caused by an upthrown fault block (Johnston 1980).

Fig. 7. Incipient rhodoids, Piopio, North Island. Thin algal coatings surround Triassic lithoclasts immediately above an angular unconformity. Lens cap is 55 mm in diameter

4.5 Piopio, North Island

Basal Triassic conglomerate which was well lithified and folded is overlain with sharp angular unconformity by rhodoid limestone of Oligocene age. The rhodoids are poorly formed consisting largely of eroded fragments of Triassic rocks surrounded by thin layers of algal coating (Fig. 7). Small neptunean dykes occur within the Triassic rocks and are filled with rhodoid sediment. Associated with the rhodoids are echinoderm fragments, benthic foraminifera and glauconite in a carbonate mud matrix. The rhodoid limestone gives way vertically to bryozoan grainstones of the Te Kuiti Group (Nelson 1978b).

Interpretation. Rhodoids formed on a rocky sea-floor where little terrigenous sediment was being deposited. The size of eroded Triassic clasts (up to 200 mm) suggests strong wave activity, and this area was probably close to shore and similar to modern coastal areas close to Wellington (Fig. 1).

5 Summary of Depositional Environment and Growth Conditions (Fig. 8)

The following points summarize the general similarities in conditions during formation of rhodoids at the various localities.

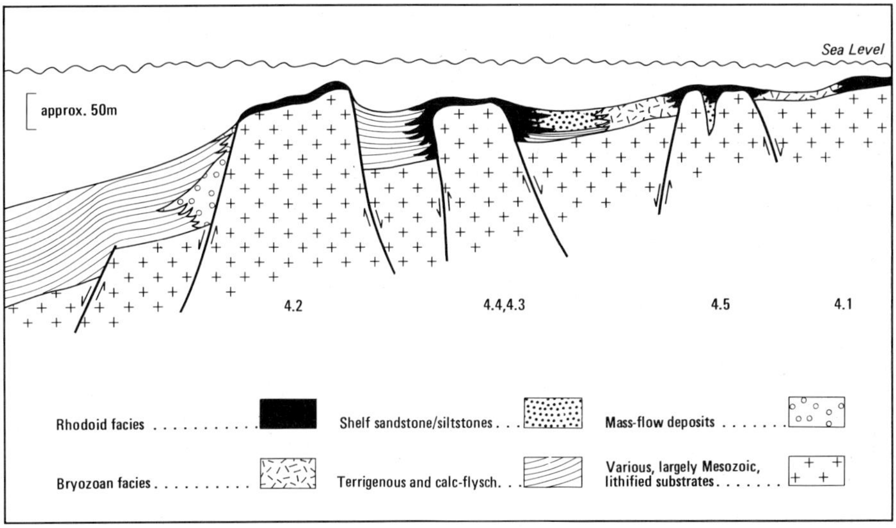

Fig. 8. Diagrammatic reconstruction of various environments of formation of New Zealand rhodoids discussed in the text. No relationship between the individual deposits is intended. The faults shown in the basement are meant to represent a variety of tectonic situations (rifting, subduction and transform faulting) and should not be viewed as a single tectonic environment

a) With the exception of Tinui, the deposits immediately overlie an erosional unconformity.
b) All deposits probably formed on upthrown fault blocks forming submarine topographic highs.
c) Autochthonous rhodolites formed as algal shoals, although in areas of low productivity the rhodolites are restricted to a few metres or less.
d) Allochthonous rhodolites were deposited in deeper peri-platform flank beds encased in dominantly terrigenous-rich sediments.
e) Associated allochems include bryozoans and the benthic foraminifera *Amphistegina* sp., with subsidiary echinoderms, molluscs and other benthic foraminifera.
f) Authigenic glauconite and phosphate occur in some samples, particularly where productivity rates were very low.
g) The rhodoids formed in shallow marine conditions, probably less than 50 m in depth, but grade into deeper shelf sediments. The restricted distribution of rhodoids reflects their sensitivity to changes in hydraulic and depth conditions. It is not possible to be specific about depth limits of any one deposit.
h) The rhodoids formed under relatively high energy conditions, but were not continuously turned. A wide variety of shapes and growth forms may occur in any one rhodolite and this suggests movement by the rare high energy event (storms, for example) rather than by regular wave or tidal activity.

6 Comparison with Modern and Ancient Rhodoid Deposits

Rhodoids have been widely recognised in modern tropical environments, particularly those associated with reef facies (Adey and MacIntyre 1973). Although widely believed to be tropical, shallow water organisms, crustose coralline algae are known to occur at depths ranging from 0 – 250 m to latitudes of 70° – 80° (Bosellini and Ginsburg 1971, Adey and MacIntyre 1973, Milliman 1977, Wray 1977, Studencki 1979). Algal nodules occur in biogenic sediments as subsidiary components on the present day New Zealand shelves (Carter 1975), although these deposits may well be relict sediments relating to periods of lower sea-level. As described above modern algal nodules also occur on rocky shores around New Zealand today. The nodules tend to be disc to spheroid shaped with abraded relatively laminar coatings up to 5 mm thick covering greywacke pebbles. No recent examples of an algal nodule/*Amphistegina*/bryozoa association in modern sediments could be found, although Carter (1975) records the presence of unnamed benthic foraminifera and bryozoa with algal nodules on New Zealand shelves. Ancient examples of similar rhodoids are numerous, but the closest similarity appears to be with the Busko-Spa limestones described by Studencki (1979), who recorded a similar faunal association. However, there are marked differences in the rhodoid type, depositional environment and associated facies, those from Busko-Spa being tropical or subtropical algal bioherms.

7 Diagenesis

As all examined occurrences of rhodolite are overlain by thick marine sequences, alteration and cementation took place on the seafloor and in submarine phreatic environments. The earliest diagenetic process is the destruction of peripheral cells by endolithic algae and fungi and sometimes by larger faunas (boring molluscs, for example). The process results in micritization of selective areas of a rhodoid, and destruction of the cellular structure. Micritized areas may also occur within a nodule, indicating periods of destruction and non-development during the nodule growth (Fig. 9). Abrasion may also inhibit the nodule development.

The initial mineralogy of algal coatings is high Mg calcite and the preservation potential of the original fabric is therefore good. The initially high cellular porosity is filled with fine acicular and blocky low Mg calcite cements, in many cases high in iron. These cements may occur as small crystals infilling original cells, and increase the specific gravity of nodules markedly. This cementation phase was probably accompanied by a loss of Mg ions from the high Mg calcite structure with no change in crystal morphology.

In rocks which have suffered a high degree of compaction and pressure solution the rhodoids act as rigid structures reflecting the stable nature of the cell walls (Fig. 10).

Rhodoid limestones contain evidence of other diagenetic alteration. Aggrading recrystallization affects rhodoids to the same extent as other allochem fragments. Several examples of iron-rich neomorphic spar, almost totally obscuring

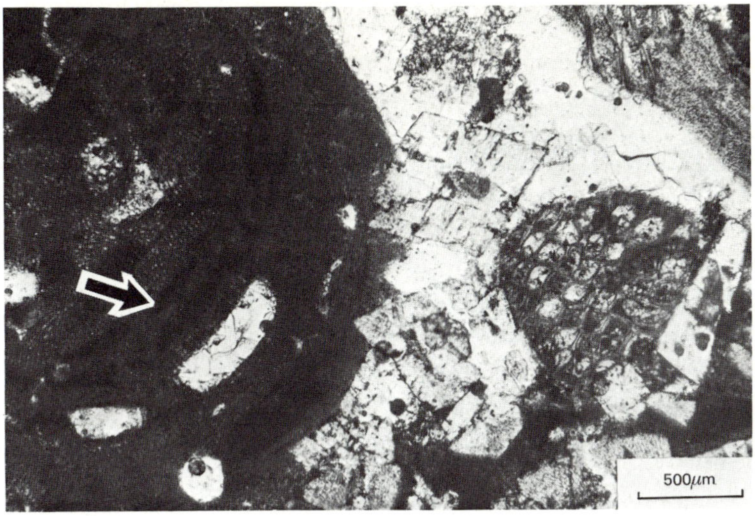

Fig. 9. Rhodoid bryozoan association with secondary calcian dolomite replacement as idiotopic crystals. Dark layers in algal nodule (*arrowed*) indicate periods of micritization by endolithic algae and fungi during static growth periods. Voids in *lithothamnium* layers are spar-filled concepticals

the original algal fabric, were observed. In partially altered rhodoids the neomorphic spar tends to affect the nucleus and neomorphism may spread from this area to other parts of the algal nodule (Fig. 10).

Dolomitization and dolomite crystal growth in coralline algae has been documented by previous authors (Schlanger 1957, Milliman 1974). True dolomite was not identified in New Zealand Tertiary rhodoids, probably because there has been insufficient time for it to form. Idiotopic calcian dolomite crystals, however, were found associated with rhodoids (Figs. 9 and 10), and were identified by staining (Dickson 1966). The molar percentage of $MgCO_3$ in these calcian dolomites, determied from 13 samples by the X.R.D. method of Goldsmith and Graf (1958), ranges from 35% – 45%. Scanning electron microprobe analysis of elemental distribution within the crystals revealed very low Fe percentages and no patterned distribution in either Mg, Na or Fe. The calcian dolomite appears to be a later stage cement replacing earlier low Mg calcite cements and allochems (Fig. 10). This suggests that the Mg cannot be directly derived from the original high Mg calcite skeleton of the rhodoids. Most of the dolomite growth occurs adjacent to formation boundaries where calcareous sediments are juxtaposed against permeable terrigenous clay-rich sediments. In some bryozoan grainstones of similar age to rhodoid limestones calcian dolomite development is associated with pressure solution seams, possibly resulting from Mg accumulation associated with accumulation of insoluble clay minerals. Therefore, the presence of calcian dolomites seems to be associated with the presence of Mg-rich clay minerals rather than original high Mg calcite skeletal fragments.

Fig. 10. Rhodoid/*Amphistegina*/echinoderm/mollusc association with well developed pressure solution suture (*arrowed a*) and idiotopic calcian dolomite crystal development. Neomorphic calcite replaces part of the algal structure (*arrowed b*) and nucleus. Note that the calcian dolomite is not affected by pressure solution

8 Conclusions

New Zealand Tertiary rhodoids are dominated by *Lithothamnium, Lithophyllum* and *Mesophyllum* and are usually associated with the benthic foraminifera *Amphistegina.* The size of nodules varies from a few millimeters to 60 mm, with rare larger growth forms. They formed in relatively shallow seas, within the photic zone and in areas of little terrigenous influx. Increasing terrigenous input choked algal growth. Most examples formed in areas of high periodic turbulence and accumulation rates were slow, allowing abrasion and boring of growing nodules. Glauconite and phosphate nodules may be associated authigenic minerals.

Rhodoids were formed of high Mg calcite and are usually well preserved, although they may be recrystallized, compressed and partially dissolved along solution boundaries. Cellular pore space is usually infilled with blocky calcite spar and sometimes with acicular drusy cements. Similar nodules are forming in coastal areas of New Zealand today in areas of high energy and low terrigenous influx.

Acknowledgements. The authors are grateful to the following for assistance in this research: Internal Research Committee, Victoria University of Wellington; Robertson Research International Limited, Llandudno, North Wales.

References

Adey WH, MacIntyre (1973) Crustose coralline algae: A re-evaluation in the geological sciences. Geol Soc Am Bull 84:883 – 904

Bosellini A, Ginsburg RN (1971) Form and internal structure of recent algal nodules (Rhodolites) from Bermuda. J Geol 79:669 – 682

Bosence DWJ (1976) Ecological studies on two unattached coralline algae from western Ireland. Palaeontology 19:365 – 395

Buchbinder B (1977) The coralline algae from the Miocene Ziqlag Formation in Israel and their environmental significance. In: Flügel E (ed) Fossil algae. Springer, Berlin Heidelberg New York, pp 279 – 285

Carter L (1975) Sedimentation on the continental terrace around New Zealand: a review. Mar Geol 19:209 – 237

Carter RM, Norris RJ (1978) Excursion to Blackmount, Waiau Basin tour guide. Geol Soc N Z Queenstown Conf, 31 p

Chapman VJ (1974) The marine algae of New Zealand, part III Rhodophyceae. 3301 Lehre Verlag von J Cramer, pp 155 – 277

Dickson JAD (1966) Carbonate identification and genesis as revealed by staining. J Sediment Petrol 36:491 – 505

Goldsmith JR, Graf DC (1958) Relations between lattice constants and composition of the Ca – Mg carbonates. Am Min 43:84 – 101

Johnson JH (1957) Geology of Saipan, Mariana Islands, Part 2. Petrology and soils. Petrography of the limestones. US Geol Surv Prof Pap 280-C:177 – 187

Johnson JH (1961) Fossil algae from Eniwetok, Funafuti and Kita-Daito-Jima. U S Geol Surv Prof Pap 260-Z:907 – 950

Johnson JH (1964) Fossil and Recent calcareous algae from Guam. U S Geol Surv Prof Pap 403-G:1 – 40

Johnston MR (1980) Geology of the Tinui-Awatoitoi District. N Z Geol Surv Bull 94:62

Kennett JP, Houtz RE, Andrews PB, Edwards AR, Gostin VA, Hajós M, Hampton M, Jenkins DG, Margolis SV, Ovenshine AT, Perch-Nielsen K (1975) Cenozoic paleoceanography in the Southwest Pacific Ocean, Antarctic glaciation, and the development of the Circum-Antarctic current. In: White SM (ed) Init Rep Deep Sea Drill Proj 29. US Gov Printing Office, Washington, pp 1155 – 1169

Lagaaij R, Gautier YV (1965) Bryozoan assemblages from marine sediments of the Rhone delta, France. Micropaleontology 11:39 – 58

Lees A, Butler AT, Scott J (1969) Marine carbonate sedimentation processes, Connemara Ireland. Reading Univ Geol Rep 2:64

Milliman JD (1974) Recent Sedimentary Carbonates. Part 1 Marine Carbonates. Springer, Berlin Heidelberg New York, 375 p

Milliman JD (1977) Role of calcareous algae in Atlantic continental margin sedimentation. In: Flügel E (ed) Fossil algae. Springer, Berlin Heidelberg New York, pp 232 – 247

Molnar P, Atwater T, Mammerick J, Smith SM (1975) Magnetic anomolies, bathymetry and the tectonic evolution of the South Pacific since the Late Cretaceous. Geophys J R Astron Soc 40:383 – 420

Nelson CS (1978a) Temperate shelf carbonate sediments in the Cenozoic of New Zealand. Sedimentology 25:737 – 771

Nelson CS (1978b) Stratigraphy and paleontology of the Oligocene Te Kuiti Group, Waitomo County, South Auckland, New Zealand. N Z J Geol Geophys 21:553 – 594

Norris RJ, Carter RM, Turnbull IM (1978) Cenozoic sedimentation in basins adjacent to a major continental transform boundary in Southern New Zealand. J Geol Soc London 135:191 – 206

Pedley HM (1976) A palaeoecological study of the Upper Coralline limestone, *Terebratula-Aphelesia* Bed (Miocene, Malta) based on bryozoan growth-form studies and brachiopod distributions. Palaeogeogr Palaeoclimatol Palaeoecol 20:209 – 234

Schlanger SO (1957) Dolomite growth in coralline algae. J Sediment Petrol 27:181 – 186

Schopf TJM (1969) Palaeoecology of ectoprocts (bryozoans). J Palaeontol 43:234 – 244

Stach LW (1936) Correlation of zoarial form with habitat. J Geol 44:60 – 66

Studencki W (1979) Sedimentation of algal limestones from Busko-Spa environs (Middle Miocene, Central Poland). Palaeogeogr Palaeoclimatol Palaeoecol 27:155 – 165

Wass RE, Conolly JR, MacIntyre RJ (1970) Bryozoan carbonate sand continuous along southern Australia. Mar Geol 9:63 – 73

Wray JL (1977) Calcareous algae. Elsevier, Amsterdam, 186 p

III.4 Early Permian Coated Grains from a Lagoonal Environment, Laborcita Formation, Sacramento Mountains, Southcentral New Mexico, U.S.A.

Donald Francis Toomey[1]

1 Introduction

The Sacramento Mountains of southcentral New Mexico (Fig. 1) outcrop at the eastern edge of the basin-and-range province, and are a fault-block range. The mountains are aligned in an approximate north-south direction, are tilted about one degree to the east, and on the west, are bounded by a gravity fault zone

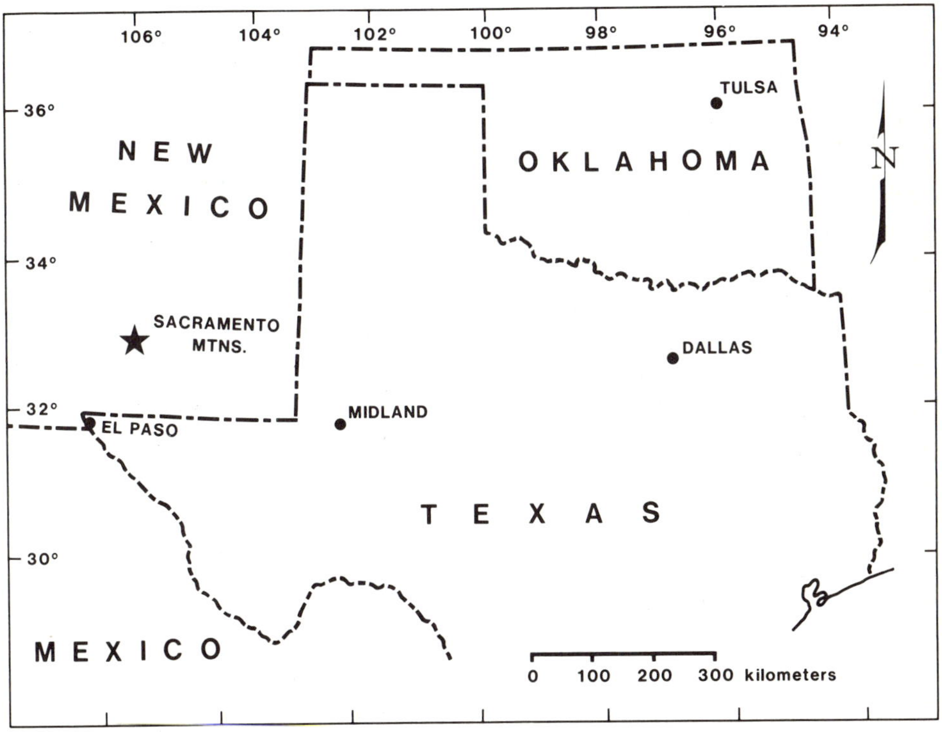

Fig. 1. Geographic location map showing the relative position of the Sacramento Mountains

1 Cities Service Company, Energy Resources Group, Box 1919, Midland, TX 79702, USA

Coated Grains (ed. by T.M. Peryt)

situated close to the present escarpment (Pray 1961). It has been estimated that the minimum displacement along the western escarpment for a distance of 29 km is on the order of 2330 m, but this decreases to about half this amount towards the northern end of the range where the Early Permian section is exposed. Most of the salient structural features were formed by major tectonic movements during latest Upper Carboniferous and Early Permian time, with only relatively little structural complication to the Permian sequence. The major uplift of the Sacramento Mountains fault-block is believed to have occurred in late Tertiary time. Significantly, Pray (1961, p. 5) reports that fault scarps seen in Recent alluvium suggest that the uplift of the mountains is still continuing.

The Sacramento Mountains are of interest in that they contain excellent exposures of biohermal structures ranging in age from Lower Carboniferous Waulsortian-type "mud mounds" (Pray 1958), Upper Carboniferous phylloid algal mounds (Toomey et al. 1977), to Early Permian mounds (Otte and Parks 1963, Toomey and Cys 1977, Mazzullo and Cys 1979).

The purpose of this contribution is to describe a unique accumulation of rather large algally coated grains from the Early Permian (Assel) Laborcita Formation of the northern Sacramento Mountains, and to discuss the paleoecologic relationships of these algally coated grains to small contemporaneous algal-sponge-brachiopod bioherms contained within a relatively thin stratigraphic interval (Toomey and Cys 1979).

2 General Stratigraphic Relationships

The coated grains described in this paper occur in the lower portion of the Laborcita Formation, near the village of Tularosa, and are exposed in the west-facing escarpment of the northern Sacramento Mountains.

The Laborcita Formation was named by Otte (1959) for a sequence of interbedded limestones, shales, sandstones, arkoses, and conglomerates exposed in the central and northern portions of the mountain range. The formation is characterized by rapid lateral and vertical cyclic facies changes, the result of sedimentological responses to spasmodic positive tectonic impulses that occurred on the Pedernal Landmass to the east. This positive element was tectonically active from Late Carboniferous to Middle Permian time, and directly influenced sedimentation during that time span. The uppermost portion of the Laborcita Formation laterally intertongues with the continental Abo Formation in an overall regressive sequence. The depositional environments reflected in the sediments of the Laborcita Formation display a response to continually changing paleoenvironmental conditions ranging from dominantly marine to dominantly terrestrial depositional sequences. At the northern end of the range, marine units are more conspicuous. Maximum thickness of the Laborcita Formation is approximately 303 m. Fusulinid microfossils have dated the entire formation as Early Permian (Wolfcamp-Assel) age (Steiner and Williams 1968).

The coated grains discussed in this paper occur on the upper surface of a limestone bed given in Otte's (1959) measured section 29. A portion of this measured

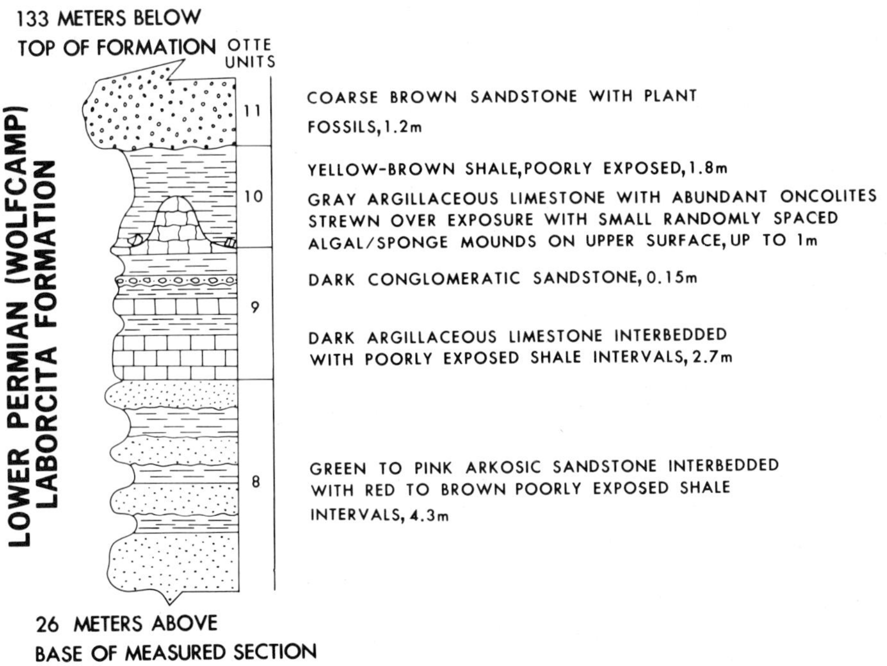

Fig. 2. A portion of Otte's measured section 29 of the Early Permian Laborcita Formation, showing the stratigraphic position of Unit 9 and the relationship of the oncolite (coated grains) horizon to contemporaneous algal-sponge mounds

section is shown in Fig. 2. It should be emphasized that facies relationships do change laterally along strike, and there is also some minor offsetting of this interval by small faults.

3 Coated Grain Horizon

3.1 Outcrop Description

As shown in Fig. 2, Otte's unit 9 consists of interbedded limestone, shale and conglomerate. The limestones are fairly well exposed and consist of dark grey fossiliferous and argillaceous types. Petrologically, they can be classified as skeletal wackestones characterized by relatively common tubular encrusting foraminifers. Some discrete algal-foraminiferal colonies also occur, along with scattered pelmatozoan and molluscan remains. In some instances this shell debris may be circumscribed by micrite envelopes that show the effect of organism boring. Locally the limestones may be burrowed, and there has been some neomorphism of the micrite matrix to neospar. Acid etching of these limestones has yielded abundant specimens of *Hyperammina,* an agglutinated foraminifer.

Overlying the clastics are small organic buildups, which are mound-shaped masses randomly distributed along the upper surface of Otte's unit 9. The height

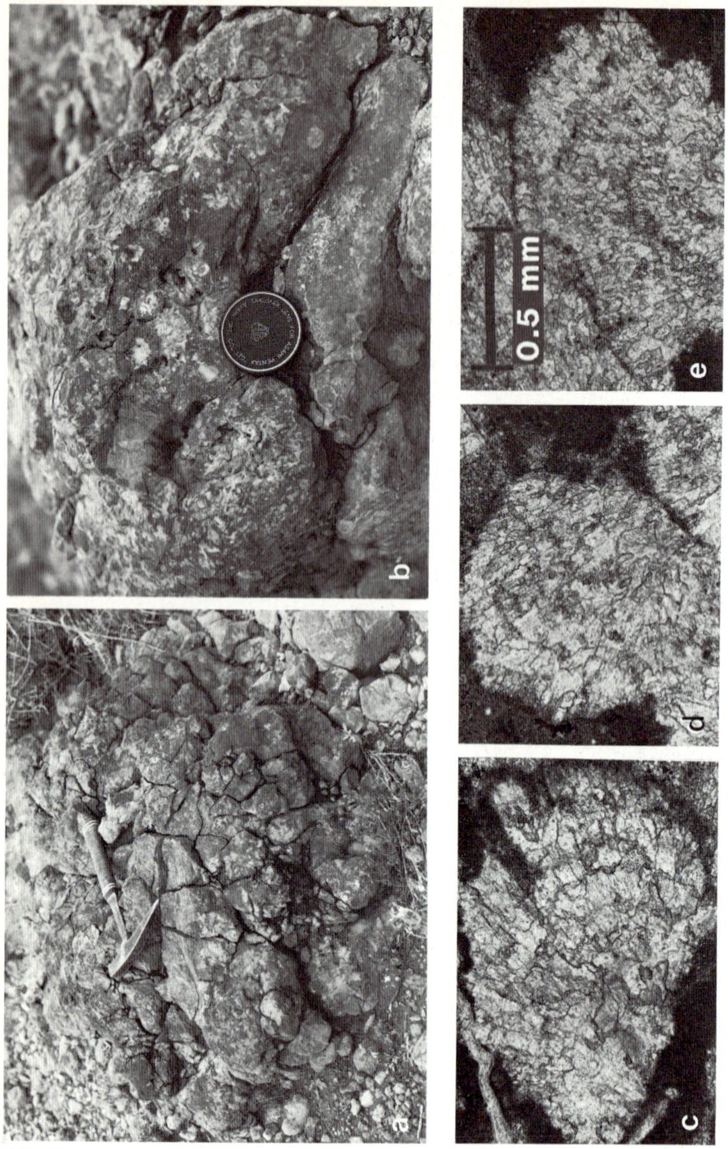

Fig. 3. **a** Outcrop photograph of a typical small Laborcita algal-sponge mound; length of hammer is 32.5 cm. **b** Close-up photograph of the top of a small Laborcita bioherm showing a representative terminal community dominated by siliceous sponges (*Heliospongia*); diameter of lens cap is 7 cm. **c – e** Thin-section photomicrographs (crossed nicols) of various cement fans of radial-fibrous crystals filling voids and/or cavities of moundrock and suggesting syndepositional marine origine for these fan cements

Fig. 4. **a** Outcrop photograph showing abundant Laborcita oncolites (coated grains) strewn over the intermound area; length of hammer 32.5 cm. **b** Close-up photograph of a relatively large coated-grain showing concentric laminated growths of *Archaeolithophyllum* surrounding a brachiopod nucleus; diameter of coin 2.1 cm. **c, d** Thin-section photomicrographs (crossed nicols) of the laminae composing a typical Laborcita coated grain. These laminae are recrystallized thalli of the red alga *Archaeolithophyllum lamellosum* Wray. Dark-colored encrustations on algal thalli are encrusting foraminifers. **e** Close-up photograph of typical Laborcita moundrock showing digitate or plumose algal-foraminiferal colonies reflecting the initial stage of mound growth. These in turn are overlain by laminar encrustations of *Archaeolithophyllum lamellosum* Wray; diameter of coin is 2.1 cm. **f** Close-up photograph of Laborcita moundrock showing discrete laminar colonial growths of *Archaeolithophyllum lamellosum* Wray separated by mud-infill between colonies. Note the superficial morphological resemblance to stacked-hemispheroid stromatolites; diameter of coin 2.1 cm

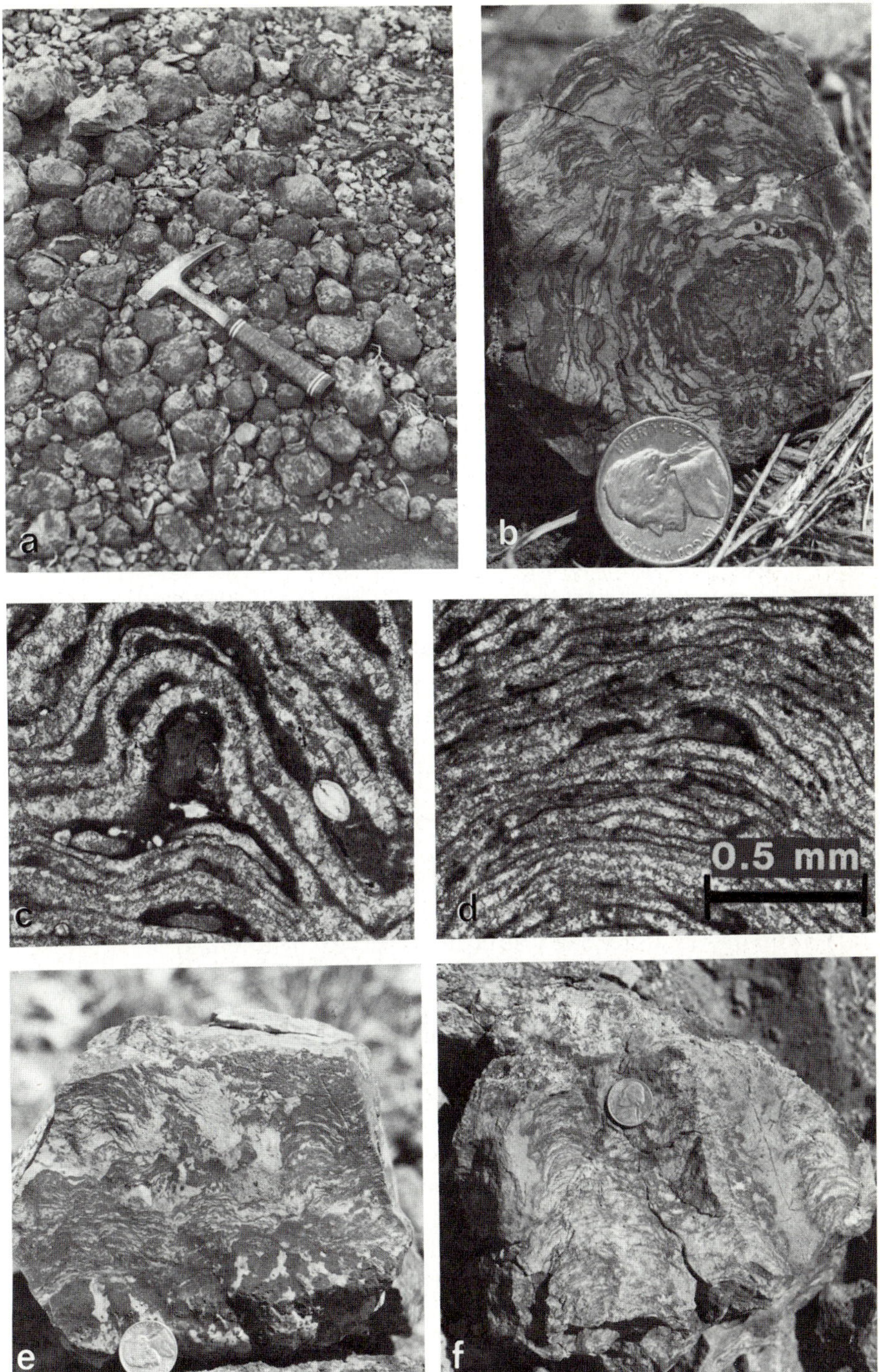
a
b
c
d
0.5 mm
e
f

of these bioherms ranges from 0.3 to 1.5 m with a diameter of from 1 to 3 m (Fig. 3a, b), and are unique because they show a well-developed vertical biotic zonation (Toomey and Cys 1979). This zonation consists of an initial pioneer community composed of small digitate or plumose-shaped colonies of algae and foraminifers (Stage 1). This is followed by a climax community that is first dominated by laminar growths of the red alga *Archaeolithophyllum* (Stage 2), then by a terminal community composed of minor *Archaeolithophyllum* and abundant erect heliosponges and brachiopods (Stage 3). Diagenetically, radial fans of probable former aragonitic cement are present in all of the moundrock units (Fig. 3c – e). This diagenetic phenomenon is thought to indicate contemporaneous submarine cementation of the bioherms.

The intermound rock consists of a discontinuous bed of dark grey limestone up to 15 cm in thickness. This horizon is distinguished by the occurrence of abundant coated grains strewn over the outcrop surface (Fig. 4a). The coated grains average from 7 to 10 cm in diameter, although some attain a diameter of up to 15 cm. These coated grains are composed of cencentric layers of the red alga *Archaeolithophyllum lamellosum* Wray (Fig. 4b – d), and as such could be aptly termed rhodoliths[2]. The nucleus, or accretionary center, of a rhodolith can be most any type of skeletal grain found within the immediate depositional environment, although brachiopods seem to be the most common algally coated nuclei. Other observed skeletal grains serving as coated-grain nuclei include sponges, gastropods, pelecypods, scaphopods, and cephalopods. Other megafossils found within this intermound horizon, though not algally coated, include very rare, small rugose corals, fragments of pinnid-type pelecypods up to 25 cm in length, and rather common trilobites.

3.2 Petrology and Diagenesis

Petrologically, the algally coated grains can be described as boundstones composed of concentric laminar thalli of algae. These concentric algal thalli may form a continuous accretionary structure (Fig. 4d), or they may in part be separated by thin layers of micrite containing scattered skeletal debris (Fig. 4c). The thalli themselves are usually strongly recrystallized to sparry calcite mosaic and most cellular microstructure has been obliterated. Examination of numerous thin sections has, however, yielded a few specimens which still retain the outlines of hypothallic cells; no perithallic cells were seen. Accordingly, all of the coating algae have been referred to *Archaeolithophyllum lamellosum* Wray.

One characteristic feature of the laborcita coated grains is the abundance of various epizoans that encrust the algal thalli. The most common epizoans are encrusting foraminifers such as *Tetrataxis* and *Calcivertella* although occasional polychaete worm tubes (*Spirorbis*) are also seen. In addition, small gastropods and various mobile foraminifers are often found enmeshed between the algal laminae in layers of micrite.

2 The term "rhodolite" commonly applied to coated grains composed of red algae is incorrect, since a rhodolite is a pyrope-almandine garnet; the correct descriptive term is rhodolith (see Binda 1973, Toomey 1975, p. 244), or – as preferred by the editor – rhodoid

Diagenetically, radial fans of probable former aragonite cement are found in small voids or cavities beneath some *Archaeolithophyllum* thalli (Fig. 3c – e). Here, the cements are believed to have grown from both the floor and roof of the voids. In some instances there are even pronounced meshworks of interlocking cement fans (Fig. 3e) filling the voids completely, and sometimes there are even algal encrustations on the fan periphery, seemingly substantiating a submarine origin for these distinctive cements.

Most of the radial fans appear to have been neomorphosed to ray-shaped crystals, and there are also even larger "loaf-shaped" crystals. In addition, some relict square-tipped fibers may be found in the radial fans.

The intermound rock, upon which the coated grains are abundantly strewn (Fig. 4a), and which appears to be coeval with the coated grains, is probably the bed from which the rhodoliths are derived from. This in situ intermound rock is mainly a burrowed skeletal wackestone containing various smaller foraminifers (*Tuberitina, Tetrataxis, Globivalvulina,* palaeotextularids, bradyinids, and nodosariids), rather conspicuous trilobite remains, and rare sponges, ostracodes, bryozoan fragments, brachiopods, gastropods, scaphopods, and echinoids. However, the most characteristic feature of this horizon is the abundance of erect colonies of algal-foraminiferal growths and encrusted skeletal grains. Typical photomicrographs prepared from thin-sections cut perpendicular to the bedding plane surface are shown on (Fig. 5b, h). These suggest that the digitate-shaped algal-foraminiferal colonies grew upon the sea floor or either upon dead heliosponges or ramose bryozoans, and alternated with *Archaeolithophyllum* thalli to form in situ organic boundstones. The digitate-shaped algal-foraminiferal colonies are very similar to the "Foraminiferen-Riffe" described from the Upper Triassic Hallstatt Limestone of Austria (Wendt 1969). As seen in thin-section, the foraminifers appear to be represented solely by the agglutinated form *Tolypammina.* However, acid residues prepared from some of the algal-foraminiferal colonies have yielded thousands of specimens of *Ammovertella inclusa* (Cushman and Waters), many specimens of *Ammodiscella* sp. cf. *A. virgilensis* Ireland, and only rare to common *Tolypammina* sp. cf. *T. nodosa* Ireland.

The matrix surrounding the algal-foraminiferal colonies is sparry and contains angular fragments of silt to fine-sand sized quartz grains, but the most characteristic feature is the abundance of varied skeletal grains circumscribed with algal-foraminiferal encrustations (Fig. 5a, c, d, g).

3.3 Paleoecology

Detailed study of the Laborcita mound/coated-grains interval suggest that this depositional sequence was laid down during a major marine transgression. This is schematically shown in Fig. 6.

Within shallow water of a probable open marine lagoon, on a substrate of burrowed skeletal wackestone, a pioneer community of plumose algal-foraminiferal colonies began to grow on small irregularities of the sea bottom. This pioneer colony (Stage 1) consisted of small colonies of probable blue-green algae

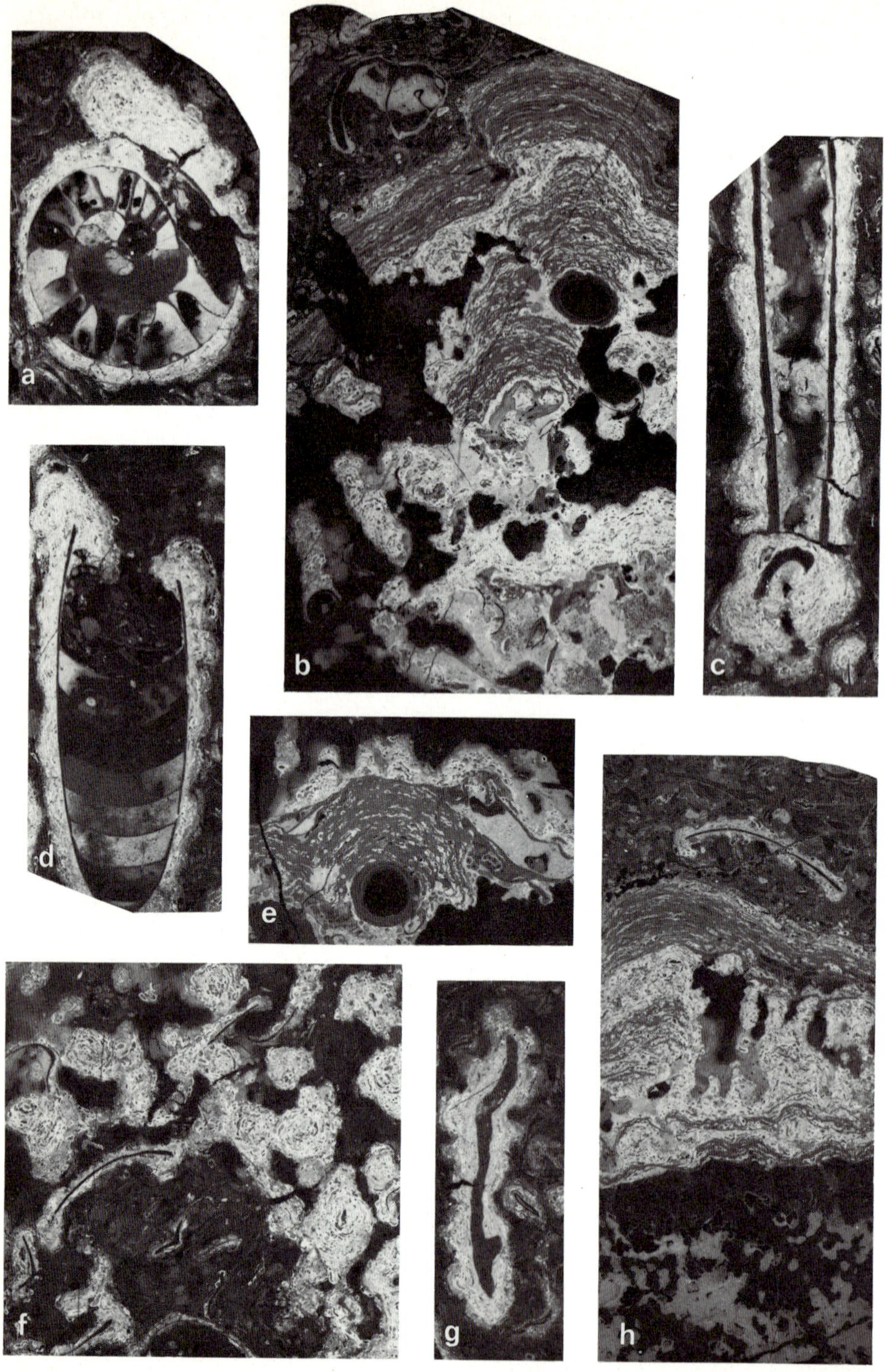
a
b
c
d
e
f
g
h

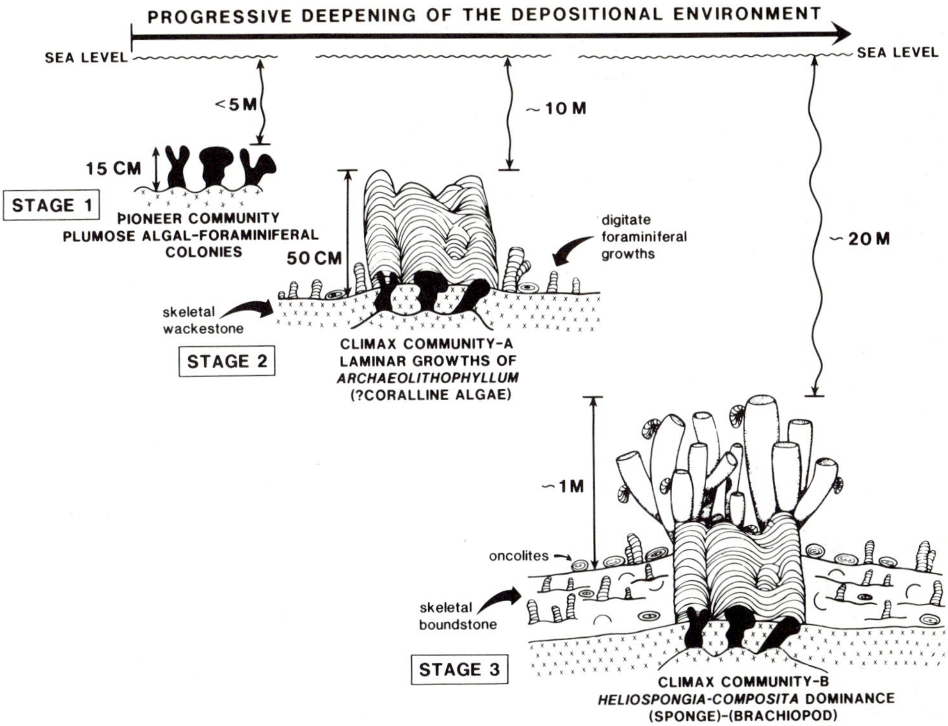

Fig. 6. Schematic diagram showing the postulated sequence in the growth of Early Permian algal-sponge-brachiopod mounds and their relationship to oncolite (coated-grain) growth and distribution

and encrusting foraminifers which formed very small boundstone growths, with not more than a few centimeters of relief. Algal-foraminiferal colonies of this type have been identified as very shallow water communities from close-by Late Paleozoic localities (Toomey et al. 1977). These distinctive rocks usually occur on the tops of phylloid algal organic buildups that have shoaled into very shallow water. Growth of the pioneer colony ceased when the lagoon deepened below the critical depth limite for optimum growth.

Fig. 5a – h. Negative prints of thin-section photomicrographs (all ×3) of various Laborcita coated-grains; **b, e, h** are perpendicular to bedding, remainder are all parallel to bedding. **a** Coiled cephalopod completely coated by algal-foraminiferal encrustations (*white-colored*). **b** Sequence showing (*from bottom to top*) algal-foraminiferal colonies (*white*), encrusted by thalli of *Archaeolithophyllum lamellosum* Wray (*dark lines*), followed by a thin layer of algal-foraminiferal encrustation, and this in turn encrusted by *Archaeolithophyllum*. Sparry matrix (*black*) on top of photograph contains a "floating" partially algal-foraminiferal encrusted gastropod. **c** Scaphopod shell encrusted by algal-foraminiferal growths. **d** Straight cephalopod (cf. *Mooreoceras* sp.) encrusted by algal-foraminiferal growths. **e** Scaphopod shell encrusted by *Archaeolithophyllum,* then later encrusted by small algal-foraminiferal digitate growths. **f** Mass of algal-foraminiferal colonies in a sparry matrix. **g** Mollusc fragment encrusted by algal-foraminiferal growth. **h** Sequence showing (*from bottom to top*) algal-foraminiferal colonies (*white*) encrusting a siliceous sponge (*Heliospongia* sp.), followed by encrustations of *Archaeolithophyllum* thalli. Sparry matrix on top of photograph contains a "floating" encrusted brachiopod fragment

The deepening of the depositional environment probably occurred rapidly, and is marked by the onset of the growth and dominance by the laminar red alga *Archaeolithophyllum lamellosum* Wray (Stage 2). Concurrently, with growth of the archaeolithophyllid colonies, whose growth form closely resembles stacked-hemispheroid stromatolites, algal growth extended out beyond the mounds. Here, in the intermound area, thalli of *Archaeolithophyllum* encrusted and circumscribed skeletal debris lying on the sea floor, thus forming the abundance of algally coated grains. The overall abundance represented within this interval would tend to suggest that this was a continuing on-going process during Stage 2 time. Along with the formation of abundant algally coated grains, numerous digitate-shaped foraminiferal growths also occupied the sea bottom and functioned as extremely effective sediment binders.

Significantly, there appears to be an absence of the mature community stage on the mounds. This stage is usually attained when diverse and abundant organisms enter the community and enrich its potential. However, the intermound area does show a rather high diversity of organisms, and it appears that this area was the center of organism diversity for "normal" organisms development and expansion.

The concluding stage of mound growth, considered as climax community-B, and represented as Stage 3, occurred when phylloid algal growth began to give way to increased abundance of erect siliceous heliosponges accompanied by abundant brachiopod growth, many of which appear to have settled on and around the larger sponge clusters. Some encrusting archaeolithophyllid algae continued to grow on and within the sponge clusters, thus adding to overall mound rigidity, along with prevassive submarine cementation. This concluding phase of mound growth, accompanied in the intermound areas with continued coated-grain formation, appears to reflect a continued deepening of the depositional environment.

The Laborcita buildups and the associated algally coated grains were killed-off by a return to rapid clastic influx and shoaling of the depositional environment.

4 Summary

Relatively large algally coated grains of Early Permian (Assel) age occur in the Laborcita Formation of the northern Sacramento Mountains, southcentral New Mexico. These coated-grains are composed of concentric thalli of the red alga *Archaeolithophyllum lamellosum* Wray and appear to be intimately related to coeval mound growth, i.e., Stage 2, when hemispherical colonies of *Archaeolithophyllum* dominated and molded these small organic buildups. A seral progression of mound evolution indicates that the coated-grains (rhodoliths) grew and developed within a rapidly deepening lagoonal depositional setting. The sequence of both mound growth and coated grain accretion ended abruptly by a return to clastic influx and overall shoaling of the depositional area caused by tectonic impulses from a nearby landmass.

Acknowledgements. Sincere thanks are due to Bill Batten of CITCO (Technical Center, Tulsa, Oklahoma) for preparation of the thin sections utilized in this study, and to John M. Cys of MAPCO (Midland, Texas) who accompanied the writer into the field and assisted in collecting samples, and to my wife Alisanne for the necessary help in the preparation of the illustrations.

References

Binda PL (1973) Form and internal structure of Recent algal nodules (rhodolites) from Bermuda: a discussion. J Geol 81:238

Mazzullo SJ, Cys JM (1979) Marine aragonite sea-floor growths and cements in Permian phylloid algal mounds, Sacramento Mountains, New Mexico. J Sediment Petrol 49:917–936

Otte, C (1959) Late Pennsylvanian and Early Permian stratigraphy of the northern Sacramento Mountains, Otero County, New Mexico. N M Bur Mines Min Res Bull 50:111

Otte C, Parks JM (1963) Fabric studies of Virgil and Wolfcamp bioherms, New Mexico. J Geol 71:381–395

Pray LC (1958) Fenestrate bryozoan core facies, Mississippian bioherms, southwestern United States. J Sediment Petrol 28:261–273

Pray LC (1961) Geology of the Sacramento Mountains escarpment, Otero County, New Mexico. N M Bur Mines Min Res Bull 35:144

Steiner MB, Williams TE (1968) Fusulinidae of the Laborcita Formation (Lower Permian), Sacramento Mountains, New Mexico. J Paleontol 42:51–60

Toomey DF (1975) Rhodoliths from the Upper Paleozoic of Kansas and the Recent – a comparison. Neues Jahrb Geol Palaeontol Monatsh 4:242–255

Toomey DF, Cys JM (1977) Spirorbid/algal stromatolites, a probable marginal marine occurrence from the Lower Permian of New Mexico, USA. Neues Jahrb Geol Palaeontol Monatsh 6:331–342

Toomey DF, Cys JM (1979) Community succession in small bioherms of algae and sponges in the Lower Permian of New Mexico. Lethaia 12:65–74

Toomey DF, Wilson JL, Rezak R (1977) Evolution of Yucca Mound complex, Late Pennsylvanian phylloid-algal buildup, Sacramento Mountains, New Mexico. Am Assoc Petrol Geol Bull 61:2115–2133

Wendt J (1969) Foraminiferen-“Riffe” im karnischen Hallstätter Kalk des Feuerkogels (Steiermark, Österreich). Palaeontol Z 43:177–193

IV Oncoids

IV.1 Oncoids: Comment to Recent Developments

Tadeusz M. Peryt[1]

Oncoids are a group of algally (red algae excepted), cyanobacterially and bacterially coated grains which originate(d) in marine and freshwater phreatic zones, although some workers prefer to use the term oncoids for all coated grains which are not ooids. Recent oncoids have been described from the lacustrine, fluvial and marine environments. The marine oncoids, occurring in the intertidal and shallow subtidal environments, are non-lithified, as opposed to hard freshwater oncoids. As noticed by Monty (1972), recent freshwater forms are closely similar to the ancient marine oncoids (see Flügel 1978, Tables 15 and 16). This in turn may be related to the replacement of porostromate oncoids by spongiostromate oncoids in marine settings in the Jurassic (see Peryt 1981, with discussion). At that time, porostromate oncoids (and mainly *Girvanella* oncoids – see Peryt 1981, Tables 1 and 2, and papers by Biddle, Bowman, Catalov, Poncet, and Tichy, this vol.) and spongiostromate oncoids existed in different environmental settings.

As discussed by Peryt (1981), *Girvanella* oncoids flourished in marine subtidal environments and may be tentatively regarded as an indicator of deposition under conditions of slow sedimentation rate. Data presented by Bowman (this vol.) give further support to this conclusion. To date, little work has been done on the morphology of *Girvanella* oncoids, but owing to the detailed study of Wright (this vol.) we are moderately acquainted with the morphology of other cyanoids (using the term introduced by Riding, this vol.), namely those built by *Ortonella, Mitcheldeania, Bevocastria* and *Garwoodia.* Their morphology is related, as evidenced by Wright (this vol.), to the degree of agitation during oncoid growth. Forms which grew in higher energy conditions are subspherical and rounded, are well laminated and have a dense fabric. Forms which grew in lower energy conditions have more irregular forms, are less well laminated and have a spongy fabric. As may be supposed from the literature, these statements apply to other porostromate oncoids, although a reservation has to be made that at least some oncoids which originated in low energy environments are subspherical and rounded; as an example the (originally porostromate, now spongiostromate – see Peryt 1981) oncoids from the Zechstein Limestone may be given (see Peryt and Wagner 1981, Table II).

1 Instytut Geologiczny, Rakowiecka 4, 00-975 Warszawa, Poland

Coated Grains (ed. by T.M. Peryt)

In the Jurassic, a general shift of oncoids into more shallow zones occurred. In the Lower Jurassic, oncoids developed in subtidal environments with a moderately high energy, such as tidal channels or back barrier (Ciarapica and Passeri, this vol.), and the Upper Jurassic oncoids, often associated with ooids, occur in even shallower environments (see Dahanayake, this vol., Gasiewicz 1983) although they may be found in deeper environments (e.g., Massari and Dieni, this vol.) where they reflect, as concluded by Massari (this vol.), very low sedimentation rate. The same seems to be the rule for the Cretaceous oncoids (Tišljar, this vol.; Carozzi et al., this vol., with discussion; Krajewski, this vol.). The morphology of spongiostromate oncoids has been the subject of extensive studies by Dahanayake, and the reader is referred to his paper in the present volume and the earlier papers quoted therein. An interesting observation has been made by Ciarapica and Passeri (this vol.) that "smooth surfaces are typical of the high energy oncoids whereas wavy surfaces with fine crenulated laminas are frequent in the muddy facies"; this closely corresponds to the situation observed in recent lacustrine oncoids (Schäfer and Stapf 1978) and in ancient fluvio-lacustrine oncoids (e.g., Anadón and Zamarreño 1981, Nickel, this vol.).

Starting from the Eocene, spongiostromate oncoids are replaced in marine settings by rhodoids (Monty 1979); sporadically they are to be found in brackish environments (cf; Richter and Sedat, this vol.).

As opposed to marine environments, the lacustrine environments have been dominated during the Phanerozoic by spongiostromate oncoids. Monty and Mas (1981, p. 119) concluded that "If we can rely on the significance of the microstructures, it can be proposed that... oncolite forming freshwater communities that we find... in Lower Cretaceous rocks persisted unaffected at least until the Pleistocene and perhaps until to-day in the same type of environment (fluvio-lacustrine)". It is worth mention that Nickel (this vol.) concluded that the oncoid microstructures are directly controlled by microbial assemblages, and that the same microstructures as reported from recent oncoids have been also described from the Paleozoic oncoids (e.g., Schäfer and Stapf 1978).

To return to the marine oncoids, the concept of tranquil environments of formation of ancient oncoids, with only intermittent turbulence, was concluded after demonstration that the coated grains occurring in shallow water lithofacies and earlier ascribed to oncoids are in fact vadoids (see Peryt, this vol., for discussion). This in turn demonstrates the need for reexamination of all pre-Jurassic shallow water facies from which the oncoids have been described. Despite the obviously different environmental settings in which oncoids and vadoids came into existence, these coated grains can co-occur due to sea level oscillations by which the oncoidal deposit may be put in the vadose zone. On the other hand, it is quite probable that both vadolites and oncolites may form the correlative layers in the sections, as both are related to the slow sedimentation rate, for which reason caution must be maintained during the extrapolation of results of studies of coated grain-bearing rocks, which too often lead to overgeneralizations as to the nature of coated grains and sedimentary environments.

The studies on ancient lacustrine oncoids stress the importance of "the French apporach" including "integration of detailed observation of morphology, internal fabric, biogenic microstructures, mineralogy and diagenesis into the geological setting" (Nickel, this vol.). This approach leads to great progress in knowledge of morphology and microstructures of lacustrine oncoids and should certainly therefore be used when studying ancient marine oncoids. In fact, some papers (e.g., by Wright and Bowman, and Dahanayake, this vol.) reflect this tendency, although the important problems of mineralogy and diagenesis are still usually ignored.

References

Anadón P, Zamarreño I (1981) Paleogene nonmarine algal deposits of the Ebro Basin, northeastern Spain. In: Monty C (ed) Phanerozoic stromatolites. Springer, Berlin Heidelberg New York, p 140–154

Flügel E (1978) Mikrofazielle Untersuchungsmethoden von Kalken. Springer, Berlin Heidelberg New York, 454 pp

Gasiewicz A (1983) Upper Oxfordian oncolites of SE Poland. N Jb Geol Paläont Mh (in press)

Monty CLV (1972) Recent algal stromatolitic deposits, Andros Island, Bahamas. Preliminary report. Geol Rundschau 61:742–783

Monty CLV (1979) Scientific reports of the Belgian expedition on the Australian Great Barrier Reef, 1967. Sedimentology: 2. Monospecific stromatolites from the Great Barrier Reef tract and their paleontological significance. Ann Soc Geol Belg 101:163–171

Monty C, Mas JR (1981) Lower Cretaceous (Wealdian) blue-green algal deposits of the Province of Valencia, Eastern Spain. In: Monty C (ed) Phanerozoic stromatolites. Springer, Berlin Heidelberg New York, p 85–120

Peryt TM (1981) Phanerozoic oncoids – an overview. Facies 4:197–214

Peryt TM, Wagner R (1981) Mikrostruktury stromatolitów i onkoidów z wapienia cechsztyńskiego Polski północno-zachodniej. Kwart Geol 25:629–650

Schäfer A, Stapf KRG (1978) Permian Saar-Nahe Basin and recent Lake Constance (Germany): two environments of lacustrine algal carbonates. In: Matter A, Tucker ME (eds) Modern and Ancient Lake Sediments. Spec Publ Intern Ass Sed 2:83–107, Oxford

IV.2 Cyanoliths (Cyanoids): Oncoids Formed by Calcified Cyanophytes

Robert Riding [1]

1 Introduction

A wide variety of calcareous nodules have been regarded as being probably formed by the activity of cyanophytes. Stromatolites too, are generally assumed to be mainly of cyanophyte origin. But it is only possible to be confident about these interpretations if the deposits contain direct internal evidence of the algae involved. In the Recent this can be provided by the presence of the living algae themselves, but in ancient material the evidence must be in the form of mineralized fossil remains or distinctive petrographic fabrics. In many cases it must be admitted that a cyanophyte origin for stromatolites and oncoids is inferred only from general similarities in form and structure between them and Recent cyanophyte mats which have trapped and bound particulate sediment. Specific evidence for the type of algae (or other microorganisms) involved is usually lacking. However, some fossil examples of stromatolites and oncoids contain convincing evidence of their cyanophyte origin due to the presence of mineralized algal remains within them. Many of the Precambrian silicified microfloras, which provide valuable information about the early history of life on Earth, occur within stromatolites. Examples include the microfloras from the Transvaal Dolomite, Gunflint Iron Formation, Belcher Group, Paradise Creek Formation, Bungle Bungle Dolomite, Beck Spring Dolomite, Bitter Springs Formation, and many others (Schopf 1977, Table 2). They confirm the cyanophytic nature of these stromatolites and have assisted interpretation of the organic controls of stromatolitic microstructure (Hofmann 1975), stromatolite evolution (Awramik et al. 1976), and stromatolite biostratigraphy (Schopf 1977). Permineralization by silica has even preserved fine details of the organic structure of the algae down to the level of the cell contents, but these silicified stromatolitic microfloras are relatively rare, and are best known from the Precambrian.

In the Phanerozoic some stromatolites and oncoids contain calcified cyanophytes. Normally, it is the sheath material which has been calcified (Fritsch 1959, p. 869) and, consequently, these fossils do not provide the same degree of detailed information about the algae as silicified specimens. Nevertheless, calcification commonly preserves features such as the size, shape and orientation of filament-

1 Department of Geology, University College, Cardiff, Great Britain

Coated Grains (ed. by T.M. Peryt)

ous cyanophytes, and these fossils are much more common than silicified microfloras. Consequently, oncoids containing calcified cyanophytes constitute an important type of algal nodule and contain considerably more information concerning the organisms responsible for their formation than do oncoids formed by trapping and binding of particulate sediment.

The purpose of this paper is to underline the proposal, first made at the Second International Symposium on Fossil Algae (Riding 1979), that we should not allow the problems of interpreting the precise identity of many oncoids to obscure the fact that some preserve a lot more information concerning their origin than others. My suggestion is that oncoids formed by calcified cyanophytes should be distinguished as a discrete category, and should be termed *cyanoliths*. Oncoids which do not contain the calcified remains of cyanophytes, and consequently whose origin, although suspected or inferred to be algal in general and cyanophyte in particular, is really uncertain should not be included in this category.

2 Cyanoliths

Cyanoliths are oncoids constructed by calcified cyanophytes (Riding 1979) and are distinguished by the preservation within them of calcified sheath material. Commonly these are tubiform (porostromate) fossils, or potential fossils in the case of Recent specimens, although filamentous cyanophytes can produce non-tubiform calcified sheaths (Riding and Voronova 1982), and calcified coccoid cyanophytes could also produce cyanoliths. Excluded from cyanoliths are nodules which are not formed primarily by cyanophytes and also oncoids formed by non-calcified cyanophytes. Consequently, the "classic" oncolite (if there is such a thing), with a spongiostrome microfabric produced essentially by the trapping and binding of particulate calcareous sediment, is *not* a cyanolith.

Cyanoliths are generally discoidal, elongate or subspherical in shape and may be flattened on one side. They form about a nucleus which may be an inorganic grain, a soft particle which has subsequently decayed, or a complete or broken skeleton. They range from less than 1 cm to 30 cm or more in size. If the nucleus is large relative to the final size of the cyanolith then it will often control the shape of the nodule. The surface ranges from smooth to irregularly nodose.

Internally, cyanoliths exhibit a combination or radial and laminar fabrics, porostromate tubes and coarse fenestral fabrics (Figs. 1 – 4) which contrast with the spongiostrome microfabrics of non-skeletal oncolites. As with any firm substrate, free or fixed, they are susceptible to both encrustation and boring by other organisms. Carboniferous oncoids, including cyanoliths, from Britain commonly contain vermiform gastropods (Burchette and Riding 1977, p. 22) (Figs. 1 and 2). Borings by polychaetes and bivalves are reported in Recent lacustrine cyanoliths by Jones and Wilkinson (1978, p. 1104). Cyanoliths may not require regular movement for their formation. Jones and Wilkinson (1978, p. 1108) argue for mainly in-place growth of the Recent Michigan marl lake cyanoliths they studied.

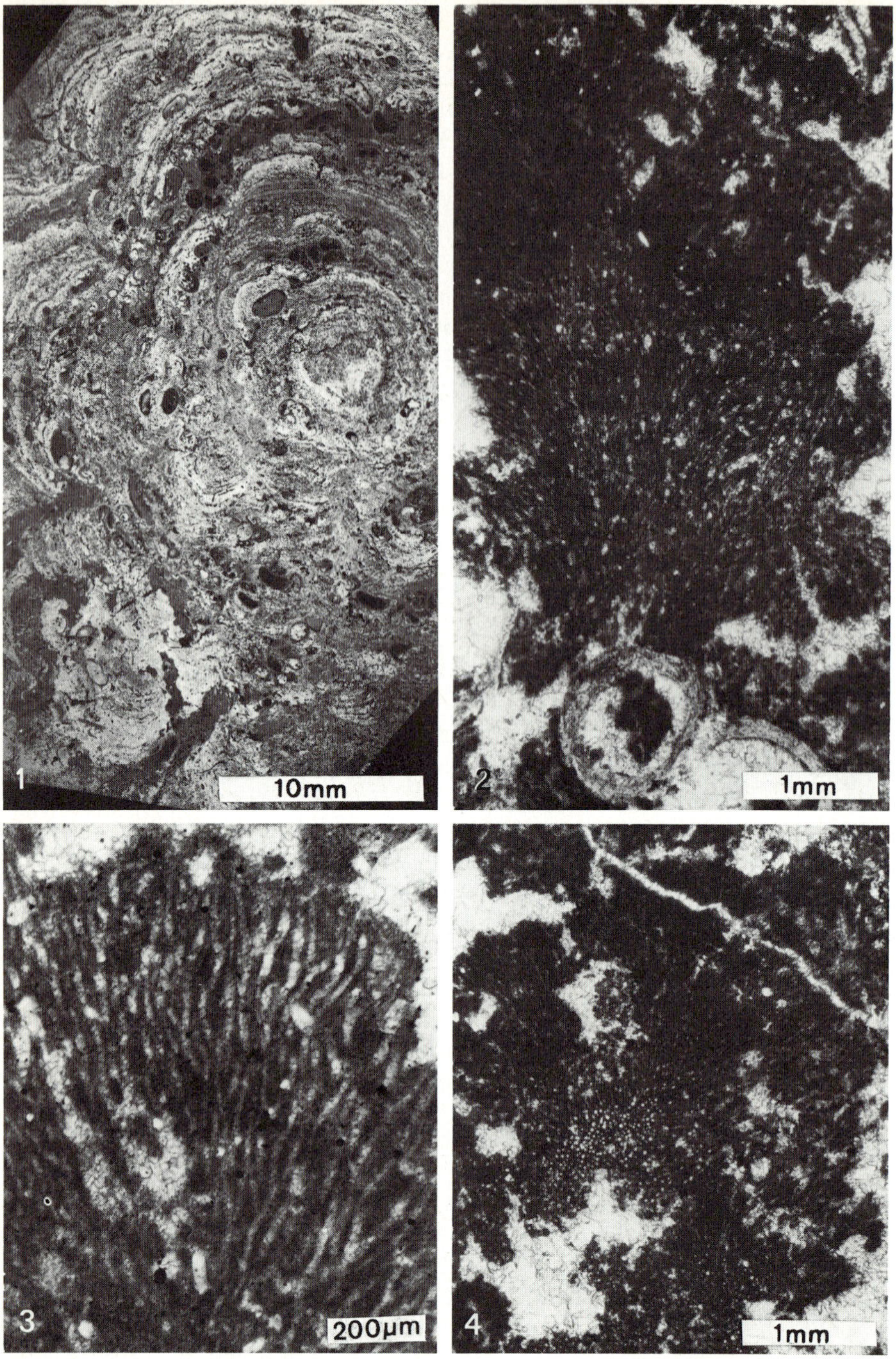
1
10mm
2
1mm
3
200μm
4
1mm

2.1 Recent Examples

Algal modules which can be classed as cyanoliths have been described from Recent freshwater lakes and rivers for many years (see Fritsch 1959, pp. 868 – 869). Lake Constance (Bodensee) in southern Germany and the Michigan "marl lakes" in the United States are well-known localities for calcareous algal nodules. Schäfer and Stapf (1978) describe oncoids (oncolites) from Untersee, the western extension of Lake Constance. They recognised two main types: smooth and hard oncoids built by *Schizothrix,* and soft and spongy oncoids built by *Phormidium, Calothrix* and *Dichothrix,* but spongy and dense layers, containing different types of algae, alternate in many specimens, the surface appearance being generated by the last algae to colonize the oncoid. *Rivularia* is an additional, but less important, constituent of these oncoid-building communities. The internal structure of the oncoids is dominated by the calcified filaments of the algae which are generally conspicuous in section, although the dense structure produced by *Schizothrix* is more obscure (Schäfer and Stapf 1978, p. 96).

Oncoids ("lacustrine pisoliths") described by Jones and Wilkinson (1978) from small marl lakes in central Michigan contain algal filaments encrusted by calcium carbonate to give the oncoids a radial internal fabric. In addition, alternations of dense layers with porous layers containing this radial fabric give a laminar component to the structure of the oncoids. Jones and Wilkinson do not specifically identify the algae responsible for forming these oncoids, but they have surface coatings of living cyanophytes. Oncoids ("water biscuits", Clarke 1900) from the northern end of Canandaigua Lake in New York State are formed by calcification of the sheath of *Schizothrix calcicola* (Riding and Voronova 1982), and also resemble the Michigan marl lake oncoids in general appearance (Fig. 5).

The cyanophytes reported to be involved in the formation of Recent oncoids which can be classed as cyanoliths mainly include *Schizothrix, Phormidium, Calothrix* and *Dichothrix* (Fritsch 1959, pp. 868 – 869; Schäfer and Stapf 1978, pp. 95 – 98; and references). These belong to the families Oscillatoriaceae and Rivulariaceae and are all filamentous forms, with the oscillatoriaceans being distinctly motile.

Fig. 1. Composite cyanolith composed of intergrowths of *Ortonella* and *Bevocastria.* Note tubes of attached gastropods and irregular lamination which becomes discontinuous locally. Lower Carboniferous, Birky Cleugh, Cumbria, England. Negative print of thin-section

Fig. 2. *Ortonella* sp., growing upon a vermiform gastropod (*lower centre*) within a cyanolith. Lower Carboniferous, Birky Cleugh, Cumbria, England

Fig. 3. *Ortonella* sp., detail of tuft to show branched filaments. Lower Carboniferous, Birky Cleugh, Cumbria, England

Fig. 4. *Ortonella* sp., tufts of filaments within a cyanolith. Note coarsely fenestral fabric produced by irregular spar-filled cavities adjacent to algal growths. Lower Carboniferous, Birky Cleugh, Cumbria, England

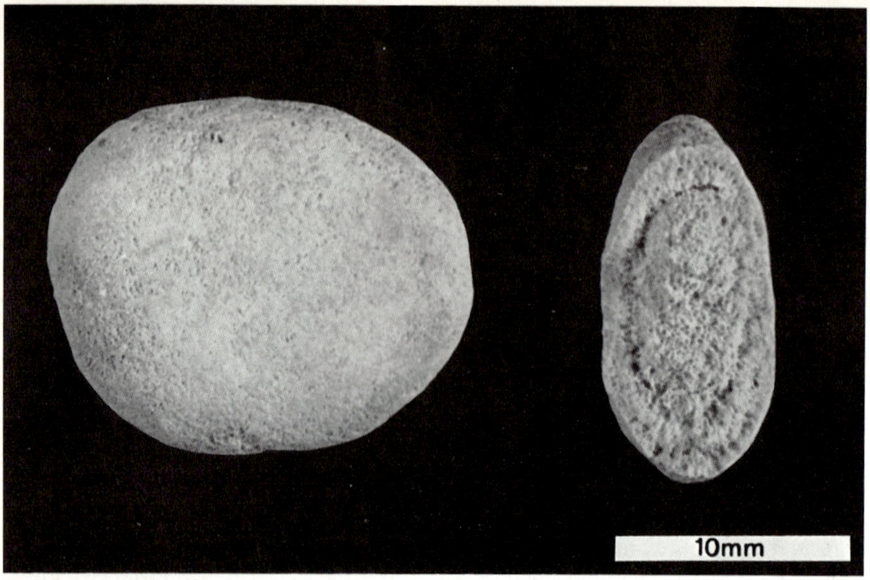

Fig. 5. Recent cyanoliths formed by *Schizothrix calcicola,* showing radial and laminar internal fabric on broken section (*right*) and smooth outer surface (*left*). Squaw Island, Canandaigua Lake, New York State, U.S.A. (see Riding and Voronova 1982)

2.2 Ancient Examples

Cyanoliths constructed by *Girvanella* and *Ortonella* are common in shallow marine facies at various times in the Phanerozoic. In fact, the type material of *Girvanella* from the middle Ordovician of south-western Scotland occurs as "loosely compacted masses" (Nicholson and Etheridge 1878, p. 23). These are a variety of oncoid, with and without distinct nucleii, which consist essentially of tangled masses of *Girvanella.* Oncoids (oncolites) also formed by *Girvanella* occur in the Upper Devonian fore-reef limestones of the Canning Basin, Western Australia (Playford et al. 1976, Fig. 11 b) as well as in numerous oncolites of different age and provenance (Peryt 1981, Table 2).

Other good examples of cyanoliths are the oncoids which occur in peritidal facies in the Lower Carboniferous on the border between England and Scotland (see Garwood 1931) (Fig. 1). *Ortonella* and *Bevocastria* are prominent components of these. *Ortonella* (Figs. 2 and 3) forms erect tufts attached to vermiform gastropods within the cyanoliths. Coarse fenestral fabrics occur due to the creation of relatively large irregular voids adjacent to the rigid tufts of calcified filaments (Figs. 2 and 4). Similar oncoids formed by *Garwoodia* Wood occur in the Lower Carboniferous marginal marine deposits of South Wales (Wright 1982), but the precise affinities of *Garwoodia* are not yet clear and if it is not a cyanophyte then these oncoids cannot be referred to as cyanoliths. Thus, the recognition of cyanoliths is dependent upon the confident interpretation of the enclosed calcified fossils.

There are numerous other potential examples of marine cyanoliths in the Palaeozoic and Mesozoic, and also of fossil freshwater cyanoliths (for example, see the references cited by Jones and Wilkinson 1978, p. 1103 and by Schäfer and Stapf 1978, p. 84). Anadón and Zamarreño (1981) describe oncoids (oncolites) from fluviatile channel deposits of the Palaeogene of the Ebro Basin, Spain. These contain erect calcified filaments similar to those formed by Recent *Phormidium, Schizothrix,* and *Calothrix/Dichothrix.*

3 Discussion

Free-lying, rounded, irregularly laminated calcareous nodules, usually no more than a few centimetres across, have been noted in Recent lacustrine and fluviatile environments for more than a century and similar bodies have been recognised in rocks since the early 1900's. They are distinguished from inorganic laminated bodies, such as oöids and cave pearls, by their size and less regular appearance, and they have been variously termed algal biscuits, lake balls, lacustrine pisoliths, oncoids, and oncolites. In addition to their general form and structure they are related to each other by their known, or assumed, origin through algal activity. They are not limited in their geological occurrence to freshwater environments, but are known also from brackish, marine and hypersaline situations.

In detail these algal nodules show a considerable variety of types. In particular, it is not difficult to distinguish those nodules which have an internal structure consisting of both radial and laminar fabrics which mould the original algal filaments (as exemplified by the Recent Canandaigua and Lower Carboniferous oncoids figured here) from those oncoids with spongiostrome microfabrics and essentially laminar structure which are presumably constructed primarily by trapping and binding of particulate sediment or degradation of calcified filaments.

Although many terms, including Linnean names (e.g., *Pycnostroma*), have been applied to algal and related nodules, very little clarity of terminology or communication has been achieved. One of the more useful terms introduced in recent years is "rhodolith" (originally rhodolite) suggested by Bosellini and Ginsburg (1971) for nodules formed by crustose coralline algae. It is also the only term currently in general use for algal nodules which explicitly refers to the type of organism involved. In a discussion of terms applied to oncoids and similar structures Bosellini and Ginsburg (1971, p. 670) comment: "These various types of algal nodules are often superficially quite similar; but, because they are formed by different algae in different ways and probably in quite different environments, it is desirable to separate them whenever possible".

Cyanoliths are distinct in internal microfabric from other algal nodules and it is appropriate to distinguish them terminologically with a separate name. Their distinctiveness arises from the calcification of the principal organisms responsible for their formation. They thus reflect their origins with much greater precision than, for example, the oncoids with spongiostrome microfabrics which, although usually regarded as the products of algal mats, cannot be assumed to be always entirely cyanophytic (Monty 1977, p. 19) and for which details of size, organization and abundance of the algae are not only obscure, but probably have been lost.

A term has no more value than the concept it represents; and several different terms can be used to embody the same idea. In cyanolith I am proposing a term for calcareous nodules known, or confidently interpreted, to be produced by cyanophytes which contain within them calcareous material directly reflecting the form of the algae. The prefix "cyano-" indicates the type of algae involved. The suffix is more a matter of taste. I favour the suffix "-lith" because it conforms to the style already established in the term rhodolith. But "-oid" would serve exactly the same purpose. So, whether cyanolith or cyanoid is preferred is less important than the concept of nodules recognisably formed by calcareous cyanophytes.

Acceptance of this term should not result in its indiscriminate use. Because it is a genetic term there will be cases where it cannot be applied due to lack of certainty about the precise nature of some oncoids. But its use where there is confidence in the cyanophyte nature of the oncoids, and where they are preserved as calcareous fossils within the oncoid, will provide a term which reflects our improved ability to distinguish subdivisions within what, hitherto, has been a confusing and heterogeneous group of organic nodules.

Because cyanoliths are defined here as being formed by *calcified* cyanophytes it is necessary to examine their internal structure in order to recognise them. They are readily distinguished from oncoids with spongiostrome fabrics formed by non-calcified cyanophytes (see Gebelein 1969, Figs. 11 and 12), but may be confused with nodules formed by the calcareous tubes of other organisms. Nodules constructed by calcareous green algae, such as *Gongrosira* (Golubić and Fischer 1975) resemble cyanoliths. They could, appropriately, be termed chloroliths!

Many ancient nodules contain porostromate and other fossils whose affinities are uncertain and possibly not algal at all. One example is *Rothpletzella* (commonly referred to as *Sphaerocodium*) which forms nodules in Devonian limestones.

Introduction of the term cyanolith helps to draw attention to a particular group of nodules by setting them apart from nodules formed by other calcareous algae and by non-algal organisms. It also emphasizes the importance of being able to recognize the nature of the fossils in nodules which have generally, but imprecisely, been described as oncoids, oncolites, lake-balls, algal nodules and algal biscuits. If the affinities of more porostromate and other nodule-forming organisms can be recognized it will be possible to more precisely define other groups of nodules upon the basis of their dominant organisms.

Acknowledgements. I am grateful to Claude Monty for reviewing an early draft of this manuscript. Paul Wright critically read the final version and made many helpful comments.

References

Anadón P, Zamarreño I (1981) Paleogene nonmarine algal deposits of the Ebro Basin, northeastern Spain. In: Monty C (ed) Phanerozoic stromatolites. Springer, Berlin Heidelberg New York, pp 140 – 154

Awramik SM, Margulis L, Barghoorn ES (1976) Evolutionary processes in the formation of stromatolites. In: Walter MR (ed) Stromatolites Dev Sedimentol 20:149 – 162

Bosellini A, Ginsburg RN (1971) Form and internal structure of Recent algal nodules (rhodolites) from Bermuda. J Geol 79:669 – 682

Burchette TP, Riding R (1977) Attached vermiform gastropods in Carboniferous marginal marine stromatolites and biostromes. Lethaia 10:17 – 28

Clarke JM (1900) The water biscuits of Squaw Island, Canandaigua Lake, N.Y. Bull NY State Mus 8 (39):195 – 198

Fritsch FE (1959) The structure and reproduction of the algae, vol II. Cambridge Univ Press, Cambridge

Garwood EJ (1931) The Tuedian beds of northern Cumberland and Roxburghshire east of the Liddel Water. Q J Geol Soc London 87:97 – 159

Gebelein CD (1969) Distribution, morphology, and accretion rate of Recent subtidal algal stromatolites, Bermuda. J Sediment Petrol 39:49 – 69

Golubić S, Fischer AG (1975) Ecology of calcareous nodules forming in the Little Connestoga Creek near Lancaster, Pennsylvania. Verh Int Verein Limnol 19:2315 – 2323

Hofmann HJ (1975) Stratiform Precambrian stromatolites, Belcher Islands, Canada: relations between silicified microfossils and microstructure. Am J Sci 275:1121 – 1132

Jones FG, Wilkinson BA (1978) Structure and growth of lacustrine pisoliths from Recent Michigan marl lakes. J Sediment Petrol 48:1103 – 1110

Monty C (1977) Evolving concepts on the nature and the ecological significance of stromatolites. In: Flügel E (ed) Fossil algae, recent results and developments. Springer, Berlin Heidelberg New York, pp 15 – 35

Nicholson HA, Etheridge Jr R (1878) A monograph of the Silurian fossils of the Girvan district of Ayrshire with special reference to those contained in the "Gray Collection", vol I, 1. Blackwood, Edinburgh

Peryt TM (1981) – Phanerozoic oncoids – an overview. Facies 4:197 – 214

Playford PE, Cockbain AE, Druce EC, Wray JL (1976) Devonian stromatolites from the Canning Basin, Western Australia. In: Walter MR (ed) Stromatolites, Dev. Sedimentol. 20:543 – 563

Riding R (1977) Skeletal stromatolites. In: Flügel E (ed) Fossil algae, recent results and developments. Springer, Berlin, Heidelberg, New York, pp 57 – 60

Riding R (1979) Cyanoliths: skeletal oncolites formed by calcified blue-green algae. 2 Symp Inter Algues Fossiles, Abstr

Riding R, Voronova L (1982) Recent freshwater oscillatoriacean analogue of the Lower Palaeozoic calcareous alga *Angulocellularia*. Lethaia 15:105 – 114

Schäfer A, Stapf KRG (1978) Permian Saar-Nahe Basin and Recent Lake Constance (Germany): two environments of lacustrine algal carbonates. Spec Publ Int Ass Sediment 2:83 – 107

Schopf JW (1977) Biostratigraphic usefulness of stromatolitic Precambrian microbiotas: a preliminary analysis. Precambrian Res. 5:143 – 173

Wright VP (this vol.) Morphogenesis of oncoids from the Lower Carboniferous Llanelly Formation of South Wales, pp 424 – 434

IV.3 Algal Micro-Reefs – Coated Grains from Freshwater Environments

J. Schneider [1], H. G. Schröder [2], and Th. Le Campion-Alsumard [3]

1 Introduction

Two types of lacustrine coated grains from an oligotrophic freshwater lake (Attersee, Austria) are described in this paper:

1. One type are the so-called "Krusten- and Furchensteine", lacustrine encrusted and furrowed stones. They have been known for more than 120 years. Schimper (1857) reported them from Lake Geneva (published by Forel 1878). Later on they were discussed more thoroughly, e.g., by Boysen-Jensen (1909), Kann (1941) and Schneider (1977). The carbonate crusts are structured like cauliflowers and build micro-"patch-reefs" on every solid substrate (inorganic and biologic). The crusts are biogenic growth structures which in addition incorporate trapped detritus and inorganically precipitated calcite (Schneider and Schröder 1980). Furrows occur on carbonate substrates beneath well developed crusts (Schneider 1977).
2. Another type of coated grains originates from encrusted macrophytes, which produce $CaCO_3$ crusts solely by themselves and/or together with epiphytic microphytes (Passarge 1904, pp. 108/109; Wetzel 1960, p. 232; Allen 1971, p. 124; Schäfer 1973; Allanson 1973; Schröder 1982).

The coated grains from the Attersee are not directly comparable with "oncoids" showing concentrically textured carbonate structures, as described, e.g., by Schöttle and Müller (1968), Schöttle (1969) or Schäfer and Stapf (1978) from the Recent Lake Constance (Germany), or by Richter et al. (1979, Fig. 3, p. 22) from the basis of Pleistocene brackish water environments of Corinth (Greece).

A critical discussion of the terminology, classification, and environmental significance of oncoids throughout the Phanerozoic is given by Peryt (1981) and also Flügel (1978, p. 109 ff., Table 13, p. 94 ff.).

In this paper we will describe the genesis and the relevance of the sedimentological and paleoecological aspects of these special types of coated grains.

---

1 Geologisch-Paläontologisches Institut, Goldschmidtstraße 3, D-3400 Göttingen, FRG
2 Deutsche Texaco AG, Industriestraße 2, D-3109 Wietze, FRG
3 Station Marine d'Endoume, Centre d'Océanographie, Rue de la Battérie des Lions, F-13007 Marseille, France

Coated Grains (ed. by T.M. Peryt)

The results presented here were obtained as part of a program about carbonate construction, decomposition and sedimentation in the Attersee; they were obtained during a joint program of the authors in July 1980.

2 Location

The Attersee is located in Austria about 40 km E of Salzburg in front of the Northern Calcareous Alps (469.2 m above Adriatic Sea, length: 20.1 km, mean broadness: 2.4 km; area: 45.9 km^2; length of coasts: 53 km). The lake basin was formed during the Würm-glaciation.

The build-up of the carbonate crusts occurs in the shallow sublittoral zone within an area of 2.8 km^2 (6.2% of the total lake area) in a water depth of 0 to 10 m. The optimal development of crusts on stones and gravels is found to be in a depth of 0.6 m (see Schröder 1982).

In the shallow nearshore subtidal areas erosion takes place by waves and currents. The coarser material from the Würm-moraine is thus exposed on the sediment surface. Therefore these stones and gravels can be used as a hardground to be settled on by calcifying microphytes. Erosion products of that erosion zone are transported downslope; they mainly derive from the biogenic production (high $CaCO_3$ contents > 75%). At the deeper subtidal lake bottom the sediments are accumulated as soft bottom which can be densely colonized by macrophytes down to a depth of approximately 10 m.

3 Carbonate Production

According to Pia (1933) we schematically differentiate the processes of carbonate construction into:

A. Abiogenic carbonate construction (e.g., sinter crusts formed by degassing of CO_2 by increasing the temperature and/or lowering the pressure when groundwaters come to the earth surface).

B. Biogenic carbonate construction
   a) physiological (e.g., changing of the micro-environment by the activity of photoautotrophic organisms, raising pH and CO_3^{2-}-concentration during assimilation of CO_2 and HCO_3^-, promoting or causing precipitation of $CaCO_3$,
   b) organic (construction of skeletal carbonates of mollusks).

In the littoral area of the Attersee the rounded and mostly flattened stones of the moraine material in some places cover an area of almost 100 m width, especially in the northern part of the lake. These gravels and stones (limestones, dolomites, sand- and silt-stones derived from the Alps as glacial till) are the substrate for the calcareous crusts, built primarily by microphytes. Between these stones

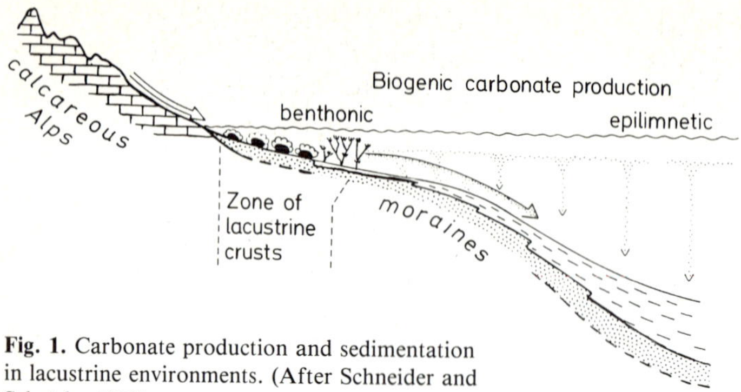

Fig. 1. Carbonate production and sedimentation in lacustrine environments. (After Schneider and Schröder 1980)

and also in greater depth, macrophytes (mostly charophytes) grow on soft sediment bottoms.

Both the hard substrates (stones) and the macrophytes are overgrown by carbonate crusts produced by microphytes. The crusts can be destroyed by storms and waves and then contribute to the sedimentation of lake marl within the littoral zones as well as marly muds in the deeper zones to the basin bottom (see Fig. 1; Schäfer 1973, Schäfer and Stapf 1978). Thus the benthic biogenic carbonate production, together with the epilimnetic biogenic decalcification, contributes to the build-up of carbonate sediments of the lake.

Epilimnetic biogenic decalcification (Minder 1922) is seasonally caused by the blooming of phytoplancton.

4 The Build-Up of Carbonate Crusts

There are two main types of "coated grains" in the littoral zone of the Attersee:

- Crusts in form of micro-patch-reefs on hard substrates like stones, gravels, shells (and also plastics etc.).
- Crusts in form of encrustations around stems of submersed macrophytes (especially of Charophytes).

Fig. 2. a "Krustenstein" from shallow sublitoral (Attersee, 0.5 m). Typical cauliflower-like surface of microphytic carbonate crust on a moraine stone. **b** Furrows (biokarst) on a limestone surface visible after removing mechanically the overgrowing crust (part of a moraine boulder from 3 m depth). **c** Cross section of an encrusted limestone with furrows ("Krusten- and Furchenstein"). The furrows was formed underneath the carbonate crust (see Fig. 5). *e* zone of endolithic microorganisms. **d** Cauliflower-like growth structure of algal micro-reef mainly built by the cyanophyte *Schizothrix*. (Organic substance removed by H_2O_2, SEM). **e** Fragment of a microphytic carbonate crust with trapped particles: Detritic grain and two coated charophyte tubes. See also Fig. 4g. (Organic substance removed by H_2O_2, SEM). **f** Cross-section through a colony of epi-/endolithic cyanophyte *Schizothrix*. *A* epilithic part with precipitated micrite; *B* carbonate substrate with destructional micrite; *C* solid limestone substrate penetrated by endolithic filaments. See also Fig. 6. (Organic structures preserved by critical point drying, SEM)

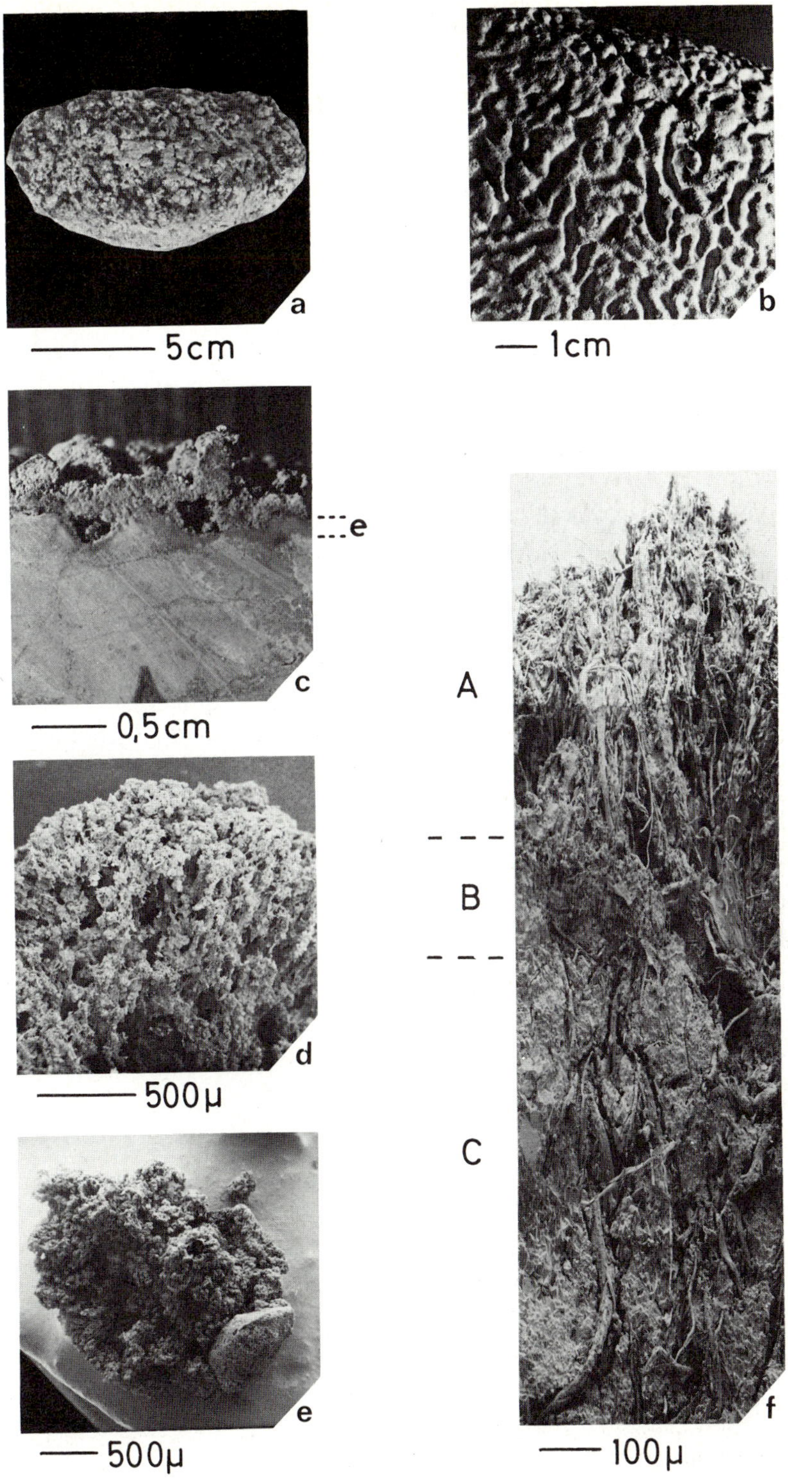
a
5cm
b
1cm
e
c
0,5cm
d
500µ
e
500µ
A
B
C
f
100µ

Different processes contribute to the build-up of these crusts (e.g., Eggleston and Dean 1976, Schneider 1977, Schäfer and Stapf 1978, Schröder 1982):

4.1 Trapping and Binding of Particles (Fig. 2e)

Sediment particles of different origin can be entrapped mechanically from suspended matter out of the lake water. They are bound by filaments of microorganisms and/or their mucus on the crust surface. Cyanophytes (Cyanobacteria, Krumbein 1979a, Ripka et al. 1979, Golubic 1979), chlorophytes, rhodophytes and also diatoms are responsible for this process. The mucus-rich diatoms and the cyanophyte *Dichothrix* play a dominant role in that trapping process.

4.2 Biogenic Carbonate Construction (see Fig. 3)

4.2.1 Carbonate Encrustation in General. In freshwater environments many epilithic cyanophytes (e.g., *Schizothrix, Scytonema*) encrust their filaments extracellularly with calcite. The calcite crystallizes as micrite around the filaments. Until now we have not yet observed whether it can be distinguished between micritic envelopes of the filaments on the one hand and a calcification of the sheath itself on the other [this, for instance, having been described by Riding (1977) from *Tolypothrix* and *Plectonema*]. During our investigations we only found micritic envelopes outside the polysaccharide sheaths of the cyanophyte *Schizothrix* (see Fig. 2d – f). The resulting structures are irregular tubes of micritic calcite. Within the group of photoautotrophic microorganisms that build up carbonate crusts, the genus *Schizothrix* plays the dominant role.

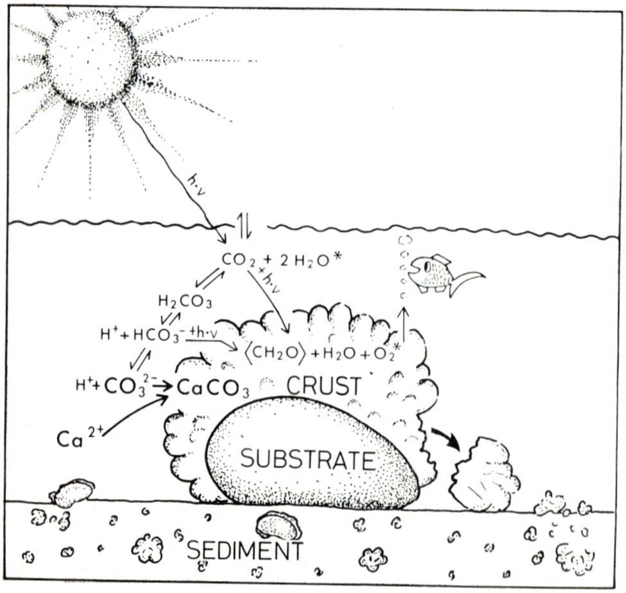

Fig. 3. Construction of lacustrine crusts and their contribution to the formation of sublittoral sediments. (After Schneider and Schröder 1980)

4.2.2 Build-Up of Specific Carbonate Structures. Organisms like the cyanophyte *Rivularia* and the chlorophyte *Chaetophora* form specific calcite structures around their radially-growing filaments within their common gel (Fig. 4a–c).

The sheaths around the cells are not calcified in freshwater, which is in contrast to two marine littoral *Rivularia* recently described by Golubic and Campbell (1981). The resulting structures of freshwater *Rivularia* are specific calcite grains and aggregates which can easily be recognized and identified as sediment particles (Schneider and Schröder 1980; see (Fig. 4d, e). According to Golubic and Campbell (1981, pp. 223/224) these specific structures may be attributed to skeletal carbonates.

We refer to Bornemann (1887), Elliott (1964), Schneider and Schröder (1980) for the ecological and taxonomical problems of calcifying *Rivularia*-like structures (*Cayeuxia, Zonotrichites*).

5 Cementation

Between the filaments of the epilithic photoautotrophic microorganisms the saturation with respect to $CaCO_3$ can be raised considerably during the assimilation of the different carbon sources (CO_2, HCO_3^-, CO_3^{2-}). The biogenically-induced high pH and high carbonate concentration in that micro-environment leads to an inorganic precipitation of $CaCO_3$ within the crust. In freshwater environments (Mg/Ca ratio in lake water $\ll 1$) calcite will be precipitated as micrite.

On and within the crusts live a great number of organisms: bacteria, fungi, gastropods, insect larvae, worms etc. All contribute in different ways to the build-up of the crusts as well as to its decomposition (see Fig. 2a–e and 5).

These carbonate crusts are a real microcosm which is comparable with a reef bicenosis with respect to the variability of species and their physiological activity. This is the reason why we designate the crusts on solid substrates as micro-reefs.

The carbonate precipitation by the processes described above leads to a gradual growth of the micro-reefs which inside become more and more stable and compact by cementation (see Fig. 5, phase 3 and 4). The micro-reefs can be partly destroyed by wave action and by bioabrasion done by gastropods, larvae, and water birds like swans, which turn the stones when searching for food. The resulting fragments may then contribute to a deposition of carbonate lake sediment as shown above.

The remaining micro-reefs achieve a patchy shape being distributed at random which might allow to address them as micro-patch-reefs.

6 Decomposition of Carbonate Substrates and Origin of Furrows

Beneath the crusts which grow on carbonate stones and gravels a more-or-less accentuated brain-like furrow pattern occurs. Ninety-four per cent of all carbonate stones and gravels in the Attersee show these furrows (Schröder 1982). Carbon-

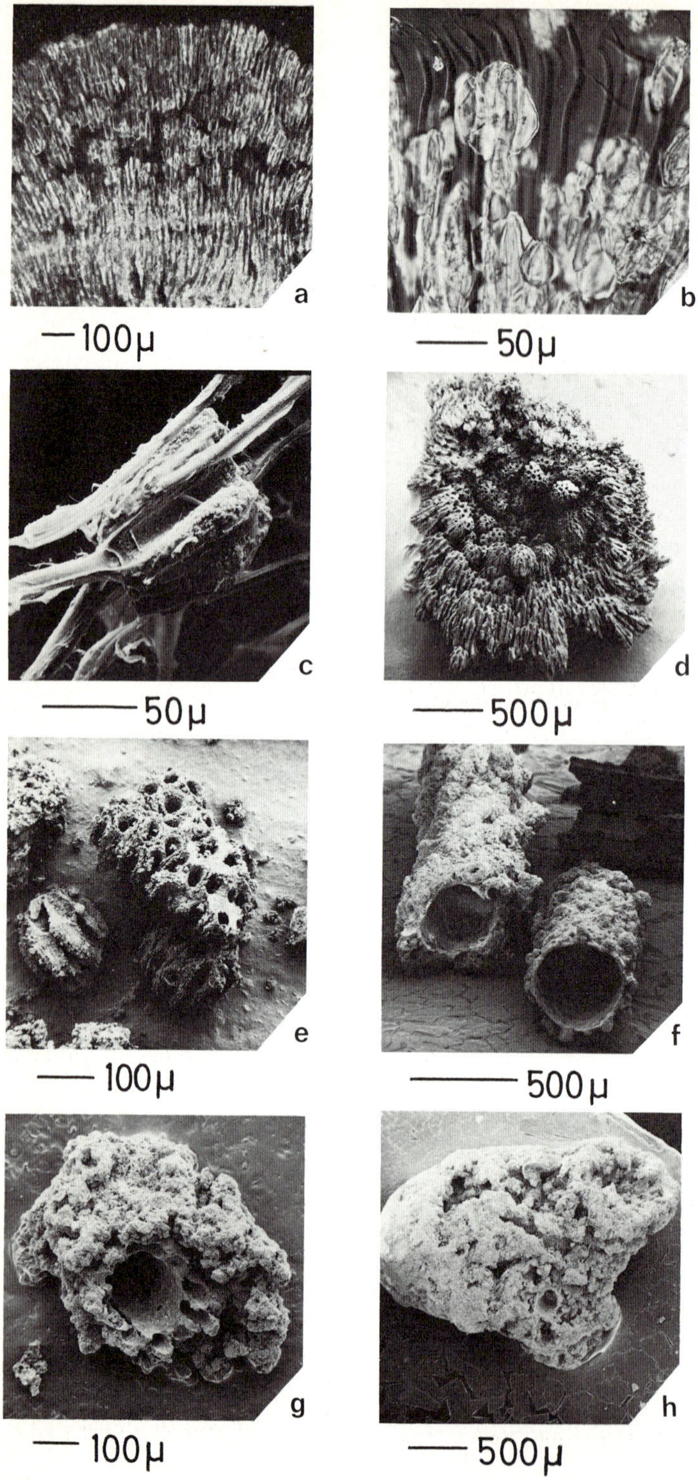
a
100µ
b
50µ
c
50µ
d
500µ
e
100µ
f
500µ
g
100µ
h
500µ

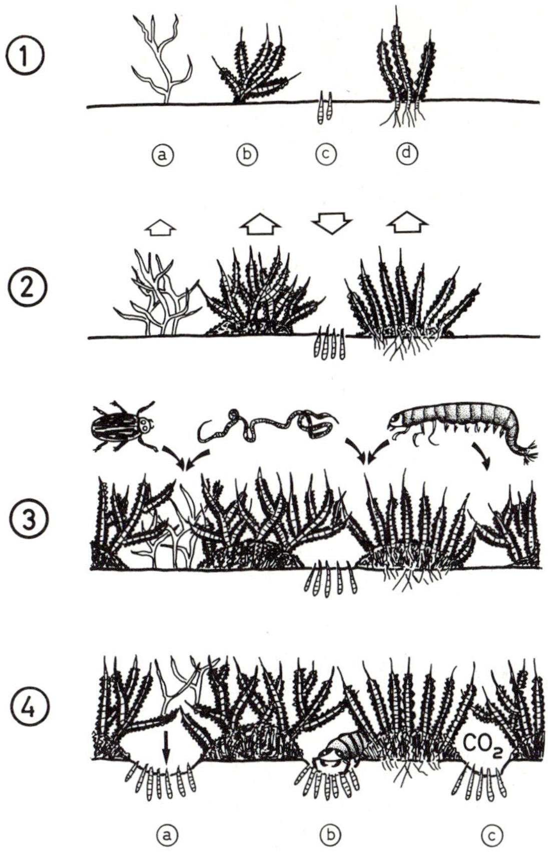

Fig. 5. Development of crusts and furrows on carbonate substrates. *1* primary colonisation; *2* different growth development; *3* invasion by different animals; *4* development of furrows (bioerosion) (Schröder 1982)

ate substrates are regularly attacked by endolithic microorganisms like fungi and cyanophytes. A schematic genetic model was already given by Schneider (1977, pp. 254/255).

Figure 5 schematically shows the different life forms of microphytes (Cyanophyceae, Chlorophyceae) on carbonate substrates, the continuous development of crusts and of furrows.

Fig. 4. a Cross section of a *Rivularia* showing the radial growth, the concentric zonation and the calcitic aggregates formed within the common gel of the living colony. (Colony cut after formaldehyde-fixation, light microscope). **b** Detail of the colony from Fig. 1 showing the calcite grains growing within the common gel around the filaments of *Rivularia* (light microscope). **c** Single calcitic grain from a *Rivularia* grown around the filaments (filaments preserved by critical point drying, SEM). **d** Calcite "skeleton" of a *Rivularia* colony (upside down). The organic substance was removed by H_2O_2, SEM). **e** Single characteristic calcite grains from a *Rivularia*-colony as to be found in the sediment fraction >63 μ. **f** Two macrophyte (charophyte) particles coated by microphyte-carbonate. *Left* type C (see Fig. 7). *Right* type B (see Fig. 7). Sediment-particles from 12 m water depth. **g** Macrophyte (charophyte) particle coated by microphyte-carbonate, type E (see Fig. 7). Sediment-particle from 0.5 m water depth. **h** Sediment particle rounded during transportation before final sedimentation. Grain built by macro- and microphyte-carbonate (see also Fig. 2e). Sediment from 12 m water depth

Phase 1: Primary colonization on the surface of a carbonate substrate by benthic microphytes in a statistical distribution pattern.

a) Epilithic forms which do not calcify (e.g., *Stigonema,* see Fig. 5, phase 1a).
b) Epilithic forms which calcify (e.g., *Scytonema,* see Fig. 5, phase 1b).
c) Endolithic forms which do not calcify (e.g., *Plectonema, Hyella, Mastigocoleus, Chlorogloea,* fungi, see Fig. 5, phase 1c).
d) Forms which live partly endolithic and partly epilithic like *Schizothrix. Schizothrix* penetrates the carbonate substrate as an euendolith (Golubic et al. 1981) down to a depth of 2.6 mm, the deepest penetration by a cyanophyte ever observed. The epilithic part of *Schizothrix* still precipitates $CaCO_3$, so contributing to the formation of a crust on the substrate surface (see Fig. 2f and Fig. 5, phase 1d).

Phase 2: The growth of calcifying epiliths (type b and d in Fig. 5, phase 1, predominantly *Schizothrix*, see Kann 1941) develops step by step.

Younger forms colonize on top of the older and more cemented calcified structures. This generates a calcification sequence which may increase the host surface by about 300%. The forms living on the surface are thus ecologically preferred with respect to light utilization.

Inside the crust a continuous cementation takes place resulting in a gradual stabilization of the micro-patch-reef (see also Fig. 6). Specific calcareous structures in the crust are formed by *Rivularia* (see Schneider and Schröder 1980, Casanova 1981).

During the growth of the crust a specific zonation develops on carbonate substrates (see Figs. 2f and 6):

A. The uppermost part of the crust consists mainly of freshly precipitated calcite (mostly micrite) and trapped particles so that the crust is still soft. The lowermost older part of the crust is more cemented and therefore more stable.

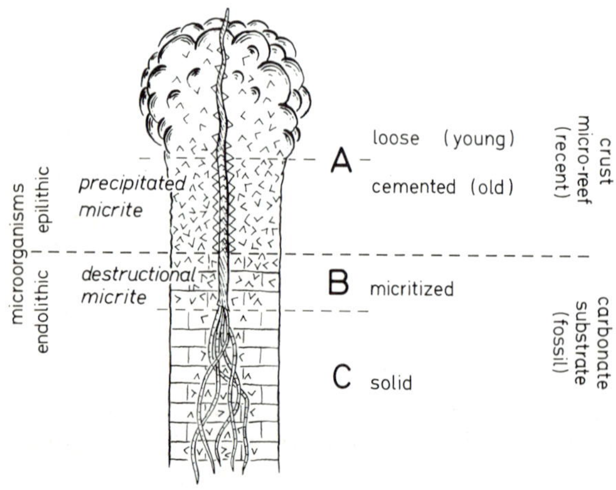

Fig. 6. Schematical micro-zonation of crusts on carbonate substrate. (Epi-/Endolith: *Schizothrix*, see also Fig. 2f)

B. The uppermost layer of the limestone substrate underneath is secondarily micritized by the intense boring activity of the endoliths similar to the genesis of micrite envelopes sensu Bathurst (1966). Thus the substrate surface is difficult to distinguish from the crust when thin sections are investigated under a light- or a scanning-electron-microscope.
C. The deeper part of the solid carbonate substrate shows its original structure and as well first hints on attacks of the endoliths (especially cyanophytes and fungi).

Kobluk and Risk (1977a, b) describe the formation of constructive- and destructive-micrite envelopes from marine shallow water environments. The authors remark that the micrite around and within algal filaments and within bore holes of endoliths is built after death of the microorganisms (very early diagenesis). In contrast to these observations from marine environments ("cementation phenomenon not related to algal life processes", Kobluk and Risk 1977b, p. 1072) our observations from freshwater environments show that the calcification takes place during life time of the organisms and is caused by the physiological activity of the photoautotrophic organisms (see also Fig. 3). A summarized overlook on problems of calcification by bacteria and algae is given by Krumbein (1979b).

Phase 3 (Fig. 5): As the crust becomes several mm thick, an invasion of animals (larvae, insects, worms, gastropods) takes place.

These animals tunnel and graze in and on the crust as well as on the micritized uppermost substrate surface. Thus the loosened substrate surface can be easily abraded by the attack of grazing organisms (mostly larvae of *Tinodes*). Thus preceding biological corrosion by endolithic microorganisms prepares the substrate for biological abrasion (for processes in marine environments see e.g., Schneider 1976, Torunski 1979, Le Campion-Alsumard 1979, Schneider and Torunski, in press).

Phase 4 (Fig. 5): The development of furrows is a consequence of complex bioerosion. The thalli of the randomly distributed micro-patch-reefs become continuously endurated by cementation. Thus light penetration is restricted beneath the thalli. As a consequence biological corrosion (Fig. 5, phase 4a) and biological abrasion (Fig. 5, phase 4b) is more intense between the micro-patch-reefs. The micritized and therefore loosened substrate surface (Fig. 6, zone B) is more intensely attacked by bioerosion and from that furrows develop between the already endurated thalli of the micro-patch-reefs. The furrows can be defined as biokarst on carbonate surfaces which develop only in combination with the formation of a crust (Schneider et al., in press; see also scheme, Fig. 1, p. 254 in Schneider 1977, see Fig. 2b).

Inorganic dissolution of carbonate can take place within crusts of several centimeter thickness. Within the resulting caverns a microbial degradation of organic material can raise the CO_2-content and lower the pH, which can lead to dissolution of the carbonate substrate (see Fig. 5, phase 4c; Golubic 1962, 1973, p. 443).

7 Coated Grains Formed by Macrophytes

In the sublittoral zone we find many macrophytes (mostly Charophytes) which build more-or-less thick carbonate crusts. These crusts are formed by a precipitation of micritic calcite during the photosynthesis of the macrophytes themselves and/or by the activity of microphytic epibionts respectively. The crust produced by the microphytic activity might be compared with the micro-patch-reefs on stones and gravels described above. The calcification results in different characteristic types of coated grains (Figs. 4f, g and 7). After the death or mechanical break down of the macrophytes parts of the crust contribute to the sublittoral carbonate sedimentation as specific coated grains (Figs. 4d – h and 7, see Schäfer 1973, Schröder 1982). These grains can easily be found in the sediment fraction of >63 μ.

8 Significance of the Trophic Water Conditions

Calcareous crusts not only occur in extremely oligotrophic lakes like Attersee. Schäfer and Stapf (1978) report an intense growth of coated grains ("Schneggli-steine") in the Untersee of lake Constance at places, where the Rhine river water is slightly polluted by waste water from a sewage plant. The Untersee is an eutrophic lake as well being rich in bicarbonate. The same was shown by Kulke and Ott (1978) from the Amper river of the Ammersee (Southern Bavaria, see as well Müller and Sigl 1977).

Kann (1940, p. 264) compared oligotrophic hardwater lakes (e.g., Lunzer See, Austria) and eutrophic hardwater lakes (e.g., Plöner See, Northern Germany) and found out that a remarkable development of calcareous crusts does not occur in all hardwater lakes. One reason could be the presence of nutrients like phosphates which are present in eutrophic waters and can act as crystal poisons to the calcite nucleation as well as organic substances, which can hinder the nucleation of $CaCO_3$ by complexing Ca ions, as known from investigations of, e.g., Suess (1970), Chave and Suess (1970), Walter and Hanor (1979a, b).

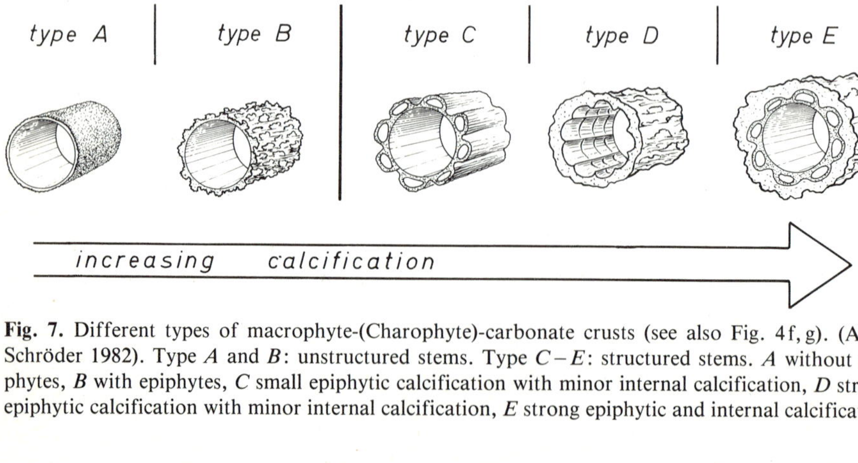

Fig. 7. Different types of macrophyte-(Charophyte)-carbonate crusts (see also Fig. 4f, g). (After Schröder 1982). Type *A* and *B*: unstructured stems. Type *C*–*E*: structured stems. *A* without epiphytes, *B* with epiphytes, *C* small epiphytic calcification with minor internal calcification, *D* strong epiphytic calcification with minor internal calcification, *E* strong epiphytic and internal calcification

These indications could be a possible, yet hypothetical explanation for the fossil record as well.

9 Geological Relevance

In freshwater sublittoral environments we find calcareous coated grains in form of complete algal micro-reefs on hard substrates as well as their destruction products and fragments of calcified macrophytes as components of the sediment.

The micro-reefs are accompanied by biokarst-furrows on the substrate surface when the crusts grow on hard carbonate substrates. The internal structures of the micro-reefs are composed of different types: spongy, dense and often fan-like structures are made by different microphytes (see, e.g., Schäfer and Stapf 1978).

The resulting coated grains (encrusted and furrowed stones and gravels as well as coated grains from macrophytes) are, with respect to their origin, indicators of:

- shallow water environments within the photic zone of lakes,
- environments of oligo- to mesotrophic hardwater lakes and streams,
- environments with primary or secondary hard substrates are restricted to non-depositional or erosional milieus,
- environments of soft bottoms with a low sedimentation rate, grown with macrophytes.

Macroid crusts and furrowed stones cannot be transported under normal conditions. Because they are autochthonous, they can be regarded as good indicators of the above listed environments.

It should be noted, however, that destruction products of crusts from micro-reefs on stones as well as coated grains from macrophytes can be transported down to the deep lake basins. As sediment particles they no longer indicate the environments of their origin.

From the fossil record corresponding carbonate crusts were described by: Bornemann (1887, Triassic equivalent of *Rivularia* on bivalve shells), Pia (1933, Permian and Tertiary), Wagner (1954, Tertiary Cyanophyte crusts on freshwater bivalves), Elliott (1964, Triassic *Rivularia*), Richter et al. (1979, Pleistocene), Reiss (1923, Tertiary), Donsimoni and Giot (1977, Oligocene), Link et al. (1978, Pliocene).

Schäfer and Stapf (1978) gave an environmental comparison of the carbonate crusts (oncoids) from Lake Constance and Permian equivalents from the Saar-Nahe basin.

Monty and Mas (1981) described oncolithic microstructures formed by different cyanophytes from lower Cretaceous freshwater environments of Eastern Spain.

10 Summary

In oligotrophic hardwater lake Attersee different types of coated grains occur:

1. Carbonate crusts on solid substrates of biological and inorganic origin.
2. Carbonate encrusted and furrowed stones on hard carbonate substrates.

The carbonate crusts are formed by trapping and binding sedimentary particles as well as by a biogenic carbonate production of photoautotrophic microorganisms (cyanophytes, chlorophytes, rhodophytes, diatoms). The furrows on hard carbonate substrates underneath the crusts are the product of an interaction of biological corrosion by endolithic microorganisms (cyanophytes, fungi) and biological abrasion by macroorganisms (Larvae, gastropods) which altogether lead to a biokarst.

Carbonate crusts around macrophytes (mostly charophytes) are the product of photosynthetic activity of the macrophytes themselves together with the life processes of photoautotrophic microorganisms living on them.

A destruction of the biogenic crusts (either by water movements and/or death of the living producers) leads to the formation of more-or-less characteristic sediment particles which contribute to lake sedimentation.

Macroid coated grains like encrusted and furrowed stones are normally autochthonous and therefore good indicators of shallow water environments within the photic zone, of environments with low sedimentation rate (or erosion), and possibly of environments of oligotrophic hard waters. Destruction products from stone crusts and from macrophytic crusts are transported to the deeper parts of the lake down to the profundal zone, no longer indicating the environment of their origin.

Acknowledgements. The investigations were partly supported by a grant of the Deutsche Forschungsgemeinschaft (Schn 16/8).

We especially thank Prof. Dr. St. Golubic and Dr. S. Campbell (Boston, Univ., USA) and Prof. Dr. E. Kann (Vienna) for many discussions and help in nomenclatorical and ecological questions. Prof. Dr. E. Flügel (Erlangen) and Dr. A. Schäfer (Bonn) we have to thank for critically reading the manuscript and giving many valuable comments. Further we acknowledge the help of B. Raufeisen for painting some figures, Dipl. Geol. H. Scholz for the SEM-photographs and Miss. K. Roberts and Prof. Dr. B. D. Erdtmann for their assistance in improving the English in the manuscript and Miss. K. Busse for typing the manuscript.

References

Allanson BR (1973) The fine structure of the periphyton of Chara sp. and Potamogeton natans from Wytham Pond, Oxford, and its significance to the macrophyte-periphyton metabolic model of R. G. Wetzel and H. L. Allen. Freshwater Biol 3:535 – 541

Allen HL (1971) Primary productivity, chemo-organotrophy, and nutrional interactions of epiphytic algae and bacteria on macrophytes in the littoral of a lake. Ecol Monogr 41:97 – 127

Bathurst RGC (1966) Boring algae, micrite envelopes and lithification of molluscan biosparite. Geol J 5:15 – 32

Bornemann JG (1887) Geologische Algenstudien. Jahrb K Pr Geol L A 1886:116 – 134

Boysen Jensen P (1909) Über Steinkorrosion an den Ufern von Furesö. Int Rev Ges Hydrobiol 2:163–173

Casanova J (1981) Etude d'un milieu stromatolitique continental. Les travertins plio-pléistocènes du Var (France). Thèse présentée à l'Université d'Aix-Marseille III, 132 pp, 30 Taf

Chave KE, Suess E (1970) Calcium carbonate saturation in seawater: Effects of dissolved organic matter. Limnol Oceanogr 15:633–637

Donsimoni M, Giot D (1977) Les calcaires concrétionnés lacustres de l'Oligocene supérieur et de l'Aquitanien de Limagne (Massif Central). Bull BRGM 2e Ser 2 Sect 1:131–169

Eggleston JR, Dean WE (1976) Freshwater stromatolitic bioherms in Green Lake, New York. In: Walter MR (ed) Stromatolites. Dev Sedimentol 20. Elsevier, Amsterdam, pp 479–488

Elliott GF (1964) Zonotrichites (calcareous algae) from the Arabian Triassic. Eclogae Geol Helv 57:567–570

Flügel, E (1978) Mikrofazielle Untersuchungsmethoden von Kalken. Springer, Berlin Heidelberg New York, 454 p

Forel FA (1878) Remarques sur la sculpture des galets des grèves des lacs Suisses. Act Soc Helv Sc Natl Bern. cited in Boysen-Jensen P (1909) p 173

Golubic S (1962) Zur Kenntnis der Kalkinkrustation und Kalkkorrosion im Seelitoral. Krustensteine und Furchensteine. Schweiz Z Hydrol 24:229–243

Golubic S (1973) The relationship between blue-green algae and carbonate deposits. The biology of blue-green algae. Blackwell, Oxford, pp 434–472

Golubic S (1979) Cyanobacteria (blue green algae) under the bacteriological code? An ecological objection. Taxon 28:387–389

Golubic S, Campbell SE (1981) Biogenically formed aragonite concretions in marine Rivularia. In: Monty C (ed) Phanerozoic stromatolites. Case histories. Springer, Berlin Heidelberg New York, pp 209–229

Golubic S, Friedmann I, Schneider J (1981) The lithobiontic ecological niche, with special reference to the microorganisms. J Sediment Petrol 51:2, 475–478

Kann E (1940) Ökologische Untersuchungen an Litoralalgen ostholsteinischer Seen. Arch Hydrobiol 37:177–269

Kann E (1941) Krustensteine in Seen. Eine vergleichende Übersicht. Arch Hydrobiol 37:504–532

Kobluk DR, Risk MJ (1977a) Calcification of exposed filaments of endolithic algae, micrite envelope formation and sediment production. J Sediment Petrol 47:2 517–528

Kobluk DR, Risk MJ (1977b) Micritization and carbonate-grain binding by endolithic algae. Am Assoc Petrol Geol Bull 61 7:1069–1082

Krumbein WE (ed) (1979a) Cyanobakterien – Bakterien oder Algen? I. Oldenburger Symposium über Cyanobakterien, 1977 Taxon Stellg Ökol, 130 p

Krumbein WE (1979b) Calcification by bacteria and algae. In: Trudinger RA, Swaine DJ (ed) Biogeochemical cycling of mineral-forming elements. Studies on environmental science, vol III. Elsevier, Amsterdam, pp 47–68

Kulke H, Ott, E (1978) Rezente und subrezente Süßwasser-Onkoide aus Oberbayern. Vortragskurzfassung GV Tagung 1978, Münster

Le Campion-Alsumard T (1979) Les cyanophycées endolithes marines. Systématique, ultrastructure, écologie et biodestruction. Oceanol Acta 1972, 2 2:143–156

Link MH, Osborne RH, Awramik S (1978) Lacustrine stromatolites and associated sediments of the Pliocene Ridge Route Formation, Ridge Basin, California. J Sediment Petrol 48 1:143–157

Minder L (1922) Über biogene Entkalkung im Zürichsee. Verh Int Ver Theor Angew Limnol 1:20–32

Monty C, Mas JR (1981) Lower Cretaceous (Wealdian) blue-green algal deposits of the Province of Valencia, Eastern Spain. In: Monty C (ed) Phanerozoic stromatolites. Case histories. Springer, Berlin Heidelberg New York, pp 209–229

Müller J, Sigl W (1977) Morphologie und rezente Sedimentation des Ammersees. N Jahrb Geol Paläontol Abh. 154 2:155–185

Passarge S (1904) Die Kalkschlammablagerungen in den Seen von Lychen, Uckermark. Jahrb Preuß Geol LA 1901 22:79–146

Peryt TM (1981) Phanerozoic oncoids – an overview. Facies 4:197–214

Pia J (1933) Die rezenten Kalksteine. Z Kristallogr Mineral Petrogr Abt B Ergänzungsband, Leipzig. 420 p

Reis OM (1923) Kalkalgen und Seesinterkalke aus dem rheinpfälzischen Tertiär. Geognost Jh 36:102–130

Richter DK, Herforth A, Ott E (1979) Pleistozäne, brackische Blaugrünalgenriffe mit Rivularia haematites auf der Perachorahalbinsel bei Korinth (Griechenland). Neues Jahrb Geol Palaeontol Abh 159 (1):14–40

Riding R (1977) Calcified Plectonema (blue-green algae) a recent example of Girvanella from Aldabra atoll. Paleontology 20 1:33–46

Rippka R, Deruelles J, Waterbury JB, Herdman M, Stanier RY (1979) Generic assignments, strain histories and properties of pure cultures of cyanobacteria. J Genet Microbiol 111:1–61

Schäfer A (1973) Zur Entstehung von Seekreide-Untersuchungen am Untersee (Bodensee). Neues Jahrb Geol Palaeontol Mh 1973:(4):216–230

Schäfer A, Stapf KRG (1978) Permian Saar-Nahe Basin and recent Lake Constance (Germany): two environments of lacustrine algal carbonates. Spec Publ Int Assoc Sediment 2:83–107

Schneider J (1976) Biological and inorganic factors in the destruction of limestone coasts. Contrib Sedimentol 6:112

Schneider J (1977) Carbonate construction and decomposition by epilithic and endolithic micro-organisms in salt- and freshwater. In: Flügel E (ed) Fossil algae. Springer, Berlin Heidelberg New York, pp 248–260

Schneider J, Schröder HG (1980) Calcification des cyanophycées et leur contribution à la genèse des sédiments calcaires. In: Humbert L (ed) Cristallisation-déformation-dissolution des carbonates, 17./18. Nov. 1980. Bordeaux, pp 421–428

Schneider J, Torunski H (1983) Biokarst on limestone coasts, morphogenesis and sediment production. Marine Ecology (in press)

Schneider J, Golubic S, Le Campion-Alsumard, T (in press) Processes of biokarst formation in freshwater and marine environments. – Int. Seminar on paleokarstic versus neokarstic processes and their economic importance. Napoli, 30/31. March 1978

Schöttle M (1969) Die Sedimente des Gnadensees. Ein Beitrag zur Sedimentbildung im Bodensee. Arch Hydrobiol Suppl 35 (3):117

Schöttle M, Müller G (1968) Recent carbonate sedimentation in the Gnadensee (Lake Constance), Germany. In: Müller G, Friedman G (ed) Recent developments in carbonate sedimentology in Central Europe. Springer, Berlin Heidelberg New York, pp 148–156

Schröder HG (1982) Biogene benthische Entkalkung als Beitrag zur Genese limnischer Sedimente. Beispiel: Attersee (Salzkammergut; Oberösterreich). Diss Univ Göttingen, 178 p

Suess E (1970) Interaction of organic compounds with calcium carbonate. I. Association phenomena and geochemical implications. Geochim Cosmochim Acta 34:157–168

Torunski H (1979) Biological erosion and its significance for the morphogenesis of limestone coasts and for nearshore sedimentation (Northern Adriatic). – Senckenbergiana Marit 11 (3/6):193–265

Wagner W (1954) Oberoligozäne Süßwasserbildungen des Mainzer Beckens bei Vendersheim in Rheinhessen. Jahresber Mitt Oberrhein Geol Verh NF 36:12–19

Walter LM, Hanor JS (1979a) Orthophosphate: Effect on the relative stability of aragonite and magnesian calcite during early diagenesis. J Sediment Petrol 49 (3):937–944

Walter LM, Hanor JS (1979b) Effect of orthophosphate on the dissolution kinetics of biogenic magnesian calcites. Geochim Cosmochim Acta 43:1377–1385

Wetzel, RG (1960) Marl encrustation on hydrophytes in several Michigan lakes. Oikos 11 (2):223–236

IV.4 Brackish-Water Oncoids Composed of Blue-Green and Red Algae from a Pleistocene Terrace Near Corinth, Greece

Detlev K. Richter and Rainer Sedat[1]

1 Introduction

According to Peryt's (1981) definition oncoids contain nuclei of any petrographical composition which are irregularly coated by cyanobacterial envelopes of the Porostroma (with visible algal filaments) or Spongiostroma type (without visible filaments) (Monty 1981). Grains and rock fragments which are coated only by coralline algae are termed "rhodoids" by Peryt (1981) and "rhodolites" by Bosellini and Ginsburg (1971). Difficulties in nomenclature arise if the grains are coated with both blue-green and red algae, as will be described in this paper. The authors of this paper are therefore inclined to use the broader definition of oncoids given by Flügel (1978), which includes any algal, foraminiferal algal, serpulid algal and micritic oncoids.

2 Stratigraphy of the Oncoid-Bearing Sediments

In the region of the Isthmus of Corinth (Fig. 1), rocks of alpidic deformation are overlain by a slightly tilted sequence of lacustrine-brackish Neogene – the "Corinthian Marl" of von Freyberg (1973) – which in its upper part contains marine layers with foraminifera and ostracods of late Pliocene to early Pleistocene age (Meulenkamp, written communication).

West of Corinth, close to the "km 88" signpost on the road Corinth-Kiato, the "Corinthian Marl" is unconformably overlain by up to 50 cm of non-marine sediments rich in the limnocardian *Cardium (Didacna) fuchsi* and with rare *Adelinella elegans* and *Theodoxus* sp. This layer contains the composite blue-green and red algal oncoids described in the next chapter. Its age is disputed in literature: Villafranchian (Gillet 1963), period between Palaeo- and Eutyrrhenian (von Freyberg 1973), or the interglacial cycle before Palaeotyrrhenian (Schröder 1975).

The overlying Quarternary sequence is mainly composed of skeletal arenites and rudites as well as oolites. In its upper part the Tyrrhenian gastropod *Natica lactea* occurs above and below a reworked soil layer (Fig. 1). In the lower part, a

1 Geologisches Institut, Ruhruniversität Bochum, Universitätsstraße 150, D-4630 Bochum-Querenburg, FRG

Coated Grains (ed. by T.M. Peryt)

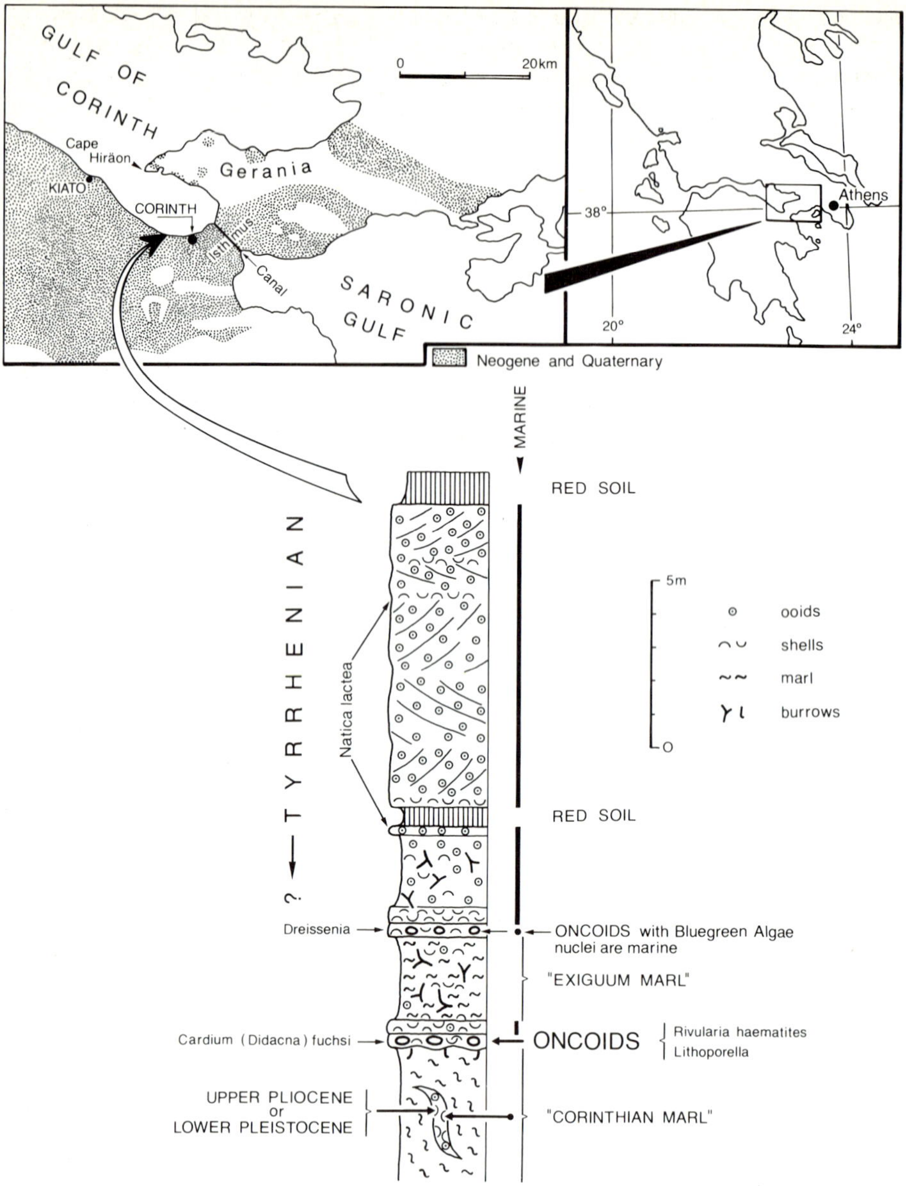

Fig. 1. Locality map and outcrop column showing the position of blue-green and red algal oncoids

(lacustrine-) brackish intercalation occurs which consists of thin-bedded pelletal marls with arenitic burrow fills, small specimens of *Cardium exiguum* and non-marine ostracods, gyrogonites and *Characaea* stalks. This "Exiguum Marl" is overlain by a layer 5 – 40 cm thick which again is lacustrine-brackish as shown by the occurrence of the pelecypode *Dreissena.* It is rich in blue-green algal crusts and oncoids which frequently envelope aragonitic stems of the coral *Cladocora cespitosa*, but also *Clamys, Pecten,* bryozoan colonies etc. These oncoid cores indicate a non-preserved marine stage between the "Exiguum Marl" and the oncoid formation.

The alternations of marine and lacustrine-brackish layers in the area around Corinth are related to eustatic sea level fluctuations during the Quarternary.

3 The Oncoids

The oncoids of the brackish layer above the "Corinthian Marl" (Fig. 1) are mainly composed of algal encrusted debris of Pleistocene algal reefs (Fig. 3 d), pebbles of Mesozoic limestones, chert, limestone breccias, as well as sandstone pebbles of Mesozoic to Neogeneous age (in decreasing frequency). These pebbles are frequently covered with boring inlets or dissolution pits (Fig. 3 c) which were formed prior to the oncoid incrustation.

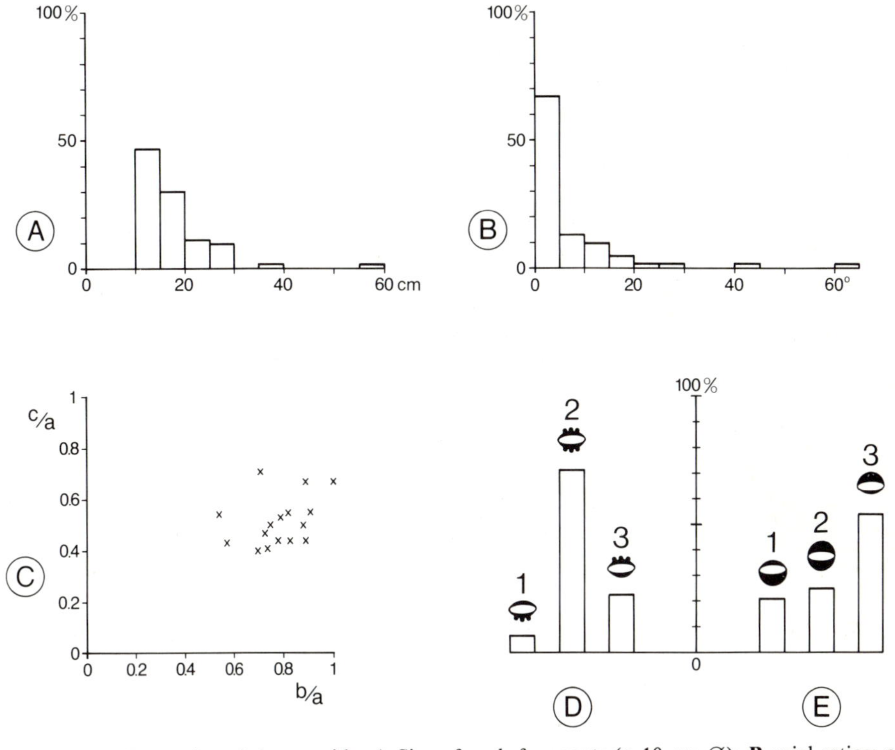

Fig. 2A–E. Properties of the oncoids. **A** Size of rock fragments (>10 cm ∅); **B** axial ratios: a (length) >b (width) > c (thickness); **C** angle between plain a/b and stratification; **D** domes at bottom only (*1*), at top and bottom (*2*), and at top only (*3*); **E** thickness of algal crusts: greater at bottom (*1*), equal at bottom and top (*2*), greater at top (*3*). Number of oncoids measured: 63 for **A**, **C**, and **D**; 16 for **B**; 24 for **E**

Being < 1 cm to 60 cm in size, the oncoids are poorly sorted. In Fig. 2, only oncoids larger than 10 cm were considered because the parameters were measured at the outcrop, while the thickness of incrustations of smaller oncoids would be visible only in thin sections. The thickness of algal crusts varies between <1 mm and 40 mm and is frequently greater on the upper surface of the oncoids (Fig. 2).

The oncoids are oval in shape and the average ratio of diameters of the three axes is 10:8:5 (Fig. 2b). The upper and lower surfaces of the oncoids are often crenulated or covered with little domes (Fig. 3a, b).

The incrustations consist of irregular µm- to mm-sized algal layers. Three types of layers can be distinguished:

a) Micritic to Microsparitic Layers. Porous layers in which only spots of algal filaments 5–50 µm thick are visible. Different species, though not determined individually, seem to be present in these layers, which may also encrust fine-grained detrital and skeletal material (e. g., ostracods).

b) Blue-green Algal Layers. Radially diverging and even branching tubules with openings 10–30 µm wide (Fig. 4a, c, d) form calcitic fans and hemispheroids $<$ 1 cm in size; they are enclosed in type (a) or type (c) (coralline algae) layers. These fans and hemispheroids consist of microcrystalline undulose calcite crystals with crystallographic c mainly parallel to the tubules. The openings are surrounded by a calcite layer, a few µm thick and poor in inclusions, which passes into syntaxial calcite rich in inclusions. Single calcite crystals may enclose several algal filaments. The colonies show a zonal increase and decrease in calcificaton with a distance of 0.2–0.6 mm. All these properties compare very well with the hemispheres of *Rivularia haematites* (De Candolle) Agardh from the Pleistocene brackish-water bioherms of the Perachora peninsula (Richter et al. 1979). Occasionally the zonal organization is missing; according to Flügel (1977, p 138) such deviations are ecologically controlled.

c) Red Algal Layers. In the corallinacean crusts of the oncoids examined no distinction can be made between hypo- and perithallus, because of the monostromate layering (i.e., composed of one layer) of cellules 10–20 µm wide and 30–70 µm thick. These figures and the shape of the thallus indicate the genus *Lithoporella*. The species could not be determined, because the conceptacles of our specimens have diameters of only 150 to 200 µm, compared with $>$ 400 µm reported by Johnson (1961, p. 51) for the three species *Lithoporella melobesioides, atlantica* and *conjuncta*. Even the earlier determination of the one-layer coralline algae from the brackish-water bioherms of the Perachora peninsula such as *Lithoporella melobesioides* Foslie (Richter et al. 1979) is questionable because of the missing data concerning the conceptacle size.

The occurrence of monostromate red algae in these oncoids is uncommon, as is the joint occurrence of red and blue-green algae. The wavy and monostromate *Lithoporella* layers in the crusts are rarely attached to one another and form laminar mats, laterally linked hemispheroids, and discrete, vertically stacked hemispheroids (Fig. 4b). The latter are connected by micritic matrix, possibly of blue-green algal origin. Unconnected pillars occur only if *Rivularia* colonies are involved together with the coralline algae.

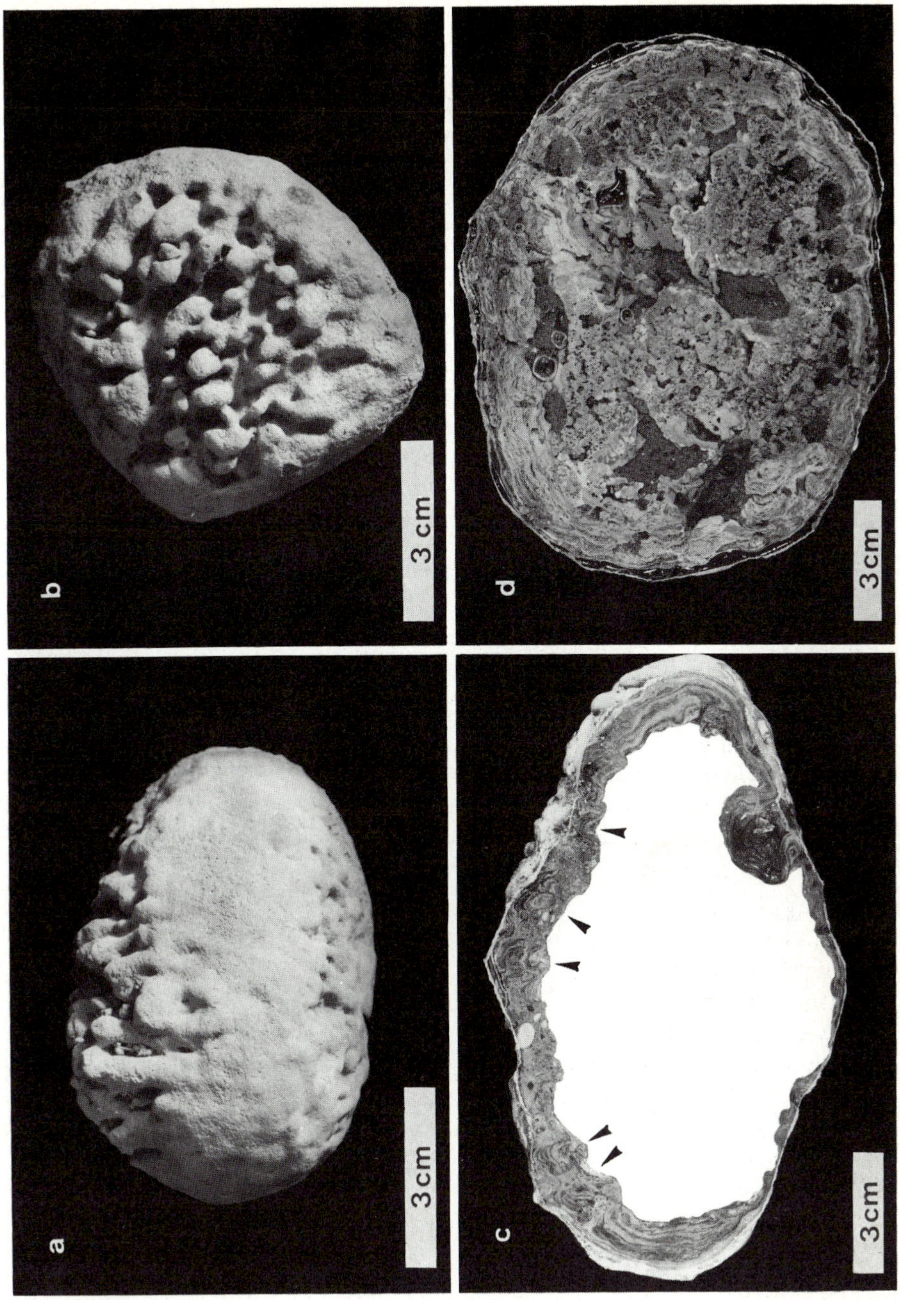

Fig. 3a – d. Surface and internal structures of typical oncoids. – **a** Lateral view. **b** Upper surface. **c** Oncoid section: Nucleus = mesozoic limestone covered with boring inlets or dissolution pits; crust = irregular blue-green and red algal layers, partly only domes (*arrow*). **d** Oncoid section: Nucleus = Pleistocene reef debris; crust = algal coating rich in *Lithoporella*-layers (*white*)

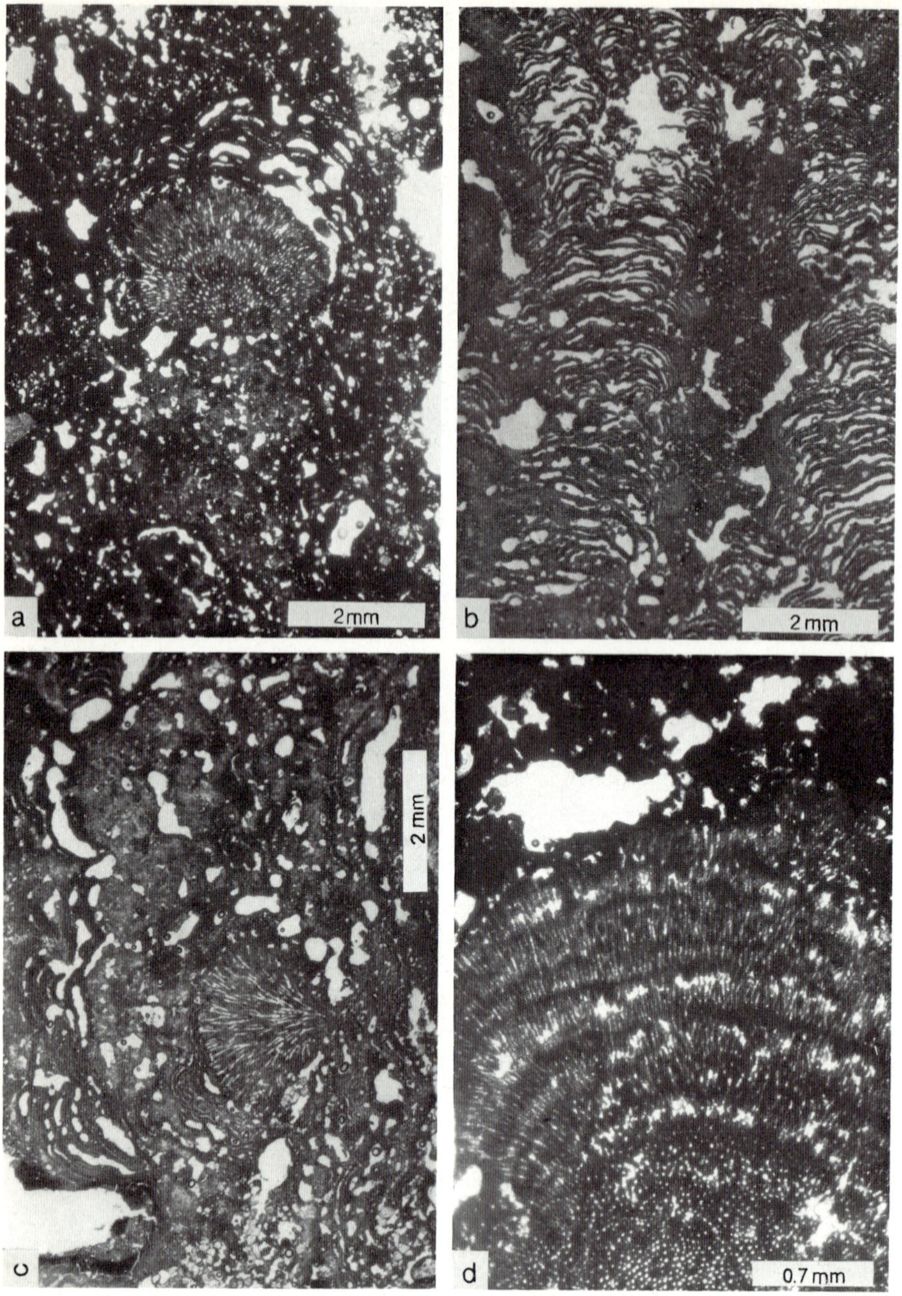

Fig. 4a – d. Details of blue-green and red algal crusts (thin sections). **a** Zoned *Rivularia haematites* incrusted by Lithoporella (vertical section). **b** Discrete, vertically stacked hemispheroids built up by *Lithoporella*. **c** Non-zoned *Rivularia haematites* incrusted by *Lithoporella* (vertical section). **d** Intensively zoned *Rivularia haematites* (vertical section)

4 The Depositional Environment

The following observations are relevant for the determination of the depositional environment of the oncoids under study:

a) *Rivularia haematites* (De Candolle) Agardh occurs at present only in lakes and rivers of limestone areas (see references in Richter et al. 1979). In modern marine sediments, only two calcareous Rivulariacea (*R. mesenterica* and *R. polyotis*) are known; these, however, use aragonite for calcification, compared with the calcitic *Rivularia haematites* (Golubic and Campbell 1981).
b) It is characteristic of coralline algae that they settle in the normal-marine shallow sea. Only very few *Corallinacea* occur in the Baltic Sea, at salinities below 20‰: *Melobesia* near the island of Fehmarn, and *Lithothamnium, Lithophyllum* and *Corallina* in the western Baltic (Pankow 1971).
c) According to the Treatise on Invertebrate Paleontology (N 591), *Cardium (Didacna) fuchsi* belongs to the group of Limnocardiacea and indicates brackish environment.
d) In the variegated, (lacustrine-) brackish Upper Pliocene "Bunte Serie" (von Freyberg 1973), about 4 km north-east of the Corinth Canal, rhodolite layers are intercalated with layers containing non-marine skeletal remains (*Melanopsis, Unio,* oogonites and stalks of Characeae, and so on). The diameter of the rhodoids (red algal oncoids) is less than 2 mm. These rhodoids are composed of an undetermined species of *Lithoporella* and are sometimes Fe-calcitic, which, according to Richter and Füchtbauer (1978), indicates that the composition was originally Mg-calcitic.
e) *Lithoporella*-type red algae occur in Palaeotyrrhenian normal-marine marl layers at Cape Hiraeon (Perachora peninsula; for the profile see Richter et al. 1979). They have also been found in non-marine marly intercalations of the "Corinthian Marl" in the Corinth Canal.
f) Modern colonies of *Rivularia haematites* in the small River Alz – the outlet of the Lake Chiem (Chiemsee) east of Munich – are composed of a calcite with 1.3 – 2.3 mol% $MgCO_3$ (Richter et al. 1979). $MgCO_3$ content in calcite of modern normal-marine coralline algae varies – according to Richter (1979) – between 11 mol% (at an average annual water temperature of 7 °C) and 18 mol% (at an average annual water temperature of 26 °C).
   In the oncoids investigated calcite of *Rivularia* layers contains 3.0 – 4.4 mol% $MgCO_3$, while *Lithoporella* layers are composed of a calcite with 2.6 – 4.0 mol% $MgCO_3$ (X-ray analyses of 10 samples each).

Whereas oncoids both with porostromate layers and isolated colonies of *Rivularia haematites* were described (e.g., from Lake Constance; Schäfer and Stapf 1978), the coexistence of oncoid crusts with *Rivularia haematites* and coralline algae with Limnocardiacea can only be interpreted in terms of a brackish environment during the formation of the Pleistocene oncoids at Corinth. The occurrence of uniserial *Lithoporella*-type coralline algae in non-marine marl intercalations of the so called "Corinthian Marl" and in Palaeotyrrhenian marine deposits at Cape Hiraeon indicates that these algae can exist on soft-sediment bottoms. The combined occurrence of *Lithoporella* and both porostromate and spongiostromate blue-green algae is not, therefore, surprising.

The increased $MgCO_3$-content in *Rivularia* layers of the oncoids investigated compared with that found in fluvial occurrences (see item f) also fits in with a brackish environment. In coronae of the sea urchin *Echinocyamus pusillus* in the marine bed overlying the oncoid layer, an original $Mg_{0.10-0.13}$-calcite was gradually replaced during diagenesis by a $Mg_{0.03-0.05}$-calcite (Richter 1979). This suggests that the original $MgCO_3$ content of calcite in *Rivularia* layers was higher and diminished by diagenetic processes. Cathodoluminescence, however, reveals a zonal growth in calcite crystals of *Rivularia haematites* in Pleistocene bioherms of the Perachora peninsula. This fabric would have been destroyed during a diagenetic replacement of an earlier high-Mg calcite by the present $Mg_{0.04-0.05}$-calcite and it thus points to the composition observed being original (Richter and Zinkernagel 1981).

The alternations of lacustrine-brackish and marine beds in the profile investigated (Fig. 1) are interpreted as results of eustatic sea level fluctuations (Richter et al. 1979). In glacial periods the sea level went down in the Mediterranean, and as a result the Gulf of Corinth was flushed between the Isthmus and a submarine ridge situated North of Patras (at present at a depth of 40 m) by freshwater; in warmer periods the Gulf again became part of the Mediterranean. Similar alternations of marine and non-marine stages in the Quarternary are suggested by Piper and Panagos (1979), even for the Gulf of Patras (West of Patras). According to Chr. Anagnostou, Athens (personal communication), faulting until Holocene time is indicated by seismic profiles North of Patras and may well be responsible for the ridge mentioned above.

References

Bosellini A, Ginsburg RN (1971) Form and internal structure of recent algal nodules (rhodolites from Bermuda). J Geol 79:669–682

Flügel E (1977) Verkalkungsmuster porostromater Algen aus dem Malm der Südlichen Frankenalb. Geol Bl Nordost-Bayern 27:131–140

Flügel E (1978) Mikrofazielle Untersuchungsmethoden von Kalken. Springer, Berlin Heidelberg New York, 454 p

Freyberg B von (1973) Geologie des Isthmus von Korinth. Erlanger Geol Abh 95:1–160

Gillet S (1963) Nouvelles données sur le gisement villafranchien de Néa-Corinthos. Practica Acad Athinon 38:400–419

Golubič S, Campbell SE (1981) Biogenically formed aragonite concretions in marine Rivularia. In: Monty CLV (ed) Phanerozoic stromatolites. Springer, Berlin Heidelberg New York, pp 209–229

Johnson JH (1961) Limestone-building algae and algal limestones. Colorado School of Mines, Boulder, 297 p

Monty CLV (1981) Spongiostromate vs. porostromate stromatolites and oncolites. In: Monty CLV (ed) Phanerozoic stomatolites. Springer, Berlin Heidelberg New York, pp 1–4

Pankow H (1971) Algenflora der Ostsee, vol I. Benthos Fischer, Jena, 419 p

Peryt TM (1981) Phanerozoic Oncoids – an Overview. Facies 4:197–214

Piper DJW, Panagos AG (1979) Surficial sediments of the gulf of Patras. Thalassographica 3:5–20

Richter DK (1979) Die Stufen der meteorisch-vadosen Umwandlung von Mg-Calcit in Calcit in rezenten bis pliozänen Biogenen Griechenlands. Neues Jahrb Geol Palaeontol Abh 158:277–333

Richter DK, Füchtbauer H (1978) Ferroan calcite replacement indicates former magnesian calcite skeletons. Sedimentology 25:843–860

Richter DK, Zinkernagel U (1981) Zur Anwendung der Kathodolumineszenz in der Karbonatpetrographie. Geol Rundsch 70

Richter DK, Herforth A, Ott E (1979) Pleistozäne, brackische Blaugrünalgenriffe mit Rivularia haematites auf der Perachorahalbinsel bei Korinth (Griechenland). Neues Jahrb Geol Palaeontol Abh. 159:14 – 40

Schäfer A, Stapf KRG (1978) Permian Saar-Nahe Basin and Recent Lake Constance (Germany): two environments of lacustrine algal carbonates. Spec Pulb Int Assoc Sedimentol 2:83 – 107

Schröder B (1975) Bemerkungen zu marinen Terrassen des Quartär im NE-Peloponnes/Griechenland. Neues Jahrb Geol Palaeont Abh 149:148 – 161

IV.5 Environmental Significance of Freshwater Oncoids, Eocene Guarga Formation, Southern Pyrenees, Spain

Enno Nickel[1]

1 Introduction

An oncolite (sic) is a stromatolite with centrifugal growth vectors, and encapsulating laminae formed around an intermittently mobile nucleus (Hofmann 1969).

Since Murray's study (1895) on "calcareous pebbles" from a pond in Michigan, ball-like oncoids (Heim 1916) have attracted the interest of many scientists. Much has been contributed to elucidate their nature as biosedimentary structures by studying marine oncoids (Walter 1976, Wray 1977). Schäfer and Stapf (1978) gave a brief list of publications on freshwater stromatolites (and oncoids).

Research on the mineralogy and geochemistry of oncoids is still scanty (Schöttle and Müller 1968, Fannin 1969, Abell et al. 1978). The elucidation of cryptalgal fabrics and their classification were the aim of the work by Hofmann (1969) and reached a culmination in the publications by Monty (1976) and by the experiments of Pentecost (1978).

Integration of detailed observation of morphology, internal fabric, biogenic microstructures, mineralogy and diagenesis into the geological setting to a great extent can be attributed to the studies of Freytet (1964, 1965, 1971, 1973), Freytet and Plaziat (1965), Bertrand-Sarfati and Fabre (1974), Boulanger and Poignant (1975), Crouzel and Lang (1975), Donsimoni et al. (1975), Van der Borch et al. (1977), Mouline (1977), Bertrand-Sarfati et al. (1978), and Lang and Segonzac (1979).

The Tremp-Graus Basin is one of the Cretaceous-Paleocene basins south of the Axial Zone of the Pyrenees. This elongate basin (Fig. 1) stretches from the eastern flank of the Mediano High, a diapiric anticline, to about 20 km east of the village of Isona. It has a total length of about 95 km and is 20 km wide on the average. During the Paleocene-Eocene about 2000 m of marine and non-marine sediments were dumped into the basin (Nijman and Nio 1975). Present geometry of the basin is the result of the last folding of the Pyrenees during the Eocene.

The stratigraphic position of the Guarga Formation within the total sediment column is shown in Fig. 2. This sediment column with a limestone member, a

1 State University of Utrecht, Institute of Earth Science, Comparative Sedimentology Division, P.O. Box 80021, 3508 TA Utrecht, Netherlands (publication no. 25)

Coated Grains (ed. by T.M. Peryt)

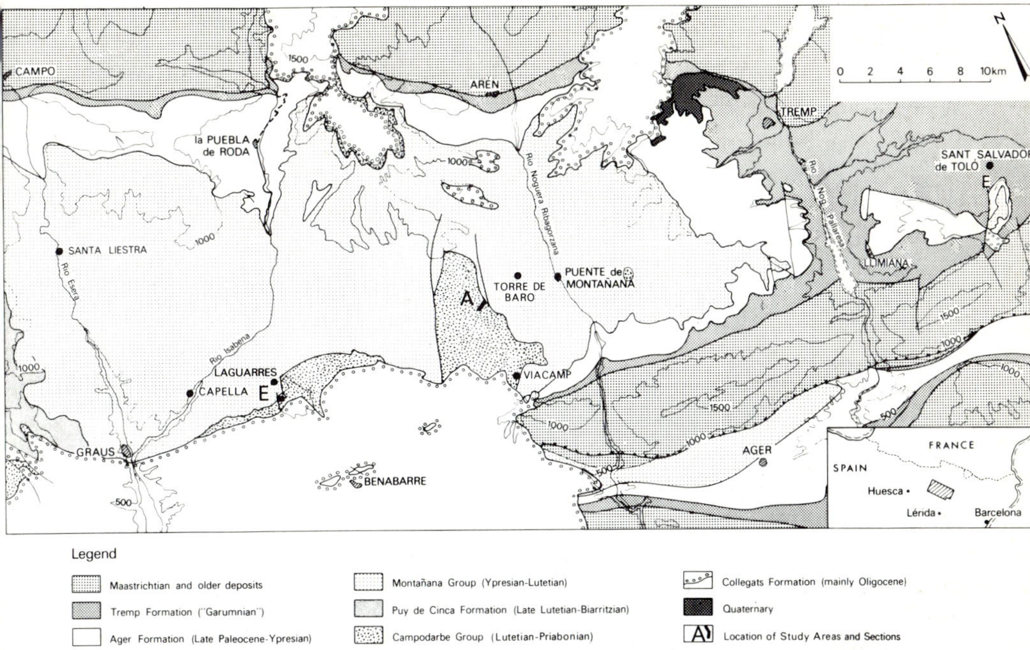

Fig. 1. Generalized geological map of the Tremp-Graus Basin, South Central Pyrenees. Measured sections are situated at A and E. (Redrawn from Gaemers 1978)

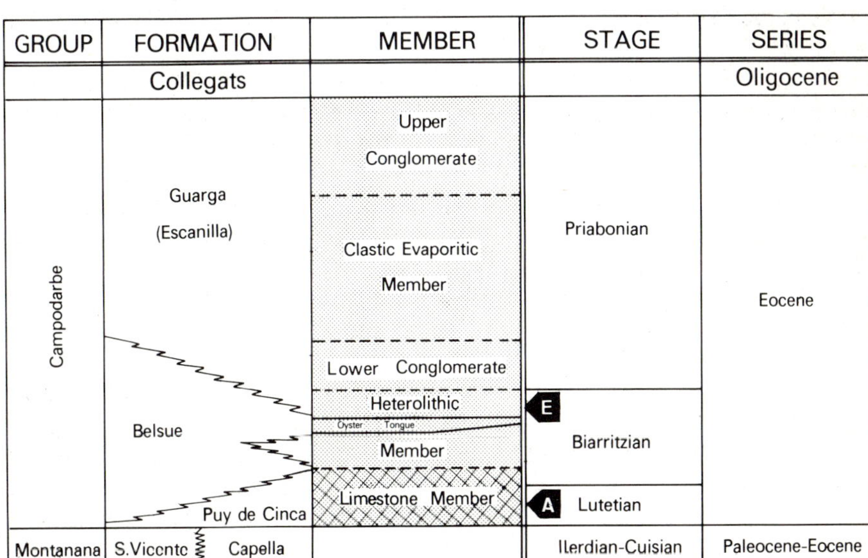

Fig. 2. Stratigraphic scheme of the Upper Eocene in the central Tremp-Graus Basin and position of samples (*A* and *E*)

heterolithic caliche-marl member, a lower conglomerate, a clastic-evaporitic member, and an upper conglomerate has its maximum thickness in the centre of the basin southeast of the village of Capella. Two outcrops were examined. The first is located west of the village of Puente de Montanana. It is a small roadcut produced by quarrying of limestones for an old limestone kiln at the road to Torre de Baró (section Ni 57-47-1, "A"). The other section (Ni 43-49-2, "E") lies on the slope beneath the road from Laguarres to Benabarre (Fig. 1). Both sections have been investigated during a large-scale study of the basal Guarga Limestones (Nickel 1980). The results form the sedimentological framework of this paper.

In this article oncoids from the Eocene Guarga Limestones in the Tremp-Graus Basin, Southern Pyrenees, Spain are described and their value for environmental interpretation is discussed.

1.1 Methods and Results

The above-mentioned sections (A and E) were measured again with more detail. The oncoid associations were classified following size, morphology, surface structure, and host sediment. An evaluation of literature revealed that classification according to these features best demonstrates the influence of a changing environment (Freytet and Plaziat 1965, Dahanayake 1978). From the observations by Donovan (1973), Mouline (1977) and Bertrand-Sarfati and Fabre (1974), it becomes evident that flat algal structures like mats, cushions, and stromatolites had to be integrated into the study because they seem to be end-members of a morphological row (see also definition of "o." by Hofmann 1969). Environmental interpretation of the Guarga-oncoids is discussed against the background of these former results.

Microstructures and cryptalgal patterns of the Guarga-oncoids were studied in sawed slabs, thin sections, and acetate peels. Mineralogy was investigated in thin sections using staining and X-ray diffractometry by the method of Goldsmith et al. (1955) and Goldsmith and Graf (1958).

2 Relation of Oncoid-Morphotypes to Lithological Sequences

The oncoids are found in two different sequences. Both are fining-upward regarding grainsize distributions. Together with this fining-upward trend lithology changes from coarse clastic to fine carbonatic in section A (Fig. 3). In section E the sediments generally are carbonates and the fining-upward trend is less accentuated (Fig. 4).

In the first case the oncoids are clearly related to a fluvial channel system. The lowermost oncoids occur on top of a polymict conglomeratic channel fill. Pebbles of all grain sizes (× mm up to several cm) are encrusted by maximally 1.5 cm thick laminated brownish-yellow (10 YR 4/2 to 10 YR 6/2) carbonate. Differential etching of these crusts by weathering reveals a pattern suggesting an alternation of dense and porous laminae. The conglomerate is capped by grey marly beds with grey to violet mottling. Two sandy lenticular intercalations within these marls show increasing carbonate content from base to op (Fig. 3). The carbonate content is due to an increasing admixture of mm-sized oncoids, oncoid debris, and carbonate cement. They are classified as sandy oncoid grainstones to oncoid packstones.

The marls above these thin beds contain several cm-thick lenses which are characterized by their light grey colour within the generally darker mottled marls. They contain more carbonate and host a collection of larger oncoids (cm-sized). The section is topped by 2 m of clean dense, micritic limestone (10 YR 5/2 – 8/2). This limestone contains the collection of largest oncoids. Laterally the carbonate wedges out into paleosoils with caliche crusts.

According to layer configuration the oncoids in this section can be assigned to three types (Logan et al. 1964, Radwanski and Szulczewski 1966, both in: Flügel 1978):

- Type C (concentrically stacked), with concentric layers molding precisely the outlines of the core;
- Type R (randomly stacked), more or less concentric layers with protruding bumps, domal overgrowths, and cauliflower structures;
- Type I (inverted stacked), oncoids with angular discordances of single layers or sets of laminae; indications of different orientation in growth direction.

Typical collections from section A are shown in Fig. 5a, b. Grainsize distribution, type classification, and position within sedimentary bodies are summarized in Table 1. The table indicates that there is a correlation between host rock lithology and degree of preservation of the oncoids. Layers including clastic material contain only oncoid fragments and/or very small whole oncoids (1 – 2 mm). Sizes of complete oncoids reach an absolute maximum in the face of section 57-47-1 between sub-sections II-7 and II-3. Towards the top the oncoids become smaller and finally disappear. The latter also holds for their distribution in SW-direction. In this area where the freshwater limestone wedges out into calcareous crusts the oncoids are relatively small (max. 0.5 cm) and have a "rusty" appearance due to sesquioxide staining. Their host sediment is a white-grey micritic nodular limestone with red mottling.

Oncoids of type C are restricted to cylindrical and stick-like forms and to small ball-shaped or chip-like individuals (0.1 cm to 1.5 cm smallest measurable diameter). The cores of these oncoids are small limestone pebbles, single pelecypod shells, or the original core is replaced by clear calcite spar, in some cases leaving an open residual cavity. The chip-like oncoids are mostly re-encrusted fragments of older oncoid crusts.

The second type R, with bumpy outer surface morphology mainly occurs among the large oncoids (2 to 15 cm max. measurable diameter). They resemble almost completely the algal nodules from Little Connestoga Creek, Pennsylvania, described by Golubić and Fischer (1975). Orientation of the long axes of these oncoids in sub-sections II-4 to II-7 indicates low-angle large-scale cross-bedding.

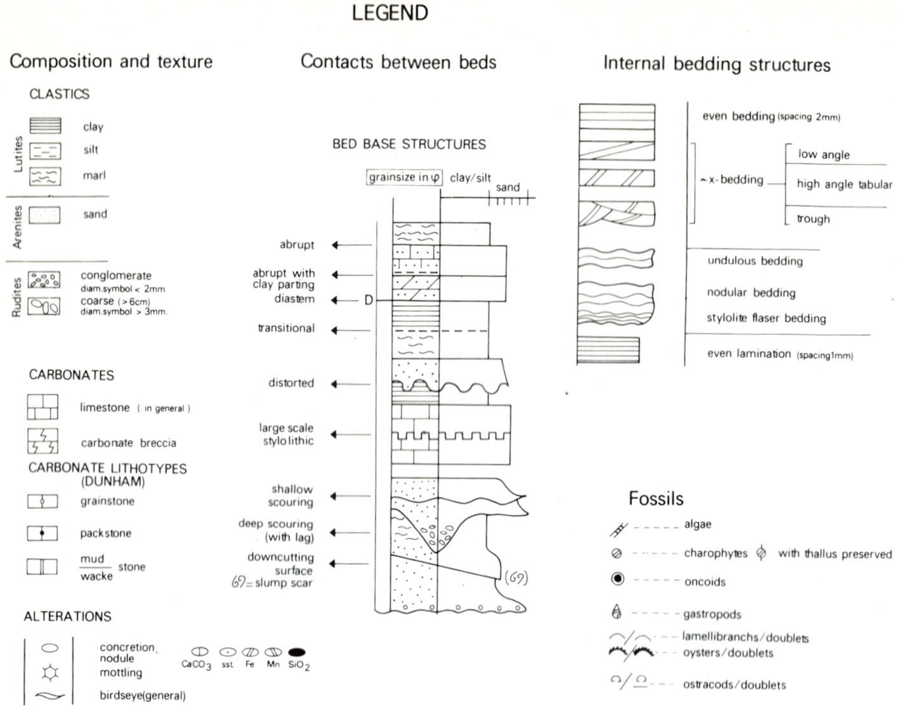

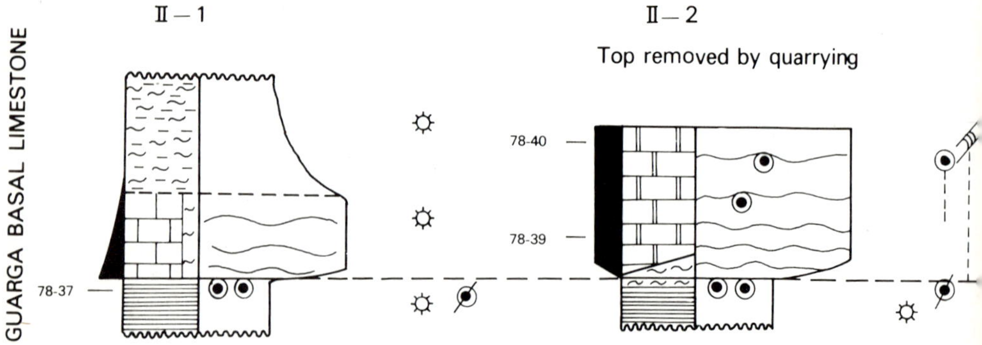

Fig. 3. Columnar sections of outcrop A (own code NI 57-47-1). Symbols in the legend are used uniformly in all litho-logs of this paper

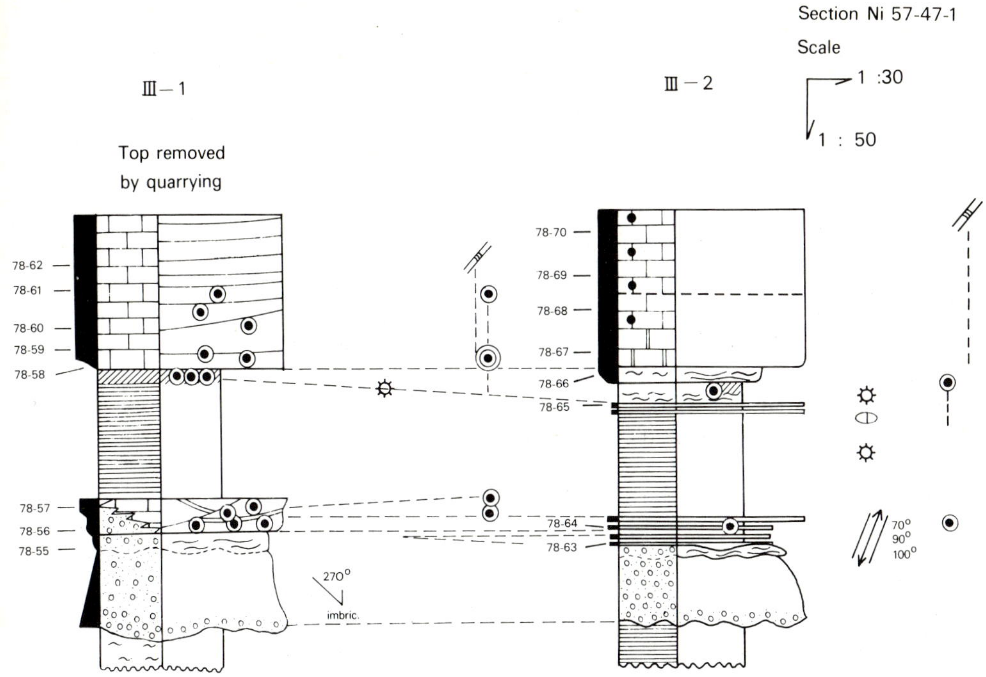
Section Ni 57-47-1
Scale
1 :30
1 : 50
III – 1
III – 2
Top removed
by quarrying
78-62
78-61
78-60
78-59
78-58
78-57
78-56
78-55
78-70
78-69
78-68
78-67
78-66
78-65
78-64
78-63
270°
imbric.
70°
90°
100°

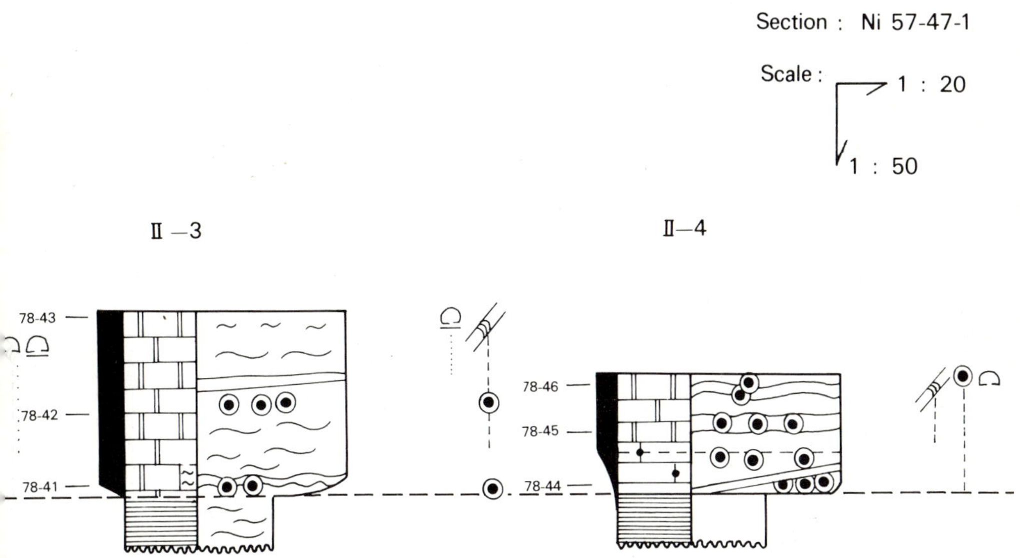
Section : Ni 57-47-1
Scale :
1 : 20
1 : 50
II – 3
II – 4
78-43
78-42
78-41
78-46
78-45
78-44

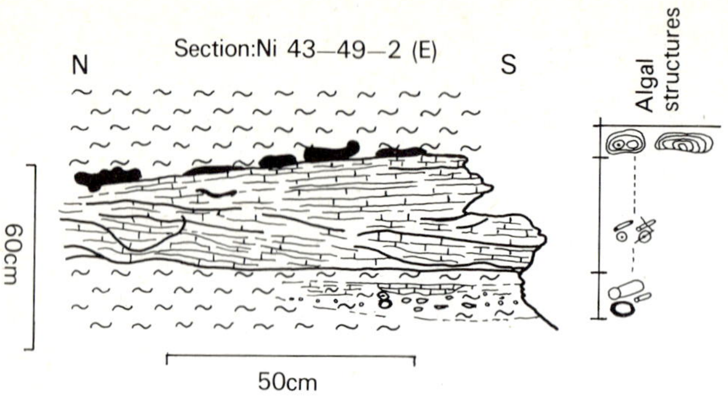

Fig. 4. Schematic drawing of outcrop E (own code NI 43-49-2)

Table 1. Sizes and layer types of the oncoids from section A. Classes according to Logan et al. and Radwanski and Szulczewski (in Flügel 1978). Italic letters mean "reworked"

Size Largest measurable diameter (cm)	Layer type	Lithology of Matrix Sediment	Section
0.5 – 1	C + *R*	Marly silt	I-1
2	C + *R*	Sandy micrite	I-2
1 – 3	C	Marl (red)	II-1
1 – 3	C + R	Marl (grey)	II-2
1	C	Micrite	
5	C + R	Micrite	II-3
0.1 (top)	C	Micrite	II-4
15 (centre)	C + R + I	Micrite	II-4
5 – 10 (base)	C + I	Micrite	II-4
0.1 (top)	C	Micrite	II-5
5 – 10 (base)	C + R + I	Micrite	II-5
0.1 (top)	C	Micrite	II-6
5 – 10 (centre)	C + R + I	Micrite	II-6
1 – 2	C + *C* + *R*	Sandy micrite	II-6
0.1 (top)	C	Micrite	II-7
5 – 10 (centre)	C + R + I	Micrite	II-7
1 – 2	C + *R*	Sandy micrite	II-7
1 – 2	C + *C* + *R*	Oncolitic sandstone	II-7
0.5 – 1	C + *C*	Micrite	III-1
0.1 – 5	C + *C*	Marl	III-1
0.1 – 2	C + *C* + *R*	Calcareous sandstone	III-1
0.1 – 2	C + R	Conglomerate	III-1
(thin crusts with respect to pebble diameter)			
0.1 – 1	C + *C*	Micrite	III-2
0.1 – 2	C + *C*	Sandy micrite	III-2

Type I is very scarce. In most cases single crusts of larger oncoids (C-type and R-type) contain internal discordances. Investigation of slabs and thin sections revealed that oncoid growth frequently startet with type-C layering, continued with type R and/or an alternation of R and C, sparsely interrupted by internal discordances assigned to type I (Fig. 6).

If compared to the first section the oncoids from section E are generally smaller. Large oncoids are sparse and only be found at the base of the section.

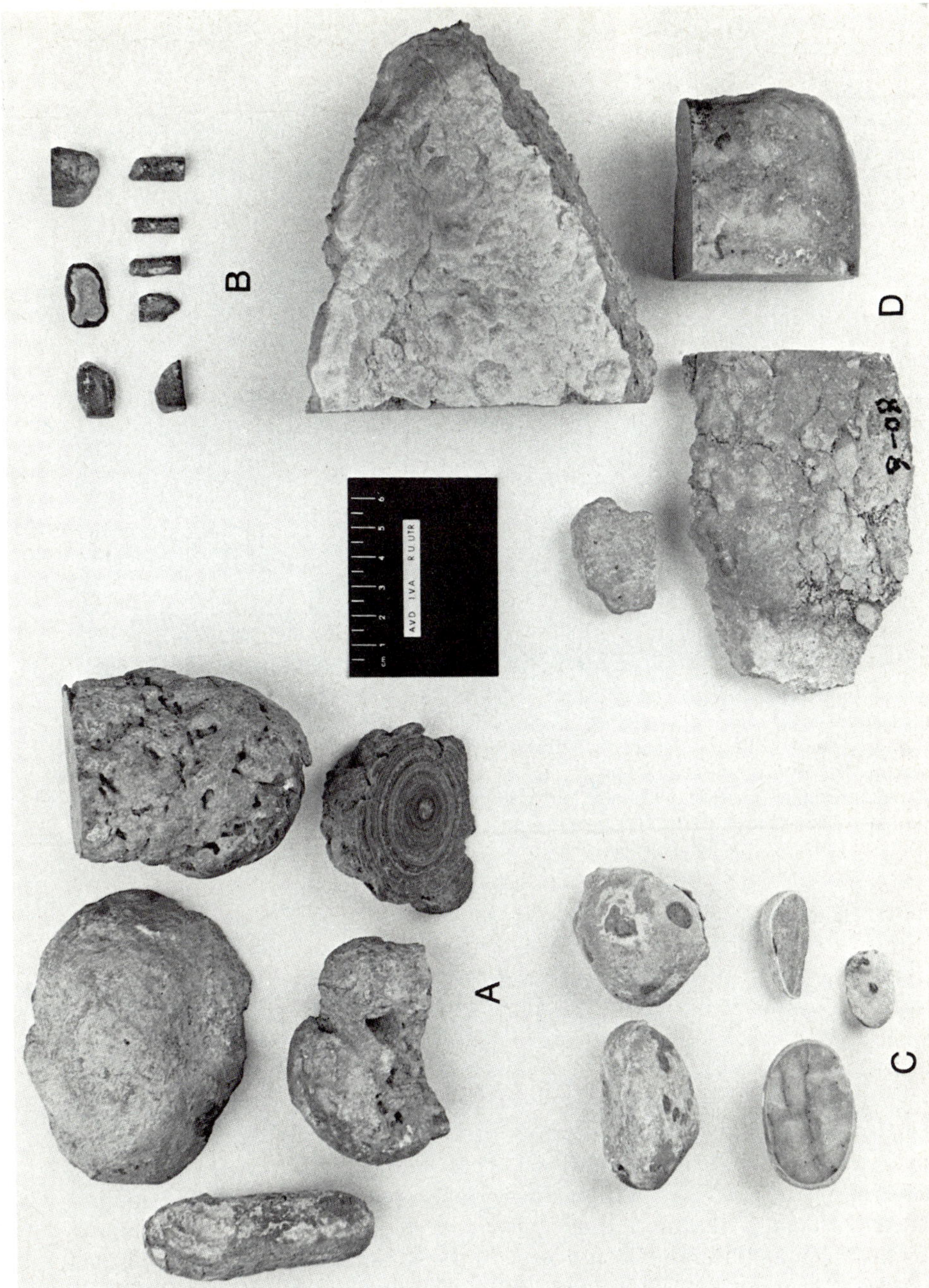

Fig. 5 A – D. Typical oncoid collections from the Guarga; **A** large individuals from section A, **B** small oncoids from section A, **C** coated pebbles from section E, **D** flat structures from section E

Fig. 6. Polished slab of large oncoid mainly composed of smooth concentric accretional laminae. Four textural zones can be distinguished: *I* core of fine oncoid hash (*x*) and caliche lithoclasts (*y*); overgrowth around this presumably early lithified core started with an inner bumpy layer of algal crusts (*a*). This zone is followed by zone *II*, smooth dense laminae. Zone *III* is an alternation of dense smooth and crinkled laminae, which was partly eroded before accretion of zone *IV*, another bumpy layer. At *v* the residues of a further laminar system can be seen

From base to top four lithological units can be recognized (Fig. 4). The basal red mottled marl contains a carbonatic lens with large oncoids, most of them as thin laminated crusts around calcareous pebbles. These crusts are up to 3 mm thick. Fully developed oncoids found together with this coated pebble collection in which the crusts are by far thicker than the cores, range in size from 1 cm to 6 cm (largest measurable diameter). All oncoids from this association have smooth surfaces (Fig. 5c, d). The elliptical or stick-like individuals (type I) suggest that they also formed around twigs, roots or other oblongue plant remains. The carbonate overlying the red basal marl is a dense whitish grey limestone made up of oncoid grainstone and wackestone. The oncoids are relatively small (0.5 mm to 5 mm) and elongate or almost round. Their grain size distribution occasionally shows coarsening-upward. The top layer of the section is characterized by isolated stromatolitic structures (algal boundstones) ranging from tens of centimeters up to 0.5 m in diameter. In contrast to the round, elliptical, or oblong oncoids with plenicinct laminae, these structures have more or less stratiform accretion of laminae. They are pancake-like, cushion-like relatively flat structures with generally convex smooth surfaces (Fig. 5c, d). These surfaces are in some cases covered by a net of straight furrows. Stromatolites with undulous, but overall convex smooth surface caught oncoid grainstone in the depressions between the bumps. The structures subsequently were covered with red silty marls. At about 5 m laterally from this section a small greyish-white marl lens 0.5 m wide and 0.1 m thick occurs within the red marl. It lies approximately at the same level as the horizon with the stromatolites above. This lens contains an assemblage of algal nodules (1 – 3 cm ∅) of fairly irregular shapes. Their surfaces are smooth or covered with small pinnacle shaped protrusions of hemispheroidal bumps (Fig. 5 c, d).

3 Microstructures

Under the microscope the more or less concentric laminae show as an alternation of dense brownish micritic and porous to fenestral layers. The dense layers are structureless to finely laminated and vary from 0.03 mm to 0.23 mm in thickness. Particle sizes are 1.5 nm to 5 nm on the average. The porous layers are 0.06 mm to 2.4 mm thick with the internal structure of clotted micrite forming a spongy layer with discrete domal, tufty, and columnar digitate protrusions (Fig. 7). These structures are vertically oriented with respect to their substrate, mostly a dense micritic layer. Their internal microstructures are superimposed tussocks. The lateral arrangement of the structures produces a system of elongate more or less vertical fenestrae.

This arrangement in a cyclic sequence of dense and porous laminae is generally found in the oncoids from section A.

A further type of microstructures which to a lesser extent was observed in the oncoids of section A but generally found in the nodules and flat structures are isolated filaments, sometimes arranged in a fan-like pattern or in isolated, more or less parallel orientation. They consist of a central clear lumen (4 nm ∅), now

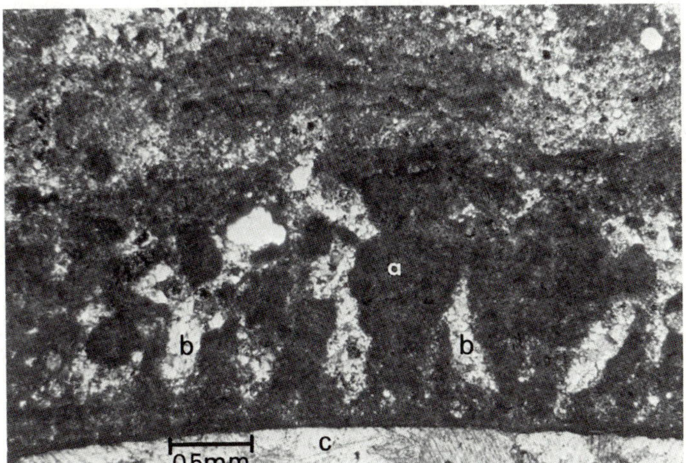

Fig. 7. Domal microstructures (*a*) with large elongate fenestrae (*b*) as crust around a conglomeratic pebble (*c*) from section A

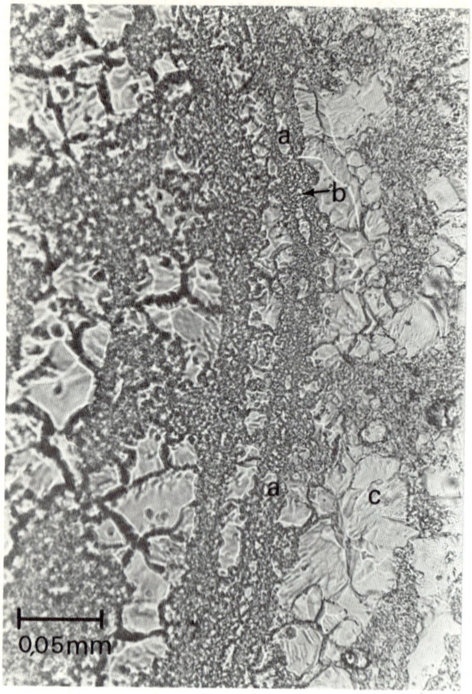

Fig. 8. Algal filaments (*a*) with microcrystalline calcite sheaths (*b*) and calcite spar cement (*c*) in a flat structure from section E. Sample 80-13, peel negative print, transmittent light

filled with microspar. This central part encrusted by a sheath of calcite crystals 2.5 nm to 5 nm ∅ (Fig. 8). The interstices between the sheaths filled with equant microspar (30 – 90 nm ∅). The filaments arranged in tufts very closely resemble the biostructures described as *Rivularia* by Schäfer and Stapf (1978), and Monty (1976). The more parallel filaments can be compared with calcified *Scytonema* (Monty 1976) or *Phormidium* (Schäfer and Stapf 1978). The process of preservation of algal filaments by calcite sheaths was studied and described by Pentecost (1978). He concluded that this process occurs very early and leads to synsedimentary lithification of the algal structures.

Rivularia-like structures and isolated filaments played the major role in the construction of the algal nodules of section E. This leads to: (a) the irregular shape of the structures, and (b) different shape of the fenestrae in the algal nodules. The latter are highly irregular in shape and are generally larger than in the oncoids of Section A. The dense micritic laminae observed in this section form more horizontal crusts and serve as substrate for the filaments and/or envelop the algal nodules as a final crust forming a relatively large flat oncoid with an algal nodule as core.

3.1 Mineralogy and Diagenesis

Analysis by X-ray diffractometry only showed low Mg-calcite with 1% to 2.9 mol % $MgCO_3$. These values lie within the ranges of typical freshwater oncoids

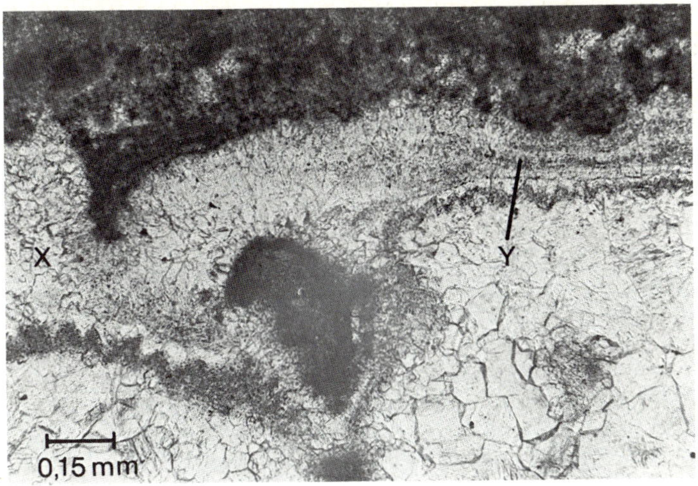

Fig. 9. Photomicrograph of the core area of an irregular oncoid. The core is dissolved and the central cavity is filled with cement A of presumably former bladed crystals as indicated by inclusions and tooth-terminations (*x*). Note dusty bands (*y*) which suggest three growth phases. Later cement B has filled the remaining cavity and replaced cement A as a pseudospar. Thin section, sample 78-74b, transmittent light

like those determined by Ordonez and Del Cura (1977) (2 mol %) and Wagner and Lamprecht (1974) (4 mol %).

According to grain sizes and crystal sizes five mineral types can be distinguished. The depositional material is thought to be dark micrite of 1.5 nm to 4 nm particle size sometimes showing clotted texture with aggregates of 90 nm diameter on the average. This micritic is the material which on one side the oncoids are made of, on the other side the interparticle matrix is made of.

The other crystals are of secondary origin. They cement original pores and fenestrae and secondary cavities or dissolution vugs. A third group of crystals documents further alteration by recrystallization leading to grain enlargement or grain diminution.

Pores and fenestrae are blocked by clear equant calcite spar (30 nm to 115 nm diameter). In several cases this clear cement seems to encase earlier cement generations which line the rims of larger pores. This is the case in oblongue oncoids wherein the original cores, pieces of wood, almost totally have been dissolved. A dusty (?) fibrous or bladed cement with acicular terminations lines the cavity walls and forms a crust around the plant remnants (Fig. 9). This encrusting cement is 0.1 mm to 0.25 mm thick and shows up to four zones of dusty inclusions. Later equant microspar has blocked the rest of the cavity and has even replaced the first cement. A second early cement are the small (2.5 nm to 5 nm) calcite crystals that form the sheaths around the algal filaments. Finally the micrite crusts on top of the oncoid grainstones in section E also have to be counted among the early cements. All the cements described hereabove are orthocements. Recrystallization of the Guarga-oncoids comprises grain diminution by secondary micrite and grain enlargement by equant microspar. The first process

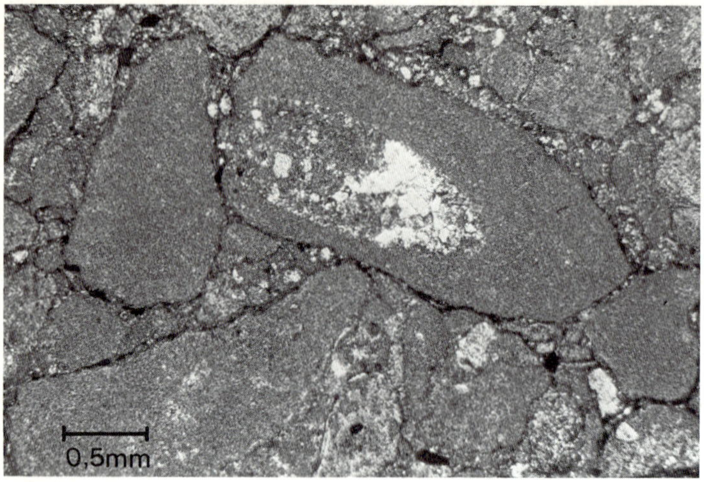

Fig. 10. Micritized oncoids in section E. Sample 80-6, peel negative print, transmittent light

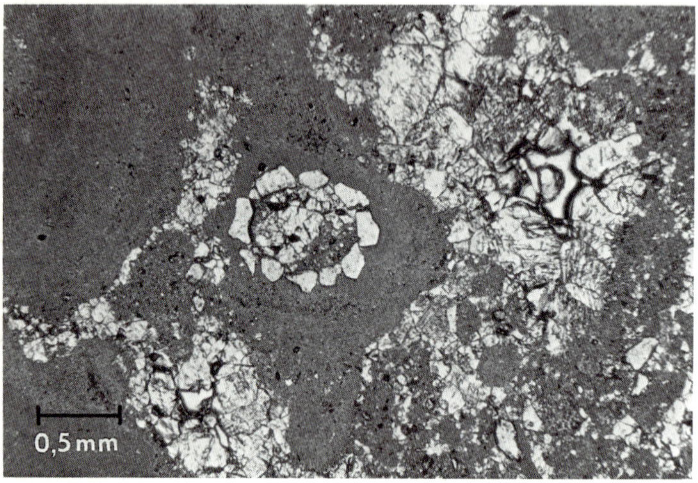

Fig. 11. Photomicrograph of internal fabric of a stromatolitic structure encorporating armoured-mud ball. Sample 79-4, section E, peel negative print, transmittent light

occurs if the oncoids are found in host rocks with indications of soil formation. This has been observed in the collection 78-74-1 from section A, sub-sections I-1 to I-2. In addition to blurring of the original structure, the oncoids show colouring by iron oxides. Micritization seems to be even stronger in section E. One gets the impression that within the oncoid grainstones micritization increases towards the top.

Shape and size of the oncoids correspond with the types in the lower part of the section, but they are internally structureless and strongly resemble peloids.

However, residual oncoids which are also micritic but show still a circular growth pattern around a, now dissolved, core sometimes with very faint lamination (Fig. 10), highly suggest that most of the grains have been former oncoids. They are therefore called "pseudopeloids" to make clear that their lack of internal structures is largely due to a non-depositional process.

There is, however, also an indication that the formation of peloids has taken place, too. The large stromatolite (Fig. 11) has encorporated several tiny armoured mud balls (0.5 – 1.5 mm ∅) which correspond to microids. These microscopically small lumps are comparable to the pseudooncolites of Dahanayake et al. (1976). Their assignation to algal or non-algal origin, however, remains uncertain in the case of the Guarga.

Recrystallization with grain enlargement produced isolated patches of clear calcite spar (15 nm to 30 nm ∅) in the dense micrite. Both, the micrite that blurs the original structures and the neomorphic spar are pseudocements.

4 Discussion

Distribution of macromorphological types from both sections (A and E) (Table 2) shows that their macroscopic features are strongly influenced by environmental changes. Their fluviatile to lacustrine origin is clearly revealed by the connection to clastic fluviolacustrine sediments and patterns. Within this regime they also depict fining-upward sequences (Reineck and Singh 1973). In section A this is seen in the following changes from base to top: (1) decrease of oncoid size, (2) decreasing amount of clastic sediment or clastic admixtures, (3) drastic change from coarse clastic to very fine carbonatic sedimentation. In section E this principle is documented in a less dramatic manner. Coated pebbles at the base, together with single large oncoids give way to finer oncoid packstones or grainstones at the top. The clastic host sediments are fine and sandstones are lacking. The principle of fining-upward here can be extended to a concept of "unattached-attached" biosedimentary structures. Like a fining-upward sequence this sequentional arrangement of biosedimentary structures reflects the decrease of transport energy in a fluvial cycle.

Mouline (1977) concluded that the requirements for oncoid and stromatolite formation in fluviolacustrine environments are low sediment load of the water, and a low dissolved charge of the water. He also postulated that growth of the algae starts at the end of high waterstands and at the beginning of new water supply to the system.

It is tempting to conclude that the distribution of oncoid-morphotypes here can be related to phases of high discharge, low discharge, abandonment, subaerial exposure, and desiccation. A rough classification of more dense, channel-bound oncoids and lacustrine oncoids of more open texture for freshwater oncoids had been made by Ordonez and Del Cura (1977). The microstructures of their oncoids support their classification. The dense oncoids have clastic material incorporated in the oncoid layers whereas the "open" structures are poor in

Table 2

Type	Size ranges	Surface ornamentation	Habit of accretion vector	Internal features	Host sediment	Sedimentary environment	Guarga sample no.
 Coated pebbles	Highly variable (1 to x cm)	Smooth to tuberculate	Spheroidal to bulbous	Nucleus	Polymict conglomerate, reddish to grey marl	Fluvial channels, bottom lags in alluvial plain depressions	78–55 78–57 78–57a 80–4
 Small oncoids, fragmented	5 to 15 mm ∅ 3.5 to 4 cm long	Smooth to slightly tuberculate	Spheroidal	Nucleus	Violet silty marls and limestones, calcareous sandstones	Shallow margins of small lakes, bottom lags of ponds and small lakes in alluvial plains, shallow channels	78–74–1 78–74 80–5 78–47 78–50 78–51
 Small oncoids	5 to 15 mm ∅ 3.5 to 4 cm long	Smooth	Spheroidal	Nucleus	Grey marls, grey calcareous sandstones, reddish grey marls, buff limestones	Shallow channels, bottom lag of alluvial ponds and lakes	78–74
 Large oncoids	< 5 mm to 6 cm ∅ < 3.5 to 15 cm long	Smooth, tuberculate, grooved to holed	Spheroidal	Nucleus	Grey calcareous sandstones, buff micritic limestones, reddish grey marls	Shallow channels, alluvial lakes (bottom-lags of alluvial ponds)	78–74 80–4
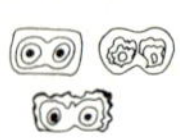 Large algal nodules	~ 10 cm ∅ 3 cm thick	Smooth to tuberculate, with furrows and rills	Nodular, stubby to crustose, tendency to horizontal accretion	Varying number of nuclei enveloped by laminae	Reddish marls	Top layer of alluvial pond	80–8 79–5
 Small algal nodules	1 to 4 cm ∅ (± round) 1 to 2 cm thick (± flat)	Smooth to tuberculate	Spheroidal to bulbous, tendency to wall formation	Combination of small nuclei and horizontal laminae	Reddish marls	Isolated very small pond in alluvial plain	80–13
 Rimstones  Composite microids	x cm plane ∅ 5 cm thick 10 cm plane ∅ 1.5 cm thick	Smooth to tuberculate with furrows and rills	Nodular, crustose, tendency to flat growth	Combination of nodular and wavy laminations	Reddish marls	Top layers of alluvial pond	80–6 80–7
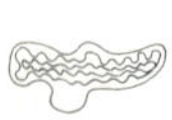 Algal stromatolites	~ 30 cm ∅ ~ 10 cm thick	Highly tuberculate to grooved	Nodular, crustose, tendency to flat growth	Combination of nodular, bulbous, and spheroidal laminae	Reddish marls		80–14 79–4

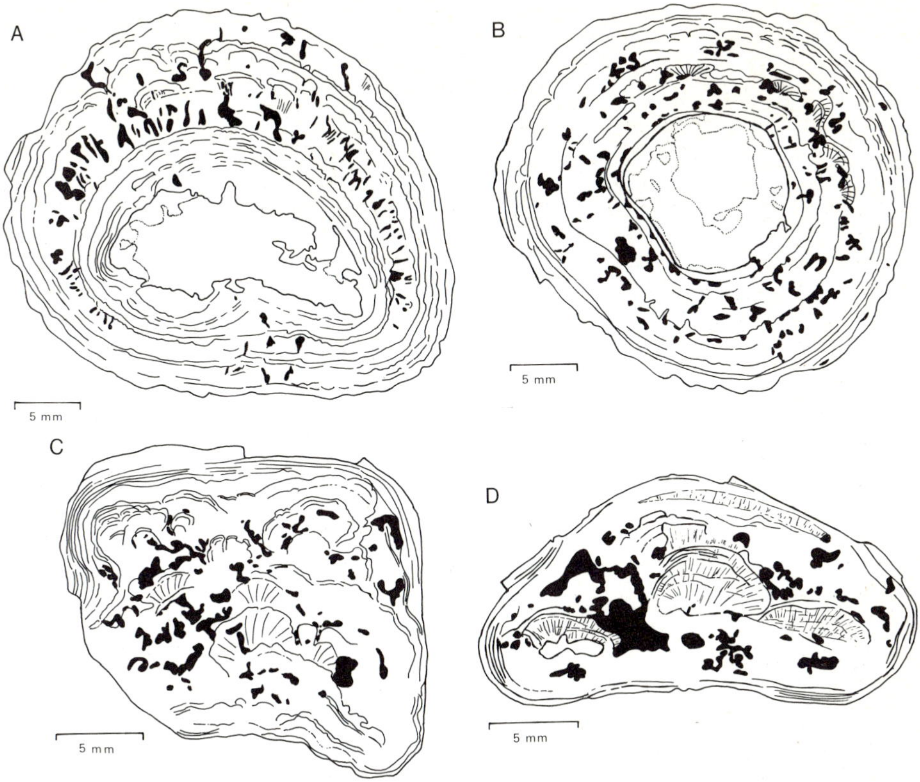

Fig. 12 A–D. Line drawings from slabs (acetate peels) of four characteristic arrangements of laminae, biomicrostructures, and fenestrae (black). **A–D** large oncoid from section A, small oncoid from section A, large oncoid from section E, algal nodule from section E

clastic material or purely carbonatic. In the case of the Guarga the oncoids of section A are grossly dense laminated, whereas even the large oncoids at the base of section E are more "open" with frequent bumpy layers, crenulated laminae, and fenestrae (see also Fig. 12). Textures in the stromatolites on top of section E are characterized by large and irregular fenestrae and pores produced by accentuated, algal growth in cushions and tufts. Core types and the relationship core-substrate to thickness and microstructure of encrustation are also the results of the response of algal growth to environmental changes.

Coated pebbles at the base of each section have fine laminated dense encrustations only a few millimeters thick. Donovan (1973) observed a positive correlation between thickness of stromatolitic coatings and grainsizes of breccias deposited in freshwater lakes of the Devonian Old Red. This relationship could not be found here. The discovery of coated pebbles in combination with algal stratiform laminae in a fluvial channel from the older Montllobat formation in this area (Ch. Atkinson, pers. comm.) (Fig. 13) demonstrates that algal growth in these systems started as soon as the water energy subsided. This is the case in especially the higher parts of the channels. Observations by Freytet and Plaziat

Fig. 13. Insitu encrustation of conglomeratic pebbles by algal laminae (*a*) from a channel fill. Eocene Montllobat Formation, Tremp-Graus Basin (donation of Ch. Atkinson, Utrecht, 1981), peel negative print

(1965) corroborate this. The channel-bound oncoid association of section A also includes a larger variety of core types, including pebbles, clams, sticks, twigs, reworked oncoid fragments, lacustrine limestone clasts, calichified lacustrine limestone extraclasts, caliche fragments, and marl clasts. This variety is a direct result of frequent erosion of alluvial substrates. In section E the role of the core as a center of encrustation is successively taken over by more or less horizontal substrates.

The occurrence of rimstone-like structures (Risacher and Eugster 1979) at the top of section E evoques the question of non-biological influences during their growth. Morphologically similar forms seem to be frequent near to the orifices of inactive hot springs (Risacher and Eugster 1979). In the case of the Guarga carbonates the structure is definitely of algal origin (algal filaments). The fact that it occurs at the top of the section fits into the scheme of a fining-upward sequence. They may represent a sediment type bound to groundwater seepages.

The final stage of the freshwater fluviolacustrine cycle would be stop of sedimentation and aerial exposure. The dense micritic crusts topping most of the large structures are highly suggestive that this has been the case in the Guarga. In how far the biosedimentary structures at the top are of subaerial origin, as was conceived by Wright (1981) for stromatolites from the Carboniferous of South Wales, cannot be decided with certainty. Their subaerial exposure, however, has to be taken seriously into consideration, because calichificaton of clearly lacustrine limestones was frequently observed in the Guarga formation (Nickel 1980).

The distribution of internal fabrics and microstructures are also controlled by sedimentation rates, algal growth, desiccation and diagenesis. Although the microstructures could not be subdivided into distinct growth zones like Bertrand-

Sarfati and Fabre (1974) did for Stephanian oncoids from Algeria, an attempt is made to classify the Guarga oncoids according to their microstructures following suggestions from Hofmann (1969) and Hoffman (1976). If we extend the approach of Hoffman (1976) to the Guarga oncoids they can be easily classified on the background of an environmental grid (Fig. 14).

The internal fabrics can be easily correlated to the morphological types of the oncoids. Dense types of section A are predominantly due to the accretion of more dense microfabrics and the structures of section E show more distinct algal colonies. The final fossil product of the former living algal meadow, mat, or mucilage, depends on type of algae, sedimentation rate, frequency of desiccation, erosion, leaching, early cementation, decay of organic matter, compaction, and late diagenesis. The microstructures observed in the Guarga oncoids follow the scheme established by Freytet and Plaziat (1965) and Pentecost (1978): (1) organic growth, (2) deposition of micrite on top of the living matter, (3) early cementation by clear calcite. The biofabrics can be classified as dense, spongy-fenestrate, and fan-like (Schäfer and Stapf 1978) (Fig. 15).

A question frequently discussed (see Riding 1977) is if the role of the algae is precipitating actively [skeletal oncolites (sic), Riding 1977] or trapping [non-skeletal oncolites (sic Riding 1977]. Pentecost (1978) clearly demonstrated that even in cases of well preserved algal filaments the average calcite production was rather due to sediment trapping than to direct precipitation during photosynthesis (1% – 5% of total calcite by algal uptake of CO_2). In the case of the

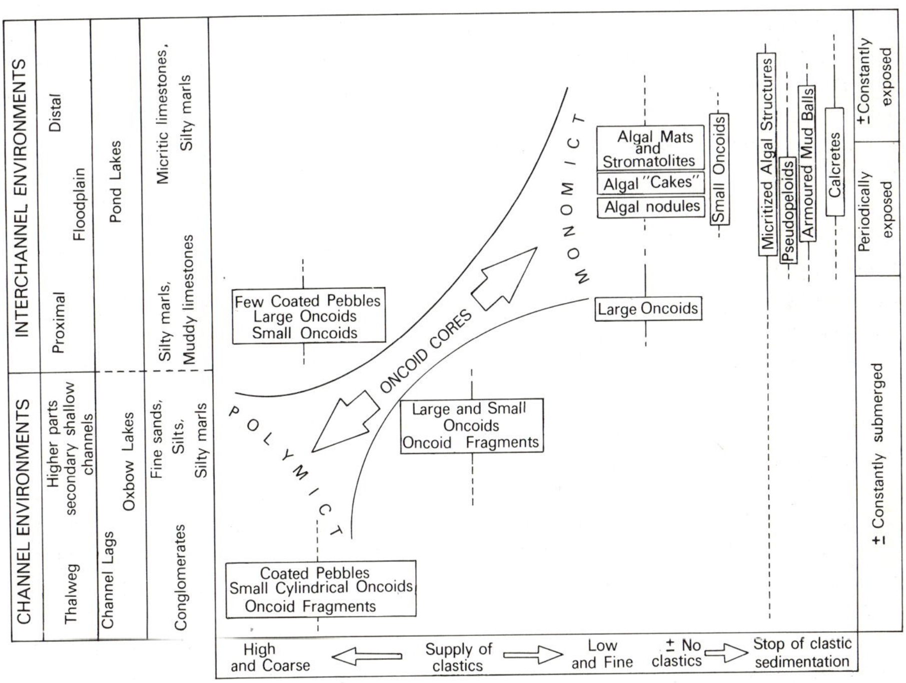

Fig. 14. Distribution of oncoid morphotypes on an environmental-sedimentary process grid

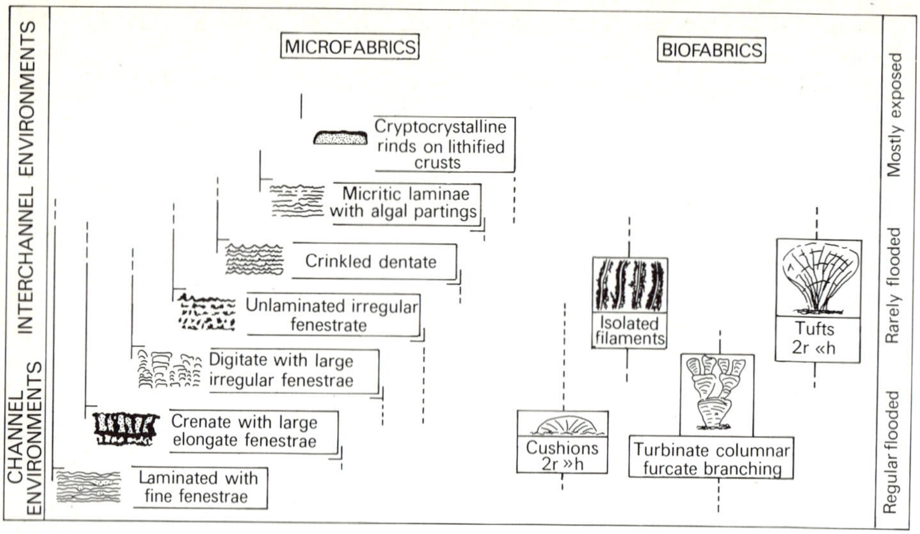

Fig. 15. Distrbution of laminae types and microbiological structures on an environmental-sedimentary process grid

Guarga oncoids one can only say that the recognizability of algal structures depends on the environment and degree of preservation. Probably the environment directly controls early cementation and preservation (Fig. 15).

This biofabric is cemented by very early calcite cement (Pentecost 1978, Schäfer and Stapf 1978, Lang and Segonzac 1979). As in other fossil examples (Bertrand-Sarfati and Fabre 1974, Donsimoni et al. 1975, Freytet and Plaziat 1975) this cement is clear microspar. This type of cementation was observed in both sections. Bertrand-Sarfati and Fabre (1974) stated that frequently micritization precedes a palisade calcite and a successive drusy calcite.

A similar sequence of palisade calcite and drusy cement was also found in the cavities of early dissolved oncoid-cores of section A (Fig. 9). The position of the micritization within the parageneses of the Guarga oncoids, however, remains uncertain.

5 Conclusions

This investigation has demonstrated that thorough analysis of morphology, texture and microfabric of oncoids may be highly indicative for freshwater environments. The evaluation of literature data showed that although most knowledge on the formation of oncoids has been achieved from research on marine oncoids there are highly promising results from recent and fossil freshwater examples. Semikhatov et al. (1979), for example, suggested oncoids have to be evaluated at five structural and textural levels

- morphology, and size of the assemblage, together with the encasing sediments,
- the individual oncoid,
- the internal laminae,
- the microstructures,
- the recognizable biota in the laminar matrix.

Based upon this scheme it was possible to conclude from the macrostructural elements of the Guarga oncoids which is their value as environmental indicators. Their microstructures are directly controlled by microbial assemblages.

The oncoids from the Guarga formation are of clearly shallow freshwater origin. The geological setting indicates that one association is related to the fluvial channel environment and that the second association belongs to the inter-channel floodplain regime. Distribution of morphological types reveals a distinct shift from closed, unattached forms in the channel regime towards open, attached structures in the floodplain. The mode of accretion seems to respond directly to water energy and sediment charge of the system. The fining-upward clastic sequences of fluvial systems are also documented in changes from

- unattached to attached algal structures,
- large unattached structures to small unattached ones,
- closed, plenicinct structures to open flat structures,
- predominantly dense laminated internal fabric to predominantly fenestrate fabric,
- preferential horizontal to vertical orientation of algal filaments.

The algal structures occur in times of low or non-clastic sedimentation and in quiet water areas. Algal growth is stopped by reactivation of channels during flood periods. This periodic flooding leads to frequent overturning of algal structures in the channel areas thus producing the plenicinct growth pattern of the oncoids. In the floodplains drowning and erosion of the algal structures was less frequent and less severe. Here the algae had maximum growth conditions. The fact that generally isochemical cores occur in the floodplain assemblage supports the concept that here the substrate was less frequently reworked. The above-mentioned systematic changes strongly suggest that the Guarga oncoids are but the unattached representatives of their stromatolitic equivalents.

The phenomenon that the algal structures are better preserved in the floodplain assemblage is probably due to very early cementation. It has been observed that micritization may easily destroy algal structures in both environments. There are strong indications that this micritization depends on aerial exposure. It therefore cannot be excluded that fluvial and floodplain oncoids were formed by the same algal associations but with accentuated growth of one or the other algal type in the respective environment. In this case the Guarga oncoid association would correspond to the concept of "bioherm series" of Krylov (1976), ... "an aggregate of the morphological modifications that could be produced by a given community of algae ..." (Semikhatov et al. 1979).

References

Abell PI, Awramik SM, Osborne RH (1978) Oxygen, and carbon isotopic variation in Pleistocene lacustrine stromatolites from Lake Turkana, Kenya. Bull Am Assoc Petrol Geol Abstr 62:488

Bertrand-Sarfati J, Fabre J (1974) The nodular stromatolites of the Autunian lacustrine deposits of the Abadla-Bechar basin (Western Algerian Sahara), Bull Hist Natl Afr N Alger 65:179–206

Bertrand-Sarfati J, Freytet P, Plaziat JC (1978) Nonmarine stromatolites from central and southern France (microstructures, algal remains); comparison with Precambrian stromatolite microstructures. Bull Am Assoc Petrol Geol Abstr 62:496

Borch van der CC, Bolton B, Warren JK (1977) Environmental setting and microstructure of subfossil lithified stromatolites associated with evaporites, Marion Lake, South Australia. Sedimentology 24:693–708

Boulanger D, Poignant AF (1975) Les nodules algaires du Miocène d'Aquitaine méridionale. Bull Centre Et Réch Sci 10:685–691

Crouzel F, Lang J (1975) Un environnement carbonaté continental: Quelques structures rayonnantes du Miocène Aquitain (France). In: Abstracts, IX° Congr Int Sedimentol T 2 Nice 1975, pp 39–43

Dahanayake K (1978) Sequentional position and environmental significance of different types of oncolites. Sediment Geol 20:301–316

Dahanayake K, Champetier Y, Hilly J (1976) Oncolithes et pseudoncolithes: Relation entre la disposition séquentionelle et la participation de Bacinella Irregularis Radioˇcic aux structures oncolithiques. Bull Soc Géol Fr 18:1665–1669

Donovan RN (1973) Basin margin deposits of the Middle Old Red Sandstone at Dirlot Caithness. Scott J Geol 9:203–211

Donsimoni M, Giot D, Lang J, Lucas G (1975) Rôle des organismes et des réactions physicochimiques dans la genèse des calcaires concrétionnés de Limagne. In: Abstracts, IX° Cong Int Sedimentol T 2, Nice 1975, pp 39–43

Fannin NGT (1969) Stromatolites from the Middle Old Red Sandstone of Western Orkney. Geol Mag 106:77–88

Flügel E (1978) Mikrofazielle Analyse von Kalken. Springer, Berlin Heidelberg New York

Freytet P (1964) Le Vitrollien des Corbières orientales: Réflexions sur la sédimentation "lacustre" nord pyrénéenne, divagation fluviatile, biorhexistasie, pédogénèse. Rév Géogr Phys Géol Dyn Sér 3 6:179–199

Freytet P (1965) Sédimentation microcyclothémique avec croûtes zonaires algues dans la calcaire de Beauce de Chaffour-Etrechy (S. et O.). Bull Soc Géol Fr Sér 7, 7:306–313

Freytet P (1971) Paléosols résiduels et paléosols alluviaux hydromorphes dans le Crétacé supérieur et l'Eocène basal en Languedoc. Rév Géogr Phys Géol Dyn Sér 2, 13:245–288

Freytet P (1973) Petrography and paleo-environment of continental carbonate deposits with particular reference to the Upper Cretaceous and Lower Eocene of Languedoc (Southern France). Sediment Geol 10:25–60

Freytet P, Plaziat JC (1965) Importance des constructions algaires dues à des Cyanophycées dans les formations continentales du Crétacé Supérieur et de l'Eocène du Languedoc. Bull Soc Géol Fr Ser 7 7:679–694

Gaemers PAM (1978) Biostratigraphy, paleoecology and paleogeography of the mainly marine Ager Formation (Late Paleocene – Early Eocene) in the Tremp Basin, South Central Pyrenees, Spain. Leidsche Geol Meded 51:151–231

Goldsmith JR, Graf DL (1958) Structural and compositional variations in some natural dolomites. J Geol 66:678–693

Goldsmith JR, Graf DL, Joensuu OI (1955) The occurrence of magnesian calcites in nature. Geochim Cosmochim Acta 7:212–230

Golubić S, Fischer AG (1975) Ecology of calcareous nodules forming in Little Connestoga Creek near Lancaster, Pennsylvania. Verh Int Ver Limnol 19:2315–2323

Heim A (1916) Monographie der Churfürsten-Matterhorn Gruppe. 3. Teil: Lithogenesis. Beitr Geol Karte Schweiz NF 20

Hofmann HJ (1969) Attributes of stromatolites. Geol Surv Can Pap 69–39, p 51

Hoffman P (1976) Stromatolite morphogenesis in Shark Bay, western Australia. In: Walter MR (ed) Stromatolites. Elsevier, Amsterdam, pp 261–271

Krylov IN (1976) Approaches to the classification of stromatolites. In: Walter MR (ed) Stromatolites. Elsevier, Amsterdam, pp 31 – 43

Lang J, Segonzac G (1979) Contribution à l'étude des algues continentales cénozoiques dans les bassins intramontagneux de Bamyan et de Yakawlang ou dans leurs environs en Afghanistan central. Les algues marins crétacées. Bull BRGM Sect IC 2:113 – 131

Monty CLV (1976) The origin and development of cryptalgal fabrics. In: Walter MR (ed) Stromatolites. Elsevier, Amsterdam, pp 193 – 249

Mouline P (1977) La place des formations stromatolitiques et oncolitiques dans l'analyse séquentionelle des sédiments fluviolacustres Tertiaires du Castrais et de l'Albigeois. Bull Soc Hist Natl Toulouse 113:132 – 151

Murray G (1895) Calcareous pebbles formed by algae. Phycol Mem 3:74 – 77

Nickel E (1980) Sedimentary petrology of continental carbonates, Eocene Guarga Formation, Tremp-Graus Basin, Southern Pyrenees, Spain. In: 1st European regional meeting of the IAS, Bochum, Germany. Abstracts pp 180 – 183

Nijman W, Nio SD (1975) Sedimentological guide to the Tertiary of the southern Pyrenees, Northern Spain – Tremp/Graus/Ainsa Basin. Reprint, IXth Int Congr Sedimentol, Nice 1975

Ordonez RH, Del Cura G (1977) Facies oncoliticas en medio continental: Application al sector de la Cuenca de Duero. Estud Geol 33:456 – 466

Pentecost A (1978) Blue green algae and freshwater carbonate deposits. Proc R Soc London Ser B 200:43 – 61

Reineck H-E, Singh IB (1973) Depositional sedimentary environments, 1st edn. Springer, Berlin Heidelberg New York

Riding R (1977) Skeletal stromatolites. In: Flügel E (ed) Fossil algae. Springer, Berlin Heidelberg New York, pp 57 – 60

Risacher F, Eugster HP (1979) Holocene pisoliths and encrustations associated with spring-fed surface pools, Pastos Grandes, Bolivia. Sedimentology 26:253 – 270

Schäfer A, Stapf RG (1978) Permian Saar-Nahe Basin and Recent Lake Constance (Germany): two environments of lacustrine algal carbonates. In: Matter A, Tucker ME (eds) Modern and ancient lake sediments. Spec Publ Int Assoc Sediment 2:83 – 107

Schöttle M, Müller G (1968) Recent carbonate sedimentation in the Gnadensee (Lake Constance), Germany. In: Müller G, Friedman GM (eds) Recent developments in carbonate sedimentology in Europe. Springer, Berlin Heidelberg New York, pp 148 – 156

Semikhatov MA, Gebelein CD, Cloud P, Awramik SM, Benmore WC (1979) Stromatolite morphogenesis – Progress and problems. Can J Earth Sci 15:992 – 1015

Wagner F, Lamprecht G (1974) Limnische Algen-Stromatolithen aus dem Perm des Saar-Nahe Beckens. Neues Jahrb Miner Monatsh 2:63 – 68

Walter MR (ed) (1976) Stromatolites. Dev Sedimentol 20. Elsevier, Amsterdam

Wray JL (1977) Calcareous algae. Elsevier, Amsterdam

Wright VP (1981) A subaerial stromatolite from the Lower Carboniferous of South Wales. Geol Mag 118:97 – 100

IV.6 Depositional Environment, Diagenesis and Reservoir Properties of Oncolitic Packstones, Macaé Formation (Albian-Cenomanian), Campos Basin, Offshore Rio de Janeiro, Brazil

Albert V. Carozzi[1], Frank U.H. Falkenhein[2], and Milton R. Franke[2]

1 Introduction and Geologic Framework

The stratigraphy of the Campos Basin (Fig. 1A, B) was controlled by the following sequence of tectonic stages characteristic of rifted continental margins (Ponte et al. 1980); pre-rift arching: denudation of Paleozoic cover (Permian-Jurassic); intracratonic rift valley: N-S collapse trough enlarged to rift valley system (late Jurassic – early Cretaceous) filled by Lagoa Feia fluvio-deltaic to lacustrine clastics; proto-oceanic gulf: restricted embayment (late Aptian) with deposition of Lagoa Feia evaporites; open marine continental margin: subsidence and seaward tilting, generation of carbonate platform consisting of offshore bars of oncolitic packstones separated by mudstones (Lower Macaé, Albian); further tilting, subsidence of platform, transgression of argillaceous pelagic mudstones and intercalated turbidites (Upper Macaé, Cenomanian) followed by basinal shales and turbidites (Carapebus Member, Campos Formation, Senonian-Eocene); renewed tilting of continental margin and rejuvenation of source areas with intense deltaic progradation (Ubatuba Member, Campos Formation (Oligocene – Pliocene).

The structural framework (Fig. 1C, D) is characterized by basement tectonics restricted to Aptian rifting and overlain by numerous dome-like structures separated by curved growth faults. The latter display a down-to-basin arrangement and disrupt only the Macaé Formation and the lower part of the Campos Formation. During the Albian-Cenomanian basinward tilting, growth faults developed with related antithetic rotation of the down-thrown blocks by gravity gliding of the limestone section over the underlying Lagoa Feia evaporites acting as a lubricant. A major result was the formation of a large SW-NE trending positive area (Fig. 2) that strongly influenced carbonate depositional patterns.

1 Department of Geology, University of Illinois at Urbana-Champaign, 254 N.H. Building, Urbana, IL, 61801, USA

2 Petróleo Brasileiro S. A. PETROBRÁS, Exploration Department (DEPEX), Rio de Janeiro, R. J., 20035, Brazil

Coated Grains (ed. by T.M. Peryt)

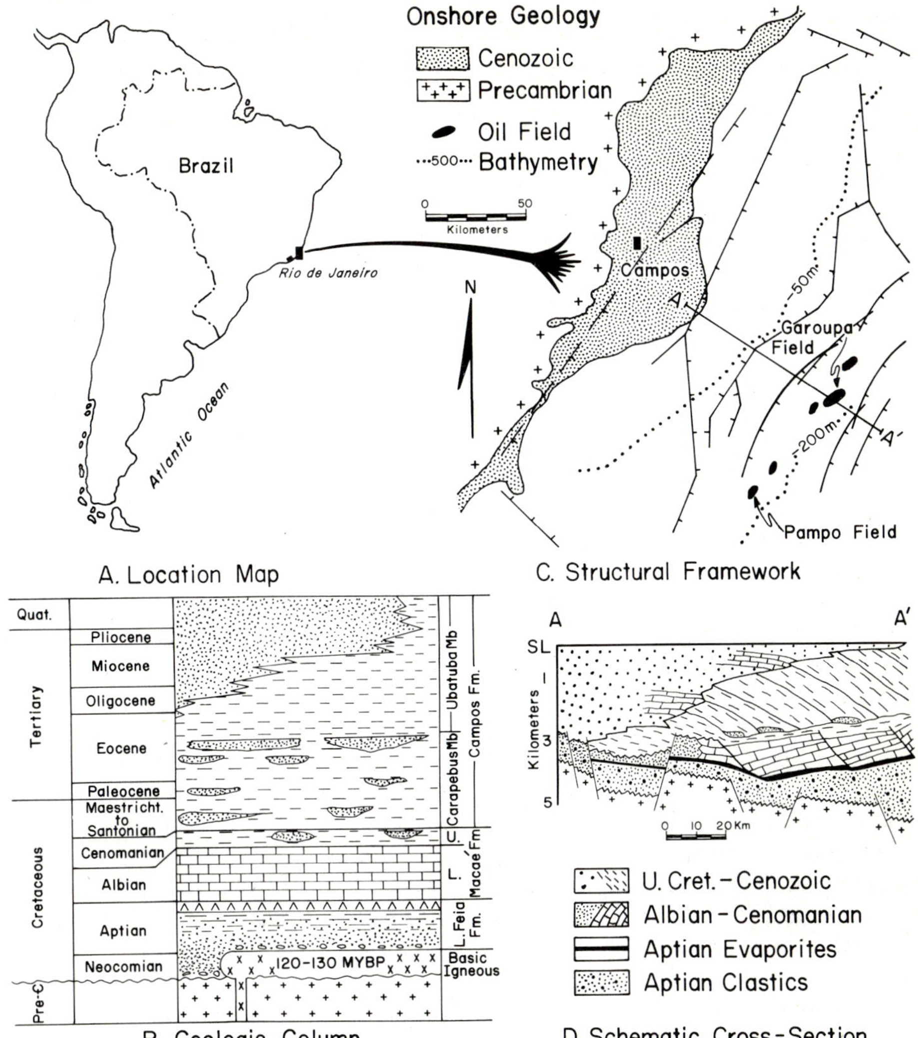

Fig. 1. Campos Basin: location, geologic column and structural framework

2 Recent Models of Formation of Oncoids

Marine oncoids are forming today in tropical shallow subtidal settings by disruption of blue-green algal mats in Bermuda (Gebelein 1969), the Bahamas (Neumann et al. 1970, Scoffin 1970, Gebelein 1976), and Bonaire (Pratt 1979).

In all these cases, algal mats occur adjacent to areas of extensive growth of marine grasses (*Thalassia*) and seaward of sandy bottoms. The seagrass beds support a dense epiphytic growth of calcareous red algae, bryozoans and foramini-

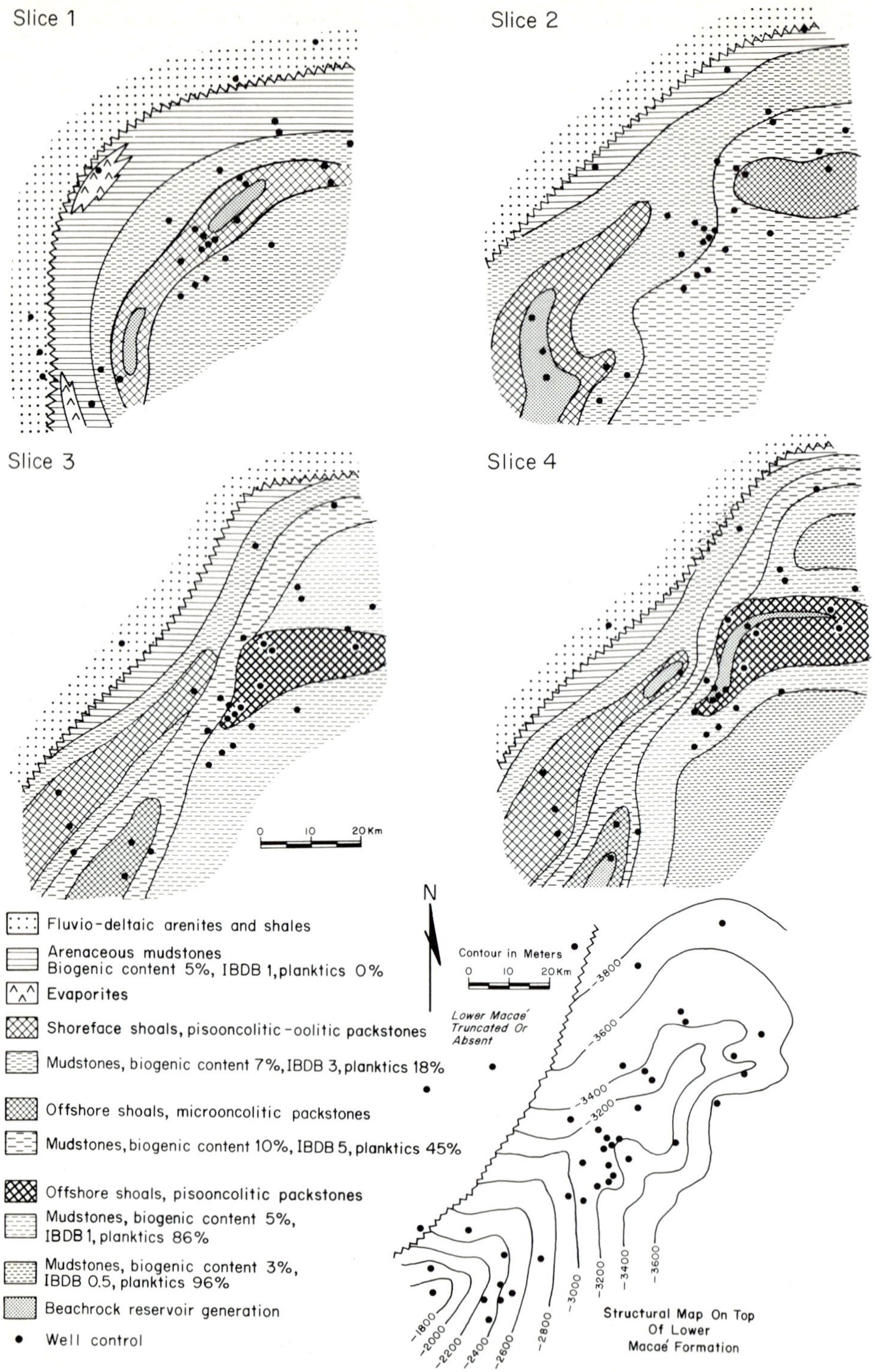

Fig. 2. Paleoenvironmental maps and structural map of Lower Macaé Formation

fers as well as numerous benthic organisms such as pelecypods and gastropods. Algal mats do not produce laminites because of environmental turbulence which continuously destroys them. Their chips, together with bioclasts from the adjacent seagrass community, become the cores of oncoids growing by means of further algal coatings and sediment binding. Final accumulation of oncoids occurs on sandy bottoms at depths of less than 1 m and accumulations of 10 to 100 oncoids m^{-2} are common.

Seagrass beds adjacent to algal mats are the critical ecologic factor. They have three main functions: (1) they provide a baffle effect which lowers the turbulence of the environment so as to protect the growth of algal mats but still allow enough periodical energy to generate oncoids from them; (2) they provide habitat and ecologic niches for the establishment of a stable and rich community (Brasier 1975, Taylor and Lewis 1970) which includes grazers and burrowers which otherwise would destroy the adjacent algal mats; (3) they generate large quantities of micrite mud from the disintegration of epibionts (Land 1970, Patriquin 1972).

3 Oncoids of the Macaé Formation

They have an irregular spheroidal to ellipsoidal shape, often deformed by compaction (Figs. 3A and 6D), and consist of pisooncoids (2 to 10 mm, in size) and microoncoids (less than 2 mm in size). Under the petrographic microscope a vague concentric layering is visible around cores of various types of bioclasts, chips of algal mats and grains of detrital quartz. Even under SEM pisooncoids no longer show filaments of blue-green algae, but only a typical internal structure of alternating porous and dense layers (Fig. 3B). The former assumed to have been originally rich in filaments represent periods of growth of the pisooncoids whereas the latter apparently devoid of filaments represent periods of interruption of growth. There are numerous instances of major interruptions of growth superimposed on the finer couplets (Fig. 3E). Natural broken surface of porous layers under SEM consists of a mosaic of micrite-size (0.5 to 4.0 μm) subhedral calcite crystals that does not give any indication of its origin (Fig. 3C), in fact it is identical to that of micrite mud. Pore casts under SEM show a spongy network of microporosity with pores smaller than 2 μm (Fig. 3D). A common feature of pisooncoids is the presence of filaments of endolithic boring algae appearing as empty flattened tubes (Fig. 3F).

4 Application of Recent Models to Macaé Oncolitic Carbonates

Open marine shelf, shallow subtidal turbulent waters and baffle systems appear to be the required conditions to generate oncolitic carbonates. They would ideally result in the juxtaposition of three distinct microfacies: mudstones, wackestones (baffle system) and oncolitic packstones, but in reality the first two cannot be easily distinguished. *Thalassia* baffle systems can only be traced back as far as the late Cretaceous, therefore it is assumed that seagrass-like communities such

Fig. 3 A – F. SEM pictures of pisooncoids. **A** Pisooncoids extracted from a friable pisooncolitic packstone. *Arrow* points to deformation by compaction. **B** Pore cast of internal structure of pisooncoid. Couplets of porous and dense concentric layers around nucleus at *left*. **C** Broken natural surface of porous layer. Micrite size (0.5 to 4.0 µm) subhedral calcite crystals with no indication of origin. **D** Pore cast of spongy microporous network. **E** Internal structure of pisooncoid showing thick nonporous concentric layers corresponding to major interruptions of growth. **F** Filament of endolithic boring alga appearing as noncalcified empty tube of organic matter inside pisooncoid

as green algae with subordinate stalked pelmatozoans and red algae played a similar role in the Lower Macaé carbonates. The effects of baffle systems is revealed by the following lithologic associations: the juxtaposition of mudstones with relatively abundant and diversified benthic bioclasts and packstones consisting only of microoncoids indicates a poorly-developed seagrass community whose ineffective baffle action allowed dispersal of bioclasts in all directions and afforded little protection to algal mats. The juxtaposition of mudstones with few bioclasts among which planktics predominate and packstones consisting of pisooncoids indicates a stable and rich seagrass community with an effective baffle action releasing bioclasts only in the direction of algal mats and thus adequately protecting them.

In order to differentiate among oncolitic packstones, an index of biotic diversity of benthics (IBDB) is used for bioclasts forming the nuclei of oncoids and for those scattered in the matrix. The index ranges from 1 to 8 by cumulative increments of one for each of the following constituents: echinoids, gastropods, molluscs, foraminifers, annelids, red algae, rudistids and hexacorals. The percentage of planktics (ostracods, foraminifers and calcispheres) relative to the total amount of bioclasts is used to differentiate among mudstones.

5 Paleoenvironmental Maps

They illustrate the depositional settings of four parachronostratigraphic slices representing basinwide shallowing tendencies (Falkenhein 1981). The datum line is the base of the Upper Macaé, a reliable chronostratigraphic surface.

Slice 1 (1000 – 700 m below datum, Fig. 2). Shallow subtidal to high intertidal shoreface shoals developed in the central part of the basin. They consisted of grain-supported arenaceous pisooncolitic to oolitic packstones with micrite matrix and sparite cement: IBDB of bioclasts in matrix: 8, of bioclasts as cores of pisooncoids and ooids: 6. The highest areas of the shoals were exposed and beachrock reservoirs generated. Adjacent mudstones: biogenic content: 7%, IBDB: 3, planktics: 18%. The baffle system of stable seagrass community is efficient.

Slice 2 (700 – 400 m below datum, Fig. 2). Migration of the shoreface shoals of slice 1 to southwest part of basin. Development in east-central area of subtidal offshore shoals. They consisted of grain-supported microoncolitic packstones with micrite matrix and rare sparite cement. The microoncoids lacked internal structure. IBDB of bioclasts in matrix: 2, of bioclasts as cores of microoncoids: 1.5. Adjacent mudstones: biogenic content: 10%, IBDB: 5, planktics: 45%. The baffle system of immature or unstable seagrass community is inefficient.

Slice 3 (400 – 200 m below datum, Fig. 2). The shoreface shoals of slice 1 persist in southwest part of basin, but without development of reservoir conditions. Southern lobe of system changed into an offshore subtidal microoncolitic shoal with adjacent relatively fossiliferous mudstones. In east-central area, the offshore subtidal microoncolitic shoals of slice 2 changed into shallow subtidal to intertidal pisooncolitic shoals. The pisooncolitic packstones with micrite matrix and sparite cement have: IBDB of bioclasts in matrix: 6, of bioclasts as cores of pisooncoids: 4–5. Beachrock reservoirs are occluded by meteoric phreatic cementation. Adjacent mudstones: biogenic content: 5%, IBDB: 1, plantics: 86%. The baffle system of stable seagrass community is efficient.

Slice 4 (200–0 below datum, Fig. 2). Final stage of conditions of slice 3. It is characterized by extensive shallowing controlled by synsedimentary tectonism. Related development of reservoirs in the offshore pisooncolitic shoals of east-central area (Garoupa field), at northern tip of shoreface pisooncolitic-oolitic shoals and in offshore microoncolitic shoals of southern part of basin (Pampo field).

6 Comparison with Other Mesozoic Oncolitic Environments

Oncolitic and oncolitic-oolitic packstones of Mesozoic carbonate shelves have been described on a worldwide basis: Dachstein of Austria (Fischer 1964), Jurassic of Paris Basin (Purser 1978) and of the Jura Mountains of Switzerland (Bolliger and Burri 1970), Jurassic Smackover of the Gulf Coast (Bishop 1968, Becher and Moore 1979), Albian Edwards Limestone of Texas (Bebout et al. 1977), and Albian-Cenomanian Regencia Formation of Espírito Santo Basin, Brazil (Tibana and Alves 1973). The Albian-Cenomanian Madiela and Catumbella Formations of Gabon and Angola basins also contain undescribed oncolitic carbonates (de Klasz 1978).

The above-mentioned examples are similar to those described here in the following critical aspects: occurrence as subtidal to intertidal offshore shoals with subordinate shoreface shoals and longshore belts, associated subtidal mudstones with very low fossil content, absence of other types of packstones and general paucity of fossils of carbonate platforms bearing large oncolite deposits.

7 Depositional-Diagenetic Sequence

The oil traps in the upper part of the Lower Macaé are mixed structural-stratigraphic resulting from closure by domes and growth faults combined with favorable depositional-diagenetic conditions. The latter are part of a sequence of ten stages of early to burial diagenesis distinguished on petrographic and cathode luminescence textures, and confirmed by stable isotope data (Franke 1981).

Low Intertidal Environment. Stage 1 (Fig. 4): deposition of initial unconsolidated sediment consisting of a framework of abraded pisooncoids with interstitial sand-size bioclasts and intraclasts set in a finer matrix interpreted as "oncoid flour" derived from the abrasion of the larger pisooncoids. Stage 2 (Figs. 4A and 5A, B): diagenetic, precipitation of thin isopachous rim cement of fibrous calcite (high magnesium and/or aragonite?).

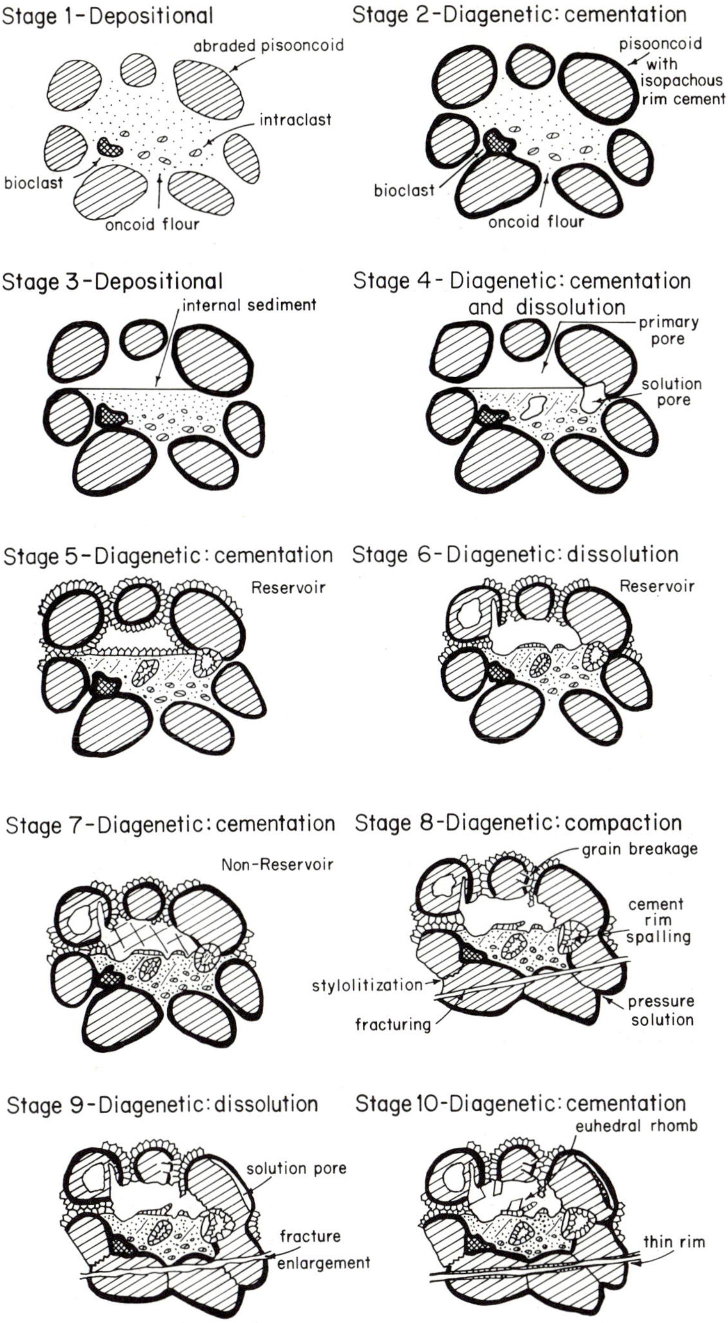

Fig. 4. Depositional-diagenetic sequence

High Intertidal Environment. Stage 3 (Figs. 4 and 5A – C): depositional, removal and reorganization of interstitial constituents into an internal sediment with geopetal features and horizontal surfaces. Energy high, predominant downward circulation of seawater, probably adjacent to a beachrock.

Beachrock Environment. Stage 4 (Figs. 4 and 5D): diagenetic, lithification of internal sediment by interparticle sparite cement followed by solution which generated secondary vuggy to channel porosity cutting across pisooncoids and internal sediment. Stage 5 (Figs. 4 and 5E): diagenetic, precipitation of rim cement of bladed calcite (high magnesium and/or aragonite?) along boundaries of all previously generated open spaces. If preserved after burial, this stage is a reservoir.

Beachrock Vadose Environment. Stage 6 (Figs. 4 and 5F – H): diagenetic, its results whenever preserved after burial correspond to the best reservoir rocks. Intense vadose dissolution after exposure of beachrock to subaerial conditions but without generation of a freshwater lens: removal of large amounts of bladed rim cement, corrosion of margins of pisooncoids, deep etching of upper surface of cemented internal sediment and local differential solution of nuclei and concentric layers of pisoooncoids. These mesopores provide a measured porosity of 25 to 30% and 200 to 400 md permeability.

Freshwater Meteoric Phreatic Environment. Stage 7 (Figs. 4 and 6A, B): diagenetic, with cementation in a freshwater lens generated upon extensive subaerial exposure. Reservoirs formed during the previous beachrock stage, including any enhancement by subsequent meteoric vadose conditions were occluded by calcite precipitation when transferred by subsidence into the meteoric phreatic zone prior to permanent burial. The phreatic calcite cement occurs as single crystals and coarse crystalline irregular to highly interlocked equant mosaic. Frequent occurrence of intense and distorted twinning indicates that cementation preceded major compaction.

Burial Environment. Stage 8 (Figs. 4 and 6C – E): compaction with grain breakage by radial microfractures, reciprocal grain interpenetration by pressure solution and stylolitization, spalling of bladed rim cement with penetration in pisooncoids, pisooncoid spalling, non-fabric selective fracturation at least in two gener-

Fig. 5A – H. Submarine to beachrock vadose diagenesis. Plane polarized pictures of thin sections impregnated with blue plastic, porosity appears in various shades of gray. **A** Submarine isopachous rim cement of fibrous calcite with traces of dissolution (*arrow*) and surrounding graded internal sediment. **B** Isopachous rim of fibrous calcite cement overlain by graded internal sediment with geopetal attitude. Note vadose dissolution effects preceding phreatic sparite cementation. **C** Horizontal surface of geopetal internal sediment. **D** Vuggy to channel secondary porosity cutting across pisooncoid (*right side*) and internal sediment. Pores rimmed by bladed calcite cement (*arrow*). **E** Rim cement of bladed calcite (*upper right*), elsewhere same calcite is cavity-filling. **F** Relic of bladed calcite rim cement on pisooncoid surface (*arrow*) from vadose dissolution before phreatic sparite cementation. **G** Reservoir rock showing vadose dissolution effects on bladed calcite rim cement (*arrow*). **H** Reservoir rock showing vadose dissolution of bladed calcite rim cement and of cortical layers of pisooncoid

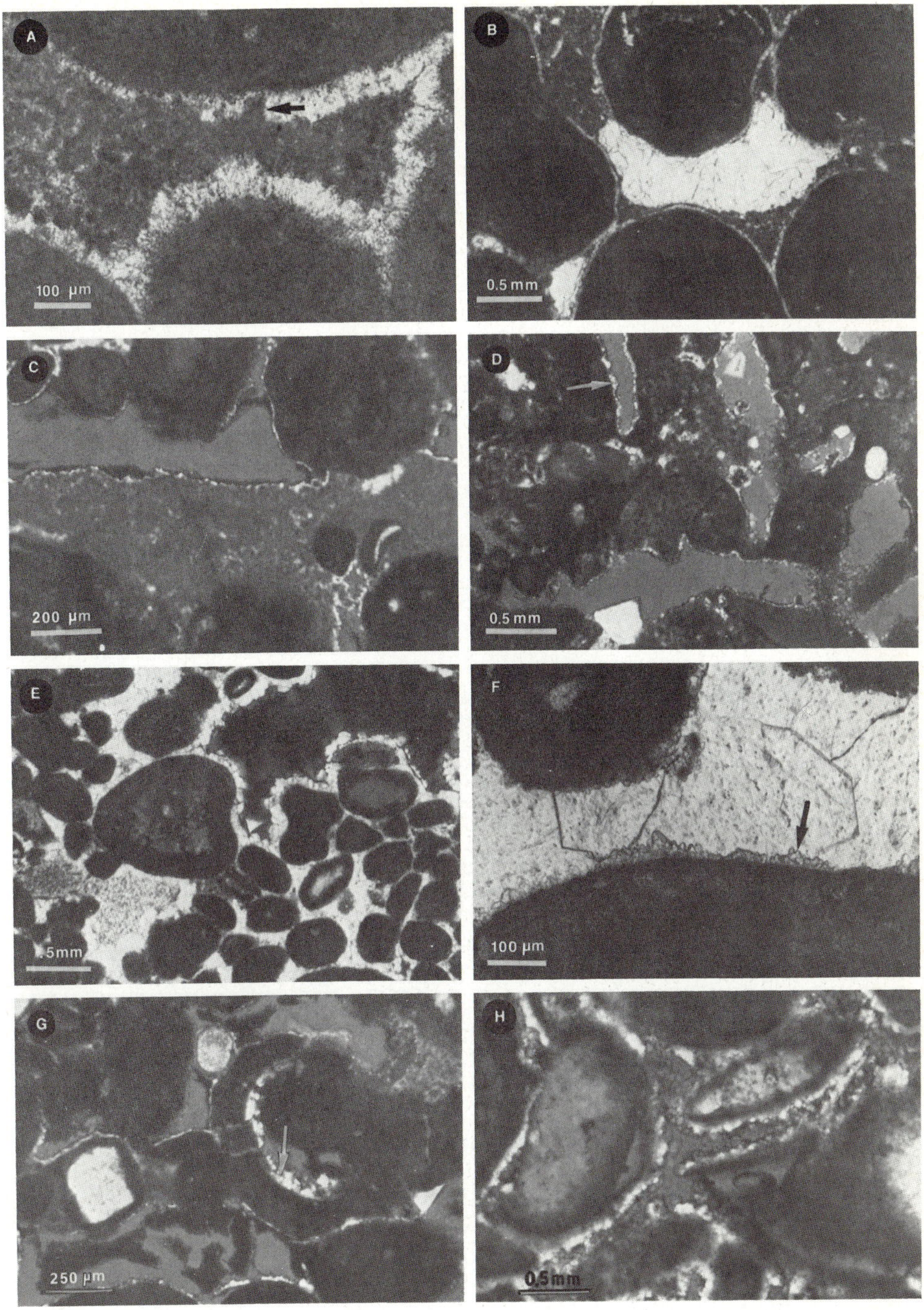
A
100 μm
B
0.5 mm
C
200 μm
D
0.5 mm
E
F
100 μm
G
250 μm
H
0.5mm

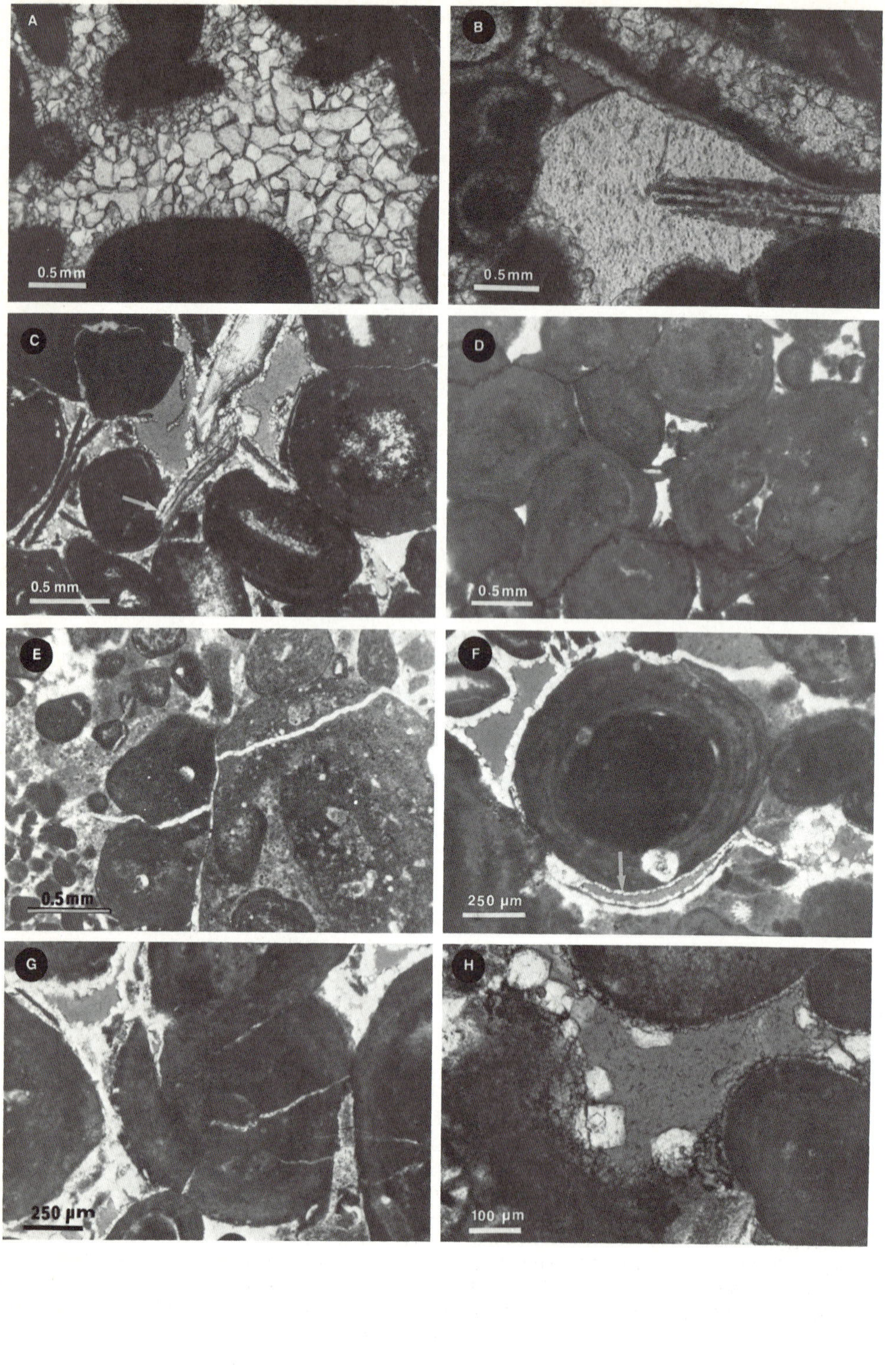
A
0.5 mm
B
0.5 mm
C
0.5 mm
D
0.5 mm
E
0.5mm
F
250 µm
G
250 µm
H
100 µm

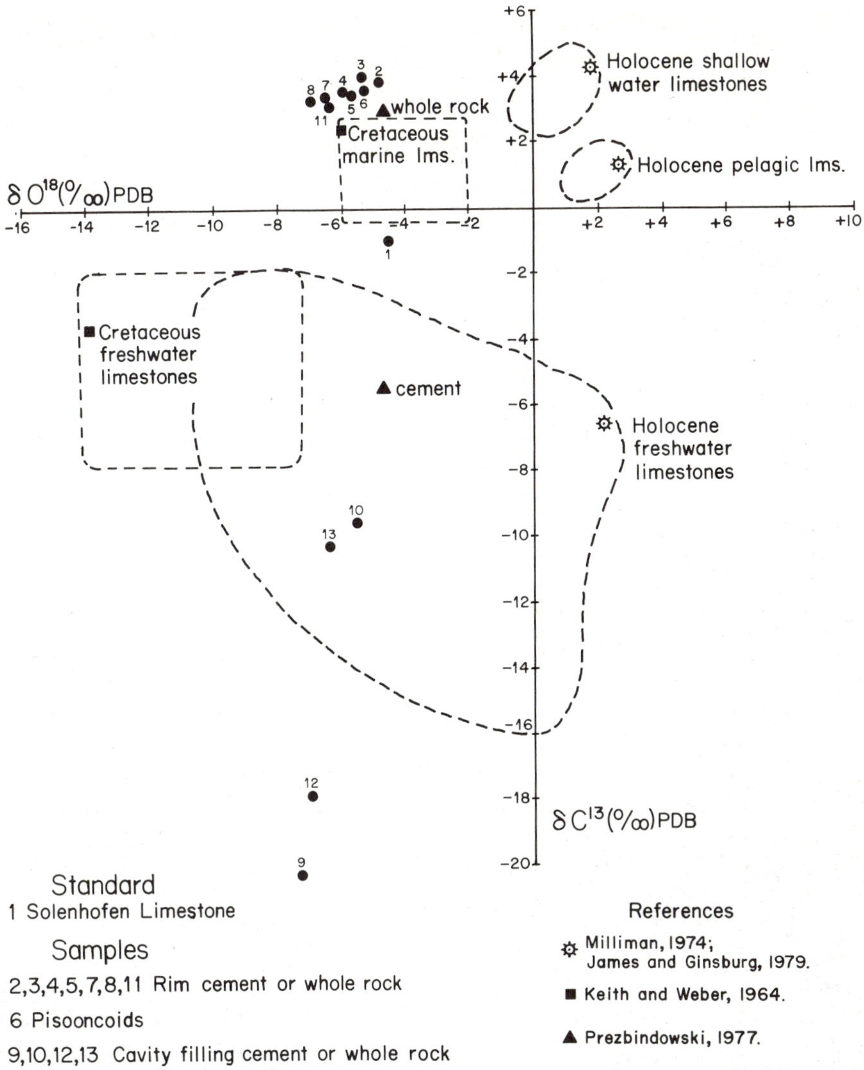

Fig. 7. Stable isotope measurements

Fig. 6 A – H. Meteoric phreatic to burial diagenesis. Plane polarized pictures of thin sections impregnated with blue plastic, porosity appears in various shades of gray. **A** Reservoir occlusion by phreatic coarsely crystalline equant mosaic of calcite. Earlier rim cement still visible. **B** Almost complete occlusion of pores by single crystal of calcite overgrowth on echinoid spine. Earlier bladed rim cement still visible. **C** Compaction effects: spalling of bladed calcite rim cement, grain breakage by microfractures, bladed rim cement penetrating pisooncoid margin (*arrow*). **D** Compaction effects: deformed pisooncoids, reciprocally penetrating pisooncoids and stylolitized contacts. **E** Compaction effects: first generation fracture cemented and second generation fracture (*trending vertical*) displacing the first and open. **F** Late dissolution pore (*arrow*) by differential solution of concentric layers of pisooncoid and showing thin rim of late calcite cement. **G** Thin rim of late calcite cement filling fractures of pisooncoid deformed by compaction. **H** Late cement as perfect rhombohedral crystals with oil staining growing on relics of earlier bladed calcite rim cement ations. Early fractures cemented, other enlarged by late dissolution. In general, pores generated during compaction remain uncemented or show a very thin rim of late calcite cement. Stage 9 (Figs. 4 and 6F): late dissolution with enlargement of all previous compaction features and generation of intraparticle porosity by differential solution of some concentric layers of pisooncoids. Stage 10 (Figs. 4 and 6F–H): late cementation is widespread but not an important porosity reducer. It occurs as a thin rim of calcite crystals and as perfect rhombohedral crystals with frequent growth lines and inclusions (oil?). This cement may be coeval with oil migration.

Other minor aspects of burial diagenesis are: aggrading neomorphism appearing as relatively large cloudy calcite crystals with diffused boundaries and abundant inclusions of unreplaced material; rare dolomite rhombs replacing matrix and cement; extremely rare silicification as overgrowth on detrital quartz grains and post-dating dolomitization; anhydritization selectively replacing pisooncoid layers and sparite cement.

8 Geochemistry of Diagenesis

The beachrock rim cement of bladed calcite and the pisooncoids gave values of δC^{13} and δO^{18} typical of marine cements, while the cavity-filling sparite cement gave very light values of δC^{13} corresponding to freshwater phreatic conditions (Fig. 7). The latter are further confirmed by a lack of cathode luminescence indicating an iron-rich, manganese-poor calcite (Meyers 1978).

9 Summary of Events and Oil Migration

The growth fault activity was responsible for the successive phases of vertical evolution displayed by the oncolitic shoals, the most evident being in the east-central part of the basin where the change of offshore shoals from microoncolitic to pisooncolitic coincided with a general high of the structural map on top of the Lower Macaé Formation. The same synsedimentary tectonic activity which localized the shoals also uplifted them into beachrock conditions, and higher into islands with extensive subaerial exposure and freshwater lenses. Subsidence of beachrocks preserved their reservoirs because no freshwater lenses had been related to them, whereas islands had their reservoirs occluded when subsidence transferred them through the phreatic environment. Reactivation of growth faults provided the necessary physical connection for the lateral migration of oil between source-beds (radioactive shale, Carapebus Member, Campos Formation) and the underlying oncolitic reservoirs.

Acknowledgements. The authors are very grateful to the Board of Directors of Petróleo Brasileiro S. A. Petrobrás for permission to publish this paper.

References

Bebout DG, Schatzinger RA, Loucks RG (1977) Porosity distribution in the Stuart City Trend, Lower Cretaceous, South Texas. In: Bebout DG, Loucks RG (eds) Cretaceous carbonates of Texas and Mexico. Bur Econ Geol Univ Texas Rep Inv 89:234–256

Becher JW, Moore CH (1979) The Walker Creek field: A Smackover diagenetic trap. In: Moore CH (ed) Geology of carbonate porosity. Am Assoc Petrol Geol Course Note Ser 11:A24–A46

Bishop WF (1968) Petrology of Upper Smackover Limestone in North Haynesville field, Clairborne Parish, Louisiana. Am Assoc Petrol Geol Bull 52:92–128

Bollinger W, Burri P (1970) Sedimentologie von Schelf-Carbonaten und Beckenablagerungen im Oxfordien des zentralen Schweizer Jura. Beitr Geol Karte Schweiz, NS 140

Brasier MD (1975) An outline history of seagrass communities. Palaeontology 18:681–702

Falkenhein FUH (1981) Carbonate microfacies and depositional evolution of the Macaé Formation (Albian-Cenomanian), Campos Basin, Brazil. Unpubl Ph D dissertation Univ Illinois, Urbana

Fischer AG (1964) The Lofer cyclothem of the Alpine Triassic. Kansas Geol Surv Bull 169:107–149

Franke MR (1981) Natural porosity, diagenetic evolution and experimental porosity development of Macaé carbonates (Albian-Cenomanian), Campos Basin, Brazil. Unpubl. Ph D Diss Univ Illinois, Urbana

Gebelein CD (1969) Distribution, morphology and accretion rate of recent subtidal algal stromatolites, Bermuda. J Sediment Petrol 39:49–69

Gebelein CD (1976) Open marine, subtidal and intertidal stromatolites (Florida, the Bahamas and Bermuda). In: Walter MR (ed) Stromatolites Dev Sedimentol. Elsevier, New York, pp 381–388

James NP, Ginsburg RN (1979) The seaward margin of Belize barrier and atoll reefs. Morphology, Sedimentology, Organism distribution and Late Quaternary history. Int Assoc Sediment Spec Publ 3. Blackwell, Oxford

Keith ML, Weber JN (1964) Carbon and oxygen isotopic composition of selected limestones and fossils. Geochim Cosmochim Acta 28:1787–1816

Klasz de I (1978) The West African sedimentary basins. In: Moullade M, Nairn AEM (eds) The Phanerozoic geology of the world II, the Mesozoic. Elsevier, Amsterdam, pp 371–399

Land LS (1970) Carbonate mud: production by epibiont growth on *Thalassia testudinum*. J Sediment Petrol 40:1361–1363

Meyers WJ (1978) Carbonate cements: their regional distribution and interpretation in Mississippian limestones of southwestern New Mexico. Sedimentology 25:371–440

Milliman JD (1974) Marine carbonates. Springer, Berlin Heidelberg New York

Neumann AC, Gebelein CD, Scoffin TP (1970) The composition, structure and erodability of subtidal mats, Abaco, Bahamas. J Sediment Petrol 40:274–297

Patriquin DG (1972) Carbonate mud production by epibionts on *Thalassia;* an estimate based on leaf growth rate data. J Sediment Petrol 42:687–689

Ponte FC, Dos Reis Fonseca J, Carozzi AV (1980) Petroleum habitats in the Mesozoic-Cenozoic of the continental margin of Brazil. In: Miall AD (ed) Facts and principles of world petroleum occurrence. Can Soc Petrol Geol Mem 6:857–886

Pratt BR (1979) Early cementation and lithification in intertidal cryptalgal structures, Boca Jewfish, Bonaire, Netherlands Antilles. J Sediment Petrol 49:379–386

Prezbindowski DR (1977) Carbon and oxygen isotopic evolution of whole rock and cements from the Stuart City Trend. In: Bebout DG, Loucks RG (eds) Cretaceous carbonates of Texas and Mexico. Bur Econ Geol Univ Texas, Rep Inv 89:201

Purser BH (1978) Early diagenesis and the preservation of porosity in Jurassic limestones. J Petrol Geol 1:83–94

Scoffin, TP (1970) The trapping and binding of subtidal carbonate sediments by marine vegetation in Bimini Lagoon, Bahamas. J Sediment Petrol 40:249–273

Taylor JD, Lewis MS (1970) The flora, fauna and sediments of the marine grass beds of Mahé, Seychelles. J Natl Hist 4:199–220

Tibana P, Alves R (1973) Sedimentação carbonática Cenomaniana-Albiana na Bacia do Espírito Santo. Anais XXVII Congr Soc Bras Geol 3:339–356

IV.7 Albian Pelagic Phosphate-Rich Macrooncoids from the Tatra Mts (Poland)

Krzysztof P. Krajewski[1]

1 Introduction

Phosphate-rich macrooncoids occur in the Albian condensed limestones in the Tatra Mts (Poland) and they are especially well developed in the Turnia Ratusz succession (Fig. 1). These condensed limestones are built of several alternating layers (5 – 40 cm thick) of calcisphaeral and foraminiferal wackestones and echinoderm wackestones. The macrooncoids, like other cryptalgal structures (mostly stromatolites) and the majority of authigenic non-carbonate mineralizations, are associated with pelagic horizons only (Fig. 1).

The Albian limestones of the Tatra Mts are lying transgressively on the Urgonian reefal substrate (Lefeld 1968) and turn upward to basinal marly deposits. Cyclic changes in sea-level during sedimentation caused the formation of sedimentary features of open shelf, inner shelf, and lagoonal environments that are mixed or directly superposed in a very short vertical profile (Krajewski 1981 b). Such transgressive-regressive cycles can also to be visible in the Turnia Ratusz succession (Fig. 1). The main features of the Albian palaeo-environment can be compared to several recent phosphogenic open shelf environments, e.g., to that of SW African shelf (Birch 1979).

The aim of this paper is to present the microstructural variability of phosphate-rich macrooncoids from the Tatra Mts, their geomicrobiological and mineralogical features, and genesis of microbial phosphate fabric. Relation to other phosphate-rich cryptalgal structures is also discussed.

2 Macrooncoids

The macrooncoids are oval and elliptical to elongated, occasionally to composite irregular. Their external surfaces are always mamillated and porous. They consist of well distinguishable nuclei and coatings strongly enriched in phosphates

1 Instytut Nauk Geologicznych PAN, Zwirki i Wigury 93, 02-089 Warszawa, Poland

Coated Grains (ed. by T.M. Peryt)

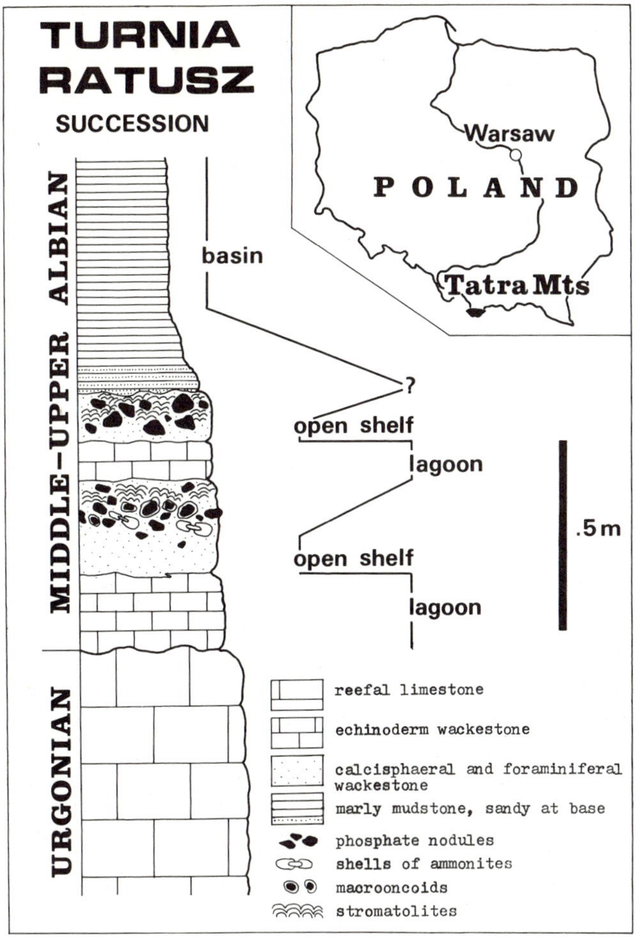

Fig. 1. Turnia Ratusz succession of the Albian deposits in the Tatra Mts and their environmental interpretation

and iron oxides which makes the macrooncoids black to dark red in colour, thus contrasting to the surrounding pale grey wackestones. The dimensions of the macrooncoids vary from 1 – 4 cm (prevailing forms) to 10 cm.

3 Nuclei

Phosphate and phosphate-ferrous nodules or their fragments occur as the nuclei of the macrooncoids. Development of phosphate and phosphate-ferrous nodules, a part of which has been then coated, was a manifold complicated process.

The conception of *hiatus concretions* (Voigt 1968) can be, in general, applied to explain the origin of the investigated nodules and nuclei. Nevertheless, hiatus concretions *sensu* Voigt usually show carbonate internal bodies (Kennedy and Klinger 1972) which have mineralized only during postlithificational stages of their formation (Kennedy and Garrison 1975). One may claim that the observed nodular matrix can be more plausibly explained by diagenetic accumulation of phosphates and iron oxides within soft carbonate sediment (non-carbonate cementation) rather than by mineralization of lithified calcareous bodies, according to the model of recent phosphate nodule formation summarized by Baturin and Bezrukov (1979).

In the Turnia Ratusz succession, the nodules occur in a single horizon within pelagic wackestone, which suggests that in the final stage of their manifold for-

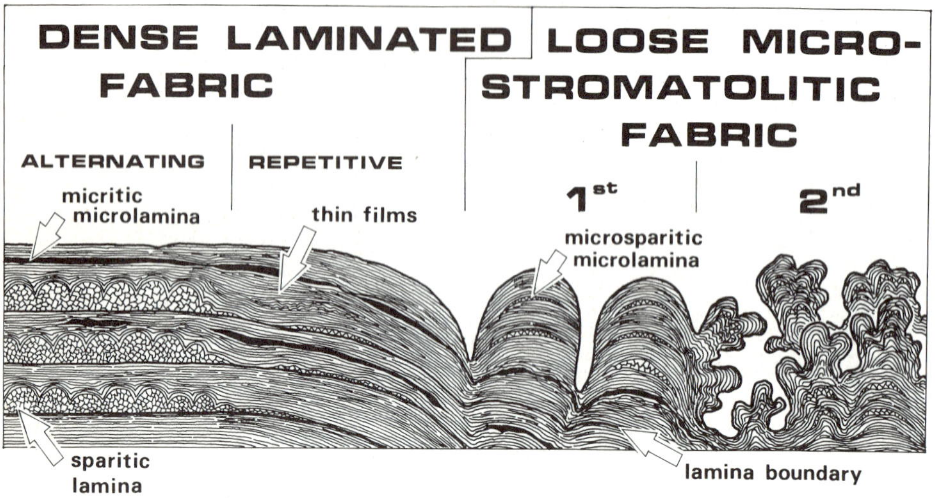

Fig. 2. Main types of fabric (I) – schematic diagram

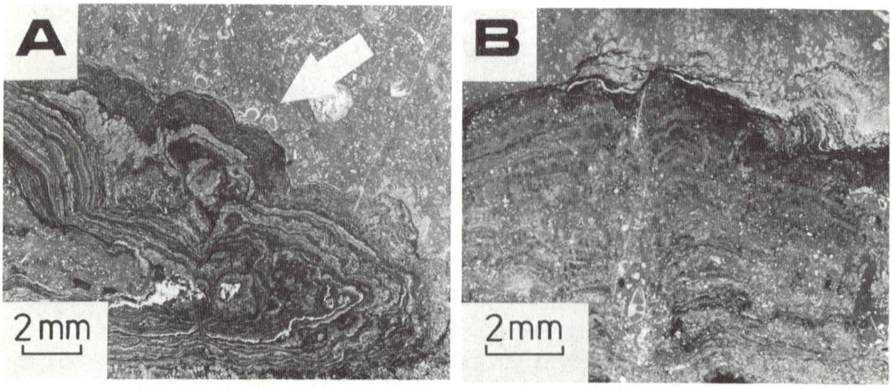

Fig. 3 A, B. Phosphate-rich macrooncoids – main types of fabric (II). **A** Oncoid coating showing dense laminated fabric (*arrowed* are foraminiferal tests incrusting the surface). **B** A part of oncoid coating showing loose microstromatolitic fabric. Thin section photographs

mation process the nodules probably concentrated at the sea-floor by winnowing. Many of the nodules exposed there were overgrown by phosphate-rich oncoid coatings.

4 Coatings

The oncoid coatings show two main types of fabric: dense laminated fabric and loose microstromatolitic one (Figs. 2 and 4). Such fabrics, if they do occur together in the same oncoid coating, may pass one to another vertically or laterally.

Fig. 4A,B. Primary fabrics (I). **A** Composite alternating lamination. **B** Microstromatolites of the first class. TLM, plane polarized light

Dense laminated fabric (Fig. 3A) may be represented by composite repetitive lamination or by composite alternating lamination (these terms will be here applied closely to Monty's, 1976, definitions). *Composite repetitive lamination* depends on a superposition of the microcrystalline laminae (several to 500 μm thick) which show internal composite microstructure. The horizontal sublaminae superposed in variable proportions can be divided as follows: (1) phosphate thin films (5 – 10 μm thick), (2) micritic microlaminae (5 – 15 μm), (3) microsparitic microlaminae (10 – 25 μ). *Composite alternating lamination* consists of an alternation of the microcrystalline laminae and the microsparitic ones.

Loose microstromatolitic fabric (Fig. 3B) consists of many juxtaposed and superposed microstromatolites growing outward within the oncoid coatings, separated by depressions, interstices or other internal spaces. Taking into account their dimensions, growth ways, and internal and microstructural proportions, the microstromatolites can be subdivided into two classes: (1) microstromatolites of the first class (500 – 1000 μm in height, 200 – 800 μm in width); (2) microstromatolites of the second class (100 – 400 μm in height, 50 – 200 μm in width).

Both dense and loose fabrics are sometimes complicated by the following additional features which occasionally occur in the oncoid coatings: (1) tests of incrusting organisms, (2) erosional surfaces, (3) fractures.

Two growth types of isolated microbiological mat, which illustrate the development of the coatings, can be distinguished: (1) growth of prostrate community, (2) growth of erect community.

Growth of prostrate community was the main growth type of microbiological mat which was composed predominantly of small cyanobacteria with horizontally elongated filaments (Fig. 6A). Associated rod-like bacteria were also common (Fig. 6B – E). Non-carbonate mineralization of a simple prostrate community resulted in the development of phosphate thin film. Moreover, rhythmical reproduction and mineralization of prostrate community caused the development of sets of thin films forming the main component of microcrystalline laminae (Fig. 5A). Reproduction of prostrate community was more or less variable in time including morphologically different disposes of thin films, sets of thin films, and their superpositions. The bloom of mat was tantamount to occupation by microorganisms of an entire coating surface or of a considerable part of it. Blooming mat had created continuous microcrystalline lamina, and in consequence, composite repetitive lamination. If the reproduction of prostrate community has stopped and started again, dormant mat reproduced only in privileged places onto a coating surface and initiated the development of microstromatolites. Under favourable conditions individual prostrate subcommunities building the microstromatolites could have joined forming thus a dense laminated fabric. On the other hand, the micritic microlaminae associated with thin films both in the dense laminated fabric and in the microstromatolites of the first class show the same primary features as have been observed in the stromatolites covering the macrooncoids (Krajewski 1981b). Such microlaminae (or laminae in stromatolites) have been developed occasionally as a result of trapping and binding of carbonate mud by living microbiological mat during high energy environmental

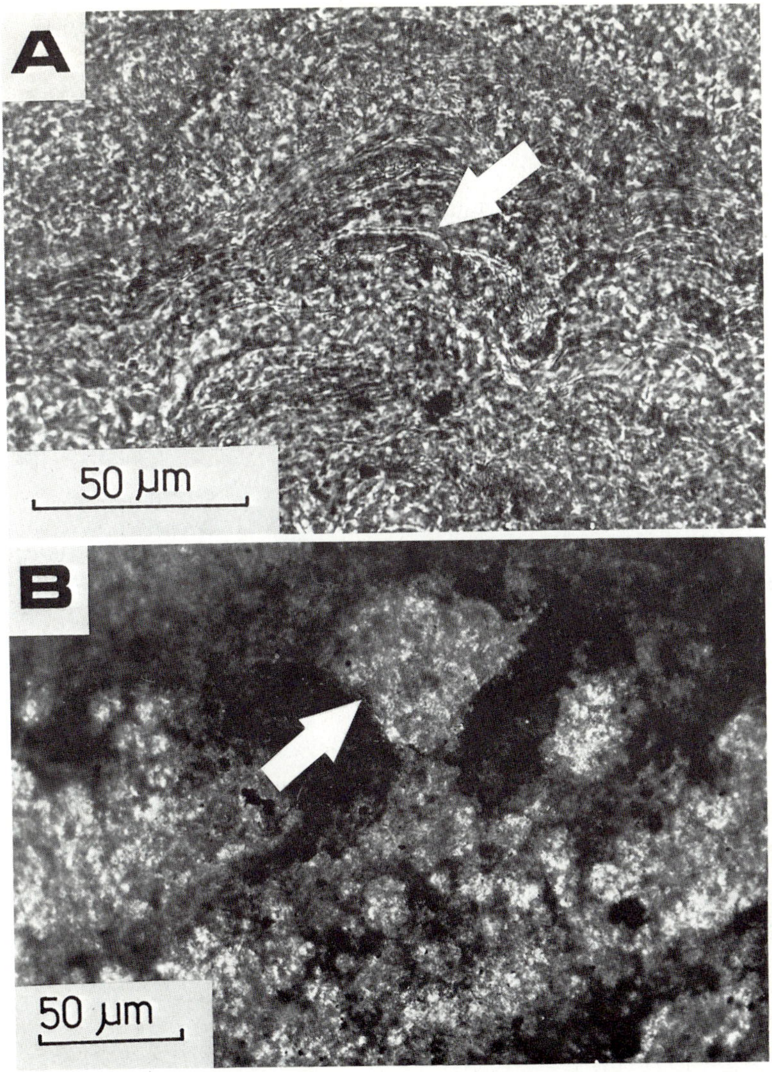

Fig. 5A, B. Primary fabrics (II). **A** Sets of phosphate thin films including remains (*arrowed*) of prostrate cyanobacterial community. **B** Fragment of microsparitic lamina including remains (*arrowed*) of erect cyanobacterial community. TLM, plane polarized light

episodes. In several cases thin-film forming prostrate community grew in association with incrusting foraminifera.

Growth of erect community was an occasional growth type of microbiological mat. Little is known of this growth type in the investigated macrooncoids. Nonetheless, the microsparitic laminae always show characteristic juxtaposed microdomes (Fig. 4A) which can be interpreted as "ghosts" after erect bundles of filamentous cyanobacteria. Microsparite that builds the microdomes, with exception

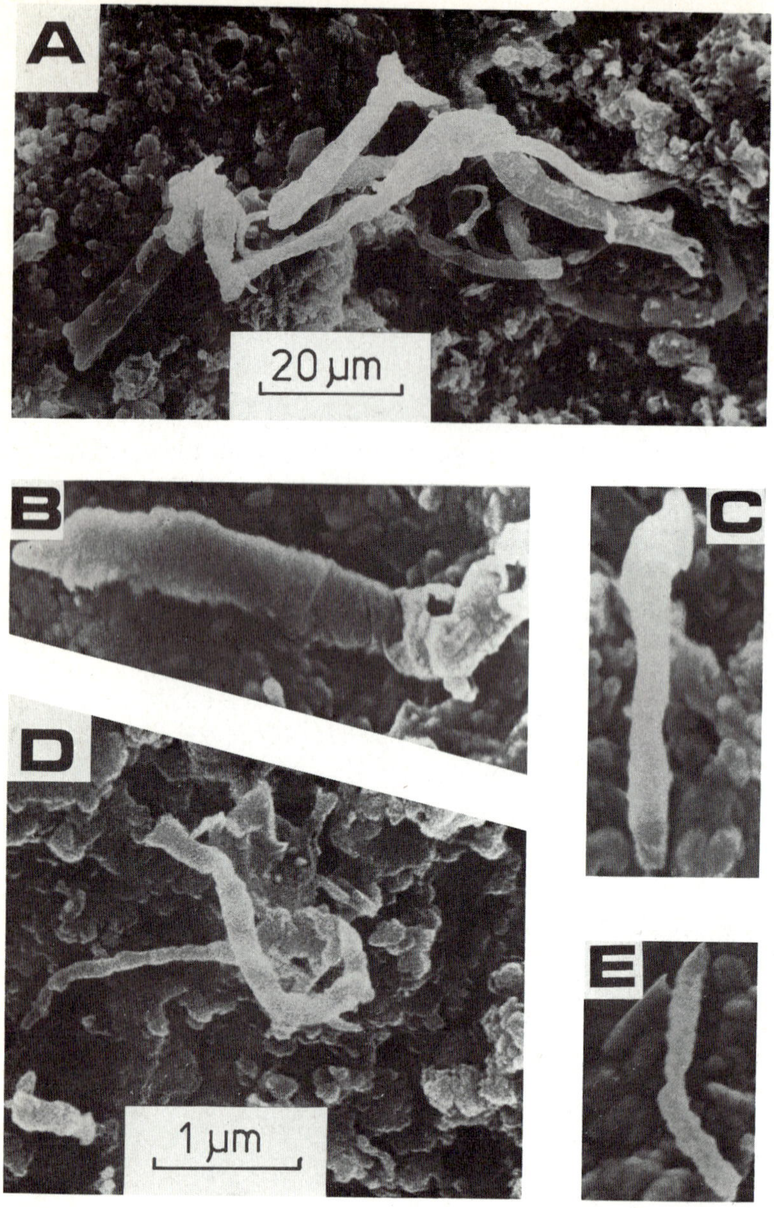

Fig. 6 A – E. Remains of prostrate community. **A** Bundle of prostrate cyanobacterial trichomes preserved as silicates and dispersed organic matter. **B – E** Rod-like bacteria "bodies" (mode of preservation as on **A**). SEM, etched preparations

of that shown on Fig. 5B, does not show any organic traces, nor radial (pseudoradial) nor radiaxial microstructures, therefore, it cannot correspond to calcified mucilageneous cyanobacterial "colonies". The origin of microsparitic microdomes can be more plausibly explained by precipitation of internal carbonate cement in fenestrae formed by oxidation of erect bundles. Microsparite building the laminae differs definitely from that one developed during the formation of secondary fabrics (see discussion on diagenesis), which suggests fenestral cementation during early diagenesis of the coatings. Episodic occurrence and lateral discontinuities of the microsparitic laminae indicate that growth of erect community was not common at least in the investigated macrooncoids. However, a part of the microsparitic laminae shows gradations to microsparitic microlaminae, the latter being interpreted as fenestral cement precipitated below detached surficial sets of thin films. Then the growth of erect community could have taken place in isolated sites below the surface of a coating and should be considered as internal growth of erect filamentous cyanobacteria. Precipitation of microsparite both in the microlaminae (detached fenestrae) and laminae (fenestrae after internal growth) was proceeding in the same early diagenetic episode, and therefore, there exists microstructural unity of such microsparite in all tested fenestrae.

It is not known whether the two distinguished growth types of microbiological mat formed the same microorganic community differentiated under fluctuating microenvironmental conditions, or separate species that developed in concurrent growth. Neither is a reason of existence of different preservation ways of such communities known. Nevertheless, similar problems are not solved from many recent examples (Monty 1976, Monty and Mas 1981).

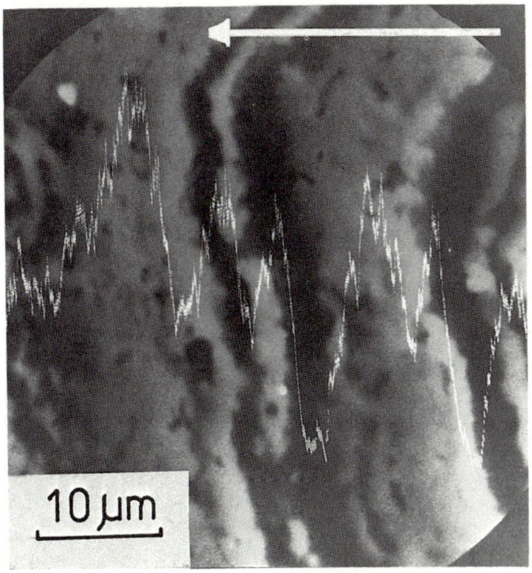

Fig. 7. Mineralogy of primary fabrics (I). Phosphate-rich thin films (*pale grey*) and microsparitic microlaminae-fenestral cement (*black*). Note the variability of intensity of phosphorus. *White arrow* shows vertical direction. Microprobe

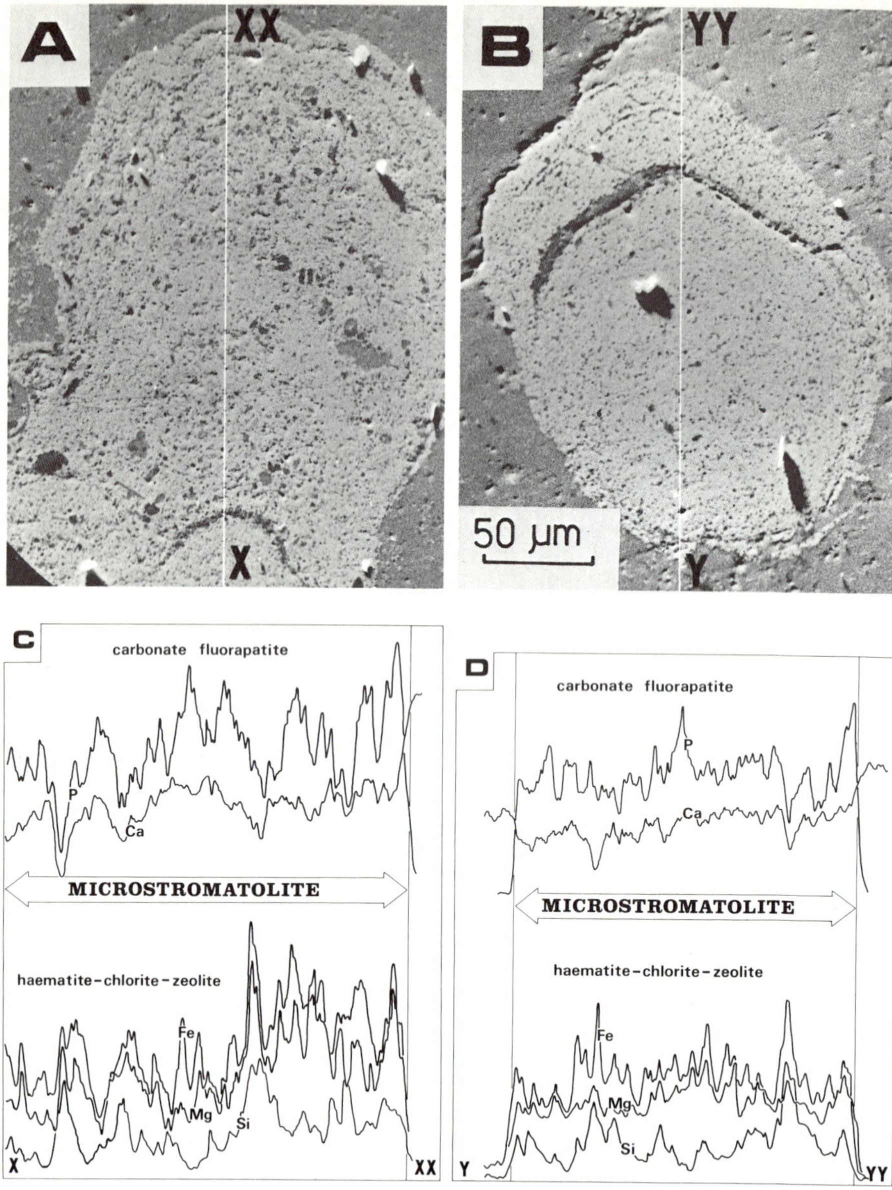

Fig. 8 A – D. Mineralogy of primary fabrics (II). **A, B** Sections of microstromatolites of the second class parallel (**A**) and transverse (**B**) to growth direction. *White lines* microprobe profiles; **C, D** Variability of intensity of studied elements (Ca, P, Si, Mg, Fe) along microprobe profiles on photos **A** and **B** respectively. Note the distinct correlations in occurrence of Ca and P (carbonate fluorapatite) and Si, Mg, Fe (dispersed impurities: haematite-chlorite-zeolite)

5 Mineralogy of the Phosphates

The investigated oncoid coatings contain abundant calcium phosphate which is limited to the thin films only and does not occur in other primary microfabrics (Figs. 7 and 8). During TLM study of thin sections, the phosphates have been observed as variable coloured homogeneous isotropic substance (collophane) and cannot be determined more precisely. However, under high magnifications sometimes small (< 1 μm) apatite crystals have been noted. Nevertheless, the majority of perceptible small crystals are in fact recrystallized impurities dispersed in the collophane, mostly haematite and chlorite. During SEM investigation the phosphates frequently show flocculous or massive ultrastructures (Fig. 9). Neither distinction of particular crystals nor observation of regular crystallographic forms is possible. On the contrary, X-ray diffraction (XRD) has revealed that the phosphate represents well-pronounced crystalline phase of carbonate fluorapatite which contains nearly 2% of carbon dioxide in apatite lattice (determined using methods by Smith and Lehr 1966, Gulbrandsen 1970).

Precipitation of the phosphates within prostrate community might depend on physiochemical activity of microorganisms that built the mat or/and was associated with semiclosed microenvironment developed directly below living mat in which decomposition of organic matter by bacteria occurred. The phosphates have been precipitated both into and onto cyanobacterial sheaths and mucilage probably due to bacteria action. Decomposition of organic matter could rise pH of the below-mat microenvironment and make an oxygen minimum in it, therefore, they were other favourable features for the precipitation of phosphate. Moreover, the inhibiting effect of magnesium/calcium ions ratio of phosphate precipitation (Martens and Harris 1970) could have been below the critical value

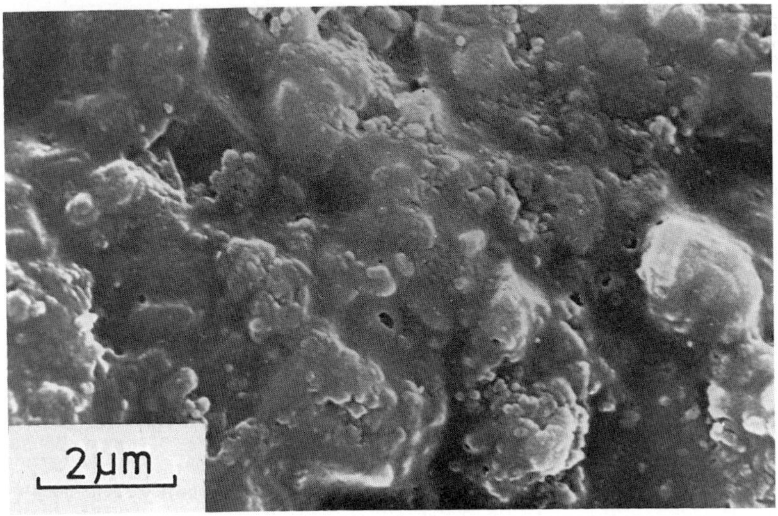

Fig. 9. Mineralogy of primary fabrics (III). Phosphates building thin films show massive ultrastructure. Crystalline phase – carbonate fluorapatite. SEM

due to adsorption of magnesium ions during formation of Mg-rich silicates (chlorite, zeolite) within the mat.

The primary nature of phosphates building thin films, however, is not equivalent to the primary nature of the apatite. Although XRD has revealed that there is well pronounced crystalline carbonate fluorapatite, SEM observations have proved the presence of variable stages of their recrystallization. The occurrence of flocculous and massive phosphate ultrastructures indicates deposition of gel-like phosphate matter rather than chemical precipitation of primary apatite crystals, according to stages of phosphate crystallization presented by Baturin and

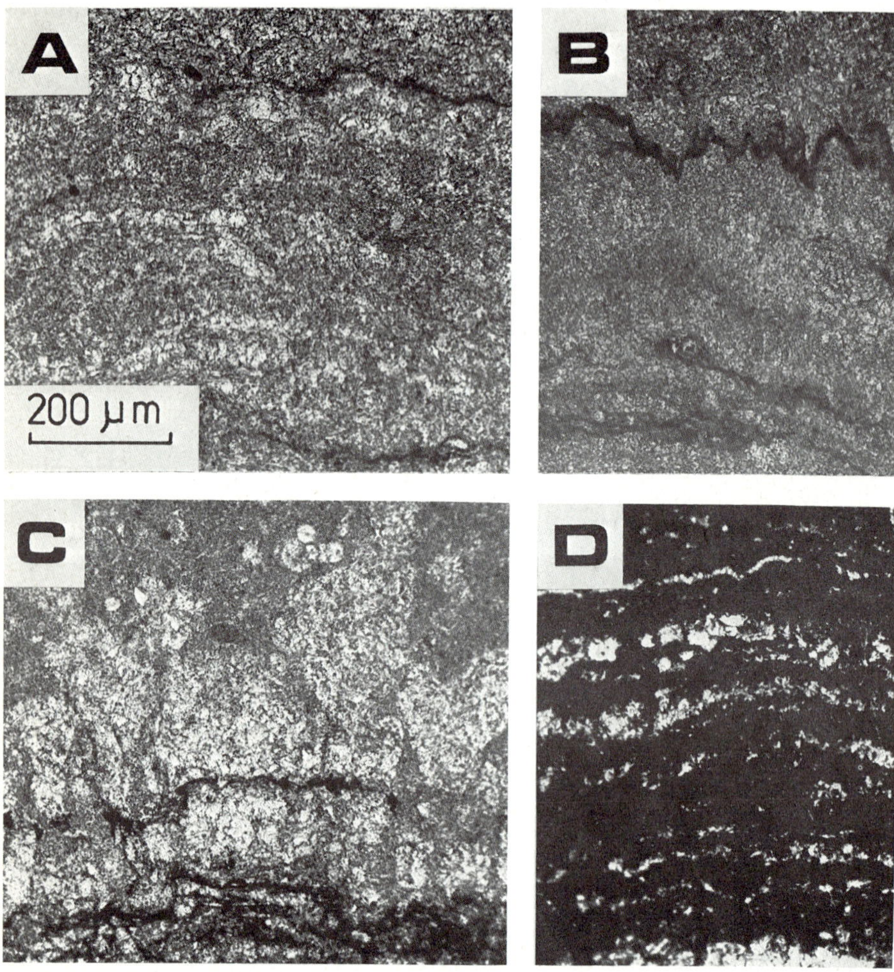

Fig. 10 A – D. Secondary fabrics. **A** Late diagenetic fine mosaic microsparite (*grey*) including remains of early diagenetic fenestral cement (*white*) and streaks of non-calcifable substances (*black*). **B** Secondary lamination consists of alternation of thick mosaic-microsparitic laminae (*grey*) and stylolite-like ones (*black*) built of iron oxides and silicates. **C** Late diagenetic microsparitization of microstromatolites. **D** Ferruginized secondary fabric; iron oxides (*black*) include remains of microsparite (*white*). Scale bar on photo **A**, for all photographs. TLM, plane polarized light

Dubinchuk (1974). Using these data, the process of flocculation of colloids instead of ionic crystallization seems to be more plausible as a mean of prostrate community mineralization.

6 Diagenesis – Development of Secondary Fabrics

The majority of tested macrooncoids shows intense diagenetic changes which caused that the fresh primary fabrics (both dense and loose) are observed relatively rarely. Early diagenetic changes of the macrooncoids, such as decomposition of organic matter, crystallization of gel-like substances, detachment of sets of thin films and fenestral cementation, are mentioned above. Long after burial, under conditions of carbonate cementation and neomorphism of surrounding wackestones, the calcification of phosphates started. Fine mosaic microsparite with crystal dimensions 5 – 20 μm and dispersed non-carbonate impurities (mostly iron oxides and silicates) developed as a result of calcification of primary phosphate fabrics (Fig. 10A, C). Calcification of the macrooncoids caused decrease of phosphate content and relative increase of non-calcifable substances which, primarily, were dispersed as impurities within the phosphates. Thus the secondary laminae composed of mosaic microsparite alternate with irregular ones enriched in non-calcifable substances. Pressure solution occurring along the boundaries of secondary laminae was responsible for the development of sharp stylolite-like non-carbonate laminae with characteristic zigzag pattern (Fig. 10B). The secondary fabrics, in places, could have been phosphatized or ferruginized by migrating phosphates and iron oxides respectively (Fig. 10D). As a collateral product of calcification of fluorapatite the fluorite inclusions were formed in the macrooncoids, according to reaction presented by Cook (1970).

7 Relation to Other Phosphate-Rich Cryptalgal Structures

Occurrence of phosphate-rich cryptalgal structures is not a common phenomenon in geological record. Such structures have been mentioned from the European Cretaceous, both in the Tethyan and epicontinental regimes, and here the example described from the Albian of the Tatra Mts is classified. Similar structures have been described from the Precambrian of India where they are common, and in occasional cases, from the Ordovician of Wales (Hofmann 1975). In the European Cretaceous phosphate-rich stromatolites occur in the Aptian-Albian condensed sequence of the Briançonnais Series (Royant et al. 1970, Rioult and Royant 1975). These structures have been formed under conditions of slow pelagic sedimentation and high biological productivity in seamount environment. A similar situation existed in the Upper Cretaceous of the Meridional Alps where phosphate-rich stromatolites have been found in association with pelagic hardgrounds (Massari and Medizza 1973). Phosphatc-rich cryptalgal structures occur also in a number of condensed sequences in epicontinental Cretaceous. Phosphates in stromatolites were regarded to be depositional primary fabrics that have developed as a result of life activity of microbiological mat (Golonka and Rajchel 1972). Division of phosphates onto biosedimentary (occurring in stromatolite columns) and inorganic occurring in colloform incrustations (Marcinowski and Szulczewski 1972) seems to be groundless. Transitions exist between both types of phosphates; they are microstructurally identical, visible as sets of thin films which include microorganic remains. Hence, both phosphate types formed in microenvironment of living microbiological mat, probably due to similar processes. Colloform phosphate incrustations correspond, indeed, to colloidal microstromatolites (Krajewski 1981a) or to microstromatolites of the secon class in the present paper. It is important to signalize that all known occurrences of phosphate-rich cryptalgal structures in the European Cretaceous were formed in pelagic or semi-pelagic phosphogenic subtidal environment with relatively low sedimentation rate and high biological productivity.

In the case of Precambrian examples, reconstruction of conditions and modes of phosphate enrichment in cryptalgal structures seems to be much more complicated. However, the cyclic development of phosphate-rich and phosphate-poor laminae indicates physiochemical rhythmicity of mineralization in microbiological mat (Banerjee 1978, and references therein). The nature of the process of such mineralization is not well known. Two alternatives can be considered here, namely coprecipitation of carbonates and phosphates within the living mat (Banerjee 1971) or progressive replacement of biologically precipitated carbonates by phosphates under conditions of active algal life (Chauhan 1979). Phosphate enriched stromatolites and oncoids have been phosphatized at the final stage of their lithification at the sea-floor (Valdiya 1972, Kumar 1977).

Acknowledgements. The author wishes to express his thanks to Dr T. M. Peryt for useful discussions and help and to Dr J. Lefeld for critical review of the manuscript.

References

Banerjee DM (1971) Precambrian stromatolitic phosphorites of Udaipur, Rajasthan, India. Geol Soc Am Bull 82:2319 – 2330

Banerjee DM (1978) Chemical rhythmicity in the Precambrian laminated phosphatic stromatolites and its bearing on the origin of algal phosphorite. Ind J Earth Sci 5:102 – 110

Baturin GN, Bezrukov PL (1979) Phosphorites on the sea floor and their origin. Mar Geol 31:317 – 332

Baturin GN, Dubinchuk VT (1974) Microstructures of Agulhas Bank phosphorites. Mar Geol 16:M63 – M70

Birch GF (1979) Phosphatic rocks on the western margin of South Africa. J Sediment Petrol 49:93 – 110

Chauhan DS (1979) Phosphate-bearing stromatolites of the Precambrian Aravalli phosphorite deposits of the Udaipur region, their environmental significance and genesis of phosphorite. Precambrian Res 8:95 – 126

Cook PJ (1970) Repeated diagenetic calcitization, phosphatization, and silification in the Phosphoria Formation. Geol Soc Am Bull 81:2107 – 2116

Golonka J, Rajchel J (1972) Upper Cretaceous stromatolites in the vicinity of Cracow. Kwart Geol 16:652 – 667 (in Polish, English summary)

Gulbrandsen RA (1970) Relation of carbon dioxide content of apatite of Phosphoria Formation to regional facies. US Geol Surv Prof Pap 700-B:9 – 13

Hofmann HJ (1975) *Bolopora* not a cryozoan, but an Ordovician phosphatic, oncolitic accretion. Geol Mag 112:523 – 526

Kennedy WJ, Garrison RE (1975) Morphology and genesis of nodular phosphates in the Cenomanian Glauconitic Marl of south-east England. Lethaia 8:339 – 360

Kennedy WJ, Klinger HC (1972) Hiatus concretions and hardground horizons in the Cretaceous of Zululand (South Africa). Palaeontol 15:539 – 549

Krajewski KP (1981a) Phosphate microstromatolites in the High-Tatric Albian limestones in the Polish Tatra Mts. Bull Acad Polon Sci Ser Sci Ter 29:175 – 183

Krajewski KP (1981b) Pelagic stromatolites in the High-Tatric Albian limestones in the Tatra Mts. Kwart Geol 25:731 – 759 (in Polish, English summary)

Kumar S (1977) Stromatolites and phosphorite in the Tirohan Limestone of Chitrakut area, District Satna, MP. Curr Sci 46:341 – 342

Lefeld J (1968) Stratigraphy and palaeogeography of the high-tatric Lower Cretaceous in the Tatra Mountains. Stud Geol Polon 24:1 – 115

Marcinowski R, Szulczewski M (1972) Condensed Cretaceous sequence with stromatolites in the Polish Jura Chain. Acta Geol Polon 22:515 – 538

Martens CS, Harris RC (1970) Inhibition of apatite precipitation in the marine environment by magnesium ions. Geochim Cosmochim Acta 34:621 – 625

Massari F, Medizza F (1973) Stratigrafia e paleogeografia del Campaniano-Maastrichtiano nelle Alpi Meridionali (con particolare riguardo agli hardgrounds della Scaglia Rossa Veneta). Mem Ist Geol Min Univ Padova 28:62

Monty C (1976) The origin and development of cryptalgal fabrics. In: Walter MR (ed) Stromatolites. Elsevier, Amsterdam, pp 193 – 249

Monty C, Mas JR (1981) Lower Cretaceous (Wealdian) blue-green algal deposits of the Province of Valencia, eastern Spain. In: Monty C (ed) Phanerozoic stromatolites. Springer, Berlin Heidelberg New York, pp 85 – 120

Rioult M, Royant G (1975) La "croute stromatolithique de Manara": dans le série brianconnaise du Monte Armetta (Alpes ligures). XI° Congr Int Sedimentol Nice, 8 p

Royant G, Rioult M, Lanteaume M (1970) Horizon stromatolithique à la base du Cretace supérieur dans le Brianconnais ligure. Bull Soc Geol Fr 7:372 – 374

Smith JP, Lehr JR (1966) An x-ray investigation of carbonate apatites. J Agric Food Chem 14:342 – 349

Valdiya KS (1972) Origin of phosphorite of the Late Precambrian Gangolihat Dolomites of Pithoragarh, Kumaun Himalaya, India. Sedimentology 19:115 – 128

Voigt E (1968) Über Hiatus-Konkretionen (dargestellt an Beispielen aus dem Lias). Geol Rundsch 58:281 – 296

IV.8 Oncoids and Stromatolites in the Rosso Ammonitico Sequences (Middle-Upper Jurassic) of the Venetian Alps, Italy

Francesco Massari [1]

1 Introduction

A specific type of oncoids, mostly found in association with stromatolitic domes in decimetric cycles, is quite common in red nodular limestones (Rosso Ammonitico) of the Trento Plateau area (Southern Alps). This was a pelagic horst block of the southern continental margin of the Tethys (Fig. 1) which evolved from a strongly subsiding Liassic carbonate platform. Thin Bajocian sequences of clean-washed *Bositra*-bearing coquinas and crinoidal sands (the *"Posidonia alpina"* beds) commonly underlie – and in places interfinger with the lowermost part of – the "Rosso Ammonitico" pelagic limestones (Sturani 1971). These

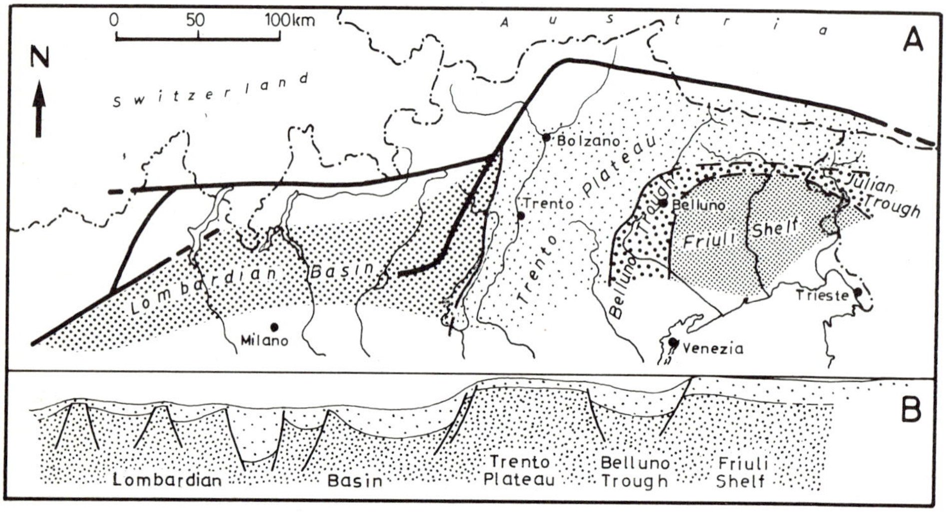

Fig. 1A, B. Paleogeography of Southern Alps during Late Jurassic. *Heavy lines* main Alpine faults bounding Southern Alps (**A**). Postulated W-E sea-floor profile during Late Jurassic. *Dense dots* pre-Jurassic rocks; *sparse dots* Jurassic sediments (**B**). (Redrawn from Gaetani 1975)

1 Istituto di Geologia, Paleontologia e Geologia Applicata dell' Università, Via Giotto 1, 35100 Padova, Italy

Coated Grains (ed. by T.M. Peryt)

form condensed and discontinuous sequences in the Trento Plateau area, and are usually represented by two members. The lower member (Rosso Ammonitico inferiore, Upper Bajocian to Middle Callovian) is often dominated by an oncolitic-stromatolitic facies and is more massive; moreover, in places it includes sparse interbeds of clean-washed and rippled microcoquinas of thin-shelled pelecypods. The upper member (Rosso Ammonitico superiore, Upper Oxfordian to Tithonian) grades upwards into Majolica limestones and shows a lower frequency of cryptalgal structures as well as the presence of sparse thin shaly or marly interbeds; it is most commonly characterized by a flaser-nodular facies in which evidence of exposure of the nodules at the depositional interface is often lacking. The two members in addition are locally separated by some meters of cherty limestones.

2 Oncoids

Oncoids consist of a simple or compound core and a layered, often polyphasic coating (Figs. 2 and 3a–d). They have subspherical or more commonly ovoidal shape. Irregular shapes may however occur when the nucleus is an angular intraformational lithoclast. Diameters range from 3.5 to 35 mm, larger diameters usually being displayed by oncoids with compound cores. In most cases the core consists of a nodule displaying meniscus-filled burrows and sometimes a microbored surface. Such nodules conceivably resulted from an early selective cementation in the shallow subsurface and were subsequently exhumed (Elmi 1978, 1981, Braga et al. 1981, Dommergues et al. 1981) and coated. The bacterial decay of organic substances associated with burrows and differences in porosity in comparison with the adjacent sediment may have played a major role in the initiation of the cementation (Kennedy 1975).

Coating of oncoids mostly consists of irregular tapering out laminae arranged in sets that usually envelope only a portion of the body (Fig. 2). Evidently growth took place only on the upper side and was episodic; it occurred during the timespan the particle rested on the bottom, being interrupted by displacement due to periodic turbulence or sediment turnover by burrowers or grazers. The resulting structure is a series of partial envelopes, more or less overlapping. In detail, the coating consists of an alternation of thicker, granular laminae, and much thinner micritic laminae (Fig. 3e). The former consist of small pellets, *Globochaete alpina* Lombard, debris of thin-shelled pelecypods, sometimes very small gastropods, foraminifera, minute echinoderm debris and other fine unrecognizable detritus. These particles are usually finer-grained than those occurring in the matrix of the oncoids and are commonly arranged with an inverse grading (Fig. 3e). In addition they are set in a matrix often affected by some degree of aggrading neomorphism which led to a slightly crystalline micrite or even microspar. The micritic laminae on the other hand are composed of dense micrite containing rare and very small-sized debris. It should be noted that the elongated particles such as thin shells of pelecypods (or fragments of them) tend to be arranged tangentially to the surface of the micritic laminae, whereas they often show an edgewise orientation within the granular laminae.

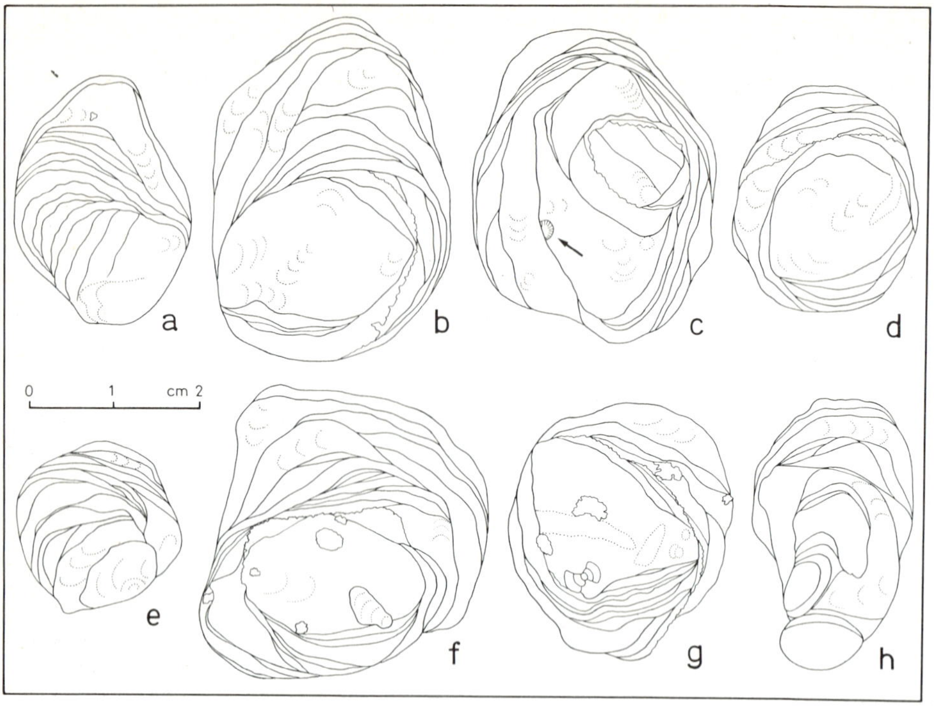

Fig. 2a–h. Oncoids from the "Rosso Ammonitico inferiore" of M. Lessini area north of Verona. Sections perpendicular to bedding. Burrowing traces indicated by *dotted lines*. Minutely *indented lines* indicate surfaces affected by bioerosion and iron-staining. *Arrow* in **c** small ahermatypic coral. Oncoid cores in most cases consist of exhumed nodules showing meniscus-filled burrows. In **c** core is an intraformational lithoclast. Cores of **f**, **g**, and **h** oncoids are compound multistage bodies resulting from aggregation, or inclusion one into another, of a number of particles, such as smaller oncoids (**h**), small nodules of burrowed limestone (**f**), or residual particles affected to a variable degree by bioerosion and iron-staining (**f** and **g**). Note presence of meniscus-filled burrows even within the coating of oncoids; **a, f, g** S. Ambrogio di Valpolicella; **b–e, h** Caprino Veronese

Fenestral fabrics are completely lacking. Degree of inheritance, regularity and thickness of the laminae generally depend on the grain size of the trapped sediment: coarser-grained varieties usually show a low degree of inheritance, with thicker and irregularly shaped laminae (Fig. 3d), whilst finer-grained varieties show more regular and thinner laminations (Fig. 3c). Pauses in the

Fig. 3a–d. Oncoids from "Rosso Ammonitico inferiore" of M. Lessini area north of Verona. Sections perpendicular to bedding. Note polyphasic character of coating. In **b** a small ammonite shell is included in core of oncoid; darker clasts are iron-stained and bioeroded residual particles. **a, d** Caprino Veronese; **b, c** S. Ambrogio di Valpolicella. **e** Detail of cryptalgal fabric of an oncoid with granular and micritic laminae. Sharp contact in *upper part* represents outer surface of oncoid. Granular lamina (in *middle*) shows a tendency to inverse grading. Caprino Veronese, "Rosso Ammonitico inferiore". **f** Thecideacean brachiopod (pedicle valve) encrusting surface of a growing oncoid and in turn coated by cryptalgal lamination. S. Ambrogio di Valpolicella, "Rosso Ammonitico inferiore"

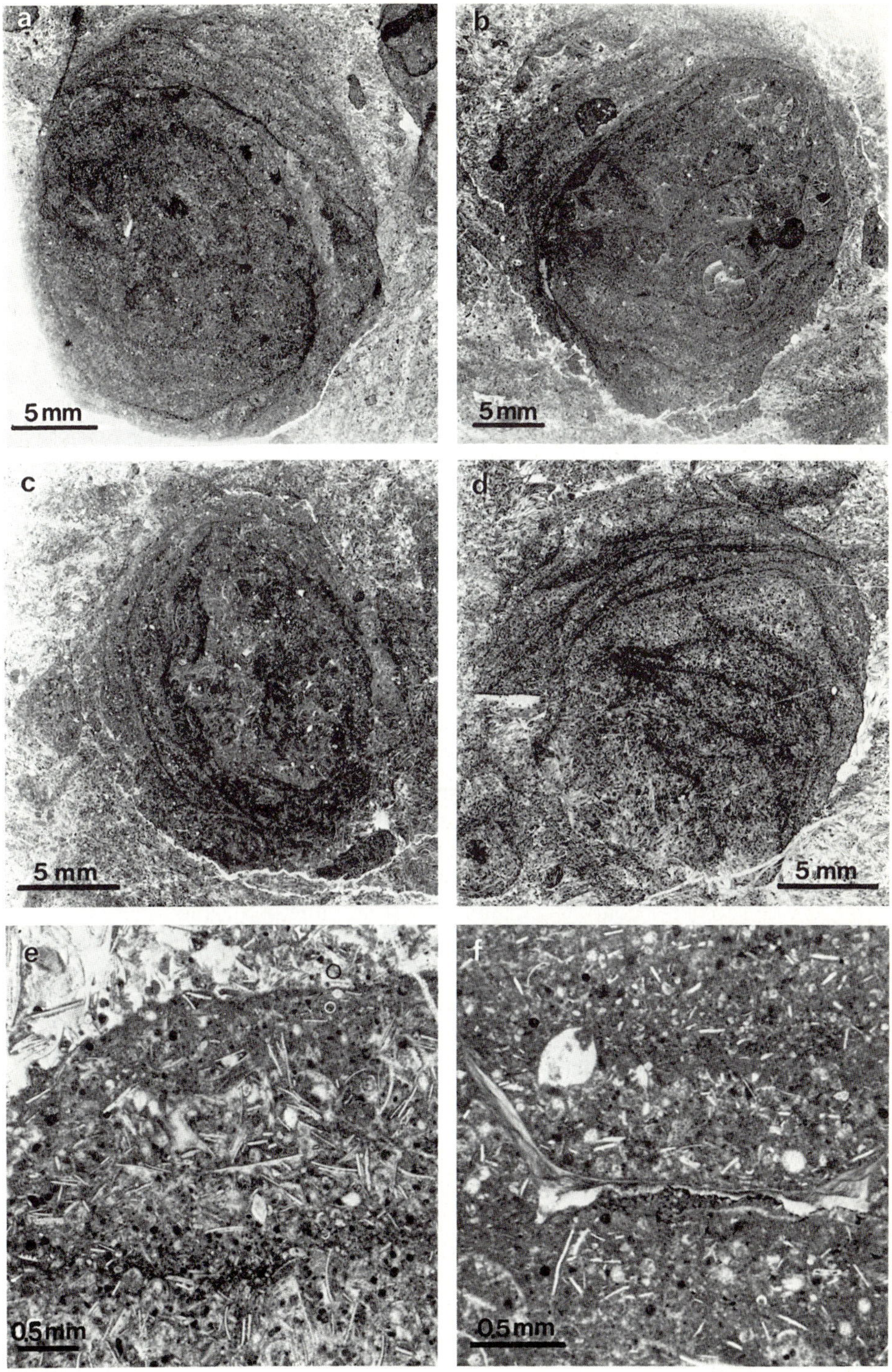
a
5 mm
b
5 mm
c
5 mm
d
5 mm
e
0.5 mm
f
0.5 mm

growth are sometimes marked by low-angle unconformities truncating the previously formed laminae and showing traces of bioerosion and iron-staining (Fig. 2b, d, f, g). Meniscus-filled burrows may occur even within the layered coating (Fig. 2). They may truncate the laminae but most commonly are developed within the granular laminae, sometimes apparently causing an enlargement of the structure from the inside.

Encrusting organisms rarely occur on the envelopes of the oncoids. These are: thecideacean brachiopods (Fig. 3f), serpulid tubes and small ahermatypic corals (Fig. 2c). Moreover very small endolithic borings of penetrative *Thallophyta* and larger borings possibly related to *Polychaeta* occur along surfaces of bioerosion, often associated with iron-staining.

3 Cryptalgal Fabric

The oncoids and stromatolites of the Venetian Rosso Ammonitico are regarded as cryptalgal structures, which originated through the sediment-binding and/or the mineral-precipitating activity of blue-greens. The growth of the layered coating was probably on average very slow and interrupted by long-lasting pauses, as suggested by the iron-stained and bio-eroded micro-unconformities occurring within the coating of some oncoids or stromatolites.

The alternation of texturally different laminae may depend on a balance between growth of blue-greens and periodic or erratic influx of detrital particles (Monty 1976). The different orientation of elongated particles within the micritic and granular laminae suggests a change from bounding to trapping mechanisms. In particular, the edgewise arrangement and inverse grading (Fig. 3e) of particles in granular laminae suggests a dimensional sorting by infiltration of the grains in between the mesh of erect filaments.

The presence of meniscus-filled burrows within the cryptalgal coating suggests that the organisms responsible for them could also behave as inner mat-dwellers and that the coating remained for some time in an uncemented state. However, early cementation of the cryptalgal fabric is suggested by (a) lack of compactional features; (b) sporadic occurrence of encrusting organisms on cryptalgal envelopes and rare boring traces; (c) presence of angular intraformational lithoclasts (mostly common in proximity to discontinuity surfaces where they often look bored and iron-stained) possibly resulting from fabric breakdown caused by the displacive force of crystallization of the growing cement; (d) absence of penetration of the stromatolitic layer by burrows existing in the overlying sediment.

4 Oncolitic-Stromatolitic Facies

The lower member and, in some intervals, also the upper member of the Venetian Rosso Ammonitico in the Trento Ridge area often display decimetric cycles. In each cycle stromatolitic domes grow on a "pavement" of particles of different

Fig. 4. Oncolitic-stromatolitic facies in cross section. Scale bar = 22 cm. "Rosso Ammonitico inferiore" from M. Lessini quarries north of Verona. Polished slab in Basilica of St. Antonio, Padua

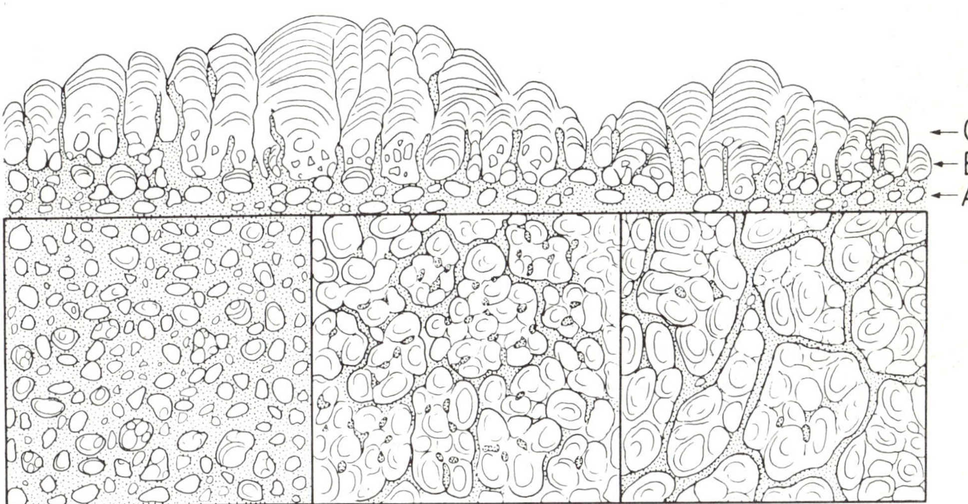

Fig. 5 A – C. Diagram illustrating stromatolitic growth on a "pavement" of oncoids and other particles. Progressive upward reduction of width of interspaces between cryptalgal structures concomitantly with dome growth is illustrated by horizontal sections at three levels of the cycle and leads in the final stage **(C)** to a massive fabric with a polygonal network of narrow interspaces separating almost contiguous dome clusters

Fig. 6. Oncolitic-stromatolitic facies showing three cycles, the lower more massive, the upper two with looser fabric. In lower cycle, a first generation of soft-stage meniscus-filled burrows are later replaced by a second generation of firm-stage burrows mostly confined to junctions of stromatolitic domes (*sparse dots*) and re-excavated in the last phase (*dense dots*). Frequency of exhumed nodules of burrowed limestone, sometimes with oncolitic coatings, should be noted throughout. Lowermost part of "Rosso Ammonitico inferiore", Caprino Veronese. Drawing from a polished slab

nature, such as oncoids, bothroids (subspherical to ovoidal bodies resulting from the aggregation of smaller particles), nodules of burrowed limestone, intraformational lithoclasts, and aggregates of irregular shape (Fig. 4). The most common organization of the stromatolitic layer consists of a cluster-like grouping of adjacent domes (Fig. 5). This cycle may correspond to the soft ground – hard ground rhythm of Hollmann (1964) and Seyfried (1981). It may record oscillatory changes in the factors controlling bottom stability such as frequency and intensity of periodic turbulence at the depositional interface, rate of sediment turnover by grazers and especially burrowers, and sedimentation rate. In the lower layer of the cycle a high rate of sediment turnover (as suggested by the frequency of meniscus-filled burrows), as well as periodic turbulence, would result in instability of the sediment-water interface, allowing periodic particle movement and preventing the binding of the bottom sediment. This may further cause exhumation of the early-cemented nodules. In the second stage of the cycle, on the other hand, lowering of periodic turbulence and possibly decrease in sedimentation rate might lead to progressive binding and stabilization of the depositional interface resulting in a drastic reduction of burrowing activity. The latter is reduced to firm-stage tubular burrows with diameters ranging from a few millimeters to little more than a centimeter, mostly confined along the dome interspaces and finally prevented from further movement by hardening of the cryptalgal fabric (Figs. 5 and 6). In a sense, the cycle is the record of the conflict between burrowing activity and growth of the cryptalgal mats, leading rhythmically to the predominance either of one or the other.

5 Depositional Setting

The presence of sparse thin layers of possibly storm-deposited microcoquinas interbedded with the oncolitic-stromatolitic facies of the Trento Plateau suggests that deposition took place at least locally within the depth of the storm wave base. Moreover, when the transition from the Trento Ridge to the adjacent Belluno Basin (Fig. 1) is analyzed, a basinward decrease in frequency of cryptalgal structures and effects of early submarine cementation can be recognized. It is indicated by the transition from the oncolitic-stromatolitic facies, through a flaser-nodular and sparsely nodular facies (with nodules mostly related to burrowing activity and little or no evidence of nodule exposure at the bottom), to non-nodular cherty facies. Approximately the same transition occurs in the vertical sequences of the Trento Plateau and, as generally agreed, reflects progressive deepening (Sturani 1971).

However, it should be stressed that in some sections of the Trento Plateau area cryptalgal structures may occur within up to the topmost layers of the Rosso Ammonitico directly overlain by the Maiolica limestones (Massari 1981). Moreover, a cryptalgal fabric has been rarely found also outside the ridge area in some condensed intervals (Late Oxfordian to Kimmeridgian in age) of sequences including carbonate turbidites and debrites in their Oxfordian portion (Casso and Claut sections, Massari 1981); these sequences occur in a setting probably corresponding to a step-faulted slope between the Friuli Platform and the Belluno Basin.

This apparently contrasting evidence seems to support the view that the cryptalgal fabric may not significantly be depth-controlled and that it may simply reflect a very low sedimentation rate in an environment affected by oscillatory changes in factors controlling bottom stability. It should be remembered that according to Monty (1971, 1977) the environmental range of cryptalgal structures may be larger than usually believed, especially if communities of blue-greens and bacteria are involved; in particular these consortia might actively grow even in the absence of light.

From this point of view, it can be thought that under similar burrowing rates the generally observed basinward decrease of the rate of early cementation, coupled with the gradual disappearance of the cryptalgal fabric, may essentially reflect basinward slowing down of the energy level at the sediment-water interface and increase of sedimentation rate. In addition, marlier lithologies predominating in the basinal areas may hamper early hardening, since it is known that beyond a certain critical threshold of clay content, early cementation is prevented. A similar transition from hard ground through nodular limestone to soft ooze has been described by Mullins et al. (1980) on Bahamian slopes and related to downslope decrease in the strength of current action; a comparable set of ridge-to-basin changes in the fossil record was described by Elmi (1978), Comas et al. (1981) and Seyfried (1981).

Acknowledgements. The author is indebted to Prof. I. Dieni for critical reading of the manuscript. This work was carried out with the financial support of the Centro di Studio per i problemi dell' Orogeno delle Alpi Orientali (Padova) (directed by Prof. B. Zanettin) of the Italian National Council for Research.

References

Braga JC, Comas MC, Delgado F, Garcia-Hernández M, Jiménez AP, Linares A, Rivas P, Vera JA (1981) The Liassic Rosso Ammonitico facies in the Subbetic zone (Spain). Genetic considerations. In: Farinacci A, Elmi S (ed) Rosso Ammonitico Symp Proc Tecnoscienza, Roma 61 – 76

Comas MC, Oloriz F, Tavera JM (1981) The red nodular limestones (Ammonitico Rosso) and associated facies: a key for settling slopes or swell areas in the Subbetic Upper Jurassic submarine topography (Southern Spain). In: Farinacci A, Elmi S (ed) Rosso Ammonitico Symp Proc Tecnoscienza, Roma 113 – 136

Dommergues JL, Elmi S, Mouterde R, Rocha RB (1981) Calcaire grumeleux du Carixien portugais. In: Farinacci A, Elmi S (ed) Rosso Ammonitico Symp Proc Tecnoscienza, Roma 199 – 206

Elmi S (1978) Bioturbation et tri mécanique, facteurs déterminants dans la genèse des calcaires "noduleux" et des "Ammonitico-Rosso". Soc Géol Fr, 6e Réun Ann Sc Terre, p 157

Elmi S (1981) Classification typologique et génétique des Ammonitico-Rosso et des faciès noduleux ou grumeleux: essai de synthèse. In: Farinacci A, Elmi S (ed) Rosso Ammonitico Symp Proc Tecnoscienza, Roma 233 – 248

Gaetani M (1975) Jurassic stratigraphy of the Southern Alps. A review. In: Squyres CH (ed) Geology of Italy. Earth Sci Soc Libyan Arab Rep, Tripoli, pp 377 – 401

Hollmann R (1964) Subsolutions-Fragmente (Zur Biostratinomie der Ammonoidea im Malm des Monte Baldo/Norditalien. Neues Jahrb Geol Palaeontol Abh 119:22 – 82

Kennedy WJ (1975) Trace fossils in carbonate rocks. In: Frey RW (ed) The study of trace fossils. Springer, New York, pp 377 – 398

Massari F (1981) Cryptalgal fabrics in the Rosso Ammonitico sequences of Venetian Alps. In: Farinacci A, Elmi S (ed) Rosso Ammonitico Symp Proc Tecnoscienza, Roma 435 – 469

Monty CLV (1971) An autoecological approach of intertidal and deep-water stromatolites. Ann Soc Géol Belg 94:265 – 276

Monty CLV (1976) The origin and development of cryptalgal fabric. In: Walter MR (ed) Stromatolites. Dev Sedimentol 20, Elsevier, Amsterdam, pp 193 – 249

Monty CLV (1977) Evolving concepts on the nature and the ecological significance of stromatolites: a review. In: Flügel E (ed) Fossil algae. Springer, Berlin Heidelberg New York, pp 15 – 35

Mullins HT, Neumann AC, Wilbur RJ, Boardman MR (1980) Nodular carbonate sediment on Bahamian slopes: possible precursors to nodular limestones. J Sediment Petrol 50:117 – 131

Seyfried H (1981) Genesis of "regressive" and "transgressive" pelagic sequences in the Tethyan Jurassic. In: Farinacci A, Elmi S (ed) Rosso Ammonitico Symp Proc Tecnoscienza, Roma 547 – 579

Sturani C (1971) Ammonites and stratigraphy of the *"Posidonia alpina"* beds of the Venetian Alps. Mem Ist Geol Min Univ Padova 28:1 – 190

IV.9 Pelagic Oncoids and Ooids in the Middle-Upper Jurassic of Eastern Sardinia

Francesco Massari and Iginio Dieni[1]

1 Geologic Setting

In eastern Sardinia a Middle to Upper Jurassic marine sequence more than 1000 m thick rests unconformably on the Palaeozoic basement and consists in the coastal area of three main lithostratigraphic units: the Dorgali Formation, consisting for the most part of dolostones (Bathonian), the ammonite-bearing Mt. Tului Limestone (Lower Callovian to Uppermost Kimmeridgian) and the Mt. Bardia Limestone (Portlandian to Berriasian). The second unit its gradually replaced westwards by oolitic and oncolitic dolostones, and is therefore incorporated in the first one (Fig. 1A, C). A transgressive trend can be recognized in the lower part of the Jurassic sequence, from Bathonian to Upper Kimmeridgian. During this time span asymmetric carbonate shelf cycles mostly consisting of upward shoaling oolite sequences were formed. They indicate the existence of a wide shallow shelf with a system of tidal oolitic bars periodically prograding seawards and separating an internal shelf lagoon largely open to marine influences from an outer shelf area. The latter during the Early Callovian-Late Kimmeridgian was the depositional setting of the Mt. Tului unit which in its lower part is represented by an oncolitic facies, mainly occurring in the more condensed and pelagic sequences.

Further onwards, from the Portlandian to the Berriasian (Mt. Bardia Limestone), a regressive trend predominates, at first with coral reef limestones and then with Purbeckian facies consisting of tidal flat and freshwater limestones.

The oncolitic facies of the Mt. Tului Limestone is best developed in the Posada-Siniscola area (Fig. 1B), where the presence of ammonites allowed to determine approximately its chronostratigraphical range. Here the lowermost part of the unit consists of a Lower Callovian condensed layer bounded by hard grounds and overlain by the typical oncolitic facies rich in ammonites; this apparently extends in age to Oxfordian and is up to 25 m thick.

2 The Oncolitic Facies

The lowermost layers of the Mt. Tului Limestone (well exposed at the base of the Posada cliff with a thickness of about 3 m) are characterized by a nodular ap-

1 Istituto di Geologia dell' Università, Via Giotto 1, Padova, Italy

Coated Grains (ed. by T.M. Peryt)

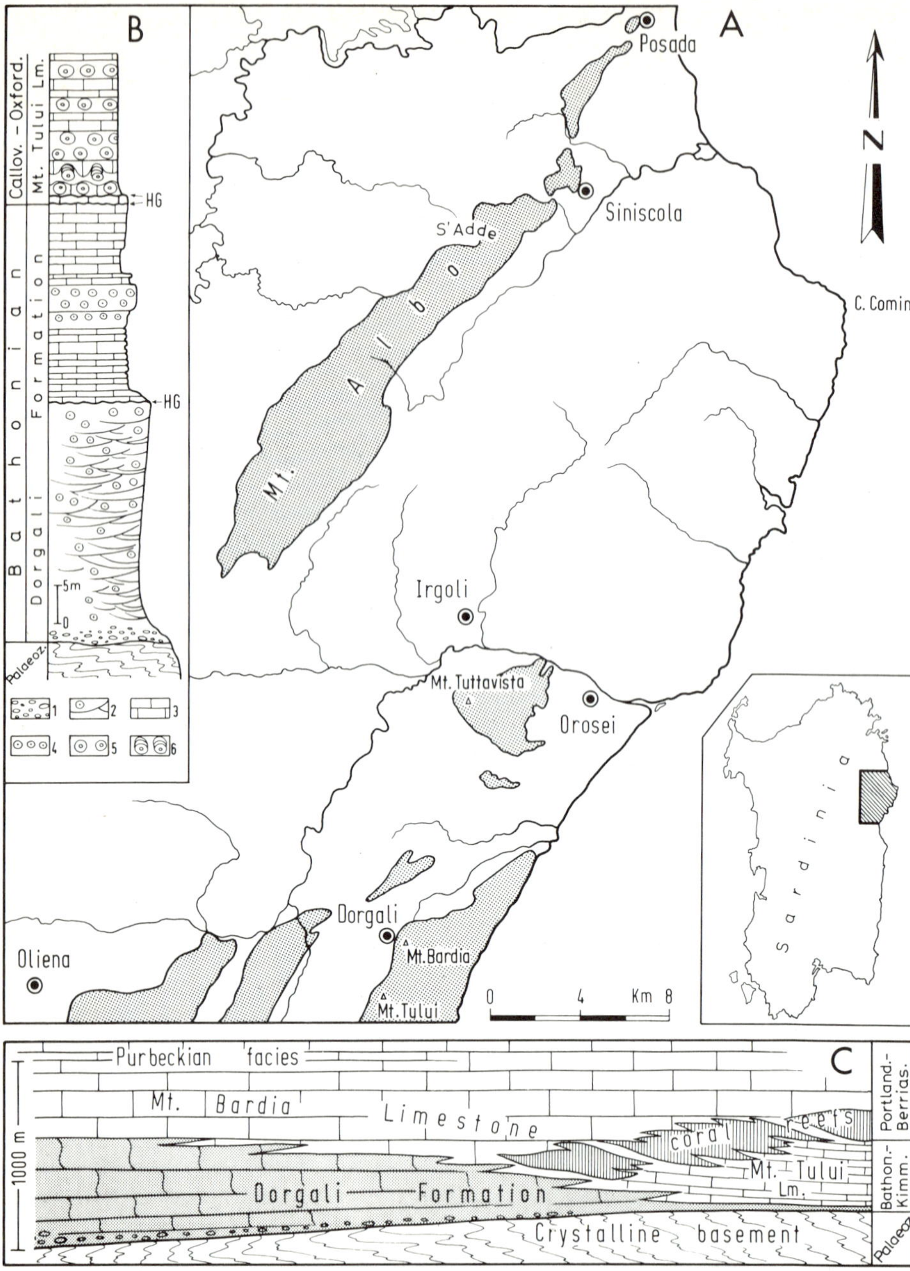

Fig. 1. A Distribution of Mesozoic (Jurassic and Cretaceous) outcrops (*dotted*) in central eastern Sardinia. **B** Jurassic sequence of the Posada cliff. *1* Conglomerates and sandstones; *2* cross-bedded calcareous and dolomitic oolites; *3* outer-shelf limestones and dolostones; *4* dolomitic oolites; *5* calcareous oncolites; *g* domical stromatolites developed on oncoids. **C** The Jurassic lithostratigraphic units of eastern Sardinia

pearance due to close packing of the oncoids which are sometimes capped by stromatolitic domes. Further upwards the oncoids are less abundant, showing nevertheless a tendency to be more or less concentrated along bands parallel to the bedding. The thicker sequences (S'Adde, WSW of Siniscola) are considerably poorer in oncoids than the condensed ones, and contain in places chert nodules resembling the burrow-fill flints of the English Chalk described by Kennedy and Garrison (1974). Toward the south-west the oncoids gradually disappear and tongues of oolitic grainstones occur more and more frequently (for instance in the Mt. Tuttavista of Orosei and Mt. Tului sections) and eventually the Mt. Tului unit merges into oolitic tidal bar sequences.

Macrofossils in the Mt. Tului Limestone include ammonites, belemnites, aptychi, brachiopods and calcitized lychniscosid sponges (the latter mostly found in correspondence with an Upper Kimmeridgian hard ground; Fig. 6h); the shells often appear minutely bored by penetrative Tallophyta. Microfossils are primarily represented by *Globochaete alpina* Lomb., calcispheres (Cadosinidae and Stomiosphaeridae), Nubeculariidae, Spirillinidae, *Globuligerina* cf. *oxfordiana* (Grig.), *Didemnoides moreti* (Dur. Delga), *Gemeridella minuta* Borza and Mišik, *Saccocoma* sp., echinoderm debris and calcitized radiolarians.

A characteristic microtexture occurring in the Mt. Tului Limestone, even in absence of the oncolitic fabric, consists of packstone to fine grainstone abounding in small micritic ooids. These bodies resemble those reported by Jenkyns (1972, 1980) in a number of Oxfordian to Tithonian sequences of the Alpine-Mediterranean domain. As in the cases studied by this author, the Sardinian ooids are almost completely micritic and show sometimes a vague concentric structure outlined by a fine-scale sparite tracery; in addition sometimes a very faint radial structure can be recognized. The core consists of several kinds of biogenic debris, among which *Globochaete alpina,* small globigerinas, cadosinas, echinoderm fragments, etc. (Fig. 5e – g). The general shape and concentric structure of these bodies is so regular that they closely mimic the geometrical perfection of true ooids (Jenkyns 1980) (Fig. 5h). Their diameters (in the 50 to 210 μ range) are decidedly smaller than those reported by Jenkyns (1972) for the Upper Jurassic pelagic "oolites" of western Sicily.

3 Oncoids

The shape of oncoids ranges from spherical/ovoidal to elongate, or lobate, the latter due in most cases to the aggregation of a certain number of smaller bodies. The diameters vary from 9 to 100 mm and most frequently lie in the 30 – 40 mm range. The nucleus may be simple or compound; in the former case it most commonly consists of a roundish nodule showing traces of meniscus-filled burrows (Fig. 2). Along an Upper Kimmeridgian hard ground a rare variety of these nodules occurs; it contains abundant spicules and fragments of the skeleton of Lychniscosida, suggesting that the sponge framework was a preferential site of the bioturbation. In other cases the core is represented by an entire or fragmentary ammonite shell (Fig. 3) or a skeletal piece of lychniscosid sponge. The

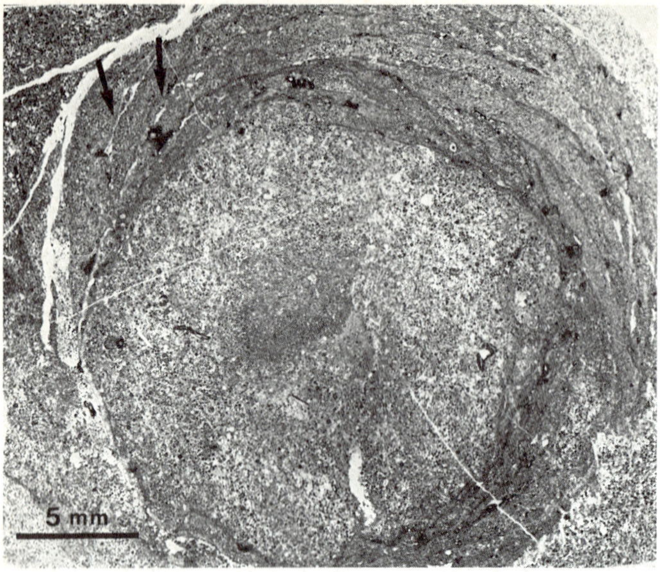

Fig. 2. Oncoid with a nodule of burrowed limestone as nucleus. *Arrows* point to shells of thecideacean brachiopods adhering to the layered envelope. The latter is moreover rich in nubecularian tests (*black spots*)

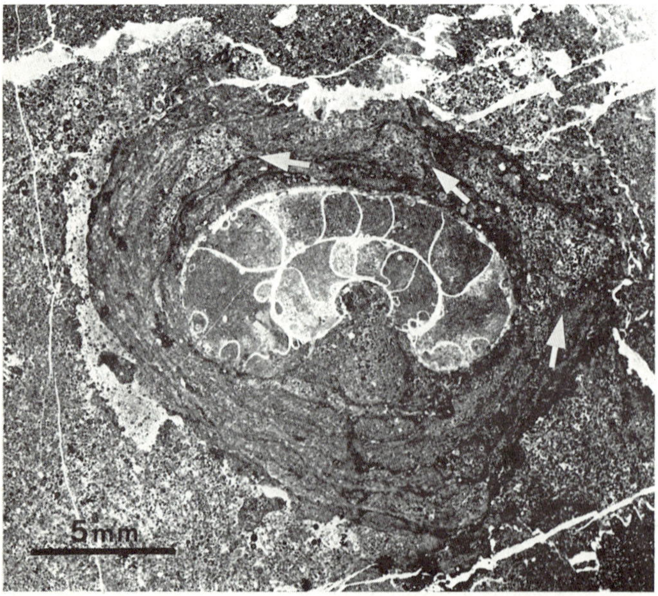

Fig. 3. Oncoid with an ammonite shell as nucleus. The coating was crossed by soft-stage burrows related to inner-mat dwellers which deformed the layered fabric causing local protrusions (*arrows*)

nucleus of some elongate oncoids consists of a tubular burrow (sometimes with geopetal vadose-like filling) probably related to a firm-stage generation following a soft-stage suite characterized by meniscus-filled burrows. Sporadic presence of microborings and of encrusting organisms around the nodules of burrowed limestones, as well as occurrence of oncolitic coating suggests an early cementation of the nodules in the shallow subsurface followed by exhumation due to periodic

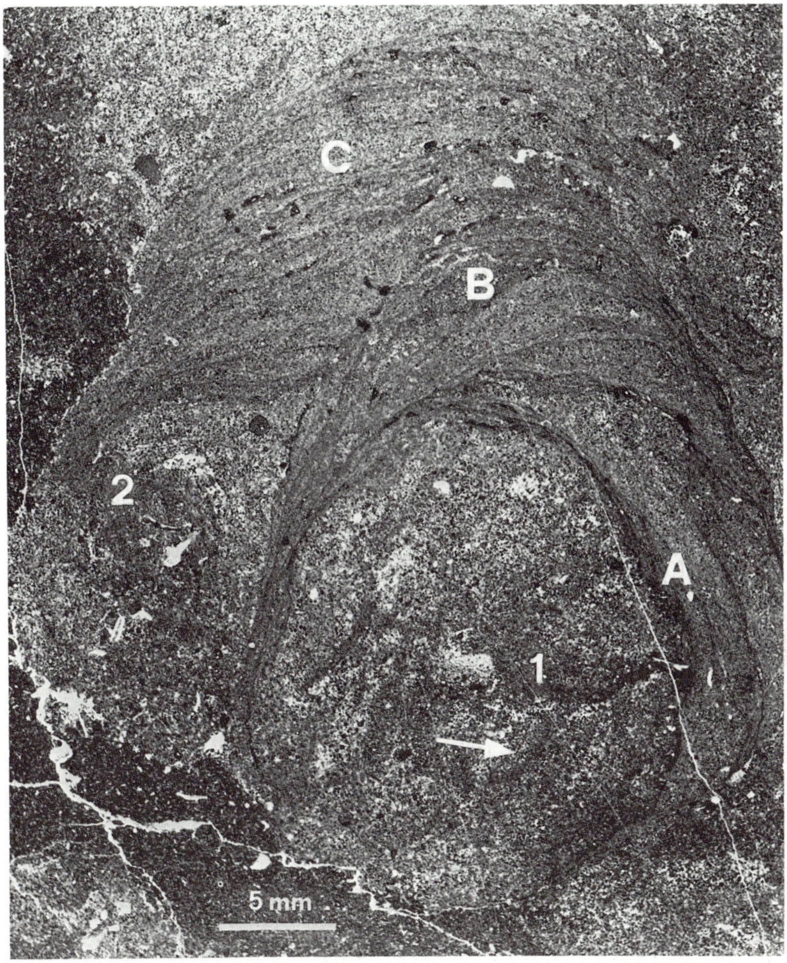

Fig. 4. Compound body displaying a multistage cryptalgal coating on nodules of burrowed limestone. The latter were affected by early cementation in the shallow subsurface prior to exhumation. Nodule *1* was at first coated by the envelope *A*, then displaced and covered by the envelope *B*. Finally nodule *2* was aggregated (possibly during a temporary burial in the shallow subsurface) and the resulting compound body was re-coated. The large dimensions attained in this stage did not allow any further displacement of the body; this was then capped by stromatolitic dome *C*, which is particularly rich in nubecularian tests. It should be noted that the coating phases *A, B, C* may in turn consist of minor subphases with slightly different orientation

turbulence or sediment turnover by burrowers or grazers (Elmi 1978, 1981; Dommergues et al. 1981).

Oncoids with compound nucleus are multistage bodies including a number of components, such as smaller oncoids, nodules of burrowed limestone or coprolites, intraformational lithoclasts, "older" residual particles showing traces of bioerosion, ammonite shells and so on. Individual components may be bound together either by oncolitic coatings, or by bioturbated limestone affected by early hardening. Repeated phases of burrowing, cementation of nodules in the shallow subsurface, exhumation and coating can sometimes be envisaged (Fig. 4). Consequently the size of particles may increase up to a maximum diameter of about 100 mm. The larger the size of particles, the more irregular and lobate is the outline, reflecting lack of mobility beyond a certain critical dimensional threshold. The most complicated history is usually displayed by oncoids associated with hard grounds or occurring within the most condensed layers.

The coating consists for the most part of an alternation of micrite and granular laminae (Fig. 5d) associated with agglomerates of *Nubecularia reicheli* Rat (Figs. 4 and 6e) occurring either as discontinuous envelopes of limited thickness or as localized irregular masses protruding outwards from the oncoid surface. The granular laminae mostly consist of pelagic ooids and secondarily of several other particles such as pellets, tiny echinoderm debris, *Globochaete alpina,* small foraminifera and some unrecognizable debris, set in a spar cement or in a matrix of micrite more or less affected by aggrading neomorphism. The particles within the granular laminae are often arranged with inverse grading (Fig. 5d), suggesting a trapping mechanism accompanied by a dimensional sorting through the mesh of a filamentous mat. The micritic laminae on the other hand are much thinner and contain only small-sized debris. The layered fabric is thought to be cryptalgal in origin and to be genetically related to blue-green communities or to consortia of blue-greens and bacteria (Monty 1976). The laminae of the cryptalgal coatings are irregular, tapering out and most commonly arranged in sets enveloping only a side of the growing body (Fig. 3). When the particle rests on the bottom, growth may take place on its upper side and continue until the body is set in motion by periodic turbulence or by sediment turnover due to burrowing and/or grazing activity; then, after some phases of partial growth, the body may appear completely enveloped by a number of discontinuous sets of laminae, more or less overlapping each other.

Fig. 5. a Part of the coating of an oncoid cortex with thecideacean brachiopods (*arrows*) adhering to the cryptalgal envelopes. **b, c** Burrows related to inner mat-dwellers: **b** Outward protrusion of the cortical part of the coating around the burrow (*upper part*) suggests excavation in a stage predating the hardening of the fabric. A small ahermatypic coral adhered to the oncoid surface after cementation took place. **c** Burrow lined by coarser particles agglutinated along the walls (?). A small attached coral is visible (*left*). **d** Layered cryptalgal fabric. The *median* granular lamina shows a distinct inverse grading. **e – g** Pelagic micritic ooids with nucleus consisting of *Globochaete alpina* Lomb. **(e)**, *Cadosina lapidosa* Vogler **(f)** and *Globuligerina* cf. *oxfordiana* (Grig.) **(g)**. **h** Pelagic micritic ooids. Note the faintly visible inner structure consisting of a very regular concentric and sometimes even radial pattern. Single ooids may be clustered together to form lumps. Numbers above scale bars are in millimeters

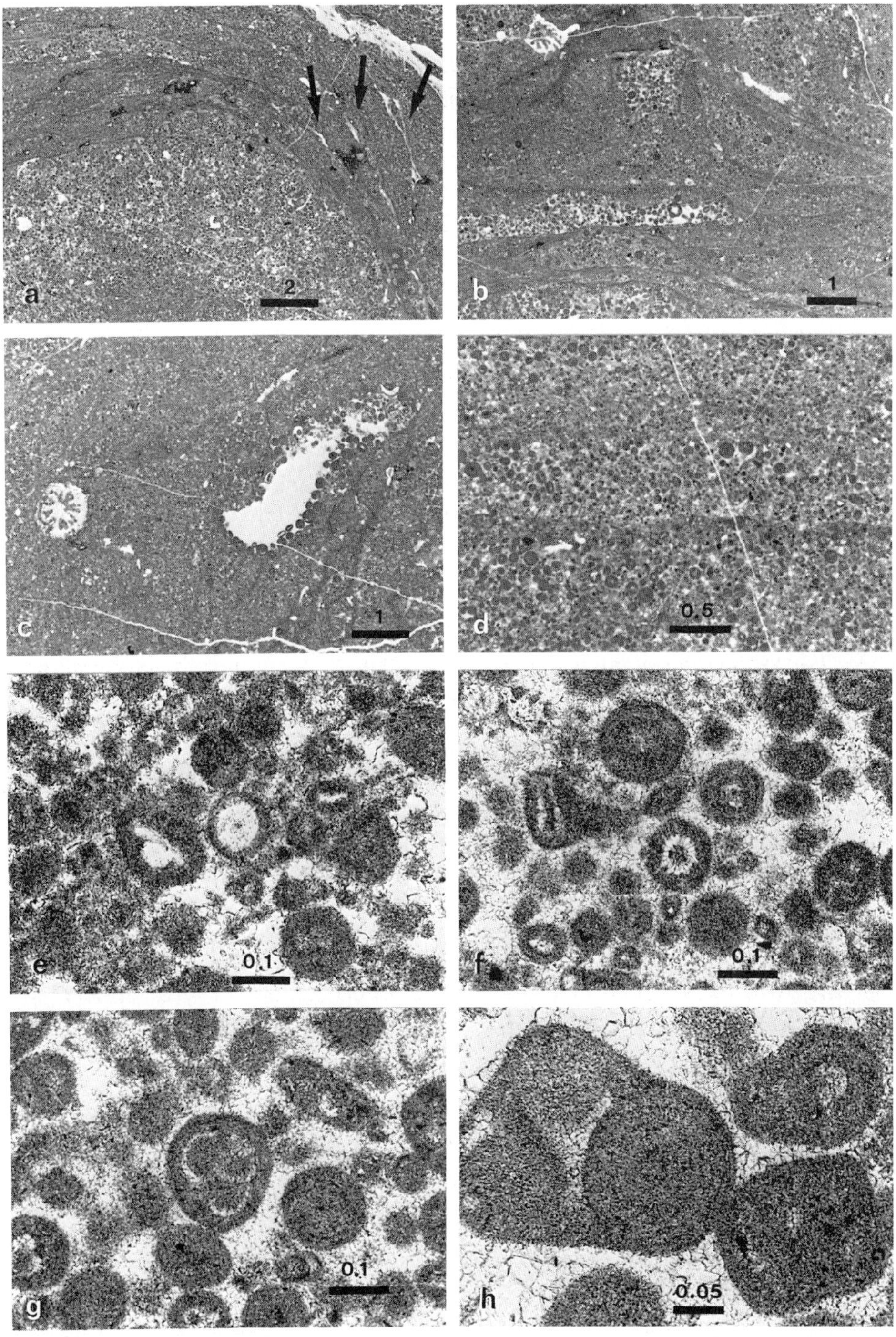
a
2
b
1
c
1
d
0.5
e
0.1
f
0.1
g
0.1
h
0.05

Burrows displaying in places meniscus-filled traces were sometimes observed even within the cryptalgal fabric (Figs. 3 and 5b, c) and are mostly developed parallel to the layering though in places they run transversely to the laminae, truncating them. Sometimes the layered envelope appears deformed (without truncation of the laminae) by the passage of inner mat-dwellers in a stage predating the hardening of the fabric. This may lead to outwards wave-like protrusions of the cortical part of the coating (Figs. 3 and 5b). Sparse borings and a variety of encrusting organisms along the envelopes (Fig. 6) indicate however an early cementation of the cryptalgal fabric.

4 The Environmental Setting

Several aspects of the examined oncolitic limestones suggest a relatively shallow water depositional environment. Among them the most significant are: (a) the local occurrence of clean-washed grainstones in addition to predominating packstones; (b) the abundance of pelagic ooids, which are considered by Jenkyns (1972) as indicative of paleodepths in the range of some tens of meters; (c) finally the presence of bryozoans and *Nubecularia* among the encrusting organisms. It should be mentioned in this regard that according to Rat (1953, 1966) the nubecularias seem to be confined to the shallow warm seas dominated by carbonate sedimentation and are mostly widespread in the laminarian zone.

Dragastan (1970, 1980) thinks that nubecularian foraminifera and blue-green algae were closely linked by a commensal relationship and that the textural characters of the fabric were controlled by intrinsic mechanisms ruling the dynamics of this ecosystem.

The facies illustrated in this paper is similar to the Upper Jurassic oncolitic-stromatolitic limestones of the Prebetic zone described by Garcia-Hernandez et al. (1981) and considered to be genetically linked to the activity of blue-green algal communities and deposited in a shallow outer shelf environment.

Acknowledgement. This work was supported by the Italian Consiglio Nazionale delle Ricerche, contr. no. 81.01898.05.

Fig. 6a–g. Encrusting organisms involved in the oncolitic coatings: **a** Small ahermatypic coral. **b** Crinoid (massive holdfast). **c** Thecideacean brachiopod (complete with both valves) in turn encrusted (*right*) by *Nubecularia reicheli* Rat. **d** *Placopsilina* sp. and (*arrow*) small bryozoan. **e** *Nubecularia reicheli* Rat. **f** Organisms *incertae sedis* ("pseudoostracods" of Samuel, Borza and Köhler 1972). **g** Tubicolous worm (*Serpula* sp.). **h** Skeletal fragment of calcitized lychniscosid sponge with cryptalgal coating. *Nubecularia reicheli* occurs both around the former and within the latter. Numbers above scale bars are in millimeters

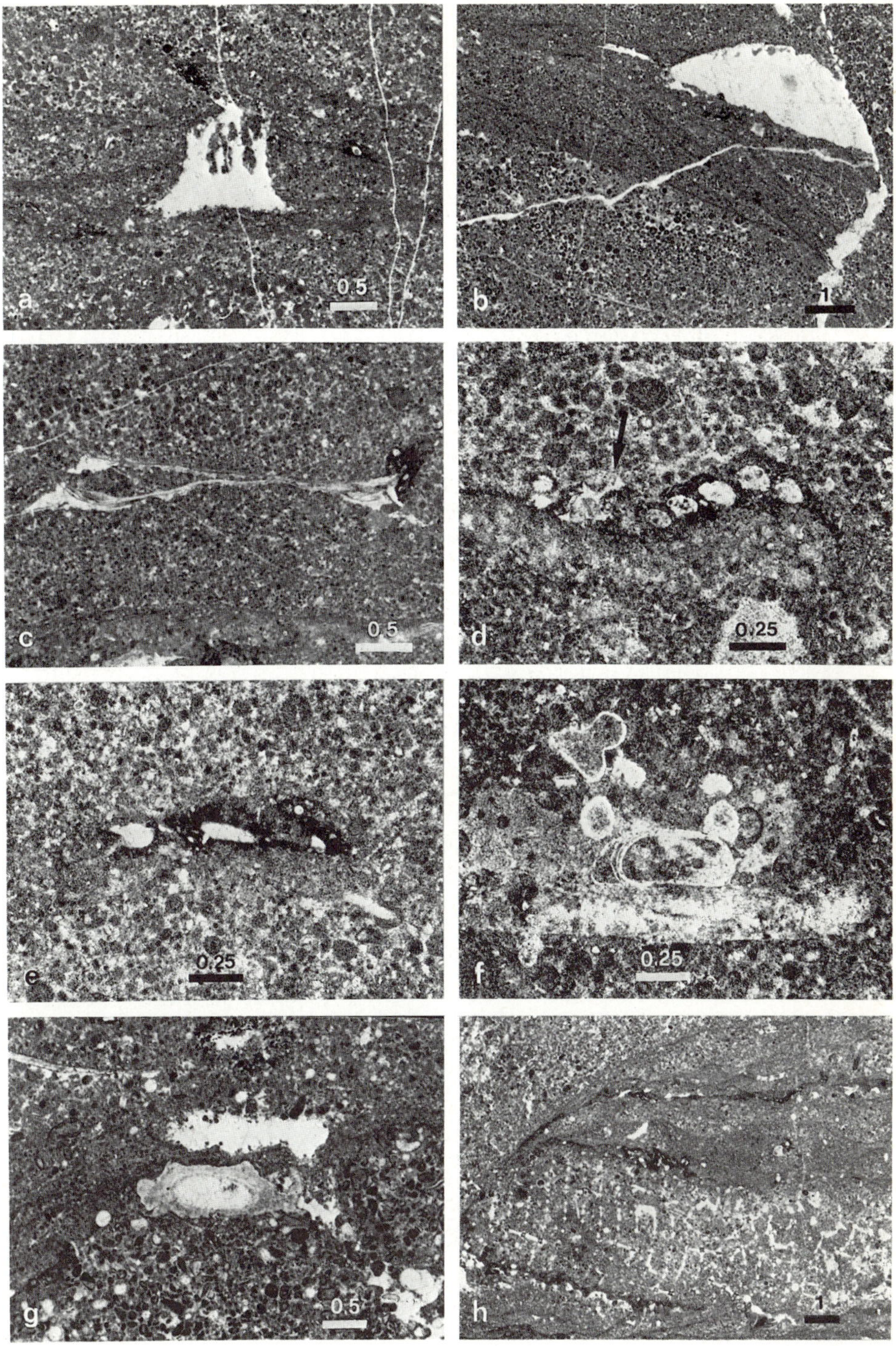
a
0.5
b
1
c
0.5
d
0.25
e
0.25
f
0.25
g
0.5
h
1

References

Dommergues JL, Elmi S, Mouterde R, Rocha RB (1981) Calcaire grumeleux du Carixien portugais. In: Farinacci A, Elmi S (ed) Rosso Ammonitico Symp Proc Tecnoscienza, Roma, pp 199–206

Dragastan O (1970) "Micro-oncolithes" dans le Jurassique supérieur de la vallée du Bicaz (Carpates orientales, Roumanie). Bull Soc Géol Fr 11:655–659

Dragastan O (1980) Calcareous algae of the Mesozoic and Tertiary of Romania. Acad Rep Social România, Bucuresti (in Rumanian)

Elmi S (1978) Bioturbation et tri mécanique, facteurs déterminants dans la genèse des calcaires "noduleux" et des "Ammonitico-Rosso". Soc Géol Fr, 6e Réun Ann Sc Terre, p 157

Elmi S (1981) Classification typologique et génétique des Ammonitico-Rosso et des facies noduleux ou grumeleux: essai de synthese. In: Farinacci A, Elmi S (ed) Rosso Ammonitico Symp Proc Tecnoscienza, Roma pp 233–249

Garcia-Hernandez M, Lopez-Garrido AC, Oloriz F (1981) Etude des calcaires noduleux du Jurassique supérieur de la zone Prébétique (Cordillères Bétiques, SE de l'Espagne). In: Farinacci A, Elmi S (ed) Rosso Ammonitico Symp Proc Tecnoscienza, Roma, pp 419–434

Jenkyns HC (1972) Pelagic "oolites" from the Tethyan Jurassic. J Geol 80:21–33

Jenkyns HC (1980) Tethys: past and present. Proc Geol Assoc 91:107–118

Kennedy WJ, Garrison E (1974) Morphology and genesis of nodular chalks and hardgrounds in the Upper Cretaceous of southern England. Sedimentology 22:311–386

Monty CLV (1976) The origin and development of cryptalgal fabric. In: Walter MR (ed) Stromatolites. Dev Sedimentol 20. Elsevier, Amsterdam, pp 193–249

Rat P (1953) Sur la découverte de calcaires pseudo-oolitiques à Nubéculaires dans le Bajocien de Bourgogne. C R Som Soc Géol Fr: 228–230

Rat P (1966) *Nubecularia reicheli* nov. sp., Foraminifère constructeur de fausses oolithes dans le Bajocien de Bourgogne. Eclogae Geol Helv 59:73–85

Samuel O, Borza K, Köhler E (1972) Microfauna and lithostratigraphy of the Paleogene and adjacent Cretaceous of the middle Vah Valley (West Carpathians). Geol Ust D Stura, Bratislava

IV.10 Depositional Environments of Some Upper Jurassic Oncoids

Kapila Dahanayake [1]

1 Introduction

Oncolites occur abundantly in the Upper Jurassic limestones of French Jura mountains, Paris Basin, Aquitain basin and also in the Spanish province of Valencia. On the basis of preliminary observations of oncolites of these regions, seven representative and well exposed oncoid-bearing carbonate sequences from the French Jura were studied in detail for stratigraphical and petrological observations (Fig. 1).

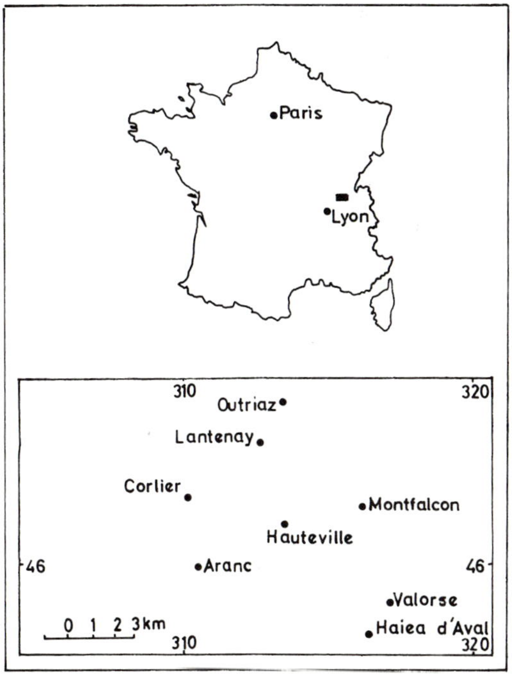

Fig. 1. Location of oncoid-bearing carbonate sequences – French Jura

1 Department of Geology, University of Sri Lanka, Peradeniya, Sri Lanka

Coated Grains (ed. by T. M. Peryt)

2 Internal Structure of Oncoids

The author employs the term "oncoid" to describe calcareous particles of recognizable algal and organic character with maximum diameters ranging from 1 mm to 5 cm and having morphologies varying from amoebodial to spherical form.

Most of the oncoids studied consist of laminations. However, some bodies do not show the laminated structure and instead, an algal mesh sometimes encrusting other laminated oncoids as well as detrital matter is observed.

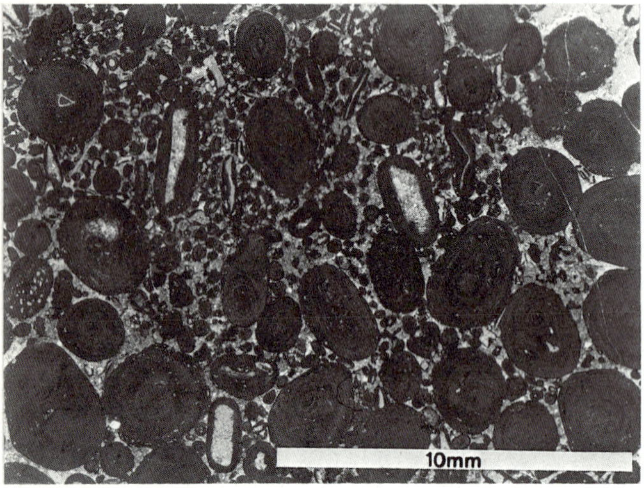

Fig. 2. Oncoids of type *I*. (Modified after Dahanayake 1977)

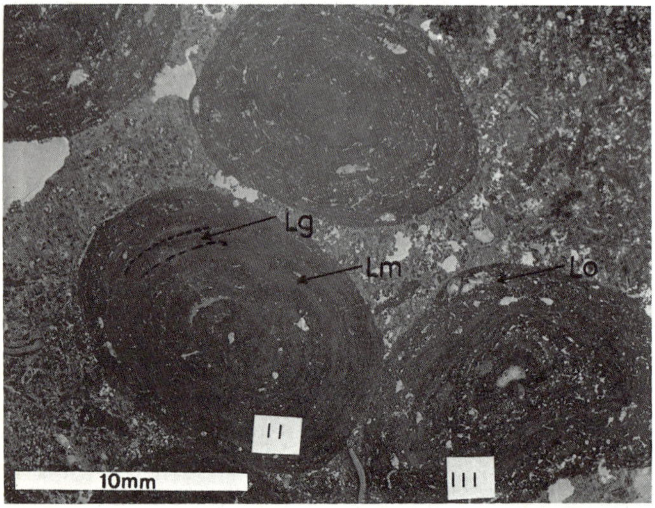

Fig. 3. Oncoids of types *II* and *III*. *Lm* Micritic laminations, *Lg* grumose laminations, *Lo* organism bearing laminations. (Modified after Dahanayake 1977)

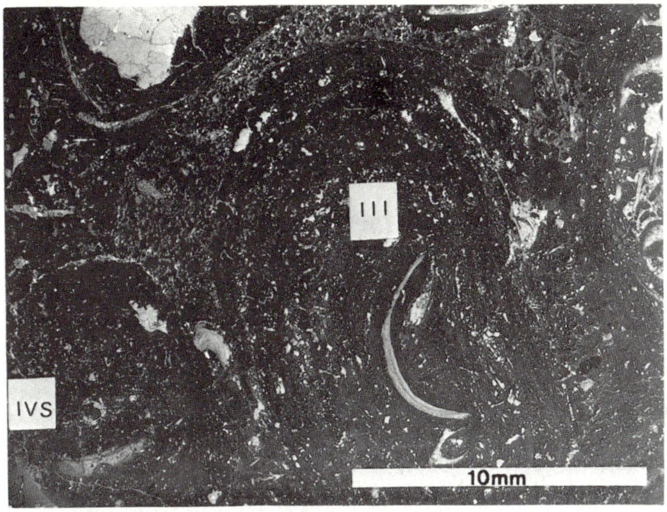

Fig. 4. Oncoids of types *III* and *IV S.* (Modified after Dahanayake 1977)

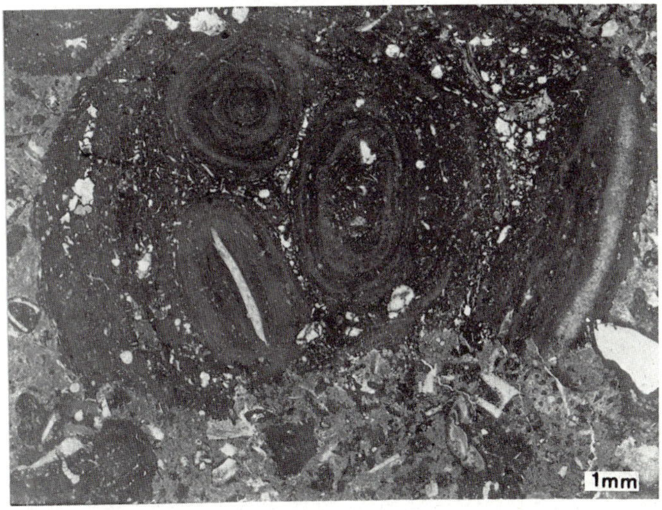

Fig. 5. Oncoid of type *IV C.* (After Dahanayake 1977)

The laminations are seen to be continuous in some oncoids whereas, they are discontinuous in others. The laminations are sometimes difficult to observe owing to obliteration by diagenetic processes such as grain diminution, aggrading metamorphism and dolomitization.

Three distinct types of laminations are recognized in the Upper Jurassic oncoids considered here. On the basis of textural and compositional variations, the following types of laminations are distinguished: (1) micritic, (2) grumose, (3) organism-bearing. The thickness ranges for the laminations are as follows: micritic laminations: 25 – 250 μm; grumose laminations: 200 – 500 μm; organism-bearing laminations: 200 μm to 1 mm. The laminations are described elsewhere (Dahanayake 1977). Depending on (a) the combinations of different types of laminations and (b) the presence or absence of laminations, four oncoid types can be recognized. Viz. I (Fig. 2), II (Fig. 3), III (Fig. 4) IV S and C (Figs. 4 and 5) (see Dahanayake 1978 for detailed discussion).

3 Sequential Position of Different Types of Oncoids

Table 1 summarizes the results obtained by detailed petrological studies of oncolites from different stratigraphic sections (Dahanayake 1978) (Fig. 6).

4 Relations of Oncoids with Other Allochems

Oncoids occur in stratigraphic relationship with other allochems such as ooids, bioclasts and pellets in sections studied in the French Jura mountains. One such section is that of Lantenay (Fig. 7) which shows three major formations consisting of several members (a, b, c...). Distinct sequences of allochems are observed within the members (Figs. 7 and 8).

4.1 Ooid Sequences

Five ooid sequences are recognized in the member (a_1) of Formation 1. Cross-bedding is observed on these oolitic sequences.

Table 1. Results from detailed petrological studies

Type of oncoid	Other allochems associated	Nature of cement/ matrix	Texture	Sequential position
I Elliptical	Oriented bioclasts Intraclasts	Sparite	*P*	Underlie type I Rounded facies
I Rounded	Pellets Rounded bioclasts	Micrite Sometimes dolomitized	*P, W, M*	Overlie type I Elliptical facies
II Rounded	Pellets Rounded bioclasts Type I oncoids	Dolomitized vuggy	*P*	1. Occur together in same facies with type I 2. Overlie type I
III Rounded	Type IV S & C oncoids (rarely) Type II oncoids (rarely)	Micrite, partly or Wholly dolomitized	*W*	1. Occur together in same facies with type II 2. Overlie Type II
IV S	Type I (rarely)	Dolomitized micrite	*M, P*	Overlie types III and IV facies
IV C	Type III oncoids	Dolomitized micrite	*W*	1. Occur together with type III facies 2. Overlie type II

P packstone, *W* wackestone, *M* mudstone

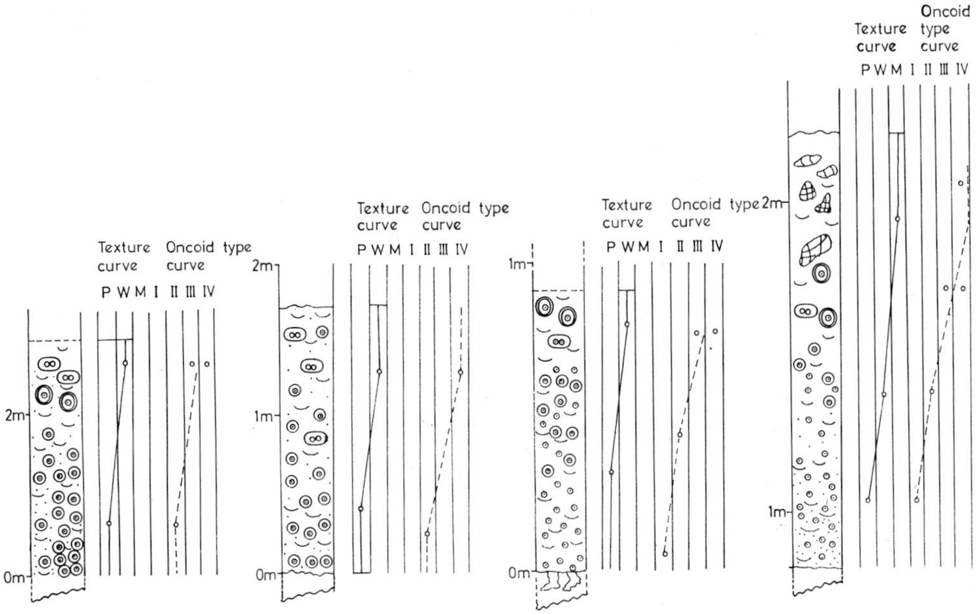

Fig. 6. Oncoid-bearing sequences showing the relative positions of different types of oncoids, French Jura

The lithofacies of all ooid sequences show a similar trend in the distribution of textures. Thus packstones and grainstones are found at the base of sequences while wackestones and mudstones occur towards the top. The grain size of allochems shows a trend decreasing from the bottom to top of the ooid sequences. At the bottom of sequences, bigger ooids are associated with bioclasts oriented parallel to stratification in sparitic cements displaying grainstone/packstone textures. Towards the top in the overlying lithofacies, smaller ooids and bioclasts appear in micritic matrices showing the wackestone and mudstone textures. In the latter texture bioturbation is common and ooids occur as pockets ("en nuage").

4.2 Bioclast-Pellet Sequences

Six distinct sequences are found in the Members (a_2) (b_2) of Formation 2. At their bases, these sequences commonly show intraclasts, bioclasts and entire organisms such as polyp corals and these are bound by a sparitic cement in grainstone/packstone textures. The pellets are rare at the base although in the overlying lithofacies they occur abundantly in bigger sizes with smaller bioclasts in micritic matrices displaying wackestone and mudstone textures.

4.3 Ooid-Pellet Sequences

Three sequences have been noted in the Member (e_2) of Formation 2. All these sequences show at their bases textures of packstone with the presence of pellets,

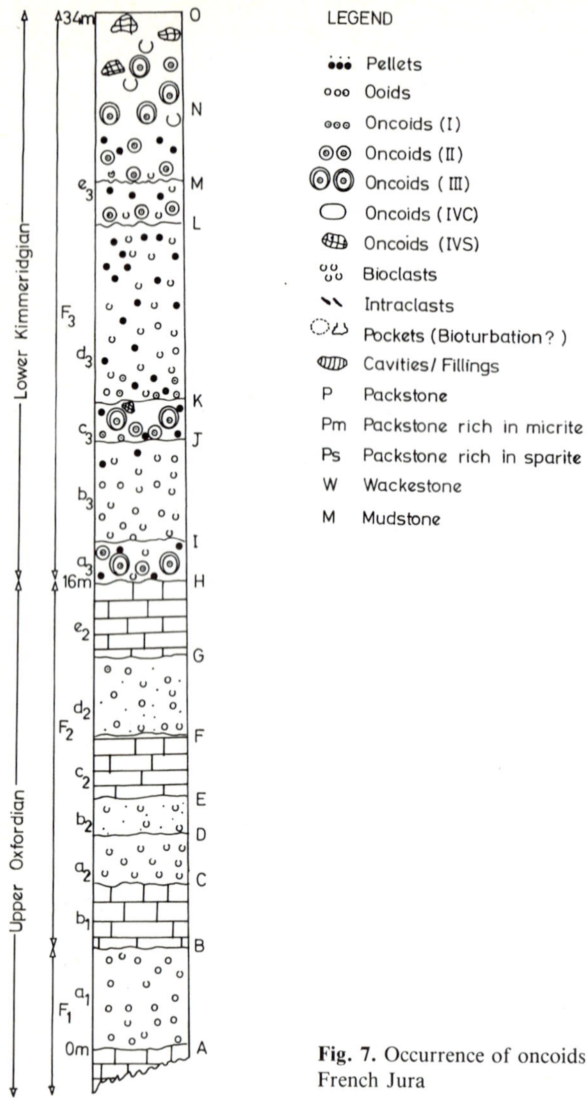

Fig. 7. Occurrence of oncoids and other allochems – Lantenay, French Jura

ooids and bioclasts together in micritic matrices. Towards the top in the overlying lithofacies these allochems show a general decreasing trend in size and they display wackestone textures. At the top, the texture is mudstone where ooids and pellets occur in pockets or "en nuage" – possibly due to bioturbation.

4.4 Oncoid Sequences

Altogether nine oncoid sequences are seen in Formation 3. In a general manner oncoid type I facies underlie type II facies and they both show packstone textures

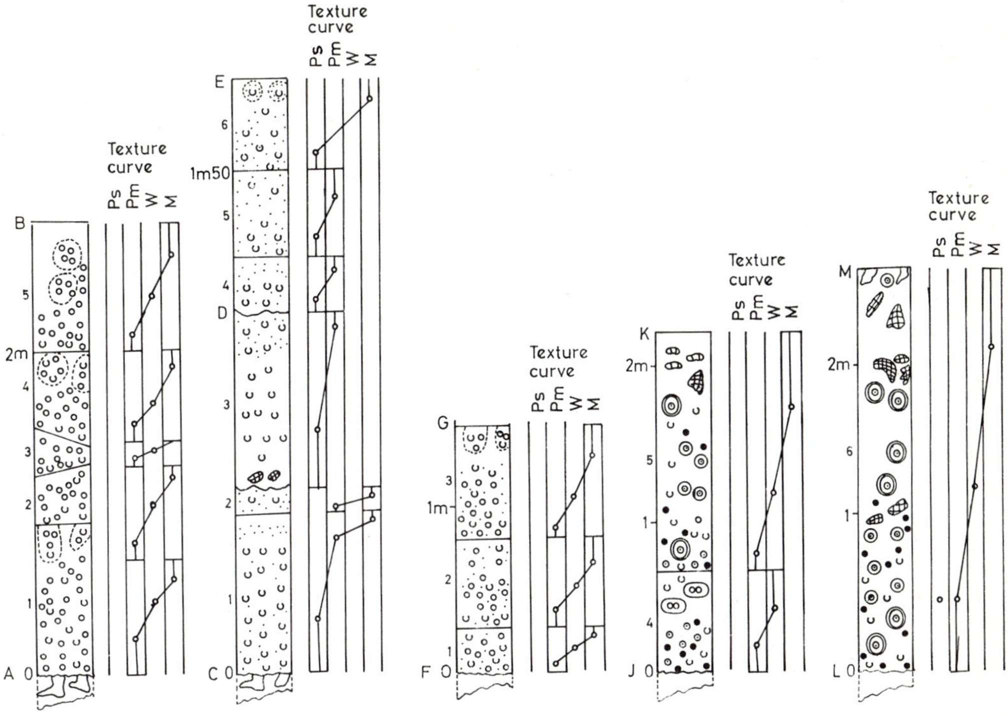

Fig. 8. Different allochem sequences from Lantenay section. *AB* ooids, *CDE* bioclasts and pellets, *FG* ooids and pellets, *JK* and *LM* oncoids

and occur at the base of sequences. Facies of type III bodies overlie those of type II in wackestone textures. Type IV facies overlie type III or II facies in a given sequence and they show wackestone/mudstone textures. Facies of type I, II and III show frequent associations of large pellets while type IV facies show the presence of relatively unbroken bioclasts, entire organisms and pellets are rare. Oncoid facies generally show effects of dolomitization and towards the top of the stratigraphic section these effects are more pronounced.

5 Relative Depositional Energies of Allochems

Presence of small-scale cross bedding with normal crystalline ooids indicates agitated depositional conditions. The association of such ooids with oriented bioclasts in sparitic cements displaying grainstone/packstone texture seems to confirm the above reasoning. On the other hand, the presence of superficial ooids and small pellets in micritic matrices with packstone/mudstone texture suggests relatively less agitated conditions for these allochems. Similar ooid-pellet associations have been observed in low energy zones of Persian Gulf (Loreau and Purser 1973). The absence of pellets in mostly sparitic coarse grained bioclasts facies at the bottom of sequences and their dominant presence towards the top in micritic matrices suggest lower energy for pellets. The association of larger pellets with oncoids in micritic matrices and the frequent dolomitization of such matrices point to low energy depositional milieu which had characterized oncoid facies.

As shown in the Lantenay section, the Upper Oxfordian period had been characterized by the deposition of ooids, intraclasts and pellets. As indicated by texture curve of allochem sequences, the depositional energies have oscillated often during the said period. The high energy-low energy P-W-M cycle may have represented such oscillations. At higher levels of energy, ooids, intraclasts and coarse bioclasts could be deposited whereas smaller pellets and bioclasts would deposit at lower energies.

As indicated by texture curves, the Lower Kimmeridgian period also had been characterized by frequent oscillations of energy perhaps with lower energy levels of longer durations as shown by thicker sequences. Dry conditions also would have prevailed thus causing dolomitization. Oncoids of type I and II which generally occur at the base of sequences perhaps indicate relatively higher energies than those of type III and IV found towards the top. The large pellets which are mostly associated with types I and II indicate that they had been deposited during the higher energy periods of the P-W-M cycles.

The allochem particles occurring in the Lantenay section can be arranged according to decreasing depositional energies as follows; ooids, and small pellets, large pellets and oncoids.

6 Depositional Environments of Oncoids

Type I oncoids with characteristic continuous micritic laminations indicate continued movement or agitated conditions as for ooids. The discontinuous organism-bearing laminations reflect short periods of non-agitation which had facilitated the growth of *Bacinella.* The oncoids of type II on the other hand show more discontinuous laminations possibly indicating many such periods of tranquility perhaps of a longer duration. The sporadic return of more agitated wet conditions could form micritic laminations on type II bodies. The dominant and continuous nature of organism-bearing laminations in type III bodies suggests relatively mild agitations occurring frequently in an otherwise calm milieu where prolific algal growth was possible. As for the oncoids of type IV, the periods of non agitation or mild agitation probably were of longer durations facilitating algal growth. Thus *Bacinella irregularis* Radoičić as well as other algae and organisms grow abundantly and *Bacinella* was capable of encrusting debris and even living organisms and also oncoids of any type in its mesh. Such growth only could have been disturbed by intermittent agitations causing fragmentation of the mesh; and ultimate depositions of type IVS and/or IVC bodies, which are thereby formed.

The petrological studies reveal that type I elliptical bodies are frequently associated with oriented bioclasts and intraclasts in a sparitic cement. The cement as well as the intraclasts point towards a high energy depositional milieu where constant flooding had removed any available mud. On comparison with modern environments such conditions exist in sub-tidal zones of a tidal flat. The fact that type I rounded oncoids are found in micritic matrices indicates relatively lower energies for them. The elliptical and rounded nature of these bodies could be attributed to the bidirectional nature of currents prevalent in intertidal channels. The type II oncoids on the other hand with their dolomitized matrices indicate a milieu favourable for dolomitization such as the upper intertidal zone or supratidal zones of a modern tidal flat. The presence of birdseye structures has also been observed in type II oncoid laminations (Dahanayake 1977). The dominant presence of pellets in type II facies attests to the shallow nature and mild energy of the milieu. The type II bodies resemble SS-R oncoid bodies described by Logan et al. (1964) in shallow intertidal milieu. The type III bodies often occur in mud-supported wackestone texture with unbroken shells and angular bioclasts showing a depositional milieu of lower energy than type I or II. The association of type III oncoids with IV C bodies in dolomitized matrices could be attributed to their deposition perhaps in supratidal depressions. Occasional flooding could bring type I and II from lower lying tidal zones to redeposit them with type III oncoids. Types I and II bodies so transported could eventually be encrusted by algal meshes during calm periods. Subsequent fragmentation and rounding during flooding could form type IV C oncoids.

References

Dahanayake K (1977) Classification of oncoids from the Upper Jurassic carbonates of the French Jura. Sediment Geol 18:337 – 353

Dahanayake K (1978) Sequential position and environmental significance of different types of oncoids. Sediment Geol 20:301 – 316

Logan BW, Rezak R, Ginsburg RN (1964) Classification and environmental significance of algal stromatolites, J. Geol 72:68 – 83

Loreau JP, Purser BH (1973) Distribution and ultrastructure of Holocene ooids in the Persian Gulf. In: Purser BH (ed) The Persian Gulf. Springer, Berlin Heidelberg New York, pp 279 – 328

IV.11 Carnian Oncolites of the Northern Calcareous Alps and the Drau Range (Austria)

Gottfried Tichy [1]

Oncolites are very characteristic in the Carnian (Middle Triassic) of the North Alpine facies. In the Northern Calcareous Alps oncolites are very abundant in the Raibl Bed. But they only occur south of the line between the localities Lech am Arlberg – Heiterwand – Wetterstein (foreland) – Guffert – Kaisergebirge – Rauschberg – Staufen – Bad Ischl – Steyerling – Windischgarsten (see also Jerz 1966, Schuler 1968, Grottenthaler 1978). As in the Drau Range oncoids occur at the beginning of the Raibl sequence. The so-called "Grenzoolithbank" or "Raibler Grenzlager" is regarded by Schneider (1953) to be the top of the Wetterstein Formation (which is also Carnian: Cordevolian in age), but it is the very base of the Raibl Member according to the author. Oncolites often occur at the transition from carbonate facies into a clastic facies. The most characteristic "Sphaerocodium Limestone" of the Northern Calcareous Alps is within the first shale layer unlike the Drau Range where this layer is missing (Jerz 1966, p. 53; Schuler 1968, p. 33). Further oncolites appear at the top of the first shale bed. There, in the Wimbach-Valley the oncolites are 7 m thick (Schuler 1968). Far to the south-east the thickness rapidly decreases.

Several oncolite beds can be found within the second shale layer. Occasionally there are also some within the Opponitz Limestone. The occurrence of oncolites in the Stubai Mesozoic shows similarities to the southern facies of the Northern Calcareous Alps.

In the Drau Range there are two distinct layers of oncolites: the "Raibl Grenzbank" between Wetterstein Limestone and Raibl Bed (Figs. 1 and 2), and at the base of the third shale horizon, the so-called "Groboolith" (Figs. 3 and 4). At the boundary between Wetterstein Limestone and Raibl Beds an obvious facies change happens. The distribution of the "Raibl Grenzbank Oolith" is strictly connected with the underlying Wetterstein Limestone of the "Bleiberg-facies" (Kraus 1969, p. 107). This oncolite bed is very thick (1.2 m) at Mežica (Yugoslavia) and so it is the same 95 km to the east, at Bleiberg (near Villach) few kilometers to the north, however, at Rubland (4 km north of Bleiberg) it is only 15 cm thick. Towards the West and the North of the Drau Range the oncolites gradually pass into a clayay-quartzitic layer rich in pyrite. The Grenzbank Oolite has abundant fossils as foraminifers, bivalves and some rare ammonites: *Car-*

1 Universität Salzburg, Institut für Geologie und Paläontologie, Akademiestraße 26, A-5020 Salzburg, Austria

Coated Grains (ed. by T.M. Peryt)

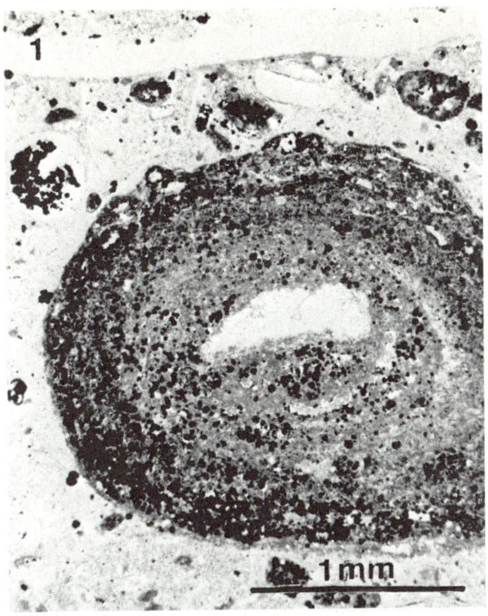

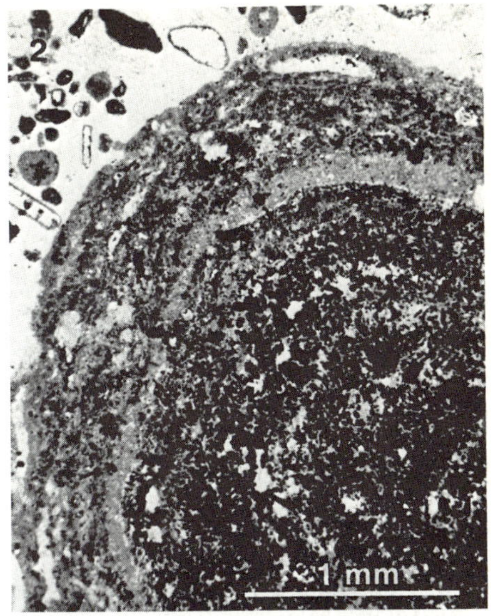

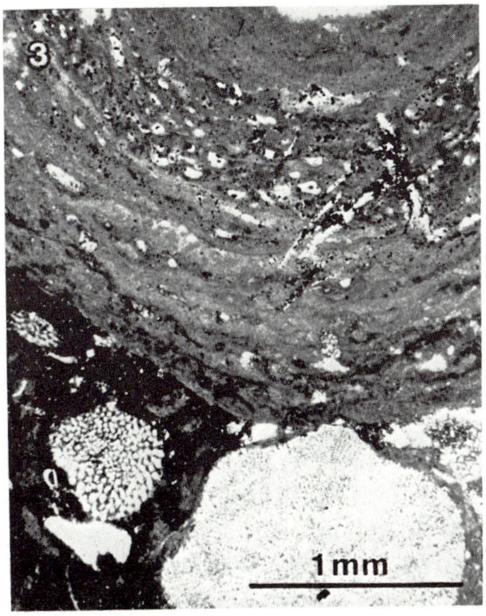

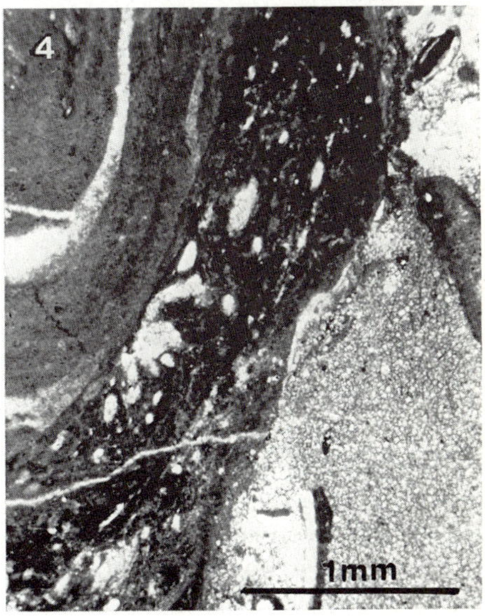

Fig. 1. Oncoid, strongly pyritized, from the bottom of the "First Shale Layer" ("Kiesoolith") of Bleiberg, Carinthia (Raibl Formation, Carnian: Upper Triassic). Sediment: oolitic grainstone

Fig. 2. Oncoid, strongly pyritized (*black*), the same locality as in Fig. 1

Fig. 3. Girvanella Oncoid, from the bottom of the "Third Shale Layer" ("Groboolith") of Bleiberg, Carinthia (Raibl Formation, Carnian: Upper Triassic). The oncoids have been called "Sphaerocodium" originally

Fig. 4. Detail of an other oncoid from the same locality as in Fig. 3

nites floridus Wulfen (Kraus 1969, Tichy 1972). The higher content of the pyrite reflects the decay of the fossils in a reducing environment where pyrite is formed.

The most typical lithologic key bed in the Carnian sequence is the oncolite bed at the base of the third shale layer (Figs. 3 and 4) which can be traced all over the Drau Range exept in the Lienz Dolomites (Kraus 1969). This layer is also very rich in fossils such as sponges, hydrozoans, bivalves (oysters), bryozoans, echinoids, and crinoids. At Dobratsch and Tschöcklnock area the oncolites are best developed and very thick (3 – 3.5 m). This thickness decreases towards the west and the north where the oncoids are mixed together with ooids. Further to the west the oncolites gradually pass into ragged resediments and in the Lienz Dolomites oncolites are not to be found. The reason why there are no oncolites in the northern parts of the Calcareous Alps or in the northern parts of the Drau Range is the great influx of terrigenous material which prevents the settlement of encrusting organisms and probably the reduced saturation of $CaCO_3$ caused by the rivers as it was pointed out by Harsch (1970). Oncolites are also missing in the Hallstatt facies and in the Central Alpine Mesozoic. In the Hallstatt facies the Raibl Member is very thick. There the cherty limestones intercalated in the thick shale indicate a deposition in deeper water as it is supposed to be in the facies with oncolites. A similar facies character exists between the Hallstatt facies (e.g., Torrener Joch) and Aflenz facies and the southernmost facies of the Northern Calcareous Alps. In the Radstadt-Tauern (Central Alpine Mesozoic) the sections show similarities with the facies at the southernmost part of the Northern Calcareous Alps as it can be studied at Hochkönig or Blünbachtal (Salzburg). Neither oncolites nor oolites can be found in the classical area of Raibl (Cave del Predil/Italy) where there is a very thick (800 m) sequence of shale.

The depositional conditions of the oncolites during Carnian time were a warm and shallow well-lit shelf sea. The rich fauna indicates an access to open water. For a short time there was a well-aerated oxygen-rich environment. When the water circulation stopped and the sea became shallower carbonate facies began. The mud banks are almost bare of fossils (only some stromatolites and very few bivalves) and dried up periodically. The rapid change in facies from Wetterstein Limestone to Raibl Beds indicates a deepening of the water and beginning of the influx of terrigenous material. The same happend at the border between the dolomite layer and the third shale layer in the Drau Range when the carbonate mud flats sank down. The influence of a well aerated and well-lit sea water established a good environment for fossils at first, the oncolites are small, but their sizes increase constantly towards the top where they became smaller again and finally vanished when the shaly facies dominated.

References

Grottenthaler W (1978) Die Raibler Schichten der Nördlichen Kalkalpen zwischen Salzach und Pyhrnpaß. Mitt Ges Geol Bergbaustud Wien 25:11–33

Harsch W (1970) Zur Sedimentologie und Paläogeographie der Raibler Schichten (Ober-Trias) im Westteil der Nördlichen Kalkalpen. Z Dtsch Geol Ges 121:253–272

Jerz H (1966) Untersuchungen über Stoffbestand, Bildungsbedingungen und Paläogeographie der Raibler Schichten zwischen Lech und Inn (Nördliche Kalkalpen). Geol Bavarica 56:1–112

Kraus O (1969) Die Raibler Schichten des Drauzuges (Südliche Kalkalpen). Lithofazielle, sedimentpetrographische und paläogeographische Untersuchungen. Jahrb Geol B A 112:82–152

Schneider HJ (1953) Lagerstättenkundliche Untersuchungen am oberen Wettersteinkalk östlich der Loisach. Diss Univ München

Schuler G (1968) Lithofazielle, sedimentologische und paläogeographische Untersuchungen in den Raibler Schichten zwischen Inn und Salzach (Nördliche Kalkalpen). Erlanger Geol Abh 71:1–60

Tichy G (1972) Beitrag zur Triasfauna von Bleiberg (Gailtaler Alpen, Kärnten) mit besonderer Berücksichtigung der Megalodontiden. Diss Philos Fak Univ Wien

IV.12 *Girvanella* Oncoids from Middle to Upper Triassic Allochthonous Boulders of the Dolomite Alps, Northern Italy

Kevin T. Biddle[1]

1 Introduction

Several types of coated grains occur within the Middle to Upper Triassic carbonate rocks of the Dolomite Alps. These structures have been mentioned by a number of authors and have been interpreted in various ways. Leonardi (1967) discussed laminated, spherical nodules, which he called Sferocodi, that were presumably formed by encrustations of the alga *Sphaerocodium*. These nodules are common in biostromal accumulations near the crests of many carbonate buildups in the Dolomites (Leonardi 1967, p. 217). Bosellini and Rossi (1974) provided several excellent illustrations of vadose pisoids from the Dolomites. In their interpretation, the vadoids occur in thin soil horizons and in areas that were subjected to Triassic subaerial exposure.

Unfortunately, much of the in-place carbonate section of the Dolomite Alps was severly altered by Late Triassic dolomitization. As a result, much of the original textural information contained in the carbonates has been lost. Many of the in-place coated grains are now only vaguely laminated forms whose origins are difficult to decipher. Unless the coated grains are correctly identified, interpretation of their depositional environments, and hence buildup histories, may be incorrect.

The purpose here is to describe a group of well preserved coated grains from the western Dolomites. These grains are primarily *Girvanella* oncoids. They are found in buildup-derived boulders that occur in Triassic basinal sedimentary sequences. Because of their basinal position, many of the boulders are still limestone and have retained most of their original textures (Biddle 1981).

2 Geologic Setting

The boulders that contain the oncoids crop out along the continuation of the Denti di Terrarossa ridge, near the southeastern edge of the Alpe di Suisi area (Fig. 1).

---

1 Exxon Production Research Company, Box 2189, Houston, TX 77001, USA

Coated Grains (ed. by T.M. Peryt)

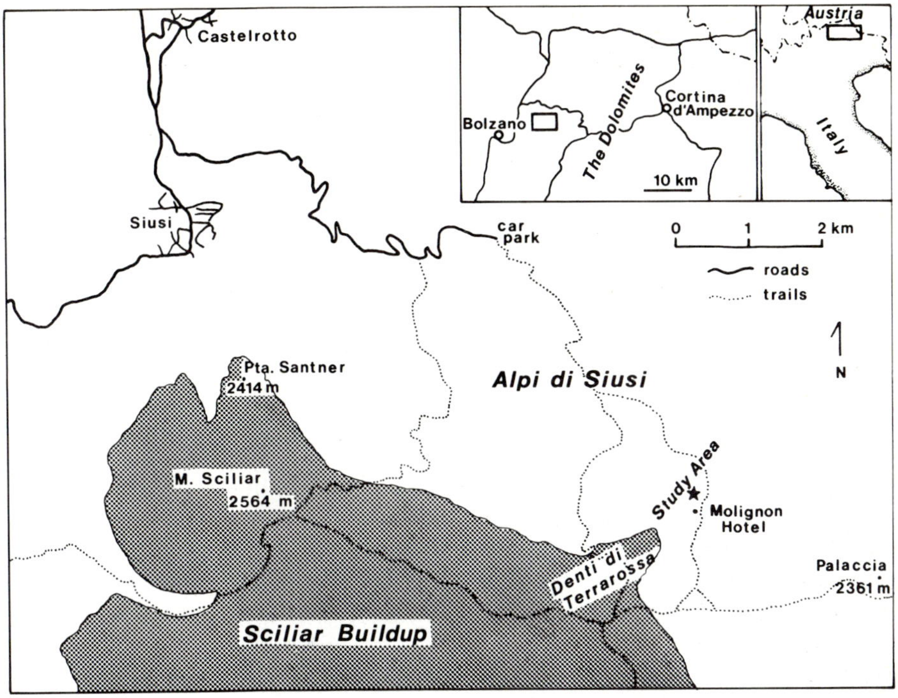

Fig. 1. Location map of the study area. Stipled pattern represents carbonate rocks of the Triassic Sciliar buildup

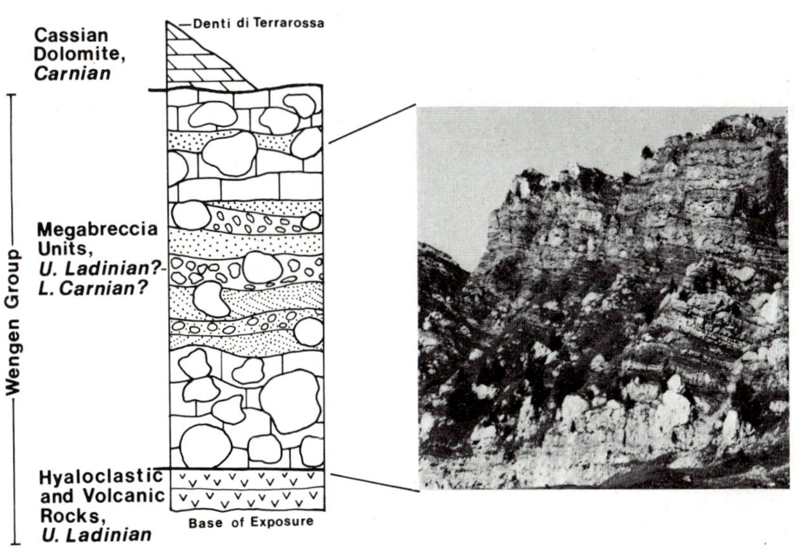

Fig. 2. Stratigraphy of the Denti di Terrarossa ridge. Nomenclature and ages are after Assereto et al. (1977), Bosellini et al. (1977) and Viel (1979). *Inset* shows bedding style of the megabreccia units. Largest boulders are about 5 m across

The ridge is formed by conglomeratic units and associated sedimentary rocks that contain the limestone boulders discussed here. These units are known informally as megabreccia (Bosellini et al. 1977). They were deposited in a basinal setting in front of the Triassic Sciliar buildup (Bosellini et al. 1977, Biddle 1979). The megabreccias sit depositionally on volcanic rocks and are overlain by prograding wedges of Cassian Dolomite (Fig. 2). The exact age of the mega-

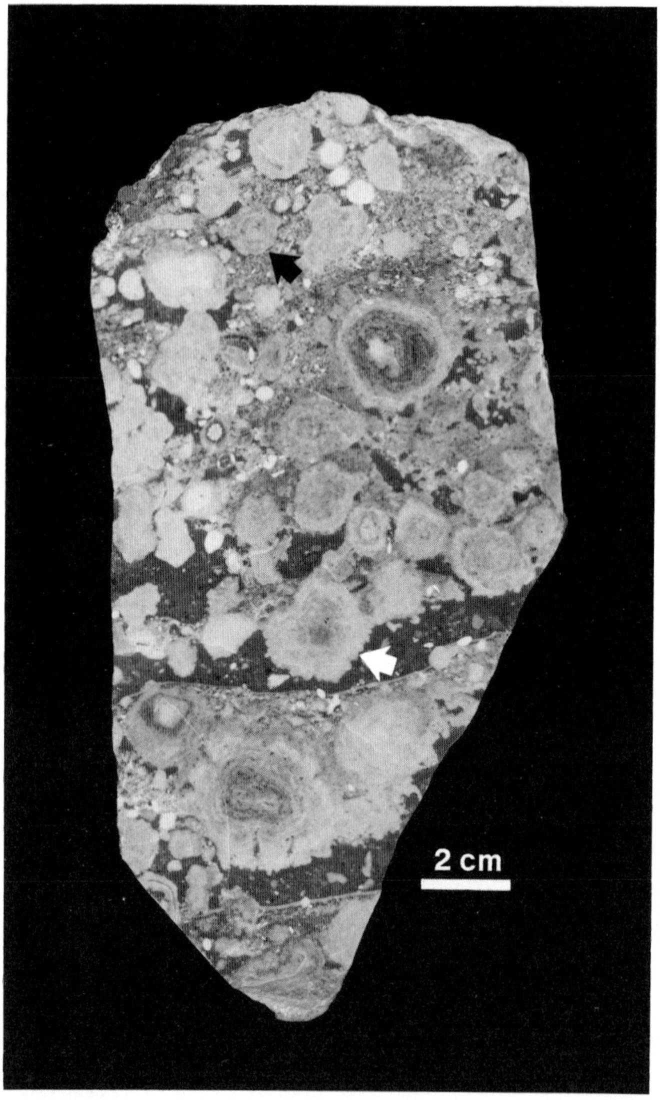

Fig. 3. Polished slab of oncolitic limestone from the Denti di Terrarossa ridge. The slab is a grainstone with large, filled solution cavities (*dark gray areas*). Where in contact with the cavities, the oncoids have irregular boundaries (*white arrow*). Oncoids not modified by solution have smoother, more regular edges (*black arrow*)

breccias is uncertain, but they are probably upper Ladinian to lower Carnian in age (Bosellini et al. 1977).

3 Lithologic Description

Several textural types of limestone boulders occur in the megabreccias, but only those that are oncolitic will be described. Oncolitic boulders are common at the base of the ridge above the Molignon hotel (Fig. 1).

All of the oncolitic boulders examined are grainstones (Fig. 3). The matrix grains are dominately micritized clasts, peloids, and abundant fossil fragments. These grains range from fine sand to granule and pebble size. The oncoids are an order of magnitude larger. In some boulders, the oncoids are tightly packed and touching (Fig. 3); in others they are scattered throughout the matrix.

Many of the oncolitic boulders contain solution cavities now filled with crystal silt, large carbonate clasts, and volcanic grains. Some of these filled cavities are quite large, and give the rock a peculiar-looking fabric (Fig. 3).

4 The Oncoids

The oncoids are formed by numerous millimeter-thick laminae, which are visible in polished slabs as slight color changes. The laminae are wavy and discontinuous, but combine to create a uniform coating around a central core.

The oncoids are either roughly spherical or elliptical (Fig. 3); those with elongate nuclei are elliptical, while those with more equant centers are spherical. Where identifiable, the nuclei are usually fossil fragments. Algal debris (*Solenopora, Cladogirvanella*), echinoid fragments, and mollusk shells are the most common nuclei.

The oncoids range in size from about 0.5 cm to 3 – 4 cm in diameter.

4.1 Microstructure

In thin-sections, the laminated structure of the oncoids is quite clear (Fig. 4).

The laminae are formed primarily by varying concentrations of flexous, micritic tubules 10 to 15 microns in diameter (Figs. 4b and 5). The size and morphology of these tubules identify them as *Girvanella*.

The *Girvanella* tubules are well preserved in laminae where they are loosely packed and surrounded by sparry calcite (Fig. 5). In areas with dense concentrations of *Girvanella*, however, individual tubules can be difficult to resolve. Occasionally, *Girvanella* can be traced into apparently structureless patches of micrite.

Two other distinct lamina types occur in the oncoids. These are micritic laminae without traces of algal filaments and micritic laminae with vertically oriented, spar-filled voids about 30 – 50 μm across. The origins of these laminae

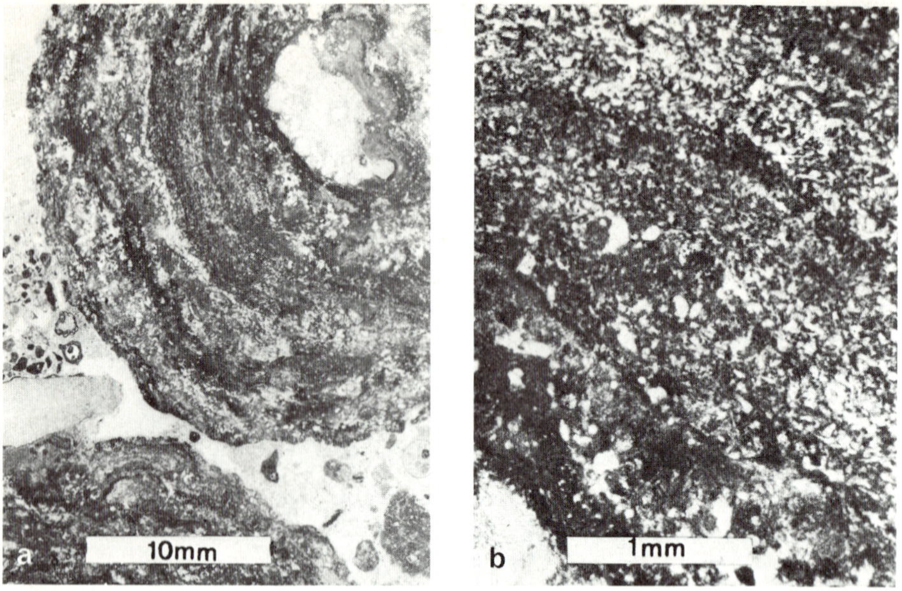

Fig. 4a, b. Microstructure of *Girvanella* oncoids from the Denti di Terrarossa ridge. **a** Large oncoid formed by numerous algal laminae coating a central fossil fragment. **b** Enlarged view of laminae shown in **a**. The laminae are defined by varying concentrations of micritic algal filaments

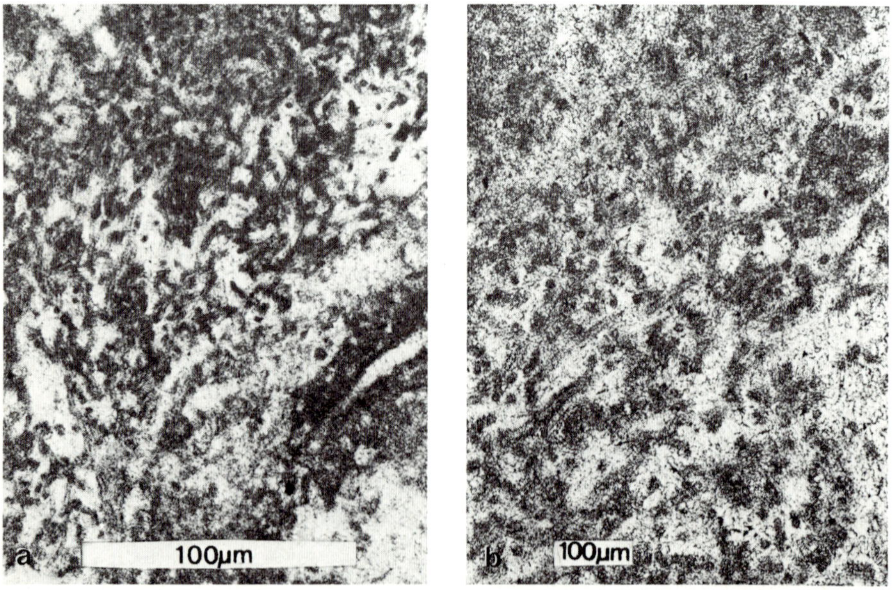

Fig. 5a, b. Well-preserved *Girvanella* from the oncolitic bounders at the northern end of the Denti di Terrarossa ridge

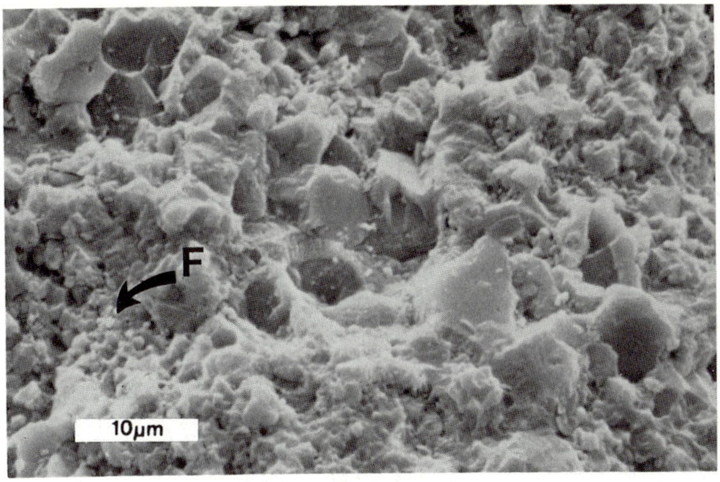

Fig. 6. SEM photograph of *Girvanella* filament and surrounding calcite microspar. The *Girvanella* tubule is poorly defined, but is formed by concentrations of micritic calcite (*F*). The microspar occurs as interlocking polyhedral grains generally less than 10 µm across

are enigmatic, but are probably related to the activities of algae. They are interlayered with well-preserved *Girvanella* laminae.

4.2 Diagenesis

The boulders that contain the oncoids have a complex diagenetic history. Many of them have been subjected to several phases of cementation, partial dissolution, conversion of original mineralogy to low-magnesium or ferroan calcite, and slight dolomitization (Biddle 1979). The oncoids, however, have not been adversely affected by these diagenetic episodes; their internal structure is generally well preserved. The oncoids are now uniformly low-magnesium calcite. The calcite occurs as micrite or as interlocking, polyhedral crystals between 5 and 10 µm across (Fig. 6). A later generation of coarser sparry calcite fills scattered solution vugs.

5 Discussion and Conclusions

The preservation of *Girvanella* in the oncoids described here is generally quite good, and appears to be related to three factors. These are: (1) possible initial composition of the *Girvanella* tubules, (2) early cementation by sparry, low-magnesium calcite, and (3) removal of the oncolitic boulders from the Triassic carbonate buildups before they were substantially dolomitized.

The initial composition of the micritic *Girvanella* walls is uncertain, but the degree of preservation and small crytal size suggest that not much alteration has taken place. It is possible that the wall-forming micrite was originally high- or low-magnesium calcite. Recent *Girvanella*-like algal filaments with calcite walls have been described by Riding (1977) and Krumbein and Potts (1978). Peryt (1980) suggested an original low-magnesium calcite composition for Triassic *Girvanella* from the Upper Muschelkalk of Germany.

The finely crystalline spar that surrounds the *Girvanella* formed early in the history of the oncoids. This cement predates the solution cavities in the boulders, which were formed during episodes of Triassic exposure. The effect of the sparry cement was to reduce the porosity and permeability of the oncoids, inhibit fluid migration through them, and decrease the opportunities for further diagenetic change.

The removal by erosion of the oncolitic boulders from the buildups, and their subsequent deposition in a basinal setting, protected them from the pervasive Late Triassic dolomitization that affected most of the buildups in the western Dolomites. Once deposited between basinal sandstones and shales, the boulders, and the *Girvanella* oncoids they contain, were immune to that dolomitization. It is the lack of dolomitization that separates the *Girvanella* oncoids of the boulders from most of the in-place coated grains of the Dolomites.

It is difficult to draw definite conclusions about the original depositional environment of the *Girvanella* oncoids. The boulders contain little information on bedding style or on the relationships between oncolitic facies and other rock types. In addition, the occurrence of *Girvanella* is not restricted to a specific depositional environment or water depth (Riding 1975). Peryt (1977) also suggested that other Triassic foraminiferal-algal oncoids may form in a variety of environments.

The compositions of many of the megabreccia boulders, however, indicate that they were derived from near the buildup margins (Biddle 1981). This suggests that the *Girvanella* oncoids may be the nondolomitized equivalents of the laminated algal nodules discussed by Leonardi (1967). If so, the occurrence of *Sphaerocodium* in the Dolomites has probably been overestimated.

In conclusion, although several types of coated grains certainly exist in the Middle to Upper Triassic section of the Dolomite Alps, diagenesis can render their interpretation difficult. The majority of well-preserved coated grains found in buildup-derived limestone boulders are *Girvanella* oncoids. Care must be taken to distinguish these, and their in-place, dolomitized equivalents, from other types of coated grains in the Dolomite Alps.

Acknowledgements. I would like to thank the personnel of the Hotel Goldknopf for making my stay in the Alpe di Siusi area a pleasant one. The manuscript was reviewed by T. M. Peryt and D. A. Yurewicz.

References

Assereto R, Brusca C, Gaetani M, Jadoul F (1977) The Pb-Zn mineralization in the Triassic of the Dolomites. Geological history and genetic interpretations. Ind Mineraria 28:267 – 402

Biddle KT (1979) Characteristics of Triassic carbonate buildups of the Dolomite Alps, Italy: Evidence from the margin-to-basin depositional system. Unpubl Ph D thesis, Rice Univ, Houston, Texas

Biddle KT (1981) The basinal Cipit boulders: Indicators of Middle to Upper Triassic buildup margins, Dolomite Alps, Italy. Riv Ital Paleontol Strat 86:779 – 794

Bosellini A, Rossi D (1974) Triassic carbonate buildups of the Dolomites, northern Italy. In: Laporte LF (ed) Reefs in space and time. Soc Econ Paleontol Miner Spec Publ 18, Tulsa, pp 209 – 233

Bosellini A, Castellarin A, Rossi PL, Simboli G, Sommavilla E (1977) Schema sedimentologico e stratigrafico per il Trias Medio della Val di Fassa ed aree circostanti (Dolomiti Central). G Geol 42:83 – 108

Krumbein WE, Potts M (1978) *Girvanella*-like structures from the Borrego Desert in Southern California. Geomicrobiol J 1:211 – 217

Leonardi P (1967) Le Dolomiti: Geologia dei Monti tra Isarco e Piave, vol I, II. Arti Grafiche Manfrini, Roverto

Peryt TM (1977) Environmental significance of foraminiferal-algae oncolites. In: Flügel E (ed) Fossil algae. Springer, Berlin Heidelberg New York, pp 61 – 65

Peryt TM (1980) Structure of *"Sphaerocodium kokeni* Wagner", a *Girvanella*-oncoid from the Upper Muschelkalk (Middle Triassic) of Wurttemberg (SW Germany): Neues Jahrb Geol Palaeontol Mh 5:293 – 302

Riding R (1975) *Girvanella* and other algae as depth indicators. Lethaia 8:173 – 179

Riding R (1977) Calcified *Plectonema* (blue-green algae), a Recent example of *Girvanella* from Aldabra Atoll. Paleontology 20:33 – 46

Viel G (1979) Lithostratigrafia ladinica: una revisione. Ricostruzione paleogeografica e paleostrutturale dell'area Dolomitico-Cadorina (Alpi Meridionali), part 1: Riv Ital Paleontol Strat 85: 85 – 125

IV.13 Triassic Oncoids from Central Balkanides (Bulgaria)

G. A. Čatalov [1]

1 Introduction

Recently an attempt has been made to modify the traditional concepts of the bathymetric and energy significance of oncoids in paleofacies interpretations (Peryt 1981). It has been pointed out that oncoids are important indicators of breaks and low rates of sedimentation as well as of positive tectonic lineaments and nearshore facies (Weiss 1969, Toomey 1974, Peryt 1977, 1981). It seems that the study of the oncoids occurring in the Ladinian-Carnian sediments of Central Balkanides may throw some light on the problem.

For the purpose of this paper two sections from the Teteven Anticlinorium (Central Fore-Balkan) and one section from the Central Balkan (Central Stara Planina) and some separate localities have been studied (Figs. 1 – 3).

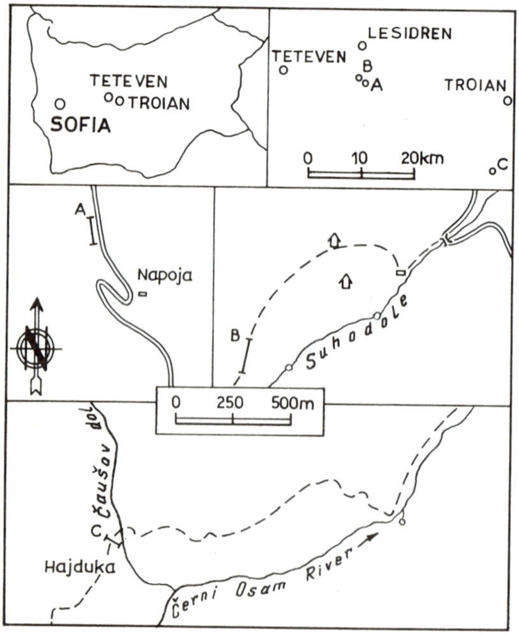

Fig. 1. Area studied

1 Geological Institute, Bulgarian Academy of Sciences, Acad. G. Boncev St., bl. 24, 1113 Sofia, Bulgaria

Coated Grains (ed. by T. M. Peryt)

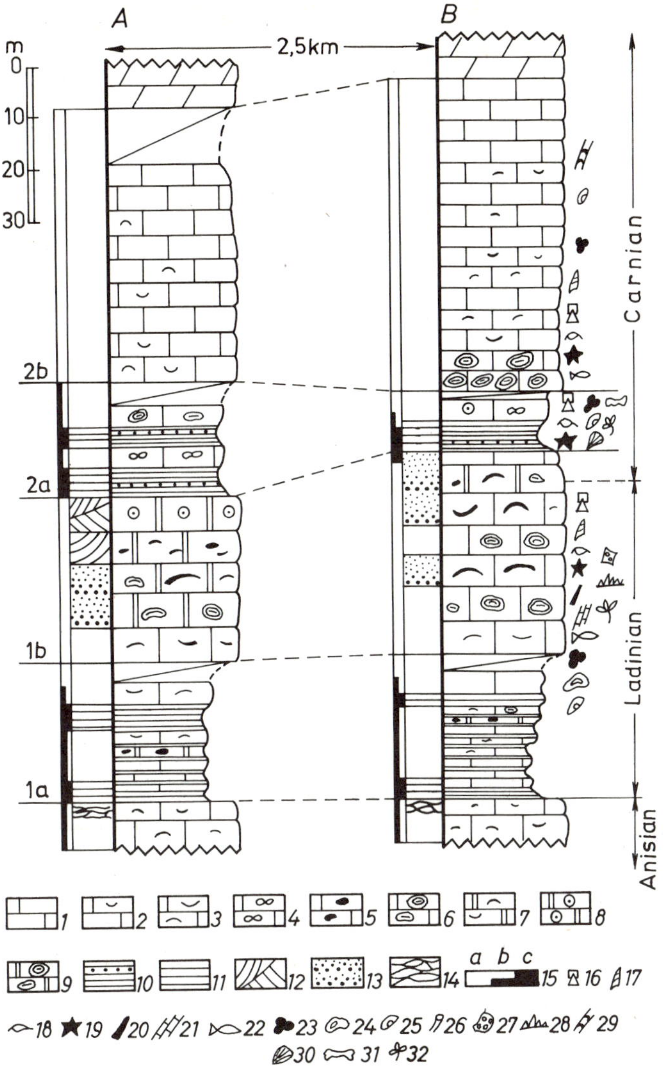

Fig. 2. Columnar sections of the Ladinian and the lower part of the Carnian sediments from Napoya (A) and Suhodole (B) (Teteven anticlinorium). Lithology: *1* micrite, *2* fossiliferous micrite, *3* biomicrite, *4* intramicrite (bahamite), *5* lithoclastic calcrudite (intraformational microconglomerate), *6* oncomicrite (superficial and normal oncoids), *7* biosparite and intrabiosparite, *8* oosparite, *9* oncosparite (superficial or embryonic and normal oncoids), *10* shale with plant remains interbedded with thin beds of siltstones and fine grained sandstones. Sedimentary structures: *11* horizontal lamination, *12* cross-stratification, *13* sorted bedding, *14* nodular structure. Colour: *15 a* light grey, *b* medium grey, *c* dark grey (*black*). Bioclasts: *16* bivalves, *17* gastropods, *18* brachiopods, *19* crinoids, *20* echinoid spines, *21* bryozoa, *22* fish remains, *23* vagile benthos foraminifers, *24* sessile calcitornellid foraminifers, *25* ostracods, *26* calcareous worms, *27* holothuroids, *28* conodonts, *29* green algae (Dasycladaceae), *30* red algae (Solenoporaceae), *31* Reptilia, *32* plant remains. Beds: *1a* lower terrigenous-limestone beds; *1b* lower limestone beds; *2a* upper terrigenous-limestone beds; *2b* upper limestone beds

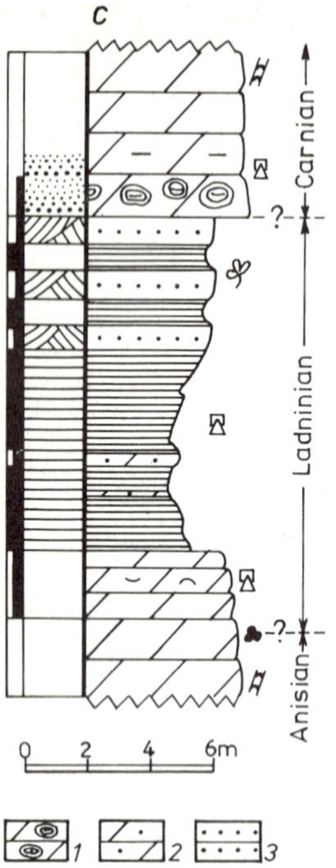

Fig. 3. Columnar section of the Ladinian and the lower parts of the Carnian sediments from Haiduka (*C*) (Central Stara Planina). Lithology: *1* fine to medium grained oncoide dolomite, *2* fine arenitic dolomite, *3* sandstones. Other symbols as in Fig. 2

2 Description of Oncolites

The oncoids studied occur in a marine dark-coloured terrigenous-carbonate formation (Ladinian-Lower Carnian) in the Teteven Anticlinorium and at the base of a marine dolomitic formation (Carnian) in the Central Balkan (Figs. 2 and 3).

In the present paper it is assumed that the oncolite contains >50% oncoids or that the latter are grain-supporting. The term oncoid-bearing rock is used when oncoids are <50% or are not grain-supporting.

Fig. 4. a Porostromate phytogenic, *Girvanella* pisooncoid in oncomicrite. Envelopes wavy, dome-like, considerably re-crystallized. 1b beds; Ladinian; occurrence Suhodole; thin section 203^{b}-193-V. **b** Porostromate phytogenic, Girvanella pisooncoid in bioncomicrite. Envelopes wavy, dome-like. Dark envelopes contain *Girvanella* tubes. 1b beds; Ladinian; occurrence Suhodole; thin section 198^{a}-188-V. **c** *Girvanella* tubes (inner diameter from 5 to 7 μm), possibly *Girvanella minuta* Wethered. 1b beds; Ladinian; occurrence Suhodole; thin section 198^{a}-188-V. **d** Porostromate zoogenic calcitornellide microoncoids in re-crystallized oncobiomicrite. Envelopes made up of sessile calcitornellide foraminifers *Planiinvoluta carinata* (Leischner). The micrite is pyritized. 1b beds; Ladinian; occurrence Suhodole; thin section 206-196-V. **e** Porostromate zoophytic, algae-calcitornellide micro-oncoid. Three types of laminae in the envelope: algal micrite, calcitornellide (*Planiinvoluta carinata*) and mixed algal-calcitornellide. 1b beds; Ladinian; occurrence Suhodole; thin section 206-196-V. **f** Paraporostromate eroded pisooncoid (allochthonous) in oncobiosparite. 1b beds; Ladinian; occurrence Suhodole; thin section 201-191-V. **g** Paraporostromate superficial microoncoid in oncoide biomicrite. The core is of crinoid. 1b beds; Ladinian; occurrence Suhodole; thin section 194-184-V

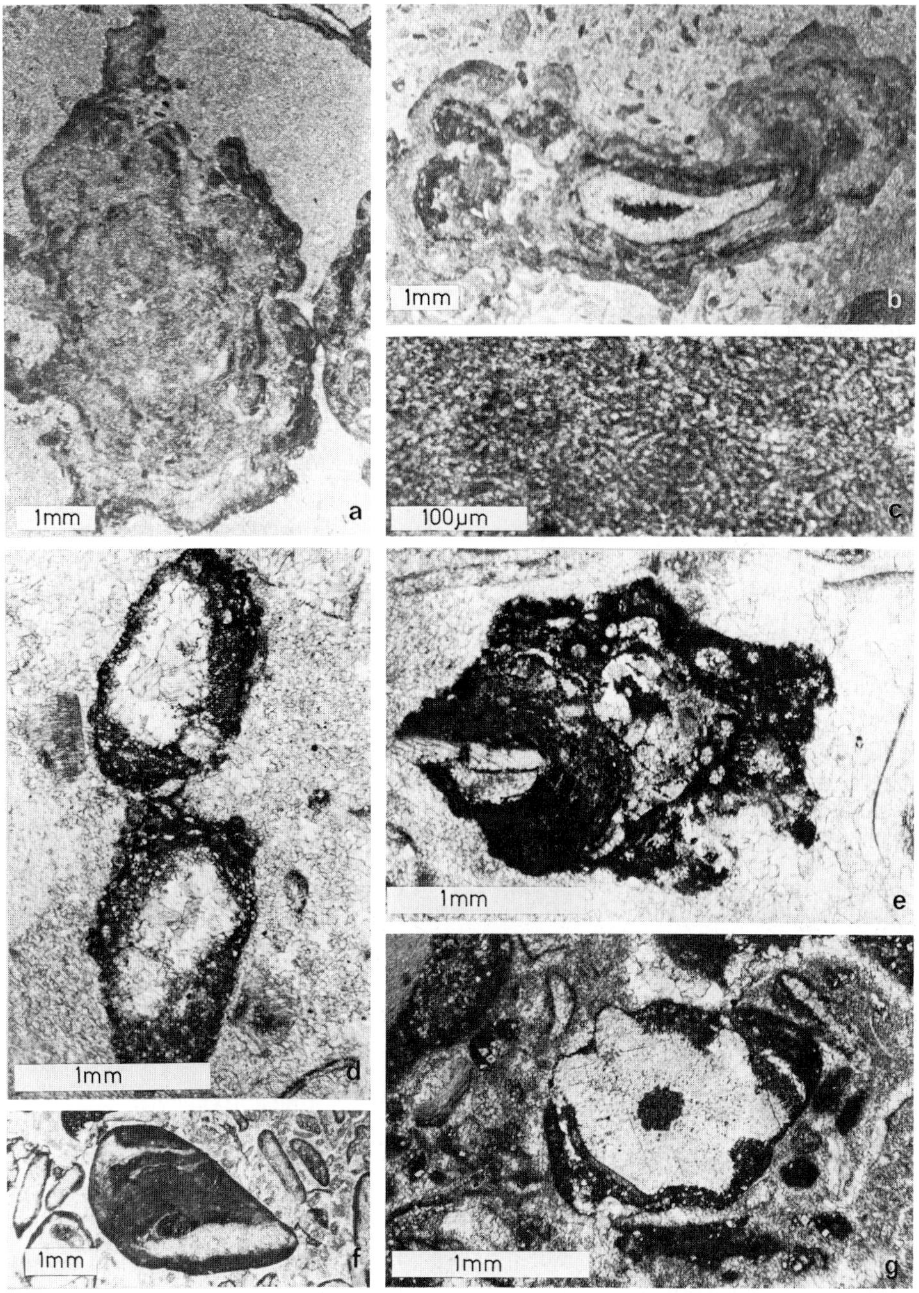
1mm
a
1mm
b
100μm
c
1mm
d
1mm
e
1mm
f
1mm
g

Oncolite rock is mainly an oncomicrite in which the content of oncoids reaches 60% – 70%. The micrite is made up of calcite grains 4 to 20 μm in size. Apart from oncoids there are sometimes bivalves, gastropods, echinoids and ostracods. The fragments are not rounded or semirounded. They are unsorted or moderately sorted. Sometimes they shown bimodal sorting but it is in most cases natural rather than hydraulic. The packing is normal. The oncobiosparites comprise 50% to 65% allochems. The species composition is close to that of the biomicrites. They are semirounded to well rounded and sorted. Eroded oncoids have been also observed (Fig. 4f). The intrasparites contain 20% – 30% oncoids but they are mainly of peloid type. The rock was formed in moderately agitated water (EI-III$_{1-3}$). The terrigenous component is insignificant (sample 36^{b-}-52-VI, Ib, 0.87% pelite).

3 Description of Oncoids

Following Monty (1979) and Peryt (1979, 1981) the oncoids have been divided into porostromate and spongiostromate. Skeletal and nonskeletal oncoids (Riding 1977) have analogous meanings. The term paraporostromate oncoids is proposed for strongly diagenetically altered initially *Girvanella* oncoids (in analogy to paramorphic minerals or paramorphism in mineralogy when minerals change their internal structure without changing their outer form and chemical composition) (Table 1).

3.1 Porostromate Oncoids

According to the organisms oncoids building them these are divided into phytogenic and zoophytogenic (Table 1).

3.1.1 Phytogenic Oncoids

The phytogenic oncoids are of restricted occurrence. Their size is from 2 – 3 mm to 2 – 3 cm (Fig. 5). Pisooncoids (2 mm to 1 cm) and small macrooncoids (1 cm to 3 cm) predominate. The normal oncoids are ellipsoidal, spheroidal or irregular. The cores are of bivalves, gastropods, brachiopods, crinoids, echinoid spines and intraclasts. The cores are usually corroded and have holes from boring organisms (most commonly of *Girvanella* and rarely bivalves and polychaetes). In some cases the core is lacking. Envelopes show more or less distinct lamination. The laminae are slightly curved or strongly domed (Fig. 4a, b). Commonly they are broken and overlap. The dark laminae are made up of algal micrite with grains 4 – 5 μm in size. The light laminae are of microspar (4 to 10 – 15 μm). The thickness of the laminae is 0.2 mm to 1 mm (Fig. 4a, b). The light and dark laminae may be considered as one pair of laminae (couplets). Sometimes they may be seen to have a gradual transition. The dark laminae show a sharp outer boundary (towards the following external light lamina) and gradual (towards the previous older light lamina) inner boundary.

Filament structures may be observed in the dark laminae. They are of two types: fine and relatively large (Fig. 6g, h). The fine filament structure is marked

Table 1. Classification of the Triassic oncoids

Microstructure	Porostromate	Made up with participation of *Girvanella, Calcitornella,* calcareous worms	
	Paraporostromate	Strongly diagenetic altered, initially porostromate	
Organic builders	Phytogenic	Mainly *Girvanella*	
	Zoogenic	Mainly *Calcitornella* ± calcareous worms	
	Zoophytogenic	*Girvanella* and *Calcitornella* ± calcareous worms	
Shape and lamination of the envelopes	Spheroidal, ovoidal	Distinctly laminated concentric	
	Spheroidal, ovoidal	Distinctly laminated wavy dome-like	
	Rounded irregular	Unclear cloud-like indistinctly layered	
	Rounded irregular	Not laminated (peloids)	
Thickness of the envelopes	Normal	Envelope bigger than nucleus	
	Superficial (embryonic)	Envelope smaller than nucleus	

by the presence of light tubes outlined by dark micrite which usually grades into the dark micrite between the tubes. The latter are made up of clear blocky calcite with grains from 2 – 3 to 5 μm. The dark calcite on the walls of the tubes and between them has dimensions of about 1 μm. The quantitative relation between tubes and the micrite between them is about 1 : 1. The tubes are straight or slightly curved and have a constant diameter. Partitions in them are lacking. Branching of tubes is not seen. Their inner diameter is 5 to 7 μm, the length is up to 70 – 80 μm. Usually they are subparallel to the laminae and rarely form irregular interlacings. These features correspond to *Girvanella minuta* Wethered (Fig. 4c, 6h), as described by Peryt (1980) and Radwanski (1968). The internal diameter of the tubes from the coarse filament structures is 20 – 25 μm, their length ranging up to 300 – 400 μm. The other features are the same. Most probably they also are of algal origin (*Girvanella* ?).

Endolithic and epilithic organisms are rare. The microboring Cyanophyceae which penetrate into different depths in the cores of the oncoids (Fig. 6b) may be referred to as endoliths. In some microholes tubes of *Girvanella* analogous to the *G. minuta* described have been observed. Epizoans are represented by tubes of calcareous worms and sessile calcitornelide foraminifers. *Nodosaria* sp. and Miliolidae gen. sp. indet have been found in the envelopes (Fig. 6b – f).

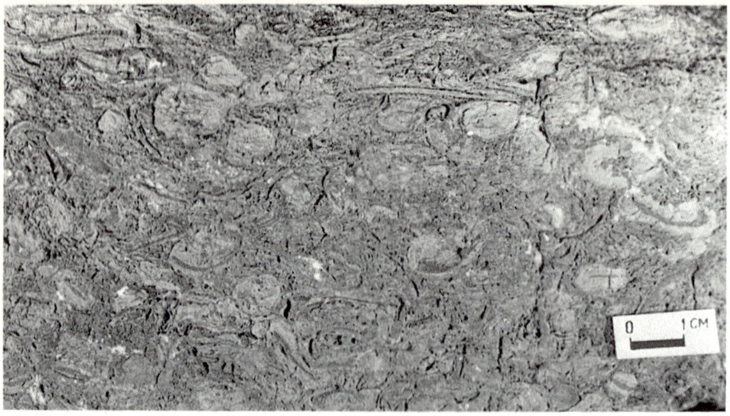

Fig. 5. Weathered surface of oncolite (2b – beds – bed 201-V, Carnian, occurrence Suhodole)

3.1.2 Zoophytogenic Oncoids

They are made up of blue-green algae and sessile clacitornelide foraminifers. The amount of plant and animal organisms must be over 10 or 20% for each component. In some cases both organism groups build up independent layers and in others they are mixed. These oncoids are rare (Fig. 4e). Their size is from 1 – 2 to 4 – 5 mm. The envelopes are indistinctly layered. Calcitornelides are represented also by *P. carinata*. The envelopes are commonly pyritized. In other cases are brown and are replaced by iron-bearing dolomite.

3.2 Zoogenic Oncoids

They are of restricted occurrence and have been found in two layers only (1a, thin section 188-173-V and 1b, section 206-196-V, Fig. 4d). Their amount is from 5% – 10% to 30% – 40%. The size ranges from 0.5 to 5 – 6 mm in diameter. In single cases they range up to 1 – 2 cm. Sizes from 1 to 3 mm predominate. The form depends upon the core which may be of bivalves, brachiopods, gastropods, crinoids or intraclasts. Superficial oncoids predominate. The envelopes show indistinct lamination. Their texture is tubular-cellular and is determined by the type of organisms. The latter are mostly sessile calcitornelide foraminifers. *Planiinvoluta carinata* (Leischner) has been identified (determination of E. Trifonova). The micrite of the walls is commonly pyritized or dolomitized (Fig. 4d).

3.3 Paraporostromate Oncoids

These oncoids are found together with typical *Girvanella* oncoids. The concept of Peryt (1980, p. 300) that such oncoids must not be referred to the spongiostromate ones only on the basis of lack of preserved *Girvanella* tubes is very reasonable.

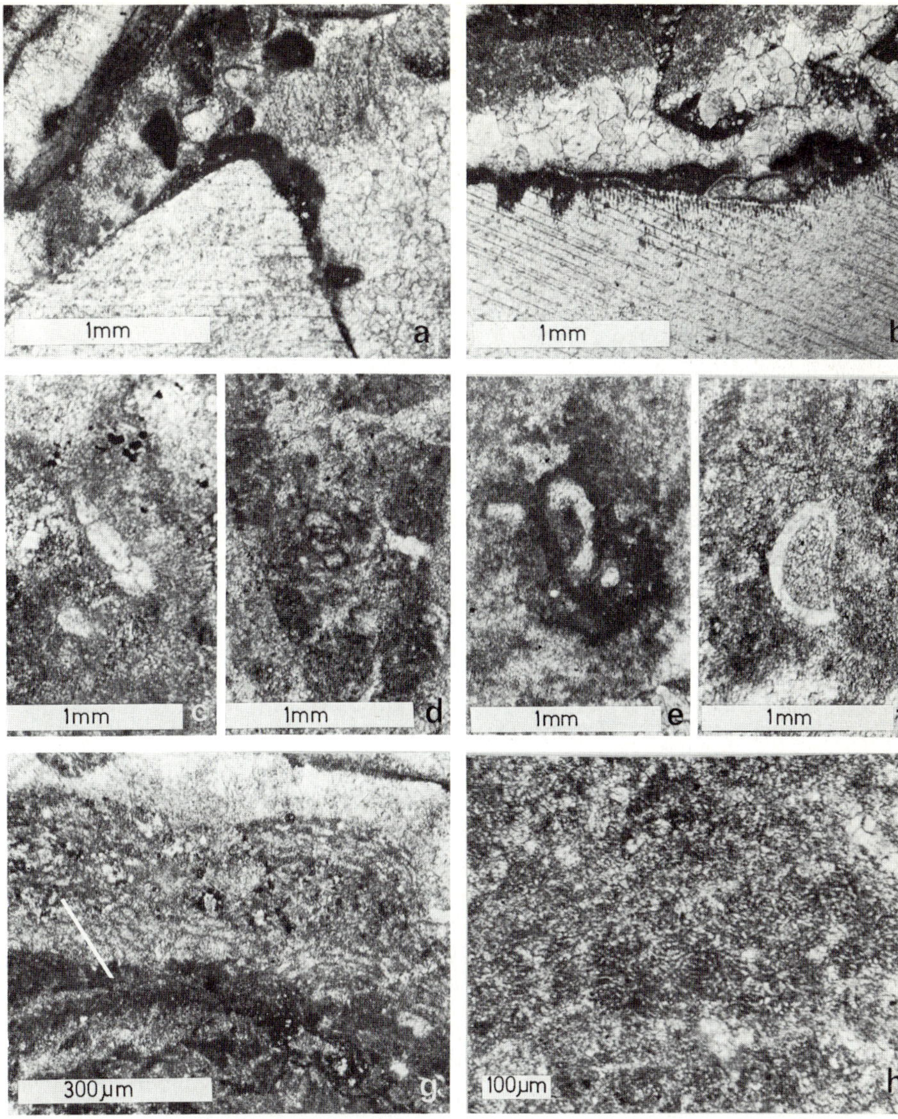

Fig. 6. a Embryonic envelope formed only of blue-green algae. 1b beds; Ladinian; occurrence Suhodole; thin section 202[a]-192-V. **b** Embryonic envelope formed both through enveloping and through the boring activity of blue-green algae. On the right – calcareous worms directly attached to the micrite cast. 1b beds; Ladinian; occurrence Suhodole; thin section 194-184-V. **c** *Nodosaria* sp. in the envelope of porostromate pisooncoid. Partial re-crystallization and dolomitization (ferro-dolomite). The rock is oncomicrite. 1b beds; Ladinian?; occurrence Suhodole; thin section 203[b]-193-V. **d** Miliolidae gen. sp. ind. in the envelope of paraporostromate peloidal pisooncoid. The rock is intramicrite. 1b beds; Ladinian; occurrence Napoya; thin section 35*-51-VI. **e** *Planiinvoluta carinata* (Leischner) (sessile calcitornellide foraminifers) in paraporostromate peloidal pisooncoid. 1b beds; Ladinian; occurrence Napoya; thin section 37[a]-53-VI. **f** Calcareous worms in the envelope of paraporostromate pisooncoid. 1b beds; Ladinian; occurrence Suhodole; thin section 194-184-V. **g** Relatively coarser tubular structure (internal diameter 20 – 25 µm) from blue-green algae forming a lamina upon a fine-tubular lamina of blue-green algae (*Girvanella*). Porostromate *Girvanella* macrooncoid in biooncomicrite. 2b beds; Carnian; occurrence Suhodole; thin section 211-201-V. **h** Fine *Girvanella* tubular structure (inner diameter 5 – 7 µm), probably *Girvanella minuta* Wethered in porostromate macrooncoid. 1b beds; Ladinian; occurrence Suhodole; thin section 198[a]-188-V

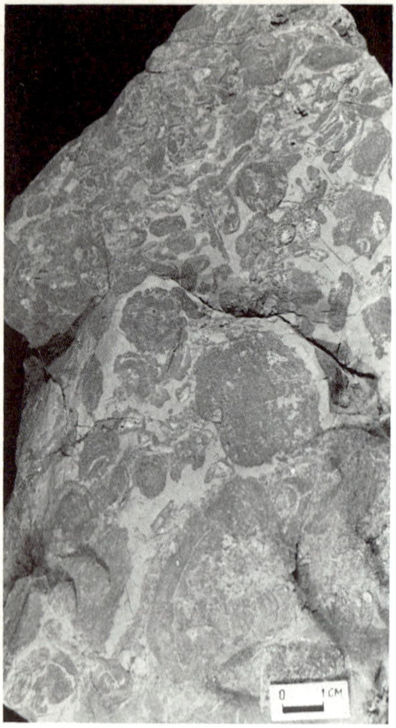

Fig. 7. Weathered surface of doloncomicrite (sorted series in a bed, occurrence Haiduka, Carnian, see Fig. 3)

The paraporostromate oncoids are of widest occurrence. This is due to intensive diagenetic alterations (re-crystallization, dolomitization, pyritization and porphyry silicification). Apart from that they are more abundant in section VI (Fig. 2) where there was a high energy environment. All of these specific features are analogous to the porostromate oncoids except the lack of preserved *Girvanella* tubes. A specific type of oncoid in high energy environment has also been observed. They are typical of 1b section VI (Fig. 2). Their size is from 0.5 to 1.5 cm and occur in the base of sorted series in the beds. These oncoids usually lack any core, have up to 2–3 indistinct laminae or are patchy (*Nubecularites* type, Maslov 1960, p. 57). Upwards in section they grade into peloids and the latter into ooids. All oncoids from dolomites in Troyan Balkan are paraporostromate oncoids. They are 2–3 mm to 2–3 cm in size. Single oncoids range up to 8–10 cm (Fig. 7).

4 Genesis of Oncolites and Oncoids

The Teteven facies of the Ladinian and the lower parts of the Carnian is very close to that of the Raibler beds of the Carnian in the Northern Calcareous Alps (Jerz 1966, Schuler 1968, Harsch 1970, Tichy, this vol.). The section in Troyan Balkan (Fig. 3). shows some similarities with the sections north of Zaile (Kalkkögel) (Jerz 1966, Fig. 11, p. 31–32).

The Teteven facies of the Ladinian and the lower parts of the Carnian was formed in a warm shallow shelf sea. The occurrence of clay-sandy sediments is related to an abrupt supply of terrigenous material. Probably this is due to epeirogenic tectonic movements mainly south of the area studied, during the Ladinian and the begining of the Carnian.

For marine oncoids a depth from 0 to 100 m is assumed (Maslow 1952, Ginsburg and Lowenstam 1958, Jerz 1966, Radwanski 1968, Weiss 1969, Harsch 1970, Peryt 1977, 1981). The typical *Girvanella* oncoids in the sections studied are associate with the richest brachiopod levels. The structural features, the composition and the organism association (brachiopods, crinoids, bivalves, *Girvanella*) show that the oncoids were formed under a constant water cover (subtidal environment). The supposed depth of these typical oncoids is about 5 – 10 to 20 – 30 m. A larger depth of several tens of meters is also permissible. The oncomicrites and the oncobiomicrites were formed at larger depth of the supposed interval while the biosparites and the oncointrasparites are shallower. A contemporary tendency towards a more deep-water environment for oncolites is quite real (Füchtbauer 1968, Taylor and Colter 1975, Peryt 1977, 1981). But the analysis of the data of Jerz (1966), Schuler (1968) and Harsch (1970) shows that oncoids are lacking in the shallowest water facies of the Raibler beds but they are also lacking in the deepest water facies. The *Sphaerocodium bornemanni* Rothpletz illustrated by Jerz (1966, p. 54, Fig. 18) is probably a *Girvanella* oncoid (possibly *Girvanella minuta* Wethered?). Mainly low energy environments are distinguished. The low energy mud bottom environment is typical of the beds richest in normal oncoids (EI is II_{1-3}, Čatalov 1972). The oncoid-bearing beds composed of oncobiosparites and oncointrasparites were formed in a high energy environment (EI is III_{2-3}). These beds show good sorting and cross stratification. In them the normal oncoids are few or single and superficial oncoids and micrite casts predominate.

Indirect and direct evidence for low rates of sedimentation or breaks in sedimentation is presented by Harsch (1970, p. 262), Bachmann and Gwinner (1971, in Peryt 1981), Peryt (1977, 1981). These data may be extended with the fact that the beds which are richest in normal oncoids occur in shale or overlay shale and siltstones (Figs. 2 and 3). Probably carbonate sedimentation, interrupted by abundant terrigenous supply, recovered very slowly. Those are the periods of mass formation of oncoids. Taking into consideration the small thickness of the Ladinian and part of the Lower Carnian sediments as compared to those of the Anisian carbonate platform a low rate of subsidence may be assumed.

The salinity of the sea water was normal. The bottom layers were well illuminated but not well aerated which is indicated by the presence of plant detritus and early diagenetic pyrite.

5 Conclusions

The following more important conclusions may be inferred: (a) the oncoids are phytogenic (*Girvanella* oncoids), zoogenic (calcitornelide oncoids) and zoophytogenic (*Girvanella*/calcitornelide oncoids); (b) in the envelopes of the oncoids *Nodosaria* sp., Miliolidae gen. sp. ind., *P. carinata* (Leischer) and calcareous worms are rarely found; (c) paraporostromate oncoids predominate;

(d) pisooncoids (0.2 to 1 cm) dominate followed by small microoncoids (from 1 cm to 3 cm); (e) the typical oncolites are made up of oncomicrites formed in low energy, slightly to moderately agitated subtidal normal sea water environment; (f) in high energy environment the oncoid-bearing beds are composed mainly of superficial oncoids and bioclasts with micrite casts; (g) in the sections stromatolites are lacking; (h) the oncomicrites are formed in relatively deep-water environment and are autochthonous while the oncobiosparites and the oncointransparites are related to relatively shallow waters and may be allochthonous; (i) the oncoids are formed under conditions of low rates of sedimentation in a sea of a slowly subsiding carbonate platform.

Acknowledgements. The author expresses his sincere appreciation to Dr. T. Peryt (Warsaw) and to Dr. C. Chrischev (Sofia) for the help and encouragement and to Dr. R. Riding (Cardiff) who read and corrected the English manuscript.

References

Čatalov GA (1972) An Attempt at energy index (EJ) analysis of the Upper Anisian, Ladinian and Carnian carbonate rocks in the Teteven anticlinorium (Bulgaria). Sediment Geol 8:159 – 175

Füchtbauer H (1968) Carbonate sedimentation and subsidence in the Zechstein basin (northern) Germany. In: Müller G, Friedman GM (eds) Recent developments in carbonate sedimentology in Central Europe. Springer, Berlin Heidelberg New York, pp 196 – 204

Ginsburg RN, Lowenstam HA (1958) The influence of marine bottom communites on the depositional environment of sediments. J Geol 66:310 – 318

Harsch W (1970) Zur Sedimentologie und Paläogeographie der Reibler Schichten (Ober-Trias) im Westteil der Nördlichen Kalkalpen. Z Dtsch Geol Ges (Hannover) 121:253 – 272

Jerz H (1966) Untersuchungen über Stoffbestand, Bildungsbedingungen und Paläogeographie der Reibler Schichten zwischen Lech und Inn (Nördliche Kalkalpen). Geol Bavarica 56:3 – 102

Maslov VP (1952) Karbonatte jelvaki organicheskogo proizhojdenie. Bull MOIP Geol 57:28 – 45

Maslov VP (1960) Stromatoliti. Dokl Geol Inst 41:1 – 168 Moskva

Monty CLV (1979) Spongiostromate vs porostromate stromatolites and oncolites (obstr.) 2 Symp Int Algues Fossiles Resumes

Peryt TM (1977) Environmental significance of foraminiferal-algal oncolites. In: Flügel E (ed) Fossil Algae. Springer, Berlin Heidelberg New York, pp 61 – 65

Peryt TM (1979) Phanerozoic oncolites – an overview (abstract) 49th Meet Paläontol Ges 1979, Oldenburg Wilhelmshaven, pp 22 – 23

Peryt TM (1980) Structure of "Sphaerocodium kokeni Wagner", a Girvanella oncoid from the Upper Muschelkalk (Middle Triassic) of Württemberg, SW Germany. Neues Jahrb Geol Paläontol Monatsh, pp 293 – 302

Peryt TM (1981) Phanerozoic oncoids an overview. Facies 4:197 – 214

Radwandski A (1968) Petrographical and sedimentological studies of the High-Tatric Rhaetic in the Tatra Mountains. Stud Geol Polon (Warsaw) 25:1 – 146

Riding R (1977) Skeletal stromatolites. In: Flügel E (ed) Fossil Algae. Springer, Berlin Heidelberg New York, pp 56 – 60

Schuler G (1968) Lithofazielle, sedimentologische und paläogeographische Untersuchungen in den Reibler Schichten zwischen Inn und Salzach (Nördliche Kalkalpen). Erlanger Geol Abh 71:1 – 60

Taylor JCM, Colter VS (1975) Zechstein of the English sector of southern North Sea basin. In: Woodland AW (ed) Petroleum and the Continental shelf of North-West Europa, vol I. Geology. Applied Science Publishers, London, pp 249 – 263

Toomey DF (1974) Algally coated grains from the Leavenwarth Limestones (U Pa Midcontinent Region USA). Neues Jahrb Geol Paläontol Monatsh, pp 175 – 191

Weiss MP (1969) Oncolites paleoecology and Laramide tectonics, central Utah. Am Assoc Petrol Geol Bull 53:1105 – 1120

IV.14 The Genesis of Algal Nodule Limestones from the Upper Carboniferous (San Emiliano Formation) of N.W. Spain

M.B.J. Bowman[1]

1 Introduction

Algal nodule limestones form a distinctive lithotype in platform carbonate successions from the Upper Carboniferous of the Cantabrian Mts., N.W. Spain (Rácz 1964, de Meijer 1971, van de Graff 1971, Bowman 1979). This paper presents an analysis of nodule bearing rocks from the Late Namurian-Westphalian A San Emiliano Formation in its type area in northern León (Fig. 1). It represents the first detailed synthesis of algal nodule limestones from the Spanish Carboniferous, discussing their genesis and palaeogeographic significance. It also discusses the relative importance of allogenic and autogenic processes in the growth and development of the nodules and their microfabrics.

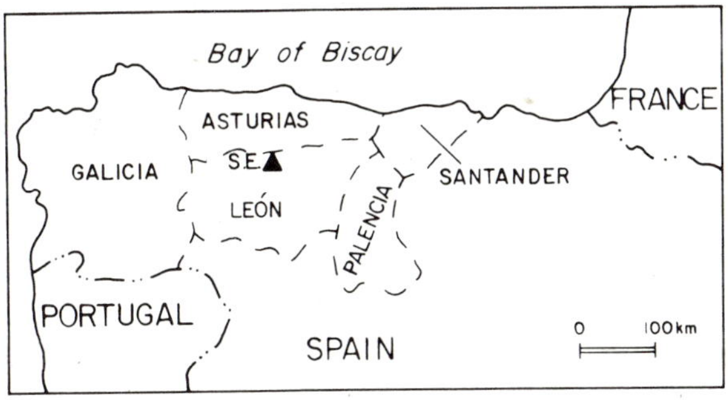

Fig. 1. Map of NW Spain showing the location of the San Emiliano area (S.E.)

2 Stratigraphy and Palaeogeographic Setting

The San Emiliano Formation comprises ca. 1800 m of alternating clastics and carbonates, overlying a thick, mainly Yeadonian carbonate sequence (Bowman 1980 and in prep.). It is divisible into three members according to variations in the proportions of clastics to carbonates (Fig. 2). To the Northeast was an extensive carbonate platform whilst to the South and West lay a rising hinterland.

1 The British Petroleum Co. Ltd., Britannic House, Moor Lane, London EC2Y 9BU, Great Britain

Coated Grains (ed. by T.M. Peryt)

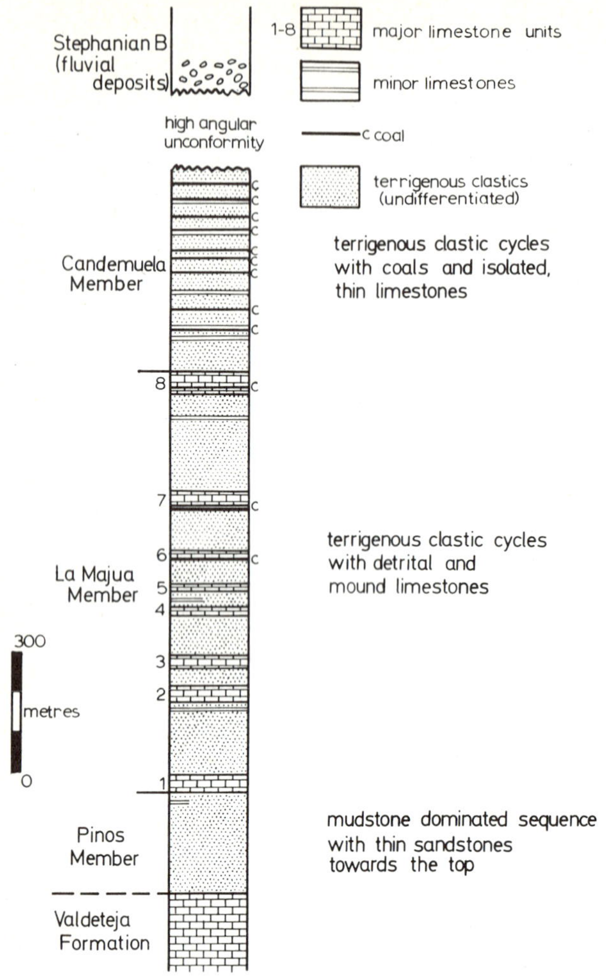

Fig. 2. The stratigraphic subdivisions of the San Emiliano Formation. (From Bowman, in prep.)

Limestones are present in all three members of the formation. They are thickest in the La Majua Member where they form eight major, 10 – 40 m thick, transgressive composite units that separate regressive, mainly deltaic intervals. The distribution and interrelationships shown by the different carbonate lithofacies suggest that they occupied broad zones aligned roughly parallel to the deltaic coastline. Facies variations between the different zones were the result of depth and energy changes, together with variations in the amount of suspended terrigenous mud within the seawater.

Algal nodule facies are common in limestones from both the La Majua and Candemuela members. They also occur in a limestone band at the top of the Pinos Member. The author has described algal nodule limestones from limestone unit 2 of the La Majua Member (Bowman 1979), but no detailed synthesis was made of all the nodules from the San Emiliano Formation.

3 Algal Nodules

The term algal nodule is used here to describe cryptalgal structures containing laminated (oncolitic) and/or non-laminated fabrics. The nodules from the San Emiliano Formation display a wide variety of micro-fabrics and growth patterns. These are described separately and their significance is discussed in the light of present knowledge of modern and ancient cryptalgal fabrics. Variations in the geometry and composition of nodules and the fossil content of the enclosing sediment, make it possible to distinguish two facies, algal nodule wackestones and oncolitic, whole fossil wackestones.

3.1 Microfabrics

Five different microfabrics have been identified within the algal nodules. These are all of organic origin. Isolated sediment particles, usually small bioclasts, may be incorporated within some examples, but discrete algal-bound detrital laminae are absent. Hence, the mechanism of nodule accretion by entrapment and binding of particulate sediment, evident in many Recent algal laminated structures, was of little importance in the development of the structures described below. This supports Monty's (1972) contention that many fossil cryptalgal laminated structures originated by in situ precipitation of carbonate sediment within algal mats.

3.1.1 Composite Micrite Laminae

These laminae are present in most algal nodules. They are composed of dark, dense homogeneous micrite which occasionally displays a clotted texture. Laminae are generally semi-continuous, varying from 0.55 mm to 5 mm thickness. Evenly spaced, dark, organic-rich bands, 10 μm thick, provide a primary microlamination within the micrite (Fig. 3a).

The laminae generally exhibit smooth to pustular outlines. However, a digitate, domal or club-shaped microstromatolitic relief may be developed in places (Fig. 3a). These are considered to result from periodic shifts in the site of active growth on a mat surface, combined with localised physical disruption of laminae during nodule movement. The degree of inheritance displayed by such structures is very variable; some persist throughout the fabric, exerting a control on the overall morphology of the nodule; others are less persistent and are rapidly overgrown by succeeding laminae.

Spar-filled fenestrae within the micrite comprise tubular burrows which disrupt or distend the primary fabric, and irregular, discordant cavities. These are probably formed during the escape of gases generated during photosynthesis or by bacterial decomposition (Monty 1976). Laminoid fenestrae are rare.

Filaments of the calcified cyanophyte *Girvanella* (Riding 1975, 1977) are often present within the micrite. Thin ($\leqslant$ 20 μm), discontinuous, sparry laminae and encrusting foraminifera occur occasionally.

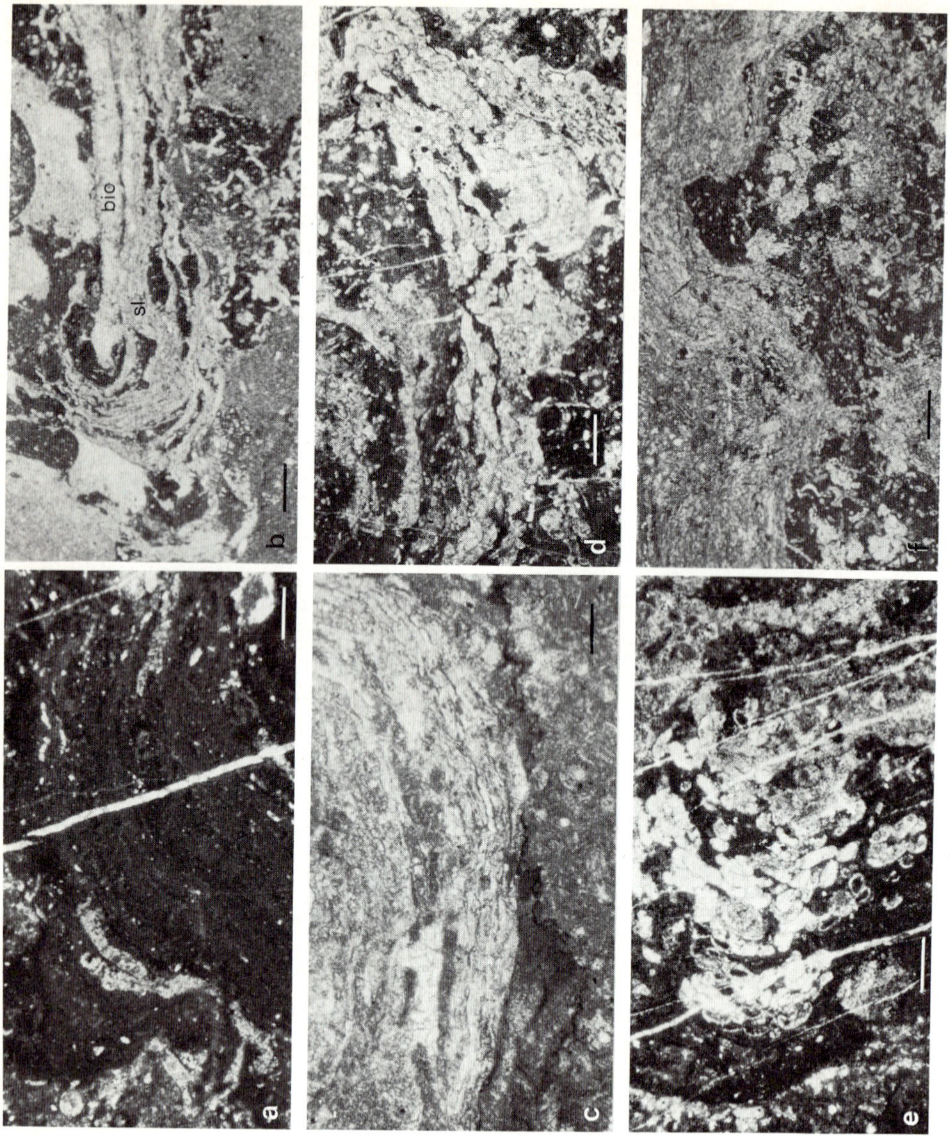

Fig. 3. a Composite micrite laminae with a club-shaped relief along the outer margin of the structure (scale bar = 250 μm). **b** Composite sparry laminae (*sl*) encrusting a recrystallised shell fragment (*bio*) (scale bar = 750 μm). **c** Detail of sparry laminae showing the characteristic stacked algal filaments (scale bar = 200 μm). **d** Discontinuities in sparry laminar fabric with disruption of some laminae as a result of bioturbation (scale bar = 250 μm). **e** Tubular foraminiferal laminae in domal micro-stromatolitic growth form (scale bar = 250 μm). **f** Clotted micrite with a well developed fenestral fabric. Note the twigs of *Donezella lutugini* Maslov and small, encrusting foraminifera within the clots (scale bar = 250 μm)

The composition and microstructure of the micrite laminae invite comparison with fabrics formed by the growth of successive mats, dominated by the gelatinous and filamentous blue-green alga *Phormidium* (Monty 1955, 1967, 1976, p. 197; Golubić and Focke 1978). The black, organic-rich band which defined individual laminae is similar to Monty's discontinuity surface separating different algal mats.

The absence of detrital sediment from the laminae and the distinctive, dense nature of the micrite preclude an origin by physical sedimentation processes. More likely the micrite resulted from biochemical precipitation during decomposition of the organic material in the mats (Dalrymple 1965, Golubić 1976).

Monty (1976) attributes the formation of purely organic biolaminated structures to conditions of diminished sedimentation. Gebelein (1976) also considers that low incidence of surface grazing organisms and potential for rapid cementation are important to the preservation of the structure. The micrite laminae described here are similar to those described by Monty (1955, 1967) from shallow, subtidal stromatolites in the Bahamas and by Leeder (1975) in oncoids from the Dinantian of northern England and southern Scotland.

3.1.2 Composite Sparry Laminae

These are common to most nodules, often occurring in regular alternation with micrite laminae. Individual sparry laminae are up to 250 μm thick and generally less continuous than the micrite laminae (Fig. 3b). They are composed of fine granular pseudospar and microspar. Most contain parallel stacks or interwoven bundles of fine filaments, 8 – 20 μm thick (Fig. 3c). Laminae commonly display a crinkled to pustular microrelief with a variable degree of morphological inheritance.

Bioturbation often disrupts the laminar fabric (Fig. 3d). Discordant and laminoid, spar-filled, gas escape cavities are also present.

Discontinuous band and clusters of sessile, tubular foraminifera are sometimes included within the fabric, together with thin micrite laminae, similar to the composite micrite laminae described above.

The composite sparry laminae are similar to the calcified layers built by *Phormidium hendersoni* Howe (formerly *Schizothrix calcicola*) in lower intertidal and subtidal stromatolites from Andros Island, Bahamas (Monty 1967, Figs. 182, 191). Similar fabrics have been figured from brackish to fresh-water algal mats (Monty 1976, Fig. 4c).

The mechanisms of calcification by blue-green algae are still poorly understood (Monty 1976). It may take place in response to a range of biological, biochemical and physical factors. The sparry laminae may have been formed by the same algal assemblage responsible for the micrite laminae but under slightly different environmental conditions.

3.1.3 Tubular Foraminiferal Laminae

These are less common than the previous two microfabrics, and generally occur as discontinuous intergrowths within other laminae. The foraminifera also occur occasionally as solitary individuals or form localised, irregular masses within nodules.

Internally, laminae consist of multiple stacks of sessile, tubular foraminifera. Tubes vary from hemispherical to flattened bean shapes, occasionally becoming irregular (Fig. 3e). The base of the tube is attached to a substrate.

Oncoids containing foraminiferal associations are present in many Upper Palaeozoic carbonate successions (Flügel 1977, Peryt 1977, and many others). They have been used as indicators of a wide range of environments from shallow agitated waters (Flügel 1977) to environments of around 30 m depth (Füchtbauer 1968, Peryt 1977). Gebelein (1976) mentions a similar foraminiferal association in Recent oncoids from shallow, high energy, subtidal areas in the Bahamas. It would seem from this that the presence of foraminiferal intergrowths within nodules is of little direct environmental significance. Their development was probably a function of biological rather than physical factors.

3.1.4 Tubular Calcareous Algal Laminae

Laminae of this type usually form the initial encrustation within a nodule or occur as thin, semi-continuous, isolated strands and bundles of tubes, embedded in dark, dense micrite. Individual tubes range from 20 µm to 50 µm in diameter. They include branching and non-branching forms. Many laminae are composed entirely of bundles of the problematic tubiform alga *Donezella lutugini* Maslov. It is unlikely that the calcareous algal laminae accreted in very agitated waters.

3.1.5 Red Algal Laminae

Discontinuous crusts displaying the characteristic cellular structure of rhodophyte algae are present in some nodules. They are most common in algal nodule wackestones associated with phylloid algal limestones.

Crusts range from smooth to "nobbly". Growth irregularities are often inherited by the succeeding cyanophyte laminae. Individual crusts are usually restricted to one side of a nodule, suggesting that colonisation took place during a period of stationary growth.

Stacheinids and phylloid algae have been observed forming laminae, but generally recrystallisation of the microstructure prevents identification of the alga present.

By analogy with present day corraline algal crusts developed on mobile substrates (Bossellini and Ginsburg 1971, Milliman 1974, 1977), the red algal laminae developed in fairly calm, clear water, probably during a period of nodule immobility. According to Adey and Macintyre (1973), Recent forms have a slow rate of growth and flourish in a moderate to low energy environment between 0 m and 160 m depth.

3.1.6 Clotted Micrite

This microfabric is typical of nodules in purer wackestone intervals. It either forms the whole nodule or alternates with other fabrics.

It is characterised by irregular to rounded masses of dark, dense micritic clots. Clots are separated by fenestrae filled with sparry calcite cement and inclined or horizontally disposed geopetal sediment. The micrite has a spongy, mottled appearance; twisted bundles of *Girvanella* filaments are sometimes present. Large patches of the enclosing sediment are often incorporated within the clotted fabric.

Individual clots range from 150 – 1000 µm and are round to ovoid. Some surround a nucleus, which is either a sessile tubular foraminifera or a *Donzella* "twig" (Fig. 3f).

In some examples this fabric is restricted to one side of a nodule, indicating a period of stationary growth; in others the fabric forms a concentric envelope around the nodule. Bioturbation may penetrate and emphasise the clotted texture.

The fenestral, clotted fabric is similar to the thrombolite fabrics of Aitken (1967). Individual clots resemble the jelly-like pustules formed by the coccoid cyanophyte, *Entophysalis* (Logan et al. 1974, Golubic 1976). Formation of the micrite within the clots was probably the result of organic processes, such as bacterially induced precipitation during decay of the cyanophyte colonies (Golubić 1976).

The nuclei present in some of the clots are not found in modern thrombolites which form by in situ aggradation of micrite, without an initial substrate. Only *Donezella* (generally *D. lutugini* Maslov) and foraminifera have been observed as organic cores: there is no obvious preference for these allochems to be encrusted by the cyanophyte colonies.

Present-day thrombolitic fabrics may form in a number of ways (Monty 1976):

1. By autochthonous growth as pustular encrustations, similar to the pustular mats in Shark Bay, Western Australia (Logan et al. 1974, Hoffman 1976a).
2. By coalescence of discrete clots or colonies, thereby creating a highly porous, spongy mass.
3. By secondary processes of corrosion and biodegradation of originally laminated cryptalgal structures.

Absence of relic laminar textures within the thrombolites precludes a secondary origin by destructive organic processes. On the other hand, the ubiquity of solitary micrite clots in other facies and the patches of matrix incorporated within the thrombolite suggest the fabric to have been formed at least partially by the mergence of originally discrete clots. Finally, the encrusting nature of the fabric suggests that in situ growth following colonisation of an unstable substrate probably also played some part in the formation of the thrombolites.

Thrombolite fabrics have been described from a wide variety of environments, from supratidal (Hoffman 1976b) through high intertidal (Logan et al. 1974, Hoffman 1976a), to shallow subtidal (Aitken 1967, Schmitt and Monninger 1977) and even basinal (Schwarzacher 1978). A shallow, subtidal origin is postulated for the fabrics described here.

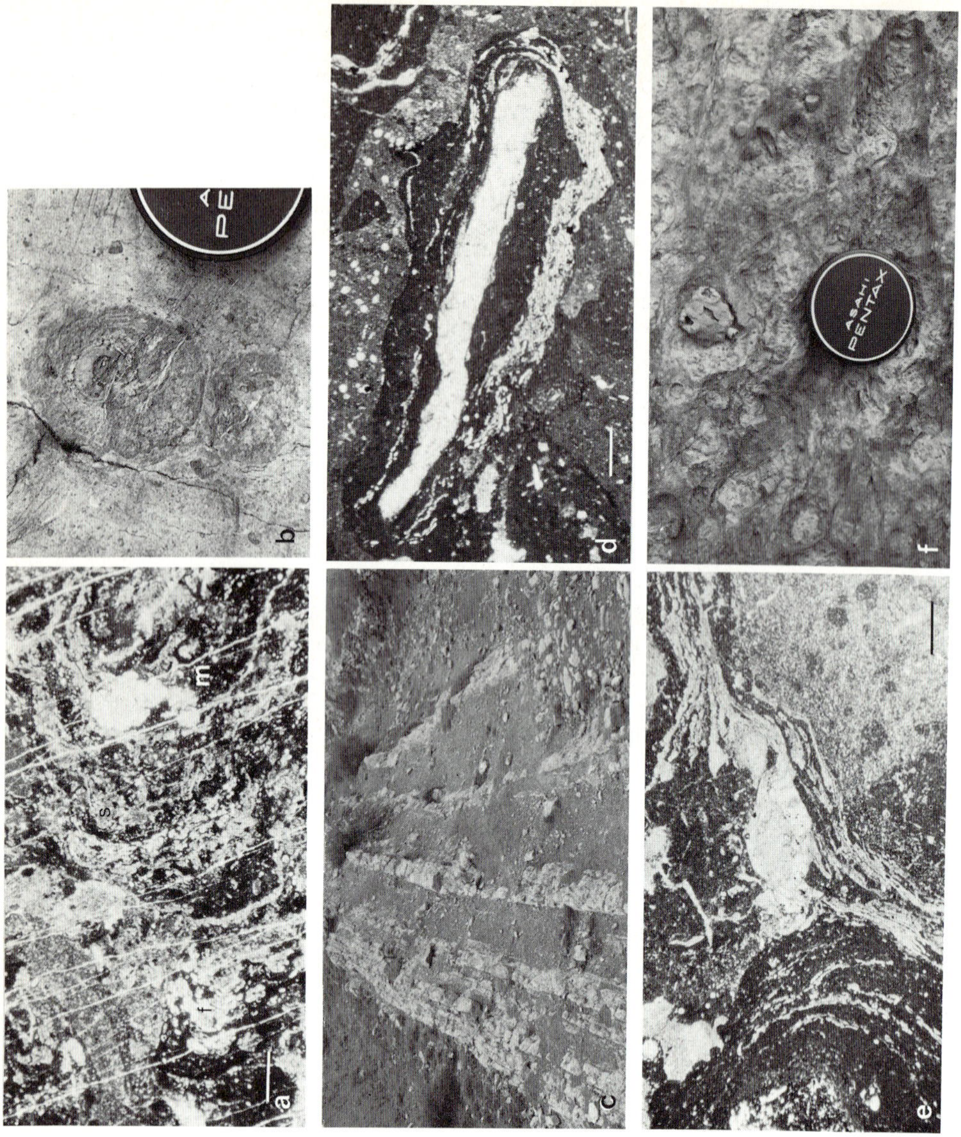

Fig. 4. a Part of a large oncoid containing composite micrite (*m*) and sparry (*s*) laminae and tubular foraminiferal laminae (*f*) (scale bar = 750 μm). **b** Small oncoids with concentric laminated fabrics from limestone unit 3 near Pinos village. **c** Interbedded oncolite, whole fossil wackestones and fossiliferous shelf mudstones near Pinos village. **d** Small oncoid showing semi-concentric micrite and sparry laminar fabrics, indicating intermittently rolling growth (scale bar = 750 μm). **e** Detail of intergrowth between two oncolites showing coalescence of laminar fabrics and incorporation of detrital sediment into the nodule structure (scale bar = 750 μm). **f** Sections of crinoid stem and calyx forming the nuclei to small oncoids in an impure lime mudstone

3.2 Nodule Microstructure

In thin section the nodules range from superficial, occasionally incomplete coats, to thick, multiple encrustations, completely enveloping the nucleus. Examples are up to 20 cm in diameter and have a rounded to flattened shape. Outer surfaces may be smooth or regular, but most examples show a pustolose or tuberculate relief, reflecting irregularities in the internal fabric.

Most nodules do not contain more than three different microfabrics, occurring in any combination (Fig. 4a). Some are true oncoids (sensu Logan et al. 1974) with only laminar fabrics present; others contain a thrombolite component. The composite sparry and micrite laminae and the thrombolite microfabrics are most common. The first encrustation surrounding the nucleus is generally micritic.

Petrographic examination often reveals a complex history of growth and intergrowth between nodules. Individual microfabrics are often discontinuous or vary greatly in thickness around a nodule, indicating periods of mainly stationary growth (Fig. 5). Mergence of nodules occurred during such periods, when fabrics were able to coalesce and intergrow. Evidence for rolling growth is provided by rarer examples with more concentric encrustations (Fig. 4b).

The nuclei of nodules are mostly bioclasts and rarely intraclasts. Endolithic *Girvanella* acts as a precursor to encrusting growth on some bioclast nuclei.

Spar-filled, non-fabric selective, fenestral porosity is characteristic of the nodules. Most fenestrae are irregular, showing no preferred orientation with respect to the microstructure. They are analogous to gas escape "birdseyes".

Ribbons of dense micrite within the enclosing sediment and the squashing of nodules against other allochems, suggest that at least the outer parts of the nodules were soft when emplaced in the sediment. However, randomly inclined geopetal sediment within some fenestrae and other cavities indicate that some form of sediment stabilisation (lithification?) took place very early during nodule accretion.

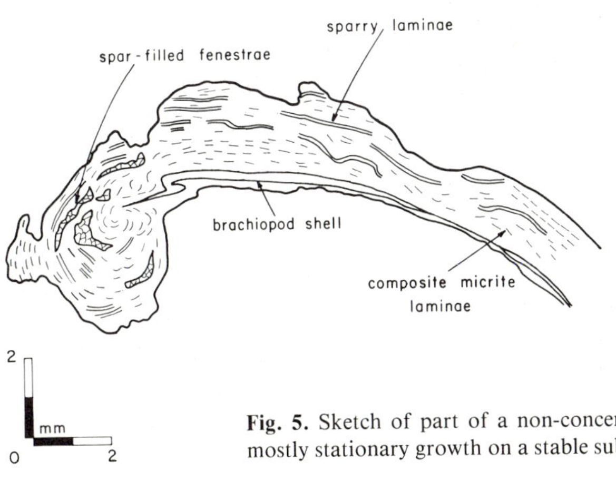

Fig. 5. Sketch of part of a non-concentric algal encrustation, indicating mostly stationary growth on a stable substrate (brachiopod shell fragment)

4 Algal Nodule Wackestone Facies

Beds attributable to this facies occur most commonly at the base of major limestone units within the La Majua Member and in close association with phylloid algal limestones. They also occur near the top of terrigenous clastic sequences and within the thin, often laterally discontinuous carbonate intervals of the Candemuela Member.

The limestone is buff to grey weathering with a mottled appearance. When fresh, it is dark grey, highly fetid and bituminous with a cryptocrystalline cement and a semi-porcellanous to muddy, mottled texture. Bedding is generally well developed but irregular and rubbly in places.

The algal nodules are supported by homogeneous micrite which contains variable amounts of terrigenous mud. Scattered silt to fine sand-sized quartz grains, occasionally with oolitic coatings, may also be present in examples found in the vicinity of clastic intervals. In places, the homogeneous micrite is recrystallised to a clotted texture.

The algal nodules are associated with abundant fragments of brachiopods, foraminifera, echinoderms, calcareous algae and bryozoa. Scattered fragments of stromatoporoids, ostracods, molluscs (mostly gastropods), corals and trilobites also occur. Most bioclasts are small and extensively micritised or bored.

Beds within terrigenous clastic intervals commonly have sharp, erosive bases and a more granular texture with scattered ooids and quartz sand grains. They contain diverse and well preserved biota, dominated by gastropods. Scattered, finely comminuted land plants may also be present.

Intraclasts occur as amorphous and eroded lumps and as micritic clots. Eroded lumps are up to 2.5 mm; they include fragments of oncoids and associated facies. Most have rounded outlines with micritised and coated margins. Micritic lumps packed with *Girvanella* filaments are also present.

The rock displays a homogeneous fabric with poor sorting and a very wide range of grain sizes. Most allochems have rounded outlines due to biogenic processes rather than mechanical abrasion.

Sparry calcite cement is found infilling the intraparticle fossil pores as well as mouldic and burrow porosity. Geopetal sediment is often present within the cavities.

Late stage pressure solution effects are often very extensive. They include thick seams and "horsetail" microstylolites. A stylobrecciated texture occurs in places.

The micrite matrix ranges from non-ferroan to ferroan calcite. The iron content becomes larger with increasing proximity to clastic intervals and with increases in the amount of terrigenous impurities within the micrite. Ferroan calcite also forms the pore-filling cement; non-ferroan calcite is present as a localised early fringing cement or as a late stage blocky pore fill.

Neomorphic alteration is very variable. In places the matrix is completely recrystallised to a ragged, fibrous calcite, whereas molluscan shells are changed to fine, granular calcite.

4.1 Interpretation

Five separate lines of evidence suggest that the algal nodules did not form in very agitated waters.

Firstly, there is only limited evidence for accretion by continuous rolling. Much growth was accomplished during stationary periods.

Secondly, the micrite matrix in which the nodules are emplaced suggests a low energy environment with no winnowing of fine material.

Thirdly, the matrix-supported wackestone texture precludes an origin for the micrite by settling of fine sediment through a self-supporting framework of nodules and other grains.

Fourthly, the development of delicate microfabrics, notably the thrombolites, indicate fairly tranquil conditions on the sea floor during nodule growth.

Fifthly, there is no evidence for prolonged high energy in most of the allochems associated with the nodules.

It thus appears that the nodules grew in a shallow, low energy, subtidal environment. The limited rolling growth on nodules was probably accomplished during intermittent higher energy events such as storm surges. The rips and sharply defined discontinuities in the internal fabric of some nodules may be considered as indicating the onset of such events, when the growths were torn from the substrate to which they had been attached. Similar effects are described by Gebelein (1969) from subtidal algal biscuits from Bermuda. Strong currents would not have been necessary to move what must have been fairly lightweight structures, composed almost entirely of organic matter.

The location of algal nodule wackestones at the base of major limestone units and within the upper parts of clastic intervals in the La Majua Member, indicates that they are related to the onset of major transgressions. These transgressive wackestones represent a transition between the terrigenous clastic belt and the landward moving, purer carbonate belt.

Thin, often laterally discontinuous, sharp-based bands interbedded with clastic sequences are considered to represent minor events at the onset of transgression, as marine conditions invaded the fringes of the active clastic belt. Skeletal debris and other allochems would be carried landwards and localised carbonate sedimentation would ensue where the seas became sufficiently clear. Sedimentation rates were probably low, small amounts of terrigenous sediment within seawater effectively inhibit carbonate sedimentation (Wilson 1975, p. 3).

5 Oncolitic, Whole Fossil Wackestone Facies

These wackestones are interbedded with fossiliferous calcareous mudstones or marls, at the top of major limestone bands within the La Majua and Candemuela members (Fig. 4c). They display a distinctive buff to yellow weathering colour and contain abundant large oncoids which give the rock a rubbly appearance at outcrop.

The oncoids are up to 25 cm in diameter and have a flattened, elliptical shape. They are embedded in a micrite matrix containing significant amounts of

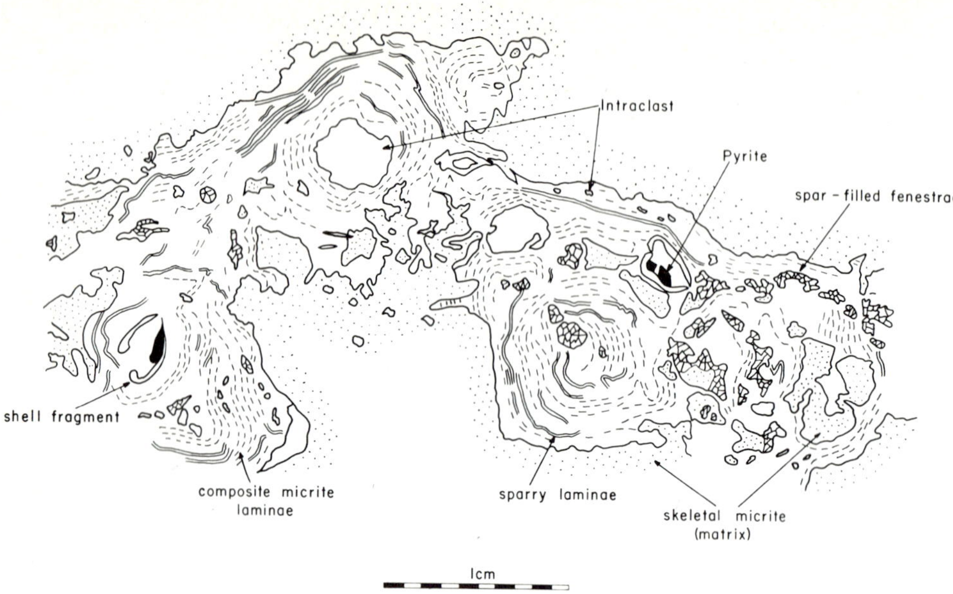

Fig. 6. Sketch of part of a composite oncoid showing complex growth history with coalescence and intergrowth of small concentric nodules

terrigenous mud. The nodules commonly have a crinkled or pustular outer surface. Composite micrite and sparry laminae form the dominant internal fabric; non-laminar fabrics are generally absent (Fig. 4d).

Complex growth histories are in evidence (Figs. 4e and 6). Initial accretion generally took place by intermittent rolling to form small ($\leqslant$30 mm), round to ovoid nodules. These then coalesced to form large ellipsoidal structures (Fig. 6). The larger nodules generally show little evidence of inversion or prolonged rolling growth, suggesting that currents were not powerful enough to move them around on the sea floor. In this context they are comparable with the subtidal algal biscuits described by Gebelein (1969) from the low energy, shallow, subtidal areas of Bermuda.

The oncoids are associated with a large, diverse and well preserved faunal assemblage, brachiopods being particularly abundant. They include various productoids, spiriferids and chonetids. Gastropods, lacy and ramose bryozoa, large fragments of crinoid stem with rarer calices (Fig. 4f) and small, delicate horn corals are also very common, together with scattered trilobites, bivalves, blastoids and echinoderms.

The oncolitic, whole fossil wackestones occupy a similar transitional position to the algal nodule wackestones, separating areas of clastic and carbonate sedimentation. However, they are associated with the commencement of regressive rather than transgressive phases (as in the algal nodule wackestones).

It is considered that conditions were more tranquil than those responsible for the algal nodule wackestones. The diverse fauna is indicative of calm, shallow conditions (Winkler Prins, pers. comm.) and its excellent preservation suggests minimal transport of the bioclasts prior to burial. This interpretation is confirmed by the comparison with Gebelein's (1969) algal biscuits, which occur only in areas subject to minimal current activity. The currents were probably sufficiently strong to move the smaller nodules, but once they attained a certain size they became stable and were able to attach themselves to the substrate.

The sedimentation rate was probably low due to terrigenous admixtures in the seawater, inhibiting carbonate sedimentation. The fossiliferous mudstones interbedded with the limestones, represent pulses of suspended material which made the sea too muddy for carbonate sedimentation to continue.

6 Conclusions

Present-day algal nodules form in a wide variety of environments, from freshwater lakes and streams to fully marine conditions. Despite this diversity they are often used as indicators of high energy, low intertidal to shallow subtidal environments in ancient carbonate sequences (following the suggestions by Logan et al. 1974). The environmental interpretation of the algal nodule wackestones at San Emiliano is derived not only from their composition and internal structure but also from the general palaeontological and sedimentological aspects of the associated sediments.

From the present study it seems that the different microfabrics described in the nodules do not reflect any marked changes in the physical environment. More likely they are the result of subtle variations in biological, sedimentological and geochemical processes within the nodule community and the environment as a whole. Monty (1955, 1967, 1976) has outlined the importance of such processes in the development of modern cryptalgal fabrics.

Nodules grew in nearshore zones adjacent to the clastic shoreline. The water was slightly polluted by terrigenous mud which probably inhibited carbonate precipitation. The environment was very shallow, generally tranquil and subtidal. Clearer, slightly deeper and more tranquil conditions existing further offshore, produced purer limestone facies associated with the nodule bearing limestones. A similar palaeolocation is inferred for many other algal nodule limestones from the Upper Carboniferous of the Cantabrian Mts.

The low energy origin for the nodules, together with the absence of algal-bound detrital laminae within the described structures, support Monty's (1972, 1974) hypothesis that many pre-Mesozoic algal nodules are more analogous to modern freshwater forms than to the intertidal and shallow subtidal forms described from present-day areas. Freshwater nodules generally grow in rather tranquil environments which experience only intermittent turbulence. The conclusions drawn from the study also support Peryt's (1981) contention that Phanerozoic (porostromiate) oncoids grew in marine subtidal settings often under conditions of reduced sedimentation.

Acknowledgements. The author wishes to thank Max Dobson and Bob Wagner for constructive criticism and NERC for financial support.

References

Adey WH, Macintyre IG (1973) Crustose coralline algae: A re-evaluation in the Geological Sciences. Bull Geol Soc Am 84:883 – 904

Aitken JD (1967) Classification and environmental significance of cryptalgal limestones and dolomites, with illustrations from the Cambrian and Ordovician of Southwestern Alberta. J Sediment Petrol 37:1163 – 1178

Bossellini A, Ginsburg RN (1971) Form and internal structure of Recent algal nodules (rhodolites) from Bermuda. J Geol 72:669 – 682

Bowman MBJ (1979) The depositional environments of a limestone unit from the San Emilian Formation (Namurian/Westphalian), Cantabrian Mts., NW Spain. Sediment Geol 24:25 – 43

Bowman MBJ (1980) The sedimentology and stratigraphy of the San Emiliano Formation and associated sequences, Carboniferous, NW Spain. Unpubl Ph D thesis, Univ Sheffield, p 316

Bowman MBJ (in prep) The stratigraphy of the San Emiliano Formation and its relationship to other Namurian/Westphalian A sequences in the Cantabrian Mts, NW Spain

Dalrymple DW (1965) Calcium carbonate deposition associated with blue-green algal mats, Baffin Bay, Texas. Univ Texas Inst Mar Sci 10:187 – 200

Flügel E (1977) Environmental models for Upper Palaeozoic benthic algal communities. In: Flügel E (ed) Fossil Algae, recent results and developments. Springer, Berlin Heidelberg New York, pp 314 – 343

Füchtbauer H (1968) Carbonate sedimentation and subsidence in the Zechstein basin (northern Germany). In: Muller G, Friedman GM (eds) Recent developments in Carbonate sedimentology in Central Europe. Springer, Berlin Heidelberg New York, pp 196 – 204

Gebelein CD (1969) Distribution, morphology and accretion rate of Recent subtidal algal stromatolites, Bermuda. J Sediment Petrol 39:49 – 69

Gebelein CD (1976) Open marine subtidal and intertidal stromatolites (Florida, the Bahamas and Bermuda). In: Walter MR (ed) Stromatolites. Developments in sedimentology, 20. Elsevier, Amsterdam, pp 381 – 388

Golubić S (1976) Organisms that build stromatolites. In: Walter MR (ed) Stromatolites. Developments in sedimentology 20. Elsevier, Amsterdam, pp 113 – 126

Golubić S, Focke JN (1978) *Phormidium hendersonni* Howe: identity and significance of a modern stromatolite building organism. J Sediment Petrol 48:751 – 764

Graff van de WJE (1971) Three Upper Carboniferous, limestone-rich, high destructive, delta systems with submarine fan deposits, Cantabrian Mountains, Spain. Leidse Geol Meded 46:157 – 235

Hoffman PC (1976a) Stromatolite morphogenesis in Shark Bay, Western Australia. In: Walter MR (ed) Stromatolites. Developments in sedimentology 20. Elsevier, Amsterdam, pp 261 – 272

Hoffman PC (1976b) Environmental diversity of Middle Precambrian Stromatolites. In Walter MR (ed) Stromatolites. Developments in sedimentology 20. Elsevier, Amsterdam, pp 599 – 612

Leeder MR (1975) Lower Border Group (Tournaisian) stromatolites from the Northumberland Basin. Scott J Geol 11:207 – 226

Logan BW, Hoffman P, Gebelein DC (1974) Algal mats, cryptalgal fabrics and structures, Hamelin Pool, Western Australia. Am Assoc Petrol Geol Mem 22:140 – 194

Meijer de JJ (1971) Carbonate petrology of algal limestones (Lois Cigüera Formation, Upper Carboniferous, León, Spain). Leidse Geol Meded 47:1 – 97

Milliman JD (1974) Marine carbonates. Springer, Berlin Heidelberg New York, p 375

Milliman JD (1977) Roles of calcareous algae in Atlantic continental margin sedimentation. In: Flügel E (ed) Fossil Algae, recent results and developments. Springer, Berlin Heidelberg New York, pp 232 – 247

Monty CLV (1955) Recent algae stromatolites in the windward lagoon, Andros Island, Bahamas. Ann Soc Geol Belg 88:269 – 276

Monty CLV (1967) Distribution and structure of Recent stromatolitic algal mats, Eastern Andros Island, Bahamas. Ann Soc Geol Belg 90:55 – 100

Monty CLV (1972) Recent algal stromatolitic deposits, Andros Island, Bahamas. Preliminary report. Geol Rundsch 61:742 – 783

Monty CLV (1974) Precambrian background and early Phanerozoic history of stromatolitic communities, an overview. Ann Soc Geol Belg 96:585 – 624

Monty CLV (1976) The origin and development of cryptal fabrics. In: Walter MR (ed) Stromatolites. Developments in sedimentology 20. Elsevier, Amsterdam, pp 193 – 249

Peryt TM (1977) Environmental significance of foraminiferal-algal oncolites. In: Flügel E (ed) Fossile Algae, recent results and developments. Springer, Berlin Heidelberg New York, pp 61 – 65

Peryt TM (1981) Phanerozoic oncoids – an overview. Facies 4:197 – 214

Raćz L (1964) Carboniferous calcareous algae and their associations in the San Emiliano and Lois Cigüera formations (prov León, NW Spain). Leidse Geol Meded 31:1 – 112

Raćz L (1965) Late Palaeozoic calcareous algae in the Pisuerga Basin (N Palencia, Spain). Leidse Geol Meded 31:241 – 260

Riding RE (1975) *Girvanella* and other algae as depth indicators. Lethaia 8:173 – 179

Riding RE (1977) Calcified *Plectonema* (blue-green algae), a recent example of *Girvanella* from Aldabra Atoll. Palaeontology 20:33 – 46

Schmitt M, Monninger W (1977) Stromatolites and thrombolites in Precambrian/Cambrian boundary beds of the Anti-Atlas, Morocco – preliminary results and developments. Springer, Berlin Heidelberg New York, pp 80 – 85

Schwarzacher W (1978) The petrology and sedimentology of the middle Glencar limestone, Sligo Basin, Northwestern Ireland – a study in bed formations. In: Bathurst RGC et al. (eds) Processes in carbonate environments (Course Notes). 1.1 – 1.4

Wilson JL (1975) Carbonate facies in geologic History. Springer, Berlin Heidelberg New York, p 471

IV.15 Morphogenesis of Oncoids in the Lower Carboniferous Llanelly Formation of South Wales

V.P. Wright[1]

1 Introduction

Oncoids are ubiquitous in the peritidal facies of the Lower Carboniferous of Britain. A study has been made of the variety of oncoids which occur in one of these peritidal units, the Llanelly Formation in South Wales, with the aim of elucidating the factors controlling oncoid morphology.

The Llanelly Formation, of probable Arundian age (Viséan), outcrops along the north eastern part of the South Wales coalfield and is one of a number of thin peritidal units within the Lower Carboniferous shelf limestones of South Wales. The detailed stratigraphy of this unit has been discussed elsewhere by the author (Wright 1981a), and consists of a varied sequence of conglomerates, sandstones, claystones and limestones of coastal floodplain and peritidal facies. The formation is subdivided into four members (Fig. 1) of which the lowermost and uppermost are of floodplain origin and the middle two are of shallow marine-peritidal origin but containing thin terrestrial horizons with calcretes and rendzina palaeosols (Wright 1980, 1981a). These two middle members contain a variety of oncoids.

2 Oncoid-Bearing Lithofacies

There are two distinct oncoid bearing lithofacies in the Llanelly Formation. Firstly oncoids occur in coarse grained, sandy, lithoclast, bioclastic grainstones containing clasts of underlying lithologies including numerous calcrete clasts; crinoid, brachiopod and orthocone fragments, foraminifera and fragments of the dasyclad *Koninckopora* and the problematic micro-organism *Uraloporella*. Such horizons are well developed at the base (the Hendre Bed of Wright 1981a) of the Cheltenham Limestone Member (Fig. 1) immediately above the truncated top of the Clydach Halt Member at Craig y Gear (British Grid reference S021,224,134); at Daren Cilau (S011,194,156); Blaen Onneu (S011,155,169) and Cwar yr Hendre (S001,099,149). This horizon can be correlated over a distance

1 Department of Earth Sciences, Open University, Walton Hall, Milton Keynes, MK7 6AA, Great Britain

Coated Grains (ed. by T.M. Peryt)

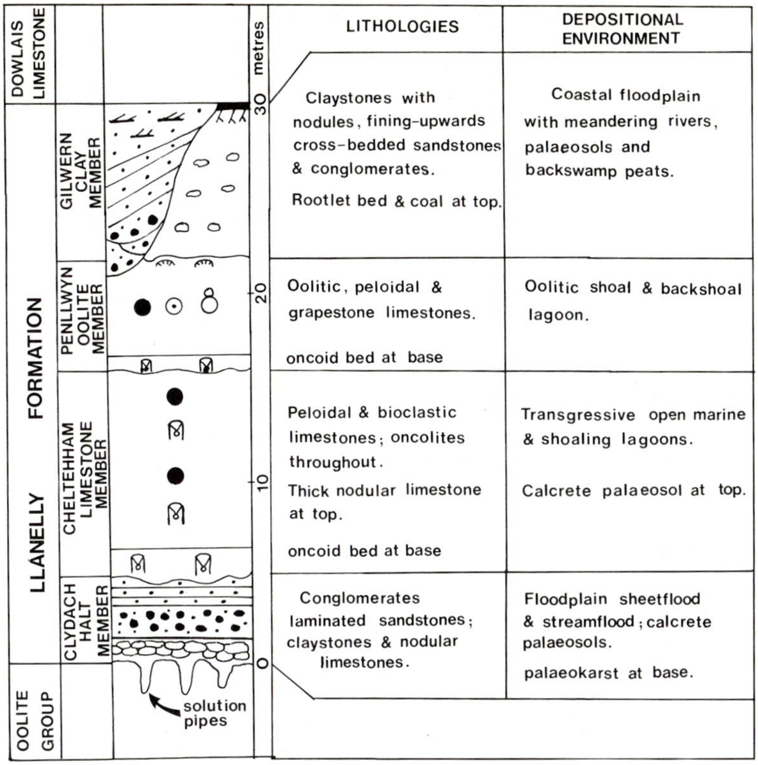

Fig. 1. Log of the Llanelly Formation

of 12.5 km along strike and reaches a maximum thickness of 50 cm. This lithofacies also occurs as a thin horizon, usually under 30 cm thick as the base of Penllwyn Oolite Member, called the *Uraloporella* Bed (Wright 1981 a), which usually has truncated a massive calcrete palaeosol at the top of the Cheltenham Limestone Member. Lastly this lithofacies also occurs within the Cheltenham Limestone, usually as very thin horizons less than 10 cm thick, but occasionally up to 40 cm. These thin horizons often occur above palaeosols or supratidal deposits and sometimes at the bases of rare fining upward cycles (see below and Wright 1981 a, b; Wright and Wright 1981).

This lithofacies is interpreted as a type of transgressive deposit, marking marine incursions over subaerial deposits, at the bases of the Cheltenham Limestone Member and the Penllwyn Oolite and at the bases of minor transgressions in the Cheltenham Limestone Member. Water depth was probably very shallow, as evidenced by the abundance of dasyclad algae. The horizon at the base of the Penllwyn Oolite has been interpreted as having formed in a schizohaline lagoon (Wright 1981 c).

The second oncoid bearing lithofacies which occurs only in the Cheltenham Limestone Member consists of a variety of peloidal, bioclastic grainstones, packstones and wackestones with a restricted fauna of ostracods and attached vermiform gastropods. This lithofacies contains rare fining upward cycles, 0.5 to 1 m thick, capped by algal laminites, fenestral limestones, gypsum pseudomorph-bearing mudstones and palaeosols. These cycles are interpreted as regressive, shoaling restricted bay facies (Wright 1981 b, Wright and Wright 1981), and these lithofacies as a whole are interpreted as having formed in a low energy, protected, very shallow restricted bay or lagoon which was periodically hypersaline or schizohaline (Wright and Wright 1981).

Thus, there are two distinct lithofacies, namely a fully marine, transgressive grainstone facies and a low energy restricted bay facies. These differences are reflected in the types of oncoid which occur.

3 Types of Oncoid

There are two types of oncoid which occur in each lithofacies and one additional type common to both.

3.1 Transgressive Grainstone Lithofacies Oncoids

Three distinct types of oncoids occur in this lithofacies.

3.1.1 Sub-Spherical Dense Oncoids. These are relatively small, dense oncoids averaging 7 – 8 mm in diameter, sub-spherical rounded, light brown oncoids (pisoncoids). They possess a distinctly laminated cortex with fine concentric laminations and show little or no asymmetry to the lamination thickness. They have a dense micritic microstructure (Fig. 2a) with only a few poorly preserved short tubes of *Ortonella* and *Mitcheldeania* (see Mamet and Roux 1975, p. 149) and *Bevocastria?* and occasionally were encrusted by vermiform gastropods. The

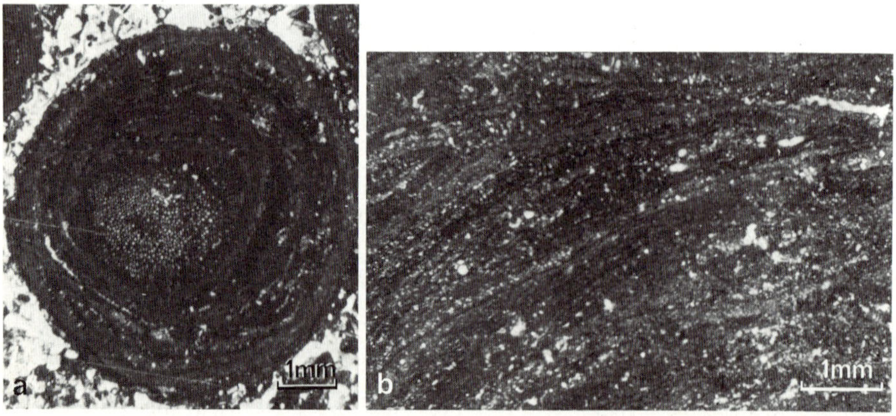

Fig. 2. a Dense oncoid from the coarse bioclastic grainstone facies (Hendre Bed). **b** Dense fabric showing microunconformities

calcite in the oncoids was precipitated not trapped and bound in place as there is an absence of micrite mud in this lithofacies and the oncoids are free of any sediment grains of the sort present in the grainstone matrix. The presence of tubes also suggests that biogenic calcification was active.

3.1.2 Large Branched Oncoids. This type of oncoid occurs only at the base of the Penllwyn Oolite in the *Uraloporella* Bed (Wright 1981 a). They reach a maximum size of 20 cm and are grey to light brown in colour and display a series of growth stages. The earliest growth stage, rarely larger than 2 – 5 cm diameter, consists of well rounded microncoids (Fig. 3) with a finely laminated cortex rather like the round dense oncoids described above but like all the oncoids at this horizon, they usually have a spar-filled nucleus making up about half the diameter, representing the site of a replaced aragonite pisoid (see Wrigth 1981 c). The laminated cortex often shows microunconformities (Fig. 2b), indicating considerable abrasion and rolling during growth. Prostromate tubes are poorly preserved consisting of a thick-tubed *Ortonella,* and rare tufts of fine tubed *Ortonella (O. kershopensis?), Bevocastria?* and the problematic organism *Stylocodium?* Attached vermiform gastropods are rare in this type of oncoid. The individual laminae contain only short *Ortonella* tubes, up to 200 μm long suggesting short periods of growth before being abraided.

The second growth stage of these macroncoids are seen in the larger macroncoids which have an inner core of the earlier growth stage surrounded, at least partly, by a cortex of irregularly branched columns up to 20 mm long and 4 mm diameter, radiating out from the core (Figs. 3 and 4a). The columns contain compact radiating bushes of the finer tubed *Ortonella* and more open elongated bushes of the thicker tubed form (Fig. 4b). The individual *Ortonella* tubes in these latter forms may reach lengths of several hundred microns. As a result of the irregular radiating, branching growth of these columns, numerous growth cavities were created within this cortex (Fig. 4a) similar to those described from modern freshwater stream oncoids by Golubic and Fischer (1975). These cavities became infilled with sediment, and contain concentrations of foraminiferas and articulated ostracods which may have inhabited these cavities (coelobites). The sides of the columns are often heavily encrusted by vermiform gastropods and also are coated by a finely laminated dark micrite (Fig. 4c). Schäfer and Stapf

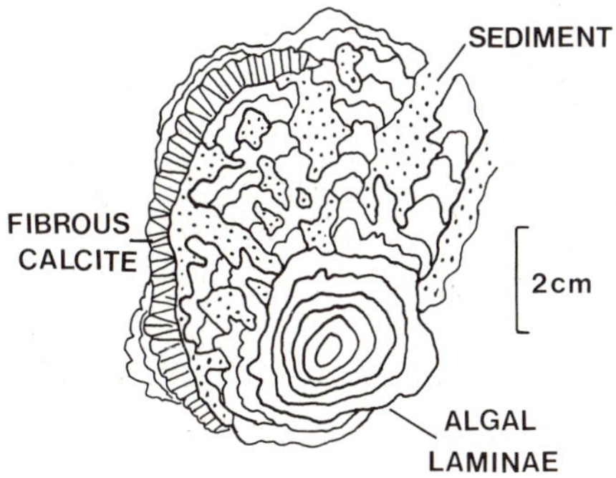

Fig. 3. Diagram drawn from a polished specimen showing one of the large branched oncoids from the base of the Penllwyn Oolite Member

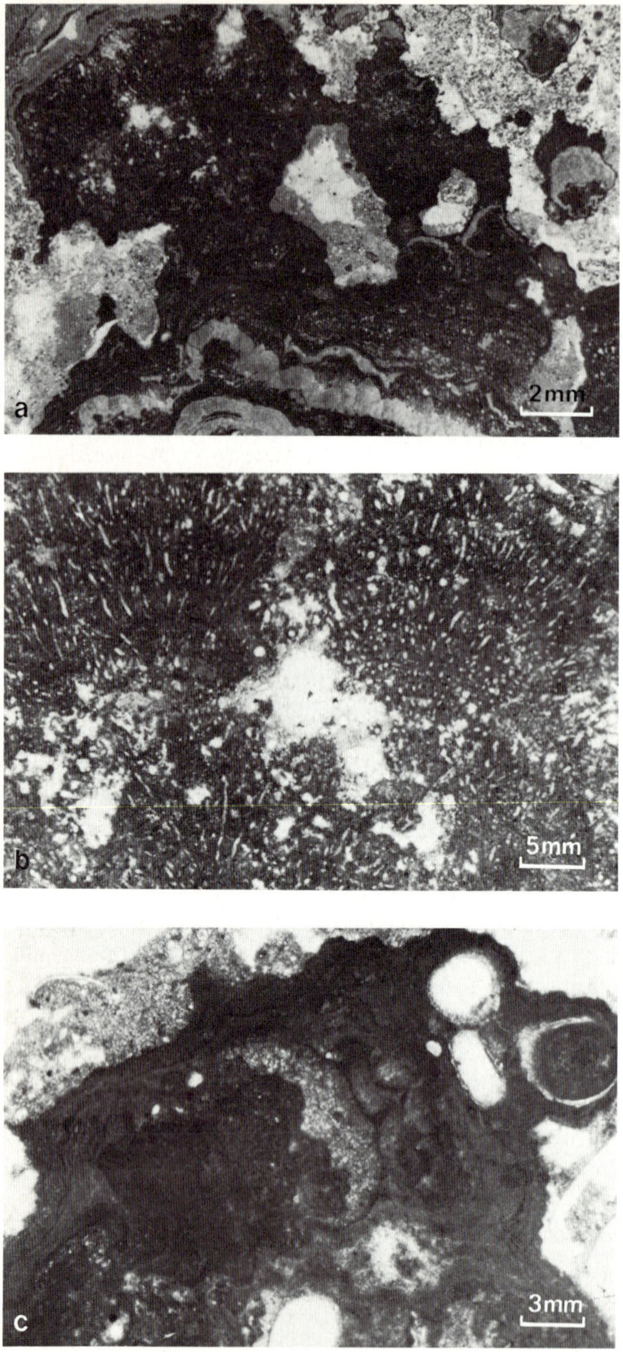

Fig. 4. a Branches composed of *Ortonella*. Note laminae of fibrous calcite at the *base* and *top left*. **b** *Ortonella* bushes. Close-up of the left hand columns from **a**. **c** Laminated coating on the sides of a column, covering a vermiform gastropod (*top right*)

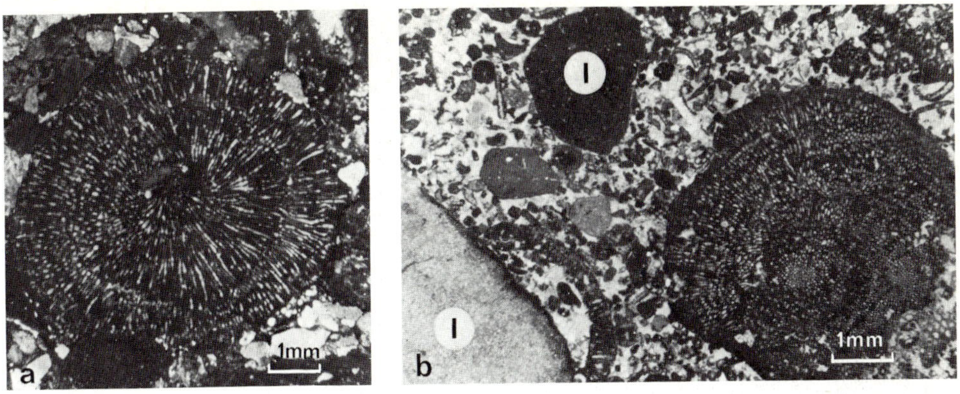

Fig. 5. a Spherical *Garwoodia* oncoid from the bioclastic grainstone facies. **b** Spherical *Garwoodia* oncoid showing evidence of breakage. Note intraclasts (*I*)

(1978, p. 96) describe similar coatings on columns on Recent oncoids from Lake Constance. The columns and branches in the Lower Carboniferous forms frequently coalesced and were then sometimes coated by prostrate filaments of *Mitcheldeania* and rarely by poorly preserved tubes of a micro-fossil resembling the *Rothpletzella-Coactilum* group. After the columns had coalesced new columns and branches developed as a result of the continuing radiating growth of *Ortonella.*

These large growth forms usually have an asymmetric form (Fig. 3) but some quite large sub-spherical forms also occur.

Interestingly at these bases of the columnar part of these oncoids, there is often a lamina heavily encrusted with vermiform gastropods which are rare in the denser, inner parts of the cortex. These oncoids as well as having spar filled nuclei have laminae of fascicular optic calcite (Fig. 3) interpreted by the author as representing replaced aragonite crusts formed during periods of hypersalinity (Wright 1981 c).

3.1.3 Spherical Garwoodia Oncoids. Associated with the round dense oncoids are spherical radiating masses of *Garwoodia,* up to 5 mm in diameter (Fig. 5 a, b). These make up about 20% of the total oncoid population. The radiating pattern of coarse tubes is very striking and differs from that seen in *Garwoodia* oncoids in the shoaling restricted by lithofacies (see below).

3.2 Restricted Bay Lithofacies Oncoids

As in the transgressive grainstone lithofacies there are three distinct oncoid types.

3.2.1 Spongy Oncoids. These are the most common types of oncoids in this lithofacies and are dark grey sub-spherical to irregular, poorly laminated forms up to 5 mm in diameter. They often display a concentric lamination showing asymmetric thicknesses. The microfabric is spongy (Fig. 6a) with poor to well

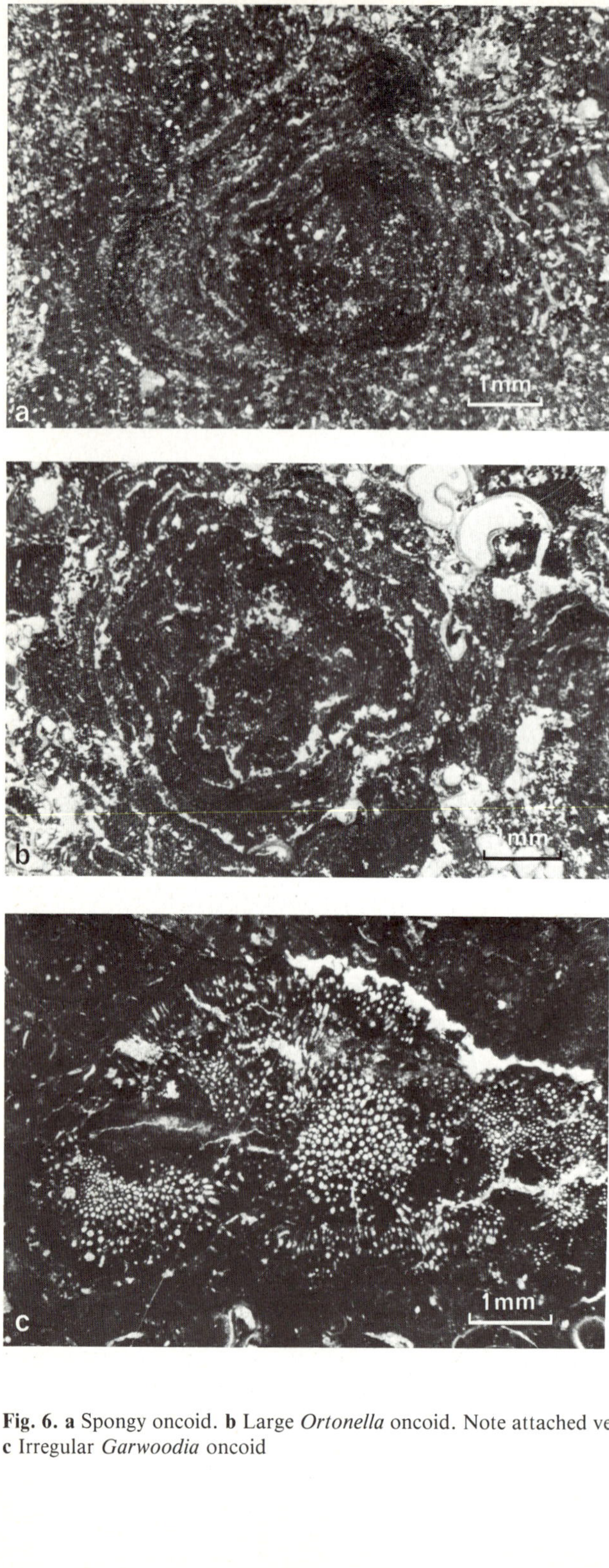

Fig. 6. **a** Spongy oncoid. **b** Large *Ortonella* oncoid. Note attached vermiform gastropod (*top right*). **c** Irregular *Garwoodia* oncoid

preserved tubes of *Ortonella, Bevocastria* (?), *Mitcheldeania* and *Girvanella.* Small tufts of *Ortonella* also occur. These pisoncoids occur in packstones, and wackestones.

3.2.2 Ortonella Macrooncoids. These oncoids have only been found at one horizon and occur in a packstone-wackestone matrix. They are up to 10 cm in diameter and have and irregular petal like outline formed of bushes of *Ortonella* (Fig. 6b). As a consequence of this dominantly bushy, radial growth pattern concentric laminations are poorly developed. The larger forms show a marked asymmetry in form. Associated with *Ortonella* are tubes of *Mitcheldeania* and well preserved *Wetheredella.* The outer laminae of the oncoids are heavily encrusted with vermiform gastropods. These oncoids coalesced as they grew and merge to form the basis for small algal-gastropod bioherms (Wright and Wright 1981).

3.2.3 Irregular Garwoodia Oncoids. A small number of oncoids built by *Garwoodia* occur in this lithofacies. They form rather irregular masses (Fig. 6c) showing marked growth asymmetry, with growth in only one or two directions unlike the spherical radial growth pattern seen in the *Garwoodia* oncoids from the bioclastic lithofacies.

4 Interpretation of Growth Forms

In the sub-spherical dense oncoids, the well laminated cortex and concentric growth pattern strongly suggest a greater degree of rolling than in the spongy oncoids. The possession of a concentrically laminated cortex has often been used as an indicator of frequent rolling (e.g., Logan et al. 1964; Dahanayake 1977) but Jones and Wilkinson (1978) from studies of recent freshwater oncoids showed that concentric laminae can form in static oncoids with the algae growing and calcifying on the sides and undersurface of the oncoid (see also Golubič and Fisher 1975). In these forms a marked asymmetry to the laminae width occurs even though each lamination is continuous around the oncoid. The lack of such asymmetry in the majority of spherical dense oncoids suggests that they were not static for long periods but were rolled during growth. This rolling was probably the cause of the short growth increments and the microunconformities in the laminations. The denser structure of these oncoids may also be related to rolling and abrasion. Schäfer and Stapf (1978, p. 101) describe hard, smooth oncoids forming in the more agitated breaker zone of Lake Constance. These oncoids had spongy layers built by bundles of calcified algal tubes which formed in quiet waters, but were abraded in the breaker zone. The hard smooth form of the oncoids Schafer and Stapf describe is the result of the preferential preservation of more resistant *Schizothrix* layers. There is no evidence that this was the case in these Lower Carboniferous oncoids but it seems likely that the absence of less resistant, bushes of calcified filaments in the higher energy forms, was due to abrasion and that the denser and more resistant laminae were preferentially preserved.

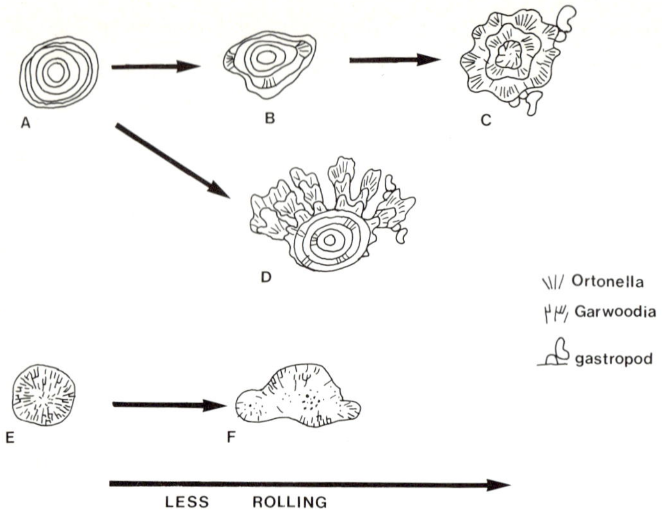

Fig. 7A – F. Summary of the types of oncoid in the Llanelly Formation. **A** dense oncoid (Fig. 2A), **B** Spongy oncoid (Fig. 6a), **C** *Ortonella* oncoids (Fig. 6b), **D** large branched oncoid (Fig. 3), **E** spherical *Garwoodia* oncoid (Fig. 5A), **F** irregular *Garwoodia* oncoid (Fig. 6c)

In the lower energy restricted bay facies the spongy fabric and the bushes of *Ortonella* would have stood a higher chance of preservation. Thus the lack of rolling and abrasion coupled with the radiating growth of the algae results in the less well laminated, sometimes asymmetric growth form. This form is best developed as the large delicate *Ortonella* oncoids (see Fig. 7) although even these forms, until quite large, do not show marked asymmetry which suggests some degree of periodic overturning.

The difference in energy levels in the two facies is also reflected in the different forms of *Garwoodia*. The forms which grew in the higher energy setting, as a result of more frequent rolling develop a strongly radial, spherical growth pattern whereas those in the lower energy restricted bay facies show asymmetry (Fig. 6c) as a result of more prolonged stationary periods during growth (Fig. 7). *Garwoodia* has much thicker walls to its tubes than *Ortonella* and was presumably more resistant to breakage.

The larger oncoids occurring at the base of the Penllwyn Oolite show both the basic growth forms, i.e. an inner, core area of well laminated, dense fabric and an outer area with bushes of *Ortonella* which occur as columns. This pattern occurs in many ancient oncoids (e.g., see Johnson 1963, p. 30). The obvious conclusion to draw here is that when the oncoids were small, under 2.5 cm in diameter, they were rolled about quite frequently, resulting in the dense, well laminated structure with numerous microunconformities. This frequent rolling also deterred encrusting vermiform gastropods. Beyond this size, the oncoids take on a radial, branched columar growth form preserving delicate calcified algal bushes. This second growth phase is most likely to have resulted from less frequent rolling. The cessation of rolling is marked in many oncoids by a surface heavily encrusted by vermiform gastropods immediately beneath the columnar growths. The change in growth does not seem to have been caused by a change in algal community because the same thick tubed *Ortonella* occurs both in the dense core and in the columns.

The change in growth might have been caused either by a change in energy level in the environment or by the oncoids having grown to such a size that they were no longer rolled about as often. If the change in growth form were due to a change in energy level, then all or most of the oncoids at that horizon should be at the same growth stage but in fact adjacent oncoids are at different stages and therefore, the second explanation seems more plausible.

Thus the change from frequently rolled to less frequently rolled results in a change from a laminar to a columnar branched form. There is a crude analogy to be drawn here with rhodoliths which also display laminar and columnar forms. The laminar forms develop in areas with frequent movement whereas the columnar and branched types form in less frequently rolled settings.

5 Oncoid Community

Oncoids, similar to the ones described in this paper, containing several porostromate genera are common in the Lower Carboniferous of Britain, e.g., see Wood (1941), Anderson (in Day 1970) and Leeder 1975. Some of these oncoids were given names, e.g, *Mitcheldeania.* Wood (1941) showed that such *Mitcheldeania* oncoids were mixtures of three algal genera. Ager (1963) considered that such *Mitcheldeania* nodules were examples of mutualism and Anderson (in Day 1970, p. 177) was also struck by the composite nature of similar algal growths and suggested some possible symbiotic associations between the algal species. The occurrence of the various constituent algal forms in monospecific masses, as oncoids or stromatolites (i.e, they were able to exist independently of one another) argues against there being such mutualistic or symbiotic associations between them.

6 Conclusions

The morphology of oncoids in the Lower Carboniferous Llanelly Formation in South Wales is related to the degree of agitation they underwent during growth (Fig. 7). Forms which grew in higher energy, transgressive grainstone lithofacies have sub-spherical and rounded forms, are well laminated and have a dense fabric. Forms which grew in lower energy, restricted bay lithofacies have more irregular forms, and less well laminated and have a spongy fabric which includes bushes of delicate porostromate tubes or consist entirely of radiating bushes of porostromates. Some oncoids change from a dense, well-laminated form to a columnar, branched form during growth which probably resulted from less frequent rolling as the oncoid grew larger and heavier.

References

Ager DV (1963) Principles of palaeoecology. McGraw Hill, New York, p 371

Dahanayake K (1977) Classification of oncoids from the Upper Jurassic carbonates of the French Jura. Sediment Geol 18:337–353

Day JBW (1970) Geology of the country around Newcastle. Mem Geol Surv UK

Golubič S, Fisher AG (1975) Ecology of calcareous nodules forming in Little Connestoga Creek near Lancaster, Pennsylvania. Verh Int Ver Limnol 19:2315–2323

Johnson JH (1963) Pennsylvanian and Permian algae. Q Colo Sch Mines 58:211 pp

Jones FG, Wilkinson BH (1978) Structure and growth of lacustrine pisoliths from Recent Michigan marl lakes. J Sediment Petrol 48:1103–1110

Leeder MR (1975) Lower Border Group (Tournaisian) Stromatolites from the Northumberland Basin. Scott J Geol 11:207–226

Logan BW, Rezak R, Ginsburg RN (1964) Classification and environmental significance of algal stromatolites. J Geol 72:68–83

Mamet B, Roux A (1975) Algues Devoniennes et Carboniferes de la Tethys Occidentale. Rev Micropaleontol 18:134–187

Schäfer A, Stapf KRG (1978) Permian Saar-Nahe basin and Recent Lake Constance (Germany): two environments of lacustrince algal carbonates. In: Matter A, Tucker ME (eds) Modern and ancient lake sediments, vol II. Spec. Publ IAS. Blackwell, Oxford, pp 83–107

Wood A (1941) The Lower Carboniferous calcareous algae *Mitcheldeania* Wethered and Garwoodia n. gen. Proc Geol Assoc 52:216–226

Wright VP (1980) Climatic fluctuation in the Lower Carboniferous. Naturwissenschaften 67:252–3

Wright VP (1981 a) The Llanelly Formation. In: Wright VP, Raven M, Burchette TP (eds) A field guide to the Lower Carboniferous Rocks naear Abergavenny. Dep Geol Univ College Wales, Cardiff, pp 36–72

Wright VP (1981 b) A subaerial stromatolite form the Lower Carboniferous of South Wales. Geol Mag 118:97–100

Wright VP (1981 c) Algal aragonite-encrusted pisoids from a Lower Carboniferous schizohaline lagoon. J Sediment Petrol 51

Wright VP, Wright EVG (1981) Palaeoecology of algal-vermiform gastropod bioherms from the Lower Carboniferous of South Wales. Neues Jahrb Geol Palaeontol Monatsh

Part V Vadoids

V.1 Vadoids

Tadeusz M. Peryt [1]

1 Definition and General Characteristics

Vadoid is a coated grain which originated in a vadose environment. To vadoids belong grains which have been (and still are) variously termed, e.g., cave pearls, fluvial pisoids, caliche pisoids, and vadose pisoids. Although commonly of pisoid size, vadoids are often of microid size and these cases were earlier called "diagenetical ooids" (Siesser 1973), "calcrete ooids" (Read 1974), "vadose ooids" (Harrison 1977) or simply "ooids" (Braithwaite 1975, Elloy and Thomas 1981, Richter, this vol.).

Because vadoids can form by a variety of processes in a range of environments from freshwater streams to hypersaline sabhka environments, from under the sky to under the rock roof or in soil, the resulting properties of vadoids are very variable; nevertheless, numerous workers note similarities between the products of different environments within the vadose zone (see Braithwaite 1979, Risacher and Eugster 1979, Chafetz and Butler 1980, Burgess, this vol.). The comprehensive characteristics of fluvial vadoids have been given by Braithwaite (1979), and of cave pearls and pisoids formed in spring-fed pools by Risacher and Eugster (1979), and the reader is referred to those papers (and literature quoted therein) for a detailed description. Folk and Chafetz (this vol.) and Chafetz and Meredith (this vol.) describe in detail travertine pisoids, and numerous papers deal with vadoids occurring in recent and ancient calcretes and peritidal facies (Table 1). Esteban (1976, Table 1), Assereto and Kendall (1977, Table 1) and Esteban and Pray (this vol., Table 1) compiled the features of vadoids occurring in those settings, but – as should be stressed – there are many controversial problems dealing with vadoids occurring in calcretes and peritidal facies, and the following text will discuss mainly these vadoids.

2 Evolving Concept of Vadoids

Vadoids were first described from continental settings and because they bear a resemblance to oncoids, the algal origin was accepted for Pliocene to Recent pisoids of the Great Plains prior to Swineford et al. (1958), who demonstrated

1 Instytut Geologiczny, Rakowiecka 4, 00–975 Warszawa, Poland

Coated Grains (ed. by T.M. Peryt)

Table 1. Vadoids in the literature (cave, travertine, fluvial and pool occurrences excluded)

Age	Provenance	Selected references
Quaternary	SW United States	Brown (1956), Chafetz and Butler (1980), Hay and Wiggins (1980)
	Yucatan	Ward (1975)
	Barbados	James (1972), Harrison (1977)
	Canary Islands	Tietz (in Flügel 1978, Plate 2/4,5)
	Spain	Calvet and Julià (this vol.)
	Algeria	Elloy and Thomas (1981)
	South Africa	Netterberg (1969), Siesser (1973), Knox (1977)
	Tanzania	Hay and Reeder (1978)
	Israel	Magaritz (this vol.)
	Persian Gulf	Purser and Loreau (1973), Scholle and Kinsman (1974)
	Aldabra (Indian Ocean)	Braithwaite (1975)
	Western Australia	Read (1974), Muir (this vol.)
	Fiji	Burgess (this vol.)
Tertiary		
Pliocene	Great Plains, USA	Swineford et al. (1958)
	South Africa	Netterberg (1969)
Oligocene	Nebraska	Schultz and Stout (1955)
Eocene	Wyoming (Green River Formation)	Surdam and Stanley (1979)
	Catalanides Range	Esteban (1973)
Paleogene	SE France	Truc (1975)
Cretaceous		
Turonian	N Israel	Magaritz (this vol.)
Lower Cretaceous	Central Texas	Amsbury (1967)
	Adriatic region, Yugoslavia	Tišljar (this vol.)
	Black Nose, western North Atlantic	Enos and Freeman (1979, reinterpreted)
Jurassic		
Lower Jurassic	Negev Desert, Israel	Goldberg and Friedman (1974)
	Taurus	de Graciansky (1972)
	Central Apennines (Calcare Massiccio Formation)	Bernoulli and Wagner (1971), Ciarapica and Passeri (this vol.)
	Trento, Italy (Calcari grigri Formation)	Castellarin and Sartori (1973)
Sinemurian-Pliensbachian	Morocco	Burri et al. (1973)
Triassic		
Upper Triassic	W Yugoslavia	this paper, Figs. 2 and 3
	Monti di Palermo, Italy	Catalano et al. (1974), Abate et al. (1977)
	Monti Aurunci (N of Naples)	Carannante et al. (1978)
	the Dolomites	Bosellini and Rossi (1974)
	Calabria	D'Argenio (1966, reinterpreted by Bernoulli and Wagner 1971)
Middle Triassic	the Dolomites (Latemar Limestone)	Gaetani et al. (1981)
	Val Brembana, Lombardy (Calcare Rosso)	Assereto and Kendall (1977), Assereto and Folk (1980)
	Gailtaler Alps (Wettersteinkalk)	Flügel (1978, Plates 2-6, 8-5)
Permian		
	N Poland (Zechstein Limestone)	Peryt and Piatkowski (1977)
	W Poland (Zechstein Limestone)	Peryt (this vol.)

Table 1 (continued)

Age	Provenance	Selected references
Permian	the Netherlands (Main Dolomite)	Clark (1980)
	Guadalupe Mountains, USA	Esteban and Pray (this vol.)
	Sacramento Mts, USA	Winchester (1972)
	Pyrenees	Nagtegaal (1969)
Carboniferous		
Namurian	Spiško-gemerské rudohorie Mts, Slovakia	Mišik (1980)
Devonian		
Middle Devonian	South-central Saskatchewan, Canada	Wardlaw and Reinson (1971)
Upper Old Red Sandstone	Scotland	Burgess (1961)
Silurian	Ohio	Kahle (1974)
Precambrian	SW China (Tongying Formation)	Cao Ruiji and Xue Yaosong (this vol.)
	N Australia (Amos Formation, McArthur Group)	Muir (this vol.)

their pedological origin. Ancient vadoids have also often been mistaken for oncoids. As stated by Dunham (1969, p. 182) "The general scarcity of identified caliche in ancient rocks is both disappointing and puzzling". From that time numerous ancient pedological carbonates have been described, and following the new Dunham interpretation of pisoids from the Permian of the Guadalupe Mts. (Fig. 1), many pisoids (and microids) earlier interpreted as oncoids have been (and are being) reinterpreted as vadoids; Figs. 2 and 3 present an example from the Upper Triassic of Yugoslavia. Duham (1969, p. 189) came to the conclusion that "the pisolite in the Upper Guadalupian is pisolitic caliche made from marine grainstone in the upper part of the vadose zone under conditions varying between those of near-surface soil and those of shallow caverns". Esteban (1976) questioned the correctness of this interpretation and suggested using the terms "vadose marine" or "vadose hypersaline" for those pisoids, to avoid the implied soil processes. He noticed that "this use of vadose retains its hydrologic connotation and is independent of water composition"; however, he distinguished between "vadose pisoliths" and caliche pisoliths (Esteban 1976). According to Esteban (1976), both types show significant differences (see his Table 1 and the discussion on pp. 2050–2052). This opinion has met with objections (Chafetz and Butler 1980), especially when it appeared that at least some "caliche pisoids" sensu Esteban resemble "hypersaline pisoids" rather than "caliche pisoids" (Burgess, this vol., see also Elloy and Thomas 1981). Perhaps the main problem of these different interpretations is that the distinction between hypersaline (or marine) "vadose pisoids" and freshwater "caliche pisoids" is somewhat overgeneralized, as also in a generally hypersaline environment phases of freshwater attack can occur. The well-known result of such an attack is dolomitization in a schizohaline environment (Folk and Siedlecka 1974), but the inventory of products of fluctuating chemical environments is, as recently noticed (Assereto and Folk 1980)

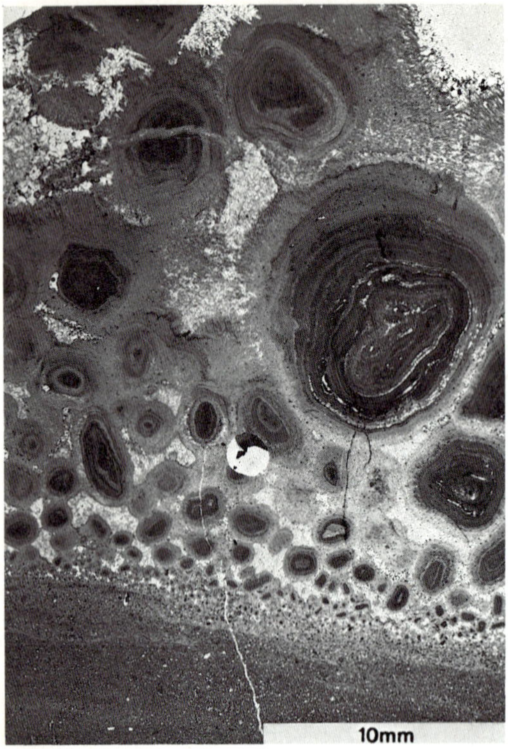

Fig. 1. Classical vadolite, described by Dunham (1969) and Esteban and Pray (this vol.), among others. Note reverse graded bedding and meniscus cement. Sample DB4063, locality Walnut Canyon; Sample furnished by D. Bernoulli (Basel)

very rich, and vadose pisoids belong to it. Therefore, because the origin of "caliche pisoids" is related to the vadose zone as in the case of "vadose pisoids", and because of the overgeneralization mentioned above, they will be described jointly in the next section.

3 Description of Vadoids

Non-skeletal grains (usually vadoid fragments, intraclasts, peloids and obscure micritic bodies) constitute the nuclei of vadoids (Figs. 1 – 7). The development of the cortex is sometimes asymmetrical and the envelopes show downward thickening (e.g., Bretz and Horberg 1949; Swineford et al. 1958; Dunham 1969, Fig. 9; Wardlaw and Reinson 1971, Fig. 12B; Assereto and Kendall 1977, Figs. 10a, 12c; Carannante et al. 1978, Fig. 8b; Cao Ruiji and Xue Yaosong, this vol., Fig. 5; Peryt, this vol., Fig. 7a, e), upward thickening (e.g., Hay and Reeder 1978), or both (e.g., Surdam and Stanley 1979; Esteban and Pray, this vol., Fig. 5; Esteban and Pray (this vol., Figs. 3A, 5D) also report lateral elongation of vadoids. In many cases there is no consistent preferred elongation of the coatings (e.g., Read 1974: 256, Harrison 1977: 141). Quite often vadoids are broken and recoated (e.g., D'Argenio 1966, Fig. 7; Dunham 1969, Fig. 5; Bernoulli and

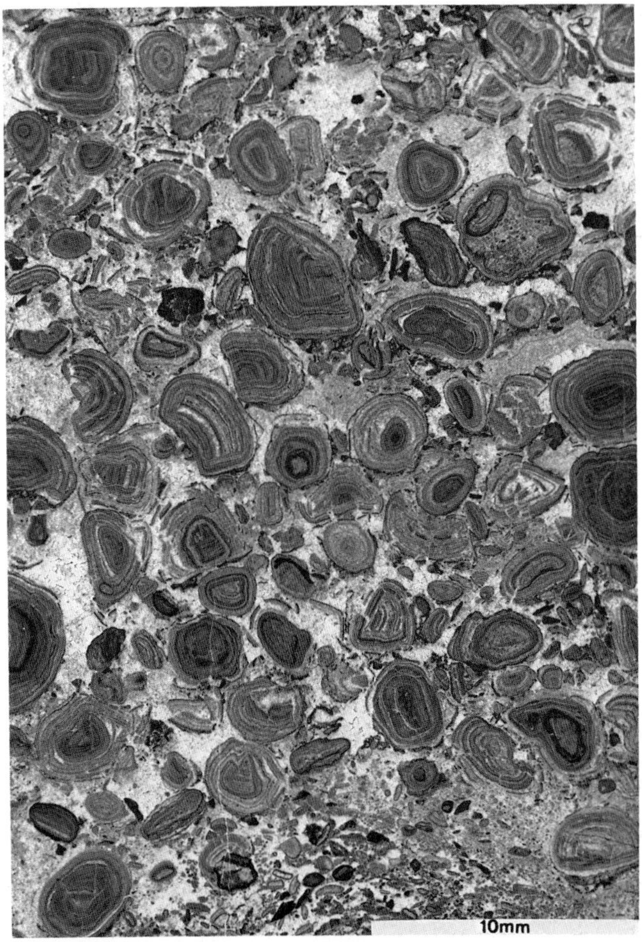

Fig. 2. Upper Triassic vadolite, Gorski Koter region near Skrad (Croatia). Sample GK838, furnished by L. Babić (Zagreb). The vadoids from the Upper Triassic dolomites of Yugoslavia have been described by Babić (1970) and interpreted as oncoids, however, they exhibit features typical of vadoids (cf. with other figures, and especially Fig. 7). Note abundance of broken and recoated forms; the repeated breaking and coating of some grains has already been noticed by Babić (1970: 13)

Wagner 1971, Figs. 6, 7; Wardlaw and Reinson 1971, Figs. 11G, 12D; de Graciansky 1972, Pl. 20, Fig. 3; Assereto and Kendall 1977, Figs. 11a, b, g; Peryt and Piatkowski 1977, Figs. 5, 6; Mišik 1980, Figs. 26, 27; Peryt and Wagner 1981, Fig. 19; Esteban and Pray, this vol., Figs. 4C, 6E; Tišljar, this vol., Fig. 4E, F; see also Figs. 2–7). Evidently episodes of fracturing alternated with episodes of vadoid growth, and the fracturing may have been caused, as concluded by Assereto and Folk (1980: 374), by desiccation fracturing and slight reworking. This reworking has been demonstrated by several workers (e.g., Castellarin and Sartori 1973; Peryt and Piatkowski 1977, Figs. 5, 6, 8). Polygonal fitting is often to be found (e.g., Dunham 1969, Fig. 9; Wardlaw and Reinson 1971, Fig. 12A; Kahle 1974, Fig. 5C; Carannante et al. 1978; Stanley and Surdam 1979, Fig. 10; Clark 1980, Fig. 14; Esteban and Pray, this vol., Fig. 5B–E; Peryt, this vol., Fig. 6c–f), and therefore this feature is thought to be diagnostic of vadoids by some workers, but this feature – resulting from vadose compaction – has also been noted in ooids (e.g., Peryt, this vol., Fig. 6a). The number of laminae in a

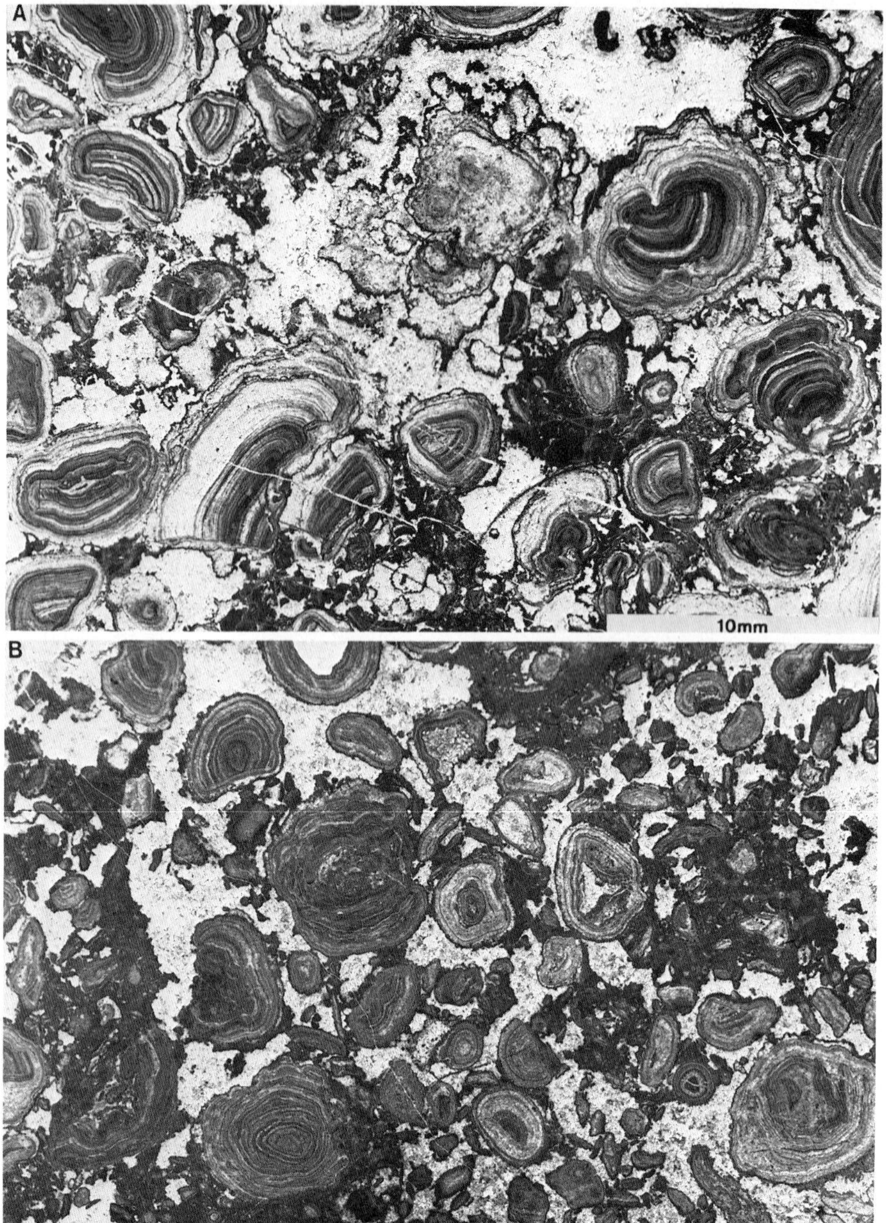

Fig. 3 A, B. Upper Triassic vadolites; see Babić (1970) for description. **A** Sample Žumb, furnished by L. Babić (Zagreb). Žumberak region near Sveta Gera (Croatia). **B** Sample GK 594, furnished by L. Babić (Zagreb). Gorski Koter near Skrad (Croatia)

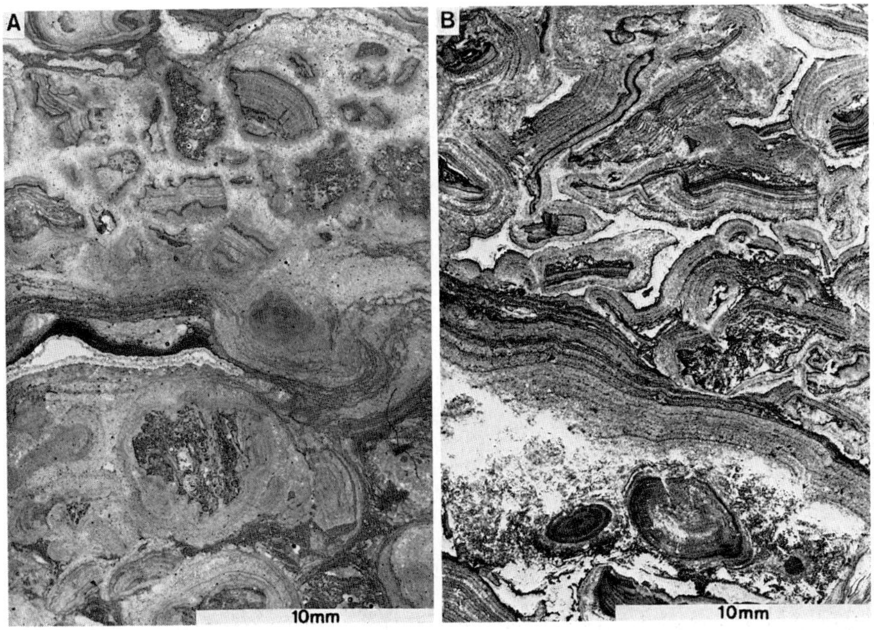

Fig. 4A, B. Middle Devonian vadolites from south-central Saskatchewan, Canada. Samples furnished by L. Fuzesy (Regina) with assistance of N. C. Wardlaw and A. C. Kendall. **A** Well 14-8-32-4W3, depth 4029 feet; Fig. 11F of Wardlaw and Reinson (1971). **B** Well 8-16-37-15W2, depth 2899 feet; Fig. 11G of Wardlaw and Reinson (1971)

cortex is very variable. According to Esteban (1976), "caliche pisoids" contain usually less than five indistinct laminae (e.g., one or two in the case of the Tarragona pisoids described by Calvet and Julia, this vol.), whereas "vadose pisoids" generally have more than ten laminae. However, numerous exceptions have been noted (e.g., few – sometimes one – laminae in "vadose pisoids" – Castellarin and Sartori 1973). The characteristic feature of some vadoids is that laminae vanish in the surrounding sediment (e.g., Castellarin and Sartori 1973; Catalano et al. 1974, Fig. 6; Peryt 1981), which clearly indicates the in situ growth in the rock. This in turn stresses the autochthonous, in many cases replacement nature of the vadoids. As Esteban and Pray (this vol.) write of the vadoids of the Capitan Reef Complex, they "formed in water under the sky, and not under a rock roof or in soil". Here it should also be stressed, however, that at least some vadoids from the paleosols formed – as did the Aldabra "concentrically zoned bodies" – "at or very close to an air/sediment interface" (Braithwaite 1975: 9).

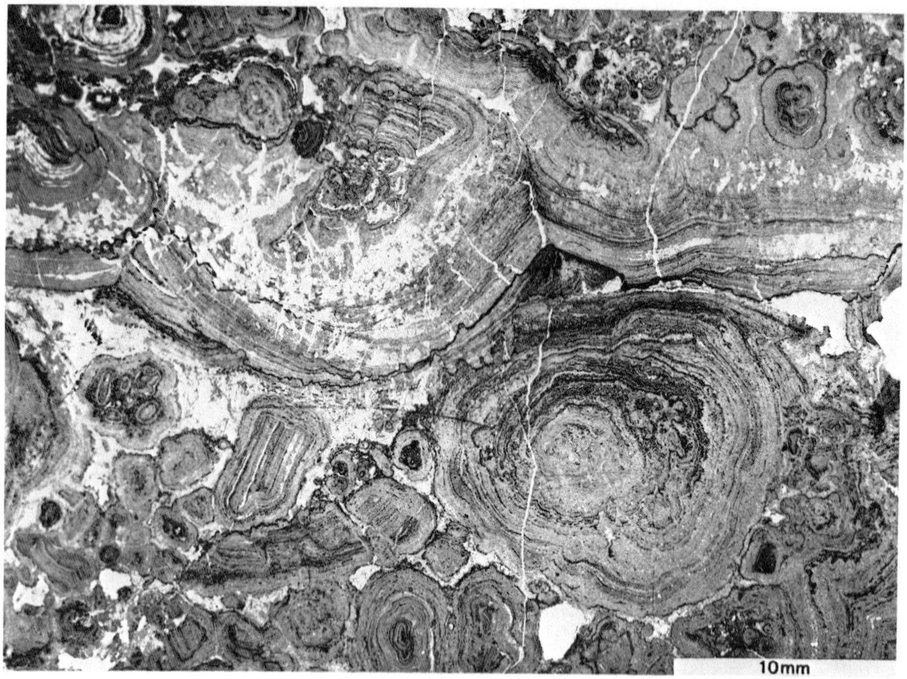

Fig. 5. Vadolite from the Zechstein Limestone of Thuringia. Sample furnished by K. Kerkmann (Gotha). Note association of vadoids (some are broken and recoated) and vadose crusts, with perched inclusions

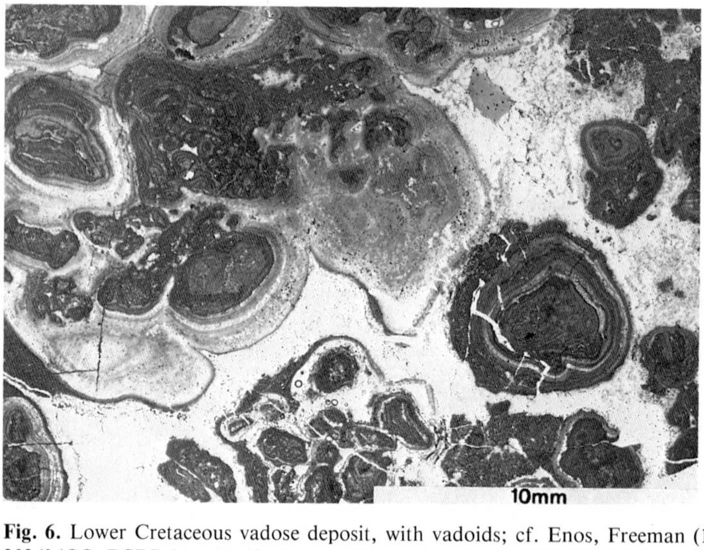

Fig. 6. Lower Cretaceous vadose deposit, with vadoids; cf. Enos, Freeman (1979). Sample DSDP 202/3/CC, DSDP Leg 44, Site 392A, Core-section-interval 22-1-108/110

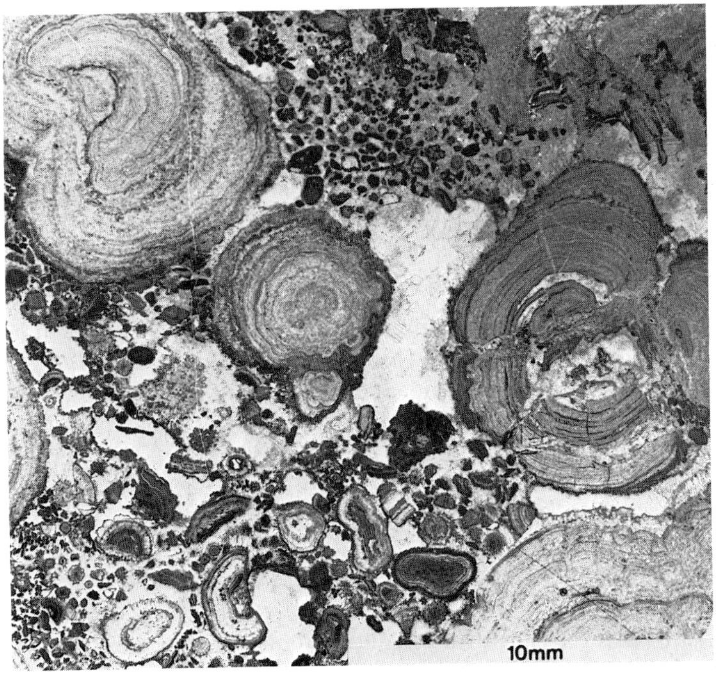

Fig. 7. Triassic vadolite from Val Brembana, Lombardy, Italy; see Assereto, Folk (1976) for detailed description. Sample VB81, furnished by RL Folk (Austin)

4 Geological Position of Vadoids

Vadoids seem to be of common occurrence in cyclic sequences. Bernoulli and Wagner (1971) noted this when studying the vadoids of the Calcare Massiccio Formation, and later many occurrences of vadoids in the Tethyan Mesozoic have been related to the loferitic (or related) cycles (e.g., Castellarin and Sartori 1973, Catalano et al. 1974, Carannante et al. 1978). As noticed by Esteban (1976: 2055), "The 'vadose-pisolite' facies might be interpreted as a depositional facies closely related to the fenestral (keystone) grain-supported facies", and this interpretation has been amplified in the recent profound paper on Capitan vadoids by Esteban and Pray (this vol.). However, there are at least several examples of vadoid development within clearly subtidal (and not shallow subtidal) deposits (e.g., Kahle 1974: 35; Peryt and Piatkowski 1977; Mišik 1980) and therefore the vadoids occurring, for example, within ooid-bearing sequences as in the Turonian of Israel (Magaritz, this vol.) may be interpreted as resulting from a significant drop of sea level as well. It is remarkable that some vadoids are related to the late stages of development of pre-evaporite basins (see Kahle 1974, Wardlaw and Reinson 1971, Peryt and Piatkowski 1977, Peryt 1981). For the Zechstein basin it has been demonstrated (Smith 1970, 1980; Peryt 1981 and this vol.) that the sea level dropped several times during the Zechstein Limestone deposition, and during the low sea levels vadoids formed (Peryt 1981, this vol.). It is supposed that the sea level changes – to some extent a prerequisite of vadoid formation – related to glacieustatic mechanism are typical of many pre-evaporite basins. Consequently, the applying of the "non-wet" hypothesis (cf. Pray 1977) affords the possibility of another approach to vadoids than that presented by Esteban (1976), as distinguishing between "caliche pisoids" and "vadose pisoids". The difference between "vadose pisolite" facies and modern caliche is that "it appears very closely controlled by specific depositional and lithologic conditions of the host rock, in contrast with the independence of modern caliche to its substrate" (Esteban 1976, p. 2055). However, assuming the sea level changes mentioned above and the situation that the vadoids developed within different substrates (as in the case of the Zechstein Limestone, within basinal mudstones and wackestones and within ooidal packstones and grainstones of the peripheral part of the basin), the controlling effect of the parent rock/deposit is problematical, like the earlier-mentioned microscopical characteristics of vadoids. As recently concluded by Chafetz and Butler (1980), the following features are useful as discriminating between "caliche pisoids" and "marine pisoids": (1) regional geological setting of the pisolite accumulation, (2) presence of fossils, (3) presence of spar crystals of triangular cross-section (indicating precipitation from freshwater), and (4) original mineralogy. In the first two points it should be kept in mind what has been written above about the effect of sea level changes; Kahle (1974: 35) notes the presence of branching corals, bryozoans, and colonial corals within vadose deposits, and this may simply reflect the vadoization of originally subtidal marine deposit, but not necessarily in hypersaline conditions. The third and fourth points of Chafetz and Butler (1980) as well as the earlier conclusions by Assereto and Folk (1976) are, in turn, very instrumental in the establishing the chemistry of the depositional and early diagenetic environments of vadoids.

5 Concluding Remarks

The general environments of formation of vadoids are: vadose freshwater environment and vadose marine environment. The connection between the genesis of given vadoids with one of these general environments is usually possible when the entire sedimentological and geological background setting is considered, but certainly the vadoids themselves need not differ. Therefore, instead of distinguishing between "caliche pisoids" and "vadose pisoids", the term "vadoid" with specific environmental precision, e.g. soil, beach, fluvial, cave, travertine or pool vadoid is suggested. All these types, although variously termed, have been described from recent and subrecent environments (Gradziński and Radomski 1967, Purser and Loreau 1973, Braithwaite 1979, Risacher and Eugster 1979, Calvet and Julià, this vol., Folk and Chafetz, this vol., accordingly, and many other papers) as well as from ancient environments [see Table 1 for soil and beach vadoids, and Tandon and Narayan (1981) and Sliwiński (1981) for other vadoids].

Acknowledgements. This study was supported by the Alexander von Humboldt-Stiftung. Samples were furnished by L. Babić, D. Bernoulli, J. C. R. Braithwaite, R. L. Folk, L. Fuzesy, K. Kerkmann, L. Passeri, B. H. Purser, P. A. Scholle, W. Sliwiński, and Deep Sea Drilling Project.

References

Abate B, Catalano R, D'Argenio B, Di Stefano P, Riccobono R (1977) Relationships of algae with depositional environments and faunal assemblages of the Panormide carbonate platform, Upper Triassic, northwestern Sicily. In: Flügel E (ed) Fossil Algae. Springer, Berlin Heidelberg New York, 301 – 313

Amsbury DL (1967) Caliche soil profiles in Lower Cretaceous rocks of Central Texas (abs). Geol Soc Amer Ann Mtg Program (Spec Pap 115), 4 – 5

Assereto RLAM, Folk RL (1976) Brick-like texture and radial rays in Triassic pisolites of Lombardy, Italy: a clue to distinguish ancient aragonitic pisolites. Sed Geol 16:205 – 222

Assereto RLAM, Folk RL (1980) Diagenetic fabrics of aragonite, calcite, and dolomite in an ancient peritidal-spelean environment: Triassic Calcare Rosso, Lombardia, Italy. J Sediment Petrol 50:371 – 394

Assereto RLAM, Kendall CGStC (1977) Nature, origin and classification of peritidal tepee structures and related breccias. Sedimentology 24:153 – 210

Babić L (1970) Sphaerocodium or onkoids from the Upper Triassic dolomite of Western Yugoslavia? Geol Vjesnik 23:11 – 19

Bernoulli D, Wagner CW (1971) Subaerial diagenesis and fossil caliche deposits in the Calcare Massiccio Formation (Lower Jurassic, Central Apennines, Italy). N Jb Geol Paläont Abh 138:135 – 149

Bosellini A, Rossi D (1974) Triassic carbonate buildups of the Dolomites, northern Italy. SEPM Spec Publ 18:209 – 233

Braithwaite CJR (1975) Petrology of palaeosols and other terrestrial sediments on Aldabra, Western Indian Ocean. Philosoph Trans Royal Soc London B 273:1 – 32

Braithwaite CJR (1979) Crystal textures of recent fluvial pisolites and laminated crystalline crusts in Dyfed, South Wales. J Sediment Petrol 49:181 – 194

Bretz JH, Horberg L (1949) Caliche in southeastern New Mexico. J Geol 57:491 – 511

Brown CN (1956) The origin of caliche on the north-eastern Llano Estacado, Texas. J Geol 64:1 – 15

Burgess CJ (this vol.) Quaternary pisoids of Lau, Fiji, pp 488 – 502

Burgess IC (1961) Fossil soils of the upper Old Red Sandstone of south Ayrshire. Geol Soc Glasgow Trans 24:138 – 153

Burri P, Du Dresnay R, Wagner CW (1973) Tepee structures and associated diagenetic features in intertidal carbonate sands. Sed Geol 9:221 – 228

Calvet F, Julia R (this vol.) Pisoids in the caliche profiles of Tarragona (N.E. Spain), pp 456 – 473

Cao Ruiji, Xue Yaosong (this vol.) Vadose pisolites of the Tongying Formation (Upper Sinian System) in SW China, pp 538 – 547

Carannante G, Carbone F, Catenacci V, Simone L (1978) I carbonati triassici dei Monti Aurunci: facies deposizionali e diagenetiche. Boll Soc Geol It 97:687 – 698

Castellarin A, Sartori R (1973) I ciclotemi carbonatici infraliassici di S. Massenza (Trento). Giorn Geologia 30:221 – 248

Catalano R, D'Argenio B, Lo Cicero G (1974) I ciclotemi triassici di Capo Rama (Monti di Palermo). Geol Romana 13:125 – 145

Chafetz HS, Butler JC (1980) Petrology of recent caliche pisolites, spherulites, and speleothem deposits from central Texas. Sedimentology 27:497 – 518

Chafetz HS, Meredith JC (this vol.) Recent travertine pisoliths (pisoids) from southeastern Idaho, U.S.A., pp 450 – 455

Ciarapica G, Passeri L (this vol.) Coated grains in contrasted environmental situations: Norian and Lower Liassic of northern Apennines, pp 576 – 586

Clark DN (1980) The sedimentology of the Zechstein 2 Carbonate Formation of eastern Drenthe, the Netherlands. Contr Sedimentology 9:131 – 165

D'Argenio B (1966) Stromatoliti triassiche della Calabria settentrionale. Boll Soc Natur Napoli 75:433–437

de Graciansky P (1972) Recherches géologiques dans le Taurus lycien. Thèse, Paris

Dunham RJ (1969) Vadose pisolite in the Capitan Reef (Permian), New Mexico and Texas. SEPM Spec Publ 14:182–191

Elloy R, Thomas G (1981) Dynamique de la genèse des croûtes calcaires (calcrètes) développées sur séries rouges pléistocènes en Algérie nord-occidentale. Contexte géomorphologique et climatique. Pétrographie et géochimie. Bull Cent Rech Explor-Prod Elf-Aquitaine 5:53–112

Enos P, Freeman T (1979) Lower Cretaceous peritidal limestones at 2,700-m depth, Blake nose, Atlantic Ocean. Geology 7:83–87

Esteban M (1973) Presencia de caliche fósil en la base del Eoceno de los Catalánides, provincias de Tarragona y Barcelona. Acta Geol Hisp 7:164–168

Esteban M (1976) Vadose pisolite and caliche. Amer Assoc Petrol Geol Bull 60:2048–2057

Esteban M, Pray LC (this vol.) Pisoids and pisolite facies (Permian), Guadalupe Mountains, New Mexico and Texas, pp 503–536

Flügel E (1978) Mikrofazielle Untersuchungsmethoden von Kalken. Springer, Berlin Heidelberg New York

Folk RL, Chafetz HS (this vol.) Pisoliths (pisoids) in Quaternary travertines of Tivoli, Italy, pp 474–487

Folk RL, Siedlecka A (1974) The schizohaline environment: its sedimentary and diagenetic fabrics as exemplified by Late Paleozoic rocks of Bear Island, Svalbard. Sed Geol 11:1–15

Gaetani M, Fois E, Jadoul F, Nicora A (1981) Nature and evolution of Middle Triassic carbonate buildups in the Dolomites (Italy). Marine Geology 44:25–57

Goldberg M, Friedman GM (1974) Paleoenvironments and paleogeographic evolution of the Jurassic System in southern Israel. Geol Survey Israel Bull 61

Gradziński R, Radomski A (1967) Pisoliths from Cuban caves. Ann Soc Geol Pologne 37:243–265

Harrison RS (1977) Caliche profiles: indicators of near-surface subaerial diagenesis, Barbados, West Indies. Bull Canad Petrol Geol 25:123–173

Hay RL, Reeder RJ (1978) Calcretes of Olduvai Gorge and the Ndolanya Beds of northern Tanzania. Sedimentology 25:649–673

Hay RL, Wiggins B (1980) Pellets, ooids, sepiolite and silica in three calcretes of the southwestern United States. Sedimentology 27:559–576

James NP (1972) Holocene and Pleistocene calcareous crusts (caliche) profiles: criteria for subaerial exposure. J Sediment Petrol 42:817–836

Kahle CF (1974) Nature and significance of Silurian rocks at Maumee quarry, Ohio. In: Kesling RV (ed) Silurian reef-evaporite relationships. Michigan Basin Geol Soc, Field Conf 1974, 31–54

Knox GJ (1977) Caliche profile formation, Saldanha Bay (South Africa). Sedimentology 24:657–674

Magaritz M (this vol.) Carbon and oxygen isotope composition of recent and ancient coated grains, pp 27–37

Mišik M (1980) Miocene sinter crusts (speleothems) and calcrete deposits from Neptunian dykes, Male Karpaty Mts. Geol zb 31:495–512

Muir MD (this vol.) A Proterozoic calcrete in the Amos Formation, McArthur Group, Northern Territory, Australia, pp 548–558

Nagtegaal PJC (1969) Microtextures in recent and fossil caliche. Leidse Geol Meded 42:131–142

Netterberg F (1969) The interpretation of some basic calcrete types. South African Archaeological Bull 24:117–122

Peryt TM (1981) Former aragonitic submarine hemispheroids associated with vadose deposits, Zechstein Limestone (Upper Permian), Fore-Sudetic area, western Poland. N Jb Geol Paläont Mh 1981:559–570

Peryt TM (this vol.) Coated grains from the Zechstein Limestone (Upper Permian) of western Poland, pp 587–598

Peryt TM, Piatkowski TS (1977) Algal-vadose pisoliths in the Zechstein Limestone (Upper Permian) of northern Poland. Sed Geol 19:275–286

Peryt TM, Wagner R (1981) Mikrostruktury stromatolitów i onkoidów z wapienia cechsztyńskiego Polski północno-zachodniej. Kwart Geol 25:629–650

Pray LC (1977) The all wet, constant sea level hypothesis of Upper Guadalupian shelf and shelf edge strata, Guadalupe Mountains, New Mexico and Texas. In: Hileman ME, Mazzullo SJ (eds) Upper

Guadalupian facies, Permian Reef Complex, Guadalupe Mountains, New Mexico and West Texas. Permian Basin Sect Publ 77-16:433

Purser BH, Loreau J-P (1973) Aragonitic, supratidal encrustations on the Trucial Coast, Persian Gulf. In: Purser BH (ed) The Persian Gulf. Springer, Berlin Heidelberg New York, pp 343 – 376

Read JF (1974) Calcrete deposits and Quaternary sediments, Edel Province, Shark Bay, Western Australia. Amer Assoc Petrol Geol Mem 22:250 – 282

Richter DK (this vol.) Calcareous ooids: A synopsis, pp 71 – 99

Risacher F, Eugster HP (1979) Holocene pisoliths and encrustations associated with spring-fed surface pools, Pastos Grandes, Bolivia. Sedimentology 26:253 – 270

Scholle PA, Kinsman DJJ (1974) Aragonitic and high-Mg calcite caliche from the Persian Gulf – a modern analog for the Permian of Texas and New Mexico. J Sediment Petrol 44:904 – 916

Schultz CB, Stout TM (1955) Classification of Oligocene sediments in Nebraska. Nebraska State Mus Bull 4:17 – 52

Siesser WG (1973) Diagenetically formed ooids and intraclasts in South African calcretes. Sedimentology 20:539 – 551

Sliwiński W (1981) Petrogeneza i warunki sedymentacji utworów weglanowych permu niecki śródsudeckiej. Unpubl Ph D thesis, univ Wrocław

Smith DB (1970) Permian and Trias. In: Johnson GAL, Hickling G (eds) The geology of Durham county. Trans Nat Hist Soc Northumberland 41:66 – 91

Smith DB (1980) The evolution of the English Zechstein basin. Contr Sedimentology 9:7 – 34

Surdam RC, Stanley KO (1979) Lacustrine sedimentation during the culminating phase of Eocene Lake Gosiute, Wyoming (Green River Formation). Geol Soc Amer Bull 90:93 – 110

Swineford A, Leonard AB, Frye JC (1958) Petrology of the Pliocene pisolitic limestone in the Great Plains. Kansas Geol Survey Bull 130:97 – 116

Tandon SK, Narayan D (1981) Calcrete conglomerate, case-hardened conglomerate and cornstone – a comparative account of pedogenic and non-pedogenic carbonates from the continental Siwalik Group, Punjab, India. Sedimentology 28:353 – 367

Tišljar J (this vol.) Coated grains facies in the Lower Cretaceous of the Outer Dinarides (Yugoslavia), pp 566 – 575

Truc G (1975) Les encroûtements carbonatés liés à la pédogenèse; rôle important des microorganisms: biocorrosion et biosynthèse de la calcite en milieu pédologique confiné. IX Congr intern Sedim, guidebook exc A2, 47 – 55

Ward WC (1975) Petrology and diagenesis of carbonate eolianites of northeastern Yucatan Peninsula, Mexico. Amer Assoc Petrol Geol Stud Geology 2:500 – 571

Wardlaw NC, Reinson GE (1971) Carbonate and evaporite deposition and diagenesis, Middle Devonian Winnipegosis and Prairie Formations of south-central Saskatchewan. Amer Assoc Petrol Geol Bull 55:1759 – 1786

Winchester PD (1972) Caliche-like limestones in the Lower Permian Laborcita Formation of the Sacramento Mountains, New Mexico (abs). Geol Soc Amer Abstr with Programs 4:707

V.2 Recent Travertine Pisoliths (Pisoids) from Southeastern Idaho, U.S.A.

Henry S. Chafetz[1] and John C. Meredith[2]

1 Introduction

Holocene-Pleistocene (?) age travertine deposits occur throughout southeastern Idaho, U.S.A., associated with the volcanic strata along the Snake River. Warm springs at some of these sites are still actively forming travertine deposits. Lithified travertine accumulations at two locations, one along Fall Creek in eastern Bonneville County, and the other along Fritz Creek in northwestern Clark County (Meredith 1980), have been investigated in detail (Fig. 1).

These deposits are considerably smaller than those in Italy (Folk and Chafetz, this volume), only having an areal extent of some hundreds of meters, and

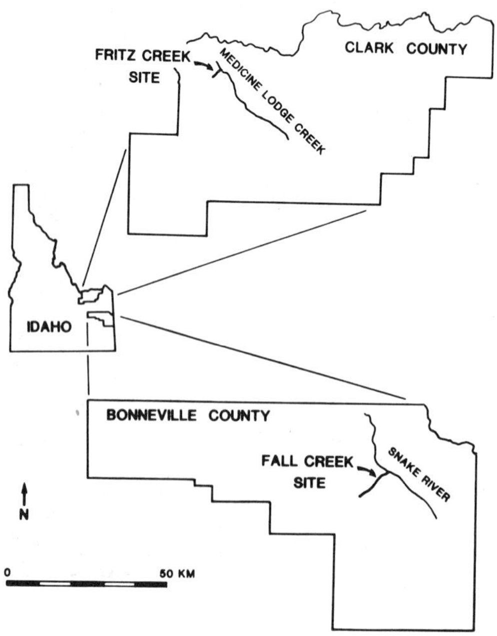

Fig. 1. Location map

1 Geology Department, University of Houston, Houston, TX 77004, USA
2 Present address: Kerr McGee Oil Company, Suite 300, 3 North Point, Houston, TX 77060, USA

Coated Grains (ed. by T.M. Peryt)

50 – 100 m in maximum thickness. Presently active warm springs form fan-shaped sloping mounds, ring-shaped cones, and benches a short way above the valley bottom (Herlinger 1981). In general, these deposits seem to be filling small, moderately steep-sided valleys by coalescence of the accumulations, primarily the large mounds. Lithologically these deposits are essentially identical to those from central Italy. However we have not, as yet, found any inorganic pisoliths in the deposits studied, only organically coated grains.

2 Bacterial Pisoliths

The two sites studied show accumulations of pisoliths formed by the encrustations of travertine intraclasts by bacterial colonies (Figs. 2 and 3). The pisoliths comprise only a minor component within the overall travertine accumulations.

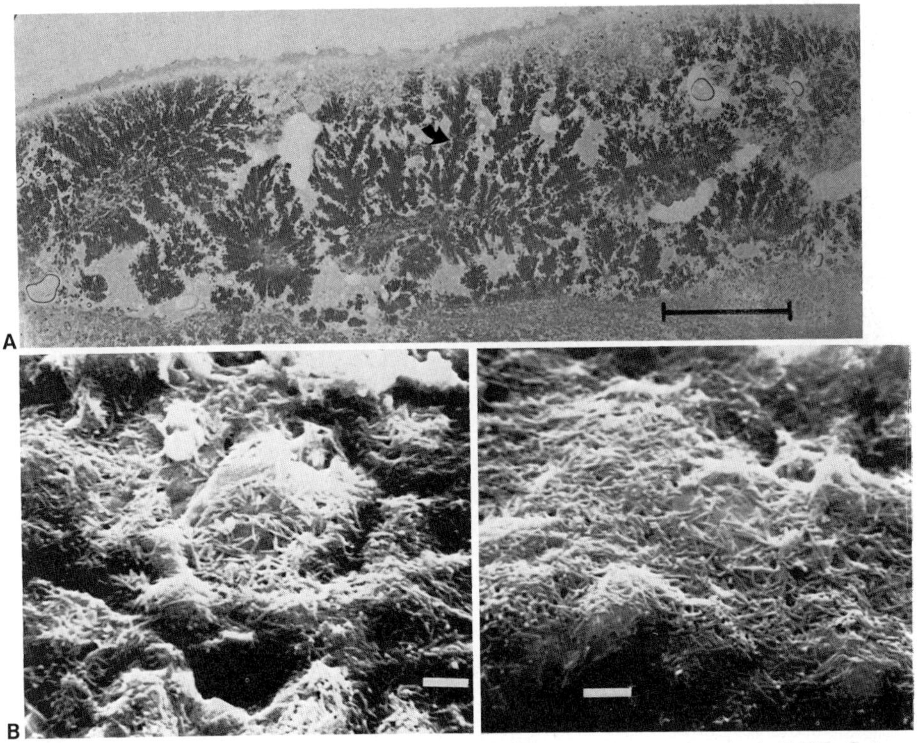

Fig. 2 A – C. Thin section (**A**) and SEM's (**B**, **C**) of pisoliths from the same host rock. **A** Photomicrograph of highly irregular bacterial pisoliths. Cortices are composed of spar encrusted branching clumps of bacteria. Pisolith in center of field displays two generations of cortical growth, the second growth layer is only present on the upper surface (arrow marks contact between layers). **B**, **C** SEM view of two samples etched in dilute HCl displaying clumps of bacteria. Scales equal 0.5 mm, 2 μm, and 2 μm for **A – C**, respectively

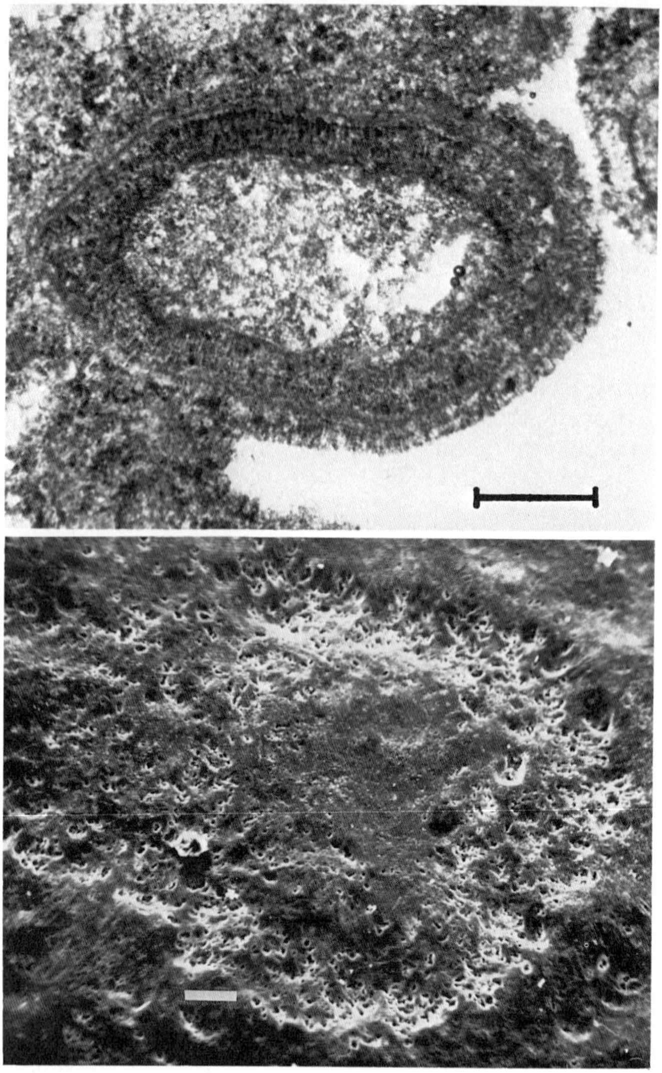

A

B

Fig. 3A, B. Thin section (**A**) and SEM (**B**) of pisoliths from the same host rock. **A** Pisolith displays regular, evenly laminated cortices. **B** Low magnification SEM view of etched pisolith shows that cortices contained an abundance of bacterial rods, now vacuoles due to diagenesis. Nuclei contained fewer bacteria as indicated by the relative abundance of holes. Observe the branching pattern, best displayed by vacuoles in the upper part of the cortices, similar to the shape of the bacterial clumps of Fig. 2A. Scales equal 0.5 mm and 100 μm for **A** and **B**, respectively

These deposits may be discontinuous and lensoidal or relatively continuous for the extent of the outcrop (generally less than 5 m). Deposits of pisoliths are in association or contact with all other lithologies seen in the travertine, but most commonly are adjacent to intraclasts, layers of "shrubs", or laminated deposits. Contacts with the surrounding deposits may by abrupt or gradational as non-coated grains grade upward into coated grains that develop thicker coatings and display an increase in overall size. The latter case results in inverse graded bedding if the nuclei are of a constant size or increase in size upward (Fig. 4).

Deposits of coated grains range from loose to densely packed. Packing may be so dense that individual coated grains are barely distinguishable except in association with rare cavities between the grains. Closely packed pisoliths may tend to conform to each other while still leaving room for spar to precipitate between the grains, or they may be so tightly packed that one hinders the growth of another and composite grains are formed.

The pisoliths range from 0.5 – 16.0 mm in diameter and average 2.0 – 3.0 mm. The nucleus is usually darker than the cortex and tends to be very pale brown to white. The nuclei range considerably in size and shape in much the same manner as the non-coated intraclasts; they are up to 15 mm long and circular to elongate or irregularly-shaped with length to width ratios usually 3 : 1 or less. Occasionally there are no apparent nuclei about which the cortex develops, resulting in a radial arrangement of shrubs.

The cortex is usually lighter-colored than the nucleus and ranges from non-existent where grains are incompletely coated, to 8 mm thick, while averaging 0.5 – 1.0 mm. The cortex may develop evenly and completely or incompletely about the nucleus, in which case it is usually best-developed on top of the pisolith (Fig. 2A). However, occasionally some other side(s) may show thicker growth. The cortex is organic and made-up of single bacterial clumps or branching forms which radiate roughly perpendicular to the nucleus. The bacteria form all the same dendritic features seen in layers of the shrubs as well as equant to highly elongate single clumps oriented perpendicular to the nucleus. There may be single shrubs growing out from the nucleus or shrubs stacked on each other that only rarely coordinate their points of origin to form even, concentric bands. The packing density of the shrubs and clumps ranges from loose to extremely dense such that individual clumps cannot be distinguished. The density of organic matter (bacterial rods) may remain constant throughout the coated grain or, not uncommonly, decrease away from the nucleus. In hand specimen those with constant density appear massive and in those of decreasing density the branching habit is evident and the margins of the pisolith are irregular.

The bacteria, around which the spar precipitated, are represented by either rod-shaped bodies or vacuoles. The rod-shaped bodies occur as isolated or semi-isolated remains or in extremely dense aggregates or clumps (Fig. 2B, C). Individual rods are commonly 0.5 – 0.8 μm long and 0.1 – 0.2 μm in diameter. The vacuoles represent the bacteria after decomposition during diagenesis. This contention is supported by the fact that rods and vacuoles are not found together, occur in the same general configuration [e.g., within the cortices of pisoliths but not in the surrounding spar (Fig. 4C)], and display the same microhabits (Figs. 2A and 3B).

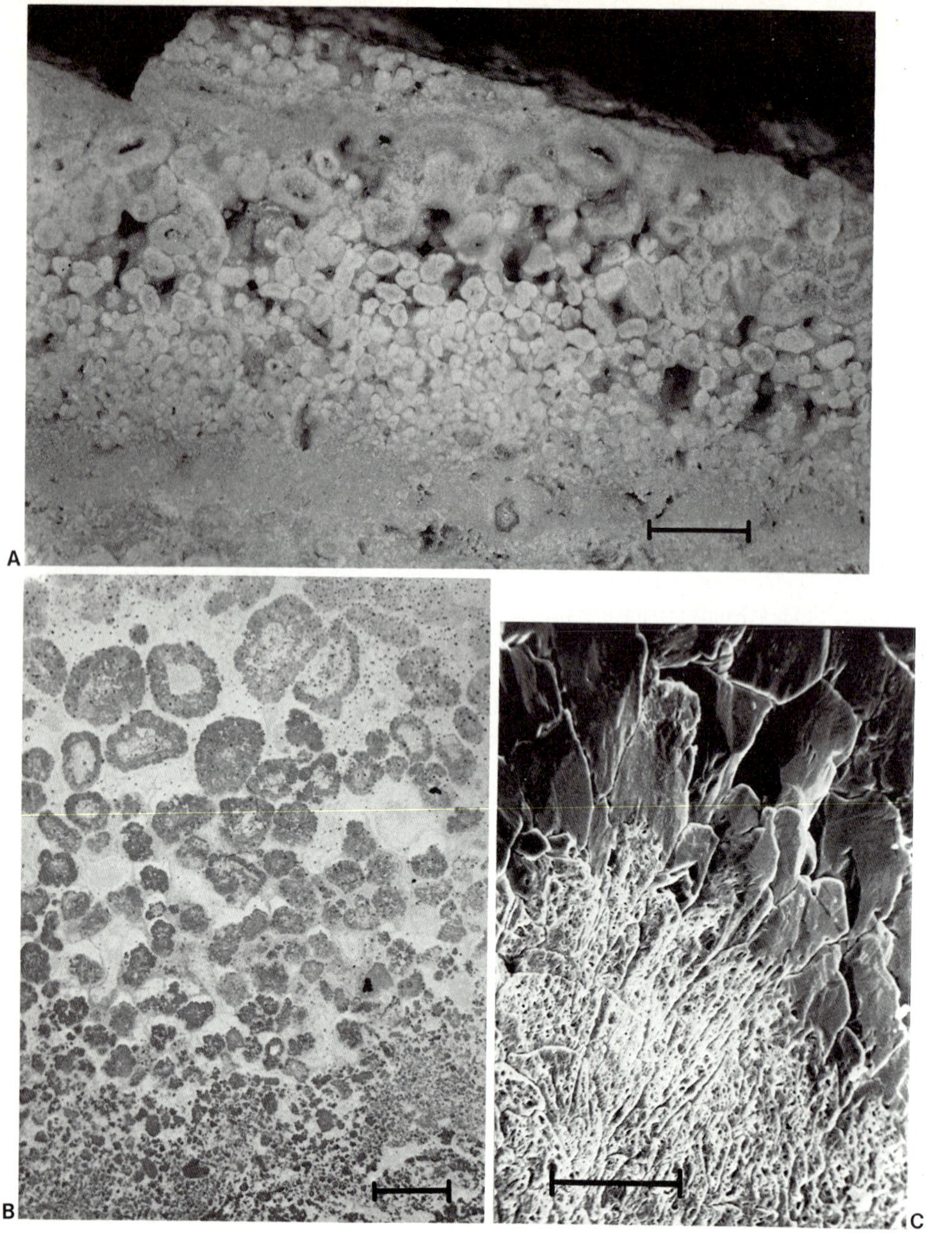

Fig. 4 A – C. Slab (**A**), thin section (**B**), and SEM view (**C**) of pisoliths from the same host rock. Etched slab (**A**) and photomicrograph (**B**) show the inversely graded layers of these pisolith accumulations. **C** Low magnification SEM view of etched border of a pisolith and adjacent cement. Observe the abundance of vacuoles in the pisolith and their total absence in the immediately adjacent cement; vacuoles are former sites of bacteria. Also observe radial pattern displayed by spar within and adjacent to the pisolith. Scales equal 1 cm, 0.25 mm, and 50 μm for **A – C**, respectively

The contact between the nucleus and cortex is usually more distinct in hand specimen than in thin section because of the color difference. The nucleus when compared to the cortex, has more poorly organized clumps of bacteria, often contains more spar, and is cemented by crystals that are more equant. Despite these contrasts it is often hard to divide the nucleus from the cortex in hand specimen where color differences do not occur and in thin section where the contact between the nucleus and cortex may be gradational.

3 Origin of the Pisoliths

The pisoliths form purely as an organic growth on intraclasts of travertine (recall that the intraclasts themselves primarily are composed of rods and clumps of bacteria embedded in calcite spar). Color differences between nucleus and cortices are due to variations in the amount of spar versus bacteria. The gradational contact between nucleus and cortex suggests that the intraclasts may not be composed of dead organic matter, but that the bacteria may be dormant over periods of time and once conditions are favorable, growth is reinitiated. This results in pisoliths in which the distinction between the nucleus and the cortex is difficult to distinguish or at least the boundary is ill-defined. At the other end of the spectrum are pisoliths that have been formed by recolonization on clasts of dead bacterial clumps in which the contact of nucleus and cortex is abrupt and color differences are distinct.

Development of a cortex completely around the grain suggests movement of the grains during growth, although Jones and Wilkinson (1978) have demonstrated that it is not necessary to move a grain in order to get complete envelopment by coatings. The presence of thicker cortices on the tops of some pisoliths plus the conformable fit of the densly packed pisoliths indicates in situ growth for some of these grains.

Acknowledgements. We gratefully acknowledge the financial support provided by the Geology Foundation, University of Houston, Sigma XI, and Getty Oil Company. In addition, we are most grateful to the Idaho Travertine Corporation for allowing access to their quarries.

References

Herlinger DL (1981) Petrology of the Fall Creek Travertines, Bonneville County, Idaho. Unpubl MS thesis, Univ Houston, Texas, p 172

Jones FG, Wilkinson BH (1978) Structure and growth of lacustrine pisoliths from recent Michigan marl lakes. J Sediment Petrol 48:1103–1110

Meredith JC (1980) Diagenesis of Holocene-Pleistocene (?) travertine deposits: Fritz Creek, Clark County and Fall Creek, Bonneville County, Idaho. Unpubl MS thesis, Univ Houston, Texas, p 263

V.3 Pisoids in the Caliche Profiles of Tarragona (N.E. Spain)

F. Calvet[1] and R. Julià[2]

1 Introduction

Caliche with pisolitic horizons have been recognized in the soils in the Mediterranean coast area of Spain from Almeria to Barcelona and Lleida. In the southern area the pisolitic horizons develop on old calchified aluvial fans as well as on volcanic rocks. In the middle and north of the Mediterranean coast they appear overlaying carbonate and metamorphic rocks.

The Tarragona-Altafulla (1°30′ long. E, 41°10′ lat. N and between 10 and 100 m high) has been chosen from this territory as study area (Fig. 1) because of its easy accessibility and also due to the large variety of profiles with carbonate differentiated horizons. More than 30 profiles (some of them younger than 1954, as C^{14} dating indicated) have been studied.

The pisoids studied show different types of the original structures and microstructures, as well as the process causing their formation. Apart from the works written by James (1972), Hay and Reeder (1978) and Hay and Wiggings (1980), very little is known about the initial microstructures of the caliches profiles.

The purposes of this chapter are: (a) To present the initial microstructures of the caliches profiles. (b) To discuss the diagenetic processes affecting the initial microstructures. (c) To interpret the origin and formation mechanism of the pisoids and the pisolitic level (peloids and matrix).

2 Geological Setting and Climate

The profiles we have studied are located in the Catalan Prelitoral depression – tectonic rift filled during the early and middle Miocene sediments – as well as on the mesozoic carbonate mountains of the Litoral chain of the Catalanides.

From the field observations we tend to infer that the substratum influence on the caliche development is almost nil, even though the existence of previous calichification etapes widespread through the region (attributed to the high Pliocene and Pleistocene) may mask the real nature of the substratum.

1 Department de Petrologia, Universitat de Barcelona, Gran Via 585, Barcelona-7, Spain
2 Instituto de Investigaciones Geológicas "Jaime Almera", Egipciacas, 15, Barcelona-1, Spain

Coated Grains (ed. by T.M. Peryt)

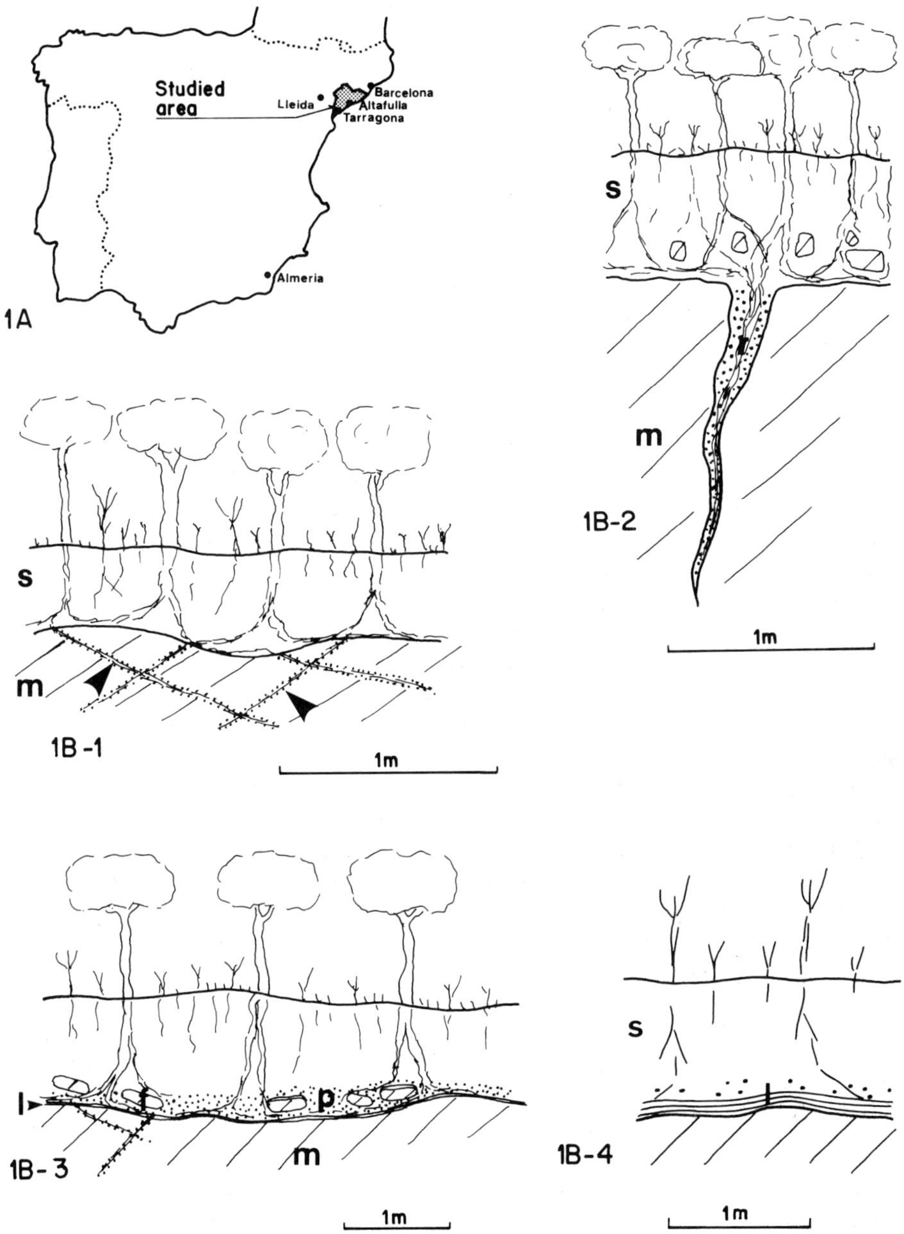

Fig. 1. A Geographic location map. **B** Several different caliche profiles of the studied sample area. *1* Subsurface micritic stringers (*arrows*) in calcareous Miocene substratum (m). *2* Vertical caliche profile in calcareous Miocene substratum (*m*). Penetration root zone (*r*). *3* Horizontal caliche profile on calcareous Miocene substratum (*m*). Soil (*s*) with pseudomycelium and calcification of some roots. *4* Horizontal caliche profile on different substrata (calcareous Miocene, calcareous Mesozoic, Pleistocene calichified alluvial-fan, etc.). Pisolitic level very little developed

The Tarragona area has an annual average rainfall of 500 to 600 mm; the annual average temperature ranges between 10 °C in January and 23 °C in August, the maximum of the maximum average is 31 °C in July. Towards the south of the Iberian Peninsula temperature generally increases and rainfall decreases except of very local areas due to the physiographic effect. The minimum peninsular pluviometric levels are in the Almeria area, they are below 200 mm; the annual average temperature is 37.7 °C in July to 0.2 °C in February.

The annual precipitation has a highly irregular distribution throughout the year. Two dry seasons (January – February and July – August) are separated by two rainy seasons (September – October and March – June) in the Tarragona area.

Besides the irregular distribution of the precipitations throughout the year we must add its stormy nature (150 mm fallen in 24 h and with a 10-year return period in Tarragona). The relative humidity in Tarragona is always high, over 55%, reaching values of 80% at 7 h during August and September.

As for its incidences in the moisture balance we must point out that it is precisely during August and September (moisture balance 0) when the relative humidity reaches the highest values; above all in areas next to the sea which compensate to a certain extent the evapotranspiration.

3 Caliche Profiles

The sequences with pedological carbonate levels show two basic geometric dispositions: development of the horizontal pisolitic level (Fig. 2a) and development of the vertical and subvertical pisolitic level. In the former we can see a downward differentiation on horizons, while in the latter there is a symmetry with respect to the penetration area of the roots with pisoids and the inserting material.

According to the distribution of the carbonate accumulations and to their textures we have recognized five levels:

Pseudomycelious Level (Fig. 2a). From 0 to 0.6 m. In the upper section of certain profiles and close to the surface white spots appear which cover roots and loamy clots. The clot colour is 7.5 YR 5/4. These spots, which have been taken from the roots, are weddellite while the ones taken from the surface of the clots are whiskers (10 – 100 μm).

Fig. 2. a Partial view of a profile caliche with horizontal development. *a* clay level with pseudomycelium and roots (*r*) with oxalate calcium mineralization; *b* pisolitic level with substratum fragments (*f*); *c* miocene substratum (*s*). *Arrows* show the laminar level. **b** Detail of the pisolitic level. *l* laminar level, *s* miocene substratum, *r* roots. **c** Spherical peloids associated with microfilaments and small roots, producing fungotufas. **d** Peloids of different morphology directly above root hair. Spherical small clots (*arrows*) and ovoid peloids (*o*). Presence of fungal filaments. **e** Ovoid peloid forming small clusters, fungal filaments. **f** Pisoid made up of a nucleus (grapestone of some peloids) and of an envelope (no laminae are observed). **g** Pisoids with different types of nuclei: peloids (*e*), substratum fragment (*f*), quartz grain (*q*), cryptocrystalline matrix (*m*)

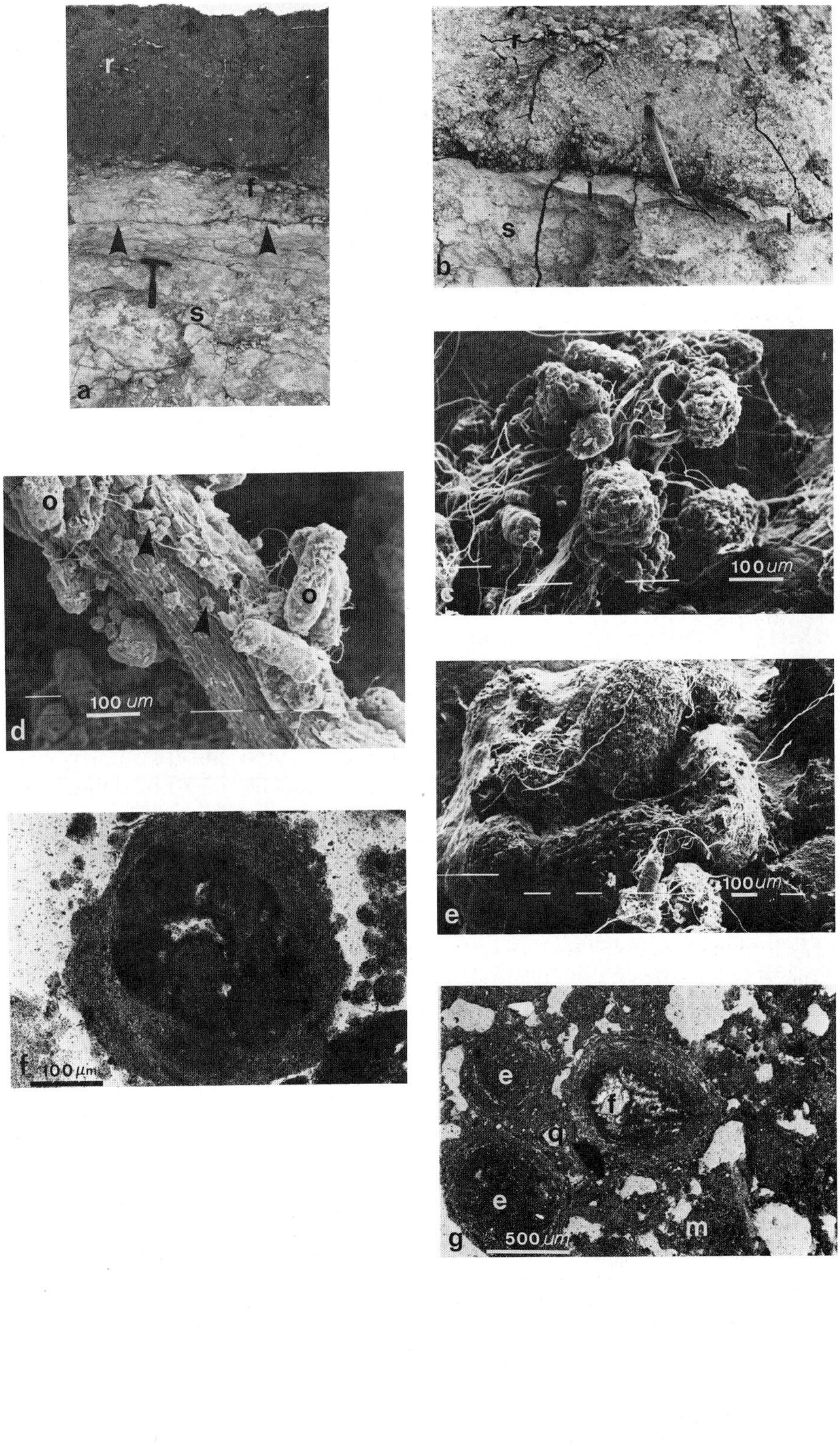
r
f
s
a
r
s
l
b
100 um
c
o
o
100 um
d
100 um
e
100 µm
f
e
f
q
e
m
500 um
g

Pisolitic Level (Fig. 2a, b). Up to 0.30 m. There are different stages of development in this level, in the areas where the maximum thickness is reached this level is made up of pisoids of a maximum diameter of 1 – 2 mm to 4 mm and dead or alive roots; the colour is greyish N 8/0 and it normally includes substratum fragments.

Cortical Laminated Level (Fig. 2a). 1 to 5 mm. Separated from the pisolitic level through a clear contact, this deposit is distinguished by the alternating 2 – 3 thin laminae which cover the stony substratum nearly uniformly. Multer and Hoffmeister (1968) name this level "thin microcrystalline rind crust". The most characteristic microstructure of this level is formed by curved tubular bodies of extended crystals 610 μm × 0.3 – 0.7 μm) of low-Mg calcite (LMC) in radial arrangement. The outer diameter ranges from 1 to 2 μm, although in some cases this internal porosity is not noticed; the length of these tubular bodies is over 100 μm.

Occlusion Laminated Level. 1 to 7 cm. Between the cortical laminated level and the overlying level, whether pisolitic or not, an other laminated level may develop not having any relation with the cortical laminated level. This new laminated level reaches maximum thickness in ponds of the substratum. The upper limit is clear although showing indentations and it becomes a root stromatolite. The lower contact is clearly overlying the cortical laminated level.

Subsurface Stringers. The stony substratum can be affected by subhorizontal to vertical laminae from 4 to 5 cm, which tend to disappear in depth or laterally. The subsurface stringers are symmetric with respect to a grey middle lamina made of roots, to the sides of this lamina a white lamina appears formed mainly by material of the substratum affected by corrosive and biocorrosive dynamic and covered by microfibres (0.5 – 2 μm), microcrystalline (1 – 3 μm) and cryptocrystalline (0.1 – 0.3 μm) anhedral crystals associated to fungic filaments and feathered whiskers.

Not all these carbonated levels appear in one and the same profile, some of them being incompatible, for instance the subsurface stringers do not appear with occlusion laminated levels. Summarizing, four types of horizontally developing profiles are found in Tarragona (Fig. 1). The vertical disposition with pisoids as a macrosubsurface stringer developed on unsteady slopes.

4 Pisolitic Level

The pisolitic level is formed by different grain types (mainly pisoids and peloids with a small amount of composed grains) and by different kinds of intragranular matrix (alveolar, peletoidal and cryptocrystalline).

4.1 Peloids

The term *peloid* is used to refer to all those components basically made up of cryptocrystalline aggregate regardless of its origin and size. Therefore this term is used only in a descriptive sense and is equivalent to ones described in caliche profiles as "pelletoid" (James 1972, Ward 1975), "calcrete pellets" (Read 1974, 1976, Adams 1980), "peloids" and "pellets" (Harrison 1977, Harrison and Steinen 1978), "pellet" (Hay and Reeder 1978, Hay and Wiggins 1980).

Description: The shape and size of the peloids is greatly variable, existing a graded variation in shape and size (from 10 – 15 μm to 1 mm). Similar size variations have been mentioned by James (1972) and Read (1976). The following kinds of peloids may be distinguished:

a) Spheric Peloids (from rounded to subrounded). The diameter range from 40 μm to 300 μm. Generally the little ones (from 40 to 100 μm) are associated with fungic filaments, making fungotufas (Fig. 2c). The fungotufas are common over and around the roots and/or root hairs.

b) Oval Peloids. They are convex towards outside, twice longer than wider. According to the size and shape the following have been differentiated: (1) from 110 to 130 μm in length and 40 – 60 μm in width; these peloids are in relationship with the spheric ones (which, in part, correspond to transversal sections of the oval shaped ones and roots at times they appear in a fan-shaped disposition. (2) 210 to 240 μm long and 90 to 120 μm wide (Fig. 2d). (3) Nearly 500 μm long and 300 μm wide (Fig. 2e), they appear forming small clusters with a certain closeness among the peloids, associated to thc root laminated levels at the bottom of the pisolitic level.

c) Irregular Peloids: show great variety of shapes, from the extended to the equidimentional ones. They can be from some dozens to 500 μm or more. Inside of these peloids ghosts of spherical and oval peloids often occur.

Microstructure: The information regarding the peloid microstructure is poor. According to Read (1974, 1976) they are formed by cryptocrystalline calcite, and according to Harrison (1977) they are formed by calcite equidimentional crystals (1 to 3 μm) randomly disposed. Hay and Wiggins (1980) suggest they are formed by calcitic micrite and sepiolite (which vary from 6 to 60%).

The microstructures found in the studied profiles are:

a) Microfibres. The microfibres range from 0.7 to 2.2 μm in length and from 0.1 to 0.2 μm in width and are randomly disposed. Main microstructure in the most cases.
b) The crypto- and microcrystalline anhedral crystalls 0.2 to several μm. In some peloids it is the dominating microstructure.
c) Clay and organic detritus.
d) Whiskers. In general the whisker crystals appear as a secondary microstructure associated to (a) and (b). In some cases, the whisker crystals appear covering the peletoidal bodies totally or partially (Fig. 3e). The whiskers rarely from the body of the peloids. They range from 5 to 10 μm in length and from 0.5 to 1.5 μm in width.

4.2 Pisoids

We understand by *pisoids* those components made up of differentiated nuclei which can be of different size, shape and nature, coated by an envelope of different thickness and disposition formed by one or more laminae and basically composed of carbonate minerals. As for the size we do not made any difference between the pisoids with a diameter smaller than 2 mm (usually called ooids) and those of a diameter bigger than 2 mm (usually called pisoids). The pisoids which have been studied, being of a pedogenetic origin may be named more precisely "caliche pisolite" (Esteban 1976) or "calcrete pisolites" (Read 1974) in order to avoid confusion with "vadose pisolites" which are of a hypersaline origin. Both these types are called "vadoids" by Peryt (this vol.).

Sizes: The pisoids measure from 250 μm to 5 mm, but in general they vary from 1 mm to 3 mm.

Nucleus: The type of the nucleus varies according to the nature of the outcrop. In general, in the vertical caliche profiles the nucleus is made up principally of grapestone of some peloids (Fig. 2f), or of only one peloid. In the latter case, it is sometimes very difficult to distinguish the nucleus from the envelope. Very often this peloid shows a slightly diffused banding as well as smaller peloid ghosts.

In the horizontal caliche profiles, with rocky hard substratum (Fig. 2a), the type of nucleus is more variable (Fig. 2g). The principal types are: substratum fragments (from 10% to 50%), one peloid (from 20% to 40%), grapestone of some peloids (from 5% to 20%) and quartz (from 2% to 10%). The proportion of each nucleus type varies according to the outcrop, and also according to its position in the outcrop.

The nucleus size varies from one pisoid to another. There exits a direct relationship between the nucleus size and the pisoid size, but not between the nucleus size and the envelope thickness. These relationship are independent of the nucleus type, although the pisoids with a nucleus made up of calcareous substratum fragments are the biggest because this nucleus type is the biggest.

When the nucleus is a detritic fragment, in general its borders are affected by corrosion, microperforation and micritization processes. Other nuclei present a recrystallization process. Similar processes have been described by Walls et al. (1975).

Fig. 3. a Whisker crystals with tangential disposition and pseudoalveolar packing. Nucleus (*n*). **b** S.E.M. view of a lamina made up of whisker crystals in a pseudoradial disposition (*r*). Lamina made up of tangential whisker crystals (*t*). Peloid (*e*), incorporated to the envelope. The nucleus is in the *lower right corner* of the photograph. **c** The lower part of the envelope is made up of the microstructure of cryptocrystalline anhedral crystals (*a*), and the upper part of whisker crystals with a tangential disposition. Microfilaments (*arrows*). Nucleus (*n*). **d** Lamina made up fundamentally of microfibres. The microfibres are disposed at random. Whisker crystals (*w*), and cryptocrystalline anhedral crystals (*arrows*). **e** Pisoids with the nucleus made up of one peloid (*e*) and of an envelope formed by only one lamina (*arrow*) made up of tangential whisker crystals. Gradual change between the envelope and the "laminae" of the alveolar texture. Patches of cryptocrystalline matrix (*m*). **f** S.E.M. view of the alveolar matrix. "Laminae" arched (*l*). Interlaminar space (*i*). Peloids (*e*). **g** Alveolar matrix made up of the "laminae" in an arched disposition (*l*). Interlaminar space (*i*). Peloids (*e*)

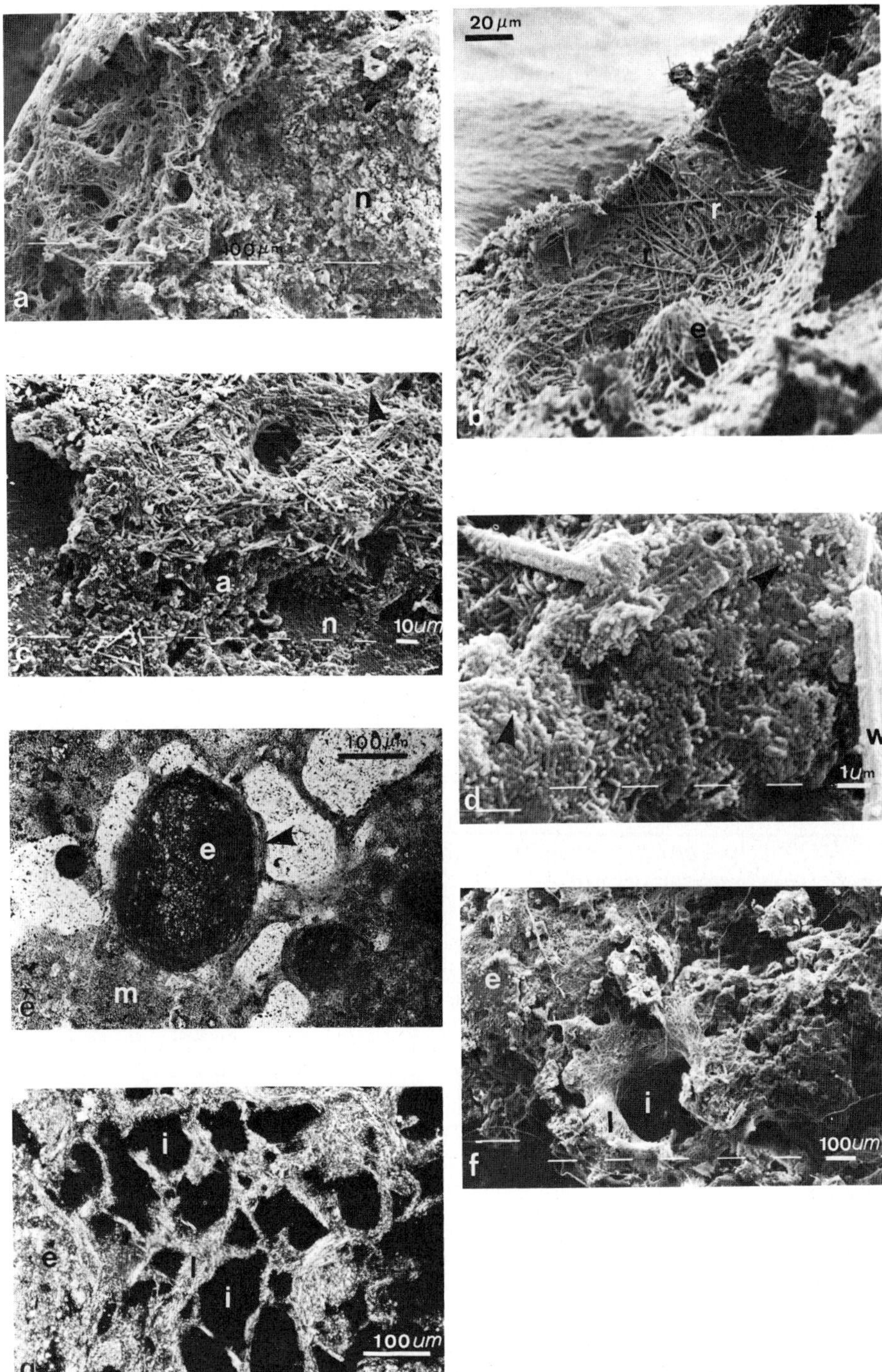
a
n
100 µm
b
20 µm
r
t
e
c
a
n
10 um
d
w
1 um
e
100 µm
m
f
e
i
l
100 um
g
i
e
l
i
100 um

Envelope: The envelope thickness is very variable, from a few microns to 500 μm or more. Rarely the envelope is bigger than the diameter of the nucleus. The envelopes are in general uniform in their thickness in the pisoids with round nuclei, although it is common to observe pisoids with their envelopes very well developed on a certain side while on the other side of the same pisoid the envelope practically does not exist (Fig. 2f).

When the nuclei have an elongated form (principally the substratum fragments) the envelope has a directional growth, and its development is thicker along the perpendicular orientation respect the major axis of the nucleus. Growths of this style have been cited by James (1972) and Walls et al. (1975); the downward and/or upward developments reported by some workers (e.g. Swineford et al. 1958, Steel 1974, Read 1976, Hay and Reeder 1978, Surdam and Stanley 1979, Hay and Wiggins 1980) have not been observed.

At times the envelopes contain peloids, quartz size silt, or some substratum fragments. Nagtegaal (1969), Harrison (1977), Hay and Wiggins (1980) have also observed this character.

The number of the laminae of the envelopes is very difficult to precise, the same has been observed by Harrison (1977) and Chafetz and Butler (1980). The laminae may present a concentric lamination, principally in the round pisoids, preserving more or less regularly their thickness. At other times, the laminae are continuous around the nucleus, but in the case of the pisoids with elongated nuclei the laminae tend to wedge out in the short side. Very often the laminae are discontinuous and truncated.

The number of laminae is relatively low (principally one or two), but can reach up to seven. This character is in concordance with the citations of the caliche pisoids (James 1972, Siesser 1973, Steel 1974, Esteban 1976, Read 1976, Harrison 1977).

The principal types of pisoids according to the laminae number and their disposition in the envelopes are:

a) Pisoids with the envelope made up of one lamina alone, which can be divided into: (1) Small-size pisoids (from 250 to 500 μm) with one lamina from 10 μm to 50 μm made up of whisker crystals in a tangential disposition (Fig. 3e). James (1972) cited similar pisoids. (2) Bigger-size pisoids with one indefinite envelope up to 500 μm (Fig. 2f).

b) Pisoids with two laminae. The first is clear and large made up principally of cryptocrystalline crystals but with a different morphology (microfibres, anhedral). The second lamina, placed over the external border of the pisoids, is sometimes discontinuous, dark and narrow (even 20 μm) made up of tangential whisker crystals. Similar envelopes have been observed by James (1972).

c) Pisoids with more than two laminae. Up to seven envelopes have been observed (Fig. 2g).

Microstructure of the laminae: The principal microstructure types observed are the following:

a) Laminae made up principally of whisker crystals. The whiskers in the laminae measure 6 – 24 μm and up to 75 μm in length and 0.6 – 2 μm in thickness. In general they are straight, but the larger crystals can be slightly curved. They present two dispositions in the laminae:

1. Tangential disposition. In general these laminae are equivalent to the dark and narrow laminae. The whisker crystals are tangential to the substratum. They may be arranged with a compact packing (Fig. 2c) (in which case the laminae measure from 10 μm to 40 μm thick), or with a pseudoalveolar packing (Fig. 3a) (in which case the laminae may measure up to 400 μm or more).

2. Pseudoradial disposition. The whisker crystals oriented at 45° respect the nucleus or the interlaminar surface are dominant (Fig. 3b). Sometimes the whiskers are packed with a certain cluster habit. Between the interwhisker porosity microfibres exit and they vary in volumetric importance according to each lamina. The secondary microstructures are crypto-microcrystalline anhedral crystals and clays.

b) Lamina made up principally of microfibres. The microfibres vary 0.5 – 1.5 μm in length and 0.1 – 0.3 μm in thickness and are rarelly curved. Their habitual disposition is at random (Fig. 3d). The secondary microstructures are whiskers, cryptocrystalline anhedral crystals and clays.

c) Laminae made up principally of crypto-microcrystalline anhedral crystals (Fig. 3c). The cryptocrystalline crystals vary from 0.05 to 1.5 μm and are equant. Sometimes they present subhedral habit. The microcrystalline crystals vary from 2 – 3 μm to 20 μm. Progressive changes have been observed between the microfibre microstructure and the cryptocrystalline anhedral microstructure and also between the latter and the microcrystalline anhedral microstructure. The secondary microstructure are whiskers, microfibres and clays. Sometimes the crypto and microcrystalline microstructures present a pseudopeletoidal aspect.

d) Laminae with calcified filaments. This lamina type is rarely observed. In this microstructure an intertwined mesh of calcified filaments is dominant. The calcified filaments are 3 – 5 μm in diameter and the crystals are perpendicular to the filament. The crystals are like the microfibres or are subhedral-anhedral cryptocrystalline equant crystals. Non-calcified filaments (being 0.8 – 1 μm in diameter), whiskers, peloids are also present.

Disposition of the different microstructure types in the envelope: From the study of the different microstructures and their real disposition in the envelope of the recent and pararecent pisoids it appears that a general rule for the disposition of the different microstructure types in the envelope does not exist. Figure 4 shows 9 real graphics of the pisoids as observed in scanning electron microscope. The conclusions from those are: (a) Each pisoid has its own microstructure distribution in the envelope. (b) The microstructure distribution is widely variable. (c) In spite of this variability the microstructures made up fundamentally of microfibres and whisker crystals are more frequent in the last lamina of the envelope, or in the outer part of the envelope when it is made up of one lamina alone. On the other hand, the microstructures composed fundamentally of crypto and microcrystalline anhedral crystals are dominant in the inner parts of the envelope. These microstructures are dominant in the recent caliche pisoids from arid zones (Almeria). Figure 5 shows in a schematic form this distribution. The envelope in the fossil pisoids is made up nearly totally of the cryptomicrocrystalline anhedral crystals.

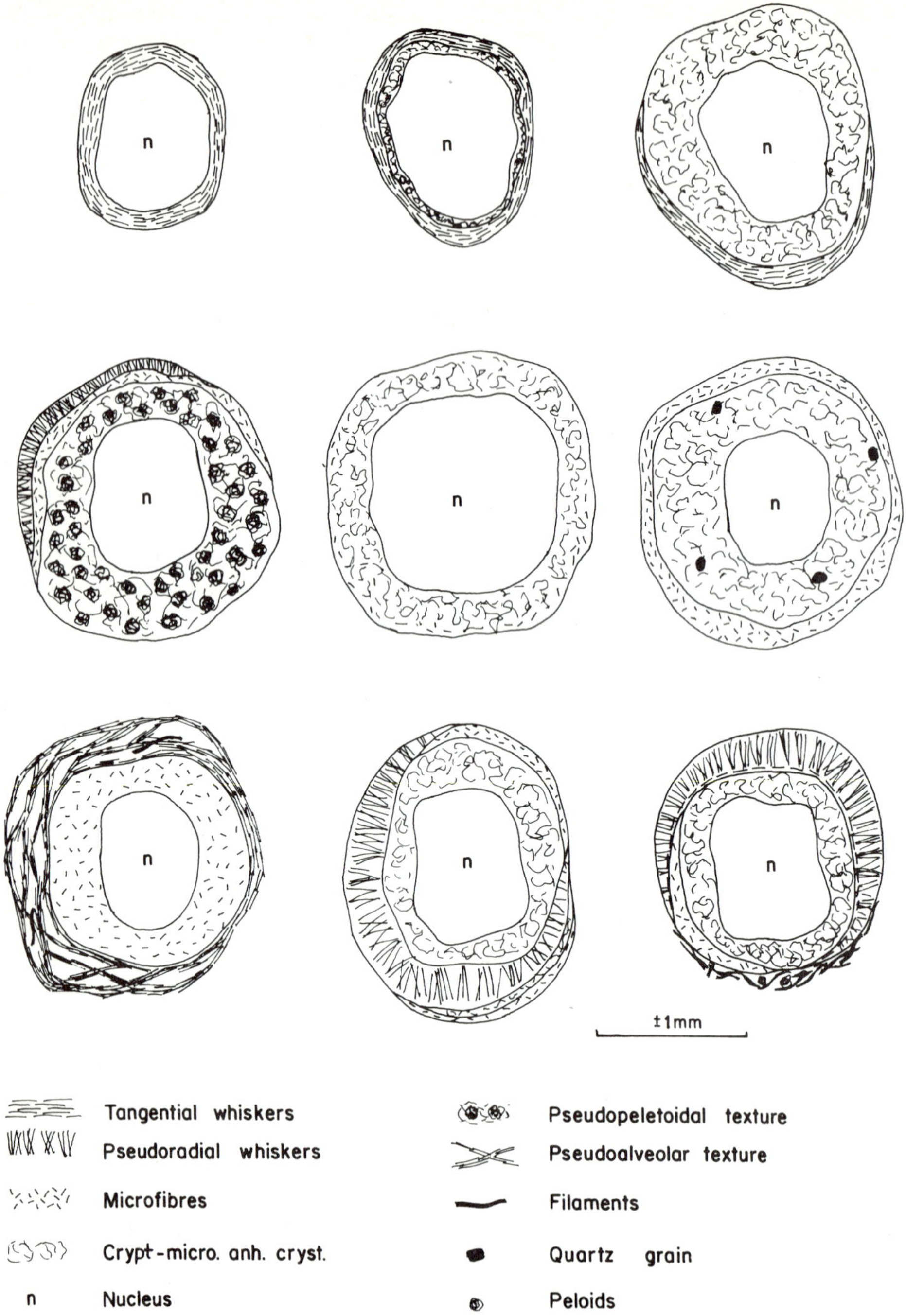

Fig. 4. Real schemes of the microstructure distribution in recent pisoids

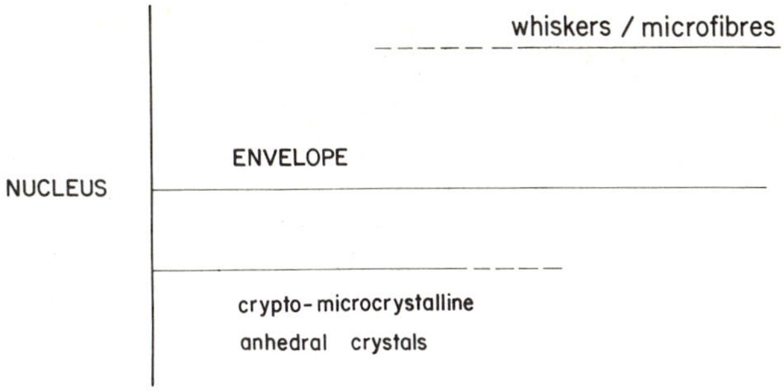

Fig. 5. Distribution of the microstructures

4.3 Matrix

The interpisoid space is occupied by different matrix types and the alveolar and peletoidal are the most important voluminously in the recent caliches, while the cryptocrystalline matrix is the principal matrix in the para-recent and fossil caliches in the Tarragona area, and in recent caliches in the Almeria area.

Alveolar matrix: Several authors have cited this texture in recent and fossil caliche (James 1972, Esteban 1974, Steinen 1974, Braithwaite 1975, Harrison 1977, Harrison and Steinen 1978, Klappa 1978, Adams 1980).

It is made up basically of "laminae" that in part cover the substratum (pisoids, peloids, etc.) (Fig. 3e) and in part form an intertwined mesh (Fig. 3f, g) in the interpisoid space. The "laminae" have an arched form. Their thickness is variable, from a few microns to 70 μm though some "laminae" can attain developments up to 300 μm.

The empty space between the intertwined mesh of the "laminae" has a variable form, principally polygonal, from equant to elongate with round borders. The mesh number of this intertwined mesh varies from a few microns to 1 mm, but in general forms a mosaic comprised between 70 μm and 400 μm. It had been cited as polygonal bubbles (Braithwaite 1975). The intralamina porosity can be progressively filled by cryptocrystalline crystals, whiskers with a random disposition, filaments (fungic, cyanobacterias, etc.).

The "laminae" are composed principally of LMC whisker crystals (Fig. 3f) from 12 to 80 μm in length and from 0.8 to 2.2 μm in thickness. The crystals are straight and some are slightly curved. The disposition of the whiskers is parallel according to the axis of the plane of the "laminae", and if these are in an arched position the whiskers turn progressively preserving their tangential disposition, while if we observe the surface of the "laminae" the whiskers are disposed at random. Similar microstructures have been cited by James (1972), Harrison (1977), Klappa (1978), Adams (1980).

The interwhisker space can be occupied by individual microfibres or forming clots, and secondary by crypto-microcrystalline anhedral crystals or clays. The "laminae" may enclose small peloids.

Pelletoidal matrix: We define the pelletoidal matrix when the interpisoid space is occupied fundamentally by peloids. It is equivalent to the "clotted micrite texture" of James 61972), Esteban (1974), Ward (1975), Hay and Reeder (1978), Hay and Wiggins (1980), to "pelletal micrite" of Scholle and Kinsman (1974), to "micrita moteada" of Esteban (1975), to "pelleted micritic matrix" of Harrison (1977), Harrison and Steinen (1978). Similar textures have been cited by Braithwaite (1975), Reeves (1976), Read (1976), etc.

Cryptocrystalline matrix: We define a cryptocrystalline matrix when the interpisoid space is occupied fundamentally by crypto-microcrystalline anhedral crystals (Fig. 2g). Some times ghosts of peloids and alveolar texture are observed.

5 Origin of Different Textures

5.1 Origin of Peloids

One of the first processes in the carbonate concentration during the development of an edaphic profile is the formation of small clots and peloids around and directly above the roots, root hairs, etc. In the descriptive part of this paper a wide morphologic variability has been seen that, in part, responds to different origins.

The small clots (from a few microns to 20 – 30 μm in diameter) are made up of organic-detritic (clays, carbonate, silica) mass. Their origin is directly related to the roots.

The peloids described as spherical can have two different origins, but they are internally related:

a) Fungotufas. According to Pomar (1976) and Pomar et al. (1975) the fungotufas are made up of organic-mineral-clay mass, as well as of rests of fungic filaments and vegetal fibres. The small clots will be the initial phases of the fungotufas origin. The agglutination mechanism is realised by organic matter and inorganic substances present in the soil (principally clays). The clots are ranged in the intertwined mesh formed by the fungic hyphae.

b) The clots may serve as a nucleus by the precipitation of different carbonate mineralogies (microfibres, whiskers and cryptocrystalline anhedral crystals) and forming a peloid by means of a centrifugal growth. Harrison (1977) cited a similar mechanism.

A large part of the ovoid peloids are of a fecal origin. According to Braithwaite (1975) they come from the fauna that colonized these levels (worms, ants, gasteropods, etc.). The calcitization mechanism of the fecal pellets in the edaphic profiles have been explained by Klappa (1978).

In the literature it is described that one of the most important origins of the peloids in the edaphic profiles is the micritization of the skeletal grains (James 1972, Read 1976, Harrison 1977, Harrison and Steinen 1978). This mechanism has not been observed in the samples studied.

In general the peloid concentration is greater or is related with the penetration root zones. Thus Harrison (1977) found more abundance of peloids (that he called pellets) in the subsurface micritic stringers and in the rhizocretion adjacent rock. Also Esteban (1974, 1975) related the mottled texture with the "channel system" or microsparite channel in Pleistocene and fossil caliches. According to both authors the peloids are round and have some tens of microns in diameter.

The "channel systems" of Esteban are in the recent profiles studied here equivalent to penetration zones more or less vertically produced by the roots. The subsurface micritic stringers are equivalent to the root penetration laminae in the substratum. Therefore it is logical that the peloids are found in textures produced by the action of the roots.

5.2 Origin of Pisoids

The origin of pisoids has been explained in different ways. For Swineford et al. (1958) it is concretionary mechanism rather than an algal origin. They relate their formation to the seasonal changes (arid climate-humid climate period). Siesser (1973) and Reeves (1976) consider that pisoids are originated through the precipitation of carbonate around a nucleus during alternating periods (dry and humid) in semiarid climates.

Ward (1975), Esteban (1974, 1975) relate the origin of pisoids with cracking processes (circumgranular cracking), shrinkage and solution. For Braithwaite (1975) the pisoids are formed through the mechanical accretion of the particles mainly clastic, he mentions the climate control in the formation of these textures. Hay and Reeder (1978) and Hay and Wiggins (1980) interpret the origin of pisoids through accretion of carbonate and detritic materials.

The observations made in the recent profiles of Tarragona allow us to state that when any component (peloid, composite grain, fragment from the substrate, etc.) starts to be coated, this coating is made through a thin cover of whisker crystals in tangential arrangement (Fig. 3e).

The main idea of the growth and development of the pisoids is that of an accretional and centrifugal style of starting in the nucleus (Fig. 6) through constructive mechanisms (the formation of different microstructures) and the progressive removal of detritical material.

The microstructure with crystal whiskers either in a tangential arrangement (predominant in the initial stages of pisoidization) or in a radial one is carried out through the precipitation of these steady LMC forms only when the conditions of formation are reached. This precipitation which we consider is controlled indirectly by fungic filaments particularly as far as the whiskers in a radial arrangement are concerned and by the remains of organic fibres of the roots when the whiskers are in a tangential arrangement.

The microstructure made up mainly of microfibres is formed by the precipitation of these crystals, this precipitation is partially controlled by filaments 1 µm in thickness (cyanobacteria, remains of plant fibrils, spheric bacteria, etc.).

These constructive mechanisms are arranged in a different way in each pisoid (Fig. 4). Thus in one part of the pisoid a microstructure may dominate while in another part a different microstructure resulting in discontinuous laminae will dominate. In other cases the envelope can be homogeneous and continuous.

This set of microstructures, which form the aggradational envelope and which are metastable (close relation microstructure-microorganism) quickly

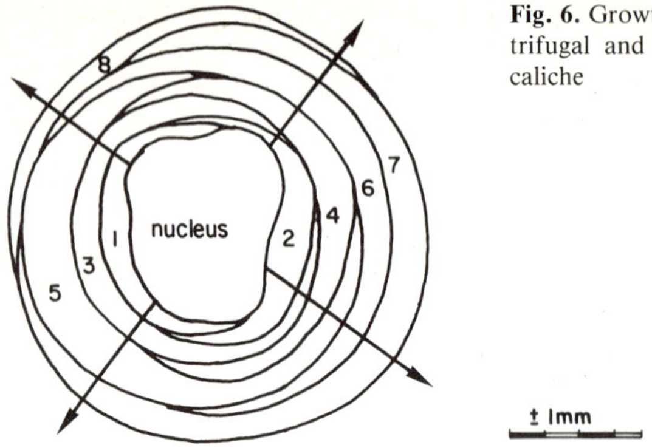

Fig. 6. Growth mechanism. The *arrows* show the centrifugal and accretional growth style of the pisoid caliche

change into more steady microstructures. Thus microfibres reorganize themselves (through an aggrading neomorphism) and become crypto-microcrystalline anhedral crystals. The whisker crystals reorganize (through a degrading neomorphism) and become crystals with a rod habit and finally crypto-microcrystalline anhedral crystals of LMC. The microfibres are much more unsteady than the crystal whiskers, that is the reason for their susceptibility with regard to the reorganization.

These mechanism of reorganization explain the existence of microstructures of the crypto-microcrystalline anhedral crystals in the inner laminae (Figs. 4 and 5), the microstructures being a result of the diagenetic transformation of the initial microstructures (microfibres and whiskers) and the true makers of the pisoid. In some pisoids and in very localised places, it is more possible to observe laminae made up of whiskers in the inner zones than laminae formed by anhedral crystals. It is possible to explain this case by means of the different susceptibility of reorganization of microfibres and whiskers.

At the same time as the processes of reorganization occur, in the outer part of the laminae or in the outer laminae, the pisoid grows aggradationally be means of whisker and microfibre formation which at the same time will be organized. While the pisoid grows aggradationally and centrifugaly it incorporates the different components found in its growth (mainly peloids and some detrital grains).

This continuous reorganization is one of the reasons for a certain homogeneity of the pisoid envelopes in the pedologic profiles. For this reason it is very difficult to observe the formed laminae, except in very uncommon conditions. This is perhaps the reason (the mechanism and the style of formation) why the pedological pisoids generally have a small number of laminae in their envelopes, comparatively with the number of laminae of the pisoids formed in other conditions (marine, cavern, etc.).

5.3 The Origin of Alveolar Texture

The mechanism of formation of the whisker walls is closely related with the rhizosphere (roots, root-hairs). When a root enters in a pedological profile, apart from the mechanic processes produced, the organic structure of the root experiences a decomposition process and its fibres remain covering the surface (the surface may be pisoids, channel walls, etc.) and also fibres remain floating in the space previously occupied by the root.

These vegetable remains (fibres) serve as substratum for the precipitation of whisker crystals and thanks to the support given by the vegetable fibres the alveolar texture is formed. The crystalls are tangentially arranged to the axis of the fibres and when these fibres are decomposed the crystal whiskers have already formed the frame of the alveolar texture.

Different author (Steinen 1974, Calvet et al. 1975, Harrison and Steinen 1978, Klappa 1978, Adams 1980) have cited the relation between the alveolar texture and the roots.

Later the alveolar texture experiences a number of diagenetic processes. The crystal whiskers that constitute the wall base experience a degrading neomorphic process and they are finally turned into a crypto-microcrystalline anhedral crystals. This process is the same as the process underwent by the laminae with whisker microstructure of the pisoids. The spaces between the walls could later remain occluded by sparitic cement of LMC, by internal sediment or by calcification of filaments.

6 Conclusions

1. The pisolitic levels are found in recent caliche profiles of the Mediterranean Iberian Peninsula (principally in N.E. Spain). The climate is semiarid and arid.
2. The pisolitic level is made up of pisoids, peloids and different matrices (alveolar, pelletoidal and crypto-crystalline) (Fig. 7).
3. The pisoids (from 1 to 3 mm) are composed of different nuclei (peloid, grapestone, rock fragment, quartz) and of envelopes containing 1 to 7 – 8 laminae.
4. The principal microstructures of the laminae are: (a) LMC whisker crystals, (b) LMC microfibres and (c) LMC crypto-microcrystalline anhedral crystals.
5. The disposition of the microstructures varies according to each lamina. The initial microstructure is the whisker and microfibres while the crypto-microcrystalline one is a neomorphic product.
6. The recent textures and microstructures are closely related to the action of the organisms and microorganisms (roots, root-hairs, fungal filaments, cyanobacterias, bacterias, rests of vegetal fibres, etc.
7. The growth development style in the pisoids is accretional and centrifugal.
8. The consequence of the recent microstructure alteration is the presence in the caliche pisoids (recent and fossil) of only few laminae in contrast to the other possible environment (marine, cavern, etc.) where the growth style is different.

Acknowledgements. The S.E.M. work has been carried out with the help of Montserrat Marsal from Metallurgy Department of Barcelona Polytechnic University and of Ramon Fontarnau from Electronic Microscopic Service of the Barcelona University.

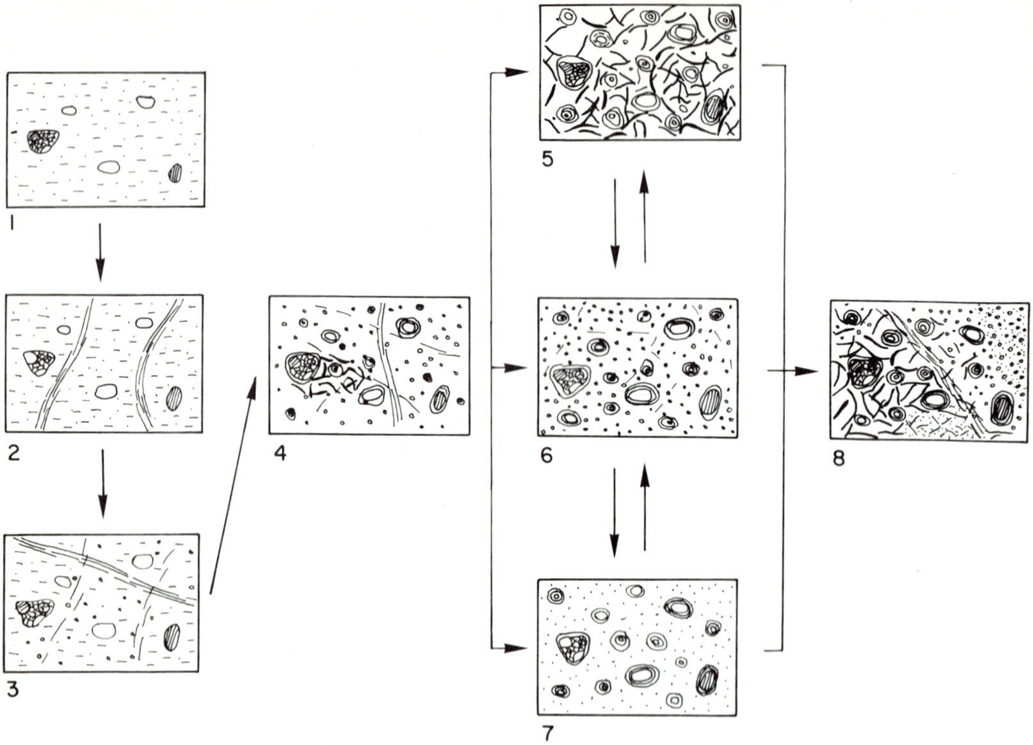

Fig. 7. Scheme of the development of the whole pisolitic level. *1* original soil with clays and substratum fragments. *2*, *3* show the initial stage of calichification of the soil by means of root penetration (*long lines*) forming the first peloids (fungotufas). *4* shows an intermediate stage of the formation of pisolitic level. Some substratum fragments and some peloids serve as nuclei to the pisoid origin. New roots may penetrate forming new peloids. The old roots undergo decomposition processes which influence the alveolar matrix formation and the pisolitic process in general. *5* – *7* show a theoretical mature pisolitic level with different matrices. *8* shows a more real pisolitic level with a mixture of different matrices

References

Adams AE (1980) Calcrete profiles in the Eyam Limestone (Carboniferous) of Derbyshire: petrology and regional significance. Sedimentology 27:651 – 660

Braithwaite CJR (1975) Petrology of paleosols and other terrestrial sediments ou Aldabra, Western Indian Ocean. Philos Trans R Soc London Ser B 273:1 – 32

Calvet F, Pomar L, Esteban M (1975) Las rizocreciones del Pleistoceno de Mallorca. Inst Invest Geol XXX, pp 35 – 60

Chafetz S, Butler JC (1980) Petrology of recent caliche pisolites, spherulites, and speleothem deposits from Central Texas. Sedimentology 27:497 – 518

Esteban M (1974) Caliche textures and "Microcodium". Suppl Boll Soc Geol It 92:105 – 125

Esteban M (1975) Características de los suelos carbonatados semiáridos: El problema del caliche y sus productos asociados. Actas Semin Int Evaluación Tierras Zonas Aridas y Semi-Aridas. America Latina Roma, pp 317 – 327

Esteban M (1976) Vadose pisolite and caliche. Am Assoc Petrol Geol Bull 60:2048 – 2057

Harrison RS (1977) Caliche profiles: indicators of near-surface subaerial diagenesis, Barbados, West Indies. Bull Can Petrol Geol 25:123 – 173

Harrison RS, Steinen RP (1978) Subaerial crust, caliche profiles, and breccia horizons: Comparison of some Holocene and Mississippian exposure surfaces, Barbados and Kentucky. Geol Soc Am Bull 89:385 – 396

Hay RL, Reeder RJ (1978) Calcretes of Olduvai Gorge and the Ndolanya Beds of northern Tanzania. Sedimentology 25:649 – 673

Hay RL, Wiggins B (1980) Pellets, ooids, sepiolite and silica in three calcretes of the southwestern United States. Sedimentology 27:559 – 576

James NP (1972) Holocene and Pleistocene calcareous crust (caliche) profiles: criteria for subaerial exposure. J Sediment Petrol 42:817 – 836

Klappa CF (1978) Morphology, composition and genesis of Quaternary calcretes from the western Mediterranean: a petrographic approach. Unpublished Ph D thesis, Univ Liverpool, p 446

Multer HG, Hoffmeister JE (1968) Subaerial laminated crusts of the Florida Keys. Geol Soc Am Bull 79:183 – 192

Nagtegaal PJC (1969) Microtextures in recent and fossil caliche. Leidse Geol Meded 42:131 – 142

Pomar L (1976) Procesos telodiagenéticos en rocas carbonatadas del litoral catalán y Baleares: Su relación con Microorganismos. Tesis Doctorado. Dep Petrología. Univ Barcelona, p 279 (unpublished)

Pomar L, Esteban M, Llimona X, Fontarnau R (1975) Acción de líquenes, Algas y Hongos en la Telodiagénesis de las Rocas Carbonatadas de la zona litoral prelitoral Catalana. Inst Invest Geol XXX:83 – 117

Read JF (1974) Calcrete deposits and Quaternary sediments, Edel Province, Shark Bay, Western Australia. In: Evolution and diagenesis of quaternary carbonate sequences, Shark Bay, Western Australia. Memoir 22, pp 250 – 282

Read JF (1976) Calcretes and their distinction from stromatolites. In: Walter MR (ed) Stromatolites, Elsevier, Amsterdam, pp 55 – 71

Reeves CC Jr (1976) Caliche, origin, classification, morphology and uses. Estacado Books, Lubbock, Texas, p 233

Scholle PA, Kinsman DJJ (1974) Aragonitic and high magnesium calcite caliche from the Persian Gulf – a modern analogue for the Permian of Texas and New Mexico. J Sediment Petrol 44:904 – 916

Siesser WG (1973) Diagenetically formed ooids and intraclasts in South African calcretes. Sedimentology 20:539 – 551

Steel RJ (1974) Cornstone (fossil caliche). Its origin, stratigraphic and sedimentological importance in the New Red Sandstone, Western Scotland. J Geol 82:351 – 369

Steinen RP (1974) Phreatic and Vadose Diagenetic Modification of Pleistocene Limestone: Petrographic observations from Subsurface of Barbados, West Indies. Am Assoc Petrol Geol Bull 58:1008 – 1024

Surdam RC, Stanley KO (1979) Lacustrine sedimentation during the culminating phase of Eocene Lake Gosiute, Wyoming (Green River Formation). Geol Soc Am Bull 90:93 – 110

Swineford A, Leonard AB, Frye JC (1958) Petrology of the pliocene pisolitic limestone in the Great Plains. State Geol Surv Kansas Bull 130:97 – 116

Walls RA, Burleigh Harris W, Nunan WE (1975) Calcareous crust (caliche) profiles and early subaerial exposure of carboniferous carbonates, northeastern Kentucky. Sedimentology 22:417 – 440

Ward WC (1975) Petrology and Diagenesis of Carbonate Eolianites of Northeastern Yucatán Peninsula, Mexico. In: Belize shelf – carbonate sediments, clastic sediments and ecology. Studies in Geology 2, pp 500 – 571

V.4 Pisoliths (Pisoids) in Quaternary Travertines of Tivoli, Italy

Robert L. Folk [1] and Henry S. Chafetz [2]

1 Geologic Setting

The "type locality" for travertine is at Tivoli, about 25 km east of Roma. In fact, the name "travertine" descends from the Latin, "Lapis Tiburtinus", referring to the stone quarried at Tibur (the ancient name for Tivoli). Quarries in the area around Tivoli have been active for two millenia, and the stone has been used for most of the famous monuments of Rome ranging from the Coliseum through St. Peter's to the modern University. Today there are a great many active quarries, and travertine from Tivoli is exported worldwide.

Travertines in the Tivoli area occur on the Pleistocene volcanic plain east of Roma, centered on the warm-spring resort of Bagni di Tivoli, and extending up to the edge of the Sabine Mountains. According to studies by d'Halloy (1812), Lyell (1829), Cohn (1864), Manfra et al. (1974, 1976), and Barbieri et al. (1979), the material was deposited from a series of warm springs, lakes, ponds, and swamps. Waters descended through the Mesozoic carbonate rocks, were heated at depth, and then rose to the surface. Springs were enjoyed by Roman emperors (e.g. Hadrian's villa was here) and are still active today. The waters carry large amounts of H_2S and $Ca(HCO_3)_2$, and the temperature is about 22 °C (Maxia 1950).

Deposits in the Tivoli area range up to 85 m thick; beds at the top are extremely soft and powdery, but those near the quarry bottoms are as hard and non-porous as any Paleozoic limestone.

2 Summary of Travertine Petrology

We feel (Folk and Chafetz 1980) that the bulk of the travertine was precipitated under the influence of bacteria, and much of it is made of bacterially-controlled calcite structures. These include warped but generally subhorizontal bacterial mats, "foam-rock" made up of 3 mm bubbles enclosed by a bacterial film, and shrub or bush-like colonies 1 – 5 cm high built by bacteria. The elementary build-

1 Geology Department, University of Texas, Austin, TX 78712, USA
2 Geology Department, University of Houston, Houston, TX 77004, USA

Coated Grains (ed. by T.M. Peryt)

ing block is a spherical bacteria-rich clump about 20 microns in diameter enclosed in a single crystal of calcite spar. The bacterial clumps are white in reflected light and brown and "turbid" in transmitted light. In the best preserved examples at 1125 × optical magnification one can see 1/2 – 1 micron dots (coccoids) and curving bacterial filaments 1/2 micron wide and up to 20 microns long. Etched samples examined with SEM show that these tiny dots and threads are now cavities from which the bacteria have disappeared. As diagenesis proceeds, the distinct dots and threads observed in some thin sections disappear and only a brown hazy spot remains. Spar crystals containing bacterial clumps become surrounded by a polycrystalline corona of radial calcite, and clumps aggregate into the mats and shrubs described above. In the travertine we have seen no evidence of any algal structures save for clumps of *Chara* (Lyell 1829).

In addition to the bacterially formed carbonates, inorganic carbonates occur in the form of large ray-crystal calcite crusts up to 40 cm thick, rhombic crystals of spar cement, and pisoliths.

3 Pisoliths

The travertines contain two kinds of pisoliths that are grossly different in field occurrence, size, shape, internal structure, and origin (Table 1). One kind is made of radially arranged bacterial colonies encased in calcite spar; the other by inorganically precipitated carbonate. Both consist entirely of low-Mg calcite, which is apparently original and has undergone no neomorphism. We have found a few examples where the two kinds of pisoliths are mixed, or where one pisolith will be composed of both types of structure but normally there is little mixture or "interbreeding" between these two distinct types of pisoliths, either in field occurrence, hand specimen appearance, microscopy, or mode of origin.

Table 1. Tabular summary

	Bacterial pisoliths	Inorganic pisoliths
Bedding	Thin layers; no cross-stratification	Thick accumulations in pockets, cross-stratification, graded bedding; some monolayers
Megascopic structure	Crudely radial; occasional crudely concentric	Few small radial sectors; always very fine concentric laminations
Microscopic structure	Radial bacterial clumps	Some radial layers of minute calcite needles, very delicate 1 – 5 μm concentric laminations
Shape	Roughly lumpy	Spherical, or smooth bumps and hollows
Interpretation	Almost entirely formed by bacteria	Almost entirely formed inorganically by rapid chemical precipitation; some adventitious bacterial clumps
	Low energy environment of rapid bacterial growth	High-energy environment of rapid chemical precipitation

4 Bacterial Pisoliths

Bacterial pisoliths can be rather easily recognized in the field (Fig. 1) by their bumpy, "cerebroid" outlines and crude radial structure. Concentric structure is generally lacking or only occurs vaguely and in the grossest sense. They are very commonly seen in slabs used for building stone. Pisoliths occur in 1 – 2 cm thick layers parallel with the bedding, ranging from 1 to 3 pisoliths thick. They do not show cross-stratification or fill pockets. They often occur interbedded with the bacterial "shrub" layers and, in fact, some have grown bacterial shrubs on their upper sides after they came to rest (Fig. 2). These in situ shrubs are often in structural continuity with branches of bacterial colonies in the pisoliths underneath. Thus, these pisoliths have formed in a depositional environment of little agitation, rich in bacterial activity, probably in semi-stagnant shallow ponds that may have been only a few centimeters deep. Indeed they may have basically formed in situ without the necessity for any rolling, as shown by their occasional well-fitted shapes (Fig. 1).

Bacterial pisoliths are generally 4 – 8 mm in diameter, with a grossly spherical outline (Fig. 3). Megascopically, they are made of crudely radial opaque whitish areas with intervening thin clearer areas, and the white areas protude as bumps on the exterior of the pisolith. Centrally sectioned ones show a broken piece of bacterial sediment (often laminated) as a nucleus.

Under the microscope, the white parts of the pisoliths are composed of radially branching bacterial shrubs encased in spar (Figs. 3 and 4) exactly like the in

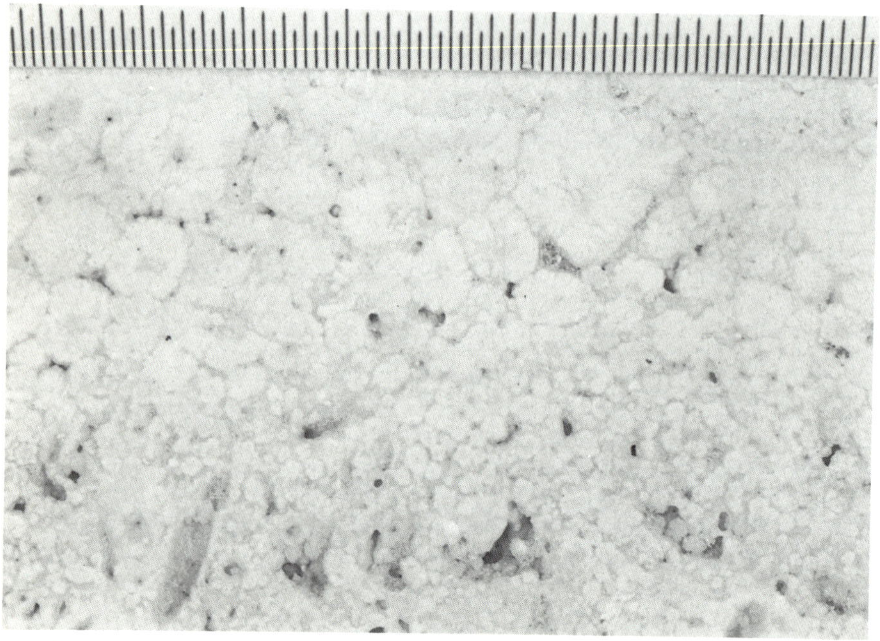

Fig. 1. Typical layer of bacterial pisoliths. Observe crude radial structure caused by bacterial shrubs, and fitted appearance indicating mainly in situ growth. Scale in millimeters

situ layers or "forests" of sub-vertical bacterial shrubs except for the orientation. The bacterial colonies in the pisoliths are made of spherical bacterial clumps 10 – 20 μm in diameter (Fig. 5)

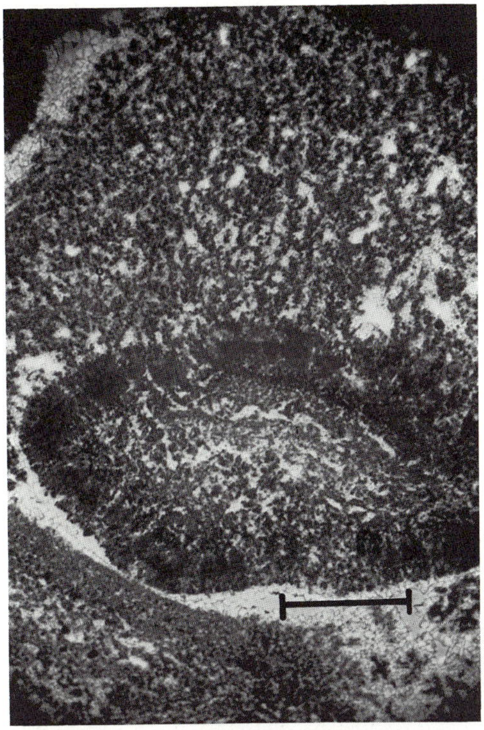

Fig. 2. Photomicrograph of bacterial pisolith with bacterial shrubs continuing growth upward after the pisolith stopped moving; scale equals 1 mm

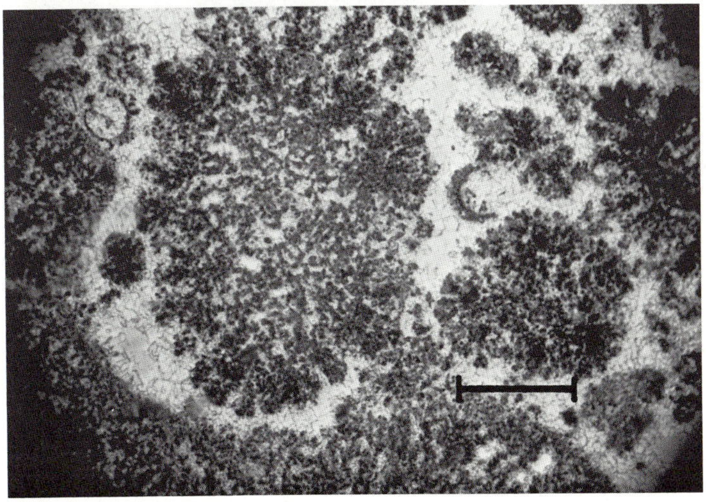

Fig. 3. Photomicrograph of pisolith with typical radial structure of bacterial shrubs; scale equals 1 mm

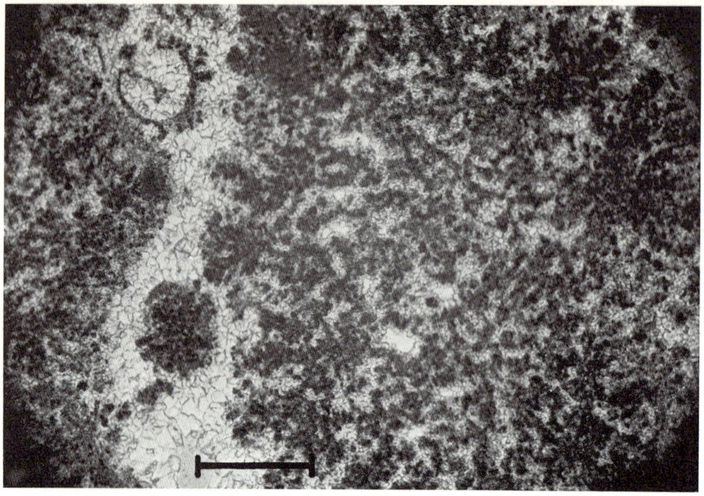

Fig. 4. Half of a bacterial pisolith (center at right edge of photomicrograph) showing radiating bacterial shrubs; scale equals 0.5 mm

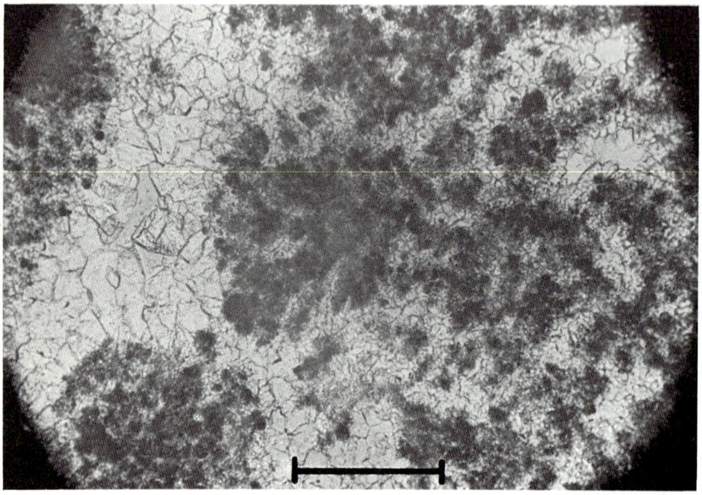

Fig. 5. Enlarged view of the edge of the pisolith shown in Fig. 4; radially branching bacterial shrubs are composed of dark nearly opaque bacterial clumps about 10 – 20 μm in diameter. Each clump is made up of a dense swarm of 1/2 μm bacterial dots (not visible in this photomicrograph). Scale equals 0.2 mm

5 Inorganic Pisoliths

The inorganic pisoliths are smoothly rounded in outline, and show extremely delicate concentric structure. Outcrops show accumulations ranging from 1 – 3 decimeters thick (usual) to as much as 2 meters thick (e.g., at Bagni di Tivoli, quarry of Alfeno Poggi, north of the one owned by Gino Poggi; Figs. 6 and 7). Because they are usually poorly cemented, they are seldom found in slabs used as building stone. These accumulations occur in pockets (Fig. 8), many with vertical to overhanging walls that sometimes meet over the top of the pisolite pocket. Many layers show cross-stratification (Fig. 7); some divergently dipping "heaps" fill pockets 1 – 3 m across (Fig. 6). The typical rock forming the slides of the pockets consists of laminated and mammilary flowstone-like crusts of radial calcite many centimeters thick (Figs. 7 and 8), also inorganically precipitated. These pisoliths have accumulated in strongly agitated environments, in surroundings that favored rapid inorganic chemical precipitation.

Pisoliths normally range from 3 – 10 mm in diameter, rarely up to 3 cm (Fig. 9). Where sectioned, the nucleus appears to be a small piece of bacterial sediment 0.3 – 0.5 mm in diameter. Often there is a central hole, whose origin we are not sure of; perhaps it is caused by solution of a poorly-cemented bacterial nucleus, or perhaps of an aragonitic core – though we have no direct evidence for the former presence of aragonite. Pisoliths are roughly spherical, but the overall shape generally corresponds with that of the nucleus, e.g., a triangular-shaped nucleus will form a large triangular-shaped pisolith, an elongate nucleus grows into an elongate pisolith, etc. (Fig. 10). Thus, they differ from "normal" ooids wherein concave surfaces tend to get filled up and corners accrete more slowly, so that the shape increases in sphericity with each additional lamination. By

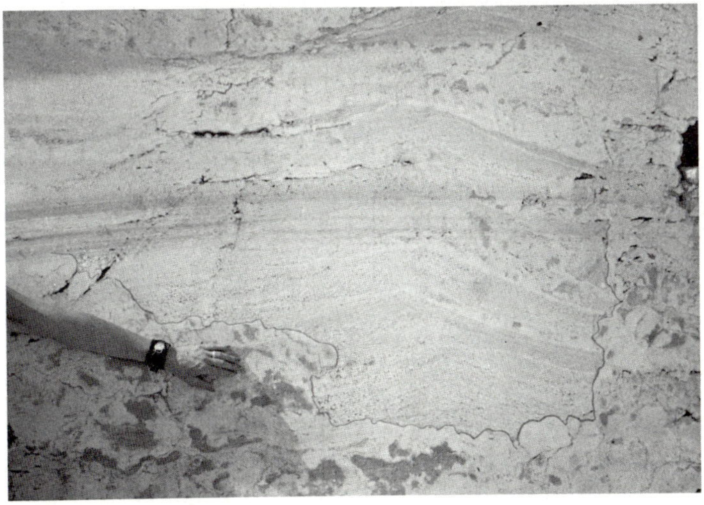

Fig. 6. Pile of bedded inorganic pisoliths over 1 m thick, filling a large pocket in calcite flowstone (outlined by black marker on quarry face). Pisoliths occur in oppositely dipping layers, thus must have formed a dune-like ridge as they filled up the pocket. Vague horizontal streaks are quarry saw-marks

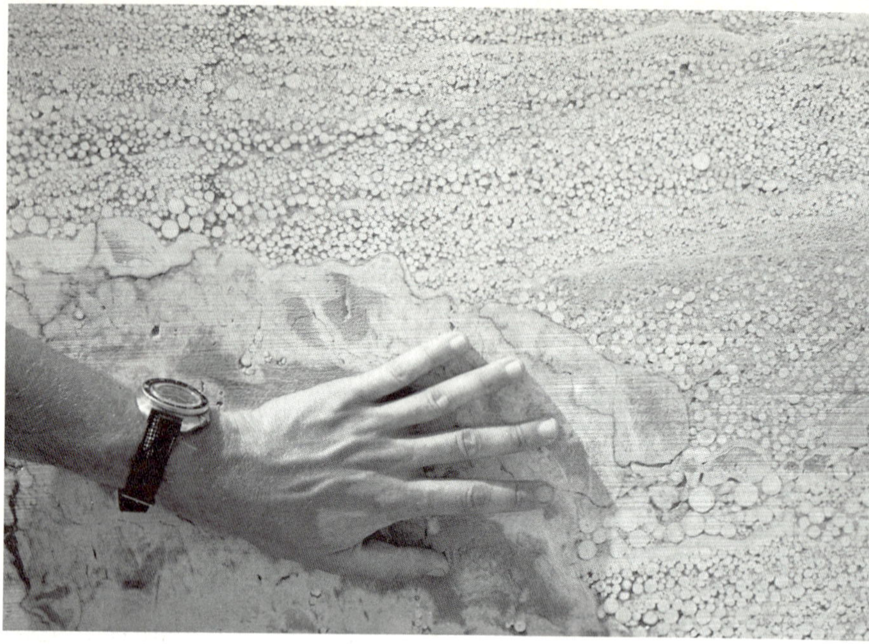

Fig. 7. Enlarged view of Fig. 6. Layers of pisoliths of varied sizes; the layer to the right of the hand shows left-dipping cross-stratification

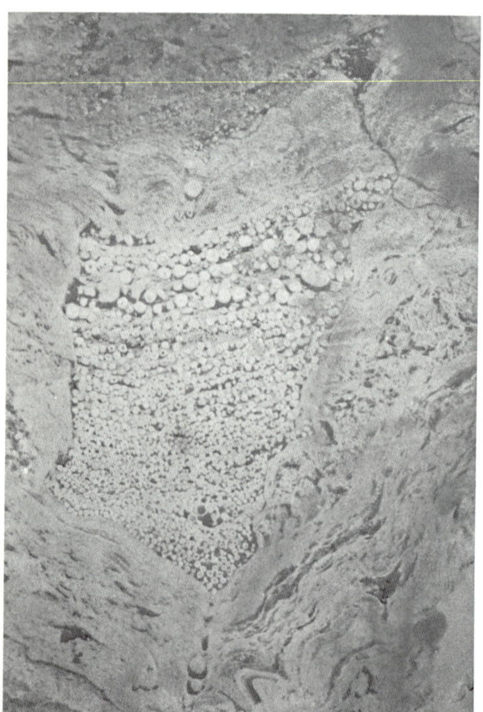

Fig. 8. Pocket of pisoliths about 30 cm thick, enclosed in calcitic flowstone layers. Observe monolayer effect of pisoliths, and porosity beneath each pisolite layer

Fig. 9. Pocket of very large inorganic pisoliths

contrast, in the travertine pisoliths, the layers added are essentially the same thickness all the way around (Fig. 11) so that the shape of the nucleus is conserved by the concentric-conformable manner of growth. Consequently many pisoliths have shallow dimples or smooth humps (Fig. 12) and often the hump of one pisolith will fit into the dimple of another (Fig. 10A). We feel this is not due to in situ growth or plastic deformation, rather is due to gentle jostling after the pisoliths basically came to rest, allowing the fitting of already-irregular, hard pisoliths to take place.

Under the microscope the inorganic pisoliths at first sight appear to consist largely of fine layers of micrite; small, thick or closely bunched bacterial clumps also occur in some of these pisoliths, but they have not controlled the structure and are only adventitious. Closer optical examination shows that most pisoliths contain some concentric layers made up of tiny length-fast radial calcite needles (Fig. 11). SEM study shows that the needle layers are approximately 0.1 mm thick and made up of a non-porous array of close-packed parallel needles about 2 – 3 μm wide (Fig. 13). Pisoliths etched briefly with weak HCl and examined under the SEM show extremely fine and regular concentric banding, appearing like a phonograph record (Fig. 14). The bands are 1 – 5 μm thick and made up of a porous, crudely radial aggregate of rodlike calcite nannocrystals ranging typically from 0.2 × 1 μm to 0.5 × 3 μm (Fig. 15). The concentric layers do not appear to have any "unconformities" or irregularities in thickness, either when viewed by SEM or optical microscope (Figs. 11 and 14), thus once formed were not later abraded. Nor have any broken pisoliths been found.

In some samples, the pisoliths occur as a series of monolayers, each layer covered by a draped saggy bacterial mat (Figs. 16 and 17). Concave-triangular pore spaces occur *beneath* the single layers of pisoliths, in a vague way resembling fenestral structure and forming a reliable geopetal indicator. We believe this to be because the pisoliths are so nearly spherical that they roll easily; it is as if one would scatter billiard balls on a table and cover them over with a soft

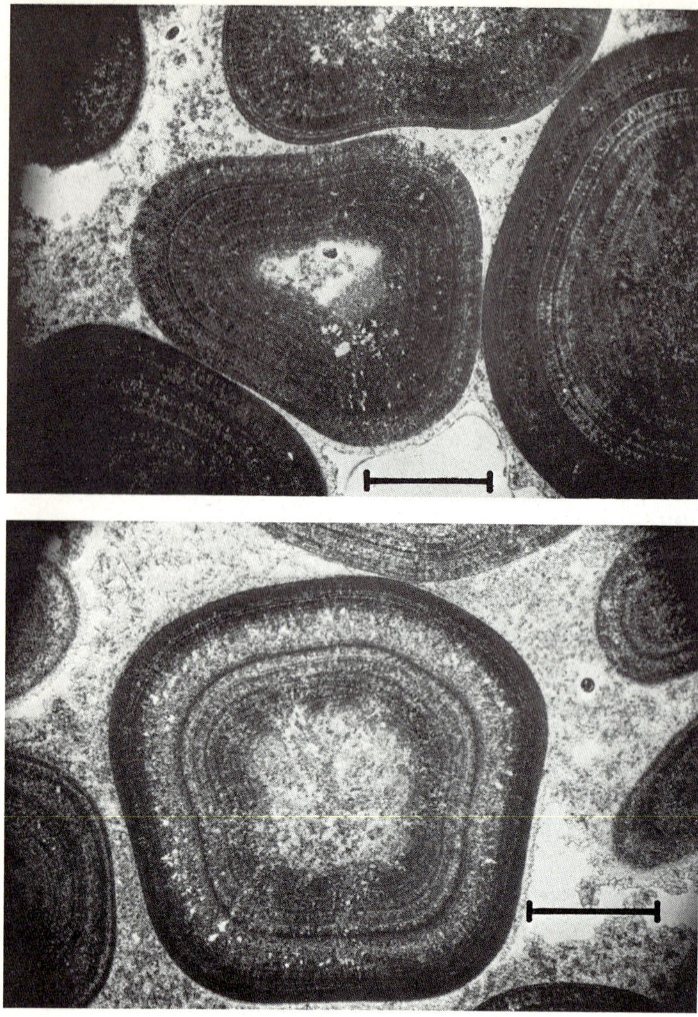

A

B

Fig. 10 A, B. Photomicrographs of inorganic pisoliths, showing conformal-concentric manner of growth such that the shape of the nucleus is simulated by the shape of the entire body. This indicates accretion at a constant rate on all sides and corners of the pisolith and shows that the indented shapes are not due to later compaction of a soft body. **A** Pisolith with a triangular nucleus, **B** with a pentagonal nucleus. The inter-pisolith material consists of delicate bacterial clumps in sparry calcite cement, only partly filling up the pore space. Scales equal to 1 mm

pancake; the "pores" would be found beneath the pisoliths. One pisolith-strewing event (perhaps a "throat-clearing" hot spring eruption) must have been followed by a quiescent interval during which time the bacterial mat formed over them and hardened, then a second pisolith-strewing event occurred, a second mat formed, and so forth.

The paradoxical thing about these pisoliths is the evidence of their formation in a highly agitated environment, contrasted with the lack of evidence of any abrasion of the layers (no unconformities) and the occurrence of *radial* calcite

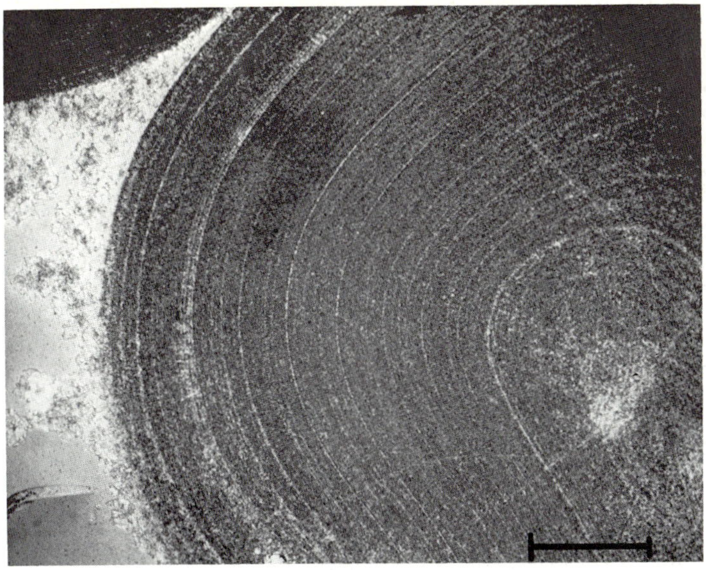

Fig. 11. Photomicrograph of extremely fine concentric banding of the inorganic pisoliths. Note continuity of layers and lack of pinch-outs or unconformities. Scale equals 0.5 mm

Fig. 12. Mass of pisoliths viewed from above, showing typical dimpled appearance

fibers. This seems to deny the idea commonly proposed for ooids that radial needles indicate low-energy environment, and that tangential needle layers (as are most common in modern aragonitic marine ooids) are formed by battering-down of aragonite needles into a felt by abrasion. It implies that the rate of chemical precipitation and growth of the calcitic pisoliths was much faster than the rate of abrasion, even when the pisoliths exceeded a centimeter in diameter. The main

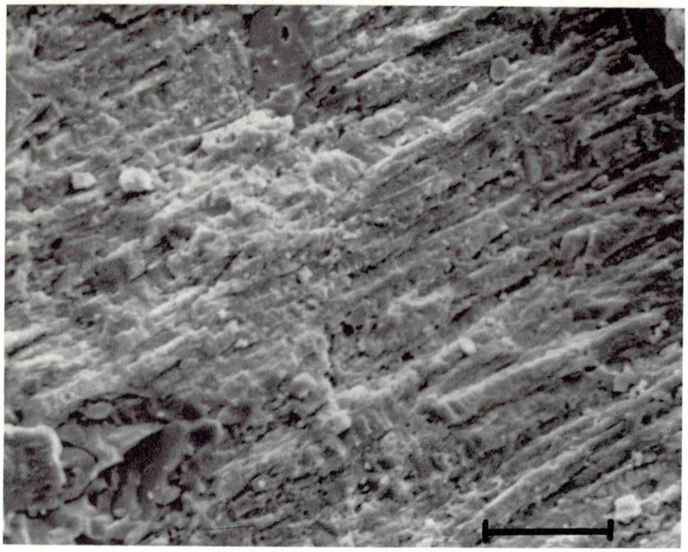

Fig. 13. SEM view of one of the layers of radial calcite crystals, illustrating their occurrence as "fasces" of tightly fitting bundled rods. Fresh fracture surface; scale equals 20 μm

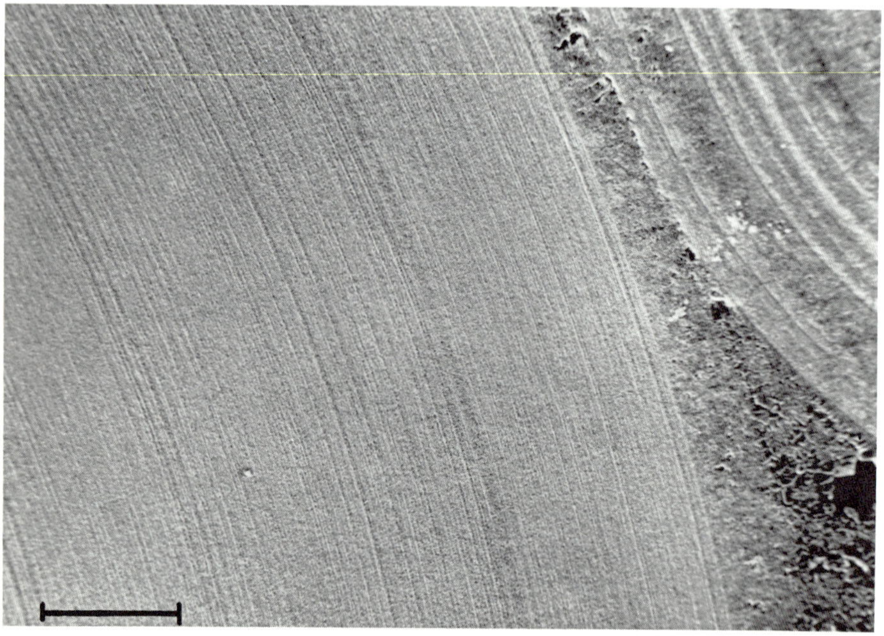

Fig. 14. SEM view of inorganic pisolith, weakly etched fracture surface (1% HCl for 1 min). "Phonograph-record" appearance due to incredibly delicate and continuous concentric lamination, illustrating extremely regular growth with no pinch-outs or unconformities. Scale equals 0.2 mm

Fig. 15. Enlarged SEM view of the "phonograph record" showing radial orientation of tiny calcite rods. Layers 2 – 10 μm thick. Lightly etched fracture surface; scale equals 10 μm

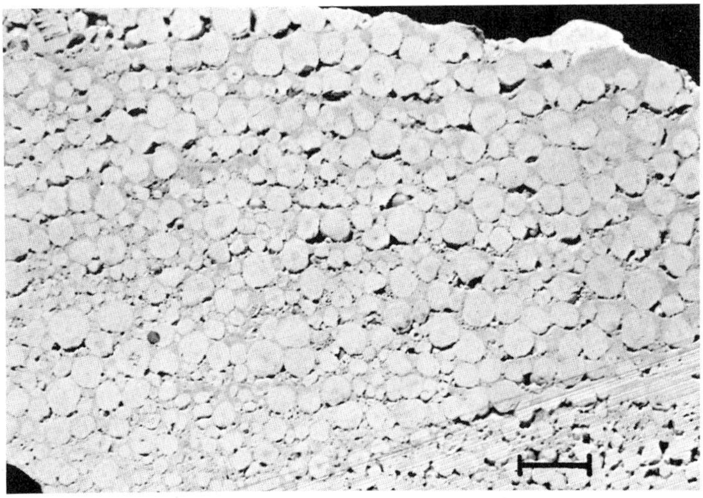

Fig. 16. Slab of pisoliths showing their occurrence in monolayers, each layer draped over by a bacterial mat. Concave-triangular pore spaces underlie the pisolith layers. Scale equals 10 mm

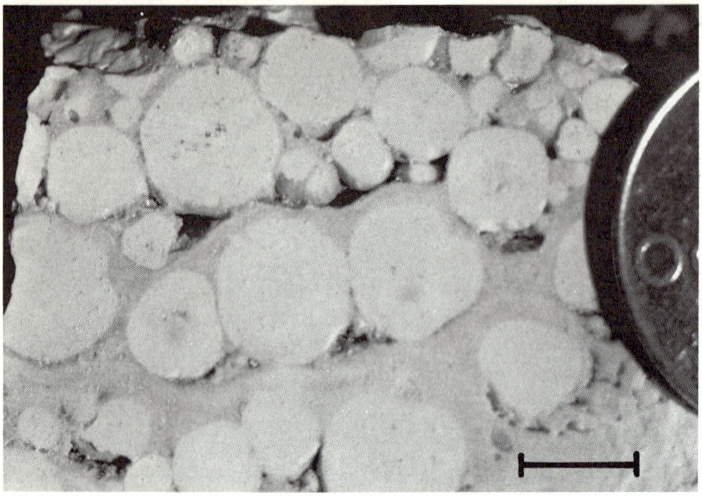

Fig. 17. Three monolayers of pisoliths with pore spaces underneath the pisoliths and a flabby bacterial mat draped on top of them. Scale equals 10 mm

reason that calcite crystals produce a radial fabric seems to be that the larger radial crystals appear to be bundled into thick, tight accumulations (like fasces) and therefore resist abrasion; further since they are parallel they may cement together by developing occult "bridging" overgrowths too small to be separately visible in the SEM. Aragonite needles are more delicate and separated, thus more easily battered into a tangential arrangement. This argument has been suggested also by Land et al. (1979) and Geno and Chafetz (1980).

Acknowledgements. It is a pleasure to acknowledge the continual logistic help in Italy given to us by Patrizia and Mario Massano; and in the field work by Nat Smith and Sandra Waisley (1979), Paolo Tortora, Pietro Zizzari and Ada de Marino (1980); and many of the quarry workers and owners in the Tivoli-Bagni region. The Geology Department of the Universita degli Studi, Roma, furnished transportation. John Pigott first tipped us off to the presence of bacteria in thin section. Hiromi Honda did the SEM work at the University of Texas. We also acknowledge the Geology Foundation, University of Texas, and the Geology Foundation, University of Houston, for providing field support, and the Owen-Coates Fund, University of Texas, and the Geology Foundation, University of Houston, for helping with publication costs.

References

Barbieri M, Masi U, Tolomeo L (1979) Origin and distribution of strontium in the travertines of Latium (Central Italy). Chem Geol 24:181 – 188

Cohn F (1864) Über die Entstehung des Travertin in den Wasserfällen von Tivoli. N Jahrb Miner Geol Palaeontol, p 580 – 610

d'Halloy JJO (1812) Sur l'existence du calcaire d'eau douce dans les département de Rome et de l'Ombrone, et dans le royaume de Wurtemberg, J Mines (Paris) 32:401 – 410

Folk RL, Chafetz HS (1980) Quaternary travertine of Tivoli (Roma) Italy: bacterially-constructed carbonate rock (abs). Geol Soc Am Ann Meet, Atlanta, 428 p

Geno KR, Chafetz HS (1980) Recent Fluvial Ooids and Pisoliths in Central Texas (abs). Geol Soc Am Ann Meet, Atlanta, 433 p

Land LS, Behrens EW, Frishman SA (1979) The ooids of Baffin Bay, Texas. J Sediment Petrol 49:1269 – 1278

Lyell C (1829) On a recent formation of freshwater limestone in Forfarshire on some recent deposits of freshwater marl, and an appendix on the gyragonite or seed-vessel of the Chara. Geol Soc London Trans Ser 2:73 – 96

Manfra L, Masi U, Turi B (1974) Effetti isotopici nella diagenesi dei travertini. Geol Rom 13:147 – 156

Manfra L, Masi U, Turi B (1976) La composizione isotopica dei travertini del Lazio. Geol Rom 15:127 – 174

Maxia C (1950) Un interessante sondaggio nel travertino di Bagni de Tivoli (acque albule fra Tivoli e Roma). Contr Sci Geol (Roma) Suppl Ric Sci 20:22 – 26

V.5 Quaternary Pisoids of Lau, Fiji

C. J. Burgess[1]

1 Introduction

Pisoids, concentrically laminated grains between 2 mm and 10 mm in diameter, originate in a variety of environments and the word pisoid is now seldom used without a qualifying genetic adjective (Dunham 1969, Kendall 1969, Scholle and Kinsman 1974, Esteban 1976, Jones and Wilkinson 1978). Attempts have been made to define the characteristics of specific pisolites (Esteban 1976, Chafetz and Butler 1980), but conflicting points still exist. This paper describes the characteristics and processes of formation of Quaternary ferruginous caliche pisovadolite from the Lau group, Fiji, and the results are compared with recent and fossil caliche pisovadolite described from elsewhere.

2 Environmental Setting

The islands and atolls of Lau lie to the east and south-east of the main islands of Fiji, between latitudes 17°S and 21°S, longitude 179°W (Fig. 1). The major islands consist of Miocene basaltic, andesitic and dacitic lavas and ashes overlain and sometimes interbedded with reef limestones and associated facies of the Futuna Limestone (Ladd and Hoffmeister 1945, D. Woodhall, in prep.). The limestones are probably of Mio-Pliocene age but have been uplifted to form salient cliffs and plateaux sometimes over 100 m high. Coral reefs and associated facies surround the islands (Burgess 1978). Temperatures average 23° – 28 °C and rainfall at Lakeba (Fig. 1) reaches 2000 mm per year (Fiji Meteorological Service 1977).

Sediments range from volcanoclastic coarse sandstones and conglomerates to pure limestone with very little non-carbonate material. The Futuna Limestone proper varies in character from a recrystallized and dolomitized coral boundstone to a terrigenous rich foraminiferal packstone. Lagoonal, reef-flat, reef-front, fore-reef talus and shelf facies have been identified. Terrigenous fragments are almost exclusively weathered fragments of the underlying Lau volcanics.

1 Robertson Research International Limited, Llandudno, North Wales, Great Britain

Coated Grains (ed. by T. M. Peryt)

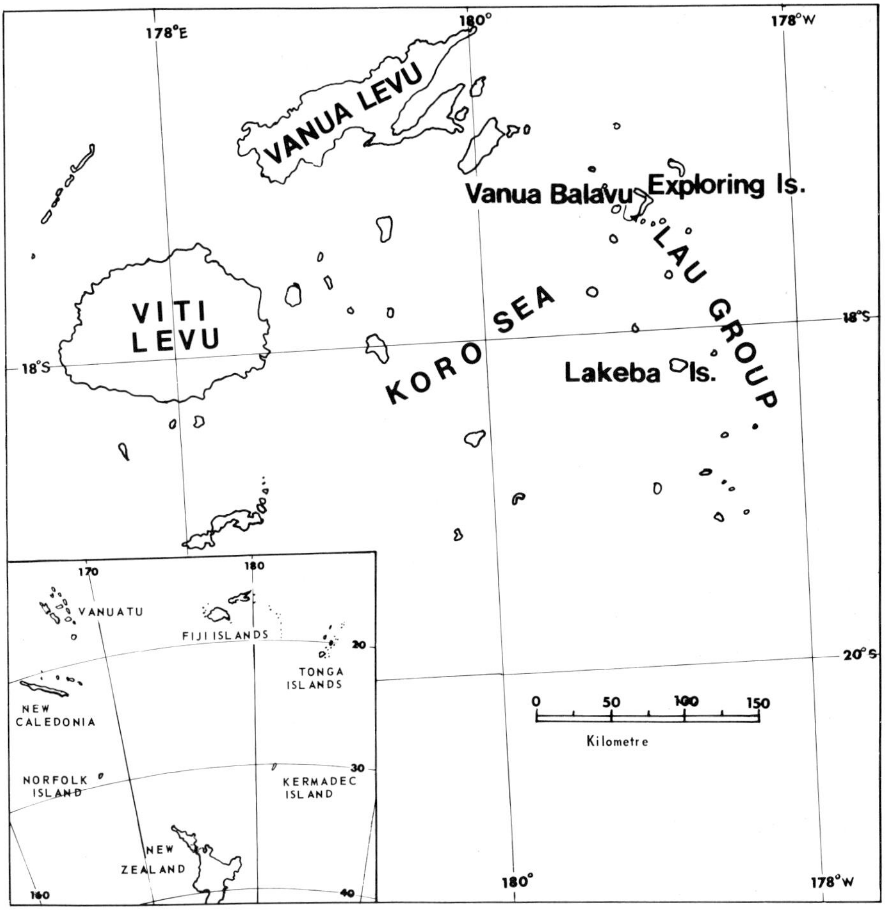

Fig. 1. Map showing the location of the Lau Group and islands on which pisoids were studied in detail

The purer Futuna Limestone is intensely weathered into jagged solution hollows, remnant knife-sharp ridges, spikes and disorientated collapsed blocks. Solution hollows, collapsed caverns and remnant travertine deposits occur throughout the limestone outcrops (Fig. 2), and drainage is entirely subterranean. Soil as a continuous layer is largely absent but survives in pockets and hollows within the limestones. Carbonate cemented red soil deposits, often with fenestral fabric, occur in exposed collapsed solution pipes and caverns. Heavy vegetation survives on all but the steepest bluffs.

Fig. 2. Exposed pisovadolite with large lithoclasts (*top right*) several metres in length filling an old solution hollow. A further solution pipe occurs *on the left*. Hammer is 260 mm in length

3 Pisoids

3.1 Distribution

Within the Lau Group pisoids occur on many of the islands where fossil reef limestones crop out, but the degree of development is variable (D. Woodhall pers. comm. 1980). This study is based on examination of deposits on Lakeba and in the Exploring Isles (Fig. 1).

Pisovadolite is found only in association with the purest Futuna Limestone, and does not develop in the thick laterites that occur on the weathered volcanics and terrigenous rich carbonate facies. Continued erosion and dissolution of the parent limestone has exposed the old subterranean drainage channels, and pisovadolite is preserved in pockets and layered crusts cemented to the parent rock. Associated with the pisovadolite are old travertine deposits, caverns and non-pisolitic lithified caliche often with well developed fenestral fabric.

3.2 Morphology

Three grain types can be recognized in the vadolite:

a) laminated pisoids,
b) unlaminated pisoids,
c) coated lithoclasts/allochems.

Laminated pisoids (Fig. 3) range in diameter from less than 1 mm (strictly speaking these are microids) to 10 mm but have an optimum development of between 3

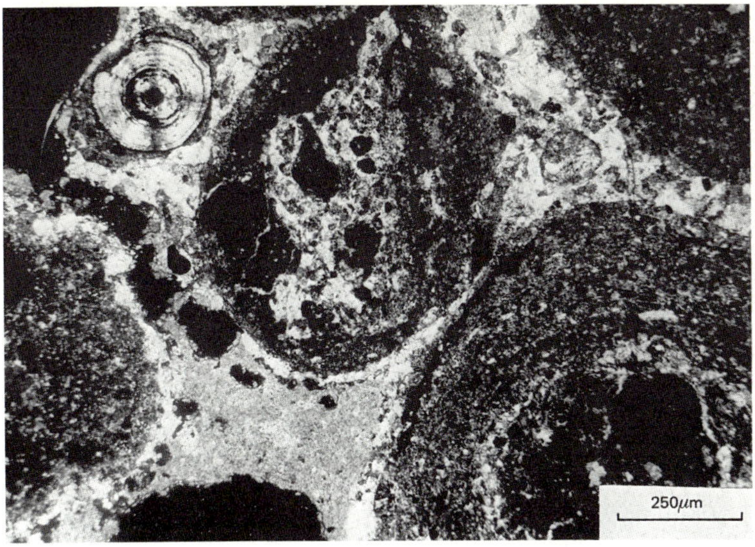

Fig. 3. Laminated pisoids surrounding a nucleus of manganese fragments and recrystallized allochems. Note how the laminae thin over projections in the nucleus. A cave pisoid occurs in the *top left*. Sparry calcite and vadose silt fill cement the pisoids

and 5 mm. The laminations are not distinct regular layers but are very irregular discontinuous coatings which are picked out by the tangential orientation of inclusions rather than layers of crystal growth. Sharp changes in mineral composition may occur and result in more distinct laminate structure. Laminated pisoids tend to be quite regular in shape, frequently spheroidal to subspheroidal, but occasionally elongate or irregular when controlled by a large non-spherical nucleus. However, in many examples the pisoids are spherical despite an irregular nucleus and laminae thin across the prominent edges (Fig. 3).

Unlaminated pisoids and microids are much less common than laminated pisoids and differ in that they show no evidence of tangential orientation of inclusions. They are spheroidal lumps, similar in shape and composition to laminated pisoids, but seldom exceeding 3 mm in size. The two types may be found adjacent to each other and the unlaminated pisoids may form a nucleus for additional laminations. Both types of pisoid may display cracking. Where a pattern is evident, the cracks tend to be concentric around part of the pisoid with a further series of radial fractures in the outer zone (Fig. 4). A number of examples also show exfoliation cracks in the outer layers. The nature and development of the fractures suggest an origin from shrinkage as a result of desiccation.

Coated lithoclasts or allochems (Fig. 5a) may also occur in association with the other grain types, but the size and space are much more variable. Lithoclasts over 3 m were recorded in a fossil cave deposit on Lakeba, but small lithoclasts may also form the nucleii of laminated pisoids or be incorporated into the laminations. Allochems are largely gastropod shells and include both terrestrial forms and also marine shells carried inland by large hermit crabs. Gastropod

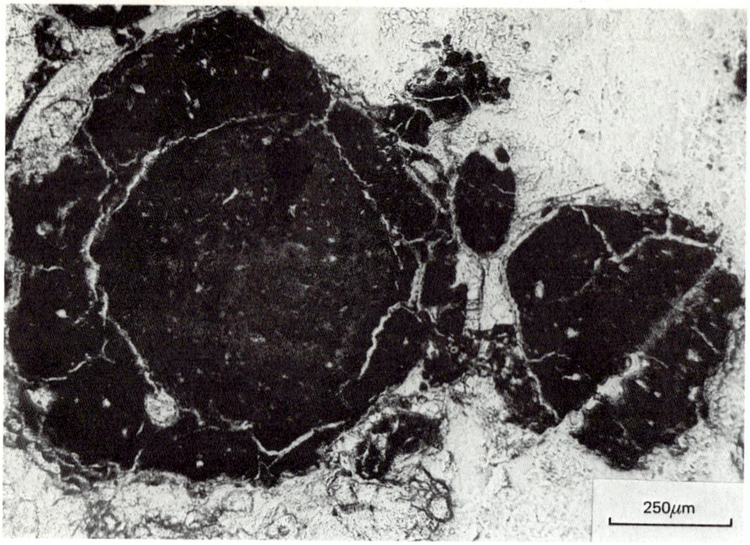

Fig. 4. Desicated, unlaminated pisoids

shell debris is a frequent constituent of lithified pisovadolite. The coating on lithoclasts is usually relatively thin (from a few microns to 2 mm) and the thickness may be very irregular filling indentations in the irregular clasts, and thinning over projections.

There is a considerable inter-relationship between these broad grain types. A small lithoclast may act as a nucleus for a laminated pisoid or a group of eroded lithified pisoids may act as a lithoclast. Nucleii are not always apparent in pisoids but where visible other pisoids, gastropod fragments and lithoclasts are the most abundant. Cave pearls, volcanic fragments and echinoid plates were also identified.

Pisovadolite deposits are usually well packed and grain supported with high porosity and permeability. Only rarely do pisoids occur in a wackestone or mudstone. Both sorting and grading of the grains may occur, but are not ubiquitous. Pisoids are frequently well sorted (Fig. 5b) but large coated lithoclasts may disrupt this sorting. Bedding is usually apparent and the deposits may be very well bedded but the lateral extent of the beds is controlled by the size and shape of the depositional pocket and beds seldom extend for more than 3 m. At the margins of the pocket the beds may become inclined, following the slope of the eroded parent rock. Graded bedding was observed in many examples but reverse graded bedding also occurs quite commonly. Inconsistencies in this pattern are reflected by poorly sorted, ungraded and unbedded deposits composed of all types of grains.

Fig. 5. a A mixture of pisoids, lithoclasts and allochems cemented by travertine type cements. Superficial sorting is evident in the upper part of the picture. Coin is 20 mm in diameter. **b** Well-sorted and well-formed vadolite. Lens cap is 55 mm diameter. **c** Coated lithoclasts and pisoids on a recent cave floor. Note the rudimentary sorting and reverse grading. Coin is 20 mm in diameter

10
cents
a
Canon
b
c

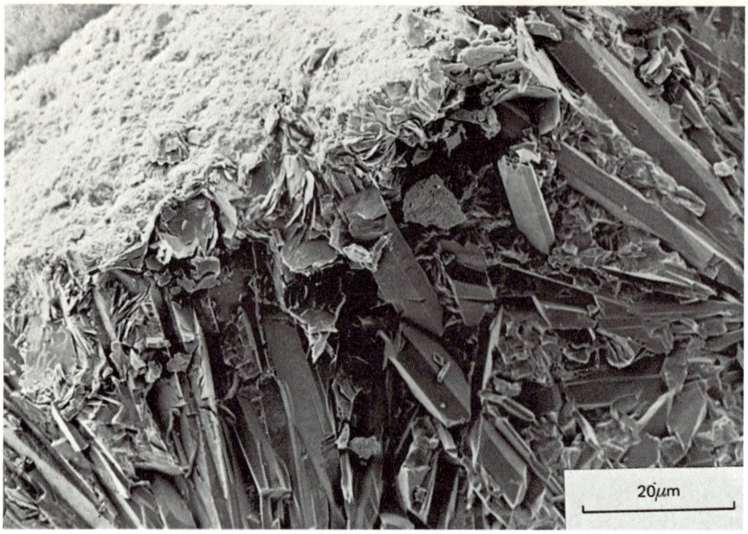

Fig. 6. Acicular calcite cements growing from pisoids and adding additional coatings, overlain by vadose deposits consisting of aluminum magnesium oxides. Scanning electron photomicrograph

3.3 Mineral Composition

The mineralogy of the grains can be divided into two categories: the nucleus and the coating. Most of the nucleii are lithoclasts derived from the parent Futuna Limestone and are thus either low Mg calcite or calcian dolomite. Dolomites were determined by staining (Dickson 1966), and thin-section evidence suggests that much of the calcite is in fact dedolomite (see: Cementation and Diagenesis). Other nucleii are of variable mineral composition. Gastropods tend to be aragonite, and eroded speleothem fragments are largely calcite. Volcanic fragments range from discrete olivine, feldspar and hornblende crystals to fragments of lavas and dacite tuff. Eroded sedimentary manganese was also observed and secondary manganese precipitation is common in pisovadolite on the southern end of Vanua Balavu (Fig. 1).

The mineralogy of the coatings is dominated by calcium carbonate and by oxides of iron, alumina and magnesium (bauxite) having a brick red colour in hand specimen and with platy appearance under the scanning electron microscope (Fig. 6). Electron microprobe analysis of the distribution of elements reflects the changing laminae of the pisoids (Fig. 7). Silicates do not occur except as detrital minerals within the coatings. Calcium carbonate (usually low Mg calcite, but high Mg calcite does occur) forms coatings and cements and the percentages of aluminum/iron/magnesium oxides to carbonate is highly variable with pure oxide pisoids at one end of the scale and almost pure carbonate pisoids at the other. On some islands the pisovadolite is very rich in phosphate and is mined.

Measurement of a cross-section of parent limestone pisovadolite samples showed considerable variation in the non-carbonate content. The unaltered Futuna Limestone contains between 0.9% and 35% non-carbonate in the form of volcanic ash and fragments. Percentages of insoluble material in pisovadolite samples vary from about 2% in recent embryonic pisoids collected from a present-day cave floor (Fig. 5c) to 36% in well cemented vadolite. Much of this material is aluminum and iron oxide and the high percentages in the cemented vadolite reflect the influx of vadose sediment, which may totally fill pores.

3.4 Cementation and Diagenesis

The Futuna pisovadolite is a product of the alteration of the parent carbonate and an account of this breakdown is pertinent to an understanding of the origin of the pisovadolite (see following section). Chemical dissolution by meteoric groundwater with high humic acid content is the most important weathering process. The dolomites of the Futuna limestones are poorly ordered and readily change to dedolomite and dissolve. The replacement of dolomite by dedolomite results in red oxide pigmentation stain around crystals and grains within the rock. A similar pigmentation may be seen in weathered limestones that were not dolomitized. Evamy (1963) noted coatings of iron-oxide pigment after dedolomitization of ferroan dolomites and attributed this to an ejection of iron from the dolomite lattice during dedolomitization. This seems an unlikely source for the Futuna oxide pigments as the dolomite is non-ferroan and the pigment ubiquitous. Accumulation of altered insoluble volcanic ferromagnesium minerals and precipitation of iron from acidic waters in a local high pH environment is another possibility.

The oxide pigments are unevenly distributed and often surround preserved resistant allochem fragments such as coralline algae and echinoderms, forming embryonic pisoids with the allochems as a nucleus (Fig. 9).

The high porosity and permeability of pisoid accumulations and their formation in the vadose zone leads to characteristic diagenetic features. Cements are predominantly blocky and acicular low magnesium calcite (Fig. 6) although isolated examples of fibrous aragonite, high magnesium calcite and gypsum also occur. The acicular form of the cements corresponds with predictions of calcite crystal morphology in environments high in Mg, although normal caliche cements are blocky and low in Mg (Folk 1974). High Mg content probably reflects calcification and solution of the parent Futuna calcian dolomites.

Many cement crystals are blunted and meniscus in form, characteristics of formation in the vadose zone (Dunham 1971). However, acicular cements crossing but not filling pores also occur. Cements may be gravitational in location (Muller 1971), or occur as distinct layers encompassing several grains. Brick-like texture and radial rays (Assereto and Folk 1976) were not observed in pisoids but similar fabrics occur in cement and travertine coatings.

High porosity results in the deposition of vadose sediment of silt and clay size. This sediment forms drapes over pisoids, floors small cavities within the vadolite and may coat already formed cements (Figs. 6 and 8), sometimes completely filling cavities. As many as five alternating layers of cement and vadose sediment were seen in one example, and alternation of cement and sediment is a common feature. Precipitation of cement and deposition of fine sediment in the pores may occur at the same time as dissolution of grains. Pisoid nucleii are

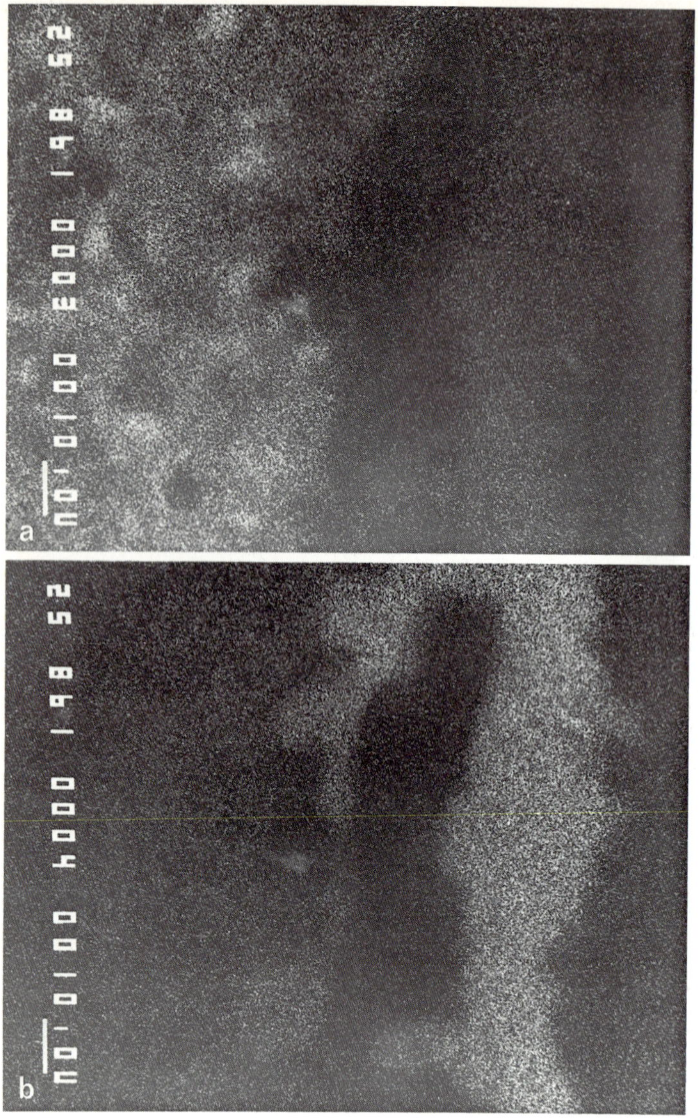

Fig. 7a–d. Electron-microprobe element distribution maps of the periphery of a pisoid. **a** Alumina, **b** Magnesium, **c** Calcium, **d** Iron. Bar in each case = 10 μm

particularly susceptible to dissolution and the void in the centre of a pisoid may itself contain cement, although not sediment. This reflects the porous but resilient nature of pisoid coatings.

The final stage in the diagenesis of pisovadolite seems to be total dissolution and reworking of grains or aggrading recrystallization. Examples of neomorphosed cements are not uncommon but total neomorphism with ghost pisoids outlined by aluminum and iron oxide pigments are rare, probably because of the continual weathering and dissolution of the sequence.

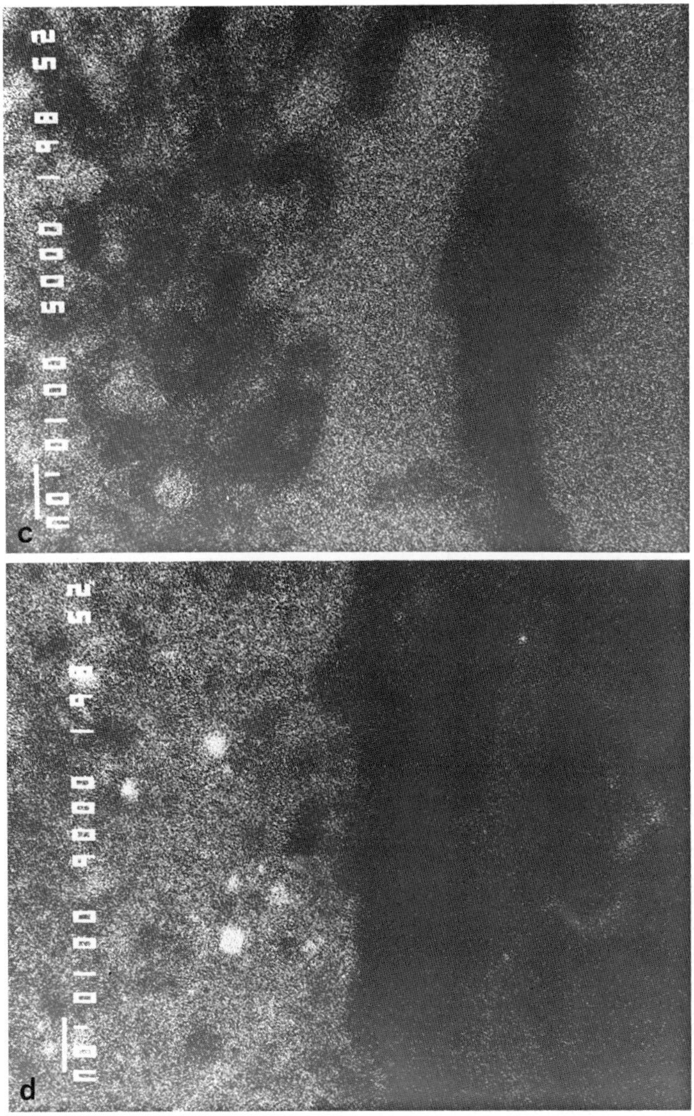

Fig. 7c, d

3.5 Origins

The genesis of the pisoids starts with dedolomitization and dissolution of the Futuna Limestone (Figs. 9 and 10). Fragments of weathered parent rock, often small allochems surrounded by iron and aluminum oxides (Fig. 9), accumulate in solution hollows, pipes and caves by the simple action of gravity (Fig. 5c). Insoluble material from the parent rock, wind-blown dust (possibly volcanic ash from Tonga), organic matter and precipitated oxides also accumulate in the ir-

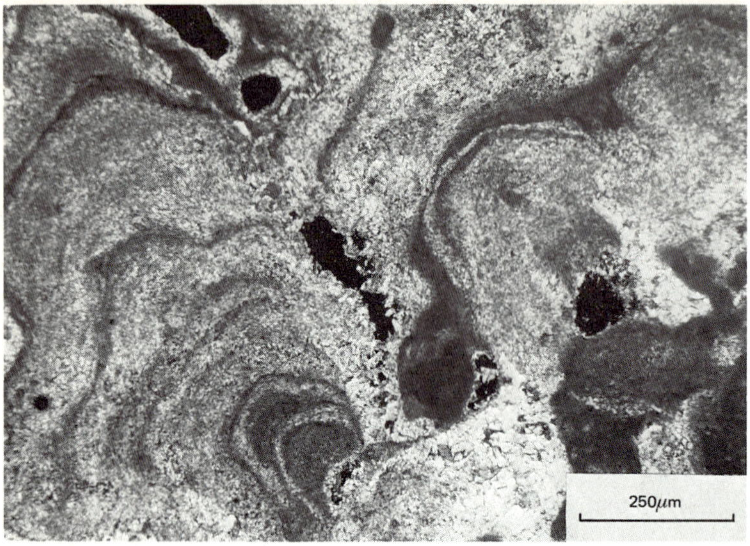

Fig. 8. Alternating calcite cement and vadose silt coatings

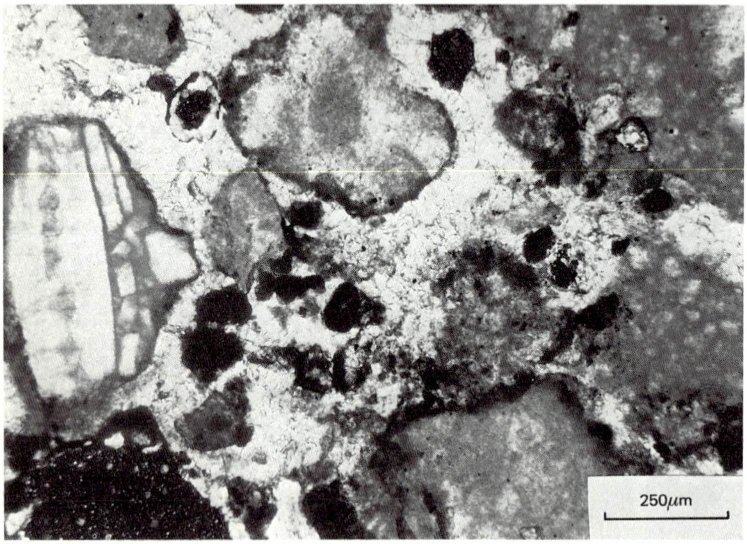

Fig. 9. Remnant allochem fragments showing precipitation of alumina/iron crust (*dark line*) during initial dissolution

regularities of the weathered limestone. These small pockets continually move downwards as more limestone is dissolved. The importance of running water in the process is debatable. Although it helps move fine-grained material downwards, the major process is thought to be gravitational fall rather than washing by moving waters. In the larger pipes lithoclasts and residues fall onto cavern

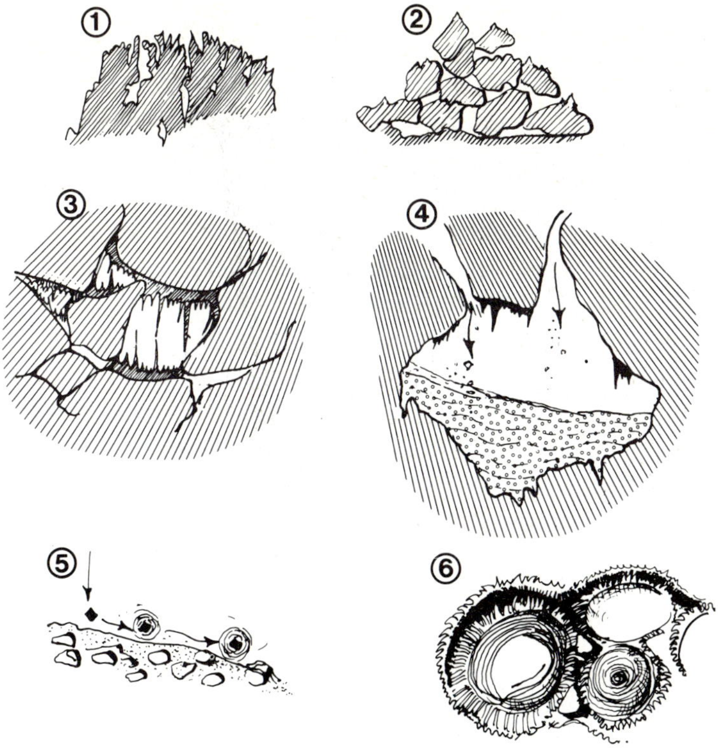

Fig. 10. Schematic cartoon of pisoid formation. *1* Dissolution of exposed limestones; *2* accumulation of disoriented blocks; *3* formation of travertine cements; *4* infill of hollows with insoluble residue, fragments and wind-blown dust; *5* "snowball" coating of grains to form laminated pisoids; *6* dissolution/cementation/re-erosion of pisoids. All processes may occur synchronously

floors and may roll. This rolling coats lithoclasts and nucleii in finer oxides and siltsized terrigenous material (a snowball effect). The coatings are naturally uneven, being thicker in hollows in the nucleus and thinner over projections. This process may explain similar features observed by Hay and Wiggins (1980). The pisoids are kept damp by the moist cave atmosphere and running water may move the grains during periods of intense rainfall, but the process is insufficient to remove coatings or abrade grains. Calcified algal filaments were observed in several pisovadoids (Fig. 11) and may help adhesion of particles to the growing grain. The process of falling and rolling also roughly sorts the grains and may account for the variable grading, bedding and sorting patterns observed. Desiccation features may occur in unlithified pisovadoids that dry and shrink as a result of atmospheric change (for example, collapse of a cavern and exposure to sunlight).

Laminated grains may be rolled and moved many times by the fall of grains from above, by the collapse of the floor on which they rest, and by running or dripping water. Unlaminated pisoids probably result from adhesion of dust as a result of wetting and drying and possibly algal activity with little or no "snowball" rolling.

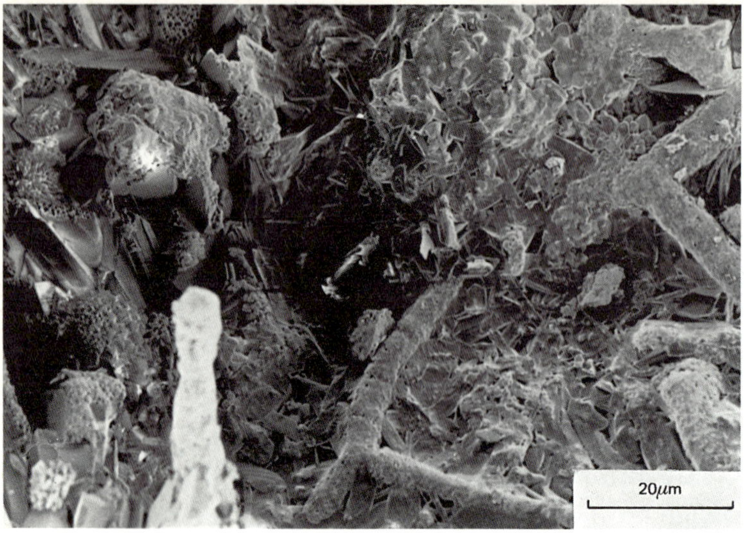

Fig. 11. Calcified algal filaments within a pisoid coating. Etchmarks are a result of preparation of the sample. Scanning electron photomicrograph

Eventual burial results in lithification by precipitation of fine grained carbonate cement from percolating groundwaters. Recent pisovadoid coatings collected from a present day cave floor were found to be compact and sufficiently hard to resist slow moving vadose waters. Cement coatings may cover several grains and further layers are added to the pisoids by this process. An alternation of cement and vadose sediment results in the in situ growth of pisoids, although because of the mineralogical differences the different growth stages are easy to see (Fig. 8). Travertine deposits may also form at this time.

Continual erosion, dissolution and collapse of the rock result in exposure and reworking of the pisovadolite, dissolution of the nucleii, fracturing and re-coating of grains, and a steady downward movement of all the insoluble material. The eventual conclusion would be the total dissolution of all the limestone and the formation of a residual layer of oxides with bauxitic pisoids.

4 Discussion and Significance of Lau Pisoids

Dunham's (1969) paper on the vadose pisolite of the Capitan reef has provided a catalyst for discussion on the different types of pisoid and their environmental significance (Loreau and Purser 1973, Chafetz and Butler 1980, and others). Two distinct types of pisoid have now been recognised from this discussion: Caliche pisoids and hypersaline pisoids (Scholle and Kinsman 1974, Assereto and Folk 1976). The problem now centres around defining the features that characterise these types of pisoid to enable more confident interpretation of fossil sequences.

Table 1. Comparison of characteristics of caliche pisovadolite

Esteban (1976)	Chafetz and Butler (1980)	This paper
Pisoids, when present, are micritic – usually with less than five indistinct laminae	Laminae evident in etched and polished slabs	Laminae vary from distinct to absent, inclusions common
Micrite and microspar abundant with clotted peloid texture	Rare	Clotted peloid texture observed but rare
Sparry cements minor, fibrous cements uncommon	Elongate sparry calcite present as gravitational cement	Rhombic calcite common, fibrous gypsum and acicular calcite locally abundant
Microspar-spar channel system circumgranular cracking	Cracking present in many pisoid types	Desiccation cracking locally abundant
Microspar replacing relic grains and caliche components	–	Aggrading neomorphism common in older well cemented pisovadolite
Rhizocretions/algal filaments common	Observed	Algal filaments observed
Independent of substratum	Developed on evaporite collapse breccia and dolomite	Only developed on limestone
No fenestral fabrics	Fenestrae present in micrite	Fenestrae present in non-vadolitic residue-rich micrite
–	Grain supported	Grain supported
–	–	Well to poorly sorted
–	–	Reverse or normal graded
Marine fossils absent	Marine fossils not observed	Marine and terrestrial gastropods common; relict marine fossils rare

The pisovadolite and associated deposits of Lau do not readily conform to either definition. They are not entirely calcareous and, in part, closely resemble lateritic or bauxitic pisoids (McFarlane 1976). However, the mineralogy, associated calcareous features and importance of gravity in the formation of the pisoids are inconsistent with known pisolaterite. Caliche is generally defined as a fine-grained chalky to well-cemented calcareous deposit formed in semi-arid temperate environments (Bretz and Horberg 1949, Blank and Tynes 1965, Reeves 1970, 1976, Goudie 1973, and others). The Lau crusts are not entirely calcareous and are forming in a tropical climate with high rainfall. Because of the island nature of the outcrop they are close to present day back-reef facies, and features such as fibrous gypsum and aragonite cements may result from the influence of wind-blown sea spray in pores. Such features are more consistent with coniatolite or hypersaline pisoids (Esteban 1976).

Despite these inconsistencies the Lau pisoids are rich in calcium carbonate and are a result of the weathering, dissolution and accumulation of residuals on a lithified outcrop in the vadose zone and are thus a type of caliche deposit. Table 1 compares the charateristics of caliche pisovadolite, summarized by Esteban

(1976) and modified by Chafetz and Butler (1980), with those identified from the Lau pisovadolite.

Many of the features of the Lau pisovadolite contradict earlier descriptions of caliche pisovadolite. The most diagnostic feature is the rudimentary lamination of grains with associated orientation of inclusions and thinning across projections in the nucleus, characteristics that result from the accretionary process.

Acknowledgements. The author is grateful to: the University Grants Committee and internal research fund, Victoria University of Wellington, for financial support in the field: D. Woodhall and other colleagues for discussion: the people of Lau for their excellent hospitality: B. Burgess for field assistance: Robertson Research International Limited for preparation of the final manuscript.

References

Assereto RAM, Folk RL (1976) Brick-like texture and radial rays in Triassic pisolites of Lombardy, Italy: a clue to distinguish ancient aragonitic pisolites. Sediment Geol 16:205 – 222

Blank HR, Tynes EW (1965) Formation of caliche in situ. Bull Geol Soc Am 76:1387 – 1392

Bretz JH, Horberg L (1949) Caliche in southeastern New Mexico. J Geol 57:491 – 511

Burgess CJ (1978) Holocene carbonate sedimentation, Lakeba and Vanua Balavu, Lau, Fiji. In: Creswell MM (ed) Lau-Tonga 1977. R Soc N Z Bull 17:191 – 209

Chafetz HS, Butler JC (1980) Petrology of recent caliche pisolities, spherulites, and speleothem deposits from Central Texas. Sedimentology 27:497 – 518

Dickson JA (1966) Carbonate identification and genesis as revealed by staining. J Sediment Petrol 36:491 – 505

Dunham RJ (1969) Vadose pisolite in the Capitan Reef (Permian), New Mexico and Texas. In: Friedman GM (ed) Depositional environments in carbonate rocks: a symposium. Soc Econ Paleontol Miner Spec Publ 14:182 – 191

Dunham RJ (1971) Meniscus cement. In: Bricker OP (ed) Carbonate cements. Studies in geology, vol 19. Hopkins Univ Press, Baltimore, Maryland, pp 297 – 300

Esteban CM (1976) Vadose pisolite and caliche. Bull Am Assoc Petrol Geol 60:2048 – 2057

Evamy BD (1963) The application of a chemical staining technique to a study of dedolomitisation. Sedimentology 2:164 – 170

Folk RL (1964) The natural history of crystalline calcium carbonate: effect of magnesium content and salinity. J Sediment Petrol 44:40 – 53

Goudie A (1973) Duricrusts in tropical and subtropical Landscapes. Clarendon Press, Oxford, 174 pp

Hay RL, Wiggins B (1980) Pellets, ooids, sepiolite and silica in three calcretes of the southwestern United States. Sedimentology 27:559 – 576

Jones FJ, Wilkinson BH (1978) Structure and growth of lacustrine pisoliths from recent Michigan marl lakes. J Sediment Petrol 48:1103 – 1110

Kendall CGStC (1969) An environmental re-interpretation of the Permian evaporite/carbonate shelf sediments of the Guadalupe Mountains. Bull Geol Soc Am 80:2503 – 2525

Ladd HS, Hoffmeister JE (1945) The geology of Lau. Bernice P Bishop Mus Bull 181

Loreau JP, Purser BH (1973) Distribution and ultrastructure of Holocene ooids in the Persian Gulf. In: Purser BH (ed) The Persian Gulf. Springer, Berlin Heidelberg New York, pp 343 – 376

McFarlane MJ (1976) Laterite and Landscape. Academic Press, London New York, 151 pp

Muller G (1971) Gravitational cement: an indicator for the vadose zone of the subaerial diagenetic environment. In: Bricker OP (ed) Carbonate cements. Studies in geology, vol. 19. Hopkins Univ Press, Baltimore, Maryland, pp 301 – 302

Reeves CC Jr (1970) Origin, classification and geological history of caliche on the southern High Plains, Texas and eastern New Mexico. J Geol 78:352 – 362

Reeves CC Jr (1976) Caliche: origin, classification, morphology and uses. Estacada books Lubbock Texas, 233 pp

Scholle P, Kinsman DJ (1974) Aragonitic and high-Mg calcite caliche from the Persian Gulf: a modern analog for the Permian of Texas and New Mexico. J Sediment Petrol 44:904 – 916

V.6 Pisoids and Pisolite Facies (Permian), Guadalupe Mountains, New Mexico and West Texas

Mateu Esteban[1] and Lloyd C. Pray[2]

1 Introduction

The abundant pisolite of the back-reef facies of the Capitan Limestone of the Guadalupe Mountain, New Mexico and Texas has long been of geologic interest. Its interpretation is important for better understanding the famous Permian Reef Complex and for interpreting pisolite, pisoids and associated facies elsewhere in the world. But major changes in prevailing concepts of the Guadalupe Mountains pisolite have occurred in the past 20 years. Consensus has not yet been reached. The purpose of this article is to summarize our current data and interpretations, emphasizing field-observable features.

Most early workers considered the pisoids were formed by algae (red algae, Ruedemann 1929; blue-green algae, Johnson 1942). Others denied or questioned their organic (algal) origin (Lang 1942, Newell et al. 1953). However, all early workers assumed the pisoids formed and were deposited in a lagoonal environment – they were free, subaqueous sedimentary particles. An interpretive revolution occurred in 1965 when, Dunham and, independently, Thomas proposed the pisoids were "soil concretions"; pisolite was a Permian caliche soil formed from a precursor marine sediment subjected to subaerial vadose diagenesis during episodic low stands of sea level. Their publications (Dunham 1969, 1972, Thomas 1968) led to wide acceptance of the "caliche" hypothesis for Guadalupe pisolite (e.g., Jacka et al. 1969, Smith 1974b) and to "caliche" and vadose interpretations for somewhat similar pisolite and pisoids elsewhere in the world (see Esteban 1976 for details). The caliche hypothesis still receives wide support.

Kendall (1969), in an important paper on the overall Permian Reef Complex, inferred a blue-green algal, seaward slope origin for many pisoids, emphasized a sedimentary clastic origin of Guadalupe pisolite and associated carbonates, and invoked a late cementation stage by phreatic and vadose water acting on supratidal to "island" areas. Nevertheless, he considered many pisoids and pisolite fabrics as "calcrete" or "caliche" in the sense of Dunham and Thomas. Purser

1 Institut Jaime Almera CSIC, c/Egipciaques 15, Barcelona-1, Spain
Present address: Amoco Production, Research Center, Box 591, Tulsa, OK 74102, USA
2 Department of Geology and Geophysics, University of Wisconsin (Madison), Madison, WI 53706 USA

Coated Grains (ed. by T.M. Peryt)

and Loreau (1973) and Shinn (1973) noted similarities in Persian Gulf pisoids to the Permian "vadose" pisoids, as did Scholle and Kinsman (1974), who stressed these analogies and suggested the need for marine-derived vadose waters to form the Permian "caliche". They inferred its formation on a marginal marine supratidal environment. Assereto and Folk (1976) and Loucks and Folk (1976) stressed the same marginal marine, supratidal environment and avoided referring to soil development or caliche.

Our joint work on Guadalupe Mountains pisolite began in 1975 after Esteban (in lectures) questioned the caliche interpretation, largely on the basis of his earlier work on ancient and modern caliches of the Mediterranean area and on his analysis of the literature on the Guadalupe Mountains pisolite (Esteban 1976). Our subsequent field studies in the Guadalupe Mountains in 1975 to 1977 led to published abstracts (Esteban and Pray 1975, 1976), and to presentation of many details of our field data and interpretations, as well as of relevant stratigraphic findings by co-workers Neese and Schwartz, in the two volumes of the 1977 symposium on Guadalupian Shelf Facies (Hileman and Mazzullo 1977, Pray and Esteban 1977). The major interpretations, herein reaffirmed, were the following:

1. Most pisoids were free sedimentary grains formed by subaqueous isopachous accretion on non-skeletal nuclei; most were deposited clastically, some show evidence of in-place growth. Some pisolitic strata show evidence of continued or later lamellar accretion of cement on primary pisoids under conditions of very shallow burial (cm to a few dm) in both phreatic to vadose environments. We believe vadose processes have been over-emphasized in the formation of the Guadalupe pisolite and relate the recognizable vadose effects largely to synsedimentary and early diagenetic processes of low supratidal flats. Deep vadose circulation (meters to tens or more meters) was not necessary to create the pisolite or associated facies. Soil products are very minor, highly localized, and post-date pisoid formation and pisolite cementation. Few pisoids formed as soil concretions or weathering products (caliche).

2. Corroboration of interpretations of Dunham and others that the predominant pisolite of the Capitan back-reef formed at a paleotopographic high (the shelf crest of Pray and Esteban 1977) of a carbonate marginal mound (Dunham 1969) that separated an evaporitic lagoon from a more open marine area.

3. Most pisoids and pisolite formed from intense precipitation from very shallow hypersaline waters in a peritidal environment that also favored development of fenestral peloidal dolomite and intraformational tepee structures. Thick pisolite development commonly occurred as parts of shoaling-upward sedimentary sequences of intertepee depressions.

4. Although the modern marginal marine environments of the south-western Persian Gulf, particularly some of the coniatolite terrane of Purser and Loreau (1973) appear somewhat analogous, a satisfying modern analog for the Permian shelf crest pisolite has not been reported.

In this article we review the major features of the shelf crest pisoids and pisolite, emphasizing field-observable features, and we infer the major processes by which the pisolite facies formed and the depositional and early diagenetic environments in which those processes operated. Our analysis includes brief discussions of the interrelationships of the shelf crest pisolite to its associated sedimentary facies, and to the intraformational tepee structures and erosion surfaces of the shelf crest. These broader relationships are relevant to better genetic understanding of the pisolite – an understanding that is still evolving.

2 Geologic Setting

The geologic setting and overall facies of the Permian Reef Complex (Fig. 1) are well known (e.g., King 1948, Newell et al. 1953, Dunham 1972, Hileman and Mazzullo 1977, Pray and Esteban 1977). Most pisolite occurs near the edge of the Northwestern Shelf in a facies belt that begins a few hundred meters or more shelfward of the Capitan (reef) Limestone and is hundreds of meters to several kilometers wide. Pisolite is cyclically interbedded with peloid grainsupportstones (packstones and grainstones) and with minor quartz sandstone. Most back-reef carbonates, including the pisolite, are pervasively dolomitized, but retain well their major depositional fabrics. Tectonic activity, except for causing repeated episodic minor subsidence, appears unimportant for facies development. Although pisolite occurs in back-reef facies equivalent to all of the Capitan Limestone, most work by ourselves and others of Guadalupe Mountain pisolite has been on strata of the younger Capitan (Yates and Tansill Formations).

Dunham (1969, 1972) interpreted the depositional profile of the outer part of the shelf as a marginal mound, with the position of the abundant pisolite being at its highest area (our shelf crest). We concur. The abundant pisolite, and associated fenestral peloid packstone and grainstone with conspicuous tepees represent a culmination facies, formed when sedimentation on the shelf crest reached its shallowest levels, a peritidal position (Fig. 2). This markedly reduced exchange

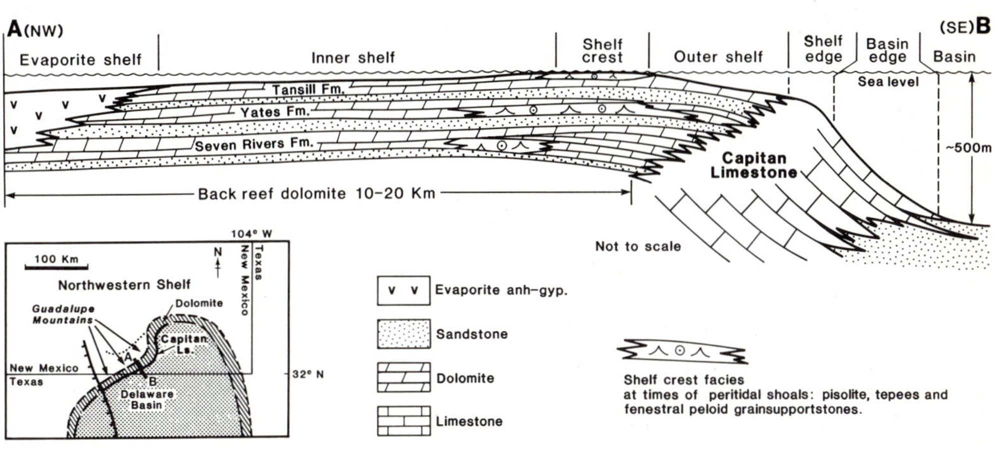

Fig. 1. Facies patterns and depositional environments of the Capitan Reef Complex. Pisolite, tepees, and fenestral carbonates of shelf crest developed only at times of peritidal shoals. Inset map (*lower left*) shows regional relationships of Northwestern Shelf (evaporites and sandstone), of the carbonate (dolomite and limestone) marginal mound and "reef", and of the Delaware Basin (sandstone)

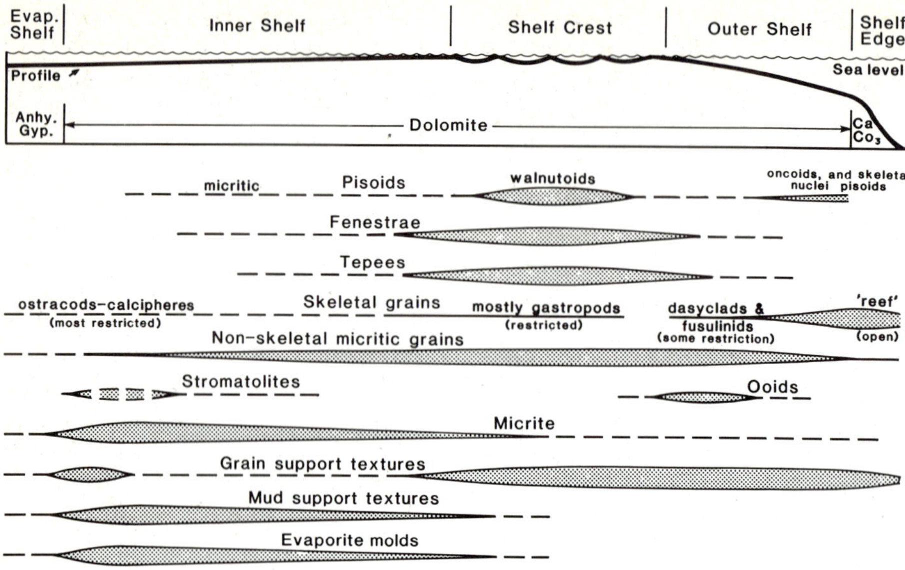

Fig. 2. Rock features and depositional profile of the shelf behind the Capitan reef in Guadalupian time

between the more evaporitic lagoon and the more open sea, creating sharp facies contrasts across the marginal mound. At times of better circulation between the open sea and the lagoonal areas (following episodic subsidence and prior to much sedimentary buildup at the shelf crest) the facies of the shelf crest appear to have been entirely subtidal, facies changes are less marked, and little pisolite formed.

3 Types of Large Coated Grains

Various types of large coated grains occur in the back-reef strata of the Capitan Reef Complex. Most are 3 – 5 mm to 1 – 2 cm in diameter. In this article we use the term pisoid for all coated grains from 2 mm to 2 – 3 cm to better fit nomenclature previous to this volume (an ooid to pisoid boundary at 2 mm) and the rock long termed pisolite in our study area. Some coated grains are smaller than

Fig. 3. A Multilaminated spheroidal pisoids, typical of shelf crest pisolite showing many thin, essentially isopachous accretionary laminae. Note similar sequence of major light and dark laminae in the outer cortex of many grains – an accretion stratigraphy. **B** Multilaminated shelf crest pisoids (walnutoids), showing outer laminae to be lighter in color, and less isopachous, thinner or locally missing along pisoid contacts (polygonal fitting). Geopetal silt in central area (peloids and quartz grains) is more abundant in upper parts of pisoids than lower (perched inclusions of Dunham). Some pisoid laminae laterally grade into this geopetal silt. Latest laminae show interpisoid linkage, and some meniscus effects. **C** Pisoids with skeletal nuclei, mostly dasyclad and mollusc fragments. Skeletal nuclei pisoids occur in grainstones of the outer shelf, and not in shelf crest pisolite. **D** Micrite-rich coated grains in micritic matrix with abundant quartz silt. Some coated grains show well-developed circumgranular cracking; others are poorly laminated aggregates of micritic peloids. Origin of micritic pisoids generally requires more data than the grain appearance

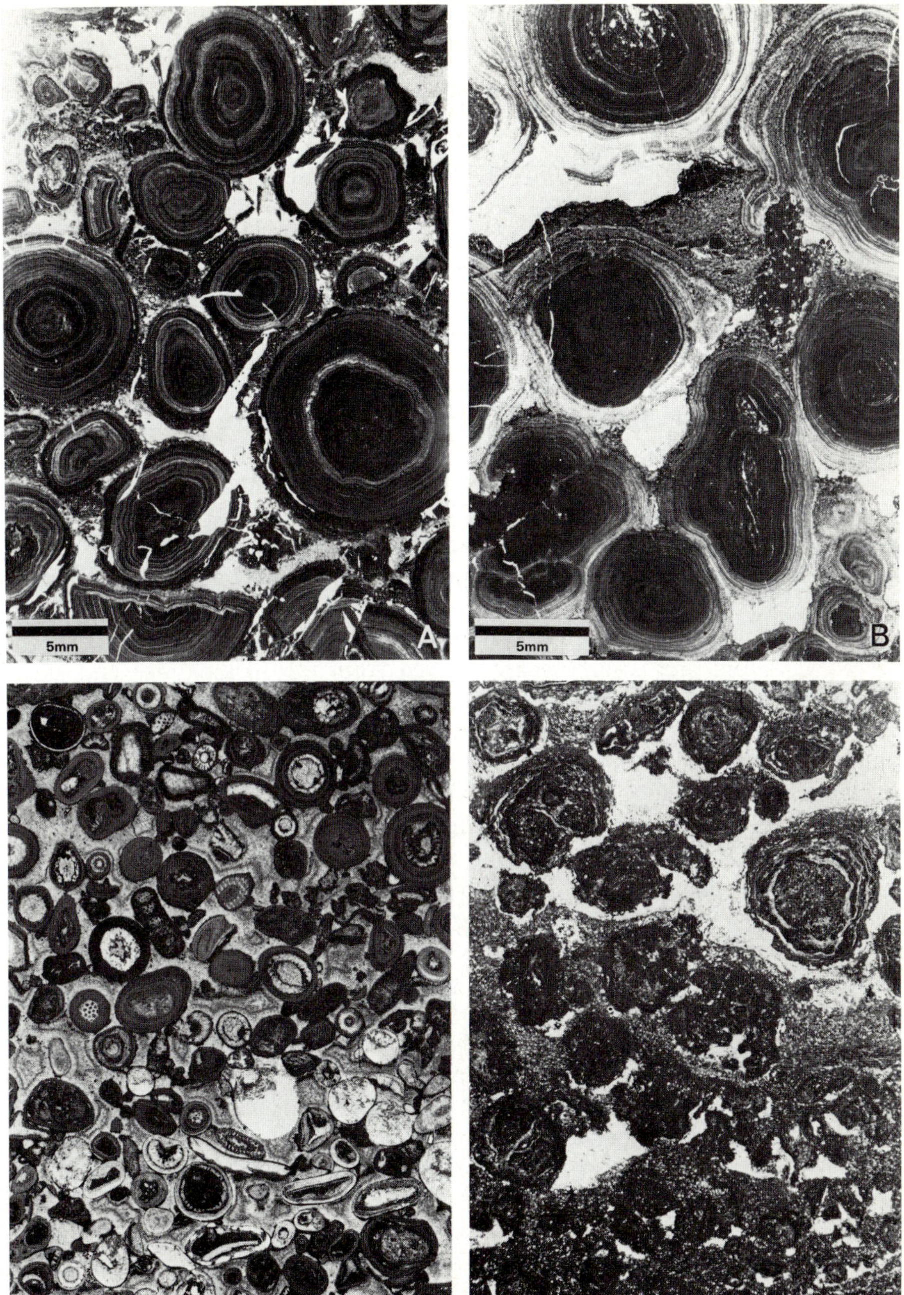
5mm
A
5mm
B
3mm
C
4mm
D

2 mm (microids or ooids); some are up to 10 cm or more across. The major types of large (+2 mm) coated grains are discussed below:

3.1 Predominant Pisoids

Most of the back-reef pisolite of the shelf crest consists almost entirely of pisoids of a distinctive type. They are well laminated, their cortex consists of many (50–100 or more) accretionary, essentially isopachous, laminae of carbonate cement (Fig. 3A, B). They have a non-skeletal nucleus, generally consisting of a peloid, if small, or a fragment of a preexisting pisoid, if 1–2 mm or larger. Some nuclei are pisolite intraclasts, others are micritic microspar bodies of obscure origin. Skeletal nuclei are virtually absent in shelf crest pisolite. We estimate this type of pisoid forms at least three-fourths of all types of back-reef pisoids. It is the type of principal concern in nearly all Guadalupe Mountain pisolite literature.

Owing to the variety of pisoid types in the back-reef strata (Fig. 4), and the number of descriptive words needed to differentiate this predominant type of back-reef pisoids from minor types that may occur in other facies positions, we here term these predominant pisoids "walnutoids", and the pisolite rock composed of them, "walnutite". Earlier we designated a characteristic sedimentary sequence involving these pisoids a "walnutite cycle" (Pray and Esteban 1977). The terms walnutoid and walnutite are derived from the striking size and shape resemblance, as initially noted by Ruedemann (1929), of most pisoids to nuts of the Texas walnut tree (*Juglans microcarpa*) which are abundant in Guadalupe Mountain canyons, such as Walnut Canyon, where most geologists have examined the Permian pisolite.

The predominant pisoids (walnutoids) are mostly 0.5 to 1.5 cm in diameter; they rarely exceed 2 to 3 cm. Typically, walnutoids are spheroidal in shape. However, irregular, polygonal and elongate shapes occur, most of which are created by coatings of broken fragments of large pisoids, or by later stages of growth when non-isopachous lamellar accretion was adapting to available interpisoid porosity. Successive laminae or groups of laminae are distinguished by contrasts of thinner, darker (in thin section), more micritic or inclusion-rich laminae with clearer, sparry (microspar) laminae. There is a tendency for inner laminae to be more isopachous, less clearly defined and more micritic and for outer laminae to be thicker, and clearer, as noted by Johnson (1942). The microspar laminae may show radial bladed and "brick-like" crystal structures (Assereto and Folk 1976). The distinction between particles (pisoids) and cement, normally easily established in carbonate rocks, is difficult and arbitrary for much walnutite, as pisoid growth was by accretion of carbonate cement laminae similar in appearance and crystal structure to laminae that later linked together individual pisoids. Outer laminae may form accretionary layers around most of a pisoid, and also may laterally link adjacent pisoids. These or others may grade laterally into sediment-rich, obscurely laminated material. Some complexities of laminae are illustrated in Fig. 3B (see also Dunham 1969, Thomas 1968).

All gradations from spar-rich (walnutoid) pisoids to more micrite-rich pisoids occur in the shelf crest area, though the former predominate. Some gradations occur within the same meter of section, in the same rock layer, or even within the same pisoid. Most walnutoids occur in rock in which they are the only grains, but many finer-textured shelf crest carbonates and some sandstones contain scattered walnutoids, or fragments of such.

3.2 Pisoids with Skeletal Nuclei

Pisoids with a nucleus of skeletal material (commonly dasycladacean algae, fusulinid, or gastropod) and with a distinct multilamellar cortex occur in more marine parts of the outer shelf (Fig. 3C). Dunham (1969) termed these "oversized marine ooliths". Their laminated cortex commonly is thin relative to the diameter of the skeletal nucleus. In thin section, the cortex resembles that of the much more abundant pisoids (walnutoids) of the shelf crest facies. Their skeletal nuclei and their facies position (Fig. 2) sets these apart from the other coated grains described herein.

3.3 Micrite-Rich Pisoids

Some pisoids of the back-reef facies are micrite-rich. The laminae are normally indistinct or obscure, relatively thick and few in number. The nucleus may be

Fig. 4A – I. Major types of shelf crest pisolite. **A** Clastic pisolite overlying peloidal sand seam. Many pisoid fragments show thin outer isopachous coatings. Grains and later interparticle cement are distinct. Sand seams show layers on left and inverse grading on right. Scale is 2 cm disc. **B** Clastic pisolite (walnutite) showing poor sorting of pisoids and some pisoid fragments. Each grain has a similar appearing, light colored, isopachous outermost coating. This, and overly close packing suggests outer coating of pisoids formed after initial clastic sedimentation. Note radial early cracking of some pisoids healed by later coating. **C** Thin section of matrix-free pisolite overlying carbonate sand. Most pisoids have angular nuclei of broken pisoids. **D** Field view of pisolite showing typical inverse grading at base of pisolite unit. Note abrupt upward size increase just above base, then more gradual size increase upward to an indefinite upper limit of the inversely graded unit (near middle of photograph). Some late stage directional elongation (mostly downward) evident on larger pisoids. Inverse graded layer overlies peloid (and siliciclastic) sand layer. **E** Inverse graded pisolite showing details of base of inversely graded unit on a pisoid-bearing peloid packstone. Pisoids show outer (light) isopachous accretions which are thicker on upper pisoids (see also Fig. 6D). Some interpisoid linkage by latest laminae. Fine-grained quartz detritus is concentrated at base of inverse graded unit. **F** Detail of the contact at base of an inverse graded pisoid unit with peloidal packstone. Contact appears sharp on right and gradational on left. Lower inverse grading involves uncoated peloids, upper grains are coated. White areas represent late solution porosity, partly filled by later cement. **G** Botryoidal pisolite. Note the laminated micritic layers partly coating three large clasts on upper side of pisolitic slab. These finely textured poorly laminated coatings, show downward, lateral and upward elongations, and formed prior to deposition of overlying skeletal-peloid grainsupportstone. Darker areas near base of pisolitic are slab are gravity cements. **H** Botryoidal pisolite, thin section view of part of a pisolite clast showing truncated composite botryoidal coatings with directional elongation. Some coatings are rich in peloids and quartz silt. Matrix (above) is a skeletal grainsupportstone with some fenestral fabric. **I** Botryoidal pisolite. Thin section view of part of a pisolite clast (center-upper left) showing truncated pisoids with directional elongations, which is itself coated by well-developed roof laminae grading into cement. Roof laminae thin into lower part of major cavity. Minor roof cements occur in the small pisoid fragments of the *lower right*. Circumgranular cracks show on edge of large pisoid of the *lower left*

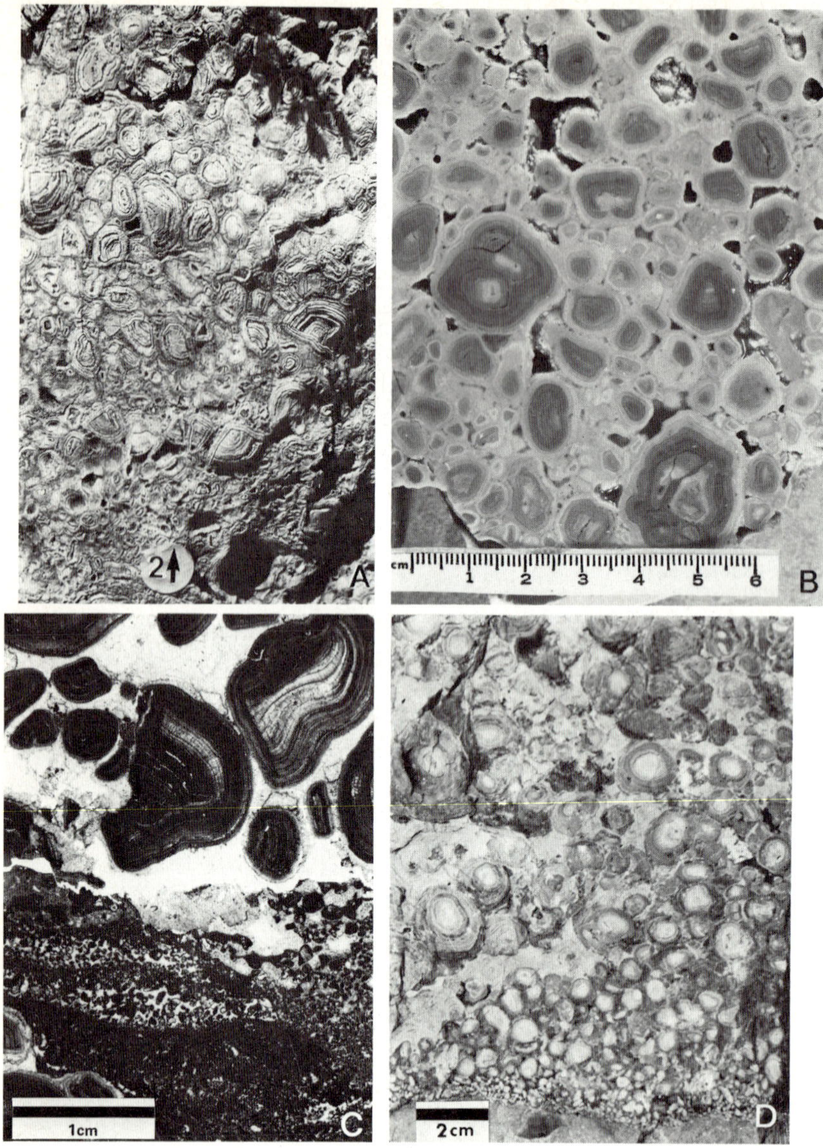

Fig. 4A – D (legend see p. 509)

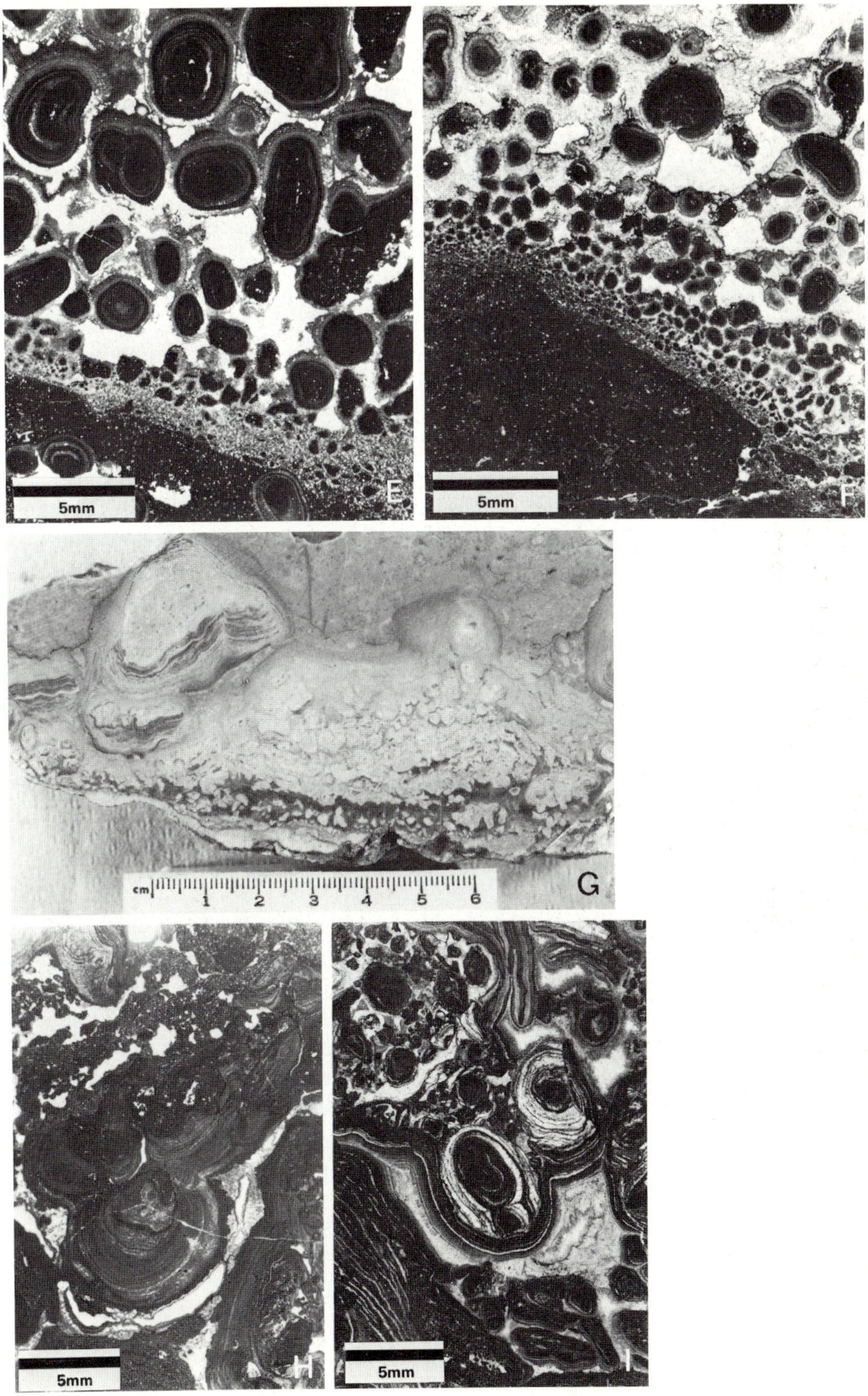

Fig. 4E–I (legend see p. 509)

sharply defined, obscure, or not discernible. Although quantitatively minor relative to the predominant pisoids (walnutoids), micrite-rich pisoids occur across much of the carbonate back-reef area. Locally they form the predominant grain type in a rock. More commonly they are scattered in finer-grained carbonate rock.

Micrite-rich pisoids of the Capitan back-reef undoubtedly have originated in various ways; for many the origin is obscure or not determinable, even if studied with a detailed knowledge of the overall geologic characteristics of the rock and its geologic setting, rather than just on the characteristics of the individual pisoid. Many micrite-rich pisoids (and macroids to 5 cm) are oncoids. These coated grains, as recognized earlier by Dunham (1968, 1972), Kendall (1969), and Tyrrell (1969) have the characteristic, rather irregular micritic laminae ascribed to binding and trapping by blue-green algae. The nucleus may be distinct or obscure. Most oncoids occur in oncoid grainstones on the outer shelf near the Capitan Limestone (Hurley 1978), but some occur scattered across the shelf crest and inner shelf facies. There are all gradations between micrite-rich pisoids (Fig. 3D) and the predominant pisoids (walnutoids). Some micrite-rich pisoids occur as grains amidst predominant walnutoids but they appear more abundant in botryoidal pisolite than walnutite. They also occur as scattered grains in peloid packstones and grainstones of the shelf crest facies. Locally, micrite-rich pisoids form discontinuous inverse graded layers several mm to a few cm thick (Fig. 4E). Some micrite-rich grains show circumgranular cracking (Swineford et al. 1958) or characteristics suggestive of micritization and may be in place of reworked soil concretions. However, we believe the scarcity of demonstrable soil profiles in the back-reef strata, and the scarcity of pisoids in the few thin profiles we have found, suggest that very few of the micrite-rich pisoids are of soil origin.

3.4 Botryoidal Macroids

Some coated grains are distinctly larger, more irregular in shape, and may be more micritic than typical walnutoids (Fig. 4G – I). These occur locally in thin, irregular layers or isolated in peloidal dolomite in shelf crest facies. Most of these show well-developed lamination. Irregular protuberances or botryoidal shapes (hence the name) are common. Many are 1 – 3 cm in diameter, but sizes to 10 cm or even larger occur. Nuclei are non-skeletal, the larger nuclei are commonly irregular intraclasts of walnutite and/or of micrite-rich pisolite or peloidal carbonate. The coated grains typically show multiple stages of fracturing, solution, and abrasion during their formation, as stressed by Dunham (1965, 1972) and Thomas (1968). Pronounced directional growths of laminae (upward, lateral, and especially downward) are common, as is direct transition from grain-coating laminae into laminated sheets that parallel or cross-cut rock fabric (Figs. 4G – I, 6A, 9F).

Table 1. Major types of shelf crest pisolite

Feature	Type of pisolite		
	Clastic	Inverse graded	Botryoidal
	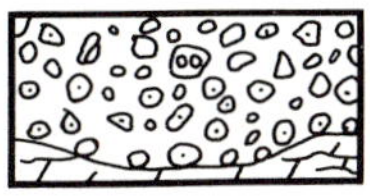	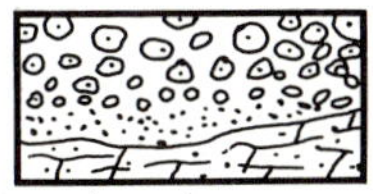	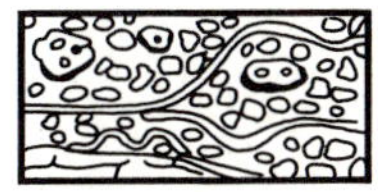
Grain type and size	Pisoids (walnutoids) 2 – 3 mm to 2 – 3 cm; minor, non-skeletal a) carbonate sand b) intraclasts to 10 – 20 cm	Pisoids (walnutoids) 2 – 3 mm to 2 – 3 cm	Pisoids (walnutoid and micritic); macroids[a] 2 – 3 cm to 10 – 20 cm; non-skeletal a) carbonate sand and silt b) pisolite intraclasts
Sorting	Well to poorly sorted (no inverse grading)	Well sorted, excellent inverse grading[a]	Erratic[a], poorly sorted, wide size range
Pisoid and macroid shape	Spheroidal to irregular, may include broken pisoid shapes[a]	Spheroidal to sub-spheroidal	Pisoids-irregular to spheroidal with downward bulging; macroids-botryoidal[a] to irregular to spheroidal; angular intraclasts[a]
Nuclei	Non-skeletal a) pisoid fragments b) obscure micritic-microspar bodies c) peloids d) pisolite intraclasts	Same as for clastic, but nuclei tend to be smaller, intraclasts uncommon	Same as for clastic, but multigeneration complex nuclei of macroids[a] suggest contemporaneous fracturing and brecciation
Lower and upper contacts	Lower-gradational to sharply erosional; upper-variable, commonly grades into finer carbonate	Lower-sharp, commonly grades into inverse graded sandsilt seam commonly overlying finer-grained matrix-rich carbonate; upper-indistinct commonly grades into clastic or, rarely, botryoidal carbonate	Boundaries gradational with finer grained carbonate above and below, or into inverse graded or clastic carbonate below; major intraformational erosion surface commonly forms upper surface
Stratal position	Intertepee depression and tepee flanks; some unrelated to tepees	Predominantly associated with intertepee depressions and lower tepee flanks	Predominantly on higher flanks of tepees and upper parts of intertepee depressions; some unrelated to tepees
Post-depositional features	Pisoids of well-sorted coarser grained pisolite may show inplace isopachous accretion and/or late stage polygonal fit and directional elongation	Same as for clastic	Same as for clastic, also multiple fractures cut cemented fabric and rocks show brecciation and internal sedimentation and cementation at early and late stages

[a] Features particularly useful for diagnosis

4 Depositional Classification of Shelf Crest Pisolite

Rocks of the shelf crest facies in which pisoids (and macroids) predominate (pisolite) display a wide variety of depositional and early diagenetic fabrics. They could be subdivided in many ways, genetically and descriptively. We find a threefold classification is practical. It is based largely on descriptive parameters, but has genetic overtones. Most shelf crest pisolite is herein referred to as Clastic Pisolite; much of it has characteristics (the individual pisoids or their aggregate) indicative or suggestive of clastic deposition of its constituent grains. The two other types are recognized by characteristics that make it readily distinguishable from the predominant or normal (clastic) shelf crest pisolite. These two are termed Inverse Graded Pisolite to recognize its distinctive and definitive feature, and Botryoidal Pisolite to recognize the common shape of many of its larger coated grains – although it has other distinctive characteristics as well. Inverse graded pisolite, termed "in-place" in our earlier publications, is used for any pisolite displaying inverse grading (termed reverse grading in Dunham's publications). The major characteristics of these three types of pisolite of the shelf crest facies are outlined on Table 1, together with sketches of typical rock fabric (see also Fig. 4). Some shelf crest pisolite lacks recognizable evidence of clastic deposition, and also lacks the distinctive characteristics of the inverse graded or botryoidal pisolite. We arbitrarily group this with the clastic category, based on our current genetic biases. Using the threefold depositional classification, most shelf pisolite is clastic; inverse graded pisolite probably forms about one-fourth and botryoidal about one-tenth of shelf crest pisolite.

Common features of all three depositional types of shelf crest pisolite are the following:

1. Nearly all grains larger than sand size (2 mm) are coated grains. Most are of the type we term walnutoids.
2. Skeletal material, whether as pisoid nuclei or as separate grains, is rare.
3. Most pisolite has little matrix material, and it consists largely of sand to silt-size peloids and siliciclastic grains. Micrite is uncommon as matrix material. Both micrite and peloids are more common in the botryoidal pisolite than the other types.
4. All three types of pisolite intergrade with each other at a cm to m scale, and all types intergrade with finer-grained peloid packstones and grainstones, with which there is abundant evidence of depositional synchroneity.
5. All types may show post-depositional modification by continued accretion of lamellar cement of type similar to that of the preexisting pisoids.
6. Evidence of pisoids replacing a precursor sediment or rock fabric is not recognized.

Dunham (1969, 1972) and Thomas (1958) focused almost entirely on pisolite whose fabrics showed in-place development features (our inverse graded and botryoidal pisolite). They did not distinguish a primary depositional fabric within rocks in which they recognized in-place (secondary) fabrics. The little clastic pisolite they reported was considered reworked from soil horizons. However,

Kendall (1969) clearly stressed the abundance of pisolite with clastic textures, recognized many pisoids as primary and differentiated their depositional fabrics from later vadose and phreatic cement fabrics.

Some key aspects in identifying and interpreting the depositional types of pisolite are briefly discussed below. Details of pisolite fabric evolution are presented in the succeeding section.

4.1 Clastic Pisolite

Most pisolite of the shelf crest occurs in irregular layers or lenses several cm to a few dm thick that lacks the distinctive features of inverse graded or botryoidal pisolite. We believe most of this was deposited clastically, as evidence of clastic deposition is clear for much of it. In some that we arbitrarily group with clastic pisolite, evidence of clastic deposition is limited or equivocal. As pisoids of clastic pisolite may show post-depositional in-place growth, careful assessment of the primary depositional fabric, and not that resulting from late stage in-place cementation, is necessary in its identification. Evidence we use for clastic deposition includes the following: (a) Fragments of pisoids (a common feature), abraded pisoids, intraclasts of pisolite, or associated rock, including (rare) sheet crack cement (Fig. 4A – C). As pisoids may fracture in-place (commonly radially), evidence of clastic deposition requires wide separation or isolation of fragments. (b) Sorting showing upward or lateral size decrease from pisolite to pisoid-bearing sand-size carbonates (Fig. 6B, C, E). (c) Pisolite with a jumbled array of pisoids of different sizes or types (Fig. 4B). (d) Relationship to basal surface: Much pisolite occurs directly above a sharp erosion surface – evidence of scouring or channeling suggests clastic deposition of the overlying pisolite (Fig. 9E). Channels as deep as 0.5 m are very rare, but evidence of channeling and scouring at much smaller scale is abundant. Particularly diagnostic evidence of clastic deposition of pisoids is the occurrence of the basal pisoids lying on the erosion surface like balls rolled onto a table, an especially unlikely geometric relationship to result from replacement processes. (e) Pisoids of differing character or pisoids with highly variable accretion stratigraphy (see Fabric Evolution section) may suggest their clastic origin. (f) Cross-bedding could be excellent evidence but we have not observed clear examples of it in shelf crest pisolite. We believe the reported pisolite cross-bedding of Kendall (1969) may have been the low-angle, seaward dipping foresets, involving skeletal nuclei pisoids of the outer shelf facies.

4.2 Inverse Graded Pisolite

The essential characteristic of this depositional type of pisolite is inverse graded bedding. This feature, first reported by Dunham (1965), in places is spectacularly developed in shelf crest pisolite (Fig. 6A). The scarcity of well-developed inverse grading in sedimentary strata and fabric features of some inverse graded pisolite requiring in-place growth of pisoids contributed significantly to Dunham's interpretation of a secondary or diagenetic origin for the pisolite. The origin of inverse grading is discussed under "Fabric Evolution"; in the shelf crest pisolite we believe depositional processes involving both depositional sorting and in-place pisoid growth were operative.

Inverse grading is readily recognizable in the field, polished slab, or thin section (Figs. 4D – F, 5A – C, F, 6A, D). Grading begins abruptly. The basal few mm to a cm or more are commonly of sand or granule size particles, upward size increase to pisoids 0.5 – 1 cm in diameter is marked over a few cm, then becomes more gradual. The upper limit of inverse grading in pisolite is indistinct. Most inverse graded beds are a few cm to 10 cm thick, locally inverse grading may persist upward for 20 or (rarely) 30 cm. All grains are coated in some units, in others, coating only occurs on grains larger than 2 – 3 mm. Spheroidal shapes and small nuclei prevail. Inverse graded pisolite rarely has matrix material between the pisoids. Broken pisolite fragments and intraclasts are uncommon, as are pisoids that are out of place vertically for their size. Most pisoids appear similar, and in many units, their "accretion stratigraphy" is similar; in some units inferred time-equivalent accretionary laminae units thicken progressively upward (Fig. 6D). Some inverse pisolite layers persist laterally for many meters, others only persist for a few dms. Inverse graded units of only sand size may be only a few cms wide. Inverse graded pisolite may merge laterally into clastic pisolite, or botryoidal pisolite or finer-grained carbonate in but a dm or less distance.

4.3 Botryoidal Pisolite

Botryoidal pisolite (Figs. 4G – I, 5A, 6A, 9F) is a distinctive minor facies of the shelf crest carbonates. Several features distinguish it from most shelf crest pisolite even though all gradations occur between it and the other two types of pisolite. The name is derived from the occurrence in this pisolite of irregular, bumpy to botryoidal growth shapes of some of the larger pisoids and the macroids, although most of the smaller and some of the larger coated grains are more smoothly spheroidal. Pisoids and macroids in botryoidal pisolite are commonly more micritic and show a greater range in sizes and poorer sorting that in other

---

Fig. 5A – F. Pisolite with directional fabrics. **A** Nest of inverse graded (*base*) to botryoidal pisolite (*top*) on flank of tepee. Nest sequence formed in concavity in peloid grainsupportstones. Upward growth of upper pisoids preceded subsequent onlap and burial by fenestral peloid sediment. Pisolite formation was synsedimentary. **B** Incipient fitted polygonal structure developed in outer laminae of inverse graded pisolite. Outermost laminae show interpisoid linkage. Note variations in sequence of light and dark laminae forming inner part of different pisoids. **C** Well-developed downward elongation and fitted polygonal structure in outer dark cement laminae of upper pisoids. Pisoids and cement in lower part of polished slab contrast with upper part and show neither elongations nor polygonal fitting. This suggests two stages of cementation or a vadose-to-phreatic contact. **D** Pisolite with coatings showing directional elongations and polygonal fitting. Directional elongation predominantly downwards, but locally also lateral and upwards. *Black areas* are residual intergranular porosity. **E** Well developed fitted polygonal structure, downward elongation and laminated interpisoid linkage cement. *White areas* are neomorphic carbonate. **F** Inverse graded, poorly laminated, micrite-rich pisoids, and pisoid fragments. Late stage, downward thickening cement (*light colored*) under many pisoids is inferred to be gravity cement of vadose origin

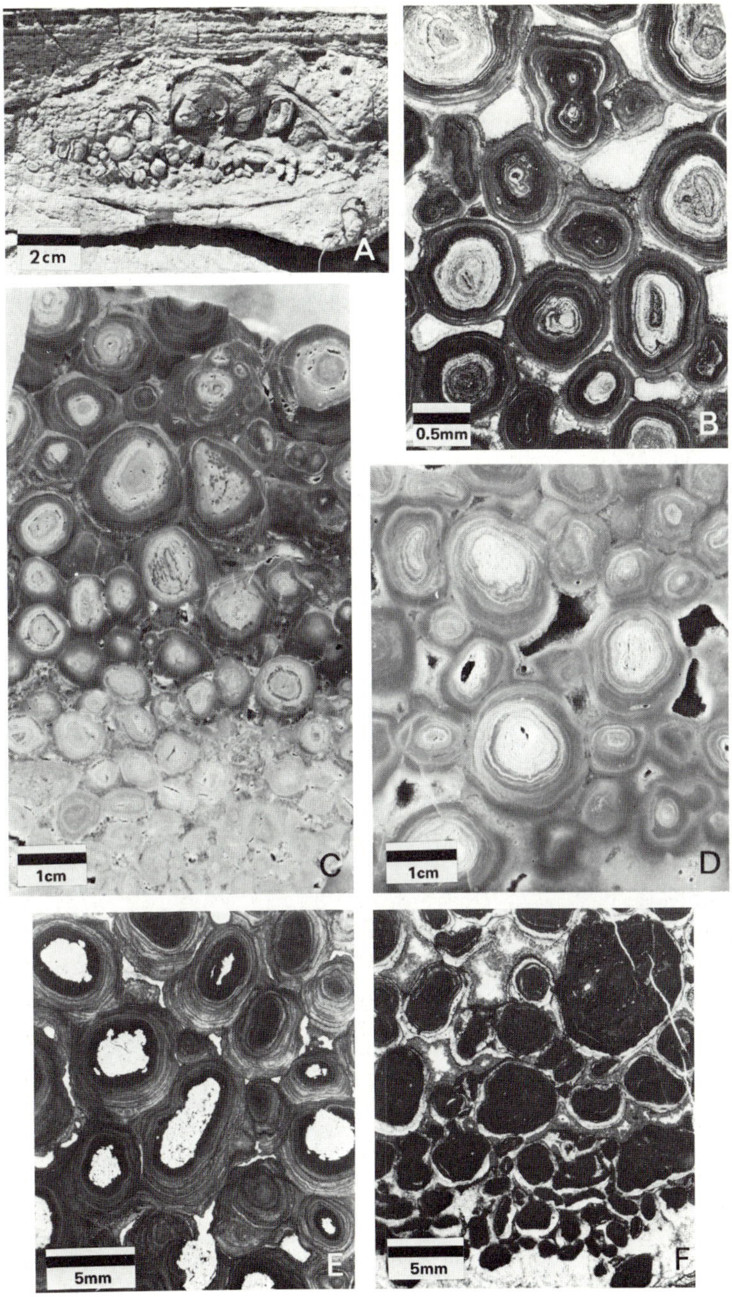
2cm
A
0.5mm
B
1cm
C
1cm
D
5mm
E
5mm
F

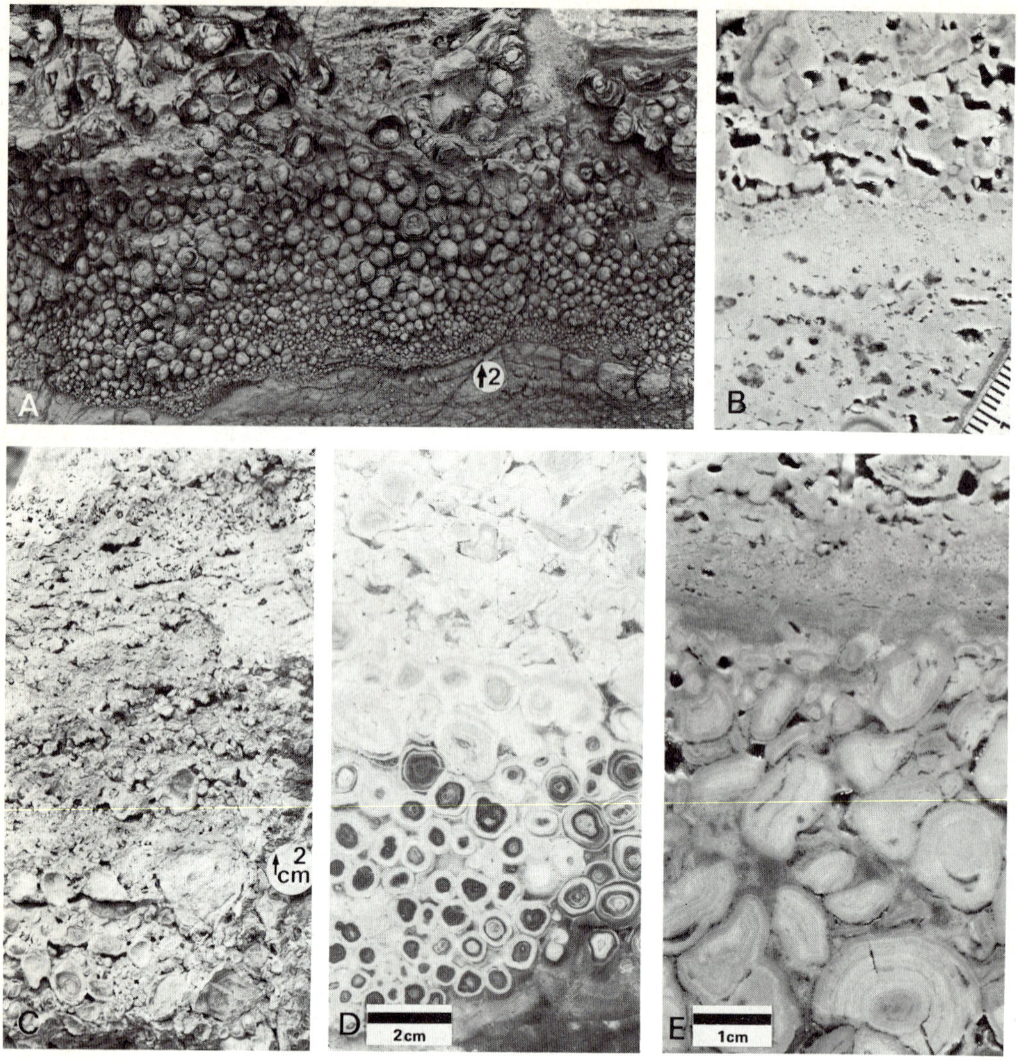

Fig. 6 A – E. Vertical sequences of pisolite. **A** Pisoid-bearing peloid grainsupportstone (*bottom*), separated by an undulose surface from a well-developed inverse graded pisolite that merges upwards into a botryoidal pisolite showing coated pisolite intraclasts, wavy laminated sheets, and directional elongations (mostly upward). Hairpin Curve, Walnut Canyon. **B** Sequence showing, in upward succession, (a) lower pisoid-bearing peloid grainsupportstone with poorly developed fenestral fabric, (b) carbonate sand seam with poorly developed inverse grading, (c) clastic pisolite with intergranular porosity grading into matrix-rich clastic pisolite. **C** Upward-fining sequence from matrix-rich clastic pisolite to pisoid-bearing floatstone with fenestral fabric. **D** Walnutite sequence showing inverse graded pisolite overlain by clastic pisolite. Cement lined sheet crack (*lower right*) cuts off lower part of inverse graded unit. Later laminae of some inverse graded pisoids show fitted polygonal structure and some late laminae link adjacent pisoids. Pisoids of lower left and center have isopachous, light-colored, outer laminae that thicken upward, suggesting that synchronous cement laminae are thicker nearer to an overlying depositional interface. Differential enlargement of loose, in-place pisoids may help form inverse grading. The substrate of the overlying clastic pisolite was a cemented crust developed on top of the inverse graded pisolite, showing synsedimentary cementation. **E** Cemented but unaltered upward fining clastic pisolite abruptly overlain by calcareous sand seam (with incipient inverse grading) which grades upwards abruptly into clastic pisolite

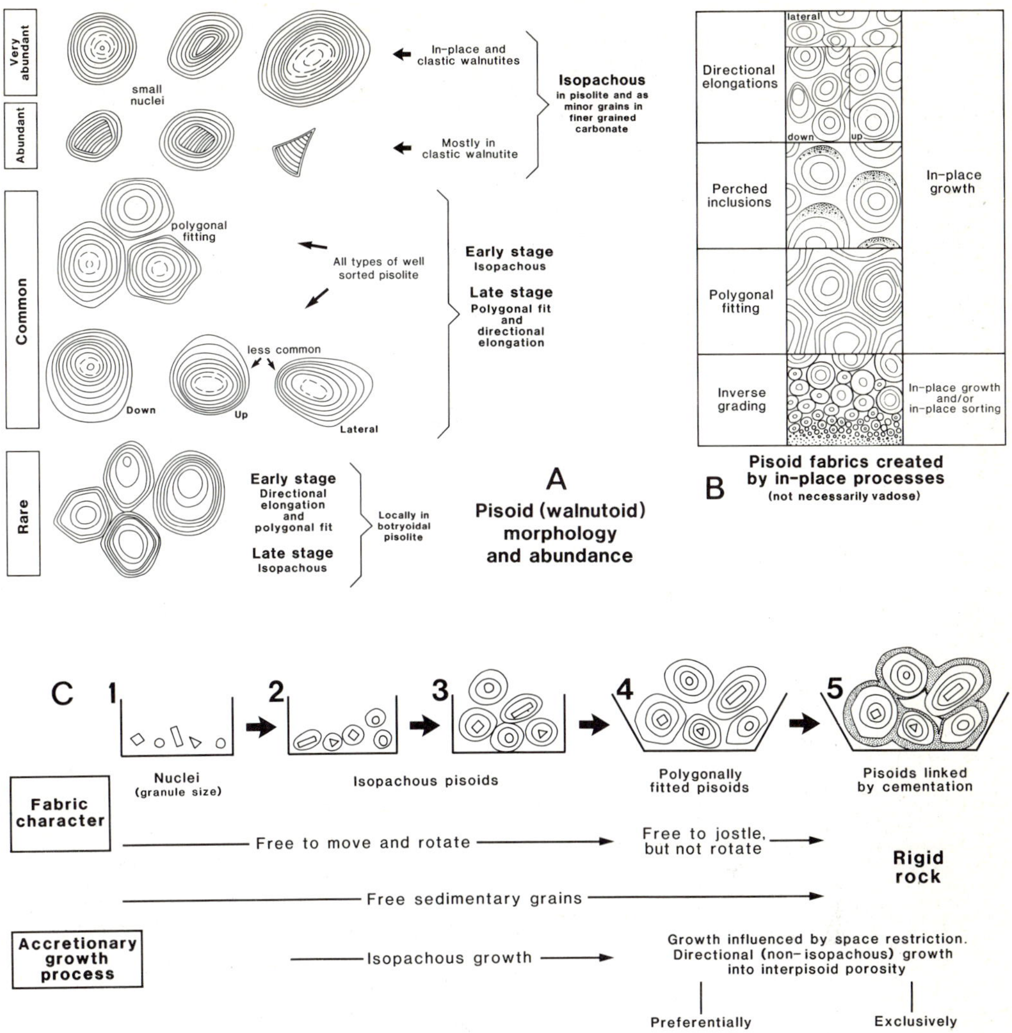

Fig. 7A – C. Fabric features of pisoids and pisolite; and concepts of their evolution. **A** Morphology and abundance of pisoids and their relationship to growth stages and to types of pisolite. **B** Fabrics of pisoids and pisolite created by in-place processes. Although these have been widely interpreted as formed by vadose processes, only directional elongation in a downward direction appears to require a vadose environment. **C** Conceptual scheme of pisolite fabric evolution in in-place cement accretion. Late stage displacive cementation (4 or 5) might cause non-tectonic fracturing

shelf crest pisolite. Individual pisoids and pisolitic intraclasts more commonly show gravity (downward elongation) cement fabrics; some of their fabrics appear to have formed at earlier stages in their growth than in the other types of pisolite. Lateral and upward growth of laminated sediment and/or cement crusts also characterize this complex rock type (Fig. 4G, 5A). Intraclasts (1 to 20 cm or larger) of brecciated pisolite and of finer grained associated carbonate are

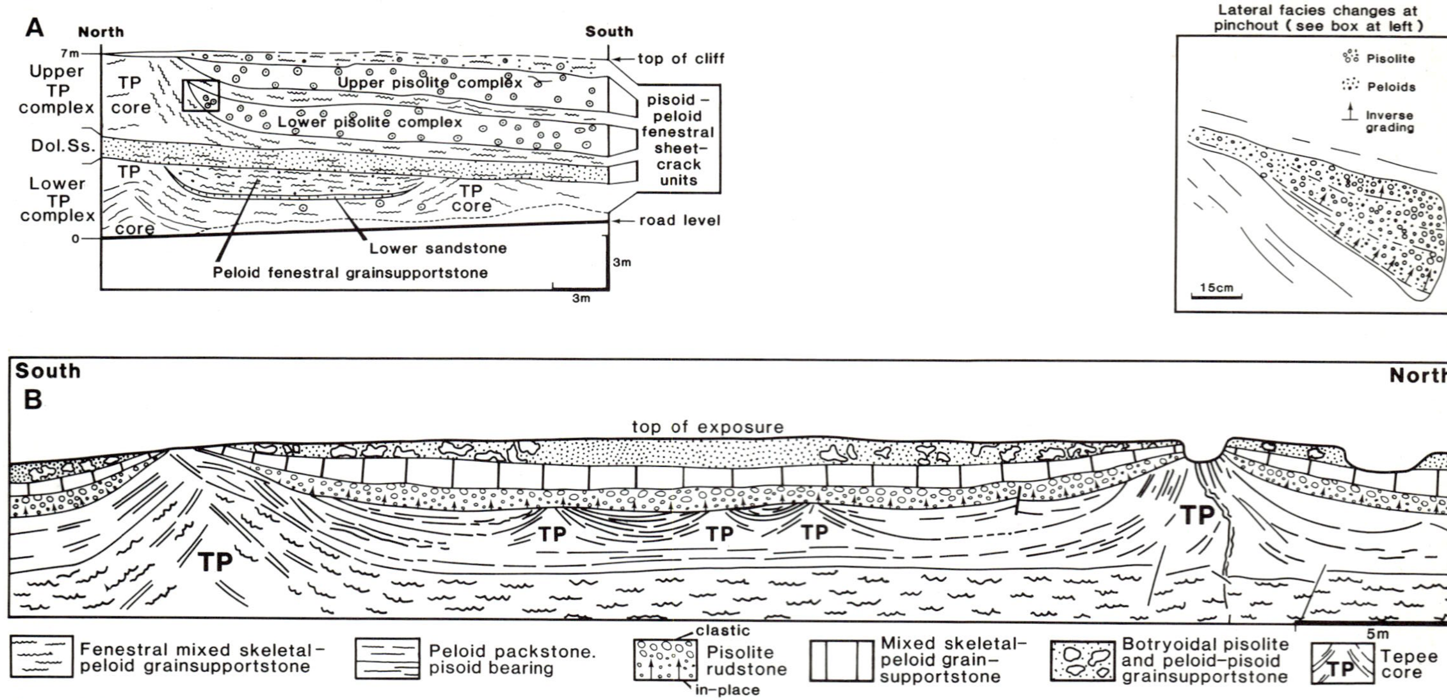

Fig. 8 A, B. Cross-section diagrams of two field localities involving pisolite and tepees. **A** Hairpin curve locality showing tepees with pisolite prominent in intertepee depression of upper tepee complex, facies variation on tepee flank (*see box on right*), and erosion surfaces truncating the tepee complexes. **B** Caverns Parking Lot locality showing relationship of clastic, inverse graded (*arrows*), and botryoidal pisolite to sedimentary filling of intertepee depression. Note that initial pisolite (inverse graded overlain by clastic) extends across entire bottom of intertepee depression and is above a major erosion surface, and that botryoidal pisolite is restricted to a late stage of sedimentation of intertepee depression and to the tepee flanks. (Diagrams adapted from Pray and Esteban 1977)

conspicuous in botryoidal pisolite, many appearing to have formed by nearly in-place processes of fragmentation. Botryoidal pisolite and its constituent coated grains show abundant evidence of multigeneration fracturing and accretion of laminated spar-rich cement. Laminated, fine sediment-rich cement also fills vertical to subhorizontal, commonly curved fractures up to 1 – 2 cm wide that transect the rock. Relationships strongly suggest that fracturing and cementation accompanied the formation of botryoidal pisolite, as a syndepositional process. Part of the rock consists of laminated, fine-grained clastic carbonate, in some places layered and in others chaotically intergrading with more pisoid-rich parts. Pockets of inverse graded pisolite (cm to dm wide) are abundant in botryoidal pisolite, and in many places inverse graded pisolite grades upward into botryoidal pisolite (Figs. 5A, 6A). Polygonal fit (Fig. 7) of coated grains is well developed in much botryoidal pisolite. The undersides of larger pisolitic intraclasts in this facies commonly show cm thick, laminated gravity cements. Only in botryoidal pisolite have we observed pisoid laminae that show directional elongation coated by later isopachous laminae. Botryoidal pisolite shows much lateral and vertical variability. This type of pisolite is most abundant in tepee areas, where it generally occurs high on tepee flanks as a final stage in the filling of an intertepee depression (Fig. 8B). In pisolite-rich strata, with or without tepee structures, botryoidal pisolite is normally in the uppermost part of the carbonate unit.

5 Fabric Evolution of Shelf Crest Pisolite

5.1 Introduction

Our observations of pisolite fabric build from the important work of Dunham, Kendall, and Thomas, who first studied details of the rock fabric of the shelf crest pisolite and appreciated its genetic complexity. Pisolite fabric evolution involves complex interweaving of depositional and diagenetic processes in an environment characterized by extensive carbonate precipitation. This precipitation strongly favored repetitive syntaxial growth from a pre-existing crystalline substrate rather than nucleation of new independent crystals, thus forming the successive accretionary laminae of the pisoids and also strongly favoring pisoid fragments and pisolite intraclasts as nuclei. This process created and enlarged the pisoids. Some grew essentially in-place, but we believe most grew and moved as independent clasts prior to their final deposition. Pisoids of some high permeability depositional assemblages may show later enlargement by continued deposition of accretionary lamellar cement. These outer laminae are commonly isopachous, but may be non-isopachous and show polygonal fitting and directional orientations. At some stage further accretionary cementation linked pisoids together into rigid pisolite. These stages of pisoid growth by lamellar accretion, both primary and early diagenetic, formed the major fabric of the present pisolite; this evolution is the focus of the following discussion, for which Fig. 7 depicts some of the relevant facts and our interpretations. Later diagenetic processes affecting the pisolite, such as porosity occlusion by equant sparry carbonate cement, solution (including stylolitization), neomorphism (largely of nuclei), and dolomitization (which was pervasive on the shelf crest) are not discussed in this article.

5.2 Isopachous and Non-Isopachous Lamellar Accretion

Shelf crest pisolite shows abundant evidence of pisoid growth and pisolite cementation by both isopachous and non-isopachous lamellar accretion. The abundance, morphology, internal structure and facies distribution of individual pisoids is pertinent to understanding pisolite fabric evolution (Fig. 7A). The most abundant pisoids (walnutoids) are spheroidal to subspheroidal and have an essentially (megascopically) isopachous lamellar structure. Locally abundant in clastic pisolite are more irregular shapes of broken pisoids, variably coated by lamellar, generally isopachous, accretions. Less abundant are pisoids showing non-isopachous accretion lamellae creating polygonal shapes and/or directional elongations which suggests pisoid growth under conditions of space restriction or directional processes of cementation, such as in stalactitic (gravity) cementation. Common directional growth features of individual pisoids of an assemblage, as stressed by Dunham, logically indicates in-place growth. The timing of such growth is critical to understanding evolution of pisolite fabric. Our observations reveal that pisoids of aggregates showing polygonal or directional effects almost invariably show an earlier stage of isopachous accretion. The importance of this is that the original pisoids need not have formed in-place, only during the later stage of directional growth are in-place processes required. The rarity of pisoids showing polygonal shapes or directional growth followed by later isopachous lamellae also is genetically significant. It suggests that once non-isopachous enlargement occurred in a pisoid assemblage, either there was no reworking, or reworking produced pisolite intraclasts rather than individual pisoids. Yet intraclasts of pisolite are common, and some form pisoid nuclei. These observations are compatible with very early and very shallow formation of the non-isopachous lamellar cement. The scarcity in most pisolite of enlargement by repeated alternations of isopachous accretion and of non-isopachous accretions, such as gravity cement, argues against repeated alternation of phreatic and vadose environments. Our analysis suggests that most pisoids of the shelf crest formed in a subaqueous environment, easiest interpreted as shallow open water. Some pisoid assemblages developed directional fabrics at the depositional interface, some developed later directional fabrics under conditions of very shallow burial in phreatic and/or vadose environments.

5.3 In-Place Pisolite Growth Fabrics

Dunham recognized the occurrence of perched inclusions, polygonal fit and directional elongation in some pisolite (Fig. 7B). and deduced that if pisoids of an assemblage showed such fit, or consistent orientations of inclusions or elongations that the pisoids had grown in-place. Dunham emphasized downward (gravity) elongation. In addition, Thomas noted locally consistent lateral or upward elongations, and preferential growth (multidirectional) into available porosity. These are all recognizable features in some shelf crest pisolite (see Figs. 4–7), and in-place growth is a logical explanation for the stage during which such directional features developed. These features and inverse grading (discussed below) were part of Dunham's and Thomas's rationale for interpreting the pisolite as created entirely by vadose processes. Esteban (1976), however, pointed out that perched inclusions, polygonal fit, and directional features, while logically requiring in-place growth, do not necessarily indicate a vadose environment for such growth. Consistent downward elongation and meniscus cements appear best interpreted as resulting from preferential wetting and evaporative effects in aereated porosity of a vadose zone. But other factors operative within subaqueous deposits on the sea floor or in a phreatic zone that might well influence pisoids growth in one or more directions. Fabrics could be the direction of sediment supply (providing either particle "feed" or particles that acted to inhibit cement accretion), the direction and amount of water movement (ion "feed"), agitation or interference effects at pisoid contacts, and importantly, simply the availability of porosity for accretionary cement. All of these could, and we believe, did complicate the directional aspects of cement fabrics of some pisolite.

Figure 7C depicts, simplistically, several stages in the evolution of pisolite fabric by the process of lamellar accretion of cement on initial nuclei present in a confined area. Isopachous accretion is inferred to occur in a saturated (sea floor or phreatic) environment when the subaqueous particles were free to rotate or adjust their position in response to their accretionary displacive enlargement in a loose grain supported assemblage. Further pisoid enlargement, if influenced by space restriction, such as the confining pressure of overlying sediment or a cemented rigid layer, could develop polygonal fit and/or directional elongations when grains could jostle, but not shift or rotate. In later stages displacive accretionary cementation might develop disruptive fabric stresses; possibly some of the abundant non-tectonic openings (Dunham 1968) in pisolite developed from expanding pisoid aggregates. Two variations common in the evolution of fabric

---

Fig. 9A–F. Erosional surfaces in shelf crest facies. **A** Truncated tepee structure developed in Hairpin Dolomite of Upper Yates Fm. showing truncated fenestral peloid dolomite overlain by laminated, peloid-rich quartz sandstone. No alteration effects are associated with this intraformational erosion surface. Walnut Canyon. **B** Major intraformational erosion surface (Top of Triplet Dolomite, Upper Yates Fm.) between overlying quartz sandstone and underlying dolomite. The uppermost carbonate (*above pencil*) is micritized. This alteration is evidence of incipient soil (caliche) processes but it did not form pisoids of the dolomite, it altered pre-existing pisoids. Walnut Canon. **C** Thin section view of sharp erosion surface developed on cemented but unaltered pisoids, overlain by peloidal grainsupportstone. **D** Thin-section detail of the alteration zone a few cm below the major intraformational erosion surface (see **C**). Alteration (soil process?) involved micritization of pre-existing pisoid fragments in peloidal dolomite rich in quartz silt. **E** Field view of erosional channel cut into layered pisoid-bearing peloid grainsupportstones. Channel is filled and overlain by clastic pisolite, in which some pisoids show post-depositional enlargement by lamellar accretions. Lower Yates Fm., Dark Canyon. **F** Inverse and botryoidal pisolite and associated peloid grainsupportstones on tilted flank of tepee. Hairpin Curve road cuts, Walnut Canyon

Fig. 9 A – C (legend see p. 523)

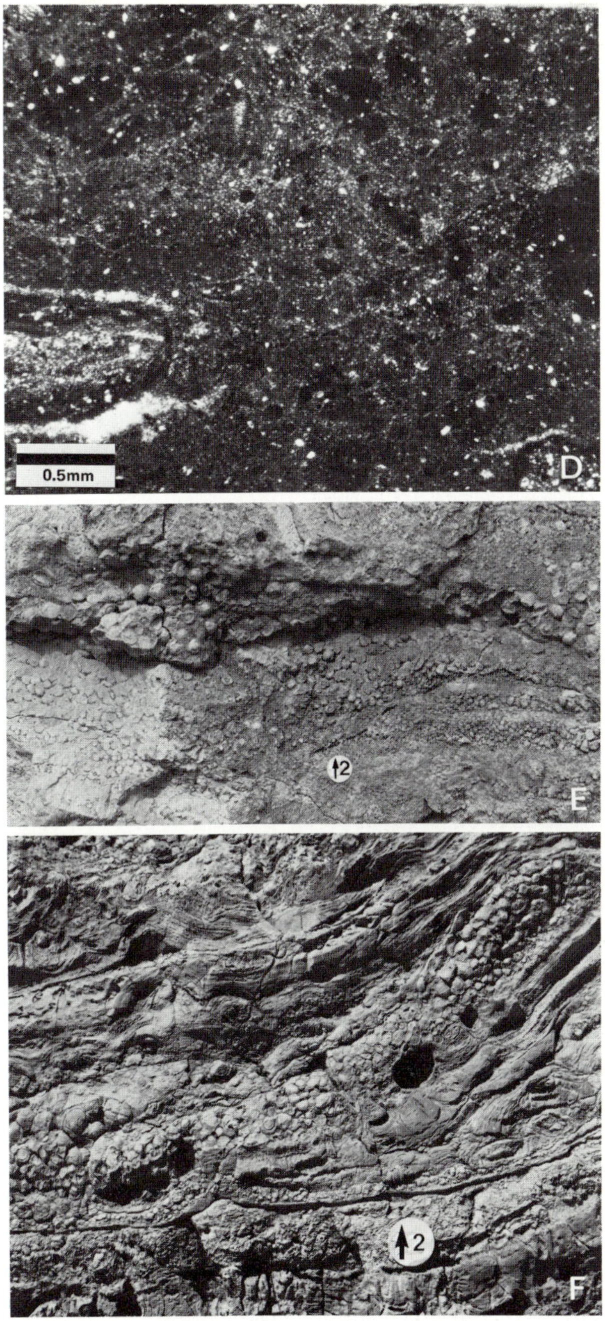

Fig. 9 D – F (legend see p. 523)

of shelf crest pisolite from that shown on Fig. 7C would involve starting with an assemblage of clastic pisoids instead of in-place small nuclei, or having accretionary cementation cease before reaching the space restrictive stage and have non-laminated cement form rigid pisolite.

Accretion stratigraphy is our term for the study and attempted correlation of a succession of accretionary laminae of different pisoids of an assemblage. It is a type of cement stratigraphy as applied to diagenetic history by Meyers (1974) and others. Successive laminae of pisoids commonly vary in color, thickness, crystal nature, inclusions, or other characteristics. Similar sequences of these distinctive accretionary laminae on different pisoids suggests time correlative accretions, somewhat comparable to tree ring or varve correlations. Correlation or lack thereof can provide valuable genetic information on a pisoid assemblage. For example, in some clastic pisolite the accretionary sequence of outer laminae may be nearly identical on most pisoids, regardless of their size or their inner laminations. If these seemingly correlatable outer sequences are continuous, unbroken and not abraded (Fig. 4B), the reasonable inference is that these outer accretions formed subsequent to their deposition as individual clasts. That is, the late accretions formed in-place. In some inverse graded pisolite (e.g., Fig. 6D) the later accretions on most pisoids are distinctive in color and crystal structure, and correlatable accretions thicken progressively toward the upper part of the graded sequence. Here accretion stratigraphy provides not only a criterion for recognizing probable in-place enlargement, but it suggests that such differential upward enlargement could be a factor in forming inverse grading. Logically such an effect might reflect increasing upward proximity to cement-providing water at a depositional interface. A somewhat related aspect to accretion stratigraphy of individual pisoids concerns possible distinctive characteristics of cement occurring in porosity between pisoids. We have observed successive layers of pisolite, commonly cm thick, that have distinctively colored cements. This provides another type of cement stratigraphy, and in some rocks of such a sequence, cements that fill remnant porosity correspond in their characteristics sequentially to a distinctive cements of the next overlying pisolite. The occurrence of this in vertically incremental units a few cm thick of pisolite, whether or not each successive unit is separated from the overlying by an erosional surface truncating pisoids and cement, strengthens the interpretation that deposition and much cementation of pisolite were penecontemporaneous processes that occurred at or near a depositional interface. The not uncommon occurrence of carbonate sand or pisoids directly on truncated underlying cement (Fig. 9C) further strengthens this interpretation.

5.4 Inverse Grading

Inverse grading, described in the section on Pisolite Classification, is a striking feature of some shelf crest pisolite. Its origin is enigmatic. We believe it could have formed by clastic sorting and/or by in-place pisoid growth, or possibly in other ways not now envisioned. We believe inverse grading formed as a primary feature of many pisolite units displaying this structure. However, post-depositional processes of enlargement of pisoids, or sorting processes such as agitation operating essentially in-place could also be factors. Inverse grading is known in some caliches, bauxites and in cave pearls, as noted by Dunham, but we discount the origin of inverse grading in shelf crest pisolite by replacement of a precursor sediment or rock through soil processes, on the basis of many geological observations cited elsewhere in this article.

Inverse grading in clastic rocks, carbonate or siliciclastic, is not common. The two closest analogies of which we are aware are provided by the mm to cm scale inverse grading reported by Clifton (1969) in sand and granule size beach and foreshore deposits, and in beach cusps as reported by Chafetz and Kocurek (1980). The latter involve skeletal material ranging up into pebble sizes in inverse graded units as thick as 30 cm with sand at the base. Lateral cusp migration forms the inverse graded units. These examples clearly indicate that local sorting processes can produce inverse grading. This may be the origin of the finer grained inverse graded units of sand and granule size, especially where particles are not coated (e.g., Fig. 4F) and it could be the origin of some inverse grading involving both finer and coarser coated grains. But occurrence of inverse graded units across a broad flat bottom of an intertepee depression, and the general absence of broken or abraded pisoids in many inverse graded units of pisolite are unfavorable for this explanation as the only, or perhaps the dominant mechanism. Extensive agitation of an assemblage of loose pisoids of a poorly sorted mixture can create an inverse grading, essentially in-place; possibly wave turbulence could create the needed agitation, but this seems an unlikely general explanation. Cave pearls, as noted by Dunham and earlier workers provide a good fabric analogy. Cave pearls occur in layers up to 20 cm thick. Some cave pearls occur in nests 10 – 30 cm across, and some display excellent inverse grading of well-rounded particles over thicknesses of 10 cm. Their formation seems to

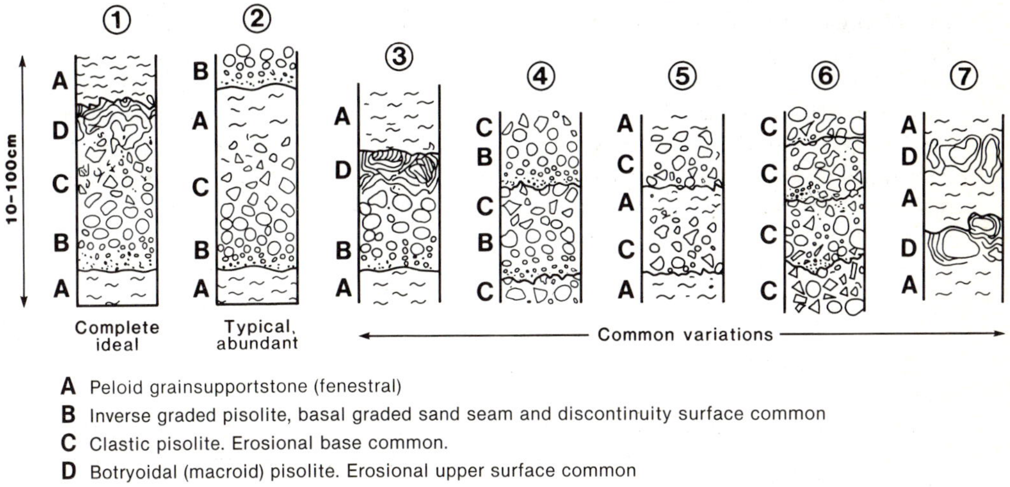

Fig. 10. Typical and common sedimentary cycles of shelf crest pisolite. The "ideal" cycle is conceptual, others are observed. Typical cycle (*2*) is termed the "walnutite cycle"

require subaqueous conditions most of the year and some amount of intermittent water agitation (flowing, or dropping into splash cups). The scale and perfection of inverse grading in some shelf crest pisolite, the absence of recognizable caverns or roofs of such, or of other speleothems, as well as the geologic setting, lead us to discount a cave pearl-cavern origin for extensive layers of inverse graded pisolite. However, the apparently requisite processes of subaqueous precipitation and agitation for cave pearl inverse grading need not be restricted to caverns, and could occur on the open sea floor, or in intertepee pools, or perhaps in small pockets formed by erosion (Fig. 5A), or related to non-tectonic fractures in an intertidal or at times supratidal environment such as we infer may have prevailed during deposition of botryoidal pisolite. Risacher and Eugster (1979) pointed out the presence of inverse grading calcitic macroids and pisoids in surface waters of a spring-fed playa and recognized some close similarities with the Permian shelf crest. Surprisingly, these pisoids occur at 4500 m in Bolivia. Inverse grading of accretionary pisolites can hardly signify a unique environment.

In-place accretionary precipitation in an aggregate of loose grains (Fig. 7C) could produce inverse grading. Mutual adjustment of enlarging pisoids during displacive growth would provide some vertical sorting; contemporary upward preferential enlargement (Fig. 6D) could be very important. Some agitation of loose pisoids could improve inverse grading and help prevent pisoid linkage that otherwise would stop growth of the expanding pisoid aggregate, but too much agitation would seem to destroy inverse grading. The common upward transition in the walnutite cycle from inverse to clastic pisolite (Fig. 10) suggests this.

What causes the selective precipitation of accretionary laminae on pisoids, rather than linking of pisoids? This vital selective placement of cement is perhaps related to factors other than agitation. We suggest the possibility that the favorable microchemical environment of precipitation was beneath an outer organic coating (algal or otherwise) on the particles.

Much research needs to be done to understand the inverse grading of the shelf crest pisolite. Investigation of details of the accretionary cements using both accretion stratigraphy and available geochemical techniques of trace elements, stable isotopes, and fluid inclusions seem to offer promise. We expect that clastic and precipitation processes logically associated with a nearly awash peritidal environment can account for the origin of the inverse grading of shelf crest pisolite, without recourse to soil-forming environments or caverns.

6 Sedimentary Sequences of Shelf Crest Pisolite

Genetic understanding of shelf crest pisolite is furthered by consideration of the stratigraphic sequences or cycles involving various types of pisolite as well as by analysis of the rock fabric and stratigraphic setting of the pisolite. Early in our field study certain vertical sequences of facies became apparent. For example, upward gradation from inverse graded pisolite into clastic pisolite was common. The converse was not. This section briefly summarizes common vertical sequences. None of them appears interpretable as a caliche profile, despite the wide variability of caliche-type soil profiles – a further reason to discount the caliche hypothesis.

Vertical sequences involving pisolite can be considered as consisting of four major lithofacies: (a) peloid grainsupportstone of either packstone and grainstone of sand to granule size range and commonly with irregular to laminated fenestral fabric, (b) inverse graded pisolite, the basal part of which commonly has inverse graded carbonate sand-granule sized particles directly overlying a discontinuity (erosion) surface, (c) clastic pisolite – grainstone commonly with an erosional base, and (d) botryoidal (macroid) pisolite – a coated grain, cement boundstone.

Figure 10 shows several sedimentary sequences or cycles recognized involving types of shelf crest pisolite. The most common upward sequence involving two types of pisolite and the finer-grained peloid carbonate is ABCA. This is generally developed in intertepee depressions. We termed this sequence the "walnutite" sequence (Pray and Esteban 1977) and interpret it as a shoaling-upward sequence (BCA) related to sedimentation following deepening, possibly an episodic shelf subsidence of a meter or so that submerged formerly intertidal flats (A). Other common sequences involving two types of pisolite are ABD [Fig. 10(3)] and BC [Fig. 10(4)]. Sequences involving but one type of pisolite are CA, CC and AD [Fig. 10(5–7)]. In many sequences the position of an erosional surface is part of the repetitive or cyclic pattern. Overall, the analysis of pisolite sequences supports many as shoaling-upward sedimentary sequences. We believe these developed in hypersaline water on the complex environment mosaic of a peritidal shelf crest, and not as rock developed by in-place soil processes from a precursor rock.

It is tempting to formulate an "idealized cycle", as shown in Fig. 10(1), involving all four facies, ABCD. Although all four facies rarely occur in direct upward progression in the field such might be inferred from the observed partial sequences, or the lateral facies of intertepee depressions and Walther's Law of Facies Succession. However the complexity of the depositional environment suggests caution. For example, the peloidal carbonates and botryoidal pisolite both occur as the highest (shoalest) of numerous sequences. Factors other than water depth are undoubtedly involved. Clastic pisolite need not be depth dependent; inverse graded and botryoidal pisolite probably involve water character as well as water depth and other factors.

In addition to the sequences involving direct upward successions of pisolite, a larger scale vertical pattern occurs in many sedimentary fills of the larger intertepee depressions. This broader scale pattern, which we termed "megasequences" (Pray and Esteban 1977) involves three stages. Each are a meter or so thick and are separated by distinct erosion surfaces. The lower or "Early" stage consists largely of peloid-fenestral carbonate and has little pisolite. The middle or "Intermediate" stage has abundant pisolite, mostly of clastic and inverse graded pisolite. The upper or "Late" stage is complex, may have any type of pisolite but botryoidal pisolite is well developed – and may be associated with more skeletal and/or burrowed peloidal carbonates. Water character and depositional topography apparently changed with time prior to the development of the major intraformational erosion surfaces that truncate most major pisolite-tepee carbonate stratal units.

7 Pisolite-Tepee Relationships, the Tepee Pool Environment

Tepees were named for sharply antiform, penecontemporaneous structures occurring in the Capitan back-reef (Adams and Frenzel 1950). Assereto and Kendall (1977) and Smith (1974a), who provides careful description and analysis of Guadalupe Mountains tepees, have recently reviewed tepee structures and their still somewhat enigmatic origin. Our interest here is in their relationship to shelf crest pisolite. Most recent workers have recognized that on the Guadalupian shelf tepees are abundant in most areas where pisolite (walnutite) is abundant (Fig. 2). The tepee ridges of up-buckled brecciated strata are commonly several dms to 1 – 2 m high; many can be deduced to enclose polygonal areas, the intertepee depressions, or as we refer to them, tepee pools. Some are as wide as 20 – 30 m. Tepees are variably truncated by intraformational erosion surfaces. Downward they die out abruptly into largely undisturbed strata. Some form stacked successions through 10 – 15 m of strata. Most well developed tepees show intense

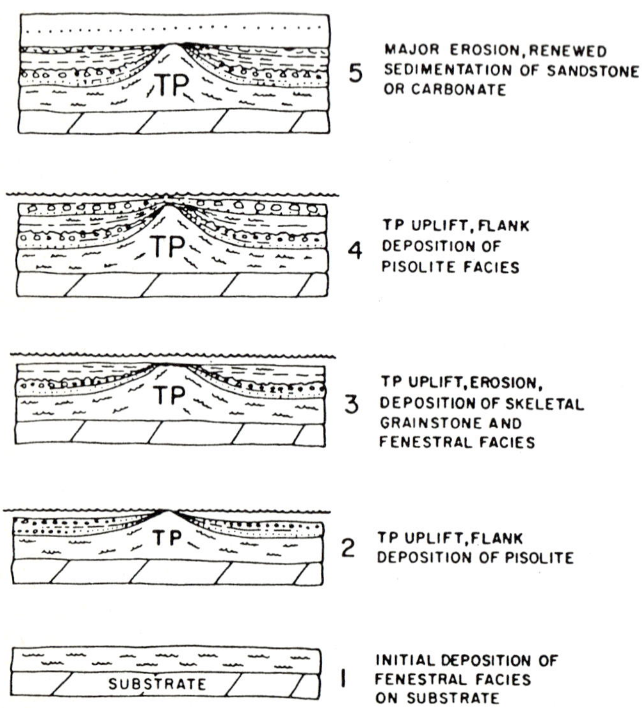

Fig. 11. Sequence of events interrelating deposition, tepee growth, and erosion. Episodic shelf edge subsidence (~ 1 m) provides depth for renewal of subaqueous sedimentation and subsequent shoaling upward cycles

fracturing that produced open sheet cracks in the ridge (core) area that diminish in size and frequency outward into the intertepee depression. Where tepees are associated with pisolite, there is an intense and pervasive cementation of the core area and of the sheet cracks of the flank and intertepee strata. The cement fillings (see Dunham 1968) typically consist of coarsely crystalline, palisade to fan arrays of 1 – 2 cm long crystals. Many blunt terminations suggest original aragonite (cf. Dunham 1969, 1972, Assereto and Folk 1976).

Pisolite formation is strongly but not invariably associated with the intertepee depressions, the tepee pool environment. However, the converse is not true, many well-developed tepees, of all scales, appear unrelated to pisolite. Salient aspects of the interrelationship of pisolite and tepees, reported in more detail in Pray and Esteban (1977) are summarized below (see also Figs. 8, 11):

1. Most tepees developed in peloidal carbonates in which pisoids are minor or absent and fenestrae are common. We infer that these sediments were deposited in shallow to temporarily emergent terrain and that cracking, probably by contraction processes, created initial polygons to depths of dms to a m or more (comparable to reports of some Persian Gulf intertidal areas reported by Evamy 1973). Subsequently expansion (caused by unknown conditions-cementation? and/or thermal expansion) favored tepee growth. Minor subsidence may have more deeply flooded intertepee depressions.

2. The thickest (up to 1 m) and best-developed pisolite and pisolite stratal sequences occur in intertepee depressions. Pisolite and associated peloidal carbonates (also locally quartz sandstone) thin toward tepee flanks, and commonly pinch out against the core (Figs. 8, 9, 11). This wedging is due more to depositional thinning, than to erosional truncation, as shown by the flattening upward fan-arrays of the sedimentary units, a geometry characteristic of flank accumulation on growing structures. Planes of fenestrae and directions of inverse grading show progressive tilting approaching the tepee crest. While much tepee growth preceded sedimentary filling of the intertepee depression, sedimentation also occurred penecontemporaneously with repeated episodes of tepee growth and with extensive carbonate precipitation of cement in the strata and associated sheet cracks.

3. Most pisolite sequences of intertepee depressions shoal upward. Pisolite facies also change toward the edges of intertepee depressions, a change particularly evident on the higher flanks (Fig. 8B). Toward the flanks inverse graded and clastic pisolite thin, and become more poorly sorted and matrix-rich. Vertical differences of 10 – 20 cm can markedly affect facies suggesting very shallow water on edges of tepee ponds. Inverse graded pisolite favors the deepest water, the botryoidal the shoalest and possibly temporarily more emergent areas (Fig. 8B).

4. Some tepee crests probably were in an intertidal to supratidal zone at times. The crests and, at times, strata of the flanks and intertepee depressions were subject to subaqueous to subaerial? erosion. At times all of the tepee area was inundated by more marine water than normally prevailed on the shelf crest, leading to deposition of packstones and grainstones of peloids and skeletal debris (albeit largely of restricted water biota), and but little pisolite.

In summary, much shelf crest pisolite was deposited in subaqueous environments of intertepee depressions at times of intensive precipitation of carbonate cement, both at the depositional interface and at very shallow depths below it. Explanations of this association can be several, some suggesting a genetic association, some not. Tepee ridges, acting as partial dikes, whether emergent or somewhat submerged, could have restricted or otherwise modified water circulation on or across the shelf crest area. This could have led to creating anomalous water (more hypersaline or at times, brackish?) in the intertepee depressions. Tepee ridges form permeable conduits for seasonal emergence of underground water in some of the marginal marine terrane in South Australia (Warren 1981) in which pisolite is forming[3]. However, emergent ground waters seem unlikely if the marginal mound hypothesis is correct, as we believe it is, for the Guadalupian shelf. Tepee ridges could be effective in inhibiting sediment transport across areas of tepees, could modify water turbulence, and could effectively trap coarse sediment. We infer that the intertepee pools themselves, whether tepee crests were emergent or not, were the likely site of formation of most of the clastic pisoids, as well as the pisoids of the inverse-graded and botryoidal pisolite. Somehow the conditions promoting pisoid formation were particularly favorable to the intertepee depressions. Were they particularly favorable to the development of organic sheaths on particles that may have promoted precipitation of lamellar cement beneath the surficial organic material? Such seems likely. It is probable that some pisolite of intertepee depressions formed from anomalous water that at times occupied the low parts of a peritidal area of tepees when there was not a genetic relationship between tepees and the water favorable for pisoid formation. Lastly, the field relationships show that some pisoids (especially those more micritic than the normal shelf crest pisoids) appear unrelated to tepees; and some strata with abundant tepees, of all scales, are not pisolitic.

8 Inferred Tidal Range

We infer that the shelf crest, during the time of tepee and pisolite development, was peritidal, involving deposition over a period of time at infratidal, intertidal, and supratidal environments. We infer that pisolite formation was primarily subaqueous. We favor shallow subtidal to intertidal depositional environments to supratidal environments for most pisolite, though the botryoidal facies may have formed on low supratidal flats, and at times tepee phenomena may have ponded water favoring pisolite formation in a supratidal position. But what was the tidal range? We do not know, but believe the shelf crest strata are most compatible with deposition in an area of a small tidal range, perhaps even at the scale of decimeters, rather than a meter or more. Three lines of evidence supporting but a limited tidal range are the following:

1. Paucity of channels within shelf crest facies. Neither field observation nor study of detailed air photographs of outcrop traces of the strata have revealed

3 Note added to page proof: Articles of Warren and Fergusson et al., in J Sediment Petrol (December 1982) show some partial analogies in a more terrestrial setting

many channels in the strata, and the observed few channels are generally less than 0.5 m in depth. We do not recognize major channels crossing the shelf crest.

2. Facies differences in pisolitic strata on the flanks of tepees appear to be sensitive to vertical differences of but a few decimeters.
3. Vertical successions of pisolite strata show abrupt differences in cementation (whether submarine, phreatic and/or late stage vadose cements) over only a cm or a few cm or dm of strata. This sensitivity to height is more compatible to a small than a large tidal range unless the site is supratidal.

9 Shelf Crest Pisolite and Erosion Surfaces

Erosion surfaces are numerous in the back-reef strata. Some are planar, others undulatory (± 1 m relief), some are smooth, others ragged; some are regionally persistent, some are only traceable within intertepee depressions or form a shallow (Fig. 9E) channel 1 or 2 m wide. Some surfaces separate major carbonate units from sandstone units, other occur within carbonate units. The location, nature, and relationship of these surfaces to the depositional and diagenetic processes of the substrate are critical for genetic interpretation of the pisolite and associated strata. In essence, the erosion surfaces testify to the very early and very near surface development of pisoids and cemented pisolite (including those showing vadose effects). Erosional surfaces, at all scales, truncate pre-existing pisolite; any associated alteration zones overprint pre-existing pisolite fabric.

In the shelf crest area, the best-developed incipient soil zone (first noted by Smith 1974b) is directly below the regionally persistent erosion surface separating the Yates Sandstone from the overlying Tansill Dolomite. Here small (< 1 cm) calcareous, non-laminated nodules replace some sandstone. The depth of alteration does not exceed 30 cm. This probable soil zone is not apparent below the Tansill in the outer shelf, evidence compatible with Dunham's marginal mound profile and our concept of but limited crestal emergence.

Some erosion surfaces overlying carbonate strata show thin alteration zones that are probably attributable to soil processes. The alteration is largely that of micritization, and, locally, reddening by iron oxides. The effects are only in the upper 10 – 20 cm recognizable soil products, including pisoids, have not developed, and the alteration overprints pre-existing pisolite fabrics, tepees, and associated strata (Fig. 9C, D). These zones may be the "case-hardening" noted by Kendall (1969). Smith (1974) stressed the large number of recognizable erosion surfaces in the cyclic back-reef strata. We agree, but do not find the extensive meter-thick zones of pisolitic and non-pisolitic calichified strata that he reported. However, some soil caliche, including micritic pisoids, has been found locally in older Permian strata of the region (see Sarg in Hileman and Mazzullo 1977, and Pray and Esteban 1977).

Many erosion surfaces that truncate pre-existing pisolitic and other carbonates show no evidence of alteration below the contact, and show sharp truncation of both pisoids and cement (Fig. 9B). Others, especially below local cm-thick seams of carbonate sand where infiltration can be observed, suggest erosion of uncemented pisoid assemblages. Ragged erosion surfaces with local relief (+1 m) occur in some tepee areas. These sharply truncate underlying cemented pisolite and finer-grained carbonates and are overlain by peloid skeletal (dasyclad-rich) grainstones, commonly with scattered cobble size intraclasts of the substrate. We detect no splash feature, organic borings, notches, or shoreline effects, as in coniatolite areas (Purser and Loreau 1973), and believe these are submarine erosion surfaces.

Two genetic associations we do note of pisolite to erosion surfaces are that botryoidal pisolite commonly terminates upward at an erosional surface, and that pisolite showing good inverse grading may occur in intertepee depressions above, not below, a distinct erosion surface (Fig. 8B). These are compatible with our interpretations that botryoidal pisolite is an occasionally awash phase of shoaling upward sediment sequences of the shelf crest, and that inverse graded pisolite forms subaqueously.

10 Comparisons to Other Coated Grains

Pisoid fabrics showing some similarity to those of shelf pisolite of the Capitan Reef Complex have been observed in Permian and other age rocks in many areas of the world (e.g., Bernoulli and Wagner 1971, Wardlaw and Reinson 1971, Burri et al. 1973, Catalano et al. 1974, Peryt and Piatowski 1977). None of these or other known occurrences provides a completely satisfactory analog, although each may show one or several of the fabrics of the shelf crest pisolite here described: such as multilaminated coatings, polygonal fit, directional elongations, perched inclusions, interpisoid linkage by pisoid laminae, botryoidal fabrics, syndepositional fracturing, and/or tepees. Unfortunately, many of these literature-stated analogies appear to have been made under the strong influence of the Permian "vadose-caliche" (soil origin) interpretation and they do not provide critical observations permitting reevaluation.

Many of the other pisolite localities (Esteban 1976) show an association with fenestral, non-skeletal, grain-supported carbonates and the pisolite occurs in a shelf environment between higher energy, more marine carbonates and facies of a restricted to evaporitic lagoon. Pisoids of most other localities appear to be more micrite than those of the Guadalupe shelf crest, and are more suggestive of an algal (oncoid) origin [e.g., Peryt and Piatkowski (1977) note the hybrid character by terming the pisoids as "algal-vadose"]. None of the pisolite shows a genetic relationship to major erosion surfaces, to continental facies, or to soil fabrics or products, so a true caliche origin seems unwarranted (Esteban 1976). The genetic association of pisolite to tepee structures has not been documented elsewhere although tepee structures occur in various pisolite localities. The gravity-cemented intraclastic breccias of the botryoidal facies have similarities to the larger-scale senile tepee breccias of northern Italy described by Assereto and Kendall (1977).

Several authors (Purser and Loreau 1973, Shinn 1973, Scholle and Kinsman 1974) have noted some similarities between the aragonite and high-Mg calcite supratidal encrustations of the Persian Gulf (coniatolite-pelagosite) and the shelf crest pisoids of the Guadalupe Mountains. The cortex of many pisoids in the two areas show similar features. Other features of the walnutite (such as inverse grading, tepee-pisoid association, non-skeletal nuclei, and low fossil) content have been reported in the Persian Gulf. Significantly the available descriptions of the coniatolites areas show no evidence of soil processes or fabric replacement of precursor rock. The botryoidal pisolite here described shows the closest analogies to the coniatolite facies. The coniatolites are associated with low coastal cliffs, stromatolitic algal mat flats, cross-bedded pisoid grainstones and to pisoids with abundant skeletal debris. These features are very rare in the Permian shelf crest pisolite.

Lang (1942), Newell et al. (1953) and Dunham (1969, 1972) cited the resemblance of cave pearls to the Permian pisoids. In addition Dunham stressed some striking fabric similarities; specifically inverse grading, polygonal fit, directional elongation and interpisoid linkage. The cave pearl analogy is especially tempting for small pockets of inverse graded pisolite within some botryoidal and clastic pisolite and does not require replacement of soil-originating processes. However, a cave (speleothem) origin for the Permian pisoids appears most unlikely. We do not recognize contemporaneous caverns, or roofs above the depositional surface of pisolite. The sheet crack openings cut existing pisolite when associated with it and other speleothems or karst features are absent (except some mm-cm scale features in botryoidal pisolite). The analogy illustrates the danger of assigning one environment to similar appearing products of carbonate precipitation. On the Permian shelf crest, most pisoids formed in water under the sky, and not under a rock roof on in soil.

Conclusions

1. Pisoids and pisolite of the shelf crest are not Permian soil (caliche), or cave pearls in karst caverns, or features formed prevailingly in vadose water regimes. Soil products (rare) are related to erosion surfaces truncating pisolite, caverns are not recognized and most vadose features are overprints on pre-existing isopachous (phreatic) accretionary fabrics.
2. The three major intergrading types of pisolites (Table 1) formed at the surface of the peritidal shelf crest. Much pisolite occurs in shoaling upwards sequences in shallow intertepee depressions. Clastic pisolite (predominant) could have been deposited anywhere in a peritidal environment. We favour subaqueous genesis for inverse graded pisolite and phreatic-vadose genesis for botryoidal pisolite.
3. The Introduction and Figs. 7 and 11 summarize many of our conclusions.

Acknowledgements. Cooperation of personnel of the U.S. National Park Service of Carlsbad Caverns National Park was essential to field research, and most helpful. All specimens are in the repository of the Department of Geology and Geophysics, University of Wisconsin, Madison. Valuable information and stimulation has been provided by University of Wisconsin graduate students conducting concurrent thesis research on Guadalupe Mountain shelf deposits. These are the following, cited together with the year of their thesis completion: Neal F. Hurley, M.S. 1978; Douglas Neese, M.S., 1978; Arthur H. Schwartz, M.S., 1981; J. Frederick Sarg, Ph.D., 1978; and Donald A. Yurewicz, Ph.D., 1976. The manuscript has been reviewed by Drs. John Hudson and T. E. Peryt, whose assistance is appreciated.

References

Adams JE, Frenzel HN (1950) Capitan Barrier Reef, Texas and New Mexico. J. Geol 58:289–312

Assereto R, Folk RL (1976) Brick-like texture and radial rays in Triassic pisolites of Lombardy, Italy: A clue to distinguish ancient aragonitic pisoids. Sed Geology 16:205–222

Assereto R, Kendall CGStC (1977) Nature, origin and classification of peritidal tepee structures and related breccias. Sedimentology 24:153–210

Bernoulli D, Wagner CW (1971) Subaerial diagenesis and fossil caliche deposits in the Calcare Massiccio Formation (Lower Jurassic, central Apennines, Italy). Neues Jahrb Geol Paläontol Abh 138:135–149

Bossellini A (1967) La tematica deposizionale della Dolomia Principale (Dolomiti e Prealpi Venete). Bol Soc Geol Ital 86:133–169

Burri PR, du Dresnay, Wagner CW (1973) Tepee structures and associated diagenetic features in intertidal carbonate sands. Sed Geol 9:221–228

Catalano R, D'Argenio B, Lo Cicero (1974) I ciclotemi triassic di Capo Rama (Monti di Palermo). Geol Romana 13:125–145

Clifton HE (1969) Beach lamination, nature and origin. Mar Geol 7:553–559

Chafetz HS, Kocurek G (1980) Genesis of inverse graded beds by beach cusp migration. Abstract with Program. Geol Soc Am 400

Dunham RJ (1965a) Vadose pisolite in the Capitan reef. Soc Econ Paleontol Mineral Permian Basin Sec Program, 10th Annl Mtg 13–14

Dunham RJ (1965b) Vadose pisolite in the Capitan reef (abs.). Am Ass Petrol Geol Bull 49:338

Dunham RJ (1968) Asymmetrically-filled veins in Capitan reef and their genetic similarity to vadose pisolite, New Mexico and Texas (abs.). Geol Soc Am Ann Mtg Program Spec Paper 121:83–84

Dunham RJ (1969) Vadose pisolite in the Capitan reef (Permian), New Mexico and Texas, In: Depositional environments in carbonate rocks – a symposium: Soc Econ Paleontol Mineral Permian Basin Sec Publication 14:182–191

Dunham RJ (1972) Capitan reef, New Mexico and Texas: facts and questions to aid interpretation and group discussion. Soc Econ Paleontol Mineral Permian Basin Sec Publication. 72-14, variously paged

Esteban M (1976) Vadose pisolite and caliche. Am Ass Petrol Geol Bull 60:2048–2057

Esteban M, Pray LC (1975) Subaqueous, syndepositional growth of in-place pisolite, Capitan reef complex (Permian), Guadalupe Mountains, New Mexico and West Texas (abs.). Geol Soc Am Abs with Programs 7:1068–1069

Esteban M, Pray LC (1976) Non-vadose origin of pisolitic facies, Capitan Reef Complex (Permian), Guadalupe Mountains, New Mexico and West Texas (abs.). Am Ass Petrol Geol Bull 60:670

Esteban M, Pray LC (1977) Origin of the pisolite facies of the shelf crest. In: Hileman ME, Mazzullo SJ (eds). Upper Guadalupian Facies. Permian Reef Complex, Guadalupe Mountains, New Mexico and West Texas. 1977 Field Conference Gdbk. Permian Basin Sec. Soc Econ Paleontol Mineral Permian Basin Sec Publication 77-16, 1:479–487

Evamy BD (1973) The precipitation of aragonite and its alteration to calcite on the Trucial Coast of the Persian Gulf. In: Purser BH (ed) The Persian Gulf. Springer, Berlin Heidelberg New York, pp 329–342

Hileman ME, Mazzullo SJ (eds) (1977) Upper Guadalupian Facies, Permian Reef Complex, Guadalupe Mountains, New Mexico and West Texas. Soc Econ Paleontol Mineral Permian Basin Sec Publication 77-16, 1:508

Hurley NS (1978) Facies Mosaic of the Lower Seven Rivers Formation (Permian) North McKittrick Canyon, Guadalupe Mountains, New Mexico. M.Sc. University of Wisconsin (Madison), 198 p

Jacka AD, Thomas CM, Beck RH, Williams KW, Harrison SC (1969) Guadalupian depositional cycles of the Delaware basin and northwest shelf. In: Elam JG (ed) Cyclic Sedimentation in the Permian Basin: West Texas Geol Soc Publication 69-56:152 – 196

Johnson JH (1942) Permian lime secreting algae from Guadalupe Mountains, New Mexico. Geol Soc Am Bull 53:195 – 226

Kendall CGStG (1969) An environmental reinterpretation of the Permian evaporite carbonate shelf sediments of the Guadalupe Mountains. Geol Soc Am Bull 80:2503 – 2526

King PB (1948) Geology of the southern Guadalupe Mountains. US Geol Survey Prof Paper 215, 183 p

Lang WB (1942) The Carlsbad dolomite and the pisolites of the Guadalupe Mountains of New Mexico (abs.). Am Ass Petrol Geol Bull 26:901

Loucks RG, Folk RL (1976) Fanlike rays of former aragonite in Permian Capitan Reef pisolite. J Sediment Petrol 46:483 – 485

Newell HG, Rigby JK, Fischer AG et al. (1953) The Permian Reef Complex of the Guadalupe Mountains Region, Texas and New Mexico. San Francisco, Freeman WH, 236 p

Peryt TM, Piatkowski TS (1977) Algae-vadose pisolites in the Zechstein Limestone (Upper Permian) of Northern Poland. Sed Geology 19:275 – 286

Pray LC, Esteban M (eds) (1977) Upper Guadalupian Facies, Permian Reef Complex, Guadalupe Mountains, New Mexico and West Texas. 1977 Field Conference Gdbk. Permian Basin Sec., Soc Econ Paleontol Mineral Permian Basin Sec Publication 17-16, 2:194

Purser BH, Loreau JP (1973) Aragonitic, supratidal encrustation on the Trucial Coast, Persian Gulf. In: Purser BH (ed) The Persian Gulf. Springer, Berlin Heidelberg New York, pp 343 – 376

Risacher F, Eugster HP (1979) Holocene pisoliths and encrustations associated with spring-fed surface pools, Pastos Grandes, Bolivia. Sedimentology 26:253 – 270

Ruedemann R (1929) Coralline algae, Guadalupe Mountains. Am Ass Petrol Geol Bull 13:1079 – 1080

Scholle PA, Kinsman DJJ (1974) Aragonitic and high-magnesian calcite caliche from the Persian Gulf – a modern analog for the Permian of Texas and New Mexico. J Sediment Petrol 44:904 – 906

Shinn EA (1973) Sedimentary accretion along the Leeward, SE coast of Qatar Peninsula, Persian Gulf. In: Purser BH (ed) The Persian Gulf. Springer, Berlin Heidelberg New York, pp 199 – 209

Smith DB (1974a) Origin of tepees in Upper Permian shelf carbonate rocks of Guadalupe Mountains, New Mexico. Am Ass Petrol Geol Bull 58:63 – 70

Smith DB (1974b) Sedimentation of upper Artesia (Guadalupian) cyclic shelf deposits of northern Guadalupe Mountains, New Mexico. Am Ass Petrol Geol Bull 58:1699 – 1730

Swineford A, Leonard AB, Frye JC (1958) Petrology of the Pliocene pisolitic limestone in the Great Plains. Kansas Geol Survey Bull 130:97 – 116

Thomas CM (1965) Origin of pisolites (abs). Am Ass Petrol Geol Bull 49:360

Thomas CM (1968) Vadose pisolites in the Guadalupe and Apache Mountains, West Texas. In: Silver BA (ed) Guidebook Permian Basin Section of the Soc Econ Paleontol Mineral Permian Basin Sec Publication 68-77:32 – 35

Tyrrell WW Jr (1969) Criteria useful in interpreting environments of unlike but time equivalent carbonate units (Tansill-Capitan-Lamar), Capitan Reef Complex, West Texas and New Mexico. In: Friedman GM (ed) Depositional Environments in Carbonate Rocks. Soc Econ Paleontol Mineral Spec Publication 14:80 – 97

Wardlaw NC, Reinson GE (1971) Carbonate and evaporite deposition and diagenesis, Middle Devonian Winnipegosis and Prairie Evaporite Formations of southcentral Saskatchewan. Am Ass Petrol Geol Bull 55:1759 – 1786

Warren JK (1981) Holocene tepees and stromatolites from South Australia and its paleohydrologic significance (abs). Am Ass Petrol Geol Bull 65:1005

V.7 Vadose Pisolites of the Tongying Formation (Upper Sinian System) in Southwest China

Cao Ruiji [1,2] and Xue Yaosong [2]

1 Introduction

The Sinian Tongying Formation is widespread in Southwest China, attaining a thickness of about one thousand meters (Fig. 1).

Many pisolitic beds, several cm to 20 cm thick, occur in of the B Submember of the Tongying Formation throughout an area of more than 130,000 km^2 (Fig. 2). The pisolites occur within cyclothems that are widely developed in the B Submember of the formation (Fig. 1). In the past, pisoids of the Tongying Formation were generally considered to be of cyanobacterial origin and were interpreted as having been produced in high energy shallow water environments within the subtidal to lower intertidal zone (Cao Ruiji et al. 1979, Tang Tianfu et al. 1980a, Yang Wanrong et al. 1978).

Recently, we examined many pisolites in the Tongying Formation and found that vadose marine processes were of primary importance for their origin. The present paper deals with the depositional environments and the paleogeographical relations of the pisolitic beds.

2 Characteristics of the Pisolitic Beds

The B Submember of the Tongying Formation is composed mainly of various cyanobacterial dolostones having a "botryoidal banded" (Fig. 3B) or "lacy" structure (Fig. 3A) according to the terminology in use by Chinese geologists. The former is made up of symmetrical botryoidal crusts, several up to about 10 cm wide and tens to hundreds of meters long; the latter consist of symmetrical striped pectinate crusts made up of needles of fibrous dolomite one to several mm wide. In the upper part of this submember, laminated dolostones usually occur which contain the remains of cyanobacteria, gas cavities, and remnants of originally gelatinous, calcified mucilage (Fig. 3C). It is inferred that the laminated dolostones were heaped up with a large number of smooth mats of mucilage secreted by cyanobacteria. Sometimes desiccation cracks are observed in them. In the middle and lower parts of this submember, another type of rock frequently

1 Formerly tsao Rui-chi

2 Nanking Institute of Geology and Paleontology, Academia Sinica, Chi-Ming-Ssu, Nanking, China

Coated Grains (ed. by T.M. Peryt)

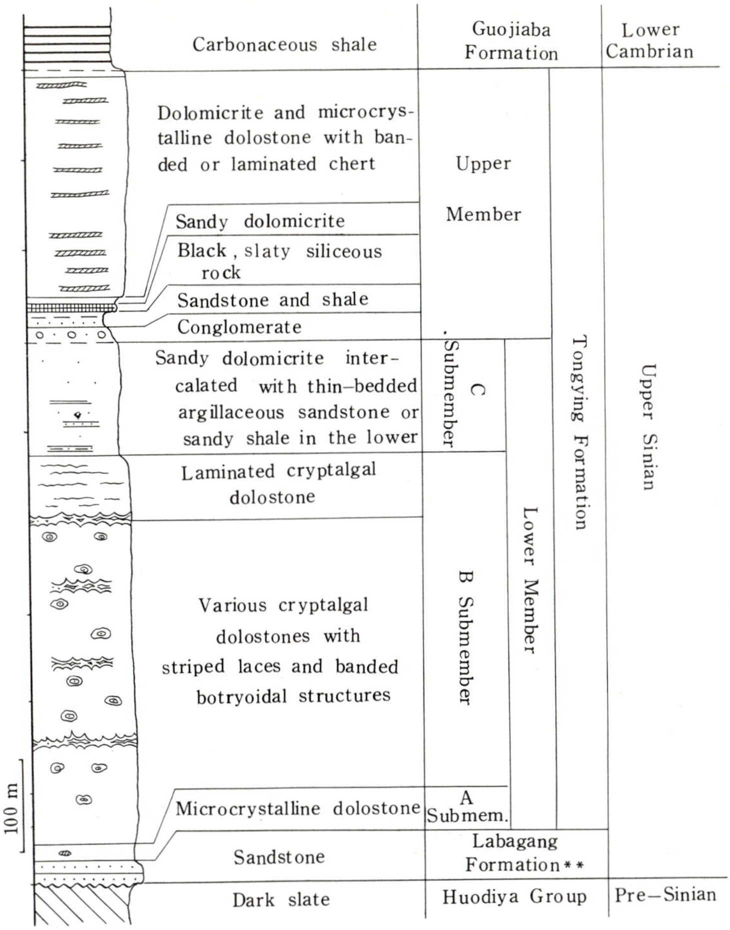

** Corresponding to the upper part of the Duoshantuo Formation

Fig. 1. Stratigraphic column of the Tongying Formation at Nanjiang, Sichan

occurs, viz., a porous carbonate with "spongy-like" structure (an irregular form of birdseye structure). These rocks contain many remains of cyanobacteria, calcified mucilage, clots and pisoids. It is inferred that they were formed from a pustular cyanobacterial mat.

The characteristics of the pisoid-bearing carbonate strata of this submember in the area studied may be summarized as follows: (1) it is homogenous and light in colour, containing >98% dolomite, <1% insoluble residue, and nearly no terrigenous clastic materials; (2) it contains abundant cyanobacterial remains but lacks stromatolitic structure (although sometimes stromatolites occur in the A Submember of the formation); (3) it is porours, and birdseye structures and botryoidal calcareous crusts and laces are commonly developed.

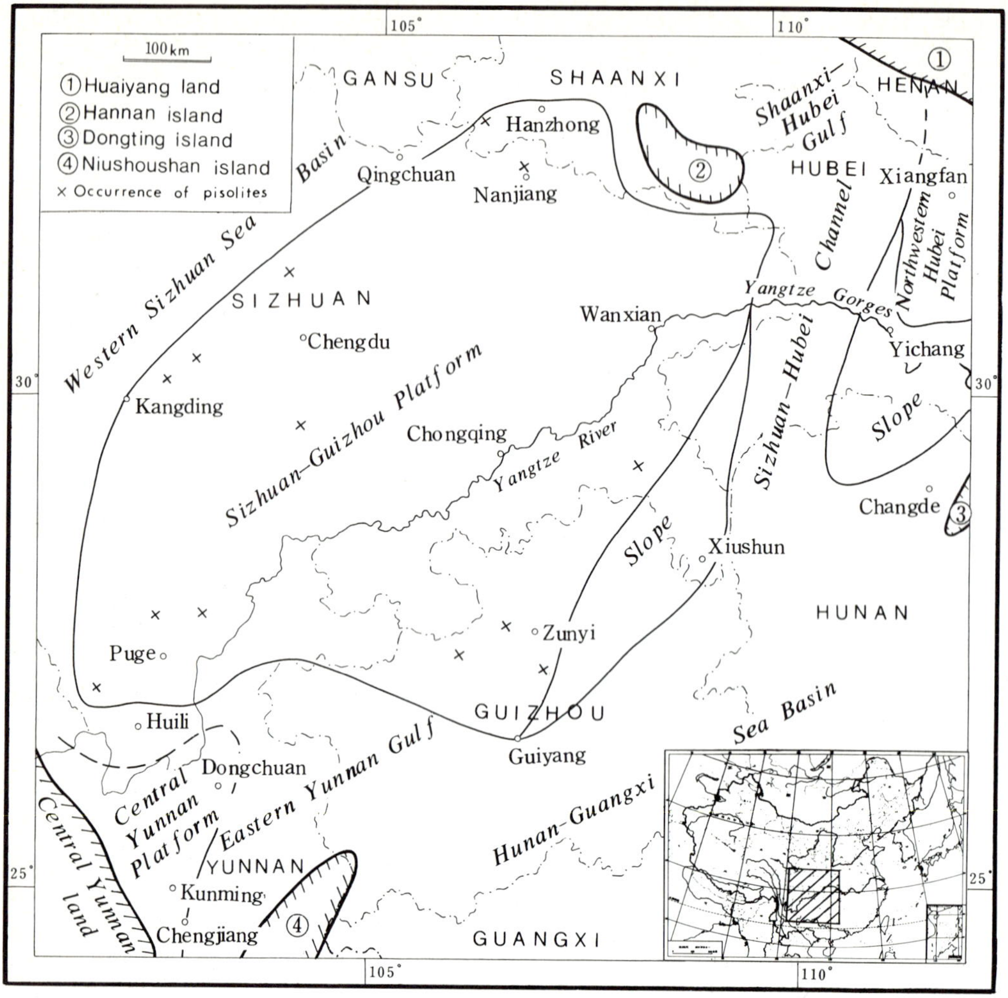

Fig. 2. Palaeogeographic map of the Tongying Formation, Southwest China. (After Tang Tianfu et al. 1980)

The pisoids of the Tongying Formation are spherical, oval or irregular in shape, 0.5 mm to 20 mm (predominantly 2 to 10 mm) in diameter. A few pisoids appear to be polygonal in shape. A typical Tongying pisoid is composed of two distinct parts, a relatively big nucleus is coated by concentric envelopes (alternating bright and dark laminae of variable thickness, Fig. 4). The nucleus is usually composed of cryptocrystalline dolomite containing cyanobacterial remains, and often contains dolomite spar (probably pseudomorphs after gypsum) and occasionally the lithoclasts of laminated cyanobacterial dolostone or dark dolomicrite.

On the base (or, rarely, on the top) of the pisoids the laminae are relatively wide and abundant, whereas on the opposite side they are closely spaced, a

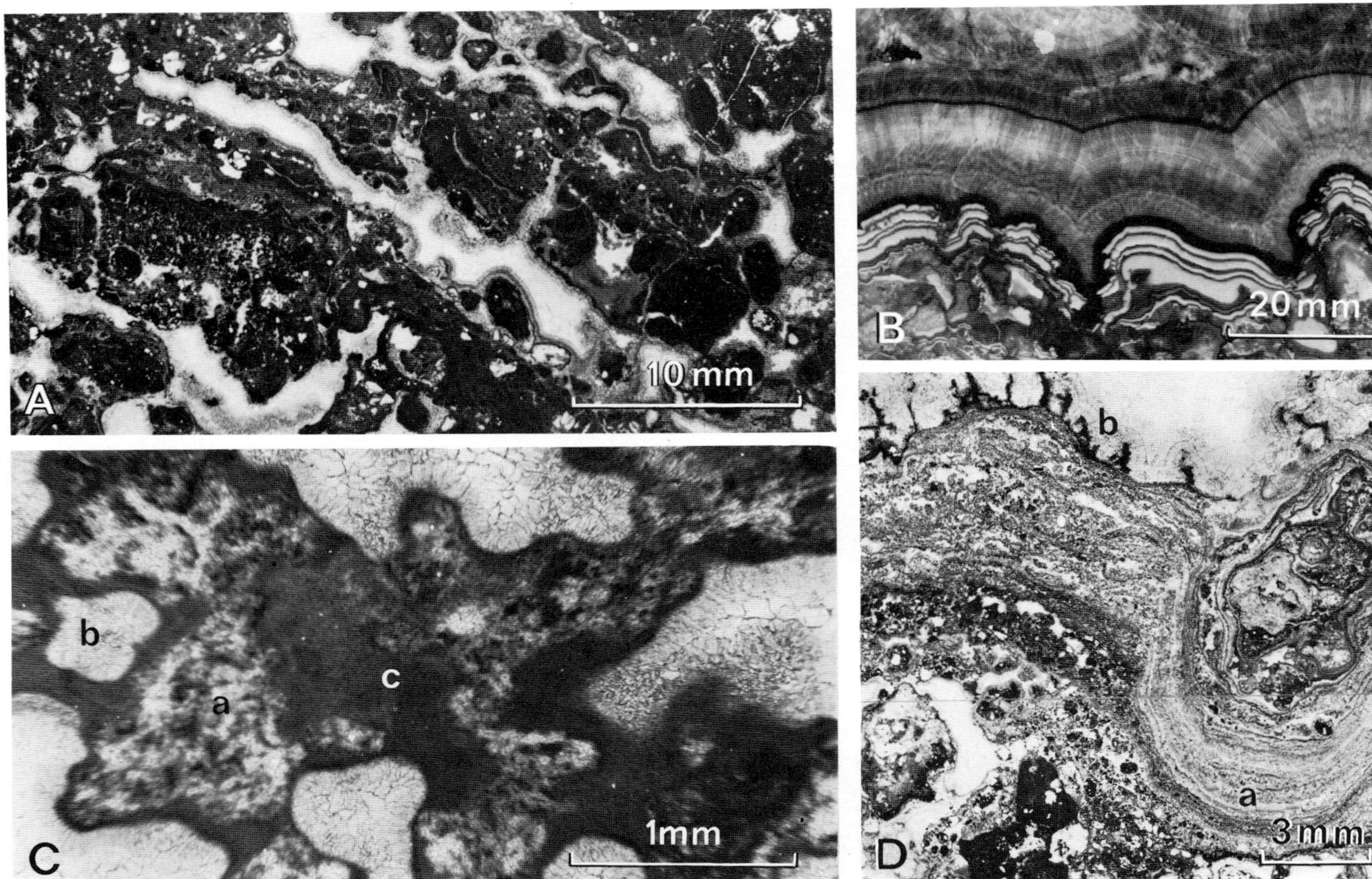

Fig. 3 A – D. Cyanobacterial carbonate microstructures in thin sections, the B Submember of the Lower Member of Tongying Formation at the Nanjiang county, Sichuan province. **A** Crusts made up of needles of fibrous dolomite. **B** The botryoidal crust. **C** The laminated dolostone containing *a* the remains of cyanobacteria, *b* gas cavities and *c* calcified mucilage. **D** Vadose laminae (*a*) and botryoidal (*b*) crust on the bottom of a large birdseye pore

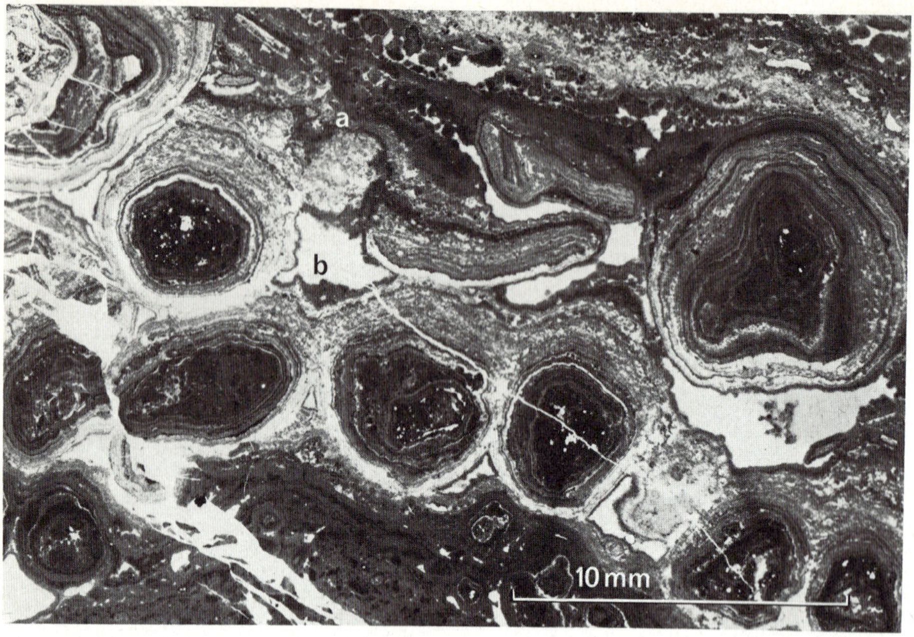

Fig. 4. The pisolitic dolostone, the B Submember of the Lower Member of Tongying Formation at Nanjiang. The outer concentric laminae of the pisoliths have a tendency to grow upwards; *a* the scouring surface and the fragments on it; *b* geotropic deposits on the bottom of the birdseye pore

feature obviously showing their direction of growth. Gravel-sized grains have never been observed in the pisolitic laminae. Dissolved traces and perched sediments (Fig. 5) occur on the upper side of some pisoids. In some cases, coatings may envelop several nuclei to form a compound pisoid (Fig. 6a), or several pisoids are surrounded by common coatings which are often semi-continuous (Fig. 6b). A few fragments of pisoids occur within the carbonate matrix, and it is occasionally observed that such fragments form the nucleus of other pisoids. All these features suggest that the pisoids were mostly formed in situ.

Chemical analyses provide evidence that the composition of the pisoids is similar to that of the botryoidal calcareous crust and of the porous carbonates.

According to these data, most of the pisoids of the formation can be interpreted as vadoids formed by a vadose marine process within porous pustular cyanobacterial mats and under hypersaline conditions during subaerial exposure. Some laminae of the pisoids are yellow in colour, containing occasional cyanobacterial remains. It is possible that these coatings were formed by vadose mucilage. Others composed of "clear" fine (10 μm) dolomite were probably formed by vadose water.

In addition, some spherical pisoids having smooth laminae which contain cyanobacterial remains, clots and silt-sized grains may be typical cyanobacterial oncoids.

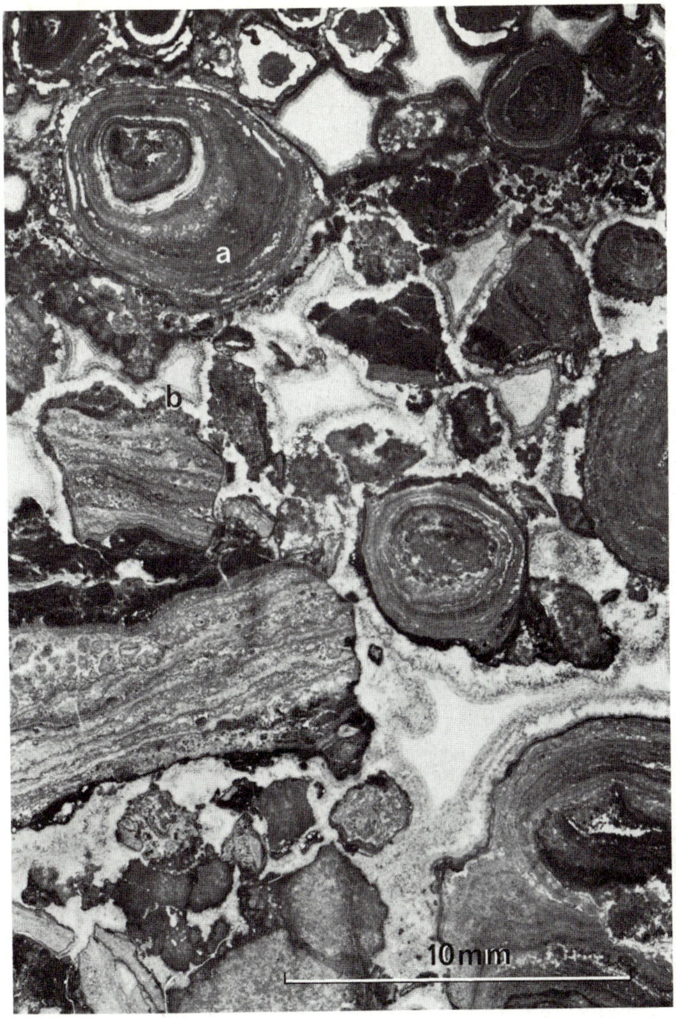

Fig. 5. Vadose pisoids growing downwards (*a*) and perched sediments on the fragments in pebble size (*b*). The B Submember of the Lower Member of Tongying Formation at Nanjiang

3 Environmental Analysis

The complete section of the B Submember is exposed in Nanjiang County, Sichuan Province, where it is 435 m thick, and composed almost entirely of rather pure cyanobacterial dolostones. The lower part consists of light gray cyanobacterial dolostones with a "spongy" structure (similar to irregular fenestral, peloid wackestone) and often intercalated with pisolitic beds, gray and blue-gray dolomicrite and cyanobacteria-lump dolostone; pisolitic beds are best developed here. The middle part of the submember consists mostly of cyano-

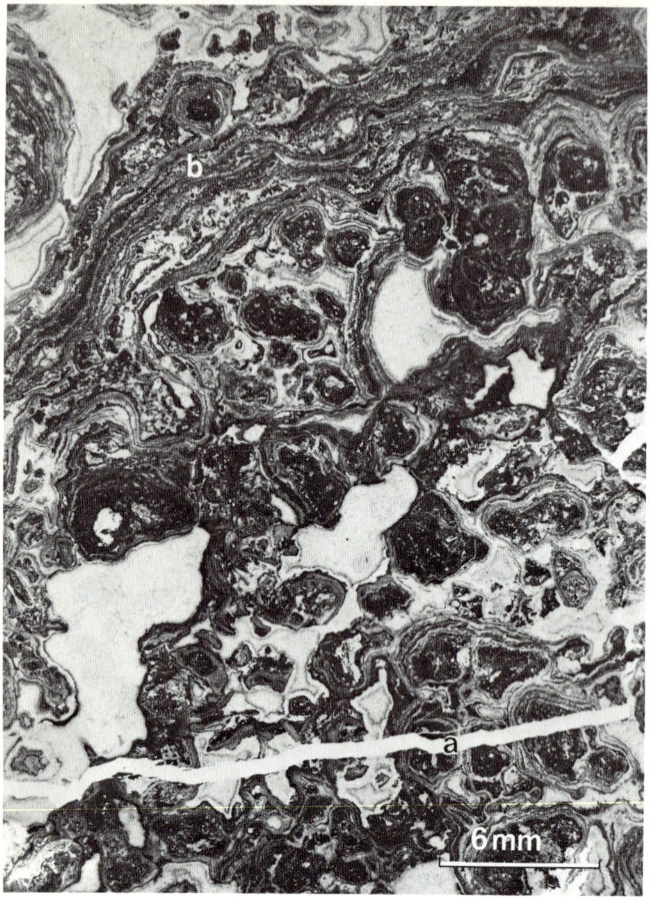

Fig. 6. Vadose pisolitic dolostone, the B Submember of the Lower Member of Tongying Formation at Nanjiang, Coatings may envelop several nuclei to form a compound pisoid (*a*) or several pisoids are surrounded by semi-continuous coatings (*b*)

bacterial dolostones with a "spongy" structure, intercalated with pisolitic beds. The upper part is composed chiefly of laminated cyanobacterial dolostones, occasionally intercalated with some pisolitic beds. Botryoidal calcareous crusts and birdseye structures are commonly developed in the B Submember. These botryoidal calcareous crusts are similar to Holocene and Pleistocene calcareous crusts described from northern Barbados (James 1972, Harrison 1977).

The pisoid-bearing carbonate strata contain no terrigenous clastic materials, but vadose traces (Fig. 7A, B) and desiccation cracks are often observed in them. Irregular desiccation cracks and pseudomorphs after gypsum can also be observed (Fig. 7C), even in the nuclei of some pisoids.

The cyanobacterial dolostones reach 400 m or more in thickness, suggesting that their deposition could in principle compensate for the subsidence of the depositional basement. However, evaporites, vadose traces and desiccation

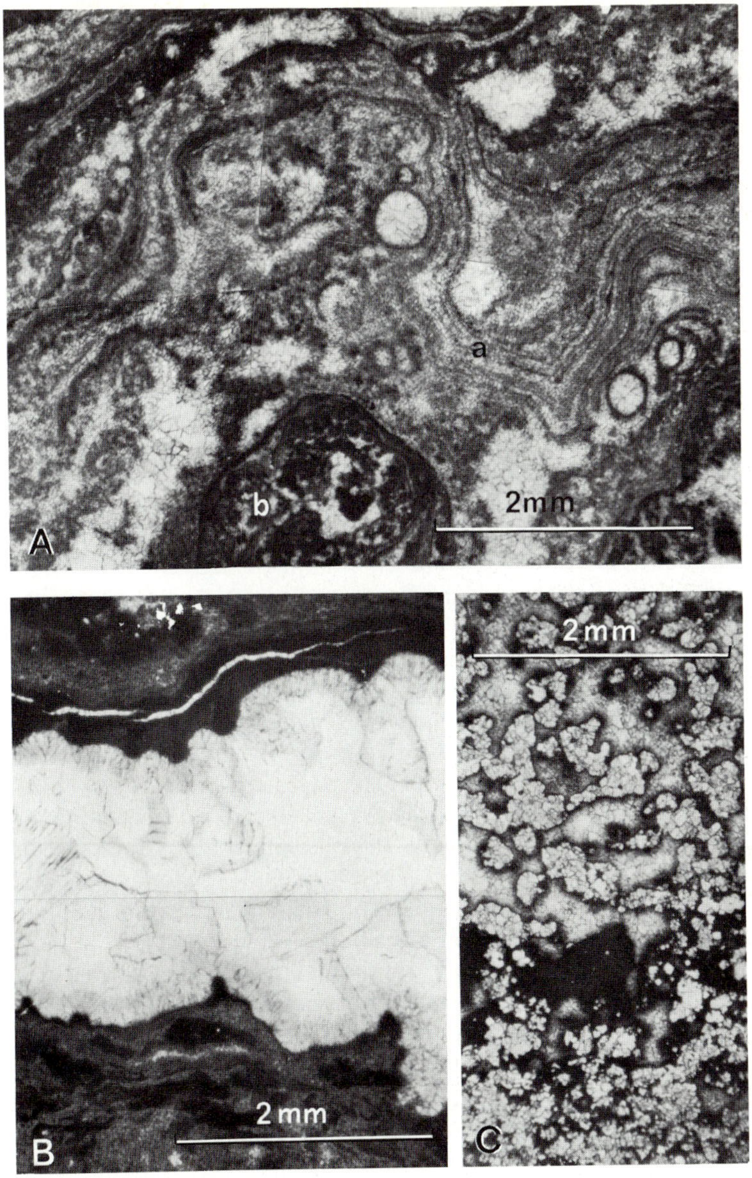

Fig. 7A–C. Microstructures of cryptalgal dolostones, the B Submember of the Lower Member of Tongying Formation at Nanjiang. **A** Vadose laminae (*a*) and vadose pisoid (*b*) in laminated cyanobacterial dolostone. **B** Vadose microstalactites and vadose microstalagmites in the top and bottom of a birdseye pore. **C** The framework of "spongy"dolostone was already replaced by spar dolomite, but two stages of cementation still are obvious in the framework porosities. The replacement took place before the cementation cracks occurring in these sediments show that the sediments were frequently exposed subaerially during deposition. It may thus be interpreted that cyanobacterial carbonate rocks at Nanjiang were formed in various environments (e.g., shallow subtidal, intertidal and supratidal). During subaerial exposure, the sediments were subjected to repeated evaporation and vadose influences during early diagenesis, and vadoids were formed. The environment in which the pisoids originated has some similarities with the Recent supratidal zone under hypersaline conditions from the Persian Gulf (Scholle and Kinsman 1974). A few of the pisoids of the Tongying Formation were probably formed due to cyanobacterial activity in shallow subtidal to lower intertidal environments.

4 Paleogeographic Distribution

Based on their study of the depositional environment of Sinian cyanobacterial carbonate rocks in south China, Tang Tianfu (1980a) consider that the cryptalgal dolostones of the Tongying Formation are only confined to structural platforms (such as the Sichuan-Guizhou, West Hubei, Nanjing-Chaoxian, and West Zhijiang Platforms). Vadose pisoids of the Tongying Formation discussed in the present paper are distributed within the Sichuan-Guizhou Platform (see Fig. 2).

In the Yangtze Gorges, mainly on the eastern side of the area here studied, the B Submember of the Tongying Formation is thinner and darker in colour (dark gray to black) and contains more silica argillaceous matter and calcite. It is made up of dolomicrite, asphaltic limestone and dark gray slaty marl with *Vendotaenia antiqua* Gnilovskaja. Botryoidal calcareous crusts, laces and birdseye structures have not been found in these carbonate rocks. Except for a few beds containing cyanobacterial pisoids intercalated with the thick dolomicrite, vadose pisoids have not been observed in this region either. These pisoids, 1 – 3 mm in diameter, are piled up closely, and associated normally with catagraphs (corresponding to them in size) that lack concentric lamination.

In eastern Yunnan Province, mainly on the southwest side of the area studied, the B Submember of the Tongying Formation is thinner and darker in colour too, but contains more terrigenous argillo-arenaceous material. It consists mainly of gray ring-like dolostone, laminated dolostone, and microcrystalline dolostone, in which evidence of vadose influence and evaporation has not been observed, therefore, only a few catagraphites occur, and pisoids are lacking.

The Yangtze Gorges were located in the paleodepression between two platforms, while eastern Yunnan Province was on the margin of the central Yunnan Platform to the northeast of central Yunnan land; in these regions, vadoids are rare.

5 Conclusions

1. The pisoids of the Tongying Formation in southwest China, previously interpreted as oncoids, were formed mostly by vadose processes. The concept of an algal nature for other Precambrian pisoids in China and elsewhere should be reexamined.
2. The presence of porous pustular cyanobacterial mats, and a hypersaline periodically subaerially exposed environment favoured the development of vadose pisolitic beds.
3. The vadose pisolitic beds are often associated with botryoidal calcareous crusts and laces, and therefore a relationship between these structures is presumed.
4. Vadose pisolitic beds seem to occur only on ancient shallow carbonate platforms which were frequently subaerially exposed.

Acknowledgements. Thanks are extended to M. Esteban, T. M. Peryt and R. Riding for their comments on the manuscript.

References

Cao Ruiji, Yang Wanrong, Zhang Junming, Yin Leiming, Li Zaiping, Zhao Wenjie (1979) The Sinian system in southwestern China. In: Nanjing institute of geology and palaeontology academia Sinica (ed) Carbonate biostratigraphy in southwestern China. Science Press, p 1 (in Chinese)

Harrison RS (1977) Caliche profiles: Indicators of near-surface subaerial diagenesis, Barbados, West Indies. Bull Can Petrol Geol 25:123 – 173

James NP (1972) Holocene and Pleistocene calcareous crust (caliche) profiles: criteria for subaerial exposure. J Sediment Petrol 42:817 – 836

Scholle PA, Kinsman DJJ (1974) Aragonitic and high-Mg calcite caliche from the Persian Gulf – A modern analog for the Permian of Texas and New Mexico. J Sediment Petrol 44(3):904 – 916

Tang Tianfu, Xue Yaoson, Yu Congliu (1980a) Characters and environmental significance of Sinian cryptalgal carbonate rocks in South China. Mon J Sci China 25:866 – 871

Tang Tianfu, Zhang Junming, Yang Wanrong, Qiu Jinyu, Zhou Yangkang (1980b) The depositional environments of the carbonate rocks and the phosphorites from Upper Sinian in S.W. China. J Stratigr 4:264 – 272 (in Chinese)

Yang Wanrong, Tang Tianfu, Zhang Junming, Qiu Jinyu, Zhou Yangkang (1978) On carbonate rocks of the Sinian Tongying Formation in Southwest China with respect to their characteristics and reservoir behavior. Geochimica 4:247 – 260 (in Chinese)

V.8 A Proterozoic Calcrete in the Amos Formation, McArthur Group, Northern Territory, Australia

M. D. Muir [1]

1 Introduction

The rocks of the McArthur Group, which includes the Amos Formation (Fig. 1), are mainly carbonates which have many features typical of deposition under shallow water, frequently hypersaline, and emergent conditions. A sabkha model has been proposed for the Mallapunyah Formation and parts of the Amelia Dolomite (Muir 1979a), and a model based on the ephemeral lakes of the South Australian Coorong has been suggested for the Yalco Formation (Muir et al. 1980). Further evidence for shallow water and emergent conditions has been described for most formations in the measured sections summarised by Jackson et al. (1978) and Plumb (1979). The persistence of such uniform depositional conditions for most of the time required to form nearly 6 km of sedimentary rocks is remarkable. The rocks of the McArthur Group extend from the Queensland border to the northern coast of Arnhem Land. The sediments can be interpreted with certainty as shallow water and emergent at least as far north as Roper River. Poor preservation in Arnhem Land makes environmental interpretations more difficult there.

2 Distribution and Stratigraphy

The rugged karstic outcrop (Fig. 2) of the Amos Formation west of the Abner Range is a striking feature of the scenery on either side of the Tablelands Highway. The heavily jointed brownish-grey dolostones are overlain with strong unconformity by the basal sandstone of the Balbirini Dolomite, and form the only carbonate basement rocks to the Balbirini Dolomite. Elsewhere, the Balbirini Dolomite and its lateral equivalents (Muir 1979b) rest upon very silicified older beds at different stratigraphic levels in the Batten and Umbolooga sub-groups. The Amos Formation has not been observed on the eastern and southern margins of the Abner Range. In fact, it is not recognised anywhere else on the Bauhinia Downs 1 : 250,000 sheet area so that most previous identifications of this formation are incorrect. Plumb and Brown (1973) redefined the Amos

1 C.R.A. Exploration Pty. Ltd., P.O. Box 656, Fyshwick, A.C.T. 2609, Australia

Coated Grains (ed. by T. M. Peryt)

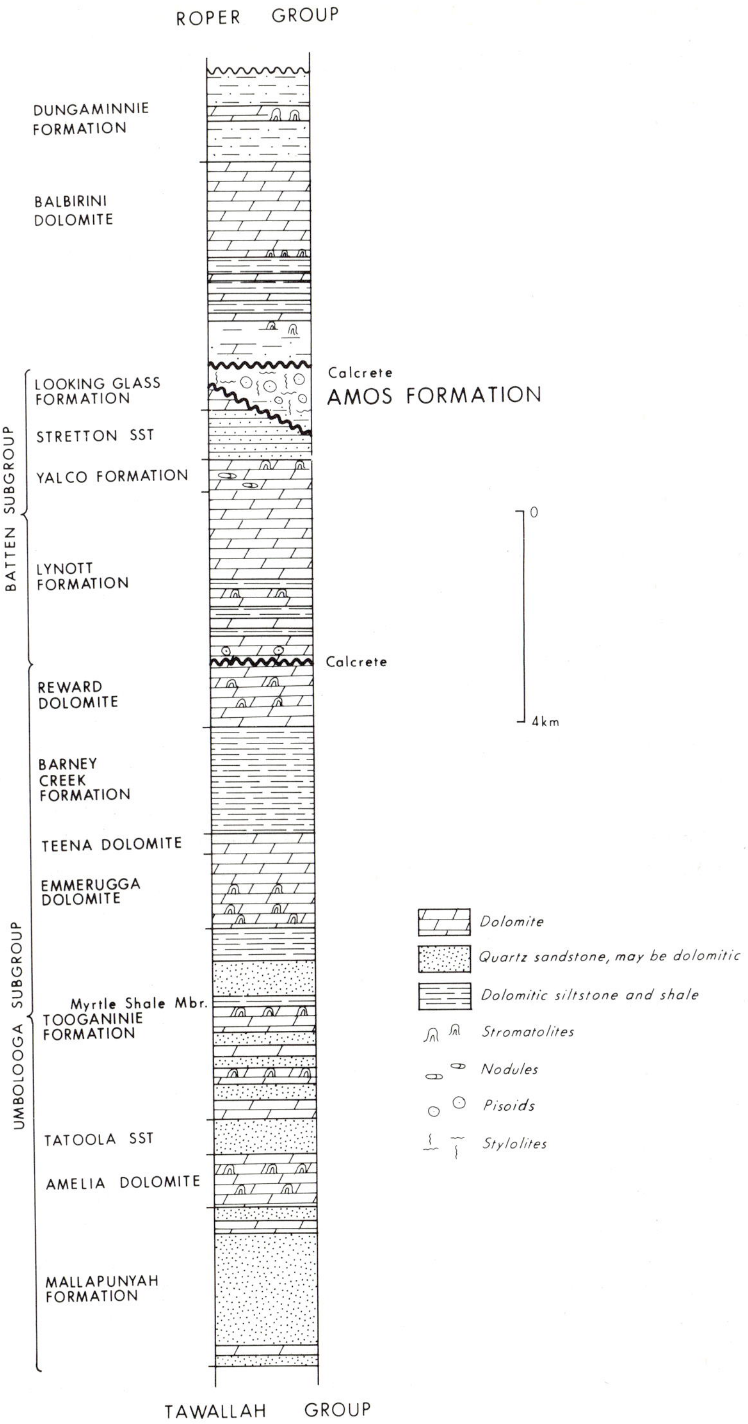

Fig. 1. Stratigraphic column of the McArthur Group showing formations mentioned in the text. The positions of calcretes which have been identified are marked

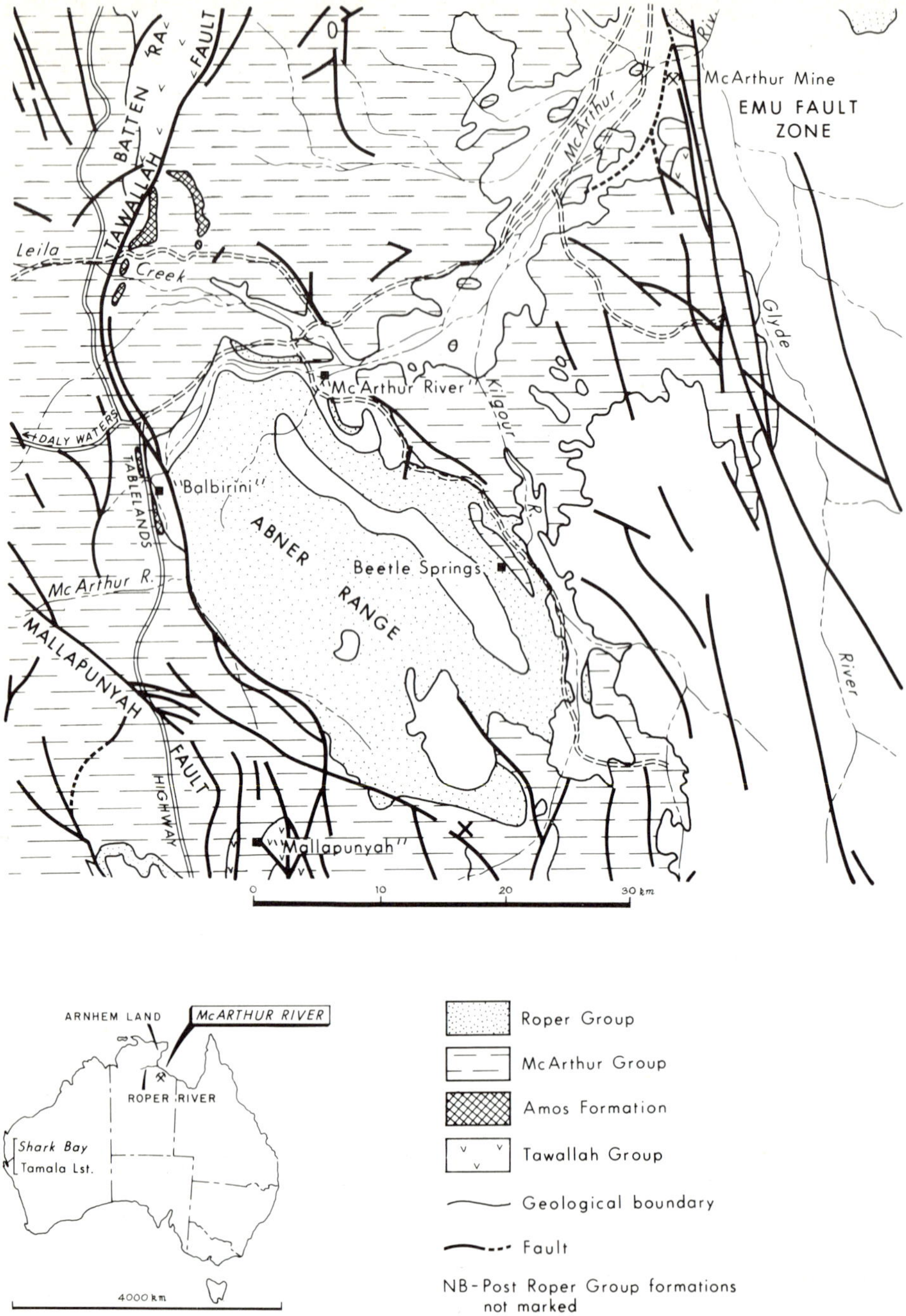

Fig. 2. Locality map

Formation, and separated it from the Myrtle Shale Member of the Tooganinie Formation with which it had been previously confused (Smith 1964). However, Plumb and Brown's definition (which has never been formalised) included the basal 63 m of the Balbirini Dolomite, above the unconformity which they did not recognise. This interval of 63 m is not now regarded as part of the Amos Formation (Jackson et al. 1978) and the Amos Formation will formally be redefined in a later publication.

The basal dolomitic siltstone of the Amos Formation overlies the Stretton Sandstone west of the Abner Range, but north of Leila Creek, it overlies the Looking Glass Formation. There is no angular unconformity between the Amos Formation and the underlying formations of the Batten Sub-Group, but its basal beds clearly cut across the stratigraphy and indicate a disconformity.

Previous work has demonstrated the large areal extent of the sub-Balbirini Dolomite unconformity (Muir 1979b, c), and its stratigraphic importance was first realised by Jackson et al. (1978). The sub-Amos Formation disconformity was previously unrecognised.

The exact age of the Amos Formation is not known. However, the Barney Creek Formation (Fig. 1) has recently been dated at 1,680 m.y. old (Page 1979, pers. comm.). This is somewhat older than previous estimates (Plumb and Derrick 1976), and compares well with the 1660 m.y. age calculated by Page (1979) for the Urquhart Shale at Mount Isa, Queensland. Therefore, the Amos Formation would be somewhat younger than this.

3 Lithological Description

The lithological description given here has been modified from that given by Plumb and Brown (1973). The most complete measured section through the Amos Formation (3 km west of the Abner Range, south of Balbirini Homestead, modified after Jackson et al. 1978) is presented as Fig. 3. The basal reddish-purple, ferruginous, micaceous, dolomitic fine sandstone and siltstone appears to follow the Stretton Sandstone conformably at this locality, with identical dip, but poor exposure prevents observation of any minor breaks in sedimentation. The Stretton Sandstone is highly silicified beneath the Amos Formation, as is the Looking Glass Formation north of Leila Creek (Fig. 2) where it directly underlies the Amos Formation.

Above these basal beds, the Amos Formation consists of alternations of thin bedded dololutite with thin bedded reddish-purple and purple dolomitic siltstone and shale (Fig. 3; 31 – 35 m interval). This is followed by a sequence of thick bedded dolarenite (45 – 60 m) with locally abundant well-rounded quartz and chert grains up to 3 mm in diameter. Carbonate flakes up to 2 cm by 3 mm are common throughout the sequence. A fine-grained white chert bed, 80 cm thick, forms a prominent marker in this dolarenite sequence. It is overlain by two cherty pinkish-orange potassium-rich mudstones (10 and 15 cm thick, respectively). Similar beds in other parts of the McArthur Group contain tuffaceous fragments and are used as marker beds ("tuffites" Jackson et al. 1978, Plumb

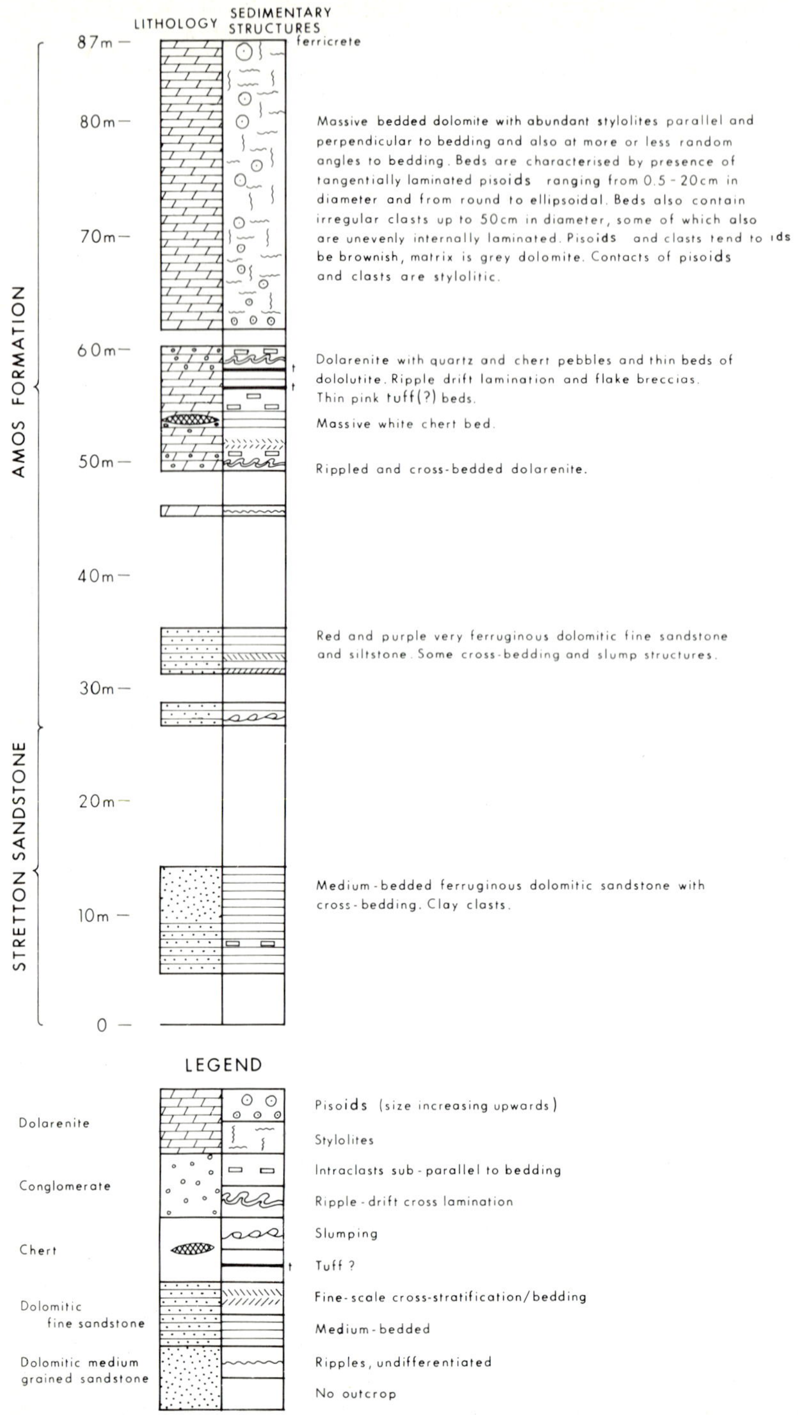

Fig. 3. Measured section Mallapunyah 77/03 for the Amos Formation. (Modified after Jackson et al. 1978)

and Brown 1973). The Amos Formation mudstones, however, contain no evidence for a tuffaceous origin, and are here regarded as potassium metasomatised mudstones, similar to those described from the Mallapunyah Formation (Muir 1979a). The source of the potassium is less obviously evaporitic than the source for the Mallapunyah Formation, but pervasive potassium metasomatism is a feature of some sediments of the McArthur Group.

Above the potassium-rich mudstones, bedding in the Amos Formation dolarenite becomes indistinct and the first pisoids appear. These are small (0.5 – 3 cm in diameter) and become more abundant and larger upwards. The largest pisoids occur at the top of the dolarenite, and these reach 25 cm in diameter. They occur in a matrix of dolarenite with a grain-size of about 1 mm. Apart from the pisoids, the other most striking feature of the upper part of the unit is the abundance of stylolites. The stylolites may be parallel to bedding, vertical, or at almost any angle. Many enclose clasts of laminated sediment, and pisoids commonly adjoin stylolites.

The uppermost part of the Amos Formation consists of a thin (2 cm maximum) fine grained ferruginous rock. This is directly overlain by the fine grained red micaceous sandstone of the basal Balbirini Dolomite.

North of Leila Creek, the Looking Glass Formation is overlain by red dolomitic mudstones of the Amos Formation identical with those which overlie the Stretton Sandstone at Balbirini. Poor exposure did not permit the measuring of a section but the pisolitic upper parts of the Amos Formation are reasonably exposed and clearly identifiable. It is capped by a thin ferricrete surface and overlain by the red micaceous fine grained sandstone of the basal Balbirini Dolomite. Between the two localities, the Amos Formation in some places appears to occupy the same stratigraphic level as the Looking Glass Formation and they were at first believed to be lateral (facies) equivalents. However, recognition of the disconformity at Balbirini suggests that this apparent lateral equivalence is better explained by the Amos Formation filling in hollows from which the Looking Glass Formation had been previously eroded.

4 Petrological Description

Samples from the upper 25 m of the Amos Formation consist of scattered dolomitic piso- and macro-ooids in a peloidal dolomitic matrix. Stylolites abound, but appear to by-pass piso-ooids rather than cut through them.

The piso-ooids consist of a finely laminated outer zone with a micritic fabric, over a core which is generally sandy dolarenite (Fig. 4).

The matrix consists of micro-ooids, with thin overgrowths, set in a micritic to sparry cement (Fig. 5). Etched and embayed sand grains also occur mixed with the peloids (Fig. 5). Some grains are quartz, and others carbonate sand. The detrital grains appear to "float" in the carbonate matrix (see Watts 1978).

All the rock samples examined are very ferruginous, and fine-grained haematite is common in all thin sections. Lamination in the Amos Formation is a product of colour banding from iron-rich and iron-poor layers. In the stylolites, the cumulate is nearly all haematite. The laminated sediment in the stylolite-bound clasts has some of the overall features of flat laminated stromatolites, but the fabric of these laminations contains abundant peloids, even in columnar forms (Fig. 6).

In pisolitic overgrowths, micro-unconformities are abundant (see Fig. 7). The angle between the microunconformable layers is variable from sample to sample. The variations in the angles depend on the orientation of earlier laminae which follow microtopography over grains, and into hollows. These features are most easily seen in thin sections.

5 Comparison with Other Calcretes

5.1 Tamala Limestone

The Holocene Tamala Limestone is a calcrete (see Read, 1976) from Western Australia. Samples were collected by the author and by Dr. P. E. Playford (GSWA) from localities south and west of Shark Bay. The Tamala Limestone cements Pleistocene dune sands and is commonly pisolitic. It is commonly ferruginous and stylolitised, and has enveloped sand grains from the dunes within its carbonate cements. Figure 8 shows the margin of a piso-vadoid from the Tamala Limestone. It compares in many respects with the piso-ooid from the Amos Formation shown in Fig. 4. Calcrete ooids (Fig. 9) and embayed sand grains (Fig. 10) from the Tamala Limestone are also identical to those from the Amos Formation (Figs. 5 and 11).

From the literature, in particular Read (1976), it appears that the massive dolomite of the Amos Formation fulfils most of the criteria used to differentiate calcretes from stromatolitic and oncolitic sediments. These distinguishing criteria which can be observed in the Amos Formation include the following (after Read 1976).

1. Calcretes develop in distinctive profiles characterised by vertical sequences of structures and fabrics. They show vadose diagenetic effects and in situ brecciation. The profile is formed by weathering processes in the vadose zone. Such weathering is inferred for the Amos Formation and the sequence from sandy dolarenite, through fractured and stylolitised recrystallised dolostones, to the pisolitic and stylolitic top of the formation (see Fig. 3) has a great many features in common with the calcrete profile at Shark Bay (Read 1976).
2. Modern calcretes follow microtopography. This is difficult to demonstrate except on a microscopic scale for the Amos Formation.
3. Modern calcretes contain micro-unconformities. These occur abundantly in the Amos Formation (Fig. 7 top right-hand corner).
4. Lamination in the Amos Formation calcretes is due to colour banding (iron staining), as it is in modern calcretes.

   Thus the Amos Formation appears to fit the majority of the criteria used to distinguish calcretes from stromatolitic laminations.

5.2 Other Calcretes

Although the bulk of the literature deals with recently formed calcretes, geologically more ancient examples are known (Peryt and Piatkowski 1977, with references, and papers in this volume).

Many of these Phanerozoic calcretes occur in association with evaporitic sequences (i.e., in particular the Upper Permian, Middle Devonian and Silurian examples), and either overlie evaporitic sequences or have developed marginally to evaporite basins. The rocks that immediately underlie the Amos Formation (Looking Glass Formation and Stretton Sandstone) are not evaporitic, but the Yalco Formation was deposited under hypersaline conditions (Muir et al. 1980) and the underlying Lynott Formation contains abundant evaporite traces. The Amos Formation was deposited on a silicified land surface of Batten Subgroup rocks, and filled minor hollows in this surface with carbonate detritus which may have been derived in part from older carbonate rocks by aeolian transport or by water transport (presumably in times of periodic flooding), or it may have been derived from inorganically formed carbonate crusts (sinter) [see Riding 1979 for a more recent example of this]. Some of the dolarenites near the top of the Amos Formation show no bedding structures and may have been unconsolidated carbonate sands similar to the Pleistocene dune sands of Western Australia in which the Tamala Limestone has developed.

The weight of the evidence appears to show convincingly that the Amos Formation can be interpreted as a calcrete. The conditions of deposition in the McArthur Group are such that it would be surprising if other calcretes were not discovered. In the Yalco Formation, which is interpreted as having formed in ephemeral hypersaline lakes (Muir et al. 1980), calcrete cement textures occur in the measured section near Ryan's Bend (MS Borroloola 77/1, Jackson et al. 1978, Muir, unpubl. data). A more interesting example, however, occurs at the base of the Lynott Formation west of the Kilgour River (locality X in Fig. 2). Here laminar calcrete, pisoids, and ooids (Fig. 12) occur in a coarse-grained gently dipping dolarenite at the base of the Lynott Formation. It overlies the erosional top of the Reward Dolomite which here is vertical and has undergone extensive karstic weathering.

6 Mineralogy

The mineralogy of the Amos Formation and Lynott Formation calcretes varies from dolomite to magnesian calcite. All the carbonates in the McArthur Group have suffered some post-Cretaceous recrystallisation, and may have been recrystallised during earlier pre-Lower Cambrian weathering episodes, or during diagenesis. Despite this, fine structures reflecting the original pedogenesis are well preserved in both calcretes. Most of the sediments of the McArthur Group are dolomite or magnesian calcite, with minor proportions of siderite, pistomesite, and magnesite. It is not surprising, therefore, that the pedogenic deposits of the Amos and Lynott Formations are magnesium-rich. Scholle and

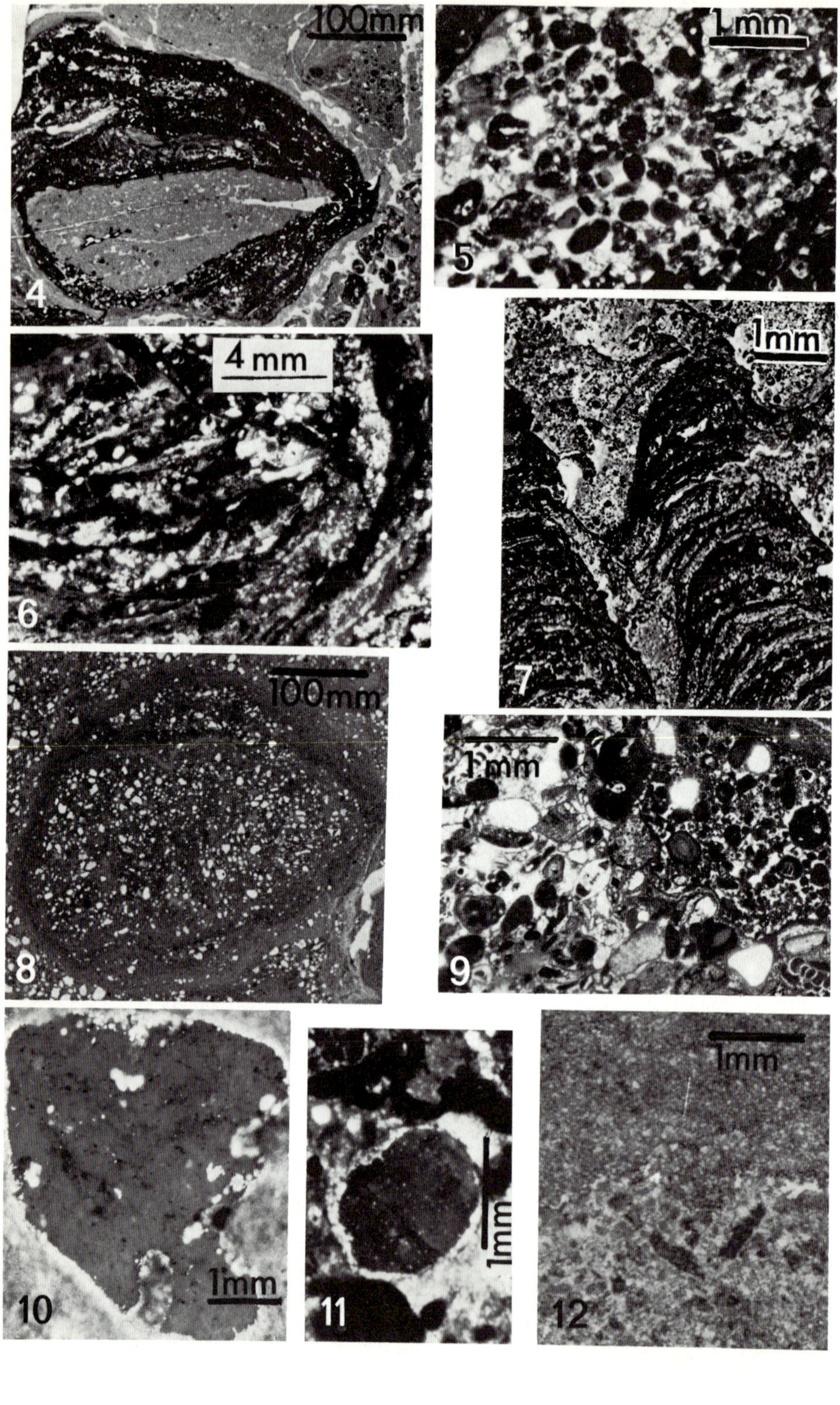
100mm
4
1mm
5
4 mm
6
1mm
7
100mm
8
1mm
9
1mm
10
1mm
11
1mm
12

Kinsman (1974) have shown that where groundwaters are hypersaline, high magnesian calcitic and dolomitic calcretes are produced in the vadose zone. There is abundant evidence in the McArthur Group for hypersaline groundwater activity (Walker et al. 1977, Muir 1979a, Muir et al. 1980) and it is reasonable to assume that hypersaline groundwater gave rise to the Amos and Lynott Formation calcretes. The well-preserved fabric of these Proterozoic calcretes is also indicative of primary precipitation as dolomite or high magnesian calcite.

7 Subaerial Exposure

Sedimentation in the McArthur Basin was controlled by approximately N-S trending faults which were active during deposition. The faults formed a number of grabens and half grabens and in many cases the upthrown block was emergent and indeed may have had considerable relief (Walker et al. 1978) during deposition of certain formations. It is likely that the vadose activity which gave rise to the Amos Formation calcrete was produced following tectonic uplift.

8 Conclusions

It thus appears that calcretes do occur in the Proterozoic rocks of the McArthur Group. Further studies and re-examination of sequences, particularly where "oncolites", "pellets", "peloids", or "ooids" have been described may indicate a large number of ancient calcretes, which have previously been unidentified.

Ancient calcretes have a number of important implications:

1. They form under very specific environmental conditions, where pre-existing rocks, or unconsolidated sands are elevated to the emergent zone and subjected to vadose diagenesis.
2. This implies sub-aerial exposure of the older sediments, and indicates unconformity or disconformity below the calcrete profile.
3. In the case of the Lynott Formation calcrete, it appears to form an unbroken sequence with the overlying sedimentary rocks, but the Amos Formation calcrete underlies a major regional unconformity.

Fig. 4. Thin section through a pisolite from the Amos Formation, north of Leila Creek

Fig. 5. Calcrete ooids and etched and embayed sand grains "floating" in a microsparite cement. Amos Formation, Leila Creek

Fig. 6. Peloids in inter-column spaces in columnar calcrete. Amos Formation, Leila Creek

Fig. 7. Microunconformity in pisolite. Amos Formation, Leila Creek

Fig. 8. Pisoid from the Holocene Tamala Limestone, Shark Bay, Western Australia

Fig. 9. Calcrete ooids, Tamala Limestone, Western Australia

Fig. 10. Embayed sand grain, Tamala Limestone, Western Australia

Fig. 11. Etched and embayed sand grain (with neomorphic quartz overgrowth). Amos Formation, Leila Creek

Fig. 12. Laminar calcrete and ooids, Lynott Formation, Kilgour River, N. T. (Locality X)

4. Hypersaline vadose activity has produced dolomitic and high-magnesian calcrete.
5. In the McArthur Group, calcretes can be taken as indication of breaks in sequence of sufficient duration for underlying beds to be subaerially exposed and weathered.

Acknowledgements. I wish to thank Dr. P. E. Playford for generously sending me samples of the Tamala Limestone and for showing me calcrete in the field in Western Australia. I am also grateful to Mr. Bryan Jones for his assistance in measuring sections through the Amos Formation and Balbirini Dolomite. Mr. M. J. Jackson made helpful comments on an early draft.

References

Jackson MJ, Muir MD, Plumb KA, Large DE, Brown MC, Armstrong KJ (1978) Field work report. McArthur basin project 1977. Bur Miner Resour Geol Geophys Rec 1978/54:1 – 53

Muir MD (1979a) A sabkha model for deposition of part of the Proterozoic McArthur Group of the Northern Territory, and implications for mineralisation. Bur Miner Resour J Geol Geophys 4:149 – 162

Muir MD (1979b) Regional unconformities. In: Plumb KA (ed) McArthur basin research, September quarter, 1979. Bur Miner Resour Geol Geophys Rec 1979/82:3 – 4

Muir MD (1979c) Amos formation and Balbirini Dolomite. In: Plumb KA (ed) McArthur basin research, March quarter, 1979. Bur Miner Resour Geol Geophys Rec 1979/44:6 – 9

Muir MD, Lock DE, Borch von der CC (1980) The Coorong-model for penecontemporaneous dolomite formation in the Middle Proterozoic McArthur Group, Northern Territory, Australia. In: Zenger DH (ed) Concepts and models of dolomitization – their intricacies and significance SEPM Spec Publ 28, pp 51 – 67

Page RW (1979) Mount Isa project. In: Geological branch summary of activities 1978. Bur Miner Resour Aust Rep 212 BMR Microform MF81:181 – 182

Peryt TM, Piatkowski TS (1977) Algal-vadose pisoliths in the Zechstein limestone (Upper Permian) of Northern Poland. Sediment Geol 19:275 – 286

Plumb KA (1979) McArthur basin research, March quarter, 1979. Bur Miner Resour Geol Geophys Rec 1979/44

Plumb KA, Brown MC (1973) Revised correlation and stratigraphic nomenclature in the Proterozoic carbonate complex of the McArthur Group, Northern Territory. Bur Miner Resour Aust Bull 139:103 – 115

Plumb KA, Derrick GM (1976) Geology of the Proterozoic rocks of the Kimberley to Mount Isa Region. In: Knight CL (ed) Economic geology of Australia and Papua New Guinea, Australas. Inst Min Metall Monogr Ser 5, I Metals, pp 217 – 252

Read JF (1976) Calcretes and their distinction from stromatolites, In: Walter MR (ed) Stromatolites. Elsevier, Amsterdam, pp 55 – 71

Riding R (1979) Origin and diagenesis of lacustrine algal bioherms at the margin of the Ries Crater, Upper Miocene, southern Germany. Sedimentology 26:645 – 680

Scholle PA, Kinsman DJJ (1974) Aragonitic and high-Mg calcite from the Persian Gulf – a modern analog for the Permian of Texas and New Mexico. J Sediment Petrol 44:904 – 916

Smith J (1964) Bauhinia Downs 1:250000 Geological Map Sheet SE53-3. Bur Miner Resour Geol Geophys

Walker RN, Muir MD, Diver WL, Williams N, Wilkins N (1977) Evidence of major sulphate evaporite deposits in the Proterozoic McArthur Group, Northern Territory, Australia. Nature (London) 265:526 – 529

Walker RN, Logan RG, Binnekamp JG (1978) Recent geological advances concerning the H.Y.C. and associated deposits, McArthur River, N.T. J Geol Soc Aust 27(7):365 – 380

Watts NL (1978) Displacive calcite: evidence from Recent and ancient calcretes. Geology 6:599 – 703

Part VI Contrasted Occurrences

VI.1 Coated Grains from the Great Barrier Reef

P.G. Flood[1]

1 Introduction

Coated grains constitute an ubiquitous component of the bioclastic carbonate sediments of the reef tract throughout the Great Barrier Reef Province (Fig. 1). They include ooids, rhodoids and a variety of encrusted grains which display a range of sizes and shapes depending upon the form of the object being coated by crustose coralline algae. The latter may contribute between 15% and 60% of the bulk of the reef top sediments.

2 Examples of Coated Grains

2.1 Ooids

Davies and Martin (1976) reported the occurrence of modern ooids in beach and channel sediments at Lizard Island. These microoids have both quartz and carbonate grains as nuclei and in many cases (see Fig. 2) the nucleation of aragonite has not occurred over the entire surface of the quartz grain. Such diverse growth forms of these ooids is suggestive of biological growth by calcareous algae on carbonate or terrigenous grains rather than chemical precipitation from normal sea water concentrations. (A similar mode of formation has been suggested by Kahle (1981) for the origin of ooids in the Pleistocene Miami Limestone, Florida Keys.) In addition a radial structure of oriented crystals is dominant in the ooid cortex. The significance of such radial fibrous ooids has been reviewed by Simone (1981).

2.2 Rhodoids

Macrorhodoids (i.e., coralline algal gravels) have been observed intertidally on the reef rim and outer reef flat of several platform reefs within the southern and northern regions of the Great Barrier Reef. These rhodoids occur as lag deposits

1 The University of New England, Geology Department, Armidale, N.S.W. 2351, Australia

Coated Grains (ed. by T.M. Peryt)

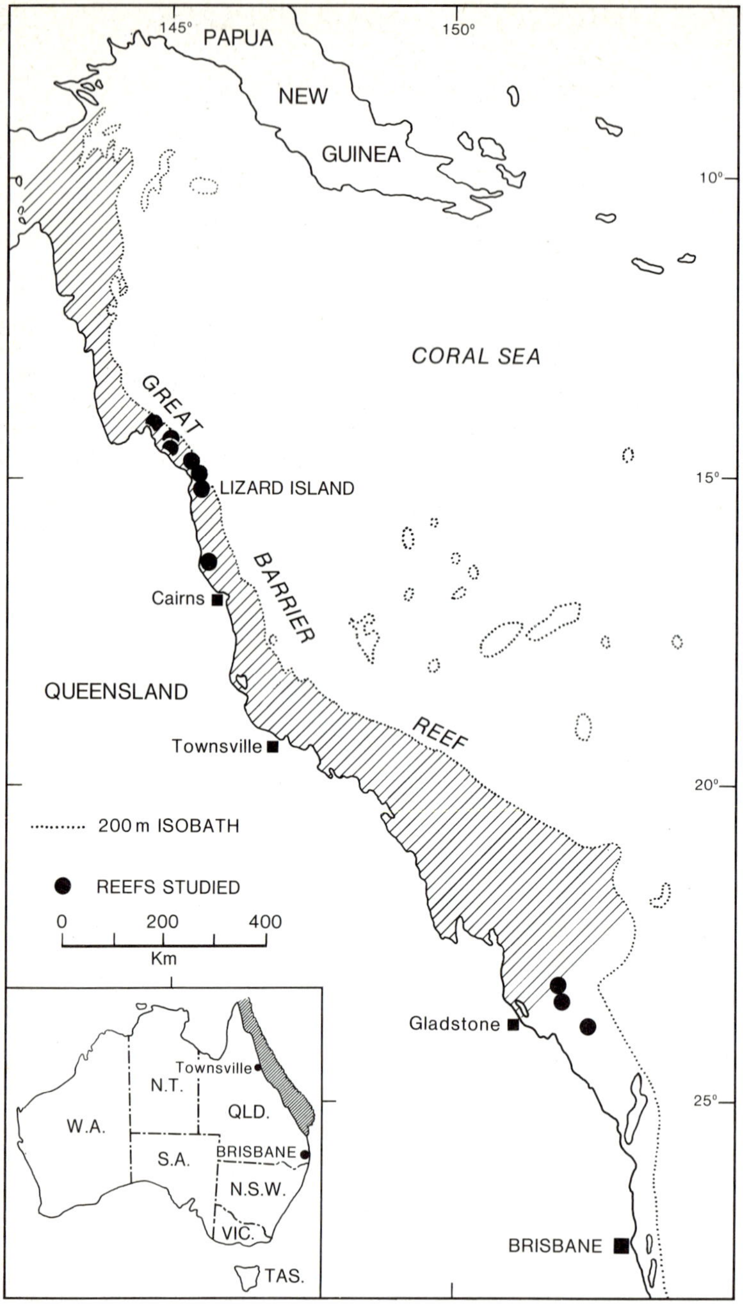

Fig. 1. The Great Barrier Reef Province occupies most of the northeastern continental shelf

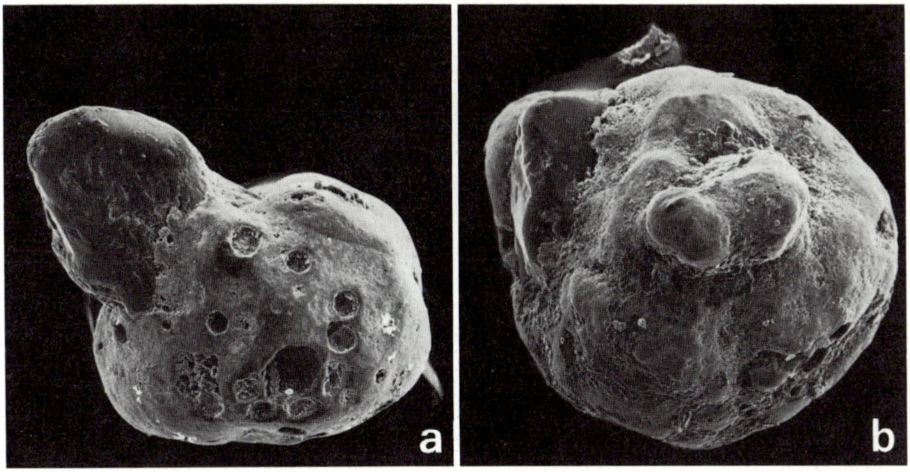

Fig. 2a, b. Ooids from the Lizard Island occurrence, northern Great Barrier Reef, described by Davies and Martin (1976). Nucleation of aragonite has not occurred over the entire quartz grain nucleus. Magnifications: **a** × 450, **b** × 520

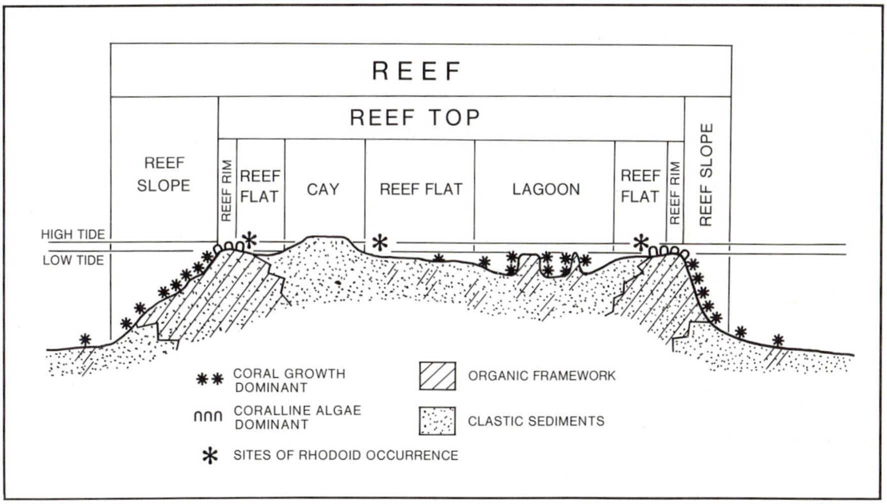

Fig. 3. Generalised cross-sectional view of a typical platform reef of the Great Barrier Reef. The location of Rhodoid occurrence is indicated

in areas subjected to relatively high energies associated with tides, waves, and wind driven currents (see Fig. 3). They display a gradation of growth forms from thin algal coatings on coral sticks to complete encrustations almost spherical in shape. Occasionally algal coatings may not cover the entire surface (see Fig. 4b). In less aggitated areas the rhodoids are less frequently subjected to movement and they commonly exhibit irregularly shaped protrusions on their upper sur-

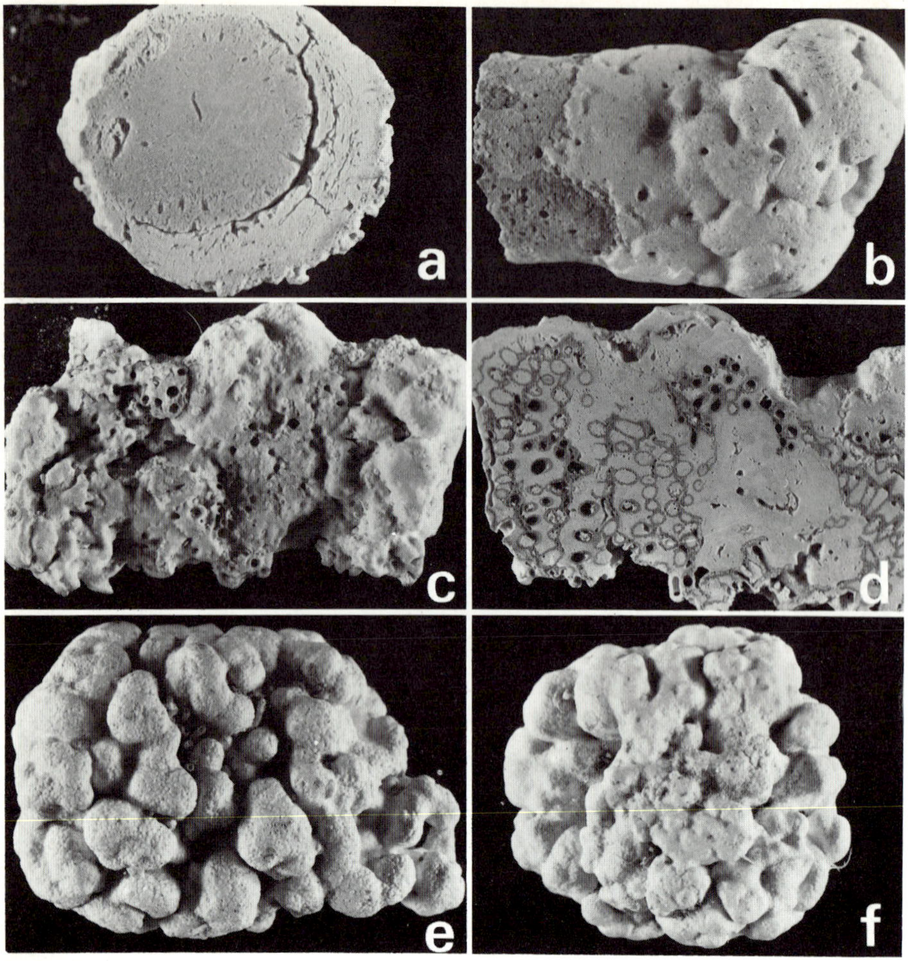

Fig. 4a–f. Rhodoids, Great Barrier Reef. **a** Sectioned rhodoid consisting of *Acropora* sp. with crustose coralline algal coating on one end. **b** Exterior of **a**. **c** Rhodoid showing both external coralline algal coating on a fragment of *Tubipora musica* (organ pipe coral). **d** Interior of **c**. **e** Typical rhodoids from intertidal surface sediment on a leeward reef flat. Magnifications: **a** × 2, **b** × 1.5, **c** × 1, **d** × 1, **e** × 2, **f** × 1.3

face. Rhodoids have been recovered from subtidal environments such as the lagoon or leeward reef slope, where they are usually buried by sediment. Such specimens commonly exhibit signs of abrasion due to transportation.

2.3 Other Coated Grains

Coated grains that are neither ooids nor rhodoids also occur in the unconsolidated reef sediments. These pisoid to macroid size grains usually exhibit thin coralline algal coatings or a red coloured coating of an encrusting foraminiferid over a varying percentage of their surface areas. No preference has been observed between the original skeletal fragment and the degree of encrustation by the coralline algae.

References

Davies PJ, Martin K (1976) Radial aragonite ooids, Lizard Island, Great Barrier Reef, Queensland, Australia. Geology 4:120–122

Kahle CF (1981) Origin of Ooids in Pleistocene Miami Limestone, Florida Keys. SEPM carbonate sedimentology and other wonderments poster display. AAPG Ann Conv, San Francisco

Simone L (1981) Ooids: A review. Earth Sci Rev 16:319–355

VI.2 Coated Grains Facies in the Lower Cretaceous of the Outer Dinarides (Yugoslavia)

Josip Tišljar[1]

1 Introduction

Coated grains are very frequent in the Mesozoic carbonate rocks of the Outer Dinarides. This chapter describes the coated grain facies in the Lower Cretaceous of the Adriatic region (Fig. 1) the aim is to characterize briefly the main types of coated grains and to evaluate their environmental significance.

The Lower Cretaceous carbonates are a part of the continuous Mesozoic carbonate sequence deposited from the Upper Triassic to the end of the Cretaceous (with some local emergences). The thickness of the Lower Cretaceous sediments varies between 750 m and 1150 m. Generally, they are shallow marine carbonates, deposited onto the great carbonate platform (for detailed discussion see: Tišljar 1978a, 1978b, Velić et al. 1979, Tišljar et al. 1981).

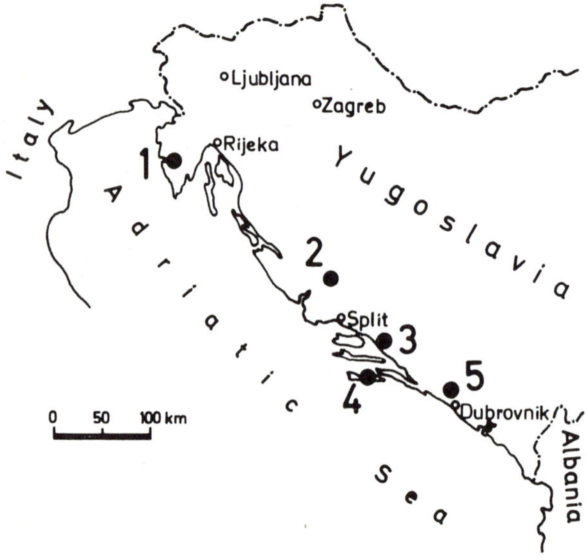

Fig. 1. Location of study area; *1* Istria, *2* Dinara mountain, *3* Biokovo mountain, *4* Korčula island, *5* environs of Dubrovnik

1 Institut for Mineralogy, Petrology and Economic Geology, Faculty of Mining, Geology and Petroleum Engineering, University of Zagreb, 41001 Zagreb, Yugoslavia

Coated Grains (ed. by T.M. Peryt)

The principal sedimentary environments and lithofacies are:

- Lithofacies A – supratidal with temporary emergences;
- Lithofacies B – shallow subtidal and intertidal;
- Lithofacies C – lagoon and back reef shoals;
- Lithofacies D – shoals with agitated water, near-reef shoals and patch reef;
- Lithofacies E – late-diagenetic dolomites.

The thickness and frequency of particular lithofacies and their vertical and lateral extension are not the same at all the localities and in all the units. Lithofacies C is most frequent. Relatively frequent is lithofacies B that is usually thin. Lithofacies A and D are less frequent, thinner and with minor lateral extension. Thick lithofacies E is frequent at certain localities or units of Beriassian, Albian and Vraconian age.

Coated grains are frequent in lithofacies C and very rare in lithofacies A. Coated grains of lithofacies C belong to oncoids, whereas coated grains of lithofacies A are vadoids.

2 Oncolite Facies

The lagoonal and back reef carbonate sediments in the Lower Cretaceous of the Outer Dinarides (lithofacies C) consist of micrites and/or biomicrites with oncoids, oncoidal intrasparites and thick oncolitic layers. The older units are dominated by oncoidal intrasparites and oncolites made of "algal ball" (Horowitz and Potter 1971) and/or "algal pellet" (Bissell and Chilingar 1967). The younger units contain macrooncoids (5 – 80 mm in diameter).

2.1 Oncoidal Intrasparites

Oncoidal intrasparites usually constitute a part of the micrite-intrasparite-oncolite cycle (Fig. 2A). Sometimes, we may find individual oncoids of "coated intraclast type" at the bottom of the oncolitic layers. The cycle begins with 0.8 – 1.3 m thick micritic layers, rarely with biomicrites, containing green algae, ostracods, gastropods, micritized benthic forams and Favreina pellets. The upper layer surface is characterized by microcracks. Above, dasycladacean micrites and oncoidal intrasparites (layer 3 in Fig. 2A) occur; these are interpreted as originating due to lowering of water energy and slow deepening, which favoured blue-green algal growth.

The total sedimentary thickness of more or less regular cycles varies from place to place from 10 to 35 m. Those cycles are well-developed in the Valangian and Hauterivian in Istria, Biokovo and Dubrovnik (loc. 1, 3, 5; Fig. 1).

Oncoidal intrasparites rarely appear outside the described cycles. They usually form thin layers in foram biosparites, intrasparites or oncolites in lithofacies C.

The main component of the oncoidal intrasparites are "coated intraclasts", micritized benthic forams, intraclasts, dasycladacean algae, gastropods and algal balls and/or algal pellets.

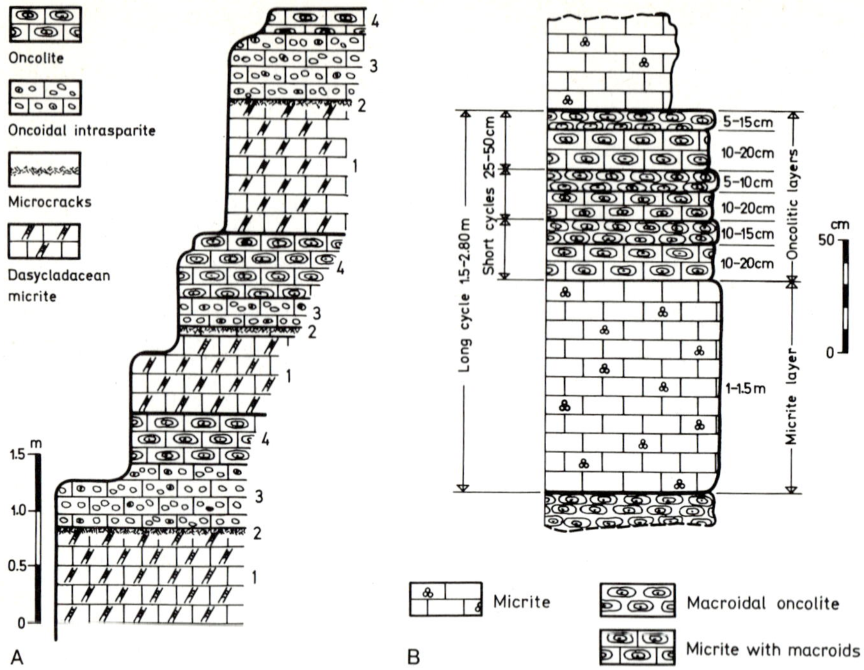

Fig. 2. A Oncoidal intrasparite and algal ball/algal pellet facies. Micrite-intrasparite-oncolite cycles in the Valangian-Hauterivian of Istria. **B** Rhythmical alternation of micrite and oncolite in the Lower Aptian of Istria

Coated intraclasts (Fig. 3A) have different dimensions (0.03 – 2 mm), degree of sorting and texture. Generally, the nuclei show micritic or clotted micritic texture, rarely, pelmicritic and biomicritic texture. They are enveloped by one, or very rarely by two micritic envelopes. The envelopes have different thickness (Fig. 3A). Grains with envelops originated by algal boring ("micrite envelopes": Bathurst 1971) are rare.

2.2 Algal Ball and Algal Pellet Oncolites

Algal ball and algal pellet oncolites may be found not only in the mentioned cycles (Fig. 2A), but very often in alternation with micrites/clotted micrites in the

Fig. 3. A Intrasparite with "coated intraclasts". Valangian-Hauterivian; the Lim channel – Istria, loc. 1. **B** Uneven sharp contact between micrite (*below*) and oncolite resulted from shallowing and short-lived emersion (evidenced by desiccation cracks and erosional surfaces) and deposition of oncolite during reflooding and formation of lagoonal environment. Tithonian – Beriassian; Quarry "Kirmenjak" – Istria, loc. 1. **C** Oncolite. "Algal ball" oncoids cemented by macrocrystalline calcite cement B and fibrous early diagenetic cement A. Tithonian – Beriassian; Poreč – Istria, loc. 1. **D** "Algal ball" with poorly preserved blue-green algal filaments. Valangian-Hauterivian; the Lim channel – Istria, loc. 1. **E** Oncolite layer. Irregularly shaped macroids with coated Bacinellas and gastropods in micritic matrix. Lower Aptian; Quarry "Kanfanar" – Istria, loc. 1 (polish slab)

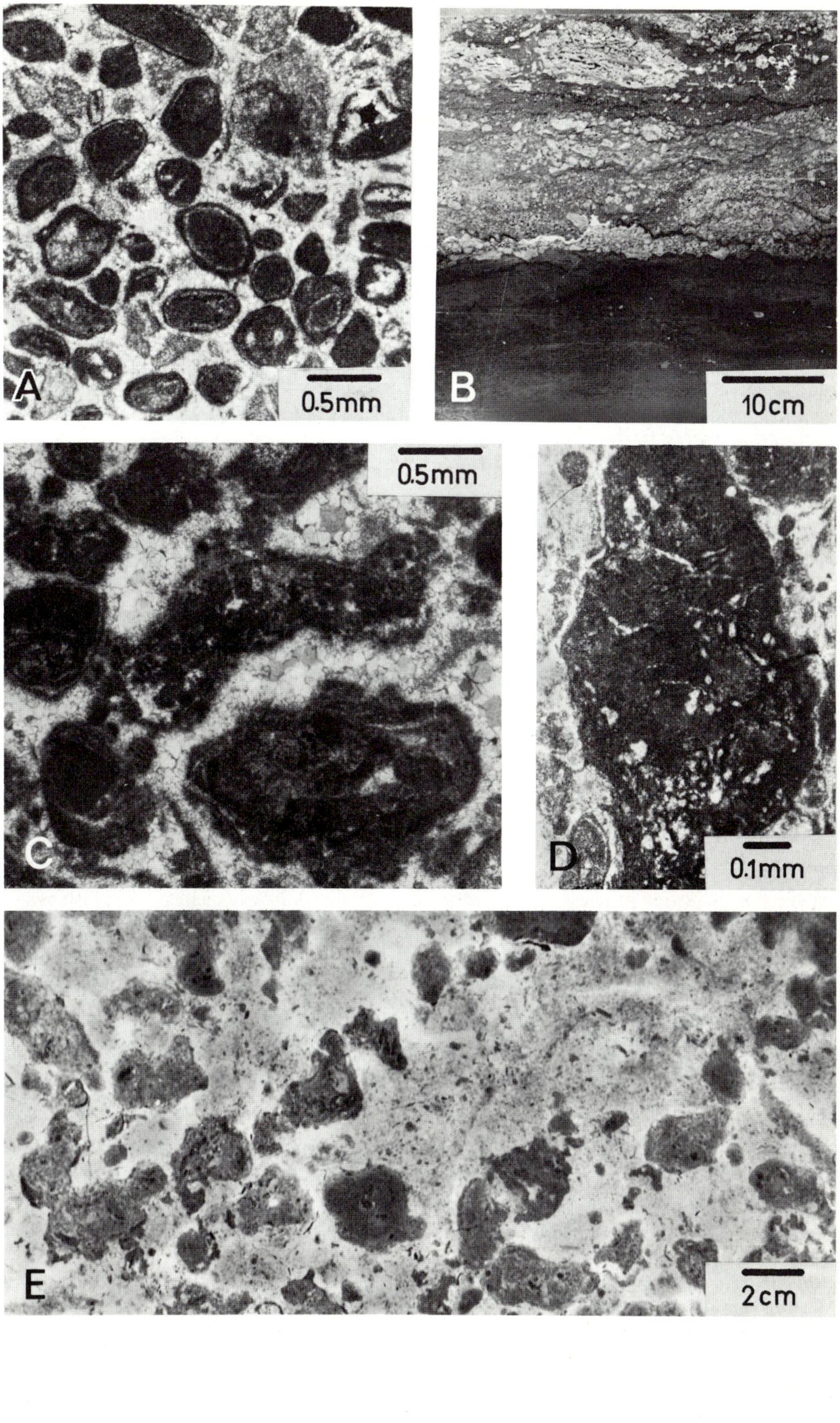
A
0.5mm
B
10cm
C
0.5mm
D
0.1mm
E
2cm

Tithonian, Beriassian, Valangian and Hauterivian in Istria, Biokovo, Korčula and Dubrovnik (loc. 1, 3, 4, 5; Fig. 1). The thickness of the sequence micrite/clotted micrite-oncolite (Fig. 3B) varies between 0.8 and 2.2 m; in the Beriassian of Istria, e.g., the total thickness of the cycles is 25 – 35 m.

Poorly sorted algal ball and algal pellet oncoids, "coated intraclasts" and peloids are the main constituents of these oncolites (Fig. 3 B, C). Tiny algal pellets (0.05 – 0.4 mm in diameter) and large algal balls (25 mm in diameter) occur together in the same thin sections. The algal balls were formed obviously by enveloping and incrustation of smaller oncoids, intraclasts, peloids, forams and gastropod fragments. There are very well-preserved blue-green algal filaments in some oncoids (Fig. 3D).

The oncolites of the facies are mostly sparitic, or less often micritic rocks. The interstices between oncoids and peloids are filled out by fibrous early-diagenetic cement A and mosaic microcrystalline cement B (Fig. 3C).

Alternation of micrite-oncolite sequences is explained similarly to the cycle from Fig. 2A. Micrites/clotted micrites originated in the subtidal-lagoonal to back reef shoal environments with tendency to gradual shallowing. The maximum shallowing coincides with the termination of micrites/clotted micrites deposition. This is evident because of erosional surfaces and desiccation cracks. They indicate early lithification and dehydration of the lower intertidal mud deposits. Algal ball/algal pellet oncoids and intraclasts were brought in from the adjacent shoals and redeposited onto the micritic layers. Later, when the oscillation ceased, convenient ecologic conditions were again established for blue-green algal growth and related formation of algal ball and algal pellet oncoids. The sedimentary conditions described were interrupted by intensive mud accumulation (new sequence).

2.3 Macroidal Oncolites

Macroidal oncolites may be found inside lithofacies C in almost all the units and at all the localities in the Lower Cretaceous of the Adriatic region. They appear as layers of great thickness alternating with micrites/biomicrites, characterizing the unit, so-called Lower Orbitolinas Limestone (approx. Lower Aptian; Velić

→

Fig. 4. A Fragment of the macroid. Green algae *Bacinella irregularis* (nucleus) with a cryptocrystalline envelope of trapped carbonate mud (*black*). Micrite matrix (above). Lower Aptian; Quarry "Lakovići" – Istria, loc. 1. **B** Oncolite composed of macroids of SS-C type ("concentrically stacked spheroids"), mosaic calcite cement and micrite matrix. Hauterivian; Biokovo mountain, loc. 3. **C** Vadolite. Intraformational erosion surface between fenestral pelmicrite (*margin bottom*) and vadolite. Crystal silt and drusy mosaic cement (*white*) flooring and filing vadoidal interstices. "Calcrete pisolites" with cryptocrystalline envelope (*white envelopes*). Beriassian; Environs of Dubrovnik, loc. 5. **D** Vadolite with geopetal fabric. Mechanically deposited (vadose) crystal silt (*bottom*) and drusy calcite cement of subaerial origin in vadoidal interstices and solutional vugs. Beriassian; Environs of Dubrovnik, loc. 5. **E** Vadolite composed of vadoids with regular concentric and radial structure. Beriassian; Environs of Dubrovnik, loc. 5. **F** Vadoids with regular concentric structure without any distinguished nucleus (*right bottom*) and broken and regenerated vadoid (*top*). The envelopes are composed of elongated tiny calcite crystals, orientated perpendicularly to vadoidal surface. Beriassian; Environs of Dubrovnik, loc. 5

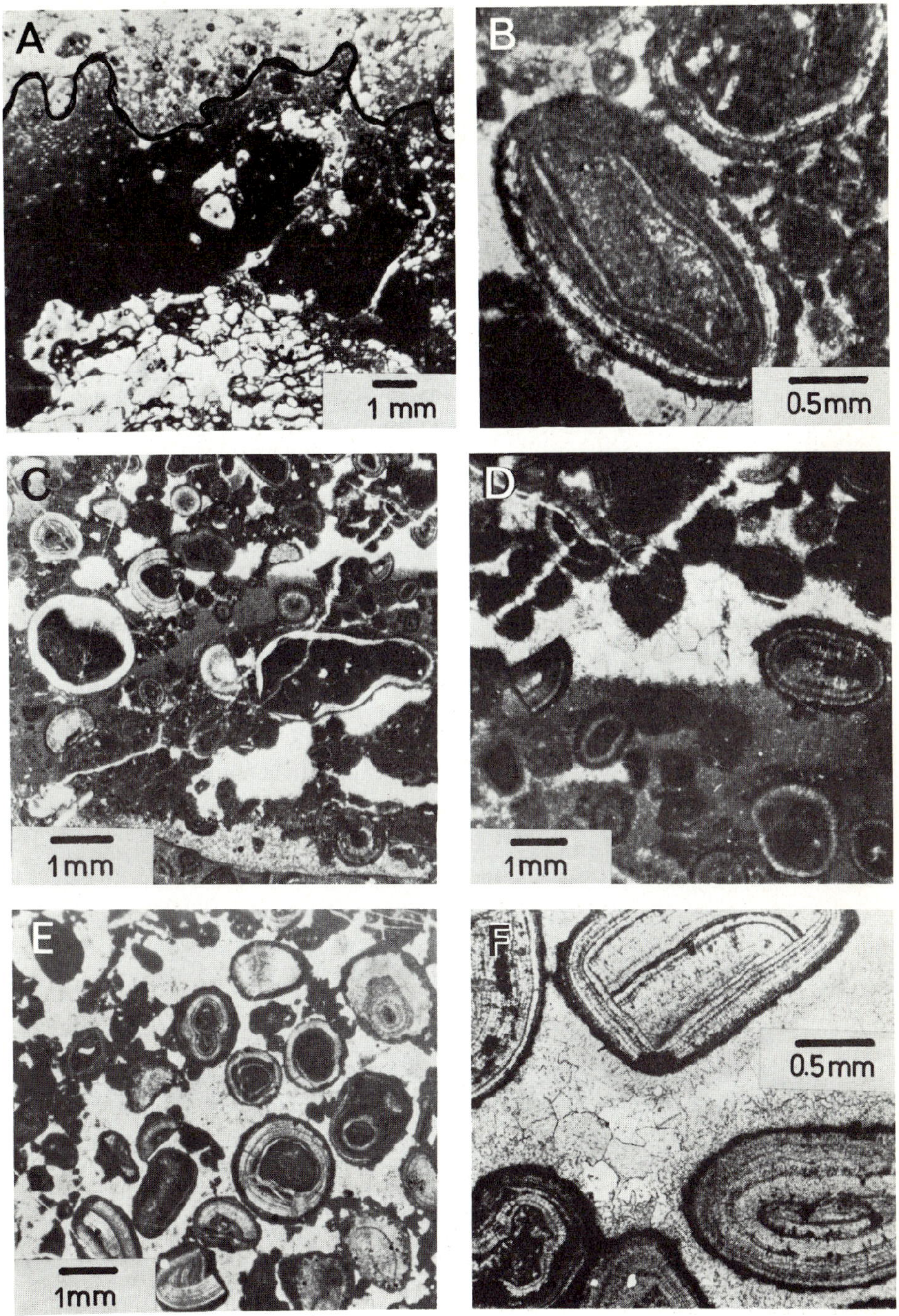
A
1 mm
B
0.5mm
C
1mm
D
1mm
E
1mm
F
0.5mm

and Sokač 1978) in the whole region of the Outer Dinarides. The limestones contain flora and fauna of restricted marine environments (green algae, blue-green algae, gastropods, ostracods, micritized benthic forams). *Sallpingoporella dinarica* Radoičić, *Selliporella dalmatica* Sokač and Velić, *Palorbitolina lenticularis* (Blumenbach), *Orbitolina (Mesorbitolina) lotzei* Schroeder and *Bacinella irregularis* Radoičić are the most frequent in these limestones. Macroidal oncolites appear also very often in the Lower Aptian of the Biokovo mountain and environs of Dubrovnik (loc. 3, 5; Fig. 1).

"The Lower Orbitolinas Limestone" consists of cyclic alternation of 0.3 to 3 m thick micritic/biomicritic and oncolitic layers. Thin layers of peloidal biosparites rarely occur. Two cycles may be distinguished (Fig. 2B).

1. Long cycle (1.5 – 3 m thick) consisting of a micritic layer (0.8 – 1.5 m) that usually contains rare macroids and several oncolitic layers (short cycles, Fig. 2B).
   The total thickness of oncolitic layers varies between 0.5 and 1.5 meters.
2. Short cycles contain several oncolitic layers. There are alternation of 10 – 20 cm thick micritic layers with macroids and 5 – 15 cm thick layers built up dominantly by macroids (Figs. 2B, 3E).

Alternation of the long and short cycles is explained by periodic changes of high and slow rate of accumulation of the carbonate mud in lagoons and/or back reef shoals. Micritic layers have been deposited at the time of high mud accumulation, and oncolitic layers in a period of changes slow-high mud accumulation.

Macroids have spherical, kidney-like or irregular clodlike shape (Fig. 3E). Diameters most often vary between 4 and 70 mm. Macroids, which built up oncolitic layers in short cycles are often fanlike distributed or overgrow each others. In this way, they form a sort of oncoidal crusts. According to macroscopic estimation on the large polished plates, volume ratio of macroids to micrites is from 1 : 2 to 4 : 1.

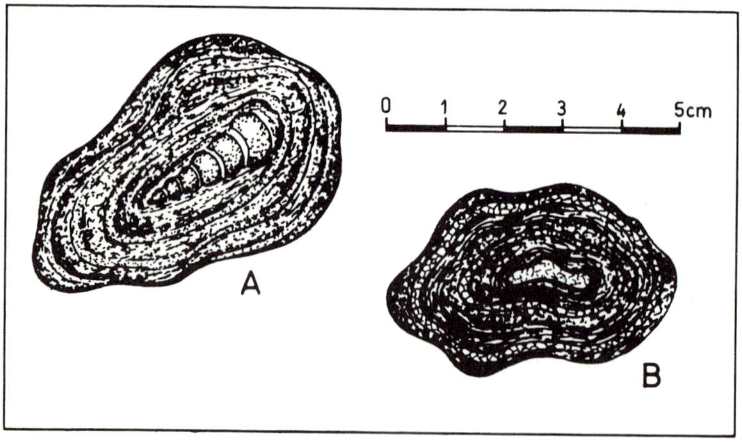

Fig. 5 A, B. Macroids from Biokovo mountain (Lower Aptian): **A** with gastropod as a nucleus, **B** with gastropod fragment as a nucleus

Bacinella (green algae) and gastropod skeletons or several incrusted forams can be seen in the most of macroidal nucles (Fig. 4A).

Irregular laminated structure is usually obvious, although some macroids have indistinct one. Macroids with cryptocrystalline envelopes, thick few millimeters up to 2 cm are frequent. Macroids of great dimension (10 – 80 mm) are the main components of oncolites in the Lower Albian of the Biokovo mountain and environs of Dubrovnik (loc. 3, 5; Fig. 1). As a rule, they have gastropod fragments as nucleus. An envelope, 3 – 70 mm thick, irregularly concretic, built up of few cryptocrystalline envelopes of changing thickness occur around the nucleus (Fig. 5). In the facies, sometimes, we may even find oncolites built up of oncoids with regularly concentric structure (Fig. 4B).

3 Vadolite Facies

Vadolite facies has been found only in the Tithonian-Beriassian in the environs of Dubrovnik and in the Hauterivian of the Biokovo mountain (loc. 3, 5; Fig. 1). Vadoids, and less often vadolitic layers, occur inside lithofacies A, which forms smaller part of peritidal and lagoonal limestones at these localities. Three sedimentary phases may be distinguished in the limestones (Fig. 6).

The limestones, their structural and textural characteristic, genesis, lithofacies and sedimentary phases were interpreted and described in detail in an earlier work (Tišljar 1980). The intertidal-supratidal phase is characterized by deposition of fenestral micrites/biomicrites, pelmicrites and LLH-stromatolites (Fig. 6). The sediments were periodically exposed to subaerial weathering in a vadose zone because of frequent, but brief, local emergences.

Vadoids, and less often vadolites, appear in few levels (Fig. 6) as irregular zones or interbeds inside the fenestral micrites and/or pelmicrites. The thickness of vadolite layers varies from 8 to 25 cm. Dimensions of vadoids vary between 0.1 and 3.5 mm (Fig. 4D – F). They show reverse graded bedding and downward elongation as do Permian "pisoliths" described by Dunham (1969). In relation to texture there are two morphologically basic types of vadoids:

1. Vadoids with regular concentric structure without any distinguished nucleus. Particular envelopes are built up of elongated tiny calcite crystals, orientated perpendicularly to vadoidal surface (Fig. 4D – F). That vadoid type is very often broken, with numerous broken parts of different dimensions observable in thin sections (Fig. 4E, F). Often the broken part is a nucleus of a new vadoid. The breaking probably resulted from fast dehydration or transportation of vadoids, particularly by repeated reflooding of vadoid deposits.

2. Vadoids containing bigger intraclast ("calcrete pisolites" – Read 1976) as nuclei and a few cryptocrystalline calcitic envelopes (Fig. 4C, right and left margin).

The first vadoid types dominate. Their interstices are filled either by crystal silt, or by drusy and gravitational cement (Fig. 4D – F).

Vadoids and thin vadolitic layers inside the limestones of lithofacies A, and inside the sediments of intertidal-supratidal phase (Fig. 6) respectively, are

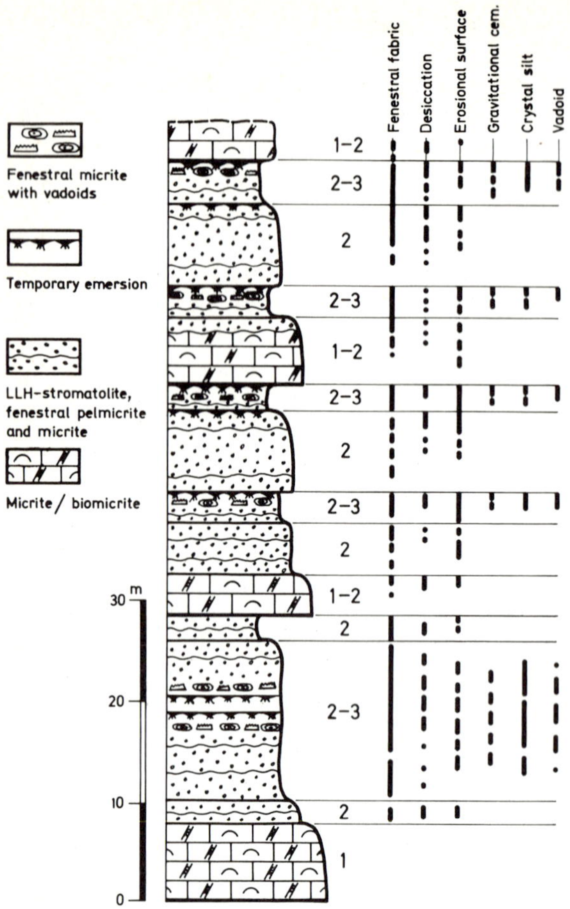

Fig. 6. Peritidal depositional phases of the Tithonian-Berriassian limestones near Dubrovnik: *1* subtidal to lagoonal phase, *2* shallow subtidal to intertidal phase, *3* intertidal to supratidal phase

evident proof of periodic and brief syndepositional emergences above sea level. Numerous solutional vugs, molds of shells, irregular fenestrae and caverns testify to the existence of humid climate and the action of meteoric water.

4 Discussion and Conclusions

The Lower Cretaceous carbonate sediments of the Outer Dinarides belong to a platform type being deposited at supratidal, intertidal, subtidal, lagoons, back reef shoals and shoals with agitated water and patch reefs. Oncoids are frequent only in lagoons and back reef shoals (lithofacies C), and vadoids in intertidal to supratidal environments (lithofacies A).

Different structural varieties and dimensions of oncoids ("coated intraclasts", algal ball/algal pellets, macroids, SS-C oncoids) and their cyclic appearance are a consequence of changes in the rate of grain movement, also in the ratio of blue-green algal growth rate and intensity of mud trapping and accumulation of deposits.

"Coated intraclasts" (Fig. 3A) and SS-C oncoids (Fig. 4B) are indicator of higher energy water environments with temporary "turning-over" of grains. Thickness and morphology of particular algal envelops is directly dependent on "turning over" interval, algal growth intensity and rate of mud trapping. Accordingly, "the grain turning over" can not have been of significant intensity, which would have led to envelope abrasion.

Algal ball/algal pellet oncoids (Fig. 3B – D) originated in restricted shallows and lagoons with lower energy water, with low rate of sediment accumulation and intensive growth of small separate blue-green algal turflets and bushes.

Macroids, which as a rule contain nuclei of *Bacinella* or gastropods (Figs. 3E, 4A, 5A, B) characterize lagoons and/or back reef shoals with still water and blue-green algal growth on the surface of skeletal remains. More or less distinctive regularity in alternation of long and short rhythmus (Fig. 2B) was caused by differing rates of mud accumulation and blue-green algal growth.

Small thickness and relatively minor lateral extension of vadoids, vadolites and crystal silt in fenestral micrites and pelmicrites, as well as differences in their vertical frequency, demonstrate that vadose conditions existed only locally, depending on the morphology of terrain and amplitude of the sea level oscillation; that they were brief relative to intertidal sedimentation, appearing often in shorter or longer time intervals. At the same time in the other regions of the Outer Dinarides (Istria, Velebit, Svilaja, Dinara, Biokovo) emergences with related smaller bauxite deposits occur.

References

Bathurst RGC (1971) Carbonate sediments and their diagenesis. In: Developments in sedimentology, vol 12. Elsevier, Amsterdam, p 620

Bissell HJ, Chilingar GV (1967) Classification of sedimentary carbonate rocks. In: Chilingar GV, Bissell HJ, Fairbridge RW (eds) Developments in sedimentology, vol 9A. Elsevier, Amsterdam, pp 87 – 167

Dunham RJ (1969) Vadose pisolite in the Capitan Reef (Permian), New Mexico and Texas. In: Friedman GM (ed) Depositional environments in carbonate rocks. SEPM Spec Publ 14:182 – 191

Horowitz AS, Potter PE (1971) Introductory petrography of fossils. Springer, Berlin, Heidelberg, New York, p 302

Read JF (1976) Calcretes and their distinction from stromatolites. In: Walter MR (ed) Stromatolites, developments in sedimentology, vol 20. Elsevier, Amsterdam, pp 35 – 71

Tišljar J (1978a) Tidal flat, lagoonal and shallow marine carbonate sediments in the Upper Jurassic and Cretaceous of Istria (Yugoslavia). Acta Geol IX/5 Book 42:159 – 194

Tišljar J (1978b) Oncolite and stromatolite limestones in the Lower Cretaceous of Istria (Croatia, Yugoslavia) (in Croatian). Geol Vjesnik 30/2:363 – 382

Tišljar J (1980) Fenestral and vadose early-diagenetic fabrics of the Tithonian and Beriassian peritidal limestones near Dubrovnik (Yugoslavia) (in Croatian). Carsus Iugosl 10/3:77 – 89

Tišljar J, Djurasek N, Sokač B, Velić I (1981) Correlation of the Lower Cretaceous Facies in the Adriatic coastal region. In: Complex oil-geological aspects for offshore and coastal Adriatic areas. Proceedings 1. Yugosl Acad Sci Arts, Zagreb (in Croatian)

Velić I, Sokač B (1978) Biostratigraphic analysis of the Jurassic and Lower Cretaceous in wider region of Ogulin, Central Croatia. Geol vjesnik 30/1:309 – 337

Velić I, Tišljar J, Sokač B (1979) Stratigraphy and depositional environments of the Lower Cretaceous in the Karst region of the Outer Dinarides (Yugoslavia). Geobios Mem Spec No 3:245 – 253

VI.3 Coated Grains in Contrasted Environmental Situations: Norian and Lower Liassic of Northern Apennines

Gloria Ciarapica and Leonsevero Passeri [1]

1 Introduction

The paleoenvironmental conditions in the Northern Apennines area during the Norian and Lower Liassic were highly different: the Norian climate was hot-arid whereas the Liassic one was hot-humid; in the Norian a large amount of sulphate evaporites precipitated, whereas no evaporites have been found in the Liassic sequences; the Norian platforms were bordered by low-energy, shallow seas, whereas the Liassic ones were built up in a pene-oceanic situation (Fig. 1).

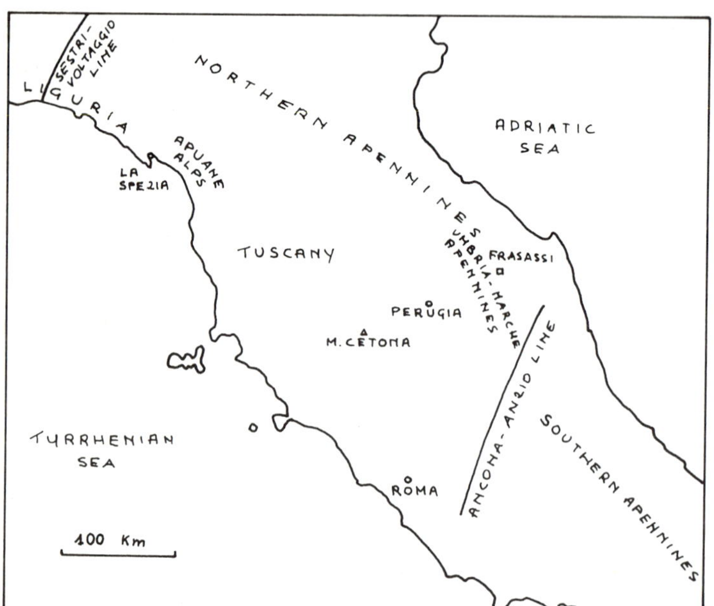

Fig. 1. Map of the Northern Apennines area

1 Istituto di Geologia dell' Università di Perugia, Piazza dell' Università, 06100 Perugia, Italy

2 Coated Grains in the Norian of the Apuane Alps

The Norian of the Apuane Alps metamorphic core is represented by the Grezzoni Formation, which overlies the Paleozoic basement and underlies the Hettangian Carrara Marble Formation (Fig. 2).

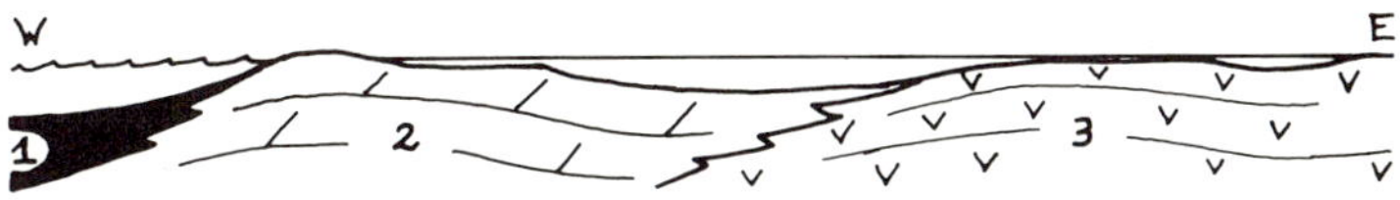

Fig. 2. Sketch of the Lower Norian paleogeography in the Northern Apennines: *1* thin bedded limestones and marls deposited in a shallow sea (*Rhaetavicula contorta* Beds of La Spezia); *2* thick bedded oolitic dolomites forming a hyperhaline barrier-lagoon complex (Grezzoni Fm. of the Apuane Alps metamorphic core, northern Tuscany); *3* sulphate evaporites interbedded with dolomites deposited in sabkhas and evaporitic coastal basins (Burano Anhydrites Fm.) (Ciarapica and Passeri 1978, Passeri 1975)

2.1 The Grezzoni Oolitic Facies

Four main oolitic facies are recognizable in the Grezzoni: (1) massive or cross-laminated beds with oolitic flat pebbles conglomerates (migrating bars, beaches and spillover fans); (2) oolitic-bioclastic flaser and wavy bedding (muddy-sandy flat); (3) oolitic-LLH stromatolitic cycles (stromatolitic sand flat); (4) oolitic rudstone (transgressive basal breccias which testifies to a synsedimentary tectonic activity) (Ciarapica and Passeri 1978) (Fig. 3).

2.1.1 Massive or Cross-Laminated Oolitic Beds. The beds, 1 or 2 m thick, show a basal erosion surface. Broken stromatolites levels and oolitic flat pebbles are fre-

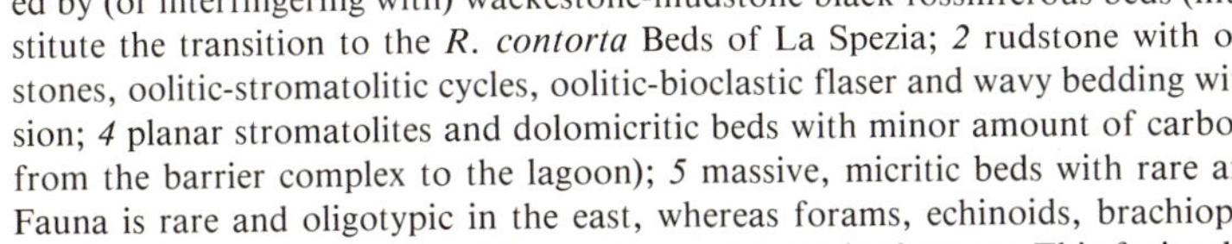

Fig. 3. The Grezzoni hyperhaline barrier-lagoon complex. *1* Oolitic rudstones and floatstones followed by (or interfingering with) wackestone-mudstone black fossiliferous beds (mud banks) which constitute the transition to the *R. contorta* Beds of La Spezia; *2* rudstone with ooids; *3* oolitic grainstones, oolitic-stromatolitic cycles, oolitic-bioclastic flaser and wavy bedding with evidences of emersion; *4* planar stromatolites and dolomicritic beds with minor amount of carbonate sand (transition from the barrier complex to the lagoon); *5* massive, micritic beds with rare and oligotypic fossils. Fauna is rare and oligotypic in the east, whereas forams, echinoids, brachiopods, pelecypods, encrusting algae and some branching corals are present in the west. This facies distribution shows the existence of a salinity threshold: hyperhaline to evaporitic conditions prevailed eastwards, whereas metahaline to normalhaline marine water predominated westwards. The Grezzoni barrier model shows some analogies with the modern Shark Bay one (Logan et al. 1974) even if the columnar stromatolites are lacking in the Grezzoni, probably because of differences in tidal range

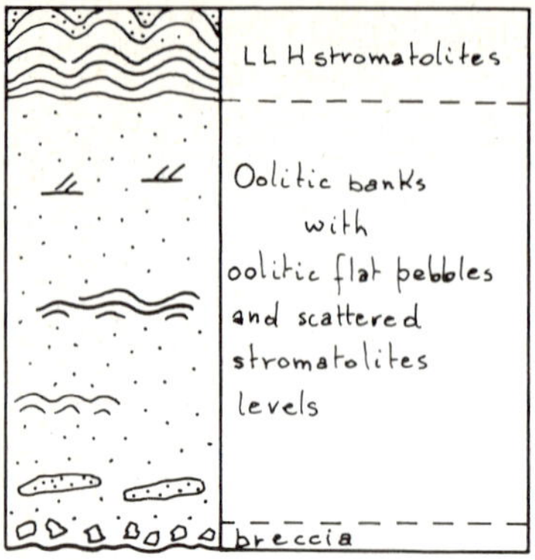

Fig. 4. Idealized oolitic-stromatolitic cycle *From bottom to top:* erosion surface with pockets filled by breccias; cross-laminated or massive oolitic grainstone with oolitic flat pebbles and scattered levels of stromatolites with a low lateral continuity (1 m); LLH-C stromatolites with oolitic interfingerings (20 – 30 cm). This cycle can be produced by migrating tidal channel

quent. Current laminations are rarely preserved. These beds are often interbedded with oolitic rudstone.

2.1.2 Oolitic-LLH Stromatolitic Cycles. This kind of cycles can be easily studied along the road to Pian della Fioba (Massa, Apuane Alps) (Fig. 4).

In modern environments the presence of LLH stromatolites and ooids is known only in the hyperhaline Shark Bay (Western Australia): in fact many environmental and morphological differences exist between the low-energy algal mats of the Florida-Bahamas platforms and the true Australian stromatolites.

2.1.3 Oolite-Bioclastic Flaser and Wavy Bedding. Well defined microcycles are lacking. Erosion surfaces mark the boundary of 2-m-thick flaser wavy bedded banks. These carbonate structures are very similar to the classic terrigenous ones: massive or cross-laminated oolitic-bioclastic sands are interbedded with lenticular micritic layers. Scattered burrows filled by carbonate sand or, alternatively, early diagenetic desiccation cracks, show that this group of structures can be produced both in subtidal and in intertidal environments with intermittent change from high to low energy.

2.1.4 Oolitic Rudstone. Oolitic breccias are very common in the lower part of the Grezzoni. Breccias and megabreccias made of oolitic or stromatolitic blocks, with early diagenetic cracks and sedimentary dykes are the most important features of the middle/lower part of the western Grezzoni (Fig. 3) (the most tectonically active zone of the barrier complex) (Fig. 5f).

2.2 Oolitic Grains and Vadoids in the Grezzoni

In the Grezzoni, the ooids show a tangential arrangement of the crystalline coating. The cores are made of bioclasts, but frequently they are of recrystallized

Fig. 5a – f. Vadoids and ooids in the Grezzoni Formation. (Apuane Alps, Norian) **a** vadoids (Vinca), note the dense coatings resistant to neomorphism; **b** vadoids (Vinca), note the broken grains and some sparry rims; **c** rough ooids (Pian della Fioba); **d** smooth ooids (Monte Grondilice); **e** oolitic flat pebbles (Monte Grondilice), note the dense rim, probably due to algal boring; **f** oolitic rudstone (Caglieglie)

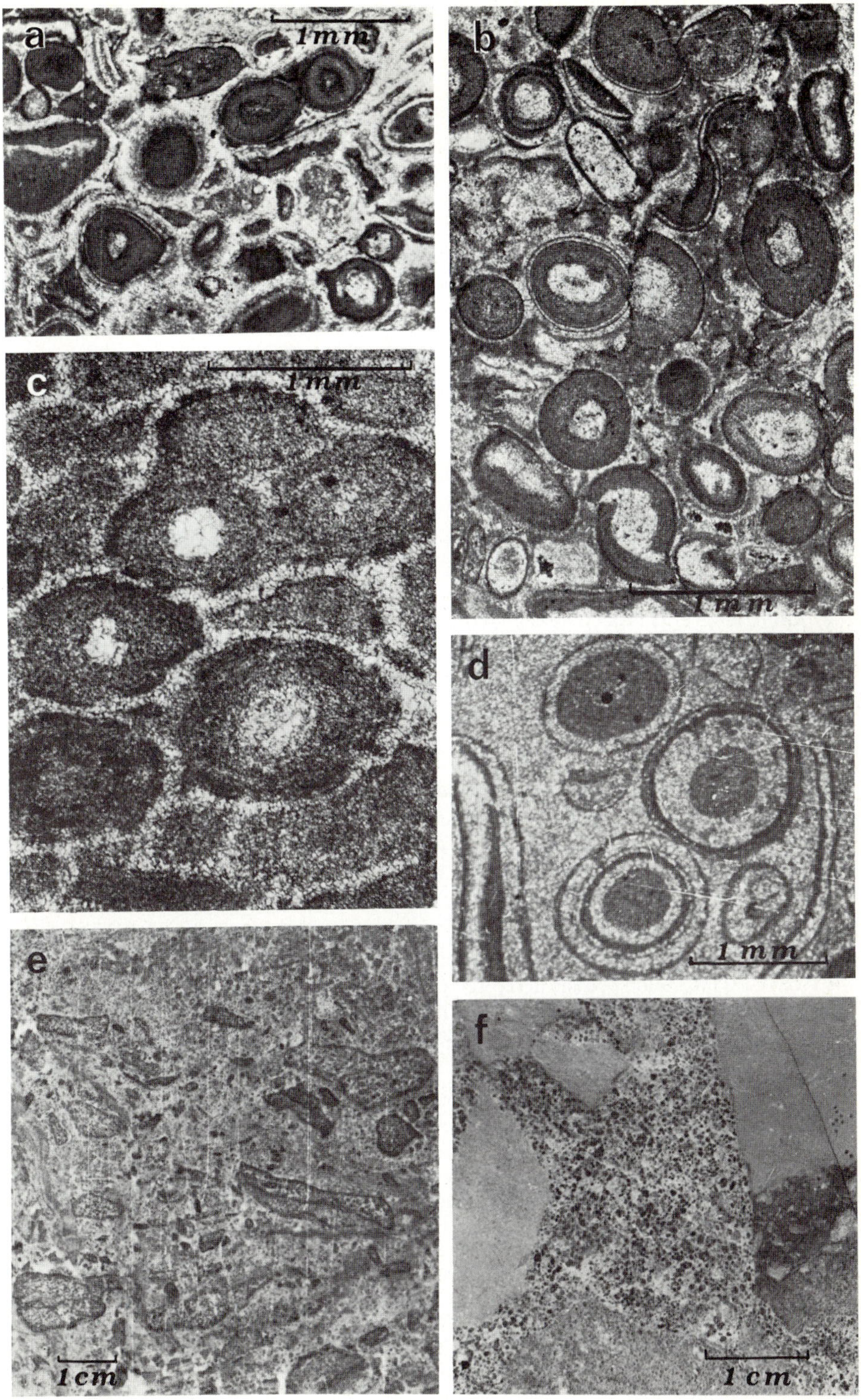
a
1mm
b
1 mm
c
1mm
d
1 mm
e
1cm
f
1 cm

micrite. The surface is often smooth (Fig. 5d), but sometimes it is rough. We cannot say if this kind of irregular surface is due to recrystallization processes or boring or irregular growth (Fig. 5c). The occurrence of lumps is also common: the external coating is made of few rings of dolomite crystals. Recrystallization often obliterated the original structures of this kind of grains.

On the contrary, the vadoids are more resistant to the recrystallization. In the Grezzoni the vadoids have a small size; their nucleus is made of bioclasts, sometimes of ooids only partially calcretized. Rarely they have been found in place: they are often reworked, so that they can have a double (calcrete-oolitic) coating (Fig. 5a, b). The calcrete coating is very dense, cryptocrystalline and it is similar to the modern Shark Bay calcrete.

3 Coated Grains in the Lower Liassic of Northern Apennines

In Northern Apennines the Lower Liassic is represented by the Calcare Massiccio Formation which is cropping out in Toscana, Umbria and Marche; in the Apuane Alps it is metamorphosed (Carrara Marble) (Fig. 6).

3.1 The Calcare Massiccio Facies and Environments

The Calcare Massiccio contains a lot of spectacular shallow water facies. The sedimentary environment is referable to a large carbonate platform with lagoon,

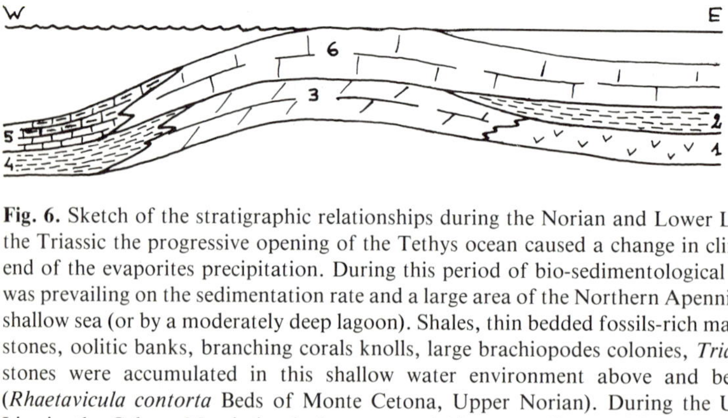

Fig. 6. Sketch of the stratigraphic relationships during the Norian and Lower Liassic. At the end of the Triassic the progressive opening of the Tethys ocean caused a change in climate, marked by the end of the evaporites precipitation. During this period of bio-sedimentological crisis the subsidence was prevailing on the sedimentation rate and a large area of the Northern Apennines was flooded by a shallow sea (or by a moderately deep lagoon). Shales, thin bedded fossils-rich marls and micritic limestones, oolitic banks, branching corals knolls, large brachiopodes colonies, *Triasina hantkeni* grainstones were accumulated in this shallow water environment above and below the wave base (*Rhaetavicula contorta* Beds of Monte Cetona, Upper Norian). During the Upper Norian-Lower Liassic, the Calcare Massiccio platform was quickly prograding on this shallow water basin. On the contrary, the La Spezia area became deeper and deeper; pelagic, thin bedded ammonitic limestones and marls were deposited at the top of the *R. contorta* Beds of La Spezia (*Schlotheimia angulata* Beds) as in the Prepiemontese units of central and western Liguria (Ciarapica and Passeri 1980a, 1980b). The end of the Calcare Massiccio sedimentation occurred at different times from west to east because of early sinking of the central Tethys area: Sinemurian Ammonitico Rosso is at the top of the Calcare Massiccio of W. Tuscany, whereas the pelagic sequence started during the Lotharingian-Carixian in Umbria-Marche. *1* Burano Anhydrites Fm; *2 Rhaetavicula contorta* Beds of Monte Cetona facies; *3* Grezzoni Fm. of Apuane Alps; *4 Rhaetavicula contorta* Beds of La Spezia; *5 S. angulata* beds; *6* Calcare Massiccio Fm. (Carrara Marble in the Apuane Alps metamorphic core)

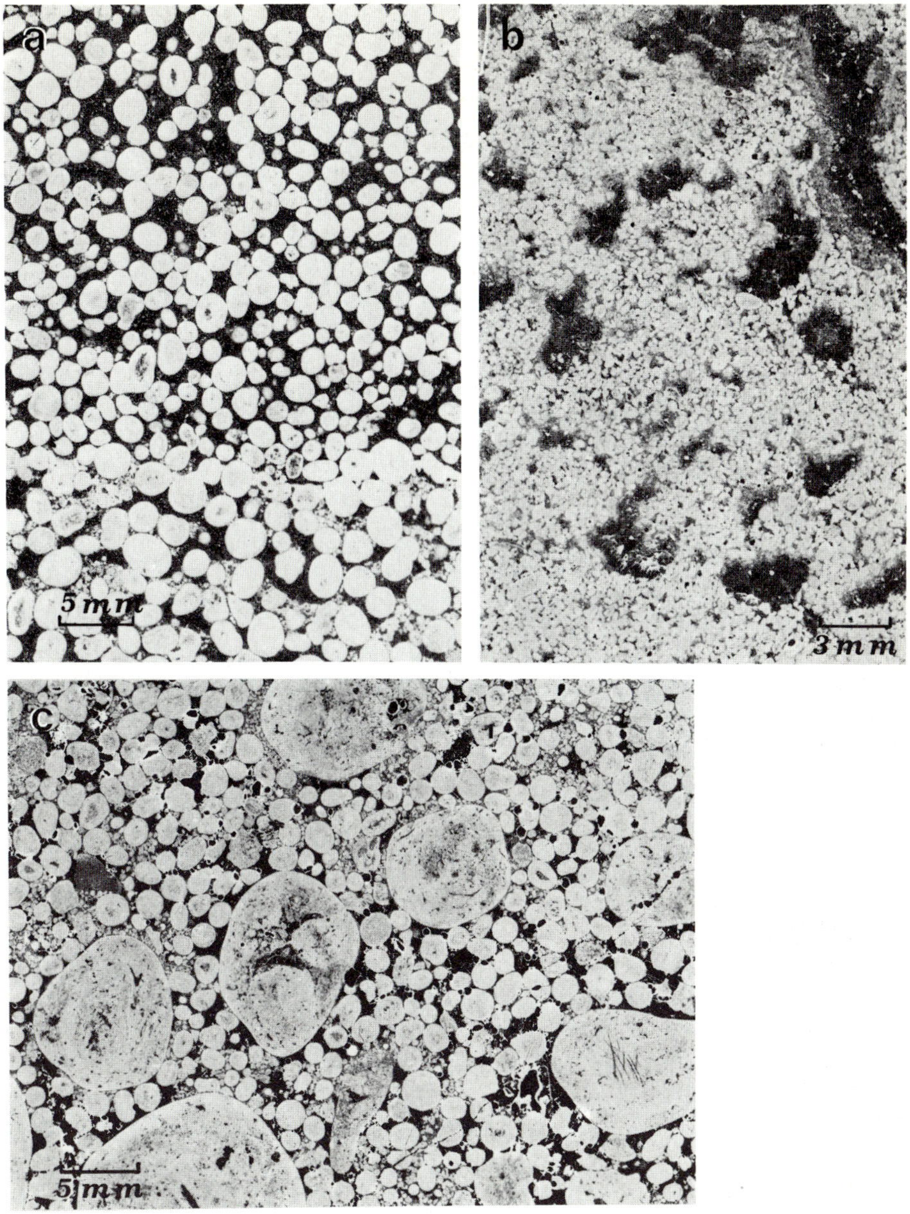

Fig. 7a – c. Ooids in the Calcare Massiccio Fm. (Lower Liassic, negative prints). **a** Oolitic grainstone (bar facies on the top of the Calcare Massiccio of Frasassi, Marche Apennines), note the very good sorting and the drusy and blocky cements; **b** oolitic grainstone with bird's eyes and microstalactitic cement (beach facies, Gualdo Tadino, Umbrian Apennines); **c** oolitic grainstone with oncoids (Monte Nerone, Marche Apennines)

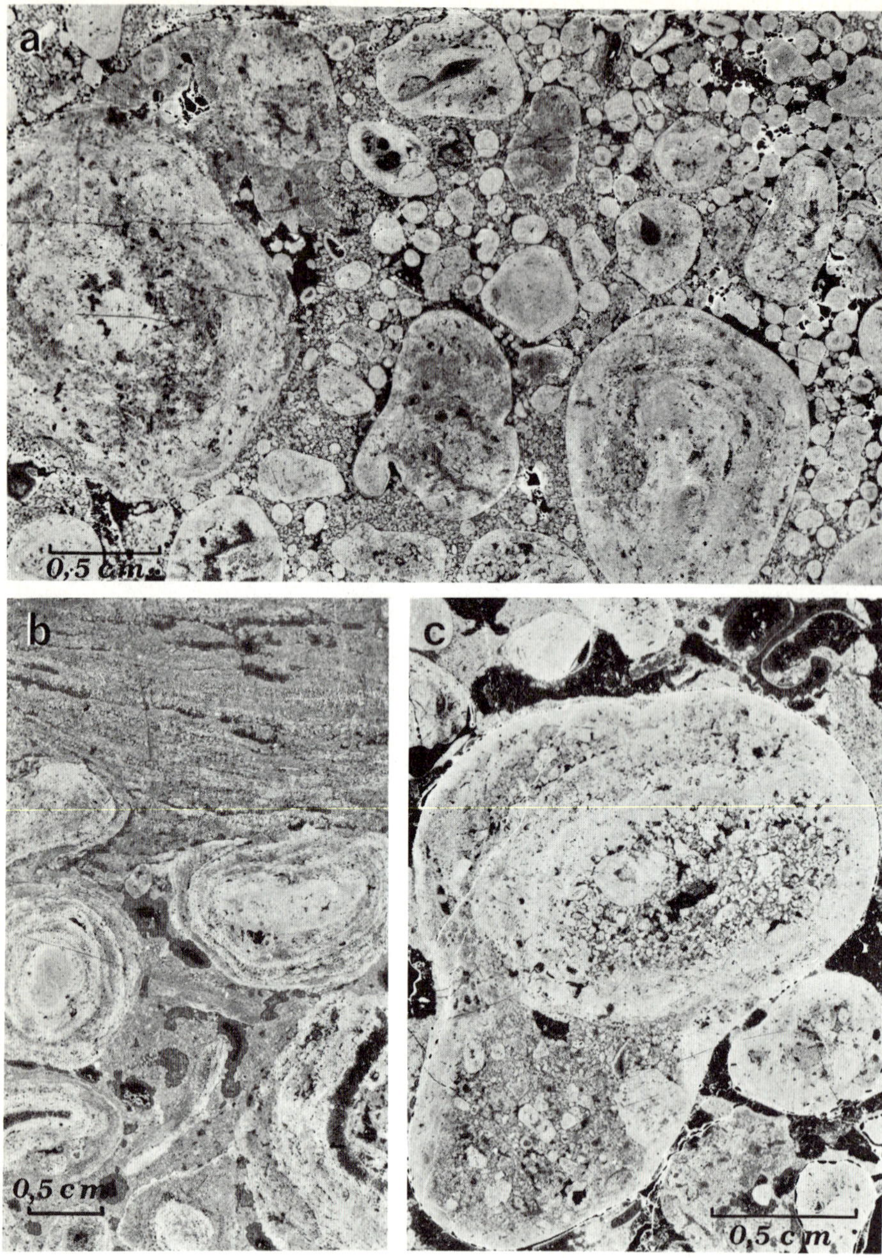

Fig. 8a – c. Oncoids in the Calcare Massiccio (negative prints). **a** Ooids and smooth oncoids grainstone (Monte Nerone); **b** oncoids in muddy matrix, followed by an intertidal low-energy algal mat with laminoid fenestrae (Frasassi); **c** smooth oncoidal lump (Monte Nerone)

muddy tidal flats, sandy beaches or sand flats, tidal channels, oolitic bars and large islands affected by strong early vadose diagenesis.

The main papers on this topic are: Boccaletti and Manetti (1972), Boccaletti et al. (1975), for the western facies, Passeri and Pialli (1972), Fazzuoli (1974a, b) for the central facies, Passeri (1971), Pialli (1971), Colacicchi et al. (1975) for the eastern ones.

3.1.1 Oolitic Facies. Many oolitic bodies are known in the Calcare Massiccio: (a) basal oolitic banks, sometimes with *Triasina hantkeni* (Monte Cetona, southern Tuscany and Ponte Arverino, southern Umbria) can represent a migrating bar deposited during the initial progradation of the Calcare Massiccio platform; (b) oolitic grainstones in western Tuscany (Elba Island), represent a part of the west-

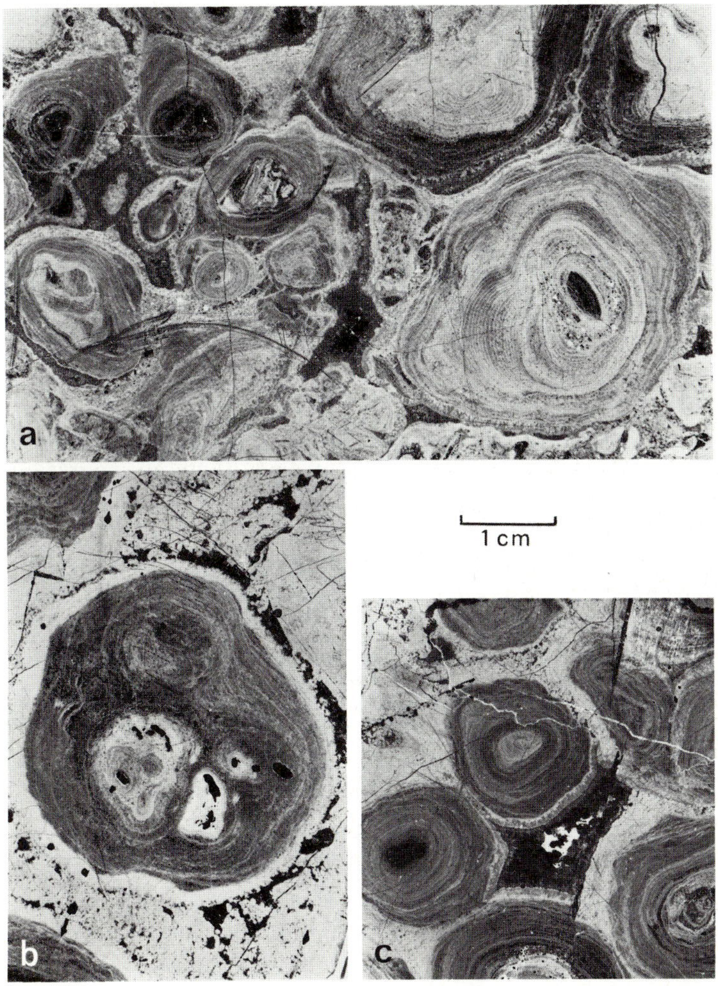

Fig. 9a – c. Pisovadoids in the Calcare Massiccio of Pioraco and Monti Sibillini ridge (southern part of the Umbria-Marche Apennines) (negative prints)

ern marginal complex, according to many authors (Boccaletti and Manetti 1972, Boccaletti et al. 1975); (c) thick oolitic banks in the upper part of many sequences of Calcare Massiccio in the Marche Apennines are thought to be related to a distensive tectonic phase and to an unequal sinking of this platform: oolitic shoals followed upward by reduced Jurassic sequences were sedimented in the raised areas (e.g. the classical section of Frasassi, Marche), while mud-supported oncoidal sediments, covered by a complete Jurassic sequence, were deposited in the depressed areas (e.g., Burano section) (Colacicchi et al. 1970, Centamore et al. 1971).

The oolitic facies (Fig. 7) makes up massive banks, dozens meters thick; bioclastic ruditic elements, lumps and intraclasts are also present; the oolitic grainstones generally show fine drusy and blocky cement.

3.1.2 Oncolite Facies. The typical oncoids are spherical in shape with a diameter 2 – 4 cm, the trapping laminae are often thick. Sometimes the oncoids are made of thin algal coating around irregularly shaped bioclasts. Oncoids can be scattered in a muddy matrix or they can be packed in oncoid-supported beds. These oncolitic beds are often followed upward by planar stromatolites as in a migrating tidal channel sequence (Fig. 8b). Oncolitic and skeletal sands, with lumps and large bioclasts, are really frequent in this facies (Figs. 7c, 8a, c).

The development of oncoids seems to be related to a subtidal environment with a moderately high energy (as a tidal channel, back barrier or tidal passes). Of course the oncoids can be easily rolled away from the site of their origin towards the near, deeper, quiet and muddy areas. Smooth surfaces are typical of the high energy oncoids whereas wavy surfaces with fine crenulated laminas are frequent in the muddy facies.

3.1.3 Vadolitic Facies. Early vadose diagenesis is very common in the Calcare Massiccio. Well developed pisovadolitic facies are known typically in the eastern and southeastern outcrops (Pioraco, Monti Sibillini) (Fig. 9) and in the western metamorphic and semimetamorphic units of Monti Pisani and Apuane Alps.

Fig. 10. Pisovadoids in the Calcare Massiccio of the Pania della Croce (Apuane Alps) (negative print)

The vadoids shows irregularly spherical shapes; their size varies from few mm to few cm; the coatings are of a very dense micrite; early diagenetic cracks are frequent and many vadoids were broken, reworked and sometimes they were encrusted again. Small or very large cavities, with clear calcite cement, are also frequent.

The metamorphic vadoids are deformed and they show an elonged shape; vadoids however are really resistent to the metamorphic processes and they can represent the best preserved elements in the fully recrystallized sequences (Fig. 10).

Table 1. Comparison between different environmental situations during the Norian and the Lower Liassic

	Grezzoni	Calcare Massiccio
Age	Norian	Lower Liassic
Climate	Hot-arid	Hot-humid
Salinity	High	Normal marine
Lateral facies	Westward: marls and limestones (*R. contorta* Beds of La Spezia) eastward: (2) marls and limestones (*R. contorta* Beds of Monte Cetona) (1) sulphate evaporites (Burano Fm.)	Westward: limestones and marls (*S. angulata* Beds) eastward: unknown
Underlying facies	Paleozoic basement or Triassic phyllites, arenites, conglomerates	Grezzoni or *R. contorta* Beds of Monte Cetona
Overlying facies	Carrara Marble (e.g. metamorphic Calcare Massiccio)	Jurassic pelagic sequence
Lithology	Early diagenetic dolomites with abundant rudstones	Limestones
Platform structures	Massive micritic banks, oolitic banks, flaser and wavy bedding, LLH stromatolites, erosion surfaces	Massive micritic banks, oolitic banks, algal mats, bird's eyes, fenestrae, neptunian dykes
Fossils	Eastward: rare and oligotypic westward: Corals, Brachiopods, Algae, Echinoids, Forams	Everywhere abundant: Algae, Forams, Pelecypods, Gastropods, Corals, Sponges
Geometry	Probably linear	Tabular, very extended
Affinities toward	Shark Bay and Persian Gulf	Caribbean platforms
Thickness of the oolitic bodies	Few meters	Several meters
Kind of ooids	Tangential, with smooth or roughly surfaces	Tangential, with smooth surfaces
Kind of oolitic facies	Rudstone-oolitic flat pebbles, oolitic banks, oolitic-LLH stromatolitic cycles	Oolitic banks with oncoids, lumps and coarse bioclasts or intraclasts
Oncoids	Unknown	Smooth or wavy surface in grainsupported or mudsupported facies
Vadoids	Microvadoids (often reworked)	Pisovadoids and macrovadoids with large cavities (sometimes reworked)

The distribution of macrovadoids in the Calcare Massiccio (in the western and eastern areas) suggests that the vadoids development has been more favoured in the marginal areas of this platform than in the inner ones.

4 Summary

Two different platforms were built in two different environmental situations during the Norian and during the Lower Liassic in Northern Apennines; consequently, the originated coated grains differ in many aspects, as summarized in Table 1.

References

Boccaletti M, Manetti P (1972) Caratteri sedimentologici del Calcare Massiccio della Toscana a Sud dell'Arno. Boll Soc Geol It 91(3):559 – 582

Boccaletti M, Fazzuoli M, Manetti P (1975) Caratteri sedimentologici del Calcare Massiccio a Nord dell'Arno. Boll Soc Geol It 94(3):377 – 405

Centamore E, Chiocchini M, Deiana G, Micarelli A, Pieruccini U (1971) Contributo alla conoscenza del Giurassico dell'Appennino umbro-marchigiano. Stud Geol Camerti 1:7 – 89

Ciarapica G, Fazzuoli M (1976) Analisi sedimentologiche di due sequenze nei Grezzoni delle Alpi Apuane. Boll Soc Geol It 95:147 – 160

Ciarapica G, Passeri L (1976) Deformazioni da fluidificazione ed evoluzione diagenetica della Formazione Evaporitica di Burano. Boll Soc Geol It 95(5):1175 – 1199

Ciarapica G, Passeri L (1978) I Grezzoni del nucleo apuano: nascita, sviluppo e morte di una piattaforma carbonatica iperalina. Boll Soc Geol It 97(4):527 – 564

Ciarapica G, Passeri L (1980a) La litostratigrafia della serie Triassica del promontorio occidentale del Golfo di La Spezia. Mem Soc Geol It (in press)

Ciarapica G, Passari L (1980b) Tentativo di ricostruzione paleogeografica a livello del Trias nella Toscana a Nord dell'Arno e sue implicazioni tettoniche. Mem Soc Geol It (in press)

Colacicchi R, Passeri L, Pialli G (1970) Nuovi dati sul Giurese umbro-marchigiano ed ipotesi per un suo inquadramento regionale. Mem Soc Geol It 9(4):839 – 874

Colacicchi R, Passeri L, Pialli G (1975) Evidences of tidal environment deposition in the Calcare Massiccio formation (Central Apennines – Lower Lias). In: Ginsburg RN (ed) Tidal deposits. Springer, Berlin Heidelberg New York, pp 345 – 353

Fazzuoli M (1974a) Facies di "laguna interna" nel Calcare Massiccio della Toscana sud-orientale. Boll Soc Geol It 93(2):369 – 396

Fazzuoli M (1974b) Caratteri sedimentologici del Calcare Massiccio nell'area della Pania di Corfino (provincia di Lucca). Boll Soc Geol It 93:735 – 752

Logan BW, Read JF, Hagan GM, Hoffman P, Brown RG, Woods PJ, Gebelein C (1974) Evolution and diagenesis of Quaternary carbonate sequences, Shark Bay, Western Australia. AAPG Mem 22:358

Passeri L (1971) Stratigrafia e sedimentologia dei calcari Giurassici di M. Cucco. Geol Rom 10:93 – 130

Passeri L (1975) L'ambiente deposizionale della Formazione Evaporitica nel quadro della paleogeografia del Norico tosco-umbro-marchigiano. Boll Soc Geol It 94(1 – 2):231 – 268

Passeri L (1979) The hypothesis of a Northern Apenninic peninsula during the Upper Triassic. Mem Soc Geol It 20:151 – 161

Passeri L, Pialli G (1972) Facies lagunari nel Calcare Massiccio dell'Umbria occidentale. Boll Soc Geol It 91(2):345 – 364

Pialli G (1971) Facies di piana cotidale nel Calcare Massiccio dell'Appennino umbro-marchigiano. Boll Soc Geol It 90(4):481 – 507

Sguazzoni G (1968) Stratigrafia dei Grezzoni del M. Grondilice (Alpi Apuane). Mem Soc Geol It 7(4):393 – 461

VI.4 Coated Grains from the Zechstein Limestone (Upper Permian) of Western Poland

Tadeusz M. Peryt [1]

1 Introduction

From the Zechstein Limestone, the carbonate member of the lowermost Zechstein evaporite cycle, the Werra, of different parts of the basin ooids, oncoids, rhodoids and vadoids have been described by various authors (e.g., Kerkmann 1966, Kostecka 1966, Füchtbauer 1968, Piekarska and Kwiatkowski 1975, Peryt 1977, 1978, 1981a, Peryt and Piatkowski 1977, Bełka 1979, Peryt and Wagner 1981, Smith, in press). Quite often the same type of coated grains occurring in the Zechstein Limestone rocks has been variously termed, depending on the general conviction of a given author as to the significance of processes leading to the origin of that type. On the other hand, usually the fact that in any given rock the grains which originated in different times in quite different environments can co-occur, has been ignored and the common practice was that the features observed in a part of coated grains occurring in a given Zechstein rock and characteristic of a certain type of coated grains, led to generalization that other grains of somewhat different features are in fact variations of that type. All these caused the picture of conditions of occurrence and formation of particular types of coated grains in the Zechstein deposits to become obscured. The aim of the present paper which is based on the author's studies in western Poland (Fig. 1), is therefore to characterize shortly the coated grains occurring in the Zechstein Limestone (Table 1), to discuss the problems of environments of their formation and primary mineralogical composition of coated grains in light of their recent analogs, showing the similarities and differences of the latter in relation to the Zechstein counterparts.

2 Geological Setting

Within the Zechstein Limestone sequences which are less than 1.0 m to more than 120 m thick, usually 40 to 60 m in the peripheral part of the basin and about 5 m in the central part (see Peryt 1978, Peryt and Ważny 1980), it is possible to distin-

1 Instytut Geologiczny, ul. Rakowiecka 4, 00-975 Warszawa, Poland

Coated Grains (ed. by T.M. Peryt)

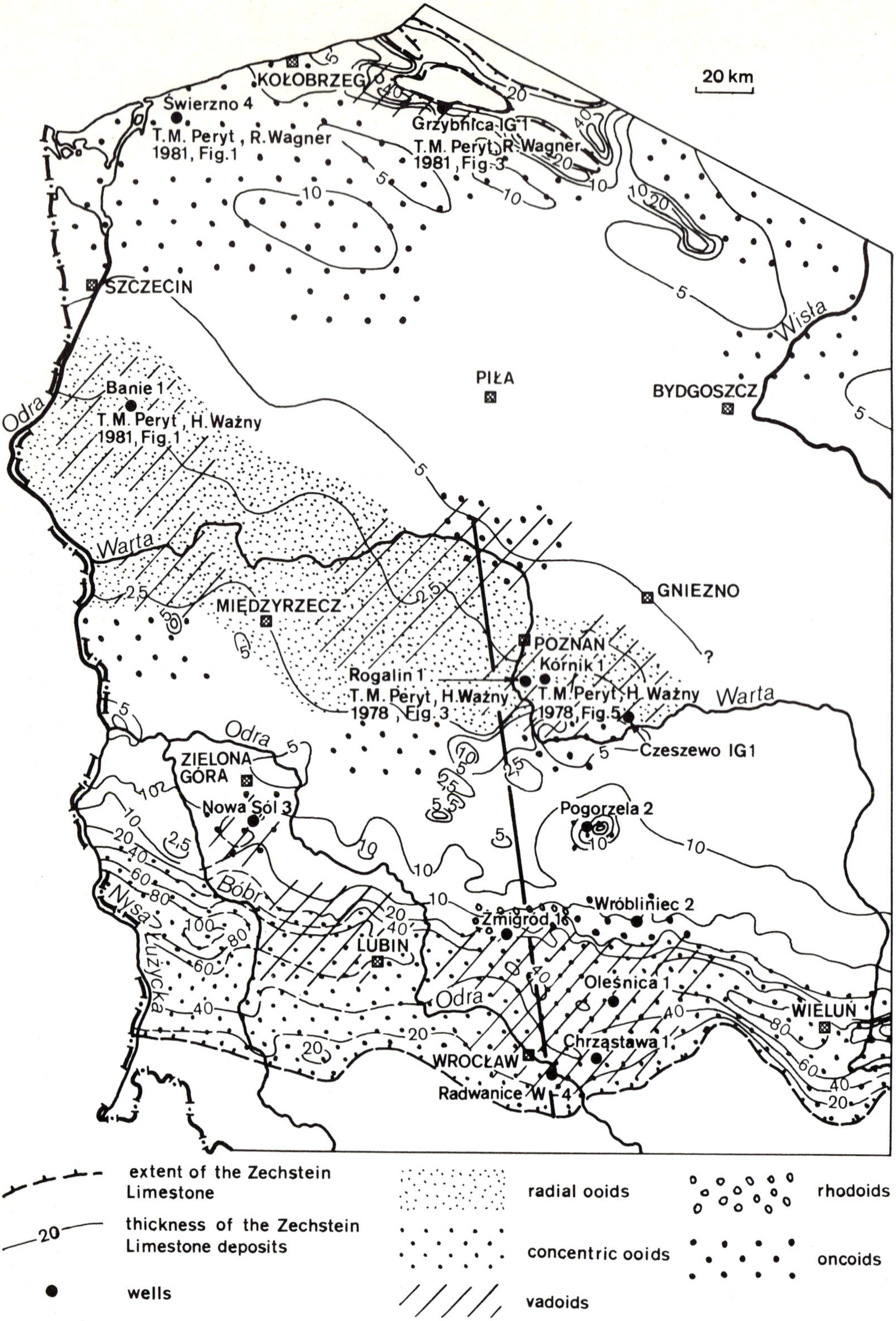

Fig. 1. Occurrence of coated grains in the Zechstein Limestone of western Poland (see referenced wells for the position of coated grains in the Zechstein Limestone sequences)

Table 1. Coated grains from the Zechstein Limestone deposits

Type	Illustration	Occurrence		Texture	Environment of formation	Mineralogy	
		Cycle	Part of the basin			Present	Primary inferred
Ooid							
Radial	Figs. 3d and 4a, b	I, II, III	Central		Quiet water environments	Calcite	High-Mg calcite
Concentric	Figs. 3a – c, 4d – f, and 6a, b, d	I, II, III	Peripheral and some shoals in central		Intertidal (and neighbouring), but often transported into other environments, usually less than 10 m deep	Dolomite	Aragonite (in pisooids also high-Mg calcite)
Oncoid	Fig. 5a – d	I, II, III	Central sporadically peripheral[a]	Spongiostromate (originally porostromate – see Peryt 1981b), locally radial texture is observed (well visible under luminoscope – Peryt 1981c, Fig. 5)	Shallow subtidal (see Peryt 1981b, Peryt and Wagner 1981)	Dolomite	High-Mg calcite (see Peryt 1981b)
Rhodoid	Fig. 4c	III	Peripheral		Upper part of subtidal zone	Dolomite	High-Mg calcite
Vadoid							
Continental (soil)	Fig. 7a, e	I, II	Peripheral (commonly), central	Very variable, what is related to the origin of vadoids in continental conditions (freshwater vadose zone) and transitional continental-marine conditions (marine vadose zone); Marine vadoids exhibit most features thought to be characteristic of that type of vadoids (see Assereto and Kendall 1977, Table 1), and continental vadoids, typical of caliche	Vadose conditions (partly hypersaline supratidal zone)	Dolomite, calcite	Low-Mg calcite
Marine (beach/coast)	Figs. 6a – f and 7b – d, f	I, II, III	(usually in the zone of condensed sequences)				Aragonite and high-Mg calcite

[a] Hitherto the term "oncoids" was used in the Zechstein (and especially in German and Polish) literature in the descriptional meaning, to determine the coated grains which do not exhibit a radial texture. From the genetical point of view, the grains described by various authors as oncoids in fact represent all types of coated grains

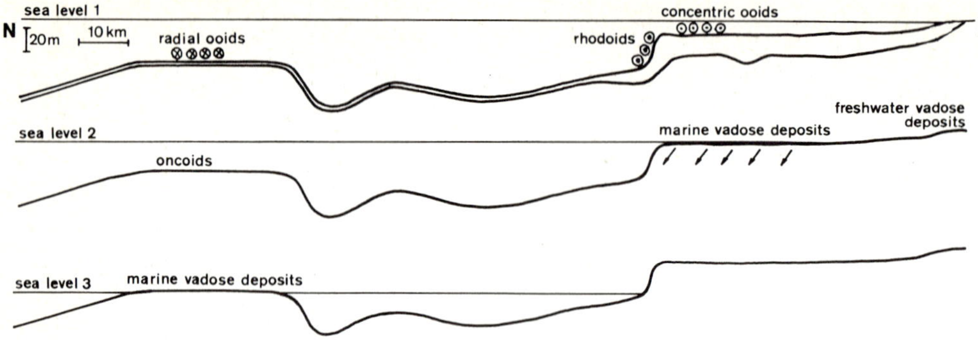

Fig. 2. Coated grains as reflecting sea level changes (line of the section is shown in Fig. 1)

guish maximally three upward-shallowing cycles which are several tens of centimeters to several tens of meters thick. The cycles are always asymmetrical, with rapid return to open sea conditions, which indicates fast deepening or low sedimentation during the transgression, and they possess the developed regressive members. The cyclicity resulted from glacieustatic changes of sea level, and this mechanism – because of the short time of duration of lowering and raising the sea level – enables the realization of the chronostratigraphical correlation; the chronohorizons mark the limits of cycles (see Peryt, in press).

Generally speaking, during the high sea levels the concentric ooids have been deposited in the peripheral part of the basin and radial ooids have formed in the shallow zones of the central part of the basin. During the low sea levels vadoids were coming into existence in the peripheral part of the basin, whereas oncoids originated in the central part. Those relations are schematically presented in Fig. 2.

3 Environments of Formation: Discussion

As presented in Table 1, the coated grains from the Zechstein Limestone originated, as may be supposed from the facies study, in the same environments in which their recent counterparts form. The pisooids which reach 6 mm in diameter and found in the well Pogorzela 2 should be discussed, as ooids of such size do not occur at present in marine environments. The size limit of ooids is 2 mm, which results from the equilibrium between precipitation on ooids and the abrasion of ooids (Bathurst 1968), and fossil ooids larger than 2 mm in diameter occur very rarely (Flügel 1978, Fig. 17). According to Radwański and Birkenmajer (1977), pisooids result from plentiful precipitation changing the water into a carbonate-milky solution, and the data as to the Pogorzela pisooids seem to support this interpretation fully. Figure 2 summarizes the data on paleogeographic conditions of formation of particular types of coated grains and explains their co-occurrence.

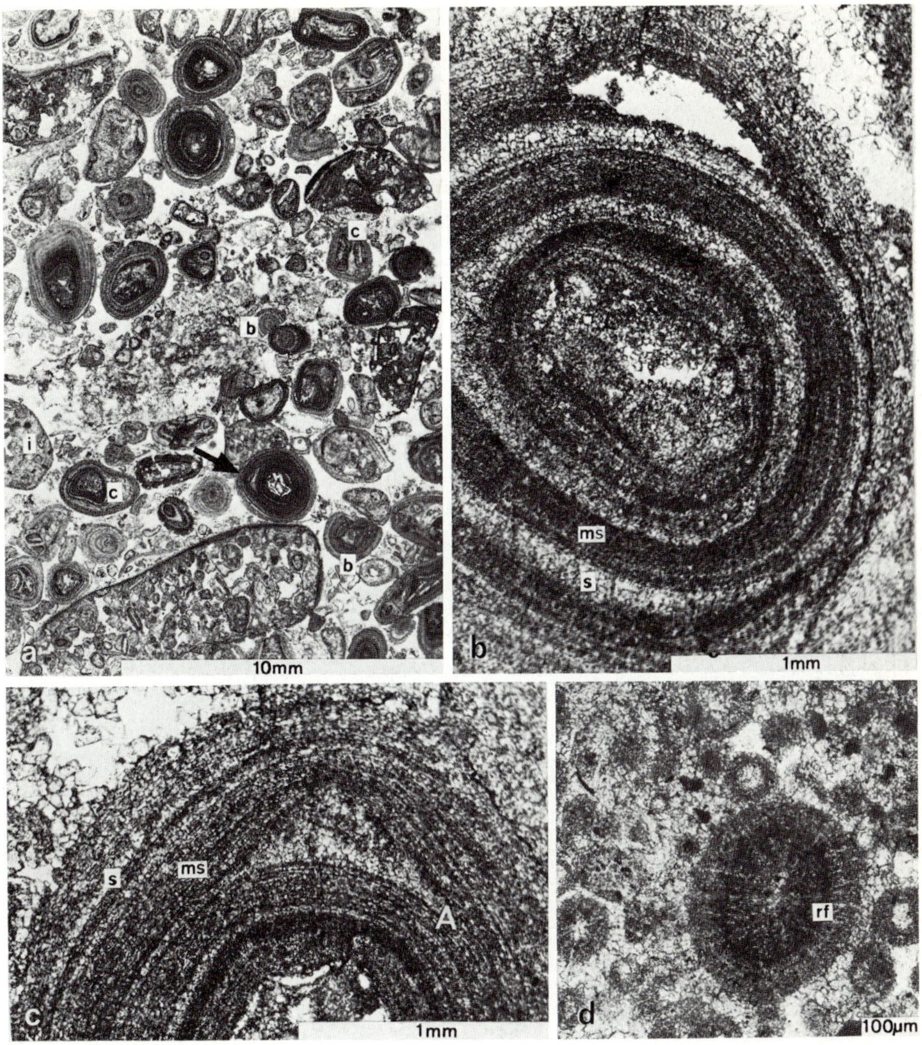

Fig. 3. **a** Loosely packed pisoolite with intraclasts (*i*). Pisooids usually simple, rarely complex (*c*) or broken (*b*). *Arrow* indicates the pisooid whose fragment is shown in **c**. Well Pogorzela 2, depth 1740.0 m. **b** Fragment of thin section presented in **a**. In cortex, the sets of microsparitic (*ms*) and sparitic (*s*) laminae are alternating. Such distinct alternation of sets of laminae as presented in photo most probably resulted from the differentiation of primary mineralogical composition. The preservation of microsparitic texture as illustrated on photo seems to indicate the recrystallization of tangentially arranged aragonite into calcite without massive dissolution (cf. Assereto and Folk 1976, Purser 1980: Fig. 154), and the sparitic texture probably originated by recrystallization of high-Mg calcite or radially arranged aragonite. **c** Fragment of **a**. Homogeneously concentric texture of pisooid. Laminae are built by microsparite (*ms*), rarely by sparite (*s*). **d** Oolite; radial texture (*rf*) of ooids is usually well visible. Well Kaleje 5, depth 3132.8 m

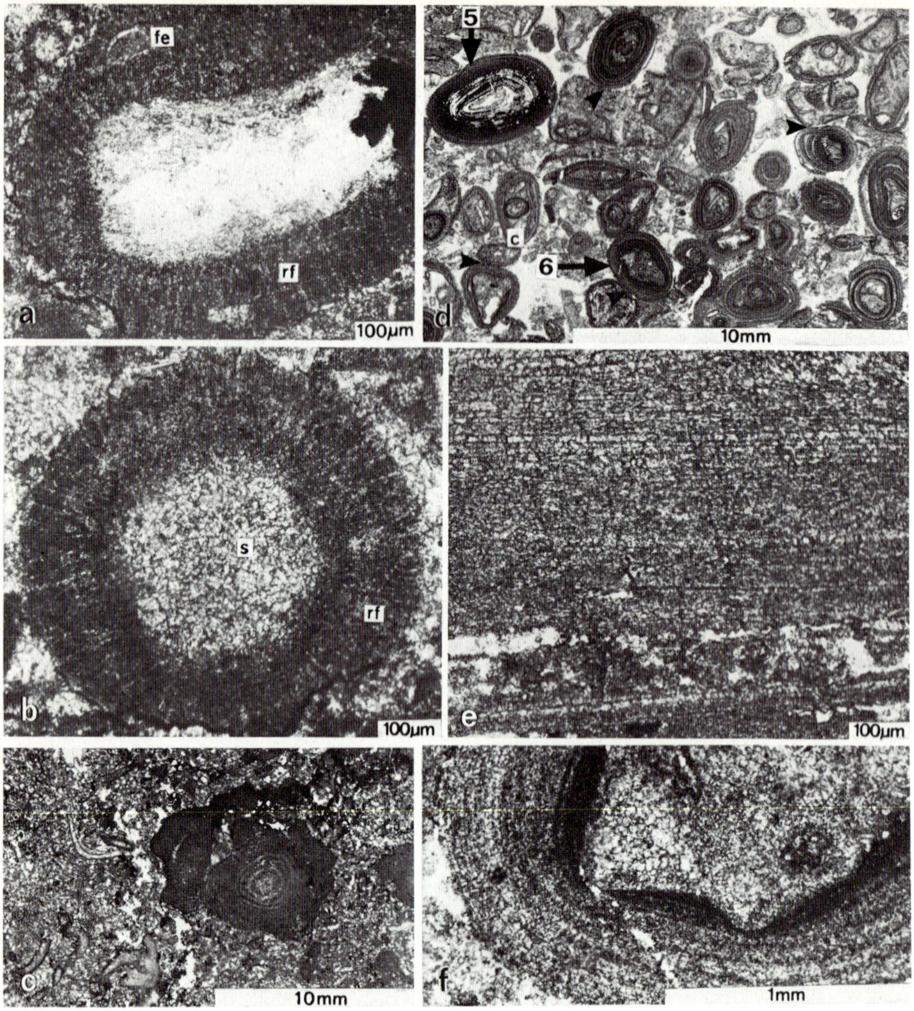

Fig. 4. a, b Radial ooids; note foraminiferal encrustation (*fe*) between the envelopes. The boundary between cortex and nucleus is usually sharp. Radial texture (*rf*) in places completely obliterated. Well Kórnik 1, depth 3491.3 m. **c** Rhodoid within bioclastic packstone/grainstone. Rhodoid texture very badly preserved which is probably connected with the very high primary content of Mg in high-Mg calcite which built those rhodoids (Late Paleozoic rhodoids were composed of calcite of variable Mg-content – Wray 1977). Well Żmigród 1, depth 1418.0 m. **d** Pisoolite more packed and better sorted than that illustrated on Fig. 3a, with intraclasts and complex pisooids (*c*). Abundant pitted contacts (*arrows*). The most common type of texture of pisoids is the alternation of sets of microsparitic and sparitic laminae. Well Pogorzela 2, depth 1738.5 m. **e** Fragment of **d**. Microstructure of pisooid in the older part of cortex (*bottom*) is as in Fig. 3b, and in the younger part as in Fig. 3c. **f** Fragment of **d**. Sparitic intraclast as nucleus of pisooid

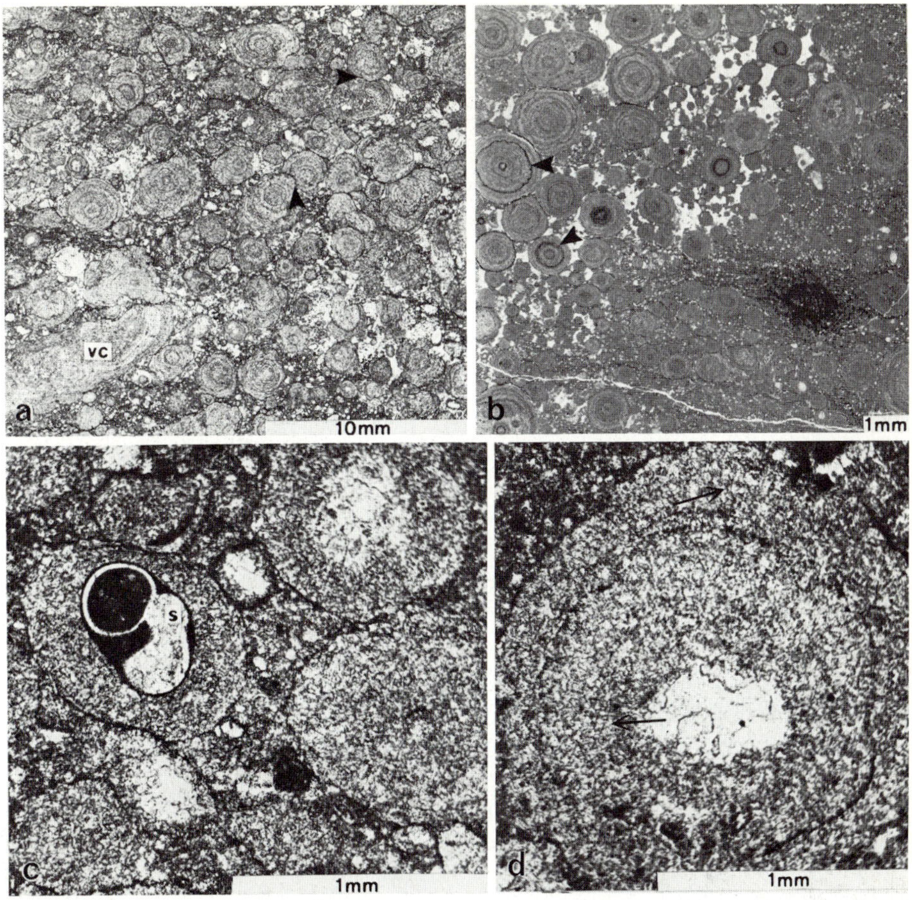

Fig. 5. a Spongiostromate oncolite with well preserved concentric texture of oncoids. Common pitted contacts (*arrows*); rare small vadose crusts (*vc*). Well Grundy Górne IG1, depth 3496.5 m. **b** Spongiostromate oncolite. The structure of cortex is faint because laminae – differing mainly in the size of crystals and colour – are little contrasted. In some oncoids (*arrows*) several phases of growth are visible. Common pitted contacts. Well Świerzno 4, depth 3151.8 m. **c, d** Spongiostromate oncolite with rudimentary preserved concentric texture of oncoids. Nuclei of oncoids often sparitic (due to neomorphism?), rarely bioclastic (gastropod shell – now sparitic – *s*). Local radial texture (*arrows*) similar to type 2 described by Monty and Mas (1981, p. 106–112, with discussion). Well Rogalin 1, depth 3191.5 m

4 Original Mineralogy: Discussion

The primary aragonitic composition of concentric ooids is indicated by oomoldic porosity which was related to dissolution in freshwater vadose and/or freatic environments (Fig. 7b). Contrary to concentric ooids, radial ooids certainly were not primarily aragonitic, as evidenced by the co-occurrence of radial ooids and oncoids of similar intensity of diagenetic processes changing the primary textures in the presence of sparitic fillings after primarily aragonitic shells in the oncoid

Table 2. Comparison of oncoids and vadoids from the Zechstein Limestone deposits of western Poland

Feature	Oncoids	Vadoids
Shape	Elliptical	Spherical to ellipsoidal
Structure	Most commonly concentric; lamination: microsparitic spongiostromate envelopes, sporadical relict radial texture	Most commonly sharply separated light and dark laminae; in places frequent interruptions in growth; sometines downward thickening of cortex; repeated brecciation in situ
Nucleus	Cryptocrystalline, rarely skeletal grain	Most commonly the fragment of deposit (also vadoids)
Size	Usually less than 2 mm	Up to 30 mm
Association	Skeletal grains, radial ooids	Other structures indicating vadose diagenesis; often lack of fossils
Mineralogical composition:		
– Present	Calcite, dolomite	Dolomite, calcite
– Primary (inferred)	High-Mg calcite	Aragonite (in marine vadose environments), calcite (in continental vadose environments)

nuclei (Fig. 5c) and their non-leaching in the vadose environment, as evidenced by their co-occurrence with calcrete (see Peryt 1981a). To date, the Fe-calcite which could speak for a primary high Mg-calcitic composition of radial ooids has not been found, but this is not unusual considering the specific conditions of replacement of high Mg-calcite by Fe-calcite (Richter and Füchtbauer 1978). The radial texture itself is characteristic of recent high Mg-calcite ooids.

5 Implications

The inferred primary mineralogy of the Zechstein coated grains (Table 1) seems to indicate that recent coated grains are their chemical analogs. This in turn means that the Mg/Ca ratio in sea water during the Zechstein Limestone deposition was very near to 5, the value characteristic of recent marine environments in

Fig. 6a – f. Vadoids strikingly similar to vadoids from Shark Bay (Read 1974) and to some vadoids from the Persian Gulf (cf. **c**, **e** and Scholle, Kinsman 1974: Figs. 5 – 7) within the ooidal deposit which underwent vadose compaction. Well Oleśnica 1, **a** depth 1595.5 m, **b** 1593.0 m, **c** 1592.5 m, **d** 1590.0 m, **e** 1596.5 m, **f** 1599.0 m. Vadose compaction happens when the carbonate sediments are subaerially exposed and flushed with freshwater (Clark 1980). The characteristic is the distinct downward intensity of compaction. The result of compaction is that the rock structure becomes packed: in the zone of grain contacts the selective dissolution occurs and the contacts become fitting, what leads to the origin of characteristically flattened vadoids (**b, d**). The presence of that texture seems to indicate the non-stable primary mineralogical composition of vadoids. Recent aragonitic and high-Mg calcitic vadoids form in vadose environments of the Persian Gulf when the hypersaline groundwater is pulled into vadose zone of the sabkha by capillary pressure. In the Persian Gulf vadose compaction has not been reported yet but it may be easily explained by the non-appearance of flushing with freshwater; furthermore, the vadoid sites are very limited in number

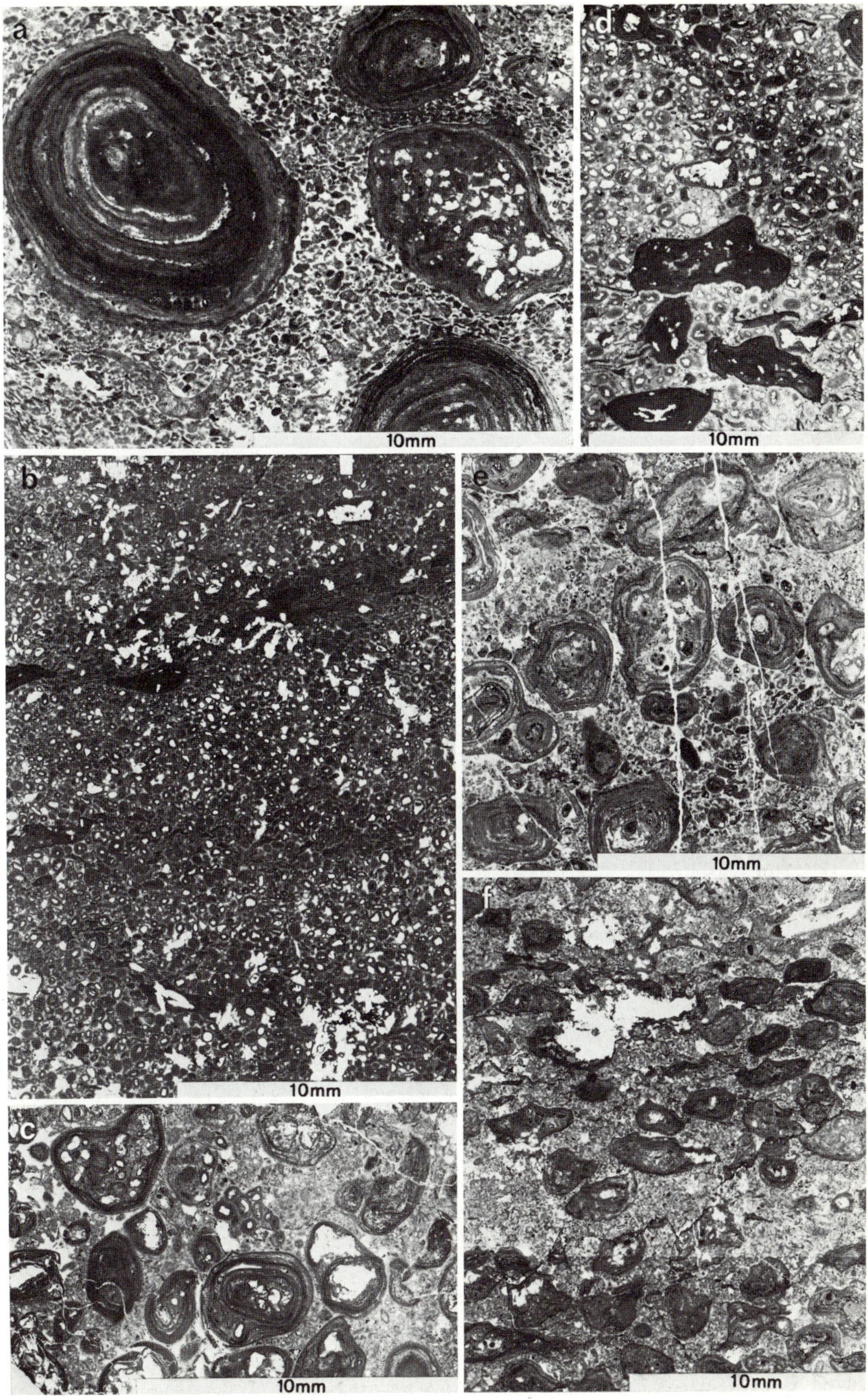
a
10mm
d
10mm
b
10mm
e
10mm
c
10mm
f
10mm

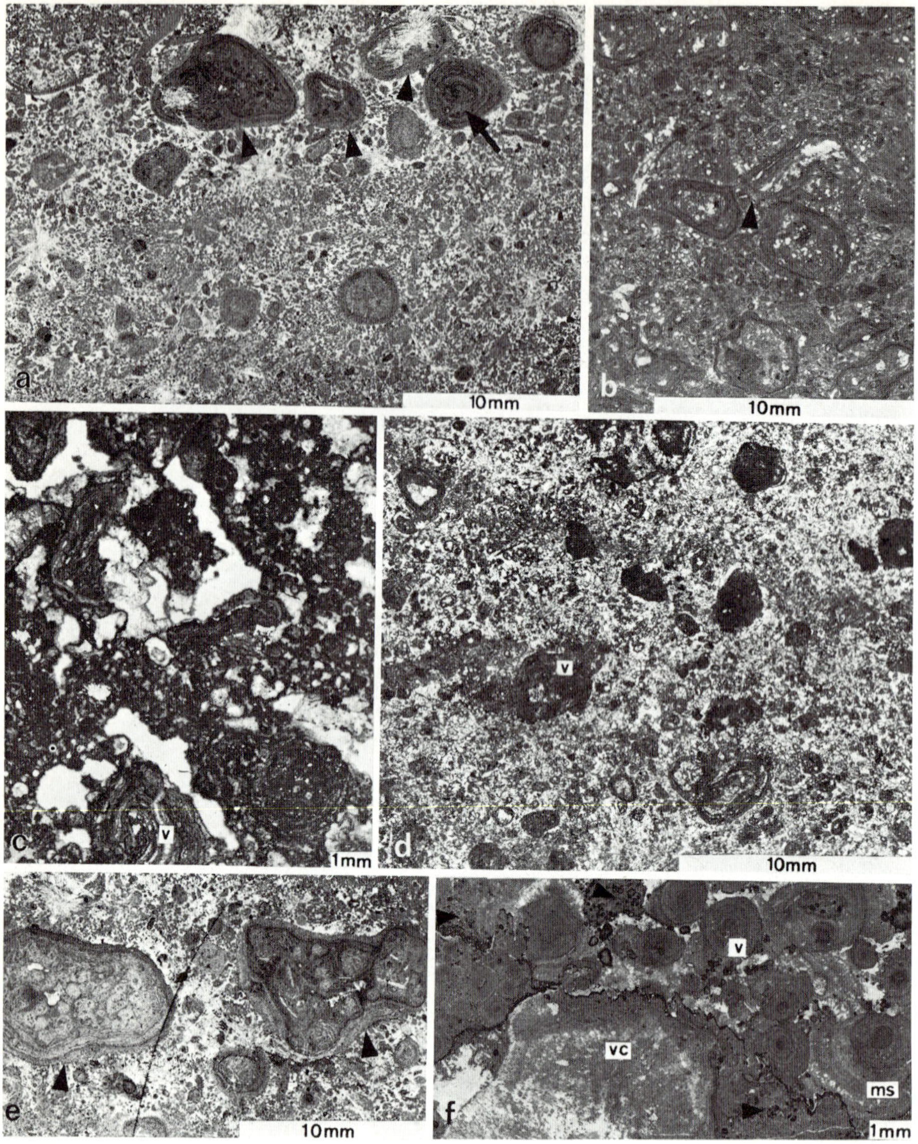

Fig. 7a–f. Vadoids occurring in caliche profile which is 28 cm thick in the well Radwanice W-4; **a** depth 1110.5 m, **e** depth 1110.4 m. Vadoids are represented by microids and pisoids of oval shape. Nuclei are formed by lithoclasts and (rarely) skeletal grains, and are commonly replaced by cortex. Cortex usually is composed of weakly individualized laminae, although quite often the lamination is very distinct. The characteristic feature of numerous pisoids (and some microids) is distinct downward thickening (*arrows*) and the presence of pockets of parent deposit between the laminae (*large arrow*). The laminae of cortex often exhibit sharp, well-fitting boundaries, similar to shrinkage fissures which have been filled with cement. The above mentioned features are diagnostic for coated grains that have been described as "calcrete ooids" and "calcrete pisoids" which originated in Quaternary near-surface vadose environments (see Peryt, this vol.) and therefore – considering also the relationship with calcrete crust – they are thought to represent vadoids which have originated in freshwater environment. Recent freshwater vadoids are composed of calcite and most probably the

which carbonates accumulate, and that in this respect the present is the analog of the past.

A commonly occurring group of coated grains in the Zechstein Limestone rocks are vadoids, earlier described by numerous workers (including the present author) as oncoids, from which they differ – as is now known – in another environment and mechanism of formation, in primary mineralogical composition and related texture dissimilarity (Table 2). Therefore, Zechstein nearshore facies should be reexamined. Oncoids in basinal facies record the slower pace of sedimentation, as earlier suggested (Peryt 1981b), especially when radial ooids co-occur with oncoids. Joint occurrence of coated grains of different origin in the Zechstein Limestone deposits and the important role played by vadoids are related to the changing environmental conditions and especially the changes of sea depth during the Zechstein Limestone deposition which resulted, as is now believed, from glacieustatic mechanism. The same mechanism must have had the same effect for sequences of coated grain types in other pre-evaporite sequences.

Acknowledgements. Most of the work was done at Institut für Geologie, Ruhr-Universität Bochum, Bochum, and Laboratoire de Geologie, Museum National d'Histoire Naturelle, Paris, and my visits there were possible due to invitations by Hans Füchtbauer and Jean-Paul Loreau, to support by the Alexander von Humboldt-Stiftung and time granted by the Instytut Geologiczny. Numerous friends and colleagues discussed the problems considered here and offered helpful suggestions, and Chris Burgess improved the language in the final manuscript. J. Modrzejewska and D. Oleksiak printed photos, and E. Lepa did the drawings.

References

Assereto RLAM, Folk RL (1976) Brick-like texture and radial rays in Triassic pisolites of Lombardy, Italy: a clue to distinguish ancient aragonite pisolites. Sed Geol 16:205 – 222

Assereto RLAM, Kendall CGStC (1977) Nature, origin and classification of peritidal tepee structures and related breccias. Sedimentology 24:153 – 210

Bathurst RGC (1968) Precipitation of ooids and other aragonite fabrics in warm seas. In: Müller G, Friedman GM (eds) Recent Developments in Carbonate Sedimentology in Central Europe. Springer, Berlin Heidelberg New York, 1 – 10

Bełka Z (1979) Shallow-water Solenoporaceae and their environmental adaptation, Upper Permian of the Holy Cross Mts. Bull Cent Rech Explor-Prod Elf-Aquitaine 3:443 – 452

Clark DN (1980) The sedimentology of the Zechstein 2 Carbonate Formation of Eastern Drenthe, the Netherlands. Contr Sedimentology 9:131 – 165

Flügel E (1978) Mikrofazielle Untersuchungsmethoden von Kalken. Springer, Berlin Heidelberg New York

above described vadoids had the same composition. **b** Ooid-vadoid grainstone. Arrows indicate the vadose compaction that is also marked, although weakly, in other places of illustrated thin section. Well Dobrzeń 1, depth 1427.9 m, **c** Vadose deposit. Vadoids (*v*) of multiphase origin, with numerous stages of breaking and regeneration. Well Wierzchowice 2, depth 1502.4 m. **d** Vadoids (*v*) within ooidal deposit. Well Czeszów 7, depth 1523.2 m. **e, f** Vadose deposits: vadoids (*v*), vadose crusts (*vc*) of distinct radial texture and microstalactitic crusts (*ms*), within only locally preserved ooidal deposit (*arrows*). Well Rogalin 1, depth 3191.3 m

Füchtbauer H (1968) Carbonate sedimentation and subsidence in the Zechstein basin (northern Germany). In: Müller G, Friedman GM (eds) Recent Developments in Carbonate Sedimentology in Central Europe. Springer, Berlin Heidelberg New York, pp 196–204

Kerkmann K (1966) Über "Oolithe" und "Stromatolithe" und die Bedeutung von Algen an der Kalksteinbildung. Wiss Z Hochsch Archit u Bauw Weimar 13:293–302

Kostecka A (1966) Litologia i sedymentacja cechsztynu synkliny gałezicko-bolechowickiej (Góry Swietokrzyskie). Prace Geol 38

Monty C, Mas JR (1981) Lower Cretaceous (Wealdian) blue-green algal deposits of the Province of Valencia, eastern Spain. In: Monty C (ed) Phanerozoic Stromatolites. Springer, Berlin Heidelberg New York, pp 85–120

Peryt TM (1977) Environmental significance of foraminiferal-algal oncolites. In: Flügel E (ed) Fossil Algae. Springer, Berlin Heidelberg New York, pp 61–65

Peryt TM (1978) Sedimentology and paleoecology of the Zechstein Limestone (Upper Permian) in the Fore-Sudetic area (western Poland). Sed Geol 20:217–243

Peryt TM (1981a) Former aragonitic submarine hemispheroids associated with vadose deposits, Zechstein Limestone (Upper Permian), Fore-Sudetic area, western Poland. N Jb Geol Paläont Mh 1981:559–570

Peryt TM (1981b) Phanerozoic oncoids: an overview. Facies 4:197–214

Peryt TM (1981c) Katodoluminescencja w badaniach skał weglanowych. Prz Geol 29:119–125

Peryt TM (in press) Chronostratigraphical and lithostratigraphical correlations of the Zechstein Limestone in Central Europe

Peryt TM, Piatkowski TS (1977) Algal-vadose pisoliths in the Zechstein Limestone (Upper Permian) of northern Poland. Sed Geol 19:275–286

Peryt TM, Wagner R (1981) Mikrostruktury stromatolitów i onkoidów z wapienia cechsztyńskiego Polski północno-zachodniej. Kwart Geol 25:629–650

Peryt TM, Ważny H (1980) Microfacies and geochemical development of the basin facies of the Zechstein Limestone (Ca1) in western Poland. Contr Sedimentology 9:279–306

Piekarska E, Kwiatkowski S (1975) Microfacial analysis of the Zechstein Limestone in the eastern part of the Peri-Baltic syneclise (Preliminary report). Acta Geol Pol 25:79–114

Purser BH (1980) Sédimentation et diagenèse des carbonates néritiques récents. Éditions Technip, Paris

Radwański A, Birkenmajer K (1977) Oolitic/pisolitic dolostones from the Late Precambrian of south Spitsbergen: their sedimentary environment and diagenesis. Acta Geol Pol 27:1–39

Read JF (1974) Calcrete deposits and Quaternary sediments, Edel Province, Shark Bay, Western Australia. Amer Assoc Petrol Geol Mem 22:250–282

Richter DK, Füchtbauer H (1978) Ferroan calcite replacement indicates former magnesian calcite skeletons. Sedimentology 25:843–860

Scholle PA, Kinsman DJJ (1974) Aragonitic and high-Mg calcite caliche from the Persian Gulf – a modern analog for the Permian of Texas and New Mexico. Jour Sed Petrol 44:904–916

Smith DB (in press) The Trow Point Bed – a deposit of Upper Permian marine oncoids, peloids and columnar stromatolites in the Zechstein of north-east England

Wray JL (1977) Late Paleozoic calcareous red algae. In: Flügel E (ed) Fossil Algae. Springer, Berlin Heidelberg New York, pp 167–176

VI.5 Unusual Algal-Crystalline Carbonate Coated Grains from the Capitan Reef (Permian, Guadalupian), New Mexico, USA

S. J. Mazzullo[1] and John M. Cys[2]

1 Introduction

An unusual type of coated-grain nodule is common throughout reef and fore-reef facies of the Capitan Limestone (Guadalupian) in New Mexico (Fig. 1). These nodules, which may attain a maximum diameter of 3.5 cm, consist internally of concentric, irregularly laminated cortices, surrounding skeletal fragment nuclei, of alternating *Archaeolithoporella* (?red alga) and fibroradial calcite replacive of marine, acicular aragonite. The occurrence within these nodules of crustiform red algae suggests their comparison to rhodoids. However, such nodules are distinct from classic modern and ancient rhodoids in terms of the absence of characteristic algal growth forms (Bosellini and Ginsburg 1971), the interpreted dual accretionary mode of cortical laminite formation (viz algal encrustation and carbonate precipitation), and mineralogy (aragonite vs high-magnesian calcite). Alternatively, the characteristic laminated cortices and radial crystalline microfabrics of these nodules suggests a certain similarity to both vadoids and coniatoids, albeit algal encrusted and solely of subtidal marine origin. This contribution describes the geologic occurrence, morphologies, internal structures, microfabrics, and origin of these peculiar coated grains.

2 General Observations

Laminated encrustations of *Archaeolithoporella* and inferred fibroradial aragonite occur in a continuum of forms of marine origin in the Capitan Limestone (Babcock 1977, Mazzullo and Cys 1978), including: (1) large, immobile cryptozoon-like heads (Fig. 2A) commonly confused for cyanobacterial stromatolites (Newell et al. 1953) and resembling coniatolites (Purser and Loreau 1973); (2) laminar crusts on bedding planes (Fig. 2B) or within reef fissures, the former bearing superficial similarity to the pelagosites of the Persian Gulf (Purser and Loreau 1973); and (3) this study: mobile (Figs. 2C, D) and semiat-

1 602 No. Baird, Suite 220, Midland, TX 79701, USA
2 Mapco Inc., Production Division – Midland District, 218 Western United Life Bldg., Midland, TX 79701, USA

Coated Grains (ed. by T.M. Peryt)

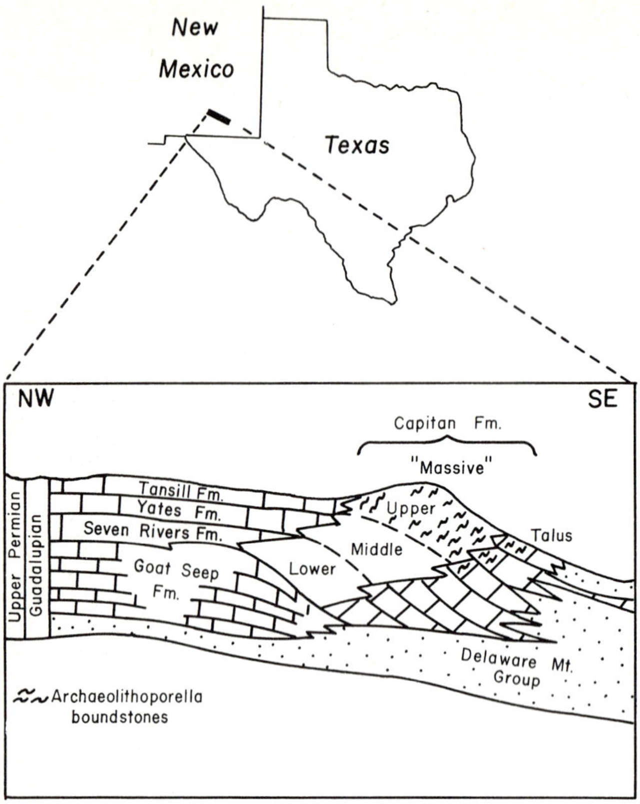

Fig. 1. Stratigraphic distribution of *Archaeolithoporella*-boundstones and pisoids in upper Capitan Fm., along a shelf-to-basin cross-section located by the *solid line* on the index map of southeastern New Mexico

tached nodules, the latter of which locally may have coalesced laterally as a result of binding by marine cements to form thick, mammillated crusts (Figs. 2E, F and 4). All these types are well-developed within the shallow subtidal, leeward-reef facies of the Capitan, where a lateral or bathymetric shoreline zonation of forms is readily apparent (Mazzullo and Cys, in prep.). However, the mobile and semi-attached nodules are now known to be ubiquitous in muddy and marine cement-rich carbonate rocks from the leeward reef to progressively deeper water, upper and middle fore-reef facies of the Capitan, in strata equivalent temporally to the back-reef Tansill and basinal Bell Canyon formations (Fig. 1). These nodules are admixed within normal marine, fossiliferous sediments, and generally comprise only a few percent of total sediment volume. Denser concentrations of such nodules do not exhibit depositional and diagenetic textures that may be indicative of peritidal or soil caliche (e.g., Dunham 1969, Scholle and Kinsman 1974, Esteban 1976).

The nodules range in diameter from a few mm to a maximum of 3.5 cm, and vary in shape from spherical to hemispheroidal (Fig. 2). In general, the larger

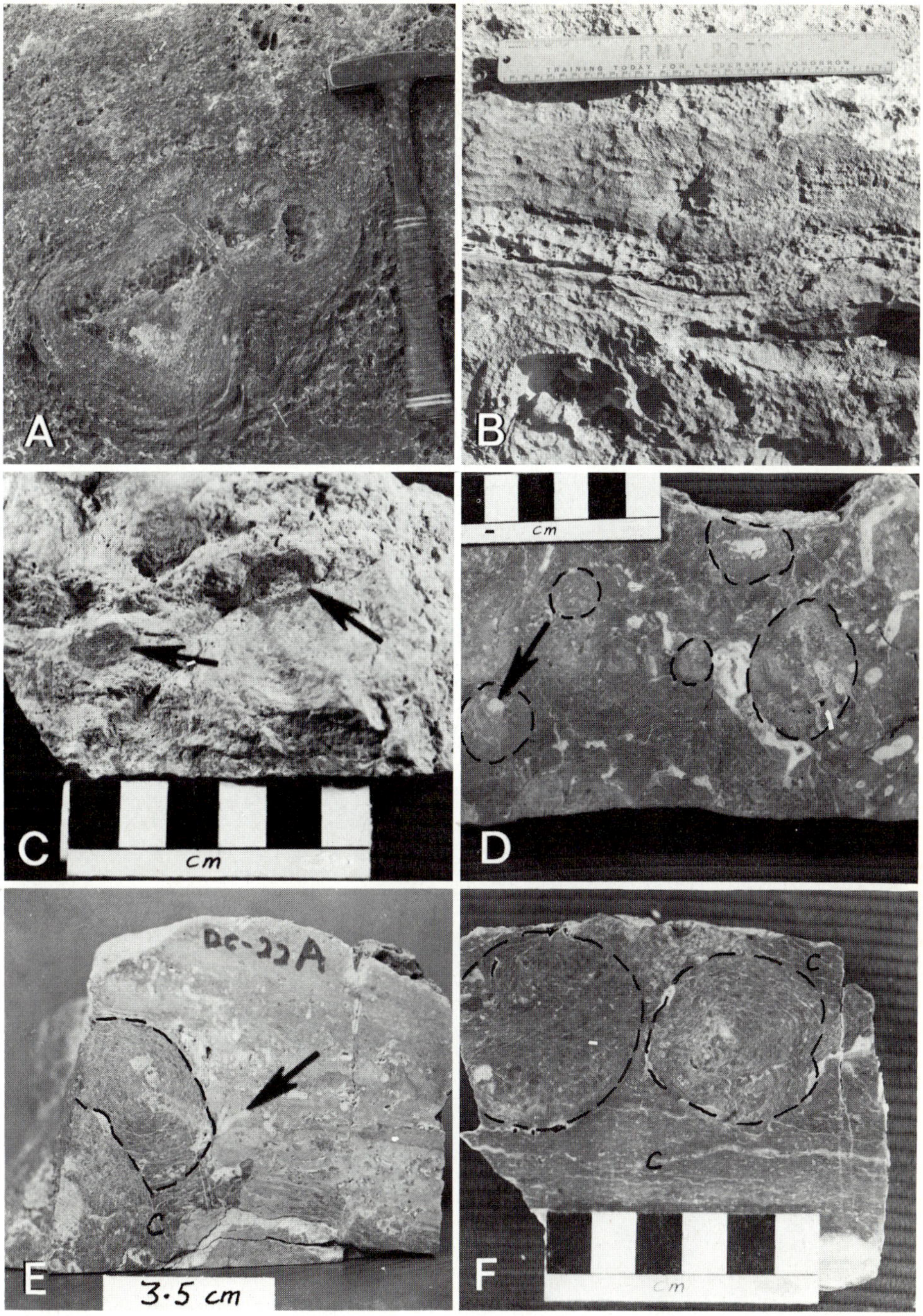

Fig. 2. A Bedding plane view of large *Archaeolithoporella*-cement "head"; reef to back-reef transition, Dark Canyon. **B** Laminar crusts of *Archaeolithoporella* aragonite boundstone, transverse view; reef to back-reef transition, Dark Canyon; lower part of ruler scaled in cm. **C** Sample of two isolated, spherical pisoids (*arrows*) with sponge nuclei; upper fore-reef, Rattlesnake Canyon. **D** Polished slab of large, partly coalesced pisoids (*outlined*), with recrystallized sponge nuclei (*arrow*); upper fore-reef, Rattlesnake Canyon. **E** Polished slab of pisoid (*outlined*) attached by marine cement (*c*) and enclosed in *Tubiphytes* (*arrow*) wackestone; reef to back-reef transition, Dark canyon. **F** Polished slab of two coalesced pisoids (*outlined*) overlain and underlain by laminar, mammillated crust (*c*); reef to back-reef transition, Dark Canyon

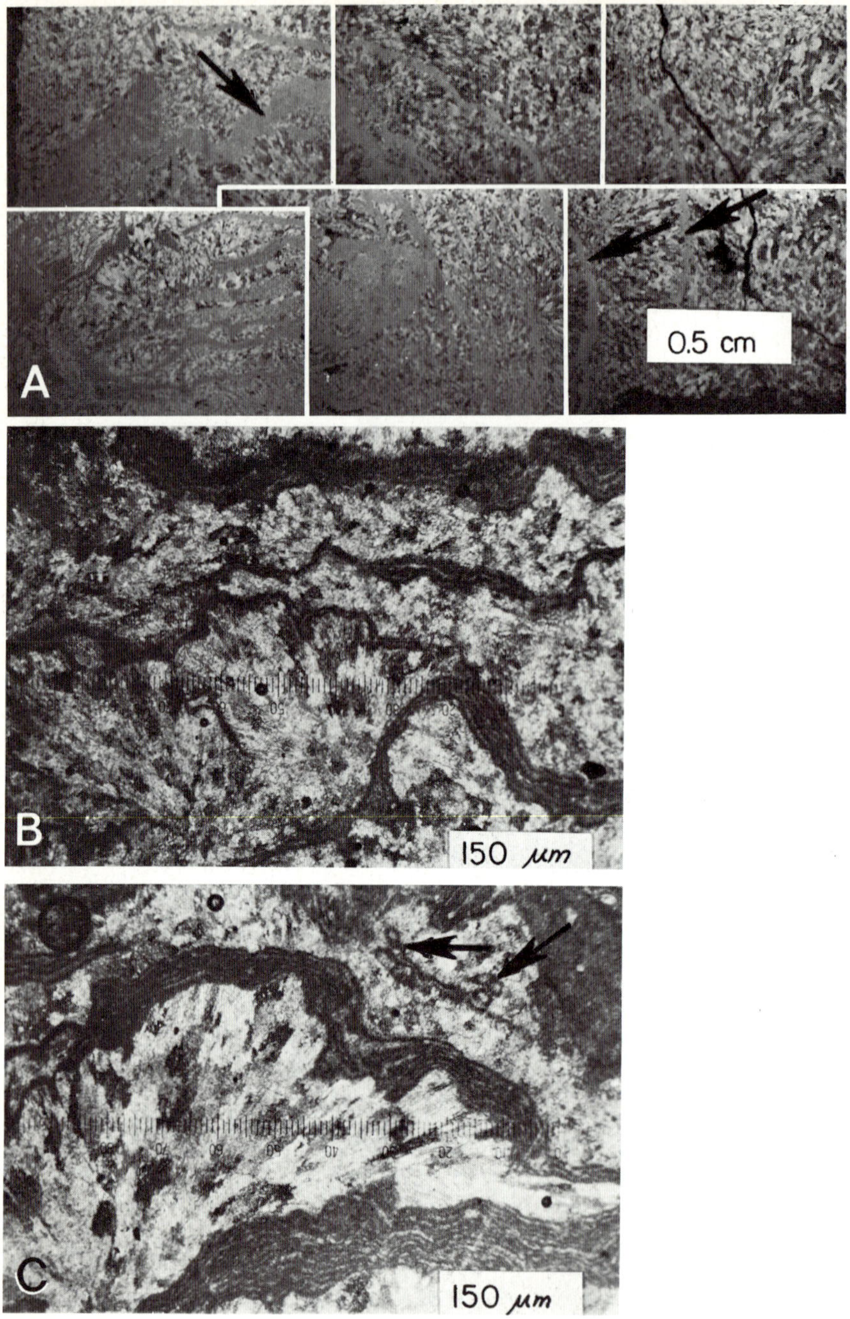

Fig. 3. A Thin-section photomicrograph of coniatoid showing *Archaeolithoporella* laminae (*arrows*) and recrystallized fibroradial calcite in cortex; nucleus out of field near lower left-hand of photo (cross-nicols). **B** Thin-section photomicrograph of coniatoid cortex illustrating *Archaeolithoporella* and fibroradial calcite. Note protuberances of calcite and crenulated laminae (cross-nicols). **C** Thin-section photomicrograph of coniatoid cortex illustrating detail of *Archaeolithoporella,* square-tipped crystals of fibroradial calcite, crenulated laminae, entrapped micritic debris (*upper right-hand corner*), and encrusting forams (*arrows*) (cross-nicols)

nodules are those hemispheroidal varieties that had become cemented to the substratum (Fig. 2E). No systematic correlation was evident between nodule size, shape, and environment of formation from reef to fore-reef facies. Rather, the potential for continual accretionary nodule growth is an inverse function of the rapidity of burial in sediment or encasement in marine cements, regardless of depositional environment. Nuclei normally include recrystallized sponge or other biotic fragments (Figs. 2 and 3) and, less commonly, small limestone intraclasts. The laminated cortices comprise approximately 70% or more of the nodules, and consist almost entirely of alternating laminae, in order of increasing volumetric abundance, of *Archaeolithoporella* and fibroradial, crystalline calcite (Fig. 3). Encrusting foraminifers, *Tubiphytes*, and entrapped peloidal debris are common in some samples (Fig. 3). although are volumetrically insignificant to cortical compositions. Cortical laminations are pronounced due to the contrast between the algal and crystalline calcite laminae (Figs. 2 and 3). However, protuberances of fibroradial calcite and variations in thickness of the associated algal encrustations result in irregular cortical laminae of wavy to crenulated appearance (Fig. 3). Successive generations of encrusting *Archaeolithoporella* are ineffective in reducing such cortical irregularities and, consequently, the outer surfaces of most of the nodules are somewhat botryoidal to mammillary (Fig. 2C–E). The cortices of the smaller spherical, mobile nodules are reasonably symmetrical and of uniform thickness (Fig. 2C, D). In contrast, the larger mobile and semi-attached nodules are characterized by asymmetrical, astropetal cortices, with resultant hemispheroidal external appearance (Fig. 2E). It is not uncommon for some of the larger nodules to exhibit a progressive internal change from spherical and symmetrical to astropetal, hemispheroidal appearance (Fig. 2E). No systemic correlation was found to exist among cortical symmetry of thickness, relative thickness of individual algal or calcite laminae, and site of nodule formation. However, there is a positive correlation between nodule growth form (e.g., cortical thickness, symmetry) and mode of formation; this aspect will be discussed later. Regardless of external morphology, symmetry, and environment of formation, the cortices of these nodules consist entirely of concentrically disposed laminae of algae and fibroradial calcite.

3 Microfabric of Cortices

3.1 *Archaeolithoporella*

The genus *Archaeolithoporella* (Virgilian-Guadalupian) in samples from the Capitan Limestone has been compared to the holotype described originally by Endo (1959) from Permian rocks of Japan by Babcock (1974, 1977); the genus occurs exclusively in marine rocks of the Capitan reef and fore-reef facies. Petrographically, samples of *Archaeolithoporella* are characterized by alternating light and dark colored calcite microlaminae, each averaging approximately 20 μm in thickness (Fig. 3). The former laminae consist of a tight microfabric of equant nannograins of 4–6 μm diameter, whereas the dark laminae are mosaics of submicron nannograins with interspersed organic matter. Within nodule cortices, encrustations of *Archaeolithoporella* may vary from a few tens of microns to several mm in thickness, but normally average approximately 90 µm. Individual laminae generally are of uniform thickness around the cortex, although some thinning of encrustations over protuberances of underlying calcite are not uncommon. There appears to be no systematic variation in the thickness of successive algal laminations within the cortices of individual nodules. However, the asymmetrical nodule types generally have thicker algal (and calcite) laminae than do the symmetrical, mobile nodules.

The lack of preserved microstructural details within specimens of *Archaeolithoporella* precludes their unequivocal placement into a specific algal family (Babcock 1977, Wray 1977). However, the genus is most commonly regarded as a problematicum, with possible affinity to the ancestral Corallinaceae (Johnson 1963, Riding 1977) or, by analogy of its habit and occurrence, to modern squamariacean red algae (Babcock 1974, Mazzullo and Cys 1978). The ubiquitous association of *Archaeolithoporella* with inferred marine aragonite, as well as the lack of biostructural detail due to recrystallization in the specimens examined, may indicate a host control on aragonite nucleation and, accordingly, an aragonitic composition (e.g., squamariacean algae) of this entity (Mazzullo and Cys 1978).

3.2 Crystalline Carbonate

Separating the *Archaeolithoporella* laminae within the nodule cortices are relatively thicker (average 300 µm) laminae of fibroradial calcite (Fig. 3). In general, these calcite laminae are thicker in the asymmetrical (hemispheroidal) than in the symmetrical (spherical) nodules, although the thicknesses of individual laminae within a given cortex are relatively uniform. The first generation of accretionary material upon a nucleus may be either algal or crystalline calcite. Successive calcite laminae are deposited on the algal substrate as an isopachous crust and/or as a series of laterally coalescent fan-druses, with radial c-axial orientations (Fig. 3). The presence of the algae does not appear to have effected the radial disposition of crystals within the calcite laminae as in the lacustrine pisoids described by Jones and Wilkinson (1978). Overlapping laminae indicative of syndepositional erosion and/or limited nodule overturning (e.g., Donahue 1969, Jones and Wilkinson 1978) are generally absent from the samples examined (not to be confused with younger symmetrical nodules that later enlarge astropetally: Fig. 2E). The bulk of the nodule cortices is composed of this crystalline calcite. Insoluble residue analyses of selected cortices and associated algal-carbonate boundstones in the Capitan Limestone yielded only 0.49 weight % (0.16% – 0.92%) organic matter, which corresponds to less than 20% by area of the cortices (Fig. 3). The syndepositional marine origin of this crystalline carbonate is indicated by the following lines of evidence: (1) multiple generations of carbonate precipitation, encrustation by marine organisms, internal sedimentation of fossiliferous peloidal debris (minor), and endolithic boring of the crystalline mosaic (although not as severe as in many Holocene rhodoids and marine hardgrounds, e.g., Alexandersson 1974, 1977, Focke and Gebelein 1978, Shinn 1969); (2) the apparent semi-lithified character of associated similar algal-cement boundstones in the Capitan (Mazzullo and Cys 1977, 1978, in prep.); and (3) the similarity of this fibroradial calcite to inferred ancient marine aragonite cements (Mazzullo and Cys 1979, Mazzullo 1980). These (Fig. 3) and other fibroradial crystalline mosaics, albeit highly recrystallized, in the Capitan and elsewhere have been interpreted by Mazzullo and Cys (1977, 1978, 1979) and Mazzullo (1980) as former marine, acicular aragonite on the basis of included relict orthorhombic fibers (Fig. 3), relatively high residual Sr^{2+} concentrations (375 – 8800 ppm), and correspondingly low $MgCO_3$ contents (0.09 – 0.95%). There is no evidence to support the contention that these crystalline mosaics resulted from the recrystallization of particulate micrite or that they represent void-filling cements of syndepositional or meteoric origin. Indeed, the coherent depositional integrity of these nodules was a consequence of the presence therein of this (former aragonite) crystalline carbonate mosaic.

4 Discussion

In terms of the presence of red algae and associated epibionts, these Permian coated-grain nodules only superficially resemble rhodoids because of: (1) the absence of laminar to columnar algal cortices as are typical of Holocene and ancient rhodoids described by Bosellini and Ginsburg (1971), Adey and Macintyre (1973), Focke and Gebelein (1978), and Toomey (1975); apparently, the laminar uniformity of these nodule cortices is inherent from the restriction of *Archaeolithoporella* to a crustiform rather than ramose habit (Babcock 1977); (2) the lack of textural features indicative of sustained endolith activity and resulting internal sedimentation and marine lithification as is so typical of Holocene rhodoids (Alexandersson 1974, 1977, Focke and Gebelein 1978); (3) the occurrence of aragonite rather than high-magnesian calcite comprising the cortices; and (4) the questionable coralline algal affinity of *Archaeolithoporella*.

The laminated cortices and abundance of fibroradial crystalline microfabrics invoke more reasonable analogy of the nodules to many Holocene and ancient peritidal pisoids, particularly vadoids, with regard to their inferred aragonite mineralogy (e.g., Shinn 1973, Assereto and Folk 1976, Esteban 1976, Loucks and Folk 1976). Accordingly, these nodules are distinct from those Holocene terrestrial vadoids whose cortices consist of interlaminated radial to bladed and unoriented (micritic) calcite (Donahue 1969, Jones and Wilkinson 1978, Braithwaite 1979, Risacher and Eugster 1979), as well as from: (1) those marine spray zone, caliche-type pisoids (vadoids) with laminae of unoriented, micritic aragonite (Scholle and Kinsman 1974); (2) presumed soil caliche vadoids (Dunham 1969). The occurence of these Permian nodules within subtidal marine rocks of the Capitan, as well as their association with marine epibionts, argues for a subtidal marine and against a peritidal origin.

These nodules are considered to be algal-encrusted coniatoids because of the observation that encrustation of allochems by aragonite in marine rocks of the

Capitan did not necessarily everywhere involve an assocication with *Archaeolithoporella* (Mazzullo and Cys 1977, 1978). Accordingly, we believe these coated grains originated as a result of alternating periods of inorganic aragonite precipitation and subsequent algal encrustation with the reservation, considering the pervasive occurrence of aragonite cements in the Capitan, that the algae were not essential in influencing ambient carbonate saturation states (e.g., that precipitation of aragonite was physicochemically induced). Although fibroradial calcite replacive of acicular aragonite occurs abundantly as inter- and intraparticle cements in the Capitan (Babcock 1977, Mazzullo and Cys 1977, 1978), the calcite laminae within the cortices of these coated grains are not cement sensu stricto. Rather, the textural evidence suggests the in situ precipitation of fibroradial aragonite around grain nuclei. We can assume, by the foregoing empirical observations, that the precipitation of aragonite laminae was more or less continuous relative to episodic algal encrustation. However, whether these algal-aragonite couplets represent a specific periodicity of growth, and the possible reasons for such alternation of events, is presently unknown. Furthermore, the reasons for pervasive aragonite rather than high-magnesian calcite precipitation, with specific reference to Holocene marine cements (e.g., Mazzullo and Cys 1979) and modern and ancient peritidal vadoids, likewise are unclear.

As with rhodoids and vadoids, variations in the external morphology and cortical structures of these nodules were indeed influenced to a great extent by frequency of movement (e.g., Bosellini and Ginsburg 1971, Donahue 1969, Focke and Gebelein 1978), although relationships between these parameters appear to have been independent of their site (e.g., depth) of formation. More important to their formation, however, was the complex competition between relative rates of cortical accretion and the growth-retardant processes of burial by sediment and/or encasement in marine cements, all of which were influenced by energy levels in the depositional environment. The natural history of these nodules appears to have involved, in many cases, a maturity-index continuum of morphological changes related to growth and nodule mobility (Fig. 4). The sphericity of

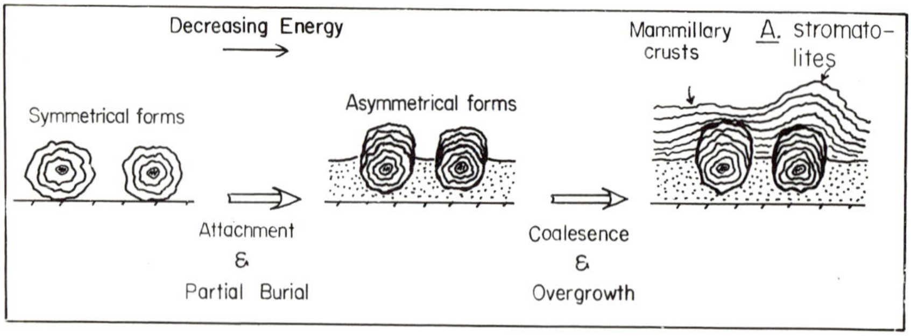

Fig. 4. Idealized schematic illustrating progressive evolution of particles and boundstones. Symmetrical forms become sedentary due to energy level decreases (relative or absolute), and are cemented to the substratum and/or buried in sediment (*stippled*). Cortical enlargement of exposed surfaces results in larger, asymmetrical (astropetal) particles. Coalescence and overgrowth of these particles by additional algal-aragonite laminae results in either superposed crusts or *Archaeolithoporella* stromatolites

the non-attached, symmetrical nodules, as well as their uniformly thick cortices (Fig. 2C, D), are features that may be suggestive of formation as a result of essentially uninterrupted accretion of laminae in an enviroment of relatively continuous nodule movement (e.g., Donahue 1969; but refer to Jones and Wilkinson 1978). However, the absence of overlapping and erosionally truncated cortical laminae enforces presently undefineable limits on maximum energy levels and degree of mobility as inferred strictly from comparative studies of particle morphology and cortical structure (refer to Jones and Wilkinson 1978). Nodules formed or transported into areas of decreased turbulence and particle movement are then subjected to an environment in which increased relative rates of interparticle marine cementation and/or sediment influx retard the continued symmetrical development of cortices (Fig. 4). Alternatively, it is possible that once spherical nodules attain a critical size, they become more sedentary and behave similarly. Such nodules commonly become attached (cemented) to the substrate (Fig. 2E), and vertical growth continues as a consequence of astropetal cortical enlargement rather than concentric accretion, resulting in asymmetrical, hemispheroidal nodules of considerably large size (maximum 3.5 cm) (Fig. 2E). Such nodules are characterized by thicker component laminae than are mobile spherical types, especially along their upper surfaces (Figs. 2E and 4). Once sedentary, adjoining nodules eventually may coalesce via interparticle marine cementation, and can be overlain by sheets of *Archaeolithoporella*-aragonite laminae to form thick mammillary crusts or *Archaeolithoporella* stromatolites (Figs. 2A, F and 4).

Assuming these nodules to be coniatoids, it is interesting that they are of subtidal marine rather than peritidal origin. Their distribution in the Capitan shows no systematic change in morphotypes across the reef depositional profile. Furthermore, correlations of particle morphology with depth and/or mobility alone are unwarranted because of the complicating effects on particle growth of concomitant particulate sedimentation and pervasive marine cementation. Accordingly, it may be an overextension of Holocene principles regarding these and other ancient coated grain occurrences and comparative morphologies as indicative of formation in specific depositional and/or diagenetic environments.

Acknowledgements. Permission to publish this paper courtesy of Union Texas Petroleum Corporation. Access to National Park and private lands granted by Carlsbad Caverns and Guadalupe Mountains National Parks Service and Mr. W. Hamann, respectively. Earlier versions of this manuscript were reviewed by A. J. Lomando and Tad Peryt.

References

Adey WH, Macintyre IG (1973) Crustose coralline algae: a re-evaluation in the geological sciences. Geol Soc Am Bull 84:883 – 904

Alexandersson ET (1974) Carbonate cementation in coralline algal nodules in the Skagerrak, North Sea: biochemical precipitation in undersaturated waters. J. Sediment Petrol 44:7 – 26

Alexandersson ET (1977) Carbonate cementation in Recent coralline algal constructions. In: Flügel E (ed) Fossil algae. Springer, Berlin Heidelberg New York, pp 261 – 269

Assereto R, Folk RL (1976) Brick-like texture and radial rays in Triassic pisolites of Lombardy, Italy: a clue to distinguish ancient aragonitic pisolites. Sediment Geol 16:205 – 222

Babcock JA (1974) Role of algae in Capitan Limestone (Permian, Guadalupian), Guadalupe Mountains, West Texas and New Mexico. PhD dissertation Univ Wisconsin, Madison, p 241

Babcock JA (1977) Calcareous algae, organic boundstones, and the genesis of the Upper Capitan Limestone (Permian, Guadalupian), Guadalupe Mts, West Texas & New Mexico. In: Hileman ME, Mazzullo SJ (eds) Upper Guadalupian facies, Permian reef complex, Guadalupe Mountains, New Mexico and West Texas. Permian Basin Sect SEPM, Publ 77-16, Midland TX, pp 3 – 44

Bosellini A, Ginsburg RN (1971) Form and internal structure of recent algal nodules (rhodolites) from Bermuda. J Geol 79:669 – 682

Braithwaite CJR (1979) Crystal textures of recent fluvial pisolites and laminated crystalline crusts in Dyfed, South Wales. J Sediment Pertrol 49:181 – 194

Donahue J (1969) Genesis of oolite and pisolite grains: an energy index. J Sediment Petrol 39:1399 – 1411

Dunham RJ (1969) Vadose pisolite in the Capitan reef (Permian), New Mexico and Texas. In: Friedman GM (ed) Depositional environments in carbonate rocks. SEPM Spec Publ 14, pp 182 – 191

Endo R (1959) Stratigraphical and paleontological studies of the Later Paleozoic calcareous algae in Japan, XIV: fossil algae from the Hyugawa Valley in the Hida Massif. Saitama Univ Sci Rep Ser B 3:177 – 217

Esteban M (1976) Vadose pisolite and caliche. Am Assoc Petrol Geol Bull 60:2048 – 2057

Focke JW, Gebelein CD (1978) Marine lithification of reef rock and rhodolites at a fore-reef slope locality (– 50 m) off Bermuda. Geol Min 57:163 – 171

Johnson JH (1963) Pennsylvanian and Permian algae. Colo Sch Mines Quat 58:211 p

Jones FG, Wilkinson BH (1978) Structure and growth of lacustrine pisoliths from Recent Michigan marl lakes. J Sediment Petrol 48:1103 – 1110

Loucks RG, Folk RL (1976) Fanlike rays of former aragonite in Permian Capitan reef pisolites. J Sediment Petrol 46:483 – 485

Mazzullo SJ (1980) Calcite pseudospar replacive of marine acicular aragonite, and implications for aragonite cement diagenesis. J Sediment Petrol 50:409 – 422

Mazzullo SJ, Cys JM (1977) Submarine cements in Permian boundstones and reef-associated rocks, Guadalupe Mountains, West Texas and southeastern New Mexico. In: Hileman ME, Mazzullo SJ (eds) Upper Guadalupian facies, Permian reef complex, Guadalupe Mountains, New Mexico and West Texas. Permian Basin Sect SEPM, Publ 77-16, Midland TX, pp 151 – 200

Mazzullo SJ, Cys JM (1978) *Archaeolithoporella*-boundstones and marine aragonite cements, Permian Capitan reef, New Mexico and Texas, USA. N Jahrb Palaeontol Mh 10:600 – 611

Mazzullo SJ, Cys JM (1979) Marine aragonite sea-floor growths and cements in Permian phylloid algal mounds, Sacramento Mountains, New Mexico. J Sediment Petrol 49:917 – 936

Newell ND et al. (1953) The Permian reef complex of the Guadalupe Mountains region, Texas and New Mexico. Freeman, San Francisco, p 236

Purser BH, Loreau J-P (1973) Aragonitic, supratidal encrustations on the Trucial Coast, Persian Gulf. In: Purser BH (ed) The Persian Gulf. Springer, Berlin Heidelberg New York, pp 343 – 376

Riding R (1977) Problems of affinitiy in Paleozoic calcareous algae. In: Flügel E (ed) Fossil algae. Springer, Berlin Heidelberg New York, pp 202 – 211

Risacher F, Eugster HP (1979) Holocene pisoliths and encrustations associated with spring-fed surface pools, Pastos Grandes, Bolivia. Sedimentology 26:253 – 270

Scholle PA, Kinsman DJJ (1974) Aragonitic and high-Mg calcite caliche from the Persian Gulf – a modern analog for the Permian of Texas and New Mexico. J Sediment Petrol 44:904 – 916

Shinn EA (1969) Submarine lithification of Holocene carbonate sediments in the Persian Gulf. Sedimentology 12:109 – 144

Shinn EA (1973) Sedimentary accretion along the leeward, SE coast of Qatar Peninsula, Persian Gulf. In: Purser BH (ed) The Persian Gulf. Springer, Berlin Heidelberg New York, pp 199 – 209

Toomey DF (1975) Rhodoliths from the Upper Paleozoic of Kansas and the Recent – a comparison. N Jahrb Palaeontol Mh 4:242 – 255

Wray JL (1977) Late Paleozoic calcareous red algae. In: Flügel E (ed) Fossil algae. Springer, Berlin Heidelberg New York, pp 167 – 176

VI.6 Contrasted Occurrence of Lower Devonian Coated Grains, Northeastern Armorican Massif, France

J. Poncet [1]

1 Introduction

In the northeastern Armorican Massif the Pont-aux-Bouchers Formation (formation d in Fig. 1), which is of Middle to Upper Siegenian age, is truncated by an erosion surface. This formation, although mainly composed of terrigenous detritus (fine sands, shales) includes a locally developed reefal horizon (reefal horizon of Baubigny; Poncet 1968). A study of the facies of this horizon has enabled the distinction of two different zones of sedimentation resulting from the development of bioherms (Poncet 1972, 1976); a zone of sedimentation lateral to the

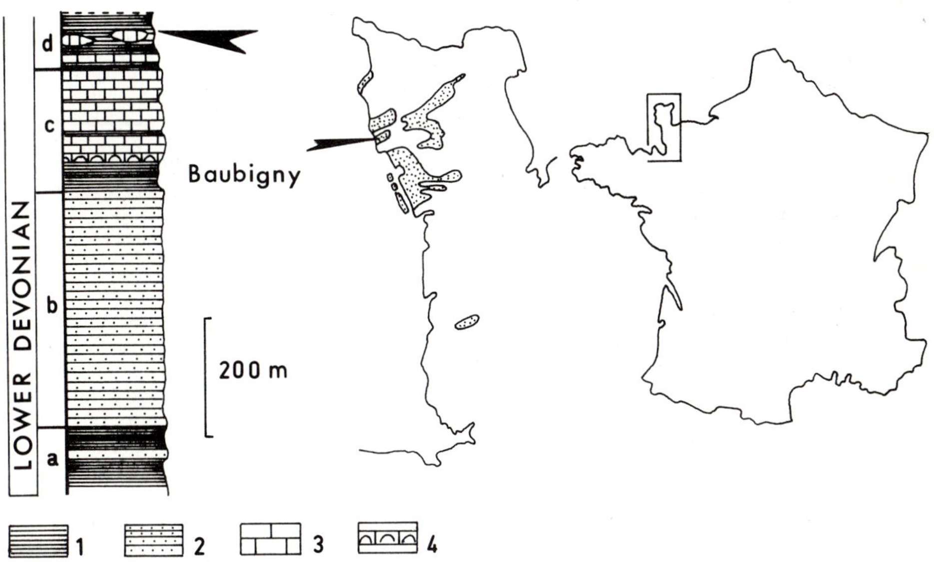

Fig. 1. Location map and stratigraphy. *Stippled:* Lower Devonian. *a* St. Germain-sur-Ay formation, *b* La Haye-du-Puits formation, *c* Néhou formation, *d* Pont-aux-Bouchers formation. *1* Shale, *2* sandstone, *3* limestone, *4* stromatolites

1 Laboratoire de Géologie armoricaine et GRECO 130007 du CNRS, Université de Caen, 14032 Caen Cedex, France

Coated Grains (ed. by T.M. Peryt)

bioherms on the one hand and a zone of back-reef sedimentation on the other. A lithologic study showed that in both zones numerous and diversified coated grains occur.

The aim of the present study is to describe these coated grains and attempt to define the characteristics of the sites of their genesis.

2 Coated Grains in the Lateral Sedimentation Zone

A calcarenitic bank, which was attached to the bioherm, has been studied. This bank consists of grainstones composed of crinoidal bioclasts and bryozoan fragments transported from the bioherm source by currents. The size of the bioclasts decreases as one moves away from the bioherm. In the distal part of this bank, ooids constitute about 60% of the sediment.

2.1 Concentric Ooids (Fig. 2)

Ooids are circular or ovoid in shape with diameters ranging between 0.350 and 0.675 mm. The internal structures consist of one to ten calcitic envelopes with a yellow to amber colour. Their thickness is not constant with thinning over irregularities in the nuclei. Nuclei are composed of crinoid and bryozoan fragments. Concentric ooids and bioclasts are cemented by large poikilotopic calcite crystals which grow syntaxially from crinoid bioclasts. Concentric ooids have a tendency to recrystallize which results in the obliteration of the envelopes.

2.2 Formation in Situ of Concentric Ooids in the Lateral Sedimentation Zone

In recent environments, concentric ooids form in environments of high turbulence and lower turbulence (Loreau 1973, Purser 1980). The ooids must remain for long periods in areas influenced by bidirectional currents. Taking all these points into account, it seems that the concentric ooids in the distal part of the lateral sedimentation zone developed in situ, in a turbulent environment (grainstone fabric) affected by bi-directional currents (cross-stratification).

Fig. 2a – f. Concentric ooids (thin section, ×80).**a** Ooid with well-individualized envelopes. Nucleus consists of a peloid. **b** Ooid with faintly individualized envelopes. Nucleus consists of a crinoid ossicle. **c** The dark micritic patches mainly localized in envelopes are probably caused by boring microorganisms (endolithic algae or fungi). Nucleus consists of a crinoid ossicle. **d** Note the variable thickness of envelopes. Nucleus consists of a bryozoan fragment. **e** Ooid with a recrystallized nucleus. The majority of envelopes except of the outermost one are replaced by neomorphic spar. **f** Ooid cemented by syntaxial calcite in optical continuity with the two crinoid grains (*right*)

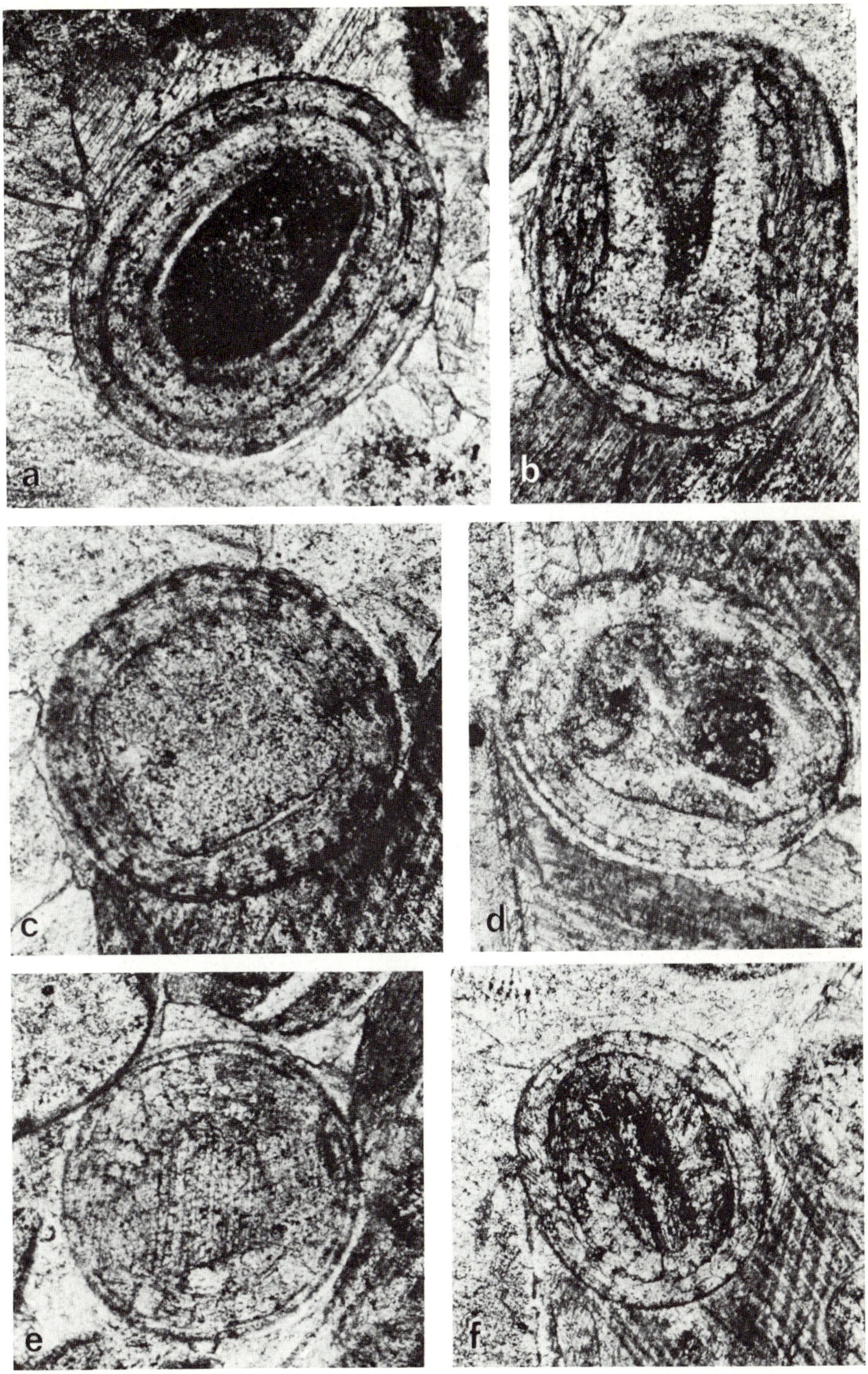

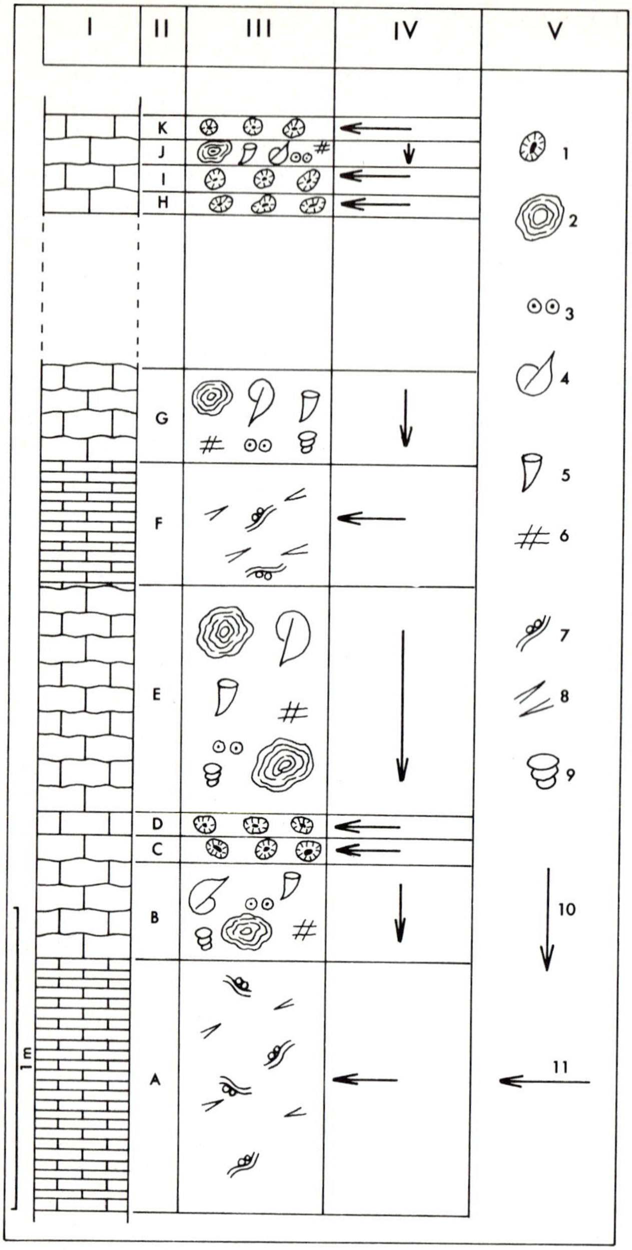

Fig. 3. Log of the back-reef section. *I* lithology, *II* beds identification, *III* organic and inorganic content, *IV* autochthonous or allochthonous character of organic and inorganic content, *V* caption. *1* radial fibrous ooid, *2* oncoid, *3* crinoid ossicle, *4* brachiopod, *5* coral (fragment), *6* bryozoan, *7* interwoven blue-green algal filaments, *8* pteropod, *9* gastropod, *10 vertical arrow* indicates the autochthonous character of organic and inorganic content, *11 horizontal arrow* indicates the allochthonous character of organic and inorganic content

3 Coated Grains in the Back-Reef Area

Sediments in the back-reef zone have been observed on a 4.30 m high section (Fig. 3). The facies in this section can be grouped into two types. Beds B, E, G, J originated in sheltered back-reef zones, while beds A, C, D, F, H, I and K contain allochthonous sediments which were brought to the back reef zone by occasional currents induced by storms and thus indicate a middle to high energy level.

3.1 Autochthonous Coated Grains in the Back-Reef Zone

The carbonate beds characteristic of back-reef sedimentation contain skeletal grains of a benthic fauna consisting of brachiopods, bryozoans, gastropods, and corals which has been slowly covered by carbonate mud. On the surface of the latter a layer of oncoids and the other coated grains, which resulted from the activity of the blue-greens, occurs.

3.1.1 Oncoids (Fig. 4). Their sizes range from 1 to 12 mm with two peaks, one at about 2.5 mm and the other at about 7.5 mm. Their shapes are diversified and assume those of the nuclei. The nuclei derive from tests of gastropods, brachiopods, fragments of bryozoans and crinoidal columnals. Encrusting blue-green algae built envelopes around these nuclei and filaments of *Girvanella* sp. may be seen in the micritic envelopes. Micrite precipitation within the mucilaginous sheath of algal filaments probably resulted from algal photosynthesis and small quartz grains were trapped in the algal envelopes. Quite often, these envelopes are punctuated by alignments of small pyrite crystals. Some oncoids have grown asymmetrically suggesting a certain amount of stability and only weak turbulence near the bottom.

3.1.2 Other Coated Grains (Fig. 4c). Are represented by tests with a micritic coating on their upper sides. In the micrite some blue-green algal filaments are visible, and it is concluded that the micritic deposit resulted from the photosynthetic activity of algal filaments, and that this is another indication of low energy levels and a subtidal algal mat.

Fig. 4. a Oncoidal wackestone-packstone. The matrix consists of a wackestone with quartz grains, fine shell debris and blebs of interwoven algal filaments (*arrow*). Oncoids are asymmetric. They consist of clear laminae (microspar) and dark laminae (micrite) in which algal filaments and trapped fine quartz grains may be seen. Nuclei consist of recrystallized tests and of bryozoans (thin section, ×12). **b** Asymmetric oncoid. Nucleus consists of a bryozoan fragment (thin section, ×26). **c** Recrystallized shell fragment with an algal micritic coating on its upper side. Note the fine quartz grains trapped in this micritic coating (thin section, ×26). **d** Oncoid with a gastropod test as nucleus (thin section, ×12)

Fig. 5a–e. Radial fibrous ooids (thin section, ×80). **a** Asymmetrical ooid. Nucleus consists of an irregular aggregate of micrite and quartz grains. **b** Asymmetrical ooid. Nucleus consists of an algal fragment (*Sphaerocodium* sp.). **c** Ooid with well-rounded shape. The micritic pillars are numerous in the outermost envelope. Note the pressure solution contact between the ooid and a bioclast in the *upper left corner*. **d** Typically asymmetrical ooid. Nucleus consists of a curved test. A tiny fragment, trapped in concavity of the test, was the site of crystal nucleation. **e** Asymmetrical ooid. Nucleus consists of a corallite fragment

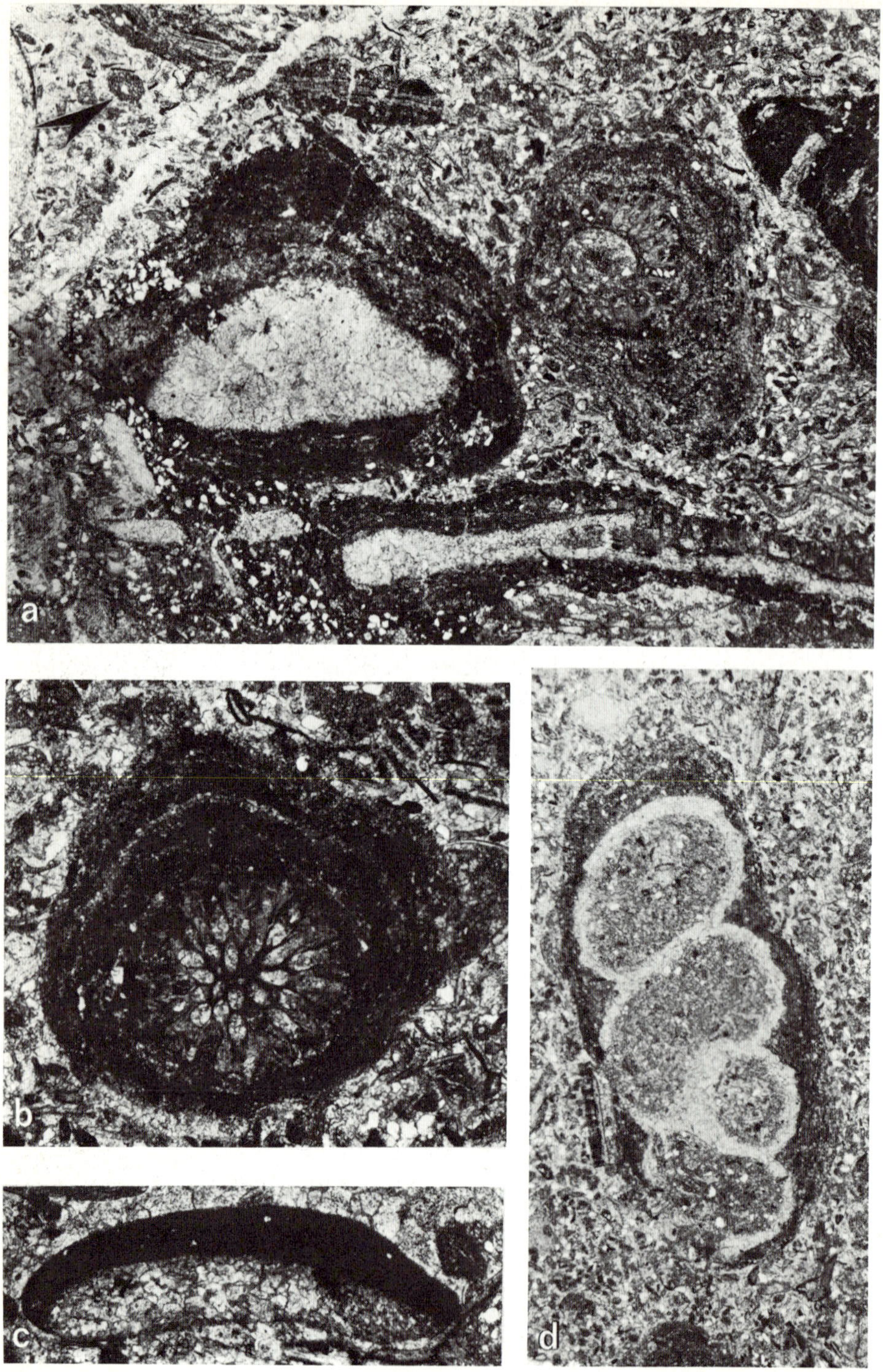

Fig. 4a – d (legend see p. 613)

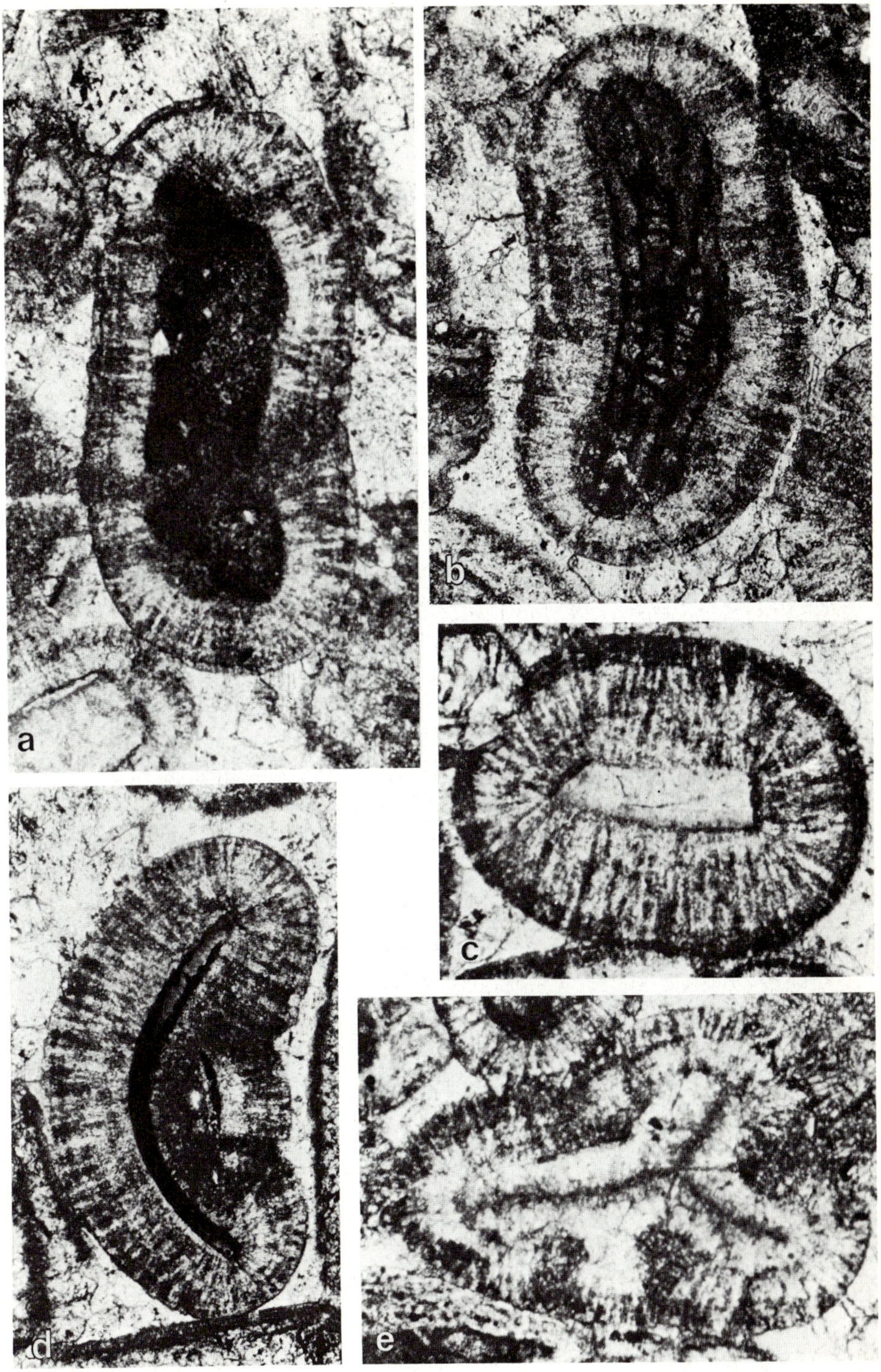

Fig. 5a – e (legend see p. 613)

3.2 Allochthonous Coated Grains Deposited in the Back-Reef Zone

Allochthonous coated grains, deposited in the back-reef zone, are to be encountered in beds C, D, H, I and K (Fig. 3) where they are mainly represented by radial fibrous ooids and bioclasts partly coated by a radial fibrous cortex.

3.2.1 Radial Fibrous Ooids (Fig. 5). These are calcitic ooids with sizes ranging from 0.345 to 0.870 mm in diameter. Their diverse shapes assume those of the nuclei. They may be perfectly circular, or they may be asymmetric with curved or elongated shapes. Some radial fibrous ooids show one or several bosses. The envelopes are faint and their number ranges from one to three, with a constant thickness. They are composed of calcitic fibres arranged perpendicular to the

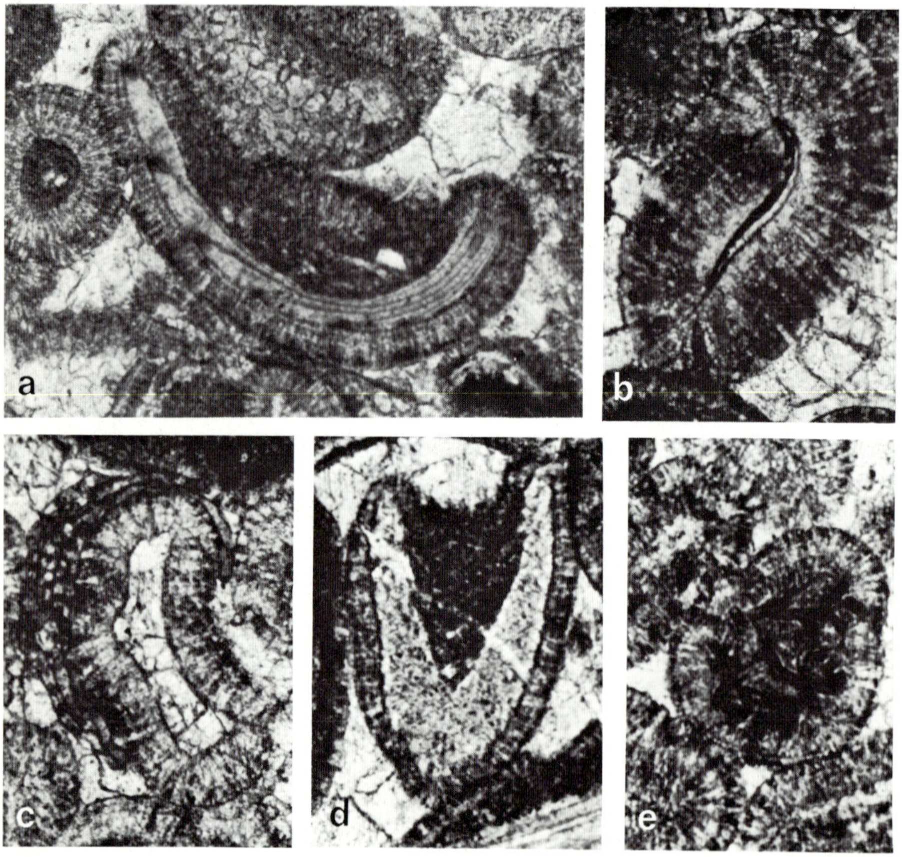

Fig. 6. a Bioclast partly coated by a radial fibrous cortex. The bioclast consists of an ostracod valve (*Cryptophyllus* sp.). In the concavity of the valve micrite and quartz grains are trapped (thin section, ×60). **b** Asymmetrical radial fibrous ooid (thin section, ×60). **c** Radial fibrous ooid encrusted by a calcareous alga (*Sphaerocodium* sp.) (thin section, ×60). **d** Bioclast (crinoid ossicle) partly coated by a radial fibrous cortex. Note the absence of radial fibrous envelopes on the micrite trapped in the concavity of the ossicle (thin section, ×70). **e** Radial fibrous ooid whose nucleus consists of a bleb of interwoven algal filaments (*Girvanella* sp.) (thin section, ×70)

nucleus. In places, the fibres are edged by dark micritic pillars, although in most cases these pillars do not protrude from the envelope. Some of them, however, cut through several envelopes. They seem to occur most commonly in the outermost envelope. Where the envelopes have been broken, the fractures tend to parallel the calcitic fibres. Nuclei consist of crinoidal fragments, algal filaments, small curved tests and peloids, often recrystallized into neomorphic spar. This recrystallization may also affect the innermost envelopes. Dark micritic patches in the ooid cortex result from algal or fungal borings. Some radial fibrous ooids are encrusted by calcareous algae (*Sphaerocodium* sp.). These radial fibrous ooids occur in an oolitic grainstone with a sparitic cement, although compaction has affected this grainstone and concavo-convex pressure-solution contacts between ooids can be observed.

3.2.2 Bioclasts Partly Coated by a Radial Fibrous Cortex (Fig. 6). These bioclasts are generally curved, and the concave hollow is filled with micrite. No coating is visible on the micrite but envelopes with radially arranged calcite fibres are well developed on the micrite-free surface of bioclasts. The micrite trapped in the concave surface of the test was not a suitable substrate for fibre nucleation. This observation is apparently inconsistent with the case of radial fibrous ooids, where crystal nucleation starts on peloid. It appears that nucleation is perhaps easier on peloids which may be more suitable substrates from a physicochemical point of view.

3.2.3 Environment of Origin of Radial Fibrous Ooids. Radial fibrous ooids are considered to form in sheltered, very shallow waters (Rusnak 1960, Freeman 1962, Loreau 1973, Davies and Martin 1976). If similar conditions are inferred for the formation of these Lower Devonian radial fibrous ooids, the following process of genesis can be proposed (Fig. 7). In a very sheltered environment (which has not as yet been identified in the reefal complex) crinoid fragments, algal fragments, ostracod tests, coral debris and peloids were the site for a radial crystal nucleation. Intermittent water movement was sufficient to move ooids and bioclasts and to prevent their cementation. During storms, ooids and bioclasts were reworked and deposited in the back-reef zone. Sessile organisms were buried by these allochthonous carbonate grains.

4 Conclusions

Sediments from the Lower Devonian reefal complex of Baubigny contain numerous and diverse coated grains. These coated grains are represented by:

- *concentric ooids* formed on a bank in a lateral position to the bioherms, under high energy conditions and influenced by bi-directional currents;
- *oncoids* formed in a back-reef zone;
- *radial fibrous ooids* and *bioclasts partly coated by a radial fibrous cortex* reworked from a low energy environment, and deposited in a back-reef zone during storms.

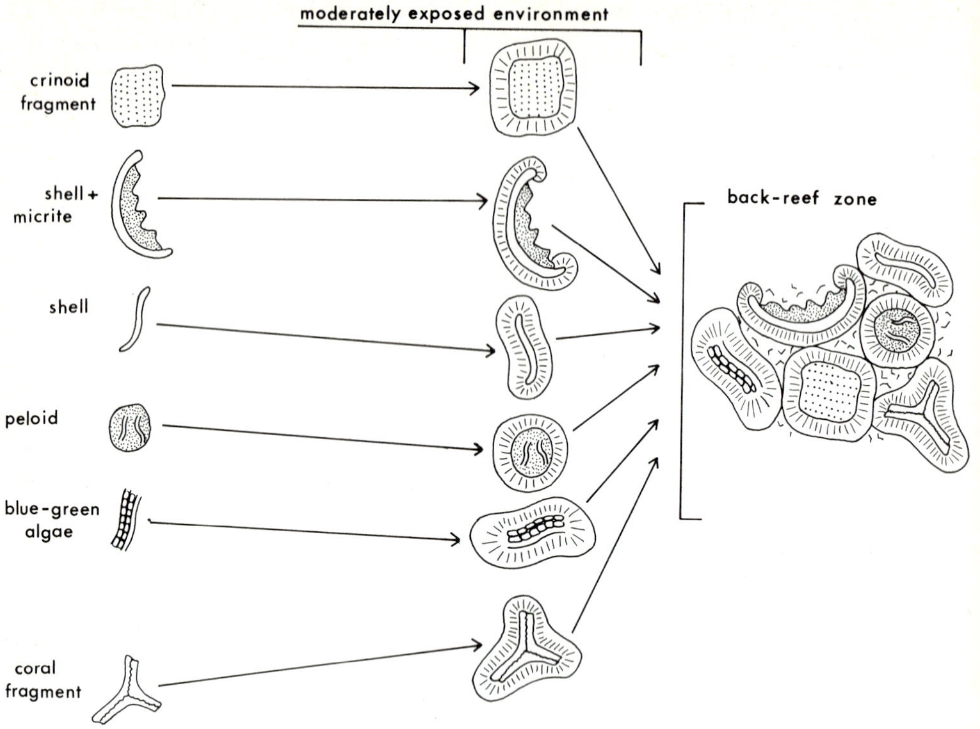

Fig. 7. A model of radial fibrous ooids formation and deposition (see text for explanation)

Acknowledgements. The author is greatly indebted to Christ Burgess for improving the English.

References

Davies PJ, Martin K (1976) Radial aragonite ooids, Lizar Island, Great Barrier Reef, Queensland, Australia. Geology 4 (2):120–122

Freeman T (1962) Quiet water oolites from Laguna Madre, Texas. J Sediment Petrol 32:475–483

Loreau JP (1973) Nouvelles observations sur la genèse et la signification des oolithes. Sci Terre 18:213–244

Poncet J (1968) Contribution à l'étude sédimentologique et stratigraphique du Dévonien de Basse-Normandie. These Caen

Poncet J (1972) Les biohermes éodévoniens de l'Horion récifal de Baubigny (Manche). Etude d'un paléomilieu. Bull BRGM IV 3:43–65

Poncet J (1976) Faciès carbonatés d'arrière-récif dans l'Eodévonien du Nord-Est du Massif armoricain (Cotentin). Bull BRGM I 1:49–68

Purser BH (1980) Sédimentation et diagénèse des carbonates néritiques récents. Technip, Paris

Rusnak GA (1960) Some observations of recent oolites. J Sediment Petrol 30:471–480

Simone L (1981) Ooids: a review. Eart Sci Rev 16:319–355

VI.7 Accretionary Lapilli and Other Spheroidal Rocks from the Archaean Swaziland Supergroup, Barberton Mountain Land, South Africa

Thomas O. Reimer [1]

1 Introduction

Since the pioneering work of Hall (1918) spheroidal or "oolitic" rocks have been described by a number of authors from the rocks of the Swaziland Supergroup (3.3 – 3.6 b.y.). Accretionary lapilli were first recognized in them by Reimer (1967). An unusually conspicuous band between the Onverwacht and Fig Tree Group, together with several similar occurrences in the Onverwacht Group, was described as accretionary lapilli by Lowe and Knauth (1978). These and other spheroidal rocks are described in this paper, and the implications of their occurrence are discussed. While the larger occurrences are described in some detail, information on the smaller ones is compiled in Table 1.

2 Onverwacht Group

From this essentially volcanic group several small occurrences of accretionary lapilli have been described. The lowermost one had been first observed by Reimer (1975a) and described in detail by Lowe and Knauth (1978) from the southwestern part of the mountain land (loc. 1 in Fig. 1). The spheroid bed lies in the upper portion of the Middle Marker, an essentially tuffaceous horizon between the Komati and Hooggenoeg Formations within a carbonate-bearing sericitic chert horizon which was interpreted as a probable silicified carbonate/evaporite sediment by these authors. The spheroids occur in poorly sorted volcanic debris in beds generally less than 10 cm thick and are themselves mostly between 1 and 2 mm in diameter, rarely up to 4 mm. They consist of a poorly sorted aggregate of fine-grained ash rarely showing nuclei and even less frequently being concentrically banded.

Prominent layers of accretionary lapilli were observed by Reimer (1967) in the middle portion of the Kromberg Formation about 1.2 km north of Geluk Beacon (loc. 4 in Fig. 1). The locality consists of several 1 – 2 m-tick bands of a fine-grained sediment on the surface of which spheroids with diameters of 5 – 15 mm are protruding, the spheroids accounting for up to 30% of the rock. These

1 Bernhard-May-Straße 43, D-6200 Wiesbaden-Biebrich

Coated Grains (ed. by T.M. Peryt)

Table 1. Speroidal rocks of the Barberton mountain land

Loc.	Stratigraphic position	Enclosing rock type	Number of beds	Thickness (m)	Lateral extent (km)	Diameter of spheres (mean/max)	Original interpretation(s)	Reference	Revised interpretation
	Onverwacht Group								
1	Middle Marker between Komati and Hooggenoeg Form.	Carbonate bearing sericitic chert	> 5	0.1 – 0.2	0.5(?)	2/4 mm	Oolites	Reimer (1975a)	Accretionary lapilli Lowe and Knauth (1978)
2	within Hooggenoeg F.	Felsic tuffs	?	?	?	as 1	Accretionary lapilli	Lowe and Knauth (1978)	
	lower part Kromberg F.	Silicified tuffs	?	?	?	as 1	Accretionary lapilli	Lowe and Knauth (1978)	
3	Kromberg F. (?)	"Greenstones"	?	?	?	?	Spherulitic greenstones	Hall (1918)	Accretionary lapilli(?)
4	Kromberg F.	Mafic tuffs	5	1 – 2	4(?)	10/15	Accretionary lapilli	Reimer (1967)	
5	?	Quartz-sericite schist	1	5(?)	?	5/10	"Quartz nodules"	Anhaeusser (1963)	Accretionary lapilli(?)
	Fig Tree Group								
6	lowest Fig Tree Group	Chert	< 50	0.05 – 1.5	40 × 100	3/10	"Some kind of volcanic rock"	Hall (1918)	Accretionary lapilli
							Siliceous oolite	Visser (1956)	Lowe and Knauth (1978)
7	above 6	Chert	1	0.1	?	3/5	Accretionary lapilli	this chapter	
8	Ngwenya F.	Silicified chert grits	1	0.15	?	6/18	Pisolites formed through coagulation of silica gel around chert fragments	Bell (1967)	

9	Ngwenya F.	Ferruginous shales	1	0.25	3	1.5/2.5	Oolites	Heinrichs and Reimer (1977)	Volcanic glass spheres from ash eruption
10	Mapepe F.	Cherts	1	1	1	1.3/3	Barytic oolite	Heinrichs and Reimer (1977)	As loc. 9, partly replaced by barytes, some round ophitic fragments
11	Mapepe F.	Shales and sand-stones	2	0.1(?)	?	1/2	Oolites	Bell (1967)	As loc. 9
12	Fig Tree G.	?	?	?	?	?	Spherulites	Visser (1956)	Ext. of one of loc. 9–11(?)
13	Zwartkoppie "Zone"	Quartz-sericite schist	"sev-eral"	0.2	?	1/1.5(?)	Spherulites	Hearn (1943)	As loc. 9
	Zwartkoppie "Zone"	Chert	"sev-eral"	?	?	3/10	Spherulites	Koen (1947)	As loc. 9
13a	Zwartkoppie "Zone"	Tuff below, gray-wacke above	1	0.4–1.6	17×30	1.7/10	Resemblance to algae pseudo-oolites biogenic	Ramsay (1963) Van Vuuren (1964) Pflug (1976)	As loc. 9, up to 65% felsic volca-nic fragments
14	lower Fig Tree G.	Carbonaceous shales	1(?)	0.3	1	1.3/2.5	As loc. 9	this chapter	
14a	upper Sheba F.	Graywackes	1	?	?	?	Similar to loc. 13a	Van Vuuren (1964)	As loc. 9(?)
15	Schoongezicht F.	Crystal tuffs	1	8–25	5	4/10	Quartz nodules in "some kind of intraformational sediment"	Viljoen (1964)	Accretionary lapilli
	Moodies Group								
16	Clutha F.	Sandy shales	1	0.1	?	1/2	Oolites	Herget (1966)	As loc. 9

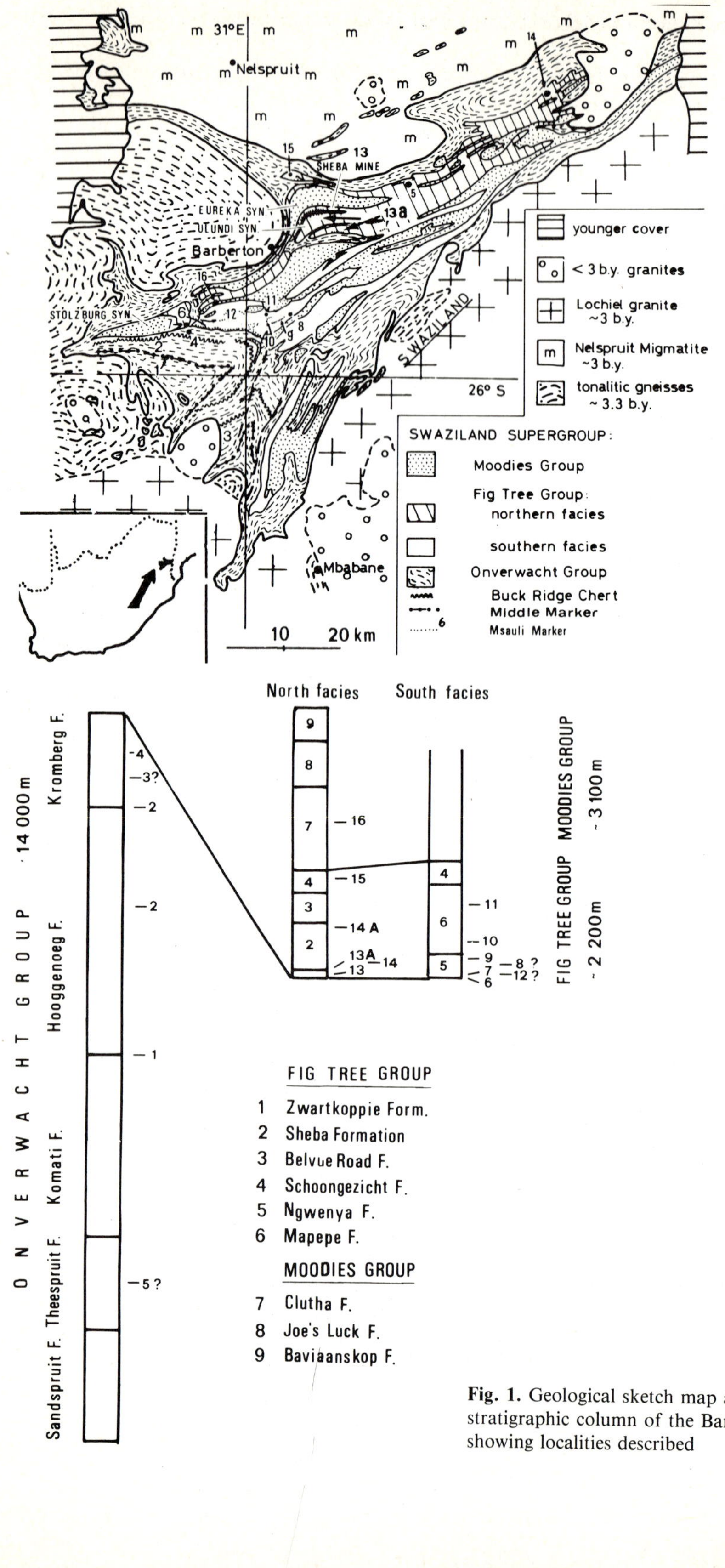

Fig. 1. Geological sketch map and standard stratigraphic column of the Barberton area, showing localities described

spheroids were apparently also observed several km farther east at an unspecified locality by Visser (1956).

3 Fig Tree Group

Based on the investigations of Reimer (1975a) and Heinrichs and Reimer (1977) two separate facies could be distinguished within this essentially sedimentary group. The northern one consists mainly of graywackes and shales and appears to be of deeper water origin. The southern one contains a highly varied lithology of shales, jaspilites, tuffs, dolomites, conglomerates, cherts, and barytes deposited in shallower and partly still water (Fig. 1).

3.1 Southern Facies

At the base of the Fig Tree Group in the southern facies (loc. 6 in Fig. 1) conspicuous 5 – 70 cm thick beds of accretionary lapilli occur within an up to 20 m thick marker band of cherts which was traced from the western part of the mountain land where it had been described by Visser (1956) and Reimer (1967) farther to the east up to the Swaziland border by Heinrichs (1980).

At detailed petrographic description was given by Lowe and Knauth (1978) who recognized it as an accretionary lapilli bed.

The spheroids themselves measure up to 10 mm in diameter, with a mean of 2 – 3 mm. In their majority, they consist of a nucleus of coarser-grained ash followed by a ring of fine-grained lithic fragments and phenocrysts and then sur-

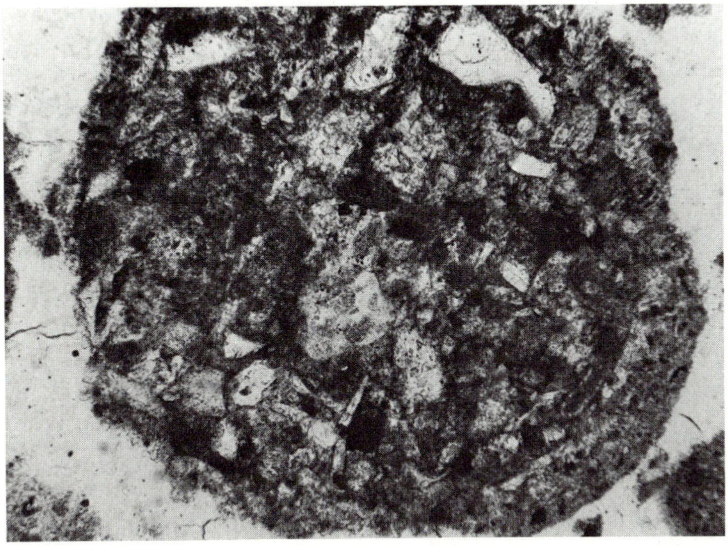

Fig. 2. Accretionary lapilli from the Msauli Marker (loc. 6), longer axis of photo 1.5 mm

rounded by several very fine-grained layers. The outer rim is usually made up of an extremely fine-grained cortex (Fig. 2). The latter appears to have been brittle on deposition while the centre still had a certain degree of plasticity. The layer surrounding the nucleus usually shows size grading from the centre outwards. A grain with a twin nucleus was described by Reimer (1967). Several spheroids were found to consist of a larger lithic fragment as a core surrounded by the very fine-grained laminae.

Lowe and Knauth (1978) investigated two layers in detail and described a peculiar type of graded bedding. In this the lower part consists only of well-sorted spheroids with gradually upwards decreasing diameters. In the higher parts gradually finer-grained ash particles make their appearance and the number of accretionary lapilli in the bed decreases. This type of graded bedding together with the sedimentary structures observed in the enclosing sediments led Lowe and Knauth (1978) to interpret the respective beds as having been deposited from volcanic ash clouds into a shallow-water to subaerial environment, probably on exposed low-energy tidal flats, without much subsequent redistribution.

In contrast to this, Heinrichs (1980), who could trace individual beds for up to 10 km along strike, favours deposition of the spheroids as original oolites with subsequent redistribution by turbidity currents from shallow water into deeper parts of the depository. It appears possible that the material was originally deposited from ash clouds stemming from several eruptions into a subaerial to subaqueous environment and that those portions which were deposited on steeper slopes were subsequently redistributed by turbidity currents. From pebbles of this characteristic rock in conglomerates of the Fig Tree and Moodies Group Reimer (1975b) reconstructed an original extent of this zone of at least 40 × 100 km. This and directions of sedimentary transport as reported by Heinrichs (1980) are shown in Fig. 3.

Heinrichs and Reimer (1977) briefly referred to a marker horizon of "oolites" in ferruginous partly tuffitic shales in the lower part of the southern Fig Tree facies (loc. 9 in Fig. 1). The spheroids are well-sorted, with a largest diameter of about 2.5 mm and a mean of about 1.5 mm. Microscopically two types of spheroids could be distinguished. Up to 75% are little deformed and consist of an outer rim of fibrous minerals growing apparently inwards from the margin in ra-

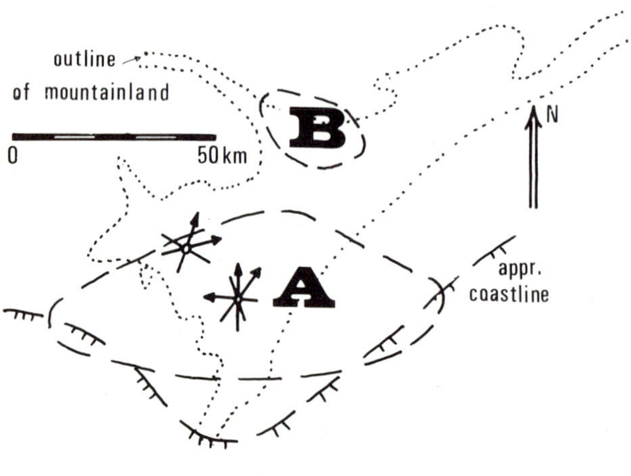

Fig. 3. Reconstructed extent of accretionary lapilli beds: *A* Msauli Marker (after Reimer 1975b), rose diagrams (after Heinrichs 1980), *arrows* from cross beds, *bars* showing orientation of ripple crests. *B* "Pseudo-oolite" layer from Zwartkoppie Zone (loc. 13a)

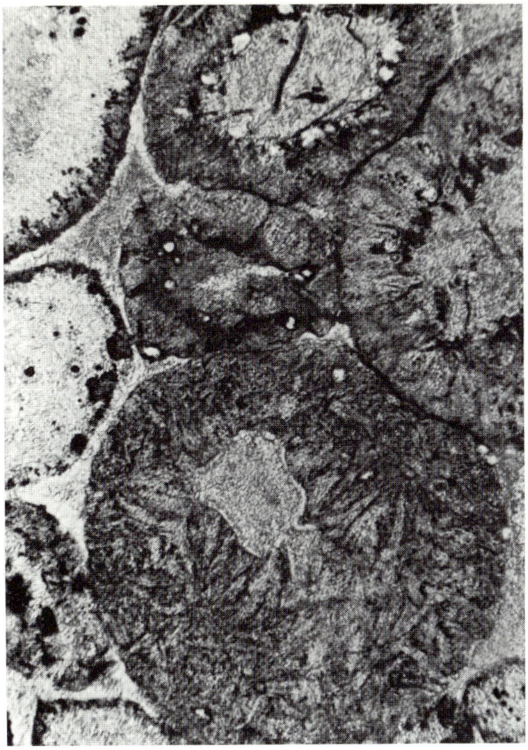

Fig. 4. Coated volcanic glass spheres from lower Fig Tree Group (loc. 9), longer axis of photo 1.5 mm

dially diverging bundles (Fig. 4). The core zone consists of irregularly bordered cherty quartz and locally macro-quartz.

The second type of spheroids consists of fine-grained translucent siliceous bodies with thin whitish rims up to 0.03 mm thick. They are usually deformed between the apparently more stable spheroids of the first type. A characteristic feature of them are perlitic cracks frequently growing outwards from a concentric crack in the centre but not extending to the rim of the spheroid (Fig. 5). The spheroid layers are usually well sorted with little recognizable detrital matrix, the interstices being filled by chert with interspersed micro-crystalline micaceous minerals.

The spheroids have a remarkable similarity to granules from hyaloclastic tuffs as described by Dimroth and Lichtblau (1979) which formed through alteration of volcanic glasses. However, in hyaloclastic tuffs the granules are usually not as perfectly spherical as the spheroids of the occurrence described here and are furthermore usually interspersed in finer-grained material. Thus, although the spheroids appear to be of similar origin, a different mode of deposition has to be sought. Their spherical nature appears to be a primary feature as shown by the relation between the perlitic cracks and their shape. They most probably formed as spheres of glass during tuff eruptions associated with low-viscosity lavas. During fall from the resulting ash cloud they and other particles were coated with fine ash like accretionary lapilli. When they were deposited in water separation

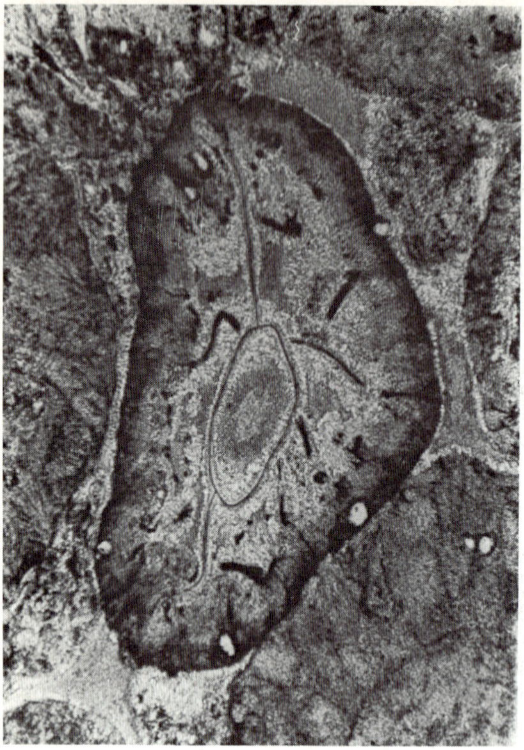

Fig. 5. Perlitic cracks in deformed volcanic glass sphere (loc. 9), longer axis of photo 1.5 mm

according to sizes occurred and they formed layers with only little admixtures of finer volcaniclastic material. The frequent lack of matrix in this and some of the other similar occurrences to be described later supports the deposition from an ash cloud rather than lateral sedimentary transport. Some subsequent redistribution has occurred. The spheroids observed would thus represent a special type of accretionary lapilli with only little coating.

Heinrichs and Reimer (1977) briefly referred to a "barytic oolite" from lenticular, up to 1 m thick and 20 m long outcrops below the sedimentary barytes occurring at the base of the middle part of the southern Fig Tree facies in the central part of the mountain land (loc. 10 in Fig. 1). The spheroids are well-sorted with a maximum diameter of about 3 mm and a mean of about 1.5 mm. Three main types of spheroids could be distinguished in this rock. Some 75% – 80% of them have a cherty core of varying size and shape, possibly representing devitrified volcanic glass, with a thin coating of sericite. The centre is frequently recrystallized to coarse quartz and contains interspersed round remnants of chert (Fig. 6). "Ghosts" of chalcedony have been observed in a few spheroids.

Some 15% – 20% of the spheroids possess a core of mostly monocrystalline barytes, which can account for up to 70% of a spheroid. The coating consists of sericite as in the first group. The sericite flakes, which can be rather coarse, in places protrude into the barytes, apparently replacing it. Instead of sericite a cherty rim may occur, and rarely the sequence from the centre outwards is

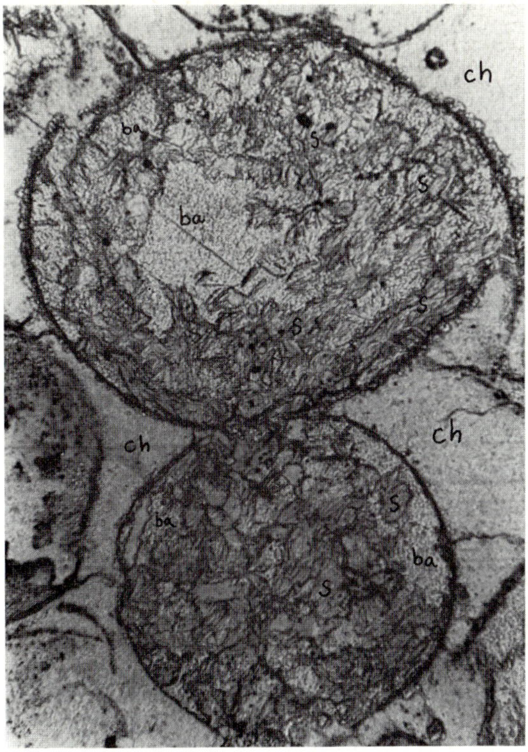

Fig. 6. Chert/baryte/sericite spheroids from middle Fig Tree Group (loc. 10), longer axis of photo 1.5 mm. *ch* chert, *ba* baryte, *s* sericite

barytes, sericite, and chert. Barytes can account for up to 20% of the rock by weight. Approximately 5% of the spheroids are well-rounded fragments of highly altered ophitic volcanics. A primary fibrous texture could still be observed in a few spheroids. Questionable glass shards as well as a few angular fragments of volcanic quartz were found.

Some of the siliceous spheroids could have been derived through erosion from the lapilli bed mentioned above (loc. 9 in Fig. 1) which stratigraphically underlies it by 150 m. Alternatively, they, as well as the fragments of the ophitic rocks, could have been derived from contemporaneous volcanic activity, in a way similar to the previously described volcanic glass spheres. The barytic nature of some of the spheroids could be attributed to selective replacement of particular zones within pre-existing spheroids, possibly of accretionary lapilli origin.

In another occurrence of spheroidal rocks, described by Bell (1967) from rocks belonging to the Mapepe Formation (middle Fig Tree Group) in the central part of the mountain land (loc. 11 in Fig. 1), the spheroids are well-sorted and closely packed, with a mean diameter of about 1 mm (Fig. 7). Original clastic matrix appears to be lacking. The spheroids consist of cherty silica intimately intergrown with sericite. Their origin is difficult to ascertain due to the limited data available. It is possible that they also represent accretionary lapilli of the type described under loc. 9 from the lower part of the southern Fig Tree facies.

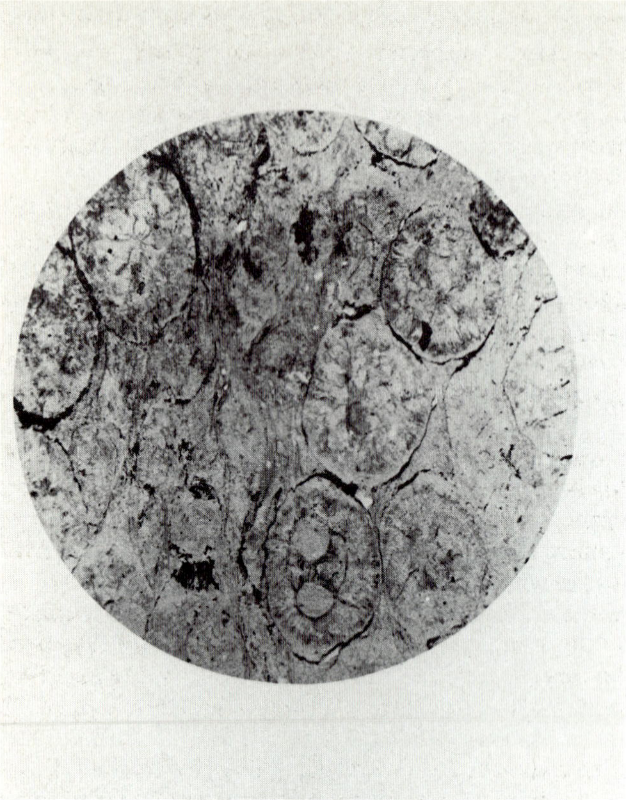

Fig. 7. Coated volcanic glass spheres (?) from middle Fig Tree Group (loc. 11). Longest grain 1.7 mm. (From Bell 1967)

In the rocks of the Schoongezicht Formation D. Lowe (pers. comm., 1982) observed several thin bands of accretionary lapilli in felsic tuffs in the western central part of the mountain land (north of loc. 12 in Fig. 1).

3.2 Northern Facies

From the rocks of the former "Zwartkoppie Zone", the base of the Fig Tree Group in the northern facies, spheroidal rocks have been described by a number of authors. The first examples were observed by Hearn (1942) in cherts of the Sheba Gold Mine (loc. 13 in Fig. 1). In quartz-sericite schists underlying thick cherts he found several sharply defined chert layers containing white concentric spheroids with diameters of 1 – 1.5 mm. They consist mainly of radiating masses of zeolites with zoisite in places. Many of these bodies contain round chert nuclei, while in some a nucleus of radiating fibrous quartz is observed. A few bodies consist of alternating shells of cherty quartz with intervening rings of zeolites. Koen (1947) described similar spheroids of 1 – 10 mm in diameter, one specimen containing a core of labradorite, from the lower part of the chert bed overlying the quartz-sericite schists. The occurrence of such large spheroids in the finely laminated cherts which to a large extent represent silicified fine-grained tuffs deposited in deeper water, militates against their having been brought in by lateral transport. Deposition from an ash cloud and an origin similar to that of the spheroids described from loc. 9 would be more feasible.

Other spheroidal rocks in the "Zwartkoppie Zone" were described in detail by Van Vuuren (1964) as the "pseudo-oolite" layer (loc. 13a). It occurs in the topmost part of the zone underlying the graywackes of the Sheba Formation. It has been located at outcrops separated by about 5 km across strike and was traced for about 10 km along strike. Ramsay (1963) has drawn attention to the supposed resemblance of some of the spheroids to primitive algae (calcisphera), while Pflug (1976) strongly advocated a biogenic origin for them as "asporogenous yeasts".

The spheroid bed on average is 0.4 m thick with a maximum of 1.6 m. It is dark green with thin yellowish-green sheared bands. It contains small lenses and thin stringers as well as small clastic slabs of black chert, some of which show thin laminations of disseminated and massive pyrites. Minor bands of disseminated pyrites occur within the spheroid bed itself.

The bed consists of rather uniformly-sized spheroids with an average diameter of about 1.7 mm, ranging from 0.5 – 5 mm. The bed shows a grain-supported texture with little detrital matrix, mostly a few isolated quartz grains replaced extensively by chert. The rock consists of 71% spheroids, 28% matrix, and 1% pyrites.

Two major types of spheroids have been recognized: The first type, which accounts for up to 65% of the spheroids macroscopically consists of greenish-gray

Fig. 8. Microfelsitic banded fragment from "pseudo-oolite" layer of Zwartkoppie Zone (loc. 13a), longer axis of photo 1.5 mm

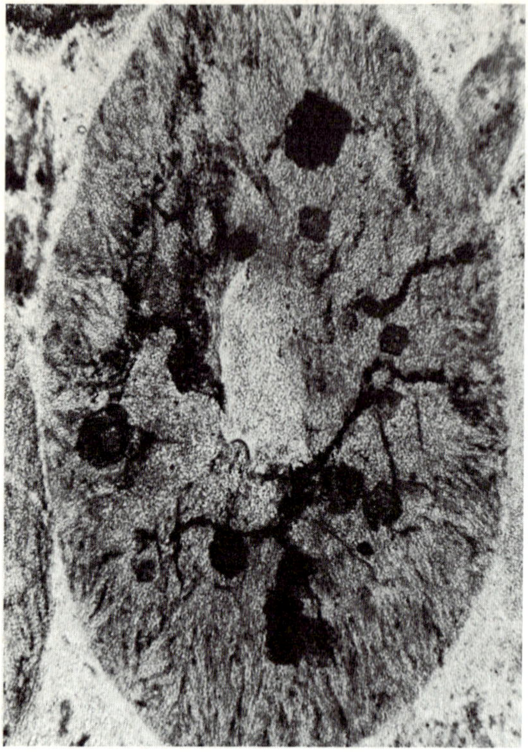

Fig. 9. Coated volcanic glass sphere with dolomite crystals (dark spots) from loc. 13a (as Fig. 8), longer axis of photo 1.5 mm

to translucent cherty material, some specimens showing thin grayish-white rims. They are usually flattened but originally were well-rounded. They consist of a microfelsitic groundmass with well-developed flow texture (Fig. 8) containing some few phenocrysts of marginally extensively replaced quartz. Some fragments of altered ophitic rocks were also observed, as well as fragments of spherulitic growth of quartz and feldspar, the latter also being replaced partly by chert.

Of the spheroids 30% – 35% consist of rings of cherty material with or without enrichment of dolomite. They mostly possess a cherty core surrounded by cherty rings containing dolomite crystals (Fig. 9). The core zone of some of the spheroids also is rich in dolomite while the rim consists of cherty quartz partly containing layers of opaques and rutile (Van Vuuren 1964). Radial growth is observed locally in the rings and in the cores spherulites of chalcedony, partly recrystallized to coarser chert or quartz, have been observed. In others chalcedony was found to grow inwards into the spheroids in radially diverging bundles. It is this type of spheroids which contains the kerogens described by Pflug (1976, pers. comm., 1981). Up to 5% of the spheroids contain round pyrite cores or consist of chrome-green chloritic aggregates which are usually highly sheared.

The origin of these spheroids is most probably similar to that of the peculiar volcanic glass spheres described above from the lower Fig Tree Group in the southern facies (loc. 9). The pronounced lack of fine-grained matrix again militates against an introduction by turbidity currents from the edge of the deposito-

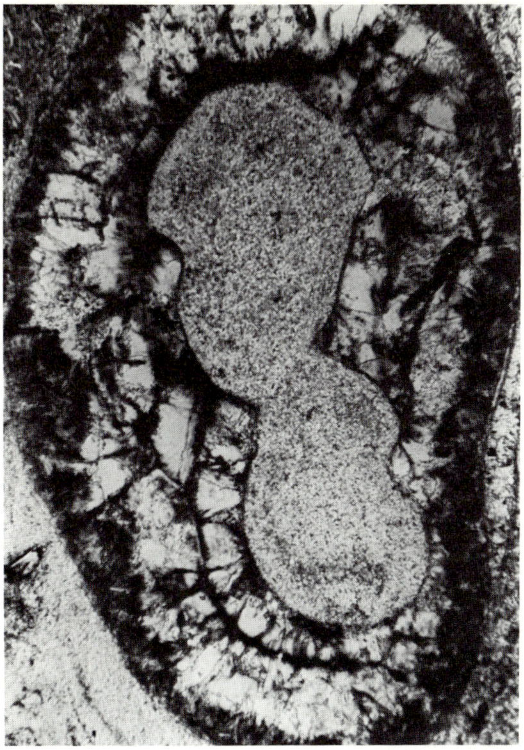

Fig. 10. Coated volcanic glass sphere from lower Fig Tree Group (loc. 14), longer axis of photo 1.5 mm

ry. Deposition from a volcanic ash cloud thus would be more feasible. The rock subsequently was intensely altered diagenetically and eventually silicified.

The bed has been deposited in deeper still water as indicated by the graywackes of the directly overlying Sheba Formation for which Reimer (1975a) reconstructed a depth of deposition of about 600 m. From the preserved outcrops an original extent of at least 17 × 30 km was reconstructed (Fig. 3). The large extent and the depth of deposition militate against a purely biogenic origin, especially in view of the volcanogenic nature of the majority of its constituents.

Another spheroid occurrence has recently been discovered in the northeastern part of the mountainland on farm M'hlati 170 JU about 2.5 km east-southeast of Malelane (loc. 14 in Fig. 1).

Macroscopically the light gray-green rock consists of densely packed spheroids with little darker-coloured siliceous matrix. In parts the rock has a grain-supported framework, but mostly contains some finegrained detrital matrix. Some 1 – 3 mm thick dark-gray shaly stringers with little or no content of spheroids were observed.

The spheroids have a median diameter of about 1.5 mm, ranging from 0.4 – 2.5 mm. In their majority they contain a gray-green, usually round cherty core, probably devitrified volcanic glass with few remnants of phenocrysts. The core is surrounded by a light-gray rim of about 0.1 mm in width, consisting of fibro-radial chalcedony associated with some fibro-radial zeolites (?). The fibrous minerals also start to grow from perlitic cracks in the centre of the spheroids. The spheroids are usually only slightly deformed, shearing having taken place mainly in the shaly interlayers. About 5% – 10% of the spheroids, usually those with very thin rims of chalcedony fibres, are highly deformed, even when in contact with undeformed spheroids with a thick rim. The latter apparently had some stabilizing influence. The spheroids are similar to those described above as coated volcanic glass spheres deposited from a volcanic ash cloud.

4 Moodies Group

The stratigraphic youngest occurrence of spheroidal rocks in the Swaziland Supergroup was found by Herget (1966) in a 10 cm thick band in sandy shales interlayered with arenites apparently belonging to the Clutha Formation which constitues the lower part of the Moodies Group in the western part of the mountain land (loc. 16 in Fig. 1). According to Eriksson (1978) the depositional environment of the enclosing rocks can be visualized as overbank flood plains in a fluvial regime.

The spheroids occur in a 10 cm thick gray-green layer within sandy shales. The individual spheroids are translucent gray or dense lightgray to yellowish with diameters of 0.5 – 2 mm. Concentric banding of the translucent and the dense material is frequently developed.

The dense non-translucent layers consists of radially arranged quartzine fibres, in places associated with chalcedony. The cherty translucent parts are made up of micro-crystalline quartz pervaded by pale-green sericite or fuchsite. They are similar to the volcanic glass spheres described above. They would thus represent locally preserved products of volcanism which at the same time in other parts of the Moodies depository led to the outflow of basaltic to andesitic lavas.

5 Implication of Data

Of the spheroidal rocks described 6 clearly consist of accretionary lapilli which formed in volcanic ash clouds by accretion of fine-grained ash around particulate or massive nuclei. Their grain size spectrum is usually rather wide, with a mean above 2 mm, the maximum observed being about 15 mm. Individual beds are up to 1 m thick. The maximum areal extent of one horizon was about 40 × 100 km as reconstructed from outcrops and indicator pebbles in conglomerates.

The data presented indicate that within the Swaziland Supergroup accretionary lapilli are more widely distributed than previously thought. However, it was estimated that only less than 0.25 km^2 of the mountain land are underlain by such rocks with a total preserved volume of possibly about 0.5 km^3. It appears unlikely that the originally produced volume of accretionary lapilli in the Swaziland Supergroup amounted to much more than about 1.5 km^3.

A total of seven occurrences consists predominantly of volcanic glass spheres partly coated with fine-grained ash during eruptions of predominantly tufface-ous material. The beds contain little finer volcaniclastic or detrital admixtures, having been deposited from an ash cloud into a body of water. The grain size spectrum of these spheroids is narrower than that of the "true" accretionary lapilli, the means always being below 2 mm, with individual grains rarely attaining sizes of up to 10 mm. The maximum thickness observed is 1.6 m and the maximum lateral extent of a particular bed was reconstructed as about 17 × 30 km from outcrops.

The majority appears to have been deposited under shallow-water to subaerial conditions. Exceptions are the specimens from the Zwartkoppie Zone in the northern Fig Tree facies (loc. 13) which were deposited in deeper water under partly euxinic conditions. Although their chemical composition is difficult to ascertain due to the intense alteration that has affected them, the majority appears to have been derived from felsic to intermediate volcanics. The only certain exception is the occurrence in the Kromberg Group (loc. 4) for which a basaltic derivation can be shown.

The small size of the occurrences described could cause such rocks to be easily overlooked in other sequences. Due to their superficial similarity with true oolites they could also easily be mistaken for the latter. Archaean rocks of Zimbabwe might serve as an example, where three occurrences of "oolites" have been described from noncarbonate environments (Mac Gregor 1937, Wiles 1957, Stowe 1968). Several bands of accretionary lapilli have recently been recognized by the author in pre-2.7 b.y. old volcanics of the Baltic Shield of Karelia (USSR). It is thus possible that a number of "oolites" described from Precambrian formations and especially from noncarbonate environments actually refer to accretionary lapilli, indicating a much more widespread distribution than so far suspected.

It is therefore premature to talk about a concentration of accretionary lapilli in the Precambrian as done by Lowe and Knauth (1978). Investigations by the author (Reimer, this vol.) suggest that the accretionary lapilli cannot be used as indicators for a higher moisture content of the atmosphere in the Precambrian. The type of accretionary lapilli observed usually in older rocks actually requires rather dry air zones around the parent ash zone for its formation. The value of accretionary lapilli lies more in their potential as time markers in fossil-free rock sequences.

Acknowledgements. This chapter presents a contribution to IGCP-Project 160 (Exogenic processes). The generous assistance of the Deutsche Forschungsgemeinschaft (DFG) for the presentation of part of the above data at the Petrozavodsk (USSR)-meeting of IGCP-Project 160 in June 1981 is gratefully acknowledged. Mr. M. Bell kindly gave permission to use Fig. 6 from his thesis, copy of which was kindly furnished by Dr. I. Saggerson (Durban, South Africa). Mr. R. Lindsay (Barberton) kindly furnished information and samples of material from loc. 14. The constructive criticism of Dr. T. Peryt (Warszawa) helped to improve the original manuscript.

References

Anhaeusser CR (1963) The geology of the Lily Syncline and portion of the Eureka Syncline between Sheba Siding and Louw's Creek Station, Barberton Mountain Land. Unpubl PhD-thesis, Witwatersrand Univ Johannesburg

Bell MC (1967) The geology of the farm Heemstede 33 and portions of the farms Schultzenhorst 31 and Mendon 32 in the Barberton Mountain Land. Unpubl MSc-thesis, Natal Univ Durban

Dimroth E, Lichtblau AP (1979) Metamorphic evolution of Archaean hyaloclastites, Noranda area, Quebec, Canada: Comparison of Archaen and Cenozoic sea floor mechanism. Can J Earth Sci: 1315 – 1340

Eriksson KA (1978) Tidal deposits from the Archaean Moodies Group, Barberton Mountain Land, South Africa. Sediment Geol 18:257 – 282

Hall AL (1918) The geology of the Barberton Gold Mining District. Geol Surv S Afr Mem 8

Hearn MG (1942) A study of the working properties of the chief gold producer of the Barberton District, Eastern Transvaal. Unpubl. PhD-thesis, Witwatersrand Univ, Johannesburg

Heinrichs TK (1980) Lithostratigraphische Untersuchungen in der Fig Tree Gruppe des Barberton Greenstone Belt im Gebiet zwischen Umsoli und Lomati. Göttinger Arb Geol Palaeontol 22

Heinrichs TK, Reimer TO (1977) A sedimentary baryte deposit from the Archaean Fig Tree Group of the Barberton Mountain Land (South Africa). Econ Geol 72:1426 – 1441

Herget G (1966) Archaische Sedimente und Eruptiva im Barberton Bergland (Transvaal – Südafrika). Neues Jahrb Mineral Abh 105:161 – 182

Koen GM (1947) Die geologie van die Sheba Gebied in die Barbertonse Distrik. Unpubl MSc-thesis, Pretoria Univ, Pretoria

Lowe DR, Knauth PR (1978) The oldest marine carbonate ooids reinterpreted as volcanic accretionary lapilli, Onverwacht Group South Africa. J Sediment Petrol 48:709 – 722

Mac Gregor AM (1937) The geology of the country around Hunter's Road Gwelo District. S Rhod Geol Surv Bull 31

Pflug H (1976) Ramsaysphera ramses n.gen.n.sp. aus den Onverwacht Schichten (Archaikum) von Südafrika. Paleontogr Abt B 158:130 – 168

Ramsay JR (1963) Structural investigations in the Barberton Mountain Land, Eastern Transvaal. Trans Geol Soc S Afr 66:353 – 401

Reimer TO (1967) Die Geologie der Stolzburg Synclinale im Barberton Bergland (Transvaal – Südafrika). Unpubl Dipl thesis, J W Goethe-Univ, Frankfurt/Main

Reimer TO (1975a) Untersuchungen über Abtragung, Sedimentation und Diagenese im frühen Präkambrium am Beispiel der Sheba Formation, Südafrika. Geol Jahrb Reihe B 17

Reimer TO (1975b) Paleogeographic significance of the oldest known oolite pebbles in the Archaean Swaziland Supergroup (South Africa). Sediment Geol 14:123 – 133

Stowe CW (1968) The geology of the country south and west of Selukwe. Rhod Geol Surv Bull 59

Van Vuuren CJ (1964) The geology of a portion of the Ulundi Syncline between Hislop's Creek and Fig Tree Creek, Barberton Mountain Land. Unpubl manuscr, Econ Geol Res Unit, Witwatersrand Univ, Johannesburg

Viljoen MJ (1964) The geology of the Lily Syncline and portions of the Eureka Syncline between the Consort Mine and Joe's Luck Siding, Barberton Mountain Land. Unpubl MSc-Thesis, Witwatersrand Univ, Johannesburg

Visser DJL (1956) The geology of the Barberton Area. Geol Surv S Afr Spec Publ 15

Wiles JW (1957) The geology of the eastern portion of the Hartley gold belt. S Rhod Geol Surv Bull 44/1

Author Index

Abate B, Catalano R, D'Argenio B, Di Stefano P, Riccobono R (1977) 438, 447
Abell PI, Awramik SM, Osborne RH (1978) 308, 327
Adams AE (1980) 461, 467, 471, 472
Adams JE, Frenzel HN (1950) 529, 535
Adey WH (1966) 228, 233, 236, 240
Adey WH (1970) 233, 236, 240
Adey WH (1973) 237, 240
Adey WH (1976) 233, 240
Adey WH, Adey PJ (1973) 227, 233, 240
Adey WH, Boykins W (in press) 228, 230, 239, 240
Adey WH, MacIntyre IG (1973) 228, 233, 234, 236, 239, 240, 243, 244, 245, 250, 254, 257, 414, 422, 605, 607
Adey WH, McKibbin DL (1970) 225, 227, 228, 233, 237, 240
Adey WH, Vassar JM (1974) 238, 239, 240
Ager DV (1963) 433
Aitken JD (1967) 415, 422
Alberstadt LP, Walker KR, Zurawski RP (1974) 197, 198
Alexandersson ET (1974) 605, 607
Alexandersson ET (1977) 605, 607
Allanson BR (1973) 284, 296
Allen HL (1971) 284, 296
Amsbury DL (1967) 438, 447
Anadón P, Zamarreño I (1981) 274, 275, 281, 283
Anderson R, Björnsson S, Blanchard DC, Gathman S, Hughes J, Jonasson S, Moore CB, Survilas HJ, Vonnegut B (1965) 59, 67
Angus JG, Raynor JB, Robson M (1979) 10, 23
Anhaeusser CR (1963) 620, 634
Assereto R, Brusca C, Gaetani M, Jadoul F (1977) 391, 397
Assereto R, Folk RL (1976) 174, 175, 440, 445, 447, 495, 500, 502, 504, 508, 530, 535, 591, 597, 605, 608
Assereto R, Folk RL (1980) 438, 441, 447
Assereto R, Kendall CGStC (1977) 437, 438, 440, 441, 447, 529, 534, 535, 589, 597
Awramik SM, Margulis L, Barghoorn ES (1976) 276, 283

Babcock JA (1974) 603, 604, 608
Babcock JA (1977) 599, 603, 604, 605, 606, 608
Babić L (1970) 441, 442, 447
Bachmann GH (1973) 93, 94, 96
Baker G, Frostick AC (1947) 148, 152
Ball MM (1967) 116, 123, 130, 132, 134, 141
Banerjee DM (1971) 356, 357
Banerjee DM (1978) 356, 357
Barbieri M, Masi U, Tolomeo L (1979) 474, 487
Barghoorn ES, Knoll AH, Dembicki AH (1977) 101, 107
Bateson JH (1965) 66, 67
Bathurst RGC (1966) 292, 296
Bathurst RGC (1968) 590, 597
Bathurst RGC (1971) 89, 90, 96, 185, 187, 568, 575
Bathurst RGC (1975) 10, 15, 17, 21, 23, 100, 101, 107, 151, 152, 203, 206, 207, 212, 213
Baturin GN, Bezrukov PL (1979) 346, 357
Baturin GN, Dobinchuk VT (1974) 354, 357
Bebout DG, Schatzinger RA, Loucks RG (1977) 336, 342
Becher JW, Moore CH (1979) 336, 342
Behrens EW, Frishman SA (1971) 29, 35
Belka Z (1979) 587, 597
Bell MC (1967) 620, 621, 627, 628, 634
Benson LV (1978) 143, 152
Berg G (1944) 91, 96
Bernoulli D, Wagner CW (1971) 438, 441, 443, 447, 533, 535
Bertrand-Sarfati J, Fabre J (1974) 308, 310, 324, 325, 326, 327
Bertrand-Sarfati J, Freytet P, Plaziat JC (1978) 308, 327
Beukes NJ (1977) 201, 207, 209, 211, 213
Beukes NJ (1978) 199, 205, 209, 211, 213
Beukes NJ (1979) 199, 209, 213
Beukes NJ (1983) 73, 96
Beyth M (1980) 163, 167

Biddle KT (1979) 392, 395, 397
Biddle KT (1981) 390, 396, 397
Biddle KT (1983) 273
Binda PL (1973) 217, 224, 264, 269
Birch GF (1979) 344, 357
Bishop WF (1968) 336, 342
Bissell HJ, Chilingar GV (1967) 567, 575
Blanc JJ (1968) 217, 224, 264, 269
Blank HR, Tynes EW (1965) 501, 502
Boardman MR (1976) 122, 128, 130
Boccaletti M, Fazzuoli M, Manetti P (1975) 583, 584, 586
Boccaletti M, Manetti P (1972) 583, 584, 586
Bodine MW, Holland DH, Borcsik M (1965) 10, 23
Boehm L, Schramm W, Rabsch U (1978) 238, 240
Bollinger W, Burri P (1970) 336, 342
Borch von der CC, Bolteon B, Warren JK (1977) 308, 327
Bordine BW (1965) 21, 24
Born SM (1972) 144, 152
Bornemann JG (1887) 289, 295, 296
Bosellini A (1967) 535
Bosellini A, Castellarin A, Rossi PL, Simboli G, Sommavilla E (1977) 391, 392, 393, 397
Bosellini A, Ginsburg RN (1971) 4, 6, 8, 217, 219, 221, 223, 224, 225, 228, 232, 236, 240, 243, 250, 254, 257, 281, 283, 299, 306, 414, 422, 599, 605, 606, 608
Bosellini A, Rossi D (1974) 390, 397, 438, 447
Bosence DWJ (1976) 217, 219, 221, 222, 224, 225, 228, 235, 238, 240, 243, 257
Bosence DWJ (1979) 227, 239, 240
Bosence DWJ (1980) 227, 237, 240
Bosence DWJ (1983) 5, 6
Bosence DWJ, Pedley HM (in press) 217, 218, 219, 221, 222, 223, 224, 234, 237, 238, 240
Bottke H, Dengler H, Finkenwirth A et al. (1969) 103, 107
Boulanger D, Poignant AF (1975) 308, 328
Bourrouilh F (1976) 174, 175
Bowman MBJ (1979) 409, 410, 422
Bowman MBJ (1980) 409, 422
Bowman MBJ (1983) 273, 275
Bowman MBJ (in prep.) 409, 422
Boysen Jensen P (1909) 284, 297
Bradley WH (1929) 103, 107
Braga JC, Comas MC, Delgado F, Garcia-Hernandez M, Jimenez AP, Linares A, Rivas P, Vera JA (1981) 360, 366
Braithwaite CJR (1975) 437, 438, 443, 447, 467, 468, 469, 472
Braithwaite CJR (1979) 437, 446, 447, 605, 608
Brand U (1981) 22, 24
Brand U, Veizer J (1980) 10, 20, 23, 24
Brand U, Veizer J (1981) 20, 24
Brand U, Veizer J (1983) 78, 96
Brasier MD (1975) 333, 342
Bretz JH, Horberg L (1949) 440, 447, 501, 502
Broecker WS, Kaufman A (1965) 143, 144, 152
Brown CN (1956) 438, 447
Bubenicek L (1971) 101, 107
Buchanan H (1970) 122, 130, 132, 134, 141
Buchbinder B (1977) 243, 257
Burchetter TP, Riding R (1977) 277, 283
Burgess CJ (1978) 488, 502
Burgess CJ (1983) 437, 438, 439, 447
Burgess IC (1961) 439, 447
Burri P, Du Dresnay R, Wagner CW (1973) 438, 447, 533, 535
Button A (1976) 100, 107, 199, 213

Cabioch J (1966) 217, 218, 222, 224, 235, 240
Cabioch J (1969) 217, 224, 227, 235, 240
Cabioch J (1972) 222, 224
Cabioch L (1968) 239, 240
Calvet F, Julia R (1983) 438, 446, 447
Calvet F, Pomar L, Esteban M (1975) 471, 472
Campos HS, Hallam A (1979) 27, 35
Cao Ruiji, Xue Yaosong (1983) 439, 440, 447
Cao Ruiji, Yang Wanrong, Zhang Junming, Yin Leiming, Li Zaiping, Zhao Wenjie (1979) 537, 546
Carannante G, Carbone F, Catenacci V, Simone L (1978) 438, 440, 441, 445, 447
Carozzi AV (1960) 72, 96, 159, 162
Carozzi AV (1961) 91, 93, 96
Carozzi AV (1962) 93, 94, 96
Carozzi AV (1963) 91, 92, 96
Carozzi AV (1964) 91, 96
Carozzi AV, Falkenhein FUH, Franke MR (1983) 274
Carozzi AV, Textoris DA (1967) 90, 96
Carter L (1975) 254, 257
Carter RM, Norris RJ (1978) 250, 257
Casanova J (1981) 292, 297
Castellarin A, Sartori R (1973) 90, 92, 96, 438, 441, 443, 445
Catalano R, D' Argenio B, Lo Cicero L (1974) 438, 443, 445, 447, 533, 535
Čatalov G (1972) 406, 408
Čatalov G (1983) 5, 6, 273
Caulet JP (1972) 227, 241
Cayeux L (1935) 71, 93, 96
Cayeux L (1970) 96
Centamore E, Chiocchini M, Deiana G, Micarelli A, Pieruccini U (1971) 584, 586
Chafetz HS, Butler JC (1980) 437, 438, 439, 446, 447, 464, 472, 488, 500, 501, 502

Chafetz HS, Kocurek G (1980) 526, 535
Chafetz HS, Meredith JC (1983) 437, 447
Chanda SH, Bhattacharyya A, Sarkar S (1977) 105, 107
Chapman VJ (1974) 245, 257
Chauhan DS (1979) 356, 357
Chave KE (1954) 86, 96
Chave KE, Suess E (1970) 294, 297
Chave KE, Wheeler BD (1965) 237, 241
Cheng Tsung-lu et al. (1977) 187
Choquette PW (1955) 92, 96
Choquette PW (1968) 32, 36
Choquette PW (1978) 73, 96, 161, 162
Ciarapica G, Fazzuoli M (1976) 586
Ciarapica G, Passeri L (1976) 577, 586
Ciarapica G, Passeri L (1978) 577, 586
Ciarapica G, Passeri L (1980) 580, 586
Ciarapica G, Passeri L (1983) 274, 438, 447
Clark DN (1980) 90, 96, 439, 441, 447, 594, 597
Clarke JM (1900) 279, 283
Clifton HE (1969) 526, 535
Cloos E (1947) 93, 96
Cloud Jr PE (1962) 29, 36
Cohn F (1864) 474, 487
Colacicchi R, Passeri L, Pialli G (1970) 584, 586
Colacicchi R, Passeri L, Pialli G (1975) 583, 586
Comas MC, Oloriz F, Tavera JM (1981) 365, 366
Conley CD (1977) 92, 93, 96, 105, 107
Cook PJ (1970) 355, 357
Craig H (1961) 27, 28, 36
Crouzel F, Lang J (1975) 308, 328

Dahanayake K (1977) 378, 379, 380, 385, 431, 433
Dahanayake K (1978) 310, 328, 380, 385
Dahanayake K (1983) 274, 275
Dahanayake K, Champetier Y, Hily J (1976) 321, 328
Dalrymple DW (1965) 413, 422
D' Argenio B (1966) 438, 440, 447
Davidson SC, McKinstry HE (1931) 148, 151, 152
Davies GR (1970) 78, 80, 96, 205, 209, 213
Davies PJ, Bubela B, Ferguson J (1978) 74, 80, 95, 96, 194, 198
Davies PJ, Martin K (1976) 15, 24, 74, 80, 96, 151, 152, 163, 167, 561, 563, 565, 617, 618
Davies PJ, Roger T (1968) 202, 213
Davis RA (1966) 90, 96
Deegan CE, Kirby R, Rae I, Floyd R (1973) 230, 241
Deffeyes KS, Lucia FJ, Weyl PK (1965) 211, 213
De Graciansky P (1972) 438, 441, 447
Deines P, Langmuir D, Harmon RS (1974) 28, 36
D'Halloy JJO (1812) 474, 487
Dickinson KA (1969) 90, 96
Dickson JAD (1966) 84, 97, 255, 257, 494, 502
Dimroth E (1976) 100, 107
Dimroth E, Chauvel J-J (1973) 101, 107
Dimroth E, Lichtblau AP (1979) 625, 634
Dodd JR (1967) 14, 15, 24
Dommergues JL, Elmi S, Mouterde R (1981) 360, 366, 372, 375
Donahue J (1965) 83, 97
Donahue J (1969) 17, 24, 83, 97, 604, 605, 606, 607, 608
Donahue J (1978) 73, 97, 161, 162
Donovan RN (1973) 310, 323, 328
Donsimoni M, Giot D (1977) 295, 297
Donsimoni M, Giot D, Lang J, Lucas G (1975) 308, 325, 328
Dragastan O (1970) 375
Dragastan O (1980) 375
Dravis J (1979) 87, 97, 121, 122, 123, 131
Dunham RJ (1965) 503, 515, 535
Dunham RJ (1969) 212, 213, 439, 440, 441, 447, 488, 500, 502, 503, 504, 505, 508, 509, 512, 514, 530, 534, 535, 573, 575, 600, 605, 608
Dunham RJ (1971) 495, 502
Dunham RJ (1972) 503, 505, 512, 514, 530, 534, 535
Dunlop JSR (1978) 67
Dupont B, Lemoalle J (1971) 13, 24

Eardley AJ (1938) 80, 93, 97, 167, 203, 213
Edmonds EA, Poole EG, Wilson V (1965) 103, 107
Eggleston JR, Dean WE (1976) 286, 297
Elliott GF (1964) 289, 295, 297
Ellis, Solander (1786) 217, 224
Elloy R, Thomas G (1981) 437, 438, 439, 448
Elmi S (1978) 360, 365, 366, 372, 375
Elmi S (1981) 360, 366, 372, 376
Endo R (1959) 603, 608
Enos P (1974) 116, 117, 131
Enos P, Freeman P (1979) 438, 444, 448
Eriksson KA (1977) 209, 213
Eriksson KA (1978) 632, 634
Eriksson KA, Truswell JF (1973) 205, 213
Esteban M (1973) 438, 448
Esteban M (1974) 467, 468, 469, 472
Esteban M (1975) 468, 469, 472
Esteban M (1976) 437, 439, 443, 445, 446, 448, 462, 464, 473, 488, 501, 502, 503, 504, 523, 534, 535, 600, 605, 608
Esteban M, Pray LC (1975) 504, 535
Esteban M, Pray LC (1976) 504, 535
Esteban M, Pray LC (1977) 535
Esteban M, Pray LC (1983) 437, 439, 440, 441, 443, 448

Eugster HD, Chou IM (1973) 101, 107
Evamy BD (1963) 495, 502
Evamy BD (1973) 78, 97, 530, 535
Evamy BD, Shearman DJ (1962) 84, 97

Fabricius F (1967) 3, 6, 75, 97
Fabricius F (1977) 4, 6, 71, 74, 77, 78, 80, 82, 97, 167
Fabricius F, Klingele H (1970) 71, 80, 81, 87, 97, 154, 157, 162, 203, 213
Falkenhein FUH (1981) 335, 342
Fannin NGT (1969) 308, 328
Farrow G, Scoffin T, Brown B, Cucci M (1979) 227, 228, 241
Fazzuoli M (1974) 583, 586
Ferguson J, Bubela B, Davies PJ (1978) 74, 80, 95, 97, 174, 175
Ferguson J, Ibe AC (1981) 173, 175
Fischer AG (1964) 336, 342
Flood PG (1983) 228, 232
Flügel E (1977) 302, 306, 414, 422
Flügel E (1978) 3, 4, 6, 8, 72, 73, 97, 273, 275, 284, 297, 299, 306, 313, 315, 328, 438, 448, 590, 597
Focke JW, Gebelein CD (1978) 44, 54, 605, 606, 608
Folk RL (1959) 3, 6
Folk RL (1974) 147, 152, 174, 175, 183, 187, 495, 502
Folk RL, Chafetz HS (1980) 474, 487
Folk RL, Chafetz HS (1983) 437, 446, 448
Folk RL, Pittman JS (1971) 90, 97
Folk RL, Siedlecka A (1974) 439, 448
Forel FA (1878) 284, 297
Fornaseri M, Grandi L (1963) 171, 175
Fornaseri M, Grandi L (1968) 171, 175
Foslie MH (1894) 217, 218, 224, 227, 228, 235, 241
Franke MR (1981) 336, 342
Frantzen W (1887) 71, 93, 97, 161, 162
Freeman T (1962) 91, 97, 100, 107, 148, 152, 203, 213, 617, 618
Freyberg von B (1973) 299, 305, 306
Freytet P (1964) 308, 328
Freytet P (1965) 308, 328
Freytet P (1971) 308, 328
Freytet P (1973) 308, 328
Freytet P, Plaziat JC (1965) 308, 310, 323, 325, 328
Friedman GM (1964) 87, 90, 97
Friedman GM, Amiel AJ, Braun M, Miller DS (1973) 80, 97, 148, 152, 163, 167
Friedman GM, Sanders JE (1978) 163, 167
Friedman I, O Neil JR (1977) 28, 36
Frishman SA, Behrens EW (1969) 81, 97
Fritsch FE (1959) 276, 279, 283
Fritz GK (1958) 5, 6
Fritz P, Katz A (1972) 10, 24
Füchtbauer H (1964) 75, 97
Füchtbauer H (1968) 406, 408, 414, 422, 587, 597
Füchtbauer H (1974) 3, 6, 89, 97
Füchtbauer H, Hardie LA (1980) 96, 97
Füchtbauer H, Müller G (1970) 61, 68

Gaemers PAM (1978) 309, 328
Gaetani M (1975) 358, 266
Gaetani M, Fois E, Jadoul F, Nicora A (1981) 438, 448
Garcia-Hernandez M, Lopez-Garrido AC, Oloriz F (1981) 375, 376
Garrett P (1971) 137, 141
Garwood EJ (1931) 280, 283
Gasiewicz A (1983) 274, 275
Gavish E, Friedman GM (1969) 32, 36, 84, 97
Gebelein CD (1969) 282, 283 331, 342, 419, 420, 421, 422
Gebelein CD (1974) 116, 117, 129, 131
Gebelein CD (1976) 208, 213, 331, 342, 414, 422
Geno KR, Chafetz HS (1980) 486, 487
Gillet S (1963) 299, 306
Ginsburg RN, Lowenstam HA (1958) 406, 408
Ginsburg RN, Shinn EA (1964) 122, 131
Goldberg M, Friedman GM (1974) 438, 448
Goldsmith JR, Graf DL (1958) 255, 257, 310, 328
Goldsmith JR, Graf DL, Heard HC (1961) 145, 152
Goldsmith JR, Graf DL, Joensuu OI (1955) 310, 328
Golonka J, Rajchel J (1972) 356, 357
Golubić S (1962) 293, 297
Golubić S (1973) 297
Golubić S (1976) 413, 415, 422
Golubić S (1979) 288, 297
Golubić S, Campbell SE (1981) 289, 297, 305, 306
Golubić S, Fischer AG (1975) 282, 283, 315, 328, 427, 431, 433
Golubić S, Focke JN (1978) 413, 422
Golubić S, Friedmann I, Schneider J (1981) 292, 297
Goodwin AM (1956) 100, 107
Goudie A (1973) 501, 502
Gradziński R, Radomski A (1967) 447, 448
Graf DL (1960) 10, 24
Graff von de WJE (1971) 409, 422
Greensmith JT (1978) 203, 213
Gross MG (1964) 27, 32, 36
Grottenthaler W (1978) 386, 389
Gruner JW (1922) 101, 107
Grunert L, Schöps W, Lübke M (1980) 61, 62, 63, 68
Gulbrandsen RA (1970) 353, 357
Gvirtzmann G, Buchbinder B (1978) 48, 54
Gygi RA (1981) 101, 103, 107

Hagan GM, Logan BW (1975) 207, 214
Hahne C, Kirchmayer M, Ottemann J (1968) 72, 73, 82, 97
Hall AL (1918) 619, 620, 634
Halley RB (1977) 78, 79, 81, 83, 97, 101, 107, 112, 113, 114, 142, 148, 151, 152, 163, 167
Halley RB, Harris PM (1979) 13, 24
Halley RB, Harris PM, Hine AC (in press) 123, 131
Hallimond AF (1925) 107
Harris PM (1979) 116, 117, 122, 128, 129, 131, 132, 133, 141
Harrison RS (1977) 437, 438, 440, 448, 461, 464, 467, 468, 469, 473, 543, 546
Harrison RS, Steinen RP (1978) 461, 467, 468, 471, 473
Harsch W (1970) 388, 389, 405, 406, 407, 408
Hay RL, Reeder RJ (1978) 438, 440, 448, 456, 461, 464, 468, 469, 473
Hay RL, Wiggins B (1980) 13, 18, 24, 79, 97, 438, 448, 456, 461, 464, 468, 469, 473, 499, 502
Hayes AO (1915) 101, 107
Hearn MG (1942) 621, 628, 634
Heim A (1916) 3, 6, 8, 308, 328
Heinrichs TK (1980) 623, 624, 634
Heinrichs TK, Reimer TO (1977) 621, 623, 624, 626, 634
Heller PL, Komar PD, Pevear DR (1980) 113, 114, 194, 198
Herget G (1966) 621, 632, 634
Herlinger DL (1981) 451, 455
Hess FL (1929) 72, 82, 97, 148, 152
Hileman ME, Mazzullo SJ (1977) 504, 505, 533, 535
Hine AC (1977) 116, 121, 122, 123, 124, 129, 131, 132, 141
Hine AC, Neumann AC (1977) 116, 122, 131
Hine AC, Wilber RJ, Bane JM, Neumann AC, Lorenson KR (1981) 116, 121, 122, 127, 131
Hine AC, Wilber RJ, Neumann AC (1981) 116, 117, 122, 123, 128, 131, 132, 141
Hoffman PC (1976) 324, 328, 415, 422
Hofmann HJ (1969) 308, 310, 324, 328
Hofmann HJ (1975) 276, 283, 356, 357
Hollmann R (1964) 363, 366
Horowitz AS, Potter PE (1971) 567, 575
Hottinger L (1977) 47, 54
Hottinger L (1981) 55
Hottinger L, Levinson G (1978) 47, 55
Hubbard DK, Ward LG, FitzGerald DM, Hine AC (1976) 116, 131
Hudson JD (1977) 27, 32, 36
Hurley NS (1978) 512, 536
Huvé H (1954) 221, 224

Ichikuni M (1973) 10, 24
Illing LV (1954) 13, 24, 77, 79, 97, 116, 131, 141, 148, 152, 154, 162, 202, 214
Imlay RW (1952) 110, 114
Irwin H, Curtis C, Coleman M (1977) 30, 34, 36

Jacka AD, Thomas CM, Beck RH, Williams KW, Harrison SC (1969) 503, 536
Jackson MJ, Muir MD, Plumb KA, Large DE, Brown MC, Armstrong KJ (1978) 547, 550, 551, 555, 557
Jackson TA, Bischoff JL (1971) 174, 175
Jacquotte P (1962) 227, 228, 234, 241
James NP (1972) 79, 97, 438, 448, 456, 461, 467, 468, 473, 543, 546
James NP, Ginsburg RN (1979) 341, 342
Jenkyns HC (1972) 369, 375, 376
Jenkyns HC (1980) 369, 376
Jerz H (1966) 386, 389, 405, 406, 408
John DM, Lieberman DL (1979) 228, 237, 241
Johnson JH (1942) 503, 508, 536
Johnson JH (1957) 245, 257
Johnson JH (1961) 228, 232, 241, 244, 257, 302, 306
Johnson JH (1962) 223, 224
Johnson JH (1963) 432, 433, 604, 608
Johnson JH (1964) 245, 257
Johnston MR (1980) 251, 257
Jones FG, Wilkinson BH (1978) 277, 279, 281, 283, 431, 455, 488, 502, 604, 605, 607, 608
Jones JC (1925) 144, 147, 152

Kahle CF (1965) 13, 14, 21, 24, 174, 175
Kahle CF (1974) 15, 24, 74, 77, 81, 93, 94, 98, 112, 115, 142, 148, 151, 152, 163, 167, 439, 441, 445, 446, 448
Kahle CF (1981) 561, 565
Kalkowsky E (1908) 3, 6, 71, 73, 83, 91, 94, 98
Kalliokoski J (1966) 107
Kann E (1940) 284, 292, 294, 297
Kann E (1941) 297
Katz A, Sass E, Starinsky A, Holland HD (1972) 10, 24
Keegan B (1974) 239, 241
Keith ML, Weber JN (1964) 27, 36, 341, 342
Kendall CGStC (1969) 488, 502, 503, 512, 514, 515, 533, 536
Kennedy WJ (1975) 360, 366
Kennedy WJ, Garrison RE (1974) 345, 357, 369, 376
Kennedy WJ, Klinger HC (1972) 345, 357
Kennett JP, Houtz RE, Andrews PB, Edwards AR, Gostin VA, Hajós M, Hampton M, Jenkins DG, Margolis SV, Overshine AT, Perch-Nielsen K (1975) 243, 257
Kerkmann K (1966) 587, 597
Kettenbrick EC, Manger WL (1971) 93, 98, 109, 115

Kimberley MM (1978) 100, 107
Kimberley MM (1979) 101, 103, 107
King PB (1948) 505, 536
Kinsman DJJ (1969) 10, 13, 18, 19, 24, 78, 98, 159, 162
Kinsman DJJ, Holland HD (1969) 10, 24
Kitano Y, Kanamori N, Domori T (1971) 10, 14, 15, 24
Kitano Y, Tokuyama A, Arakaki T (1979) 11, 24
Kjellman FR (1883) 227, 241
Klappa CF (1978) 8, 467, 468, 471, 473
Klasz de I (1978) 336, 342
Klein de GV (1970) 208, 214
Klinker J, Reiss Z, Kropach C, Levanon I, Harpaz H, Halicz E (1976) 47, 55
Klinker J, Reiss Z, Kropach C, Levanon I, Harpaz H, Shapiro Y (1978) 47, 55
Knacke O, Pohl H (1959) 60, 68
Knatz H (1966) 82, 98
Knewtson SL, Hubert JF (1969) 87, 90, 98
Knox GJ (1977) 438, 448
Kobluk DR, Risk MJ (1977) 293, 297
Koen GM (1947) 621, 628, 634
Konishi K (1975) 228, 232, 237, 241
Kornicker LS, Boyd DW (1962) 233, 241
Kostecka A (1966) 587, 597
Krajewski KP (1981) 344, 348, 356, 357
Krajewski KP (1983) 274
Kraus O (1969) 386, 388, 389
Krumbein WE (1979) 288, 293, 297
Krumbein WE, Potts M (1978) 396, 397
Krylov IN (1976) 327, 328
Kulke H, Ott E (1978) 294, 297
Kumar S (1977) 356, 357

Laborel J (1961) 228, 241
Ladd HS, Hoffmeister JE (1945) 488, 502
Lagaaij R, Gautier YV (1965) 247, 257
Lahann RW (1978) 147, 153
Lahann RW, Siebert RM (1981) 10, 24
Land LS (1970) 7, 8, 27, 36, 333, 342
Land LS (1980) 10, 24
Land LS, Behrens EW, Frishman SA (1979) 13, 15, 24, 29, 30, 36, 81, 83, 94, 98, 109, 115, 148, 151, 152, 153, 154, 162, 167, 168, 486, 487
Land LS, Goreau TF (1970) 32, 36
Land LS, Hoops GK (1973) 10, 24
Lang J, Segonzac G (1979) 308, 325, 328
Lang WB (1942) 503, 534, 536
Le Campion-Alsumard T (1979) 293, 297
Leclaire L (1972) 230, 241
Leeder MR (1975) 413, 422, 433
Lees A, Butler AT, Scott J (1969) 248, 257
Lefeld J (1968) 344, 357
Leighton MW, Pendexter C (1962) 4, 6
Lemoine P (1910) 217, 218, 224, 227, 228, 234, 235, 241
Lemoine P (1939) 223, 224
Leonardi P (1967) 390, 396, 397
Lindholm RC, Finkelman RB (1972) 85, 98
Link MH, Osborne RH, Awramik S (1978) 295, 297
Lippmann F (1973) 174, 175
Littler M (1973) 232, 241
Lloyd RM (1966) 28, 36
Locker S (1980) 122, 128, 131
Logan BW (1961) 207, 209, 214
Logan BW, Harding JL, Ahr WM, Williams JD, Snead RG (1969) 78, 80, 98, 228, 230, 233, 237, 241
Logan BW, Hoffman P, Gebelein CD (1974) 415, 417, 421, 422
Logan BW, Read JF, Hagan GM, Hoffman P, Brown RG, Woods PJ, Gebelein C (1974) 577, 586
Logan BW, Rezak R, Ginsburg RB (1964) 207, 214, 385, 431, 434
Longman MW (1980) 182, 187
Loreau J-P (1970) 71, 98, 154, 157, 162
Loreau J-P (1973) 610, 617, 618
Loreau J-P (1982) 23, 24
Loreau J-P, Purser BH (1973) 13, 15, 18, 24, 77, 78, 79, 80, 82, 98, 148, 151, 153, 157, 162, 167, 168, 186, 187, 194, 198, 201, 202, 203, 205, 209, 214, 383, 385, 500, 502
Lorens RB (1981) 11, 24
Loucks RG, Folk RL (1976) 504, 536, 605, 608
Lowe DR, Knauth PR (1978) 56, 65, 66, 68, 619, 620, 623, 624, 633, 634
Lowe DR, Knauth PR (1979) 56, 68
Lowenstam H, Epstein S (1957) 17, 24, 29, 30, 36
Lyell Ch (1829) 474, 475, 487
Lyell Ch (1855) 3, 6, 71, 72, 98

Mac Gregor AM (1937) 633, 634
MacIntyre IG, Milliman JD (1970) 228, 230, 241
Mackenzie FT, Pigott JD (1981) 15, 25, 109, 115
Mackin JH, Coombs HA (1945) 82, 98
Magaritz M (1973) 35, 36
Magaritz M (1974) 30, 31, 36
Magaritz M (1975) 30, 32, 36
Magaritz M (1983) 438, 445, 448
Magaritz M, Gavish E, Bakler N, Kafri U (1979) 32, 36
Magaritz M, Kaufman A, Yaalon DH (1981) 30, 31, 36
Mamet B, Roux A (1975) 426, 434
Manetti P, Turi A (1973) 171, 175

Manfra L, Masi U, Turi B (1974) 474, 487
Manfra L, Masi U, Turi B (1976) 474, 487
Marcinowski R, Szulczewski M (1972) 356, 357
Markun CD, Randazzo AF (1980) 100, 101, 107
Marshall JF, Davies PJ (1975) 13, 15, 25, 72, 82, 98, 148, 151, 153
Martens CS, Harris RC (1970) 353, 357
Martin L (1971) 228, 230, 241
Maslov VP (1952) 406, 408
Maslov VP (1960) 405, 408
Massari F (1981) 365, 366
Massari F (1983) 274
Massari F, Dieni I (1983) 274
Massari F, Medizza F (1973) 356, 357
Matyja BA (1978) 5, 6
Maxia C (1950) 474, 487
Mazzullo SJ (1977) 87, 90, 98
Mazzullo SJ (1980) 605, 608
Mazzullo SJ, Cys JM (1977) 605, 606, 608
Mazzullo SJ, Cys JM (1978) 599, 604, 605, 606, 608
Mazzullo SJ, Cys JM (1979) 260, 269, 605, 606, 608
McCrea JM (1950) 27, 36
McFarlane MJ (1976) 501, 502
McGannon Jr DE (1975) 15, 25, 82, 98
McIntyre WL (1963) 9, 10, 25
McMaster RW, Conovar JT (1966) 217, 224, 228, 239, 241
Meijer de JJ (1971) 409, 422
Meldau R (1964) 60, 68
Meredith JC (1980) 450, 455
Meyers WJ (1974) 523
Meyers WJ (1978) 341, 343
Meyers WJ (1980) 20, 25
Meyers WJ, Lohmann KC (1978) 95, 98
Michard G (1968) 10, 25
Milliman JD (1967) 13, 18, 25
Milliman JD (1974) 13, 14, 15, 17, 18, 19, 25, 78, 98, 243, 244, 245, 255, 257, 341, 343, 414, 422
Milliman JD (1977) 243, 254, 257, 414, 422
Milliman JD, Barretto HT (1975) 72, 81, 98, 148, 151, 153
Minder L (1922) 286, 297
Mišik M (1980) 439, 441, 445, 448
Mitterer RM (1968) 74, 95, 98, 172, 175
Mitterer RM (1969) 172, 175
Mitterer RM (1971) 172, 175
Mitterer RM (1972) 74, 95, 98, 172, 175
Moberley R (1968) 237, 238, 241
Möller P, Rajagopolan G, Germann K (1976) 10, 25
Moguedet G (1973) 81, 98
Molnar P, Atwater T, Mammerick J, Smith SM (1975) 243, 257
Montaggioni LF (1979) 228, 230, 232, 233, 241
Monty CLV (1965) 413, 421, 422
Monty CLV (1967) 413, 421, 422
Monty CLV (1971) 365, 366
Monty CLV (1972) 273, 275, 411, 421, 422
Monty CLV (1974) 422
Monty CLV (1976) 308, 317, 318, 328, 347, 351, 357, 362, 366, 372, 376, 411, 413, 415, 421, 422
Monty CLV (1977) 282, 283, 365, 366
Monty CLV (1979) 274, 275, 401, 408
Monty C (1981) 299, 306
Monty C, Mas JR (1981) 274, 275, 295, 297, 351, 357, 593, 597
Moore CH, Druckmann Y (1981) 30, 34, 36
Moore JG, Peck DL (1962) 57, 58, 65, 66, 68
Morrow DW, Mayers IR (1978) 10, 25
Mouline P (1977) 308, 310, 321, 328
Müller G (1968) 19, 25
Müller G (1971) 495, 502
Müller G, Irion G, Förstner U (1972) 147, 153
Müller J, Sigl W (1977) 294, 297
Muir MD (1979) 547, 550, 552, 556, 557
Muir MD (1983) 438, 439, 448
Muir MD, Lock DE, Borch von der CC (1980) 547, 555, 556, 557
Mullins HT, Neumann AC, Wilbur RJ, Boardman MR (1980) 365, 366
Multer HG, Hoffmeister JE (1968) 460, 473
Murray G (1895) 308, 328
Murray JW (1954) 17, 25

Nagtegaal PJC (1969) 439, 448, 464, 473
Nahon D, Carozzi AV, Parron C (1980) 100, 107
Neev D, Emery KO (1967) 163, 168
Nelson CS (1978) 243, 252, 257
Netterberg F (1969) 438, 448
Neumann AC (1971) 129, 131
Neumann AC, Gebelein CD, Scoffin TP (1970) 121, 131, 331, 343
Newell ND, Purdy EG, Imbrie J (1960) 13, 25, 72, 74, 78, 82, 86, 87, 94, 98, 148, 153, 154, 159, 162, 202, 207, 214
Newell ND, Rigby K (1957) 122, 123, 131, 209, 214
Newell ND, Rigby JK, Fischer AG et al (1953) 503, 505, 534, 536, 599, 608
Nicholson HA, Etheridge Jr R (1878) 280, 289
Nickel E (1980) 310, 324, 328
Nickel E (1983) 274, 275
Nijman W, Nio SD (1975) 308, 329
Nissen HU (1964) 93, 98
Nissenbaum A (1977) 163, 168
Norris RJ, Carter RM, Turnbull IM (1978) 250, 257

Off T (1963) 123, 131
Ohde S, Kitano Y (1980) 11, 25
Ordonez RH, Del Cura G (1977) 318, 321, 329
Oti M, Müller G (1979) 77, 81, 94, 98, 157, 162
Otte C (1959) 260, 269
Otte C, Parks JM (1963) 260, 269

Page RW (1979) 550, 557
Pallas (1766) 217, 224
Palmer MS (1979) 116, 117, 122, 128, 131, 132, 141
Pankow H (1971) 305, 306
Parker RH, Curray JR (1956) 228, 241
Passarge S (1904) 284, 297
Passeri L (1971) 583, 586
Passeri L (1975) 577, 586
Passeri L (1979) 586
Passeri L, Pialli G (1972) 583, 586
Patriquin DG (1972) 333, 343
Patriquin DG (1975) 121, 125, 131
Payton CE (1966) 90, 98
Pedley HM (1976) 247, 257
Pedro G, Carmouse JP, Velde B (1978) 9, 17, 18, 25
Pentecost A (1978) 308, 318, 325, 329
Pérès JM (1961) 49, 55
Pérès JM (1967) 227, 241
Pérès JM, Picard J (1964) 41, 55
Perkins RD, Enos P (1968) 121, 131
Peryt TM (1977) 5, 6, 396, 397, 398, 406, 407, 408, 414, 422, 587, 597
Peryt TM (1978) 587, 598
Peryt TM (1979) 401, 408
Peryt TM (1980) 396, 397, 403, 405, 408
Peryt TM (1981) 5, 6, 8, 273, 275, 280, 283, 284, 297, 299, 306, 398, 401, 406, 407, 408, 421, 422, 443, 445, 448, 587, 589, 594, 597, 598
Peryt TM (1983) 4, 5, 6, 7, 8, 9, 217, 224, 274, 438, 440, 441, 445, 462, 596
Peryt TM (in press) 590, 598
Peryt TM, Piatkowski TS (1977) 438, 441, 445, 448, 533, 534, 536, 555, 557, 587, 598
Peryt TM, Wagner R (1981) 273, 275, 441, 448, 587, 589, 598
Peryt TM, Wazny H (1980) 587, 598
Pettijohn FJ (1975) 90, 98
Pflug H (1976) 621, 629, 630, 634
Pia J (1927) 3, 6
Pia J (1933) 285, 295, 297
Pialli G (1971) 583, 586
Piekarska E, Kwiatkowski S (1975) 587, 598
Pigott J, Mackenzie FT (1980) 22, 25
Pingitore NE (1978) 20, 25
Pingitore NE (1982) 20, 25
Piper DJW, Panagos AG (1979) 306
Playford PE, Cockbain AE, Druce EC, Wray JL (1976) 280, 283
Plinius C (1979) 56
Plumb KA (1979) 547, 557
Plumb KA, Brown MC (1973) 547, 550, 552, 557
Plumb KA, Derrick GM (1976) 550, 557
Pollock J (1928) 232, 241
Pomar L (1976) 468, 473
Pomar L, Esteban M, Llimona X, Fontarnau R (1975) 468, 473
Poncet J (1968) 609, 618
Poncet J (1972) 609, 618
Poncet J (1976) 609, 618
Poncet J (1983) 273
Ponte FC, Dos Reis Fonseca J, Carozzi AV (1980) 330, 343
Popp BN, Wilkinson BH (1983) 81, 82, 98
Pratt BR (1979) 331, 343
Pray LC (1958) 260, 269
Pray LC (1961) 260, 269
Pray LC (1977) 445, 448
Pray LC, Esteban M (1977) 504, 505, 508, 520, 528, 529, 530, 533, 536
Prezbindowski DR (1977) 341, 343
Purdy EG (1963) 116, 117, 122, 123, 131, 134, 141, 148, 153
Purser BH (1978) 336, 343
Purser BH (1980) 3, 4, 6, 591, 598, 610, 618
Purser BH, Evans G (1973) 196, 198
Purser BH, Loreau J-P (1973) 16, 25, 78, 98, 438, 446, 448, 503, 504, 533, 534, 536, 599, 608
Purser BH, Seibold E (1973) 211, 214

Rácz L (1965) 409, 422
Radwański A (1968) 3, 6, 403, 406, 408
Radwański A, Birkenmajer K (1977) 73, 86, 91, 92, 93, 98, 590, 598
Raiswell R (1976) 34, 36
Ramsay JR (1963) 621, 629, 634
Rao CM (1978) 13, 25
Rat P (1953) 375, 376
Rat P (1966) 375, 376
Read JF (1974) 79, 83, 98, 437, 438, 440, 448, 461, 462, 473, 594, 598
Read JF (1976) 461, 464, 468, 473, 554, 555, 557, 573, 575
Reeves CC Jr (1970) 501, 502
Reeves CC Jr (1976) 468, 469, 473, 501, 502
Reijers TJA (1972) 188, 191, 195, 198
Reijers TJA (1974) 188, 198
Reijers TJA (1980) 188, 197, 198
Reimer TO (1967) 619, 620, 623, 624, 634
Reimer TO (1975) 65, 68, 619, 620, 623, 624, 631, 634
Reimer TO (1983) 67, 68, 633
Reineck H-E, Singh IB (1973) 321, 329
Reis OM (1923) 295, 297
Reiss Z (1977) 47, 55
Richter DK (1972) 84, 99
Richter DK (1974) 84, 99
Richter DK (1976) 76, 77, 78, 99, 157, 160, 162, 306
Richter DK (1979) 7, 8, 13, 25, 78, 82, 84, 99

Richter DK (1980) 72, 73, 82, 84, 85, 86, 99
Richter DK (1983) 3, 4, 6, 7, 8, 159, 162, 437, 449
Richter DK, Besenecker H (1983) 7, 72, 73, 79, 87, 89, 91, 99
Richter DK, Füchtbauer H (1978) 10, 25, 72, 84, 86, 93, 95, 99, 305, 306, 594, 598
Richter DK, Herforth A, Ott E (1979) 284, 295, 298, 302, 305, 306, 307
Richter DK, Sedat R (1983) 274
Richter DK, Zinkernagel U (1981) 84, 86, 99, 306, 307
Riding R (1975) 396, 397, 411, 422
Riding R (1977) 283, 288, 298, 325, 329, 396, 397, 401, 408, 411, 422, 604, 608
Riding R (1979) 277, 283, 555, 557
Riding R (1983) 5, 6, 273
Riding R, Voronova L (in prep.) 277, 279, 280, 283
Riedl R (1966) 42, 51, 55
Rioult M, Royant G (1975) 356, 357
Rippka R, Deruelles J, Waterburg JB, Herdman M, Stanier RY (1979) 288, 298
Risacher F, Eugster HP (1979) 324, 329, 437, 446, 449, 605, 608
Rohrlich V (1974) 100, 108
Royant G, Rioult M, Lanteaume M (1970) 356, 357
Rusnak GA (1960) 81, 99, 148, 151, 153, 194, 198, 203, 214, 617, 618
Russell IC (1885) 142, 144, 153

Salomons W, Goudie A, Mook WG (1978) 30, 31, 36
Samuel O, Borza K, Köhler E (1972) 375, 376
Sandberg PA (1975) 15, 22, 25, 71, 72, 81, 83, 84, 99, 109, 112, 113, 115, 142, 148, 151, 153, 163, 168
Sannemann D, Zimdars J, Plein F (1978) 75, 99
Sarkar B (1973) 93, 99
Schäfer A (1973) 284, 286, 294, 298
Schäfer A, Stapf KRG (1978) 274, 275, 279, 281, 283, 284, 286, 294, 295, 298, 305, 307, 308, 317, 318, 325, 329, 427, 431, 434
Schlanger SO (1957) 255, 257
Schmitt M, Monninger W (1977) 415, 422
Schneider HJ (1953) 386, 389
Schneider J (1976) 293, 298
Schneider J (1977) 284, 286, 291, 293, 298
Schneider J, Golubić S, Le Campion-Alsumard T (in press) 298
Schneider J, Schröder HG (1980) 284, 285, 288, 289, 292, 298
Schneider J, Torunski H (in prep) 293, 298
Schneidermann N, Sandberg PA (1971) 145, 153
Schoen R (1962) 101, 108
Schöttle M (1969) 284, 298
Schöttle M, Müller G (1968) 284, 298, 308, 329
Scholle PA, Kinsman DJJ (1974) 7, 8, 16, 19, 25, 438, 449, 468, 488, 500, 502, 504, 534, 536, 545, 546, 556, 557, 594, 598, 600, 605, 608
Schopf JW (1977) 276, 283
Schopf TJM (1969) 247, 257
Schröder B (1975) 299, 307
Schröder HG (1982) 284, 285, 286, 289, 291, 294, 298
Schroeder JH (1973) 7, 8
Schubert H (1979) 58, 59, 60, 68
Schuler G (1968) 3, 6, 75, 99, 386, 389, 405, 406, 408
Schultz CB, Stout TM (1955) 438, 449
Schwarzacher W (1978) 415, 422
Scoffin TP (1970) 121, 131, 331, 343
Scudeler Baccelle L (1980) 171, 174, 175
Self S, Sparks RSJ (1979) 56, 57, 58, 65, 68
Semikhatov MA, Gebelein CD, Cloud P, Awramik SM, Benmore WC (1979) 4, 6, 326, 327, 329
Seyfried H (1981) 363, 365, 366
Shearman DJ, Twyman J, Zand Karimi M (1970) 71, 78, 99, 101, 108
Shinn EA (1969) 605, 608
Shinn EA (1973) 503, 534, 536, 605, 608
Shoji R, Folk RL (1964) 71, 99, 203, 214
Siehl A, Thein J (1978) 100, 108
Siesser WG (1972) 228, 230, 241
Siesser WG (1973) 79, 99, 437, 438, 449, 464, 469, 473
Simone L (1974) 71, 83, 99
Simone L (1981) 72, 83, 99, 113, 115, 561, 565, 618
Simons DB, Richardson EV, Mordin CF (1965) 208, 214
Sliwinski W (1981) 446, 449
Smith DB (1970) 445, 449
Smith DB (1974) 503, 529, 533, 536
Smith DB (1980) 445, 449
Smith DB (in press) 587, 598
Smith J (1964) 550, 557
Smith JP, Lehr JR (1966) 353, 357
Sneed ED, Folk RL (1958) 218, 219, 224
Sorby HC (1879) 7, 8, 15, 25, 71, 72, 73, 74, 78, 79, 82, 83, 89, 90, 99, 142, 153, 154, 161, 162
Squazzoni G (1968) 586
Stach LW (1936) 247, 257
Steel RJ (1974) 464, 473
Stehli FG, Hower J (1961) 21, 25
Steneck R (1978) 236, 237, 241
Steneck R, Adey WH (1976) 238, 241
Steinen RP (1974) 467, 471, 473
Steiner MB, Williams TE (1968) 260, 269
Stowe CW (1968) 633, 634
Studencki W (1979) 243, 254, 258

Sturani C (1971) 358, 365, 366
Suess E (1970) 294, 298
Suess E, Fütterer D (1972) 13, 15, 16, 25, 80, 99, 174, 175
Summerhayes CP, Milliman JD, Briggs SR, Bell AG, Hogan C (1976) 227, 228, 241
Surdam RC, Stanley KO (1979) 438, 440, 441, 449, 464, 473
Swanson HE, Fuyat RK, Ugrinic GM (1954) 159, 162
Swett K (1965) 90, 99
Swineford A, Leonard AB, Frye JC (1958) 437, 438, 440, 449, 464, 469, 473, 512, 536

Taft WH, Arrington F, Maimovitz A, MacDonald C, Woolheater C (1968) 121, 131
Tan FC, Hudson JD (1974) 27, 30, 32, 34, 36
Tandon SK, Narayan D (1981) 446, 449
Tang Tianfu, Xue Yaoson, Yu Congliu (1980) 537, 545, 546
Tang Tianfu, Zhang Junming, Yang Wanrong, Qiu Jinyu, Zhou Yangkang (1980) 539, 546
Taylor HP Jr (1978) 33, 36
Taylor JCM, Colter VS (1975) 406, 408
Taylor JD, Lewis MS (1970) 333, 343
Tazieff H (1978) 61, 68
Teichert C (1970) 71, 99
Thomas CM (1965) 503, 536
Thomas CM (1968) 503, 508, 512, 514, 536
Thompson CW (1877) 228, 241
Thorarinsson S (1968) 63, 68
Thorarinsson S (1969) 58, 68
Tibana P, Alves R (1973) 336, 343
Tichy G (1972) 388, 389
Tichy G (1983) 273, 405
Till R (1970) 13, 18, 25
Tišljar J (1978) 566, 575
Tišljar J (1980) 573, 575
Tišljar J (1983) 274, 438, 441, 449
Tišljar J, Durasek N, Sokač B, Velić I (1981) 566, 575
Toomey DF (1974) 398, 408
Toomey DF (1975) 217, 224, 228, 230, 237, 242, 264, 269, 605, 608
Toomey DF, Cys JM (1977) 260, 269
Toomey DF, Cys JM (1979) 260, 264, 269
Toomey DF, Wilson JL, Rezak R (1977) 260, 267, 269
Torunski H (1979) 293, 298
Trendall AF (1965) 58, 65, 68
Trichet J (1968) 173, 175
Truc G (1975) 438, 449
Truswell JF, Eriksson KA (1975) 212, 214
Turekian KK (1972) 11, 14, 20, 25
Turmel R, Swanson RW (1976) 232, 242
Tyrrell WW Jr (1969) 512, 536

Urey HC, Lowenstam HA, Epstein S, McKinney CR (1951) 27, 36
Usdowski HE (1962) 83, 94, 99

Valdija KS (1972) 356, 357
Van Vuuren CJ (1964) 621, 629, 630, 634
Veizer J (1974) 10, 25
Veizer J (1977) 21, 25
Veizer J (1978) 10, 17, 25, 26
Veizer J, Demovič R (1973) 171, 175
Veizer J, Lemieux J, Jones B, Gibling MR, Savalle J (1977) 10, 26
Velić I, Sokač B (1978) 572, 575
Velić I, Tišljar J, Sokač B (1979) 566, 575
Viel G (1979) 391, 397
Viljoen MJ (1964) 621, 634
Visser DJL (1956) 67, 68, 620, 621, 623, 634
Visser JNJ, Grobler NJ (1972) 199, 214
Vital DA (1948) 142, 153, 163, 168
Voigt E (1968) 345, 357

Wagner F, Lamprecht G (1974) 318, 329
Wagner W (1954) 295, 298
Walker GPL (1971) 64, 65, 66, 68
Walker RN, Muir MD, Diver WL, Williams N, Wilkins N (1977) 556, 557
Walker RN, Logan RG, Binnekamp JG (1978) 556, 557
Walls RA, Harris WB, Nunan WE (1975) 462, 464, 473
Walls RA, Mountjoy EW, Fritz P (1979) 32, 36
Walter LM, Hanor JS (1979) 294, 298
Walter MR (1976) 308, 329
Walther J (1888) 99
Wanless HR (1981) 125, 131
Ward WC (1975) 438, 449, 461, 468, 469, 473
Wardlaw NC, Reinson GE (1971) 439, 440, 441, 443, 445, 449, 533, 536
Warren JK (1981) 531, 536
Wass RE, Conolly JR, MacIntyre RJ (1970) 247
Watts NL (1978) 554, 557
Weber JN (1967) 29, 37
Weber JN, Deines P, Weber PH, Baker PA (1976) 27, 28, 37
Weber von Bosse A, Foslie M (1904) 228, 232, 242
Weiler Y, Sass E, Zak I (1974) 9, 26
Weiss MP (1969) 398, 406, 408
Wendt J (1969) 265, 269
Wetzel RG (1960) 284, 298
Whelam TH, Roberts HH (1973) 30, 37
Wherry ET (1916) 91, 99
White AF (1977) 11, 26
White AF (1978) 10, 26
Whitehead HC, Feth JH (1961) 144, 153
Wilber RJ (1981) 116, 122, 131
Wiles JW (1957) 633, 634
Wilkinson BH (1979) 72, 83, 84, 99
Wilkinson BH, Landing E (1978) 71, 83, 93, 99, 103, 108, 110, 115

Wilkinson BH, Popp BN, Owen RM (1980) 83, 99, 142, 148, 153
Wilson JL (1975) 129, 131, 419, 422
Winchester PD (1972) 439, 449
Winland HD (1969) 10, 26
Winland HD (1971) 134, 141
Winland HD, Matthews RK (1974) 116, 121, 131
Winterer EL, Bosellini A (1980) 169, 175
Wolf KH (1960) 3, 6
Wolf KH (1965) 4, 6
Wood A (1941) 433, 434
Wray JL (1977) 254, 308, 329, 592, 598, 604, 608
Wright RP (1971) 110, 115
Wright VP (1980) 424, 434
Wright VP (1981) 324, 329, 424, 425, 426, 427, 429, 434
Wright VP (1983) 273, 275, 280, 283
Wright VP, Wright EVG (1981) 425, 426, 431, 434
Yang Wanrong, Tang Tianfu, Zhang Junming, Qiu Jinyu, Zhou Yangkang (1978) 537, 546

Zanevald JS (1958) 228, 233, 242
Zeller EJ, Wray JL (1956) 13, 16, 26
Zemann J (1969) 9, 10, 26
Zeng Yun-fu, Lee Nan Ho, Huang Yang-zohn (1983) 90, 99
Zohary T, Reiss Z, Hottinger L (1980) 47, 55

Subject Index

Abrasion 49, 113, 114, 145, 203, 223, 236, 254, 293, 296, 336, 418, 427, 483, 485, 486, 512, 564, 575, 590
Accretion 113, 151, 316, 324, 327, 411, 417, 419, 420, 431, 469, 504, 508, 515, 521 – 523, 526, 528, 607
Accretionary lapilli 56 – 58, 61, 63 – 67, 619, 623, 626 – 628, 632, 633
Accumulation rate 129
Acid
  Amino 172, 174
  Humic 15, 23, 495
  Organic 15
Adhesion 499
Adriatic Sea 285
Africa 31, 237
  South 65 – 67, 73, 199, 619
Agglomeration 56, 58, 105
Aggradation 211, 415
Aggregation 174, 369
Agitation 15, 19, 74, 78, 80, 91, 113, 114, 139, 151, 194, 273, 384, 433, 477, 526 – 528
Albian 330, 336, 344, 355, 567, 573
Algae 3, 5, 15, 19, 133, 167, 203, 264, 275, 279, 281, 293, 327, 395, 431
  Blue-green (= cyanophytes, = cyanobacteria) 5, 29, 75, 94, 265, 276, 277, 280 – 282, 286, 288, 289, 291 – 293, 296, 299, 302, 306, 321, 333, 348, 349, 356, 375, 403, 411, 413 – 415, 469, 471, 503, 512, 537, 538, 542, 572
    Anacystis nidulans 173
    Bevocastria 273, 280, 426, 429
    Calothrix 279
    Coccoid 277, 435
    Dichothrix 279, 288
    Garwoodia 273, 280, 429, 431, 432
    Girvanella 273, 280, 390, 393, 395, 396, 401 – 403, 405 – 407, 417, 418, 429
    Gongrosira 282
    Mitcheldeania 273, 426, 429, 430, 433
    Ortonella 273, 280, 426, 427, 429 – 432
    Phormidium 279, 318, 413
    Plectonema 288
    Rivularia 279, 289, 295, 302, 305, 317, 318
    Rothpletzella 282, 429
    Schizothrix 279, 288, 431
    Scytonema 288, 317
    Sphaerocodium 282, 390, 406, 616
    Tolypothrix 288
  Green (= chlorophytes) 15, 94, 282, 288, 289, 296, 535, 567, 572, 573
    Charophytes 286, 294, 296
    Dasycladaceans 424, 425, 509, 567
  Red (= rhodophytes) 5, 15, 243, 249, 264, 268, 288, 296, 302, 305, 331, 333, 414, 503, 599, 604, 605
    Coralline 4, 47, 49, 51, 217, 225, 227, 233, 234, 254, 255, 302, 305, 495, 561, 565
Algal bioherm 254, 431
Algal boring 101, 568
Algal carbonate 144
Algal incrustation 44
Algal limestone 243, 244, 409, 414, 418
Algal mat 19, 34, 121, 128, 163, 203, 204, 281, 325, 331, 333, 335, 411, 413, 534
Algal mound 260
Algeria 324
Allochem 101, 184, 246, 248, 249, 251, 253, 254, 380 – 384, 401, 415, 417, 419, 490, 495, 497, 605
Alps
  Apuane 577
  Dolomite 390, 396
  Northern Calcareous 285, 386, 405
  Southern 358
  Swiss 3
Amalgamation 212
Amazon shelf 72, 81, 148
Anhydrite 155, 182, 183
Anhydritization 339
Apatite 103, 353
Apennines 174
  Northern 576, 580, 586
Aragonite 9 – 12, 14 – 17, 19 – 23, 29, 32, 34, 71, 73, 78, 81, 86, 89, 92, 94, 95, 109, 134, 142, 147, 148, 154, 157, 160, 163 – 165, 167, 174, 203, 265, 305, 336, 339, 427, 429, 480, 486, 494, 495, 501, 530, 534, 561, 599, 604 – 607
  Stoichiometric 159
Archean 56, 65 – 67, 100, 102, 619, 633
Ash cloud 59 – 67, 629, 631

Ash fall 56
Assimilation 286, 289
Atlantic 74, 172, 228, 233, 235, 237
Atmosphere 66, 67
Australia 13, 66, 82, 83, 163, 547
  Great Barrier Reef 13, 72, 80, 82, 92, 95, 148, 561
  South 531
  Western 58, 67, 280, 554, 555, 578
Austria 73, 265, 284, 285, 294, 336, 386
Azores 57

Back-reef 205, 207, 232, 503, 508, 509, 512, 533, 567, 572, 610, 613, 618
Bacteria 5, 8, 54, 94, 173, 289, 293, 348, 353, 365, 372, 453, 455, 469, 471, 474, 475
  Sulphate-reducing 34
Bacterial mat 474, 482
Bahamas 7, 12, 14, 17, 18, 71, 74–79, 87, 94, 116, 122, 129, 132, 133, 138, 154, 157, 159, 171, 173, 174, 202, 206, 209, 331, 413, 414
  Andros Island 103, 105
  Bimini Lagoon 12, 18
  Great Bahama Bank 12, 18, 29, 30, 78, 122, 125, 128, 207
  Joulters Cay 12, 123, 128, 133, 170, 171, 208
  Little Bahama Bank 122–124, 128
Banding 112–114
Bank 116, 122, 130, 249
Bar 133, 139, 141, 183, 185, 577, 583
  Oolitic 182, 367
  Tidal 128, 134, 139, 369
Barbados 79, 543
Barrier 183
Basin 144, 198, 250, 295, 587
Bauxite 100, 494, 526, 575
Beach 144, 145, 147, 177, 182, 184–186, 245, 446, 526, 561, 577, 583
Beach rock 16, 335, 337, 339, 341, 342
Bedform 121, 123, 125, 128–139, 207
  Large-scale 116
  Relict 124, 129
Bermuda 44, 228, 232, 236, 331, 419, 420
Binding 277, 281, 288, 296, 333, 348, 362, 364, 411, 512, 600
Bioabrasion 289
Bioclast 180, 333, 335, 336, 380, 382, 383, 385, 420, 578, 584, 610, 617
Biocrystallite 71
Biodegradation 435
Bioerosion 230, 293, 360, 362
Bioherm 197, 264, 327, 609, 610, 618
Biokarst 293, 295, 296
Bioturbation 369, 381, 382, 413, 415
Birdseye 385, 417, 538, 543, 545
Bivalve 239, 277, 295
Bonaire 233, 331
Boring 167, 261, 277, 301, 375, 533, 580, 616
  Endolithic 362, 604
Boundstone 264, 528, 605
Brachiopod 196, 246, 249, 264, 420, 613
Branching 222, 223, 235
Brazil 81, 330, 336
Breaking 573
Breccia
  Basal 577
  Dolomite 208
  Encapsulated 209
  Intraclastic 534
  Oolitic 578
  Supratidal 209
Brecciation 554
Brine 17
Britain 227, 433
Bryozoans 41, 47, 51, 228, 232, 239, 245, 247, 249, 253, 254, 331, 420
Buildup 261, 268, 390, 392, 396, 506
Bulgaria 398
Burial 27, 105, 110, 336, 339, 355, 421, 504, 522, 603, 606
Burrowing 133, 372
Burrows 101, 103, 360, 362, 371, 578

Calcification 238, 276, 279, 281, 288, 292–294, 305, 355, 413, 427, 471, 495
Calcispheres 369
Calcite 10, 11, 73, 83, 102, 145, 148, 160, 174, 176, 180, 182, 185, 203, 284, 286, 289, 292, 294, 302, 305, 306, 341, 414, 455, 461, 474, 480, 545, 585, 599, 603, 605, 606
  Ferroan 72, 73, 95, 395, 418, 594
  High-Mg 9–11, 13, 14, 16, 19–23, 32, 72, 80, 109, 113, 167, 183, 254–256, 306, 336, 339, 494, 495, 534, 594, 599, 605, 606
  Low-Mg 9, 17, 19–21, 23, 31, 32, 79, 81, 86, 109, 142, 147, 148, 151, 171, 174, 254, 255, 318, 395, 396, 460, 467, 471, 475, 494, 495
Calcitization 87, 89, 109, 159–161, 468
Calcrete 8, 30, 31, 79, 83, 424, 425, 437, 503, 547, 554–557, 580, 594
Caliche 16, 23, 30, 31, 77, 94, 95, 310, 313, 323, 437, 439, 440, 443, 446, 456, 458, 461, 462, 464, 465, 467, 469, 471, 490, 500–504, 526, 528, 534, 600, 605
Calichification 324, 456
Cambrian 21, 86, 87, 89–92, 113
Canada 102, 105
Canary Islands 228, 239
Capillary pressure 59–61, 63, 66
Capitan Reef Complex 212, 533
Carbonate platform 169, 188, 194, 196, 197, 199, 205, 207, 209, 211, 213, 330, 336, 358, 407, 409, 546, 566, 580
Carbonate supersaturation 147
Carboniferous 84, 85, 87, 90, 260, 277, 280, 281, 324, 409, 424, 428, 431, 433
Carlsbad (Czechoslovakia) 7, 71, 73, 77, 79, 81, 94, 154, 161
Carpathians 171, 174
Carribean 230
Case-hardening 533

Catagraph 545
Cathodoluminescence 72, 73, 77, 84, 86, 306, 336, 341
Cave 7, 16, 17, 23, 32, 82, 94, 151, 437, 446, 495, 497, 500, 526, 527, 534
  Dripstone 82, 83
Cavern 439, 470, 471, 489, 490, 498, 499, 527, 574
Cavity 265, 313, 318, 319, 326, 341, 393, 396, 411, 413, 417, 418, 427, 475, 481, 483, 495, 537, 585
Celestite 159, 182, 183
Cement 74, 78, 84, 86, 87, 89, 95, 170, 171, 180, 183, 211, 255, 264, 312, 381, 383, 418, 471, 486, 495, 496, 500, 504, 508, 514, 521, 523, 526, 528, 529, 531, 554, 555, 585, 606, 616
  Blocky 191, 256, 495
  Drusy 191, 256, 573
  Fibrous 165
  Gravitational 89, 516, 521, 573
  Interparticle 85
  Marine 341, 600, 603, 606
  Meteoric 151, 532
  Phreatic 532
  Subaqueous 152
Cementation 20, 87, 89, 93, 170, 174, 182, 212, 254, 264, 268, 289, 292, 293, 326, 327, 336, 346, 355, 360, 365, 372, 395, 413, 494, 515, 522, 523, 529, 530, 532, 607, 618
  Early 325, 362, 371, 375
  Incipient 121
  Intraparticle 86
  Late 32, 503
  Stalactitic 522
Cenomanian 330, 336
Cenozoic 83, 151, 243, 250
Cephalopod 264
Chalcedony 101, 626, 630–632
Chamosite 101, 103
Channel 128, 133, 139, 141, 196, 205, 207, 208, 236, 515, 532, 561
  Discharge 232
  Drainage 490
  Fluvial 310, 323
  Tidal 133, 139, 141, 182, 185, 202, 209, 213, 232, 274, 385, 583, 584
Chert 100, 102, 369, 628, 630
Chertification 211–213
China 90, 176, 537, 545, 546
Chlorite 353
Clay minerals 255
Clotted texture 318, 411
Coast 41, 184, 228
Coastline 203, 410
Coating 15, 494
Coelobite 427
Cohesion 61, 63, 65
Cohesive force 59, 60
Collapse 105, 499, 500
Collophane 353
Colombia 103
Colonization 197, 221, 239, 292, 414
  Sea-grass 121
Community 268, 333, 348, 351, 365
Compaction 20, 92, 93, 103, 105, 110, 170, 180, 182, 201, 211, 213, 249, 254, 325, 333, 339, 616
  Vadose 441
Condensation 60
Coniatolite 501, 504, 533, 534, 599
Coral 28, 47, 49, 197, 264, 381, 617
Corrosion 339, 415, 462
  Biological 293, 296
Cortex 4, 5, 7, 57, 66, 72–74, 76, 78, 80, 93, 109, 151, 154, 159, 163, 165, 178, 182, 185, 191, 193–196, 198, 201, 203, 211–213, 222, 426, 427, 429, 431, 440, 443, 453, 506, 534, 604, 613, 624
Cracking 491
  Circumgranular 469, 512
  Contraction 530
Cretaceous 21, 274, 295, 308, 330, 333, 355, 566, 570, 574
Crinoid 110, 191, 420, 617
Cross-bedding 196, 380
Crust 77–80, 154, 155, 159, 163, 164, 221–223, 227, 228, 230, 234, 236–239, 284–286, 288, 291–296, 300, 302, 305, 310, 313, 315, 318, 319, 324, 429, 480, 516, 537, 538, 541, 543, 545, 546, 555, 599, 600, 604
Crustacean 239
Cryptalgal 205, 207, 308, 310, 344, 359, 362, 364, 365, 372, 375, 411, 415, 421, 545
Crystal silt 573, 575
Crystallite 74, 77, 78, 80, 103, 154
Crystallization 60, 362
Crystallization rate 14
Current 208, 222, 235, 248, 251, 285, 365, 385, 420, 421, 563, 610, 613
  Bidirectional 610, 618
  Bottom 235
  Oscillatory 129
  Surface 129
  Tidal 129, 205, 208, 227
  Traction 243
  Turbidity 624, 630
Cycle 205–207, 362, 367, 426, 528, 529, 567, 570, 572, 587, 590
  Carbonate-evaporite 177
  Fluvial 321
  Fluviolacustrine 324
  Loferitic 445
  Oolite-grainstone 116
  Progradational 209
  Stratification 144
  Transgressive-regressive 344
  Walnutite 508
Cyclic sequence 317
Cyclicity 129, 590

Dead Sea 9, 163, 167
Decalcification 286
Decay 5, 30, 325, 360, 388, 415
Decomposition 173, 284, 289, 411, 413, 453, 471
Dedolomite 494, 495
Dedolomitization 495, 497
Deformation 92, 481
  Plastic 185
Degassing 285
Degradation 281
  Microbial 293
Dehydration 570, 573
Denmark 90
Density 59
Desiccation 321, 324, 325, 441, 491, 499, 537, 543, 570, 578

Devonian 280, 282, 323, 555, 609, 617, 618
Diagenesis 27, 35, 66, 74, 84, 86, 87, 89, 91, 94, 95, 159, 174, 180, 199, 264, 274, 275, 308, 324, 336, 351, 395, 396, 453, 475, 494, 496, 503, 556, 583, 584
  Early 293, 545
  Eggshell 93
Dinantian 73, 77, 413
Dipole 60
Discontinuity 112, 351, 419
Dissolution 17, 21, 30, 60, 87, 89, 91–93, 103, 110, 159, 161, 182, 183, 293, 301, 318, 339, 395, 490, 495–497, 500, 501, 593
Distortion 93, 180
Dolomite 20, 73, 75, 86, 95, 182, 186, 201, 203, 208, 211, 255, 495, 504, 512, 538, 539, 556, 567, 580
Dolomitization 183, 185, 201, 211, 213, 255, 339, 379, 383, 384, 390, 395, 439, 521
Drainage system 185

Echinoderm 77, 84, 86, 246, 253
Echinoid 78, 249
Ecological succession 223, 240
Edge
  Bank 128, 130
  Platform 116, 207, 209, 211, 243
Edgewise conglomerate 205
Electrostatic attraction 59–61, 63
Electrostatic forces 65
Emergence 533, 566, 573, 575
Encrustation 3, 265, 277, 286, 288, 321, 323, 390, 414, 415, 417, 451, 534, 563, 565, 599, 604, 606
Encruster 75
Endobiont 42
Endolith 292, 293, 605
England 90, 105, 280, 413
Endolithic algae 17, 254
Enlargement 521, 522, 526
Envelope 610, 616
Environment
  Anoxic 73
  Basinal 415
  Brackish 85, 86, 284, 305, 306
  Coastal 245
  Depositional 74, 94, 139, 183, 185, 195, 264, 268, 305, 375, 603, 606, 632
  Evaporitic 101
  Fluvial 273, 281, 327
  Freshwater 288, 289, 293, 295, 326, 339, 437, 446, 593
  High-energy 78
  Hypersaline 15, 16, 163, 167, 437, 546
  Infratidal 532
  Intertidal 182, 183, 186, 273, 415, 421, 527, 532, 545, 574, 578
  Lacustrine 273, 281
  Lagoonal 4
  Marine 4, 15, 85, 154, 167, 273, 293, 446, 504
  Meteoric 21, 339
  Mixed water 21
  Mudflat 209
  Nearshore 144
  Peritidal 504, 528, 532
  Phreatic 5, 339, 342, 522, 593
  Platform 123
  Pool 529
  Reducing 72, 84
  Restricted 86
  Schizohaline 439
  Subaerial 624
  Subtidal 38, 182, 184, 186, 207, 273, 415, 421, 532, 545, 564, 578, 584
  Supratidal 28, 415, 504, 527, 532, 545, 574
  Swamp 34
  Vadose 5, 8, 339, 437, 446, 522, 523, 593
Eocene 21, 103, 274, 308, 310, 330
Epibiont 294, 333, 605
Epilith 292
Epizoan 264, 403
Equilibrium
  Chemical 14, 16, 17
  Inorganic 14, 17, 152
  Isotopic 31
  Partitioning 15
Erosion 41, 48, 155, 185, 234, 285, 323, 325, 327, 396, 490, 500, 505, 515, 528, 529, 531–534, 577, 604, 627
Eruption 56, 58, 59, 65
Europe 3, 31
Evaporation 28, 31, 143, 185, 545
Evapotranspiration 458
Evaporite basin 445, 555
Evaporites 67, 330, 543, 576
Exfoliation 491
Exhumation 371, 372
Expansion 530
Exposure 103, 105, 234, 321, 324, 327, 339, 342, 359, 365, 390, 396, 500, 541, 545, 556
Extraclast 323

Fenestrae 318, 319, 321, 351, 411, 414, 417, 530, 531, 574
Fenestral fabric 277, 360, 489, 528
Ferricrete 552
Ferruginization 101, 103
Fiji 488
Fitting 481
Flat
  Intertidal 528
  Mud 207, 209, 212, 213, 388
  Reef 232, 488
  Sand 133, 134, 139, 141, 577, 583
  Supratidal 205, 504, 532
  Tidal 203, 367, 385, 583, 624
Flooding 116, 128, 130, 139, 327, 385, 555
Floodplain 424
Florida 12, 230, 561, 578
Flow
  Gravity 211
  Storm-generated 129
  Tidal channel 139
Flow regime 208
Flowstone 480
Fluid inclusion 10, 14, 528
Fluid migration 396
Foraminifera 75, 77, 110, 178, 180, 184, 185, 228, 239, 246, 248, 251–253, 264, 265, 331, 375, 403, 405, 415, 427
  Encrusting 230, 232, 261, 264, 411, 603
  Nubecularia 372, 375
Fore-reef 44, 197, 230, 232, 280, 488, 599, 603
Fracturation 339

Fracturing 441, 500, 512, 529, 534
Fragmentation 516
France 52, 227, 609
Fringing reef 47, 127, 232
Fungi 182, 228, 254, 289, 291, 293, 296
Fungotufa 461

Gastropod 19, 84, 87, 95, 180, 185, 239, 251, 264, 289, 420, 573
 Vermiform 277, 280, 426, 429, 431, 432
Geopetal features 337
Geopetal sediment 415, 417, 418
Geopetal structure 91
Germany 19, 71, 75, 77, 87, 89–91, 94, 161, 279, 284, 294, 396
Glauconite 248, 251–253, 256
Goethite 15, 101, 103
Gorgonian 39, 42, 44, 51
Grading 65, 155
 Inverse (= reverse) 65, 362, 372, 514–516, 526–528, 531, 533
Grain diminution 319, 379
Grain enlargement 319, 321
Granulation 58, 60, 61, 63
Granule 167
Grapestone 116, 128–130, 462, 471
Great Salt Lake (Utah) 13, 16, 74, 77, 78, 80, 81, 83, 91, 93–95, 112, 154, 163, 172
Greece 8, 73, 76–80, 87, 89, 92, 93, 95, 284, 299
Greenalite 101, 102, 105
Guatemala 57
Gulf of Aqaba 80
Gulf of Mexico 34, 228, 237
Gypsum 101, 155, 239, 495

Hailstone 54, 58
Halite 16
Halotite 9
Hardground 49, 285, 363, 367, 369, 605
Hawaii 57, 58, 232
Hematite 101–103, 191, 353, 554
Hiatus concretion 345
Holocene 12, 15, 16, 18, 21, 22, 31, 76, 78, 79, 82, 95, 113, 116, 121, 128, 130, 132, 142, 148, 237, 306, 450, 543, 554, 604–607
Hot spring 5, 7, 71–73, 77, 79, 81, 87, 91, 154, 324, 482
Hurricane 121
Hydrocarbons 173
Hydrosphere 67

Ignimbrite 61
Impact 179
Inclusion 73, 81, 82, 89, 171, 203, 302, 319, 339, 491, 502, 508, 522, 523, 534, see also Fluid inclusions
Incrustation 301, 302
India 13, 31, 355
Indian Ocean 233
Interlamination 101
Internal sediment 339, 471
Interstitial fluid 84
Intraclast 110, 178, 180, 185, 222, 336, 384, 440, 451, 453, 455, 508, 512, 515, 516, 521, 522, 567, 568, 570, 574, 575, 603
Inversion 81, 420
Ireland 234, 235
Ironstone 101, 103
Isolator 59
Israel 30–32, 34, 35, 445
Italy 39, 90, 169, 359, 388, 390, 450, 474, 534

Jurassic 21, 30, 34, 77, 83, 84, 86, 90, 103, 105, 110, 170, 171, 174, 273, 274, 330, 336, 358, 367, 375, 377, 584

Karelia (USSR) 67, 633
Karst 630
Kerogen 630
Keystone 445
Kinsman's distribution coefficient 159
Klypeite 79

Labradorite 629
Lag deposit 206
Lagoon 128, 130, 152, 177, 184, 187, 203, 233, 265, 367, 504, 506, 534, 564, 567, 572, 575, 580
 Distal 116
 Platform 207, 209
 Schizohaline 425, 426
Lake 83, 91, 142, 144, 148, 285, 286, 288, 289, 295, 296, 305, 474
 Brackish 142
 Eutrophic 294
 Freshwater 279, 284, 323
 Hypersaline 152, 163, 555
 Marl 277
 Mesotrophic 295
 Oligotrophic 294–296
 Saline 82, 186
Lake Chad 13, 16–18
Lake Constance 19, 279, 284, 294, 295, 305, 428, 431
Lamina 72–74, 78, 112, 159, 202, 207, 225, 308, 311, 316–318, 321, 326, 349, 375, 393, 402, 403, 405, 411, 417, 420, 421, 427, 429, 430, 443, 456, 465, 467, 468, 470, 494, 508, 512, 516, 521, 527, 534, 539, 541, 584, 603, 604, 607, 624
 Algal 248, 264, 414
 Monomineralic 147
Lamination 73–75, 77, 81, 94, 105, 161, 320, 348, 360, 378–380, 384, 385, 405, 426, 430, 464, 481, 490, 502, 512, 545, 554, 555, 603, 604
 Algal 104
 Cryptalgal 205, 207, 208
 Current 578
 Horizontal 201
 Wrinkled 204
Laminite 426, 599
Laterite 490
Lattice 9–11, 14, 21, 159, 495
Leaching 181, 187, 594
Limestone 17, 30, 60, 71, 86, 95, 101, 103, 105, 129, 139, 155, 169, 171, 174, 183, 185, 186, 187, 201, 211, 213, 249–252, 260, 261, 280, 282
Lithification 32, 318, 339, 356, 417, 500, 570, 605
Lithoclast 359, 362, 372, 490–492, 498, 499, 539
Lithosome 129
Louisiana 30
Luminescence zonation 77
Lump 321, 418, 584

Macrophyte 284–286, 294–296
Maerl 227, 237
Magnesite 555
Magnetite 101
Maledivas 42
Malta 221
Mariana Is. 232
Marl 83
Marsh 207
Mascarene Is. 230, 232
Matrix 20, 54, 74, 80, 81, 102, 103, 134, 174, 251, 261, 265, 318, 326, 335, 336, 346, 360, 372, 393, 415, 418, 419, 427, 430, 456, 460, 467, 468, 514, 516, 531, 541, 552, 554, 584, 626, 627, 631
Mediterranean 38, 49, 52, 221, 227, 228, 230, 234, 306, 471, 504
Melting 61
Mesopore 399
Mesozoic 3, 72, 95, 225, 281, 301, 336, 385, 388, 421, 445, 474, 566
Metazoan 101
Micrite 249, 261, 264, 288, 289, 292, 293, 318, 321, 325, 333, 360, 372, 393, 395, 401–403, 405, 407, 413, 414, 418, 427, 481, 509, 512, 514, 534, 570, 572, 585, 605, 613, 617
Micritic envelope 261, 288, 292, 293, 568, 613
Micritization 17, 82, 87, 89, 134, 151, 170, 254, 319, 326, 327, 462, 468, 512, 533
Microboring 371
Microfacies 333
Microflora 276
Microlamina 348, 351, 603
Microlamination 411
Microorganisms 3, 7, 94, 95, 203, 293, 348
  Endolithic 74, 82, 86, 276, 288, 291
  Photoautotrophic 296
Microperforation 462
Microphyte 284–286, 291, 292, 295
Micropore 95
Microporosity 77–81, 87, 95, 159, 333
Microspar 317, 319, 326, 360, 402, 413, 508
Microsparite 349, 351, 469
Microstromatolite 348, 356
Microstylolite 418
Microunconformity 362, 427, 431, 432, 554, 555
Mine 72, 73, 82
Mineralization 103, 230, 344, 346, 348, 356
Mineralogical stabilization 20, 21
Miocene 221, 237, 238, 250
Mississippian 21, 90, 93–95, 212
Missouri 16
Mixing 144
Moisture 59–61, 63–65, 67
Mollusk 77, 228, 232, 246, 251, 253, 254, 286
Mold 77, 574
Mottling 105
Mound 265, 268, 506, 533
Mud ball 320

Nannograin 7, 81, 112, 113, 151, 170, 174, 603, 604
Neogene 81, 82, 142
Neomorphic aggradation 211
Neomorphic replacement 151
Neomorphism 255, 261, 475, 496, 521
  Aggrading 339, 360, 372, 379, 470
  Degrading 470
Neospar 261
Nevada 81, 82, 142
New Mexico 259, 265, 503, 599
New Zealand 58, 65, 243–245, 249, 254–256
Nontronite 17, 19
North America 142, 288
North Sea 30, 34
Norway 234
Nucleation 15, 65, 294, 521, 561, 604, 617, 618
Nucleus 3, 15, 17, 57, 60, 61, 63–66, 74, 78, 87, 89, 91–94, 110, 112, 134, 145, 152, 164, 167, 178, 180, 184, 191, 194, 201, 248, 255, 264, 277, 299, 308, 335, 339, 344, 345, 359, 369, 371, 372, 393, 415, 417, 427, 453, 455, 464, 469, 471, 477, 480, 481, 492, 494, 495, 499, 502, 504, 508, 509, 512, 516, 521, 522, 539, 541, 543, 561, 568, 573, 580, 594, 599, 603, 610, 613, 616, 619, 624, 628, 632

Ocean 141
Oil and gas accumulation 176
Oil migration 339
Oil trap 336
Oligocene 243, 250, 252, 295
Oomold 89, 91–93, 95
Ordovician 90, 280, 355
Organic framework 188, 197
Organic mat 29
Organic matter 5, 7, 19, 28, 30, 74, 101, 172, 174, 325, 353, 355, 419, 453, 455, 468, 497, 604
Organic mucilage 78
Organisms
  Benthic 228
  Boring 41, 402
  Encrusting 362, 375, 388
  Grazing 293
  Incrusting 41, 348
  Photoautotrophic 286, 289, 293, 296
Orthocement 319
Ostracod 110, 180, 265, 617
Oversaturation 31
Oxidation 101, 104, 105, 351

Paleocene 308
Paleoecology 243
Paleogene 281
Paleosol 443
Paleozoic 22, 72, 83, 87, 89, 95, 96, 274, 281, 330, 414, 474
Paramorphism 401
Partition
  Coefficient 10, 19–21, 23
  Isotopic 28
Patch reef 207, 209, 233, 284, 286, 292–294, 574
Pedocrete 7, 8
Pelagosite 534, 599
Pelecypod 19, 180, 185, 264
Pellet 33, 56, 110, 128, 129, 178, 180, 184, 185, 360, 380, 382, 383, 461, 469, 567

Peloid 9, 17, 23, 87, 128, 134, 141, 180, 182, 320, 401, 440, 456, 460, 464, 468, 469, 471, 505, 514, 528, 531, 542, 554, 570, 616, 617
Pennsylvanian 21, 22, 90
Percolation 182
Perforation 170
Peri-platform 130, 253
Permeability 65, 177, 183, 396, 492, 495, 521
Permian 75, 90, 260, 265, 295, 439, 503, 504, 508, 533, 534, 555, 573, 587, 599, 603, 605
Permineralization 276
Persian Gulf 13, 16–18, 74, 78–80, 94, 95, 171, 172, 196, 201–203, 205, 209, 211, 383, 503, 504, 530, 534, 545, 599
  Abu Dhabi 13, 173
  Trucial Coast 78, 186, 209
Phanerozoic 21, 56, 109, 113, 274, 276, 280, 284, 421
Philippines 57
Phosphate 253, 256, 294, 344–346, 348, 353, 354, 494
Photosynthesis 294, 325, 411, 613
Phytoplancton 286
Pinnacle 144
Pistomeite 556
Pleistocene 20, 21, 30–32, 74, 81, 82, 87, 91, 92, 94, 95, 139, 142, 174, 274, 284, 295, 299, 301, 305, 450, 456, 469, 474, 543, 554, 555, 561
Pliocene 295, 299, 305, 437, 456
Poland 75, 344, 587
Polychaete 239, 277
Polygonal fit 521–523, 534
Pond 474, 477
Pool 154, 155, 446
  Sea-marginal 80, 163
  Spring-fed 473
Pore 78, 180, 182, 187, 256, 318, 319, 333, 339, 418, 482, 495, 496
Pore fluid 60
Porosity 62, 77, 177, 183, 254, 360, 396, 415, 418, 460, 465, 467, 492, 495, 508, 521–523, 526
  Channel 339
  Intraparticle 339
  Oomoldic 74, 87, 593
  Vuggy 339
Precambrian 56, 66, 67, 73, 86, 91, 92, 105, 276, 355, 546, 633
Precipitation 17, 28, 31, 73, 74, 87, 94, 105, 143, 159, 194, 203, 204, 286, 289, 294, 325, 336, 339, 351, 411, 415, 421, 446, 469, 471, 480, 485, 494, 496, 500, 504, 521, 527, 528, 531, 534, 556, 561, 590, 599, 606, 613
  Biochemical 413
  Biological 27
  Chemical 27
  Direct 19, 23
  Inorganic 14, 23
  Internal 87
  Interstitial 87
Preservation 66, 73, 86, 325, 413, 431
Pressure-solution 93, 182, 254, 255, 339, 355, 418, 617
Productivity rate 253
Progradation 206, 209, 583
Progression 268, 529
Proterozoic 58, 547, 556
Pseudocement 321
Pseudomorphs after gypsum 34, 539, 543
Pseudospar 413
Pyrite 386, 388, 613, 629
Pyritization 405

Quatar 11
Quartz 84, 103, 145, 182, 191, 199, 630
Quaternary 8, 73, 79, 80, 84, 85, 87, 103, 105, 299, 301, 306, 474, 488

Rainwater 31
Recolonization 455
Recrystallization 20, 21, 30, 32, 33, 81, 86, 93, 201, 203, 212, 319, 321, 405, 414, 462, 556, 580, 604, 605, 616
  Aggrading 254, 496
Red Sea 38, 42, 44, 45, 74, 94, 95, 163, 172
Reef 47, 205, 207, 209, 211, 213, 230, 232, 234, 254, 289, 295, 301, 367, 488, 490, 503, 561, 599, 600, 603, 607
  Platform 47, 48, 561
Regression 176, 187
Rendzina 424
Replacement 87, 101, 103, 159, 183, 306, 443, 515, 534
Reservoir 187, 335, 339, 342
Residence time 53
Reworking 145, 441, 496, 500, 522
Rhizocretion 469
Rimstone 324
Rimstone dam 83
Rod 7, 71, 77–79, 95, 154, 157, 160, 167, 453
Root 458, 460, 461, 468, 469, 471
Ryukyu Is. 232

Sabkha 16, 437, 547
Salinity 15, 95, 167, 182
Sardinia 171, 174, 367
Scaphopod 264
Scotland 30, 34, 90, 227, 280, 413
Sea floor 38, 129
Sea grass 133, 232, 331, 333, 335, 336
Sea level curve 139
Sea urchin 77, 78, 306
Seawater 10, 15–17, 20, 23, 28, 78, 80, 95, 96, 337
Sediment binder 268
Sediment sliding 93
Sedimentary dyke 578
Sedimentation rate 53, 54, 243, 249, 273, 274, 295, 324, 325, 356, 363–365, 419, 421
Sepiolite 461
Sericite 626
Serpentinite 80
Serpulid 51, 232, 239
Shark Bay 78, 207, 209, 415, 555, 578, 580
Shelf 134, 198, 243, 333, 336, 344, 367, 375, 488, 505, 512, 514–516, 522, 528, 529, 531, 534
Shelf break 133
Shoal 133, 139, 182, 183, 185, 335, 336, 342
  Active 121

Shoal
  Algal 253
  Ooid 132
  Oolite 129, 202, 205–207, 209, 211, 213
  Sand 123, 128
Shoreline 144, 145, 250, 421, 533, 600
Shrinkage 469, 491
Sicily 171, 174, 369
Siderite 101, 103, 105, 556
Silicate 101
Silicification 201, 211, 213, 339, 405
Silurian 21, 89, 90, 555
Sinian 537
Sinter 82, 285, 555
Sintering 61
Soft substrate 54
Soil 29–32, 299, 319, 390, 439, 446, 456, 468, 503, 504, 512, 514, 528, 529, 533, 534, 600
Solubility 93, 145
Solution 10, 17, 74, 86, 95, 159, 160, 256, 339, 395, 396, 469, 489, 495, 512, 521
Solution collapse 104, 105
Sorting 72, 139, 492, 499, 515, 526, 568
Source-bed 342
Spain 90, 188, 281, 295, 308, 409, 456, 471
Spar 313, 319, 321, 372, 393, 396, 411, 417, 427, 446, 453, 455, 475, 477, 539, 616
Sparite 335, 336, 339, 341
Spelean 148, 151
Speleothem 72, 494, 527, 534
Spitsbergen 73, 86, 91, 92
Splash cup 17, 83, 527
Sponge 228, 232, 264, 265, 268, 603
Spring 144, 155
Stabilization 121, 232, 364, 417
Stalagmite 17
Storm 127, 163, 237, 250, 253, 365, 419, 613, 618
Storm deposit 41, 54, 205
Stratification 139, 144, 381, 477
Stream 295
Stromatolite 4, 100, 102, 201, 205, 207, 209, 211, 268, 276, 308, 310, 316, 320, 321, 324, 344, 348, 355, 362, 388, 407, 413, 433, 460, 538, 554, 573, 577, 584, 599, 607
Stromatoporoid 197
Strontianite 159
Strontium 19, 21, 23, 78, 79, 82, 159, 161, 171, 174
Stylolite 552, 554
Stylolitization 329, 521
Subdelta 207
Subsidence 185, 330, 339, 342, 505, 506, 528
Subsurface 132
Supersaturation 74, 147
Suspension 49, 194
Swamp 29, 30, 474
Sweden 89, 90
Switzerland 336
Synaeresis 179

Tepee 504, 505, 521, 529–534
Tephra 58
Terrace
  Coastal 87
  Marine 73, 77, 79
Tertiary 142, 245, 255, 256, 260, 295
Tethys 170, 358
Texas 11, 12, 336, 503
  Baffin Bay 11, 12, 29, 30, 34, 81, 83, 94, 95, 113, 148, 163
Thallus 221–223, 230, 233, 235, 264, 265, 293
Thermocline 39, 42, 47
Thermodynamic stability 20
Thrombolite 415, 417, 419
Tidal delta 186, 196, 202, 209
Tide 123, 128, 185, 563
Timor 232
Trace elements 9–23
Transformation 23, 73, 84, 86, 93, 173, 183
Transgression 148, 176, 265, 330, 419, 425, 590
Transport 66, 116, 121–123, 127, 128, 130, 141, 194, 223, 321, 421, 531, 555, 626, 629
Trapping 249, 277, 281, 288, 296, 325, 348, 372, 511, 574
Travertine 5, 437, 446, 450, 451, 453, 455, 474, 475, 481, 489, 490, 495, 500
Triassic 73, 86, 87, 89–93, 96, 154, 161, 176, 186, 265, 295, 386, 390, 395, 396, 398
Trilobite 264, 265
Truncation 533
Tufa 144, 145
Tuff 66, 629
Turbidite 330, 365
Turbulence 65, 74, 129, 130, 155, 256, 274, 333, 360, 363, 372, 607
Turkey 79, 81, 87, 91, 154
Turonian 32, 34, 35, 445
Tyrrhenian 77

Unconformity 79, 252, 360, 481, 483
  Angular 144
  Intraformational 197

Vacuole 453
Vadoization 446
Van der Waals-forces 60, 61, 65
Virginia 91
Void 134, 265, 280, 393
Volcano 58
Vug 182, 318, 395, 574

Wales 280, 324, 355, 424
Walther's Law of Facies Succession 529
Water
  Brackish 144
  Capillary 86
  Continental 11
  Environmental 152
  Fresh 11, 19
  Hypersaline 163, 504, 528
  Interstitial 34, 86
  Lake 19, 23, 144, 147, 288, 289
  Marine 504
  Meteoric 20, 28, 32, 182, 183, 185, 187, 212
  Natural 11
  Saline 185
  Spring 144, 147
  Surface 144
  Vadose 500, 504, 541
Water energy 54
Waulsortian 260
Wave base 51
Weathering 100,105, 501, 504, 554–556, 573
Wetting 523
Winnowing 129

Worm 289
  Serpulid 228, 246

Yeast 629
Yucatan 78
Yugoslavia 566

Zechstein 75, 87, 89, 273, 445, 446, 587, 590, 594, 597
Zimbabwe 66, 633

Zone
  Breaker 431
  Euphotic 38
  Freshwater 180, 182, 273
  Intertidal 185, 207, 227, 385, 531, 537
  Littoral 239, 286
  Mixing 147, 182
  Nearshore 147, 421
  Photic 53, 239, 256, 295, 296
  Phreatic 5, 7, 16, 160, 161, 180, 273, 523
  Saline 182
  Spray 605
  Sublittoral 294
  Subtidal 185, 537
  Supralittoral 239, 285, 385, 531, 545
  Vadose 5, 7, 16, 274, 437, 439, 440, 495, 501, 523, 554, 556, 573

Cyclic and Event Stratification

Editors: **G. Einsele, A. Seilacher**
1982. 180 figures. XIV, 536 pages
ISBN 3-540-11373-8

Stratification is a fundamental property of sedimentary rocks which is used in stratigraphic correlation, structural and economic geology. This book focuses on two major processes that produce pronounced bedding in the marine realm:

- cyclic or periodic gradual changes of environmental conditions, and
- rare and unpredictable events (storm wave-induced tempestites; turbidites).

These processes affect a great variety of sediments including black shales and carbonates, where primary futures can be masked or overprinted by later diagenesis. Periodic phenomena with potential time information are represented by some limestone-marl successions.

The 44 contributions by authors from many countries range from detailed descriptions of characteristic stratification phenomena, sedimentary and biostratinomic structures and faunal responses to articles presenting general conceptual models and interpretations of the depositional environment. Unique in its summary of the state-of-the-art in this interdisciplinary field, *Cyclic and Event Stratification* is of interest to all those concerned with bedded sequences.

Contents: Limestone-Marl Rhythms and Climate-controlled Facies Changes. - Event Stratification: Calcareous and Quartz-Sandy Tempestites. Other Events Deposits. - Cyclicity and Event Stratification in Black Shales. - Summary.

E. Flügel

Microfacies Analysis of Limestones

Translated from the German by K. Christenson

1982. 53 plates, 78 figures. 58 tables. XIV, 633 pages.
ISBN 3-540-11269-3

This revised and expanded translation of the successful German original introduces the reader to the concept and methodology of microfacies analysis. Concentrating primarily on carbonate rocks and sediments, it covers the recognition and evaluation of microfacies characteristics, fossils in thin-sections, facies diagnosis and models, and the interrelationships between microfacies and geochemical criteria. The text is supplemented by numerous experiences, carefully selected thin-section photographs and an extensive bibliography.

From the reviews of the German edition: "This book offers a longneeded, concise overview ot the extensive, widely disseminated information on limestone petrography, petrology (including some aspects of palaeontology), and environmental reconstructions, without an undue overlap with existing publications on sedimentary carbonates. ...During the detailed examination of the publication I caught myself describing the mode of presentation with complimentary superlatives as it was a pleasure to observe the point-by-point layout of the data within the text, well supported by many descriptive, genetic/interpretive and comparative tables and diagrams, thus neatly summarizing the information, all of which is supplemented by conceptual models... The publisher is to be congratulated for the presentation of this book – the printing of the text, figures and photographs are excellent – and Dr. Flügel for having assembled the widely distributed data..."

Earth Science Review

"The book is beautifully produced, and photomicrographs and scanning electron micrographs are superbly clear..." *Geological Magazine*

Springer-Verlag
Berlin
Heidelberg
New York